CAMBRIDGE LIBRARY COLLECTION

Books of enduring scholarly value

Polar Exploration

This series includes accounts, by eye-witnesses and contemporaries, of early expeditions to the Arctic and the Antarctic. Huge resources were invested in such endeavours, particularly the search for the North-West Passage, which, if successful, promised enormous strategic and commercial rewards. Cartographers and scientists travelled with many of the expeditions, and their work made important contributions to earth sciences, climatology, botany and zoology. They also brought back anthropological information about the indigenous peoples of the Arctic region and the southern fringes of the American continent. The series further includes dramatic and poignant accounts of the harsh realities of working in extreme conditions and utter isolation in bygone centuries.

An Account of the Petrological, Botanical, and Zoological Collection Made in Kerguelen's Land and Rodriguez During the Transit of Venus Expeditions 1874–75

The Kerguelen Islands, known also as the Desolation Islands, lie in the extreme south of the Indian Ocean. By the late nineteenth century they were still relatively unexplored, but they represented a fascinating puzzle: although the Islands were four thousand miles away from South America, they shared the same species of flora. Rodrigues, an island off the coast of Madagascar, was also a point of increasing interest among naturalists. While most archipelagic islands then discovered were volcanic, explorers noted that the caves of Rodrigues were formed of limestone, and that most local species were not indigenous. Like the Kerguelen Islands, they provided some of the first clear evidence that modern sea-levels were much altered from those of prehistory. Naturalists visited both locations as part of the expeditions to study the transit of Venus in 1874. Originally published in 1879, this collection of essays is a comprehensive catalogue of their findings.

Cambridge University Press has long been a pioneer in the reissuing of out-of-print titles from its own backlist, producing digital reprints of books that are still sought after by scholars and students but could not be reprinted economically using traditional technology. The Cambridge Library Collection extends this activity to a wider range of books which are still of importance to researchers and professionals, either for the source material they contain, or as landmarks in the history of their academic discipline.

Drawing from the world-renowned collections in the Cambridge University Library and other partner libraries, and guided by the advice of experts in each subject area, Cambridge University Press is using state-of-the-art scanning machines in its own Printing House to capture the content of each book selected for inclusion. The files are processed to give a consistently clear, crisp image, and the books finished to the high quality standard for which the Press is recognised around the world. The latest print-on-demand technology ensures that the books will remain available indefinitely, and that orders for single or multiple copies can quickly be supplied.

The Cambridge Library Collection brings back to life books of enduring scholarly value (including out-of-copyright works originally issued by other publishers) across a wide range of disciplines in the humanities and social sciences and in science and technology.

An Account of the Petrological, Botanical, and Zoological Collection Made in Kerguelen's Land and Rodriguez During the Transit of Venus Expeditions 1874–75

Alfred Edwin Eaton

CAMBRIDGE UNIVERSITY PRESS

Cambridge, New York, Melbourne, Madrid, Cape Town,
Singapore, São Paolo, Delhi, Mexico City

Published in the United States of America by Cambridge University Press, New York

www.cambridge.org
Information on this title: www.cambridge.org/9781108050098

© in this compilation Cambridge University Press 2012

This edition first published 1879
This digitally printed version 2012

ISBN 978-1-108-05009-8 Paperback

This book reproduces the text of the original edition. The content and language reflect
the beliefs, practices and terminology of their time, and have not been updated.

Cambridge University Press wishes to make clear that the book, unless originally published
by Cambridge, is not being republished by, in association or collaboration with, or
with the endorsement or approval of, the original publisher or its successors in title.

The original edition of this book contains a number of colour plates,
which have been reproduced in black and white. Colour versions of these
images can be found online at www.cambridge.org/9781108050098

PHILOSOPHICAL TRANSACTIONS

OF THE

ROYAL SOCIETY

OF

LONDON.

VOL. 168.
(EXTRA VOLUME.)

LONDON:
SOLD BY HARRISON AND SONS, ST. MARTIN'S LANE, W.C.
AND ALL BOOKSELLERS.

AN ACCOUNT

OF THE

PETROLOGICAL, BOTANICAL, AND ZOOLOGICAL COLLECTIONS

MADE IN

KERGUELEN'S LAND AND RODRIGUEZ

DURING THE

TRANSIT OF VENUS EXPEDITIONS,

CARRIED OUT BY ORDER OF HER MAJESTY'S GOVERNMENT

IN THE YEARS 1874–75.

LONDON:
PRINTED BY GEORGE EDWARD EYRE AND WILLIAM SPOTTISWOODE,
PRINTERS TO THE QUEEN'S MOST EXCELLENT MAJESTY.

MDCCCLXXIX.

PREFACE.

When, in the year 1874, Her Majesty's Government determined to despatch several expeditions to observe the Transit of Venus, the Council of the Royal Society resolved to request the Treasury to attach naturalists to those destined for Kerguelen's Land and Rodriguez, two of the least explored and most inaccessible oceanic islands of the southern hemisphere; and a Committee, consisting of Sir J. D. Hooker, Professor Huxley, and Mr. P. L. Sclater, was appointed to prepare the application, which was laid before Her Majesty's Government in the following terms:—

"It is an unexplained fact in the physical history of our globe, that all known oceanic archipelagos distant from the great continents, with the sole exceptions of the Seychelles and of a solitary islet of the Mascarene group (which islet is Rodriguez), are of volcanic origin. According to the meagre accounts hitherto published, Rodriguez consists of granite overlaid with limestone and other recent rocks, in the caves of which have been found the remains of recently extinct birds of a very singular structure. These facts, taken together with what is known of the Natural History of the volcanic islets of Mauritius and Bourbon to the west of Rodriguez and of the granitic archipelago of the Seychelles to the north of it, render an investigation of its natural products a matter of exceptional scientific interest, which, if properly carried out, cannot fail to be productive of most important results.

"As regards Kerguelen's Land, this large island (100 by 50 miles) was last visited in 1840 by the Antarctic Expedition under Sir James Ross, in midwinter only, when it was found to contain a scanty Flora of flowering plants, some of which belong to entirely new types, and an extraordinary profusion of marine animals and plants of the greatest interest, many of them being representatives of North-temperate and Arctic forms of life.

"H.M.S. 'Challenger' will no doubt visit Kerguelen's Land, and collect largely; but it is evident that many years would be required to obtain even a fair representation of its marine products; and though we are not prepared to say that the scientific objects to be obtained by a naturalist's visit to Kerguelen's Land are of equal importance to those which Rodriguez will yield, we cannot but regard it as in every respect most desirable that the rare opportunity of sending a collector to Kerguelen's Land should not be lost."

Her Majesty's Government acceded to the request preferred by the President and Council; and subsequently, on their recommendation, the Treasury sanctioned the appointment of four naturalists, three to Rodriguez and one to Kerguelen's Land. Those selected for the work in Rodriguez were Dr. I. B. Balfour, who was charged with the duties of botanist and geologist; Mr. George Gulliver, who was directed to investigate the Fauna generally; and Mr. H. H. Slater, whose special duties consisted in the exploration of caves and in collecting the remains of extinct animals. The naturalist attached to the Kerguelen's Land expedition was the Rev. A. E. Eaton, who was well qualified to bring to bear his experience of the Arctic fauna and flora upon those existing in the southern hemisphere under parallel physical conditions.

The collections and observations made by these naturalists fulfilled the expectations of the Council. A Committee appointed to consider the best means of rendering the collections serviceable to science, recommended that they should be entrusted to competent persons for examination and description, and that their reports should be published as a separate volume of the Philosophical Transactions. This recommendation was adopted by the Council, who requested Sir J. D. Hooker and Dr. Günther to undertake the editing of the work.

With regard to the specimens, the Council directed that complete sets should be reserved for the National Collections, and that the remainder should be distributed to the Museum of the Royal College of Surgeons in London, the Government Museums of Natural History in Edinburgh and Dublin, the University Museums of Oxford and Cambridge, and other Institutions.

By Order of the President and Council.

Burlington House,
   March 1879.

G. G. STOKES, *Secretaries,*
T. H. HUXLEY, *Royal Society.*

TABLE OF CONTENTS.

THE COLLECTIONS FROM KERGUELEN ISLAND.

INTRODUCTORY NOTES. *By the* REV. A. E. EATON.
   I. *The Physical Features of Kerguelen Island* . . . . page 1
   II. *Recent visits of Naturalists to Kerguelen Island* 4

BOTANY.
   Observations on the Botany of Kerguelen Island. By J. D. HOOKER, P.R.S.   5
   Flowering Plants, Ferns, Lycopodiaceæ, and Characeæ. By J. D. HOOKER, P.R.S. (*Pls. I., II.*) 17
   Musci. By W. MITTEN, *A.L.S.* (*Pl. III., Figs.* 1–5) 24
   Hepaticæ. By W. MITTEN, *A.L.S.* (*Pl. III., Figs.* 6–11) 40
   Lichens. By the REV. J. M. CROMBIE, *F.L.S.* 46
   Marine Algæ. By G. DICKIE, *A.M., M.D.* (*Pl. V., Fig* 3) 53
   Freshwater Algæ. By HR. P. F. REINSCH. (*Pls. IV. and V., Figs.* 1, 2)   65
   Fungi. By the REV. M. J. BERKELEY, *M.A., F.L.S.* 93

ZOOLOGY.
   Seals and Cetaceans. By W. H. FLOWER, *F.R.S.* 95
   Birds. By R. B. SHARPE, *F.L.S.* (*Pls. VI.–VIII.*) 101
   Eggs of Birds. By H. SAUNDERS, *F.L.S.* 163
   Fishes. By DR. A. GÜNTHER, *F.R.S.* 166
   Mollusca. By E. A. SMITH, *F.Z.S.* (*Pl. IX.*) 167
   Polyzoa. By G. BUSK, *F.R.S.* (*Pl. X.*) 193
   Crustacea. By E. J. MIERS, *F.L.S.* (*Pl. XI.*) 200
   Entomostraca. By G. S. BRADY, *M.D., F.L.S.* (*Pl. XII.*) 215
   Arachnida. By the REV. O. P. CAMBRIDGE. (*Pl. XIII.*) 219

Insecta.

  Observations on the Insects of Kerguelen Island. By the Rev. A. E. Eaton, M.A. . . . . page 228

    Coleoptera. By C. O. Waterhouse. (*Pl. XIV.*) . . . 230
    Lepidoptera. By the Rev. A. E. Eaton, M.A. 235
    Diptera. By G. H. Verrall. (*Pl. XIV.*) 238
    Neuroptera. By the Rev. A. E. Eaton, M.A. 248
    Collembola. By Sir J. Lubbock, F.R.S. (*Pl. XIII.*) . . 249
    Mallophaga. By C. Giebel. (*Pl. XIV.*) 250
  Marine Annelida. By W. MacIntosh, F.R.S. (*Pl. XV.*) . . 258
  Terrestrial Annelida. By E. Ray Lankester, F.R.S. . . . 264
  Echinodermata. By E. A. Smith, F.L.S. (*Pl. XVI.-XVII.*) . . 270
  Actinozoa. By the Rev. A. E. Eaton, M.A. 281
  Hydroida. By Prof. Allman, M.D., P.L.S. (*Pl. XVIII.*) . . 282
  Spongiida. By H. J. Carter, F.R.S. 286

THE COLLECTIONS FROM RODRIGUEZ.

Introductory Notes.
  I. *The Physical Features of Rodriguez.* By Is. B. Balfour, Sc.D. . . 280
  II. *Reports of Proceedings of the Naturalists* 293

Petrology. By N. S. Maskelyne, F.R.S. 296

Botany.
  Introductory Remarks. By Is. B. Balfour 306
  Flowering Plants and Ferns. By Is. B. Balfour. (*Pls. XIX.-XXXVII.*) 326
  Musci. By W. Mitten, A.L.S. (*Pls. XXXVII., XXXVIII., A. B.*) . 388
  Hepaticæ. By W. Mitten, A.L.S. (*Pls. XXXVIII., C. D.-XL.*) . . 396
  Lichens. By the Rev. J. M. Crombie, F.L.S. 402
  Fungi. By the Rev. M. J. Berkeley, M.A., F.L.S. 413
  Algæ. By G. Dickie, M.D., F.L.S. 415

Zoology.
  Extinct Fauna:
    Observations on the Bone Caves of Rodriguez. By H. H. Slater, B.A. 420
    Birds. By Dr. A. Günther, F.R.S., and E. Newton, M.A., F.L.S. (*Pls. XLI.-XLIII.*) 423
    On the Osteology of the Solitaire (Pezophaps solitaria). By E. Newton, M.A., F.L.S., and J. W. Clark, M.A. (*Pls. XLIV.-L.*) . . . 438
    Reptiles. By Dr. A. Günther, F.R.S. 452

Recent Fauna:

Mammalia. By G. E. Dobson, *M.A.* . . . . page 457
Birds. By R. B. Sharpe, *F.L.S.* 459
Reptiles. By Dr. A. Günther, *F.R.S.* 470
Fishes. By Dr. A. Günther, *F.R.S.* 470
Mollusca. By E. A. Smith, *F.Z.S.* (*Pl. LI.*) 473
Crustacea. By E. J. Miers, *F.L.S.* 485
Myriopoda and Arachnida. By A. G. Butler, *F.L.S.* (*Pl. LII.*) . . 497
Coleoptera. By C. O. Waterhouse. (*Pl. LIII.*) 510
Hymenoptera, Diptera, and Neuroptera. By F. Smith 534
Lepidoptera. By A. G. Butler, *F.L.S.* 541
Orthoptera and Hemiptera. By A. G. Butler, *F.L.S.* (*Pl. LIV.*) . 545
Annelida. By Prof. Ed. Grube 554
Turbellaria. By G. Gulliver, *B.A.* (*Pl. LV.*) 557
Echinodermata. By E. A. Smith, *F.Z.S.* (*Pl. LI.*) 564
Corals. By Dr. F. Brüggemann 569

THE COLLECTIONS

FROM

KERGUELEN ISLAND.

INTRODUCTORY NOTES.

By the REV. A. E. EATON, M.A., *Naturalist to the Expedition.*

I. *The Physical Features of Kerguelen Island.*

KERGUELEN Island is little else than a succession of hills and mountains formed almost exclusively of volcanic rock. In its greatest diameters it measures about 80 miles from N.W. to S.E., and 70 from N.E. to S.W., but no part of the interior is farther than 10 or 12 miles from the sea, for the coast on all sides is exceedingly intricate and abounds in large inlets and narrow fiords, which run far inland at frequent intervals between ranges of precipitous hills. A district of considerable extent in the midst of the island is occupied by snowfields, whence glaciers descend east and west towards the sea. This district is bounded on the S.E. by a series of snow-clad mountains and lofty hills extending in a curve northwards from Mount Ross almost across the island. The area intermediate between this range, Royal Sound, and the Mount Crozier hills, and likewise the islands in Swains' Bay and Royal Sound, contain series of interrupted ridges, the majority of which are crowned and terraced horizontally with basalt more or less amygdaloidal, and rarely exceed 600 feet in altitude. Their south-eastern and eastern slopes are generally very favourable for vegetation, being sheltered from the prevailing winds; but scarcely any of them attain to heights sufficient to be suitable for the growth of plants restricted to elevated sites in this island. The Mount Crozier hills, and those on the opposite side of the sound, are upwards of 2,000 or 3,000 feet high. Their summits are covered with snow until late in the season. Most of them are massive and rather simple in contour, but near Mount Crozier a few of the peaks are singularly picturesque, bristling with pinnacles, needles, and castellated towers of rock, some of which are visible from the southward, but not many. A similar departure from the prevailing type of Kerguelen Island scenery is noticeable among the hills near Sprightly Bay, and the aiguilles on the south side of Mount Ross. An active volcano is reported to exist in the neighbourhood of Bonfire Beach, and at one or two places on the same side of the island it is said that there are hot springs, resorted to by the sea elephants for the purpose of recreation at certain seasons of the year. A cold mineral spring oozes forth at the head of a small patch of boggy ground very near the western terminus of Swains' Haulover. It lies within a stone's throw of the sea, in a line with the nearest end of one of the nameless islands in the bay and the apex of the semi-pyramidal extremity of a high precipitous hill, which constitutes a conspicuous landmark on the opposite shore. The water is free from smell, but has a pronounced mineral flavour, seemingly of alum.

On the steep slopes bordering Royal Sound, near Swains' Haulover to the S.E., two or three small patches of yellow clay are visible. The clay appeared to include nothing but small fragments of volcanic rock, similar in composition to the masses *in situ* around it. A little farther on, along the same shore, opposite the south-eastern point of Seal Island, one of the hills is intersected by a nearly vertical wall of trap conspicuous from the intervening channel.

The coal beds found by Dr. Hooker at Christmas Harbour and Cumberland Bay, and a stratum at the first-named place, containing silicified trunks of trees, are almost the only formations in the island that are not distinctly of igneous origin. Some limestone, however, is stated to occur at Foundry Branch, and after our departure from the country a cast of a fossil conchifer was shown to me by Mr. J. Stone, Assistant Surgeon to H.M.S. "Supply," which was given to him by one of the sealers, who had picked it up "somewhere near Thumb Peak."

The whole island, exclusive of the snowfields already mentioned, abounds in freshwater lakes and pools on the hills and lower ground. Counting large and small together, there were about two dozen of them within a two and a half mile radius of the chief English observatory; two of these were upwards of a mile in length, but the majority of them were much smaller. Beyond this limit, however, some lakes of considerable size existed. They appeared to be uninhabited by fish or by *Neuroptera*, but a small Entomostracon (*Centropages*) was exceedingly plentiful in many of them.

Between the eastern base of the Mount Crozier hills and the coast a broad tract of low-lying land slopes very gently towards the sea. It is saturated with the drainage of the adjacent heights, and with the excessive precipitation upon it of rain and snow, resulting from its situation immediately to leeward of them. Some portions of it are said to be unsafe to persons attempting to walk over it, either through being undermined by subterranean streamlets, or on account of the existence of dangerous bogs. Streams descending from the steep declivities of the hills wear by degrees narrow trenches in the stiff clayey soil of the low ground, whose sides, in course of time collapsing by their own weight, meet and grow together in an arch above, being kept by roots of plants (especially *Azorella*) from crumbling to pieces. In the tunnel thus formed the work of excavation proceeds; the hidden channel is enlarged until it acquires the dimensions of a chasm, thinly roofed over by a layer of living plants, continuous with the surface of the firm adjacent ground, but easily broken through, into whose dark cavernous entrance the stream in mid career tumbles abruptly to a depth of many feet, sooner or later to reissue at a lower level. Years pass by; the chasm becomes wider, until at length it is disclosed by the subsidence of the yielding vault into the deep cavity beneath. Dr. Kidder states that some of the pits excavated in this manner by streams in the district in question are upwards of 30 feet in depth; the tunnels seen by me were of much smaller dimensions. The sides of such trenches and the entrances to the tunnels frequently offer favourable sites for the more delicate species of *Hepaticæ*.

Walking over the island is at certain localities very unsafe, in consequence of the frequent occurrence of mud-holes, the surface of which is precisely similar in appearance (or very nearly so) to many surrounding patches of firm bare soil. Holes of this description are produced by underground springs deep enough down to permit the surface to remain tolerably dry, while they keep the earth below in the consistency of liquid mortar, the water that issues from the source escaping by percolation through the soil. They are sometimes dangerously deep. The Antarctic Expedition, in the course of an overland journey in the northern part of the island, found their average depth on the western coast to be up to a man's waist.

In places where the ground is soft and boggy (as is frequently the case on gentle slopes and in the valleys) it is well to take notice of the kind of plants growing upon it. Patches of incoherent *Hepaticæ* and tremulous sheets of moss are best avoided. Stunted *Azorella* is somewhat uncertain, but is generally firm. Spaces overgrown with *Acæna*, as a rule, may also be traversed with safety. But wherever the surface waves freely, whatever may be growing there, it is advisable to be cautious.

So serious are the obstacles to travelling on foot in Kerguelen Island presented by fiord, lake, bog, and terraced hill, that some parts of it are absolutely inaccessible by land from districts immediately adjoining them, while others can be reached only by following the devious windings of a complicated maze. Measurements of the map afford no very trustworthy bases for estimates of the time needed for a journey of a given length in a certain direction.

The climate of Kerguelen is tempestuous, chilly, and wet. Days perfectly calm are of extremely rare occurrence, gales, or at least strong breezes, being almost constant. The wind is usually westerly, and cold; but now and then, in the south of the island, it comes on to blow from the E. and S.E., and this change is attended with a rise of temperature. In a treeless country, such as "Desolation," the violence of a gale is recognised not by ravages committed by it, but by transient evidences of its force exhibited at the time of its progress. Waterfalls descending from the cliffs are intercepted in mid air, and driven backwards in continuous clouds of spray to the hills, from whence they leap down, leaving the stream bed empty below; and similar clouds swept up from lakelets on the mountain side roll rapidly away to leeward, like vapour from a cauldron. But perhaps the strangest sights may be observed at such times from the shelter of a cliff, for there the air resounds with the gale, which is not felt, while eddies launched from the summit plunge with tremendous force upon the sea, scour the surface into foam, and hurtle the water to and fro with fitful violence. Sudden squalls and "willy-waughs" from the hills, such as are usually met with off mountainous coasts, prevail in many parts of the island, and cause boat navigation to be attended with considerable risk. But the unconcentrated force of the usual wind is the greatest obstacle to boating. In Observatory Bay, when the wind was off shore, it was often impossible for three or four days together for anyone to land from the "Volage" and "Supply," though their distance from the landing place was only 300 or 400 yards to leeward.

The range of temperature throughout the year does not appear to be excessive, the highest readings of the thermometer in summer being under 70° F., and the lowest in winter seldom being less than 32°. At Christmas Harbour it did not descend below 27° during the stay of the "Erebus." Before the English expedition arrived at the island the Americans, early in September, found the temperature one night to be as low as 18°. In the warmer months the readings are not often much higher than 55° or 56°, or much lower than 42°, on the eastern side of the island. The western coast is, however, much more bleak than the other, as is evidenced by the belt of herbage adjacent to the sea attaining a lower altitude there than it does on the eastern side of the island. On exposed parts of the S.W. shore the hill sides are conspicuously green to a height of hardly more than 150 to 200 feet (as it seemed to the eye); but in Royal Sound, and on the lee side of the island, the upper limit of verdure is between 500 and 700 feet above sea-level.

On the high hills and mountains fog is very prevalent, but districts of only slight elevation are comparatively free from it, and enjoy more frequent sunshine than other portions of the country. While the English were in Royal Sound the glimmer of distant lightning was seen one night in the direction of Mount Ross; but this was a very exceptional occurrence. Such displays of the aurora as were observed were not remarkable for their brilliancy

II. *Recent Visits of Naturalists to Kerguelen Island.*

Kerguelen Island has been visited within recent times by five scientific expeditions,—the Antarctic under Sir James Ross, the Challenger under Sir George Nares, and three Transit of Venus' expeditions.

Christmas Harbour was selected for head-quarters by the Antarctic Expedition, which remained there from the 15th of May until the 20th of July 1840. It is a deep inlet on the east coast at the north end of the mainland, bounded on each side by steep terraced hills more than 1,000 feet high, and terminated by a sandy beach at the foot of a gentle slope of moderate altitude. At the entrance it measures a mile across, but half way up it is abruptly narrowed to a quarter of its former breadth. Lying lengthwise in the direction of the prevailing winds it is open to every gale that sweeps over the neck of low land at its head; and consequently it would seem to be not the best site in the island that could be chosen for the prosecution of researches in natural history. In fact a collector stationed here would have good cause to utter Caliban's remonstrance:—

". . . . . " here you sty me
" In this hard rock, whiles you do keep from me
" The rest of the island." . .

For it apparently is quite cut off by precipices from the country immediately to the south of it,— an obvious disadvantage. Notwithstanding these drawbacks, however, many interesting specimens can be obtained here. A small tarn, not far from the sandy beach, contains a peculiar species of *Ranunculus* (discovered by Mr. Moseley). At the foot of a huge mass of dark volcanic rock that is conspicuous on the south side of the harbour, between the summit of the hill and the mouth of the bay, is fossil wood in abundance. In a small bay on the same side, close to the Arch Rock at the entrance to the harbour, and 30 feet above the sea, and again in a little cave in shale near the centre of the small bay formed by C. François on the opposite shore, coal beds crop out. Besides these, the cliffs and hill sides in the vicinage of the harbour are resorted to by two or three species of birds that are not found in more sheltered localities.

The collections made by Dr. Hooker and his colleagues in this neighbourhood appear to have been very complete for the season during which they were formed, and the area investigated by them. They included the following numbers of species: —15 birds, 18 fishes, 4 insects, 4 Crustacea (exclusive of some *Entomostraca*), an earthworm, and about 6 marine Annelida, 7 or 8 Mollusca, 2 or more Polyzoa, a Holothurian and 2 or 3 star-fish, a few Actinozoa, 2 or 3 sponges; 16 Phanerogams, 3 ferns, 1 or 2 Characeæ, 25 Musci, 10 Hepaticæ, 1 Fungus, 19, with 2 doubtful, species, and 2 named forms of Lichens, and 57, with 1 doubtful, species of Algæ (viz., 38, and 1 doubtful, marine with 6 Diatoms, and 13 freshwater Algæ). These amount in all to about 66 animals and 135 plants.*

The Challenger Expedition stayed at Kerguelen Island from the 7th to the 31st January 1874. Three days seem to have been spent at Christmas Harbour, one at Fuller's Harbour, eight at Betsy Cove, and three at Three Island Harbour; and at various times and places along the lee side of the island dredging was successfully carried on. Lists of the plants (which were all collected by Mr. Moseley, one of the

* Dr. Hooker's manuscript journal, which is full of very interesting and valuable matter, has been most kindly placed at my disposal.

Naturalists of the expedition) have been published in the Journal of the Linnæan Society for October 1874 and March 1876; and some information concerning the zoology has been given by Sir C. Wyville Thomson in "Good Words," 1874, November, pp. 743–751; and December, pp. 814–821.* From these accounts it may be gathered that *inter alia* the following numbers of species were procured by this expedition:—3 seals, 18 birds, 12 insects; 23 Phanerogams, 3 ferns and Lycopodia, 28 Musci, 12 Hepaticæ, 4 Fungi, 13 and 1 named form of Lichens, 13 marine and 1 freshwater species of Algæ, besides 23 Diatoms. The exact localities of the plants do not appear to have been recorded.

The German Transit of Venus and Surveying Expedition arrived at the island in H.I.M.S. "Gazelle," and established an astronomical observatory in the midst of the small graveyard at Betsy Cove, which is described by Sir C. Wyville Thomson, *op. cit.* p. 815. No complete account of the natural history specimens collected by this expedition has been published as yet. The cove is favourably situated for a naturalist's station. To the south and west are the Mount Crozier hills, affording much variety of altitude and exposure. On the east and south-east a broad tract of low marshy land is extended. Probably the situation is not well adapted for marine species that require very sheltered water; but others which haunt the open coast are found there in plenty. Dredging was conducted from the "Gazelle" in the neighbouring sea; and it may be anticipated that the results obtained by this means will prove to be the most important part of the collection. Considerable attention was bestowed upon the vertebrata, careful etchings being prepared of many species. Large series of specimens of birds in different stages of development, and the skins and skeletons of three species of seals, are other specialities of this collection. It is almost to be regretted that this district should have been worked over twice at the same season in consecutive years, when so much of the island is *terra incognita*.

The American Transit of Venus Expedition occupied a position on a slope near Molloy Point in Royal Sound, close to the commencement of the rocky beach which extends from thence to the Prince of Wales' Foreland. Speaking roughly, the country adjoining this position is similar to the district in the neighbourhood of Betsy Cove, differing chiefly in its distance from the open sea and its situation on the opposite side of the Mount Crozier hills. The expedition remained at Molloy Point from the beginning of September until the commencement of the second week in January Dr. Kidder, the surgeon at the astronomical station, studied the zoology and botany of the locality, and obtained many interesting observations of the habits of birds, together with a large collection of skins and eggs. The results obtained by him are recorded in Nos. 2 and 3 of the Bulletin of the United States

* This publication contains a detailed account of the past history of the island, and of its leading physical features, together with a summary of the proceedings of the expedition while there, and notices of interesting Natural History incidents.

National Museum, Washington, 1876. They include the following species:—2 mammals (one a seal); 20 birds; 3 or 4 fishes; upwards of 12 insects, besides 3 Arachnida; 7 Crustacea; 3 or more Annelida; 14 Mollusca; several Tunicata and Polyzoa, and 4 Echinodermata; 17 Phanerogams; 6 ferns and Lycopodia; 28 Musci; 18 Lichens, and 22 marine Algæ, total 69 or more species of animals, and 91 of plants.

The English Transit of Venus Expedition sailed from Simon's Bay on the 18th of September in H.M.S.S. "Volage" and "Supply." After a stormy passage the vessels joined company at the rendezvous at Three Island Harbour in Royal Sound on the 11th of October. Leaving this in the afternoon of the following day, they came to anchor in Observatory Bay in the morning of the 13th, whence they took their final departure on the 27th of February 1875. Collections were made at the following places while the expedition was at the island:—Cat Island, in the afternoon of October 11th; Observatory Bay and the adjoining district, to a distance of five miles inland, during most of October, November, December, and February; isthmus at the head of Carpenter's Cove, January 4th-9th; Thumb Peak, December 7th noon till 10th morning; vicinage of second astronomical station and hills near Swain's Haulover, October 27th, afternoon December 22nd-24th; shores and islands of Swain's Bay, January 15-30th.

The character of the localities just mentioned has been sufficiently described above in the general account of the island. The "isthmus at the head of Carpenter's Cove" is a high pass which divides the Mount Crozier hills from those to the westward of them. The summit and upper portion of this pass is referred to as a locality, in some of the botanical reports, as a "hill N.W. of Mount Crozier." A large bog at the foot of the northern declivity of the pass is described as a "bog near Vulcan Cove," and a waterfall descending the hills a little to the east of it is similarly specified as a "waterfall near Vulcan Cove"; but their real position is doubtful, and may be nearer the promontory enclosing an inlet adjacent to the entrance of Foundry Branch than to the cove in question.

The collections represent the fauna and flora of the inland and sheltered portions of the island from altitudes of 500 or 600 feet above the sea down to the depth of about 10 fathoms along the coast, and are as a rule rather deficient in duplicate specimens. The groups which are most poorly represented are those which at the time of collecting were known to have been the special subjects of investigation by other naturalists in the island, *e.g.* mammals, birds, and the Phanerogamic plants. In such groups, with the exception of certain particular species, it was considered that the interests of science would be best consulted by restricting the series of examples to minimum proportions, in order that more time might be devoted to the acquisition and preservation of specimens in less popular branches of natural history. The number of species that came under observation were as follows:—3 (besides 1 undetermined) native mammals and 2 or 3 introduced; 22 (with 1 or 2 doubtful) birds and eggs of 16 species; 4 (with 1 doubtful) species of fishes; 21 (and 5 undetermined) native insects, besides 1 introduced; 5 (and 2 undetermined) terrestrial Arachnida; upwards of 12 Crustacea; 8 Annelida; 25 Mollusca; 3 Tunicata, and 26 Polyzoa; 13 Echinodermata; 2 Actinozoa; 7 Hydroida; 8 sponges; 20 native (besides 1 introduced) species of flowering plants; 6 ferns and Lycopodia; 1 Characeæ; 43 Musci; 15 Hepaticæ; 5 Fungi; 51 or 52 species, besides 9 named forms of Lichens; 54 marine and 81 freshwater Algæ, in all about 170 species of animals, and 277 of plants.

Preliminary notices and descriptions of new species in the collections were published in the Annals and Magazine of Natural History for 1875 and 1876, the Entomologists Monthly Magazine for August 1875 and August 1876, the Journal of Botany for February 1876, and the Journal of the Linnean Society (Botany) for July 1876.

In such a country as Kerguelen Island no one without assistance could execute a fairly complete survey of the botany and zoology of an area lying within the radius of a day's walk from his head quarters in less than a month or six weeks; and then it would hold good only for that period of the year at which it was made. Much, therefore, has to be allowed for before the absence of a species from a collection can be attributed without question to its non-existence in the district in which that collection was formed; while on the other hand, should the physical conditions presented by the district be unfavourable to its occurrence, its absence does not indicate deficiency in point of completeness in the collection.

BOTANY.

Observations on the Botany of Kerguelen Island. *By J. D. Hooker, P.R.S.*

The history of the botany of Kerguelen Island (also called Kerguelen's Land, and Desolation Island), previous to the visit of the Rev Mr. Eaton, the last and most complete explorer of its flora, is a very brief one. It commences with the visit of Capt. Cook during his third voyage, in the narrative of which the vegetation of the island is thus described by Mr. Anderson, the surgeon of the "Resolution:" "Perhaps no " place hitherto discovered in either hemisphere, under the same parallel of latitude, " affords so scanty a field for the naturalist as this barren spot. The verdure which " appears, when at a little distance from the shore, would flatter one with the expec- " tation of meeting with some herbage; but in this we were much deceived. For " on landing we discovered that this lively colour was occasioned only by one small " plant, not much unlike some sorts of *Saxifrage*, which grows in large spreading " tufts, to a considerable way up the hills." Mr. Anderson proceeds then to give some particulars of this plant (*Azorella Selago*, Hk. f.), of the cabbage (*Pringlea antiscorbutica*, Br.), of two small plants found in boggy places, which were eaten as salad, one "almost like garden cress and very fiery" (probably *Ranunculus crassipes*, Hk. f.), the other very mild and "having not only male and female, but what bota- " nists call *androgynous* plants" (? *Callitriche*). He adds to these a coarse grass (*Poa Cookii*, Hk. f.), and a smaller sort which is rarer (probably *Deschampsia antarctica*, Hk.); a sort of goose-grass (? *Cotula plumosa*, Hk. f.), and another small plant much like it (this I do not recognise). "In short," he says, "the whole " catalogue of plants does not exceed 16 or 18, including some sorts of moss and a " beautiful Lichen" (*Neuropogon Taylori*, Hk. f.) "which grows higher upon the " rocks than the rest of the vegetable productions. Nor is there the least appear- " ance of a shrub in the whole country."

The date of Cook's visit was the summer of 1776, and the specimens obtained by Mr. Anderson were deposited in Sir Joseph Banks' Herbarium, which subsequently became the property of the nation, and is preserved in the British Museum. Not having been poisoned, all the Kerguelen Island plants were, when I examined them in 1843, much injured by insects, and many were entirely destroyed.

From 1776 till 1840, when the Antarctic Expedition under Capt. (afterwards Admiral Sir James) Ross, anchored in Christmas Harbour, Kerguelen Island is not known to have been visited by any ship of war, or by the Discovery or Surveying ships of any nation, though it had become the frequent resort of English and

American sealers. During the stay of the above-named expedition all the plants enumerated by Anderson as found by him in mid-summer were refound in mid-winter, together with many more, amounting to nearly 150, of which 18 were flowering plants; the other large classes being mosses and Hepaticæ 35, Lichens 25, and Algæ 51. These have all been described in the botany of the voyage (Flora Antarctica, Part II., 1847).

The next visit of naturalists to Kerguelen's Land was that of the "Challenger" Expedition in January and February 1874, when Mr. Moseley collected most diligently, both in Christmas Harbour and on the east coast 60 to 70 miles south-east of it. He found 23 flowering plants in all, including three European weeds, all annuals and doubtless imported by sealing parties (*Cerastium triviale*, *Poa pratensis* and *annua*), and three species not in the collections of the Antarctic Expedition (two *Ranunculi* and an *Uncinia*). He also procured flowering specimens of the two endemic genera *Pringlea* and *Lyallia*, and made large accessions to the cryptogamic flora, especially from the southern localities visited. Mr. Moseley had also the good fortune to land upon Marion Island, 1,650 miles to the west of Kerguelen Island; and on Yong Island (of the Heard group), about 120 miles to the south-east of it, neither of which had been previously visited by any naturalists, and in both of which he found some of the most peculiar of the Kerguelen plants.

Mr. Eaton arrived at Kerguelen Island with the Transit of Venus Expedition early in October 1874, and left towards the end of February 1875, during which time he collected diligently, chiefly at Royal Sound, Swains' Bay, and Observatory Bay. He obtained nearly all the flowering plants of previous explorers, and added very largely in the Cryptogams, especially to the Algæ.

Nearly contemporaneous with Mr. Eaton's visit was that of the American Transit Expedition, on which Dr. Kidder was the naturalist. He arrived in September 1874 and left in January of the following year, having explored some of the same localities as Mr. Eaton. His collections, amounting to about 90 species, are described in the bulletin of the U. S. National Museum, No. 3, issued in 1876 by the Government Printing Office of Washington. The flowering plants and ferns are revised by Prof. A. Gray; the mosses are described by Thos. P. James; the Lichens by Prof. E. Tuckerman, and the Algæ by Dr. W. G. Farlow. Except amongst the Lichens, there are very few novelties. Dr. Kidder adds a list of seven plants from the Crozets, all identical with Kerguelen Island species.*

The botanical results of the German Transit Expedition to Kerguelen Island are not yet published.

The three small archipelagos of Kerguelen Island (including the Heard Islands), Marion and Prince Edward's Islands, and the Crozets, are individually and collectively the most barren tracts on the Globe, whether in their own latitude or in

---

* He also mentions "a small vine with blue flowers growing amongst scoriæ," of which no specimens were collected. This is probably some endemic plant unknown to botanists.

any higher one, except such as lie within the Antarctic Circle itself; for no land even within the N. Polar area presents so impoverished a vegetation.

The chief interest attached to the flora of these archipelagos lies in the indication it affords of their being, in all probability, the remains of a much larger land area, which, though peopled with plants mainly from the southern extreme of S. America, 4,000 miles to the westward, possessed an endemic flora of its own, which included forest trees of considerable dimensions. Before, however, proceeding to discuss the relationships of their floras, I shall describe that of the largest and the only one that is at all well known.

As pointed out in the "Flora Antarctica," the prevalent features of the vegetation of this island as then known were Fuegian; one species of flowering plant alone, of those that are not peculiar to the island, being characteristic of any other flora, namely, the *Cotula plumosa*, which is found elsewhere only in the Auckland and Campbell Islands, south of New Zealand. More recent collections have confirmed and even strengthened this Fuegian affinity, for of the three additional flowering plants procured by subsequent explorers, one is Fuegian (*Ranunculus trullifolius*), another (*R. Moseleyi*) is closely allied to a Fuegian species, and the third one, *Uncinia compacta*, is a native of the mountains of New Zealand and Tasmania, and this is so nearly allied to a Fuegian species that it may prove to be a form of a plant common to all high southern latitudes.

Not only has a further knowledge of the Kerguelen Island flora strengthened its known affinities with the Fuegian, but recent discoveries in the latter flora have done so too; some of the Kerguelen's grasses especially proving to be more closely allied to Fuegian species than was suspected. The discovery of the flowers of the endemic Kerguelen genus *Lyallia* is another instance of this affinity. In the Flora Antarctica, judging from the fruit alone, the flowers being unknown, this remarkable plant was provisionally placed in *Portulaceæ*, its resemblance in habit and foliage to the andine genus *Pycnophyllum* being indicated. Complete specimens collected by Moseley prove its close relationship to the latter genus, in juxta-position with which it had indeed been placed in the Genera Plantarum, where both had been referred correctly to *Caryophylleæ*.

The elements of the Phænogamic flora of Kerguelen Island may be thus classified:—

1 Endemic genus, which has no near ally—*Pringlea antiscorbutica*.

1 Endemic genus allied to an Andean one—*Lyallia kerguelensis*.

6 Endemic species allied to American congeners—*Ranunculus crassipes* and *Moseleyi, Colobanthus kerguelensis, Acæna affinis, Poa Cookii, Festuca kerguelensis*.

5 species common to Fuegia but not found elsewhere: *Ranunculus trullifolius, Azorella Selago, Galium antarcticum, Festuca erecta, Deschampsia antarctica*.

6 species common to America, and also to New Zealand and the islands south of it. *Tillæa moschata, Montia fontana,* Callitriche obtusangula,* Limosella aquatica,* Juncus scheuzerioides, Agrostis Magellanica.* (Most of these are aquatic or marsh plants, and those marked with an asterisk are also European, and very widely dispersed.)

2 species found elsewhere but not in Fuegia, *Cotula plumosa,* common to Lord Auckland's group and Campbell's Island south of New Zealand, and *Uncinia compacta,* a native of the mountains of Tasmania and New Zealand.

This American affinity of the Kerguelen Island flora thus clearly established by its flowering plants is very strongly manifested by its Cryptogams, amongst which, however, the only evidence of migration from South Africa occurs. This is the case of *Polypodium vulgare,* a widely distributed fern in the north temperate zone, but known in the southern only from the Cape Colony, Marion, and Kerguelen Islands; what is further curious respecting it is, that the Kerguelen Island individuals are referable to a variety with pellucid veins, hitherto known only from the Sandwich Islands.

As to the local grouping of the Kerguelen Island plants, that of the Phænogams is not altogether in harmony with the Cryptogams, the former seeming to be by far the most ubiquitously dispersed of the two groups.

All the plants hitherto collected have been from two areas, one, Christmas Harbour, in the extreme north, extending about five miles either way; the other, considerably larger, occupies the south-east coast, and following it extends for about 40 miles. The distance between these areas is about 60 miles in a N.W. and S.E. direction. Of the Phænogamic plants, 19 were found in the northern area, nearly every one of which was also found in the south-eastern one, where but two additional species were collected; whereas of the 150 Cryptogams found in the northern area, a large proportion were not found in the south-eastern, where, however, nearly four times the number of species were obtained. Again, whilst but one fern was found in the north, four occur in the south-east. Of 35 *Musci* and *Hepaticæ* collected at Christmas Harbour by the Antarctic Expedition, hardly half were found at Swain's Bay, Betsy Cove, or Royal Sound, which localities yielded about 80 additional species. Nearly 50 marine Algæ were collected at Christmas Harbour, of which 18 did not occur in the south-eastern coasts, where upwards of 30 additional species were obtained. In the case of the *Lichens,* the discrepancy is still more marked, but this is possibly more apparent than real, and is to be attributed in part to the difficulty of defining the species and recognizing them from descriptions; and in part to the difficulties caused by the irreconcilable views of Lichenologists as to the limits of the species of this order.

Whatever other causes there may be for this anomalous distribution, one, no doubt, is the nature of the Christmas Harbour area. This is almost occupied by transverse valleys that run east and west completely across the north tip of the island, from sea to sea, are bounded by hills 1,200 feet high, and are perennially swept by terrific blasts from the westward. There are, hence, no shelter on land for the terrestrial flora, and no quiet bays for the proper development of a varied marine vegetation; facts which may very well account for the paucity of Cryptogams in Christmas Harbour, but not for the presence there of nearly all the flowering plants of the island. Turning again to the south-eastern area, its more sheltered valleys and land-locked harbours favour not only a greater development of Cryptogams, but also a far greater luxuriance of the Phænogams than obtains in Christmas Harbour; which last fact renders the absence of additional species of Phænogams to the south-eastward all the more remarkable.

The question remains, granting that the great majority of the Phænogams of Kerguelen Island are derived from South America, how was their transport effected? Though this question cannot be satisfactorily answered by a reference to the facilities for distant transport possessed by the fruiting organs of the Kerguelen Island plants, it is only proper to refer to these organs in some detail. Obviously, regarding the whole flora, the plants with the most minute seeds or spores and the water-plants are the most widely distributed. Under these categories come—1. The Fungi, of which all but 2 of the 8 species found are widely dispersed over the globe. 2. The marine Algæ, of which only 8 out of the 74 are peculiar to the island. 3. The fresh water Algæ, of which 28 out of 80 are regarded as endemic. 4. The aquatic and marsh Phænogams, 8 in all, of which 6 are widely dispersed.

Of the Phænogams, whether aquatic, marsh, or terrestial, none have appliances for wide dispersion except the hooked style of the *Ranunculus,* the reversed barbs of the *Acæna* (a most powerful aid), and the hooked organ attached to the fruit of *Uncinia,* also a very adequate aid. None of the others have any aid to dispersion, though they have small seeds or fruits.

Turning to the natural agents of dispersion, winds are no doubt the most powerful, and sufficient to account for the transport of the Cryptogamic spores; these, almost throughout the year, blow from Fuegia to Kerguelen Island, and in the opposite direction only for very short periods, but appear quite insufficient to transport seeds over 4,000 miles. Oceanic currents have, doubtless, brought the marine Algæ; but the transport of the seeds of the freshwater plants, of the grasses, and of the two plants with hooked and barbed appendages to the fruit, is not apparent in the case of a country that has no land birds but an endemic one (the Chionis), and of which the water birds come to land only or chiefly at the breeding season, and this after long periods of oceanic life in a most tempestuous ocean. Even supposing that the sea birds which habitually breed in Kerguelen Island did visit Fuegia between the periods of incubation, it is difficult to imagine that any seeds that had adhered to their beaks, feet, or bodies on leaving the latter country would not have been removed by the buffets of winds and waves over upwards of 4,000 miles of ocean.

The supposition that more land formerly existed along the parallels between

Fuegia and Kerguelen Island, possibly in the form of islands, remains as the forlorn hope of the botanical geographer. By such stepping stones the land birds, so numerous in the Falkland Islands (which lie in the direction of such hypothetical islands), and of which the vegetation is identical with that of colder South America, might, favoured by the prevalent westerly gales, have passed from thence to Kerguelen Island, having adhering to them fruits and seeds. The absence of such birds from the present Avi-fauna of Kerguelen Island offers no obstacle to such a speculation, as such immigrants would on arrival speedily be destroyed by the predatory gull and petrels of the island.

Various phenomena, of very different relative value and nature, but common to the three archipelagos, Kerguelen, the Crozets, and Marion, favour the supposition of these all having been peopled with land plants from South America by means of intermediate tracts of land that have now disappeared; in other words, that these islands constitute the wrecks of either an ancient continent or an archipelago which formerly extended further westwards, and that their present vegetation consists of the waifs and strays of a mainly Fuegian flora, together with a few survivals of an endemic one.

The extreme southern point of South America, from lat. 52–54° and long. 70° W. comprising Fuegia, is deflected to the eastward. Following its general direction, the Falkland Islands group is the first land met with (in long. 60° W.); its vegetation is comparatively rich and exclusively Fuegian; it has, no doubt, been brought mainly by the land and freshwater birds which abound there, and are identical with Fuegian ones. South Georgia is the next land met with to the eastward, in long. 35° W. and 54° S.; of its vegetation nothing is known except for the scanty observations recorded in Cook's voyage, which indicate its botanical identity with the Fuegia.

Of Bouvet Island, the assumed position of which is long. 5° E. and 54° S., nothing is known; it was searched for in vain by the Antarctic Expedition in 1843. Marion Island is 37° E. and 46° S., and the Crozets, in 48° E. and 47° S., are respectively about 1,650 and 1,200 miles west of Kerguelen Island, and there is no land intermediate between them. Now, from such specimens as have been obtained of the vegetation of the first of these islands by Mr. Moseley,* it appears to be almost identical with that of Kerguelen Island; that is, to be Fuegian with the addition of some of the peculiar Kerguelen Island types,† and the same remark applies to the Crozets,‡ facts from which Mr. Moseley has drawn identically the same conclusions as those to which I had arrived thirty-five years previously from a consideration of the Kerguelen Island flora alone. He says, speaking of Marion Island (Linn. Journ. XV.,

---

* Journ. Linn. Soc. XIV., 387, and XV., 484.

† Marion Island contains several Fuegian species not hitherto found in Kerguelen Island, namely, *Ranunculus biternatus*, *Hymenophyllum tunbridgense*, and probably a *Hierochloe* (the scented grass mentioned by Moseley), together with a Cape fern *Aspidium mohrioides* and an *Asplenium*.

§ See Kidder in Bull. U.S. Nat. Mus. No. 3, p. 31.

485), "the occurrence of *Pringlea* on the island, as also on the Crozets and Kerguelen
" Island, point to an ancient land connection between these islands, which the
" antiquity and extent of denudation of the lavas would appear to bear out. It is
" difficult to see how such seeds as those of *Pringlea* could have been transported
" from one island to another by birds; and the seeds seem to be remarkably
" perishable; besides the distinctness of the genus points to a former wide extension
" of land on which its progenitors became developed. The existence of fossil tree
" trunks in the Crozets and Kerguelen Island points to similar conditions.

In the Flora Antarctica, I say, p. 220, referring to the time required for the formation of the innumerable superimposed beds of volcanic rocks, as observed by me in Kerguelen's Land, and for the growths and destructions of successive forest vegetations that once clothed the island, and are now imbedded in strata at great depths, that this time is sufficient " for the destruction of a large body of land
" to the northward of it, of which St. Paul's and Amsterdam Island may be the
" only remains; or for the subsidence of a chain of mountains running east and
" west, of which Prince Edward's Island, Marion, and the Crozets, are the exposed
" peaks." And, at p. 240, when discussing the structural peculiarities of the *Pringlea*, I say, " However loth we may be to concede to any of our vegetable pro-
" ductions an antiquity greater than another, or to this island (Kerguelen) a posi-
" tion to other lands wholly different from that it now presents, the most casual
" inspection of the land where this plant now grows will force one of the two
" following conclusions upon the mind, either that it was created after the extinc-
" tion of the now buried and for ever lost vegetation, or that it spread over the
" island from another and neighbouring region, where it was undisturbed during
" the devastation of this, but of whose existence no indication remains."*

It remains to indicate the faint traces of relationship which the Kerguelen Island vegetation presents with those of a few other spots of land in a lower latitude, and that might be supposed to share some of its peculiarities. Of these the nearest are Amsterdam and St. Paul's Islands, the names of which are often transposed in our best maps (even in the Admiralty South Polar Chart of 1839). They lie about 800

---

* These ideas, suggesting the hypothesis that the existing distribution of plants is dependent on former geographical relations of land and sea, suggested themselves to me during my visit to Kerguelen Island in 1840. The first attempt to apply similar views in extenso to the conditions of a botanically well-known country was in the late Professor Edward Forbes' paper " on the distribution of endemic plants, more espe-
" cially those of the British Islands, considered with regard to geological changes." " Brit. Assoc. Reports
" for 1845." It had, however, been previously enunciated by Lyell, who thus accounted for the identity of the Sicilian animals and plants with those of the surrounding Mediterranean shores.

He supposes these to have " migrated from pre-existing lands, just as the plants and animals of the
" Phlægrean fields have colonised Monte Nuovo since that mountain was thrown up in the 16th century," and further on he says, " we are brought therefore to admit the curious result, that the flora and fauna of
" the Val di Noto, and some other mountain regions of Sicily, are of higher antiquity than the country
" itself, having not only flourished before the lands were raised from the deep, but even before they were
" deposited beneath the waters." Principles of Geology, Ed. v. iii., p. 444, &c.

miles to the N.E. of Kerguelen Island, in 78° E. long.; the northernmost, Amsterdam Island, is nearly on the 38th and St. Paul's on the 39th parallel of latitude, so they both are very little south of the latitude of the Cape of Good Hope.

I have brought together, in a paper published in the Journal of the Linnæan Society (vol. xiv. p. 474), all the little that was then known of the flora of these islands, which, like Kerguelen, are volcanic.

Their scanty vegetation is on the whole more temperate than antarctic, and approximates to that of S. Africa in containing such genera as *Phylica*, *Spartina*, and *Danthonia*. Their fern flora is very interesting; one fern only is common to Kerguelen (*Lomaria alpina*), one (*Nephrodium antarcticum*) is peculiar, though allied to a Mauritian species, and two others (*Blechnum australe* and *Asplenium furcatum*) are natives of the Cape and other countries; but what is most singular is, that neither the *Polypodium vulgare* nor *Aspidium mohrioides* have been found in either island, though the former is common to the Cape, Marion Island, and Kerguelen's Land, and the latter to the two first of these localities.

Tristan d'Acunha, in 12° W. long. and 37° S. lat., and the adjacent islets called Nightingale and Inaccessible, all nearly in the latitude of Amsterdam Island and the Cape of Good Hope, are the only other islands whose vegetation demands a passing notice here.* Their flora is essentially Fuegian, with an admixture of Cape genera, but with none of those characteristics of Kerguelen Island. Of Cape types, it contains a *Pelargonium* and an abundance of both the *Phylica* and *Spartina* of Amsterdam Island, together with species of *Oxalis* and *Hydrocotyle*. The Fuegian and Falkland Island plants of Tristan d'Acunha and its islets, which have not hitherto been found in the islands south and east of them, are however more numerous than are the Cape genera even, and include *Cardamine hirsuta*, *Nertera depressa*, *Empetrum nigrum*, var. *rubrum*, *Lagenophora Commersoniana*, and *Apium australe*; and it contains besides the strictly American genus *Chevreulia*. Two land birds, both peculiar, are common in the Tristan group, and they possess a water hen, which has a representative in Africa and S. America. I am not aware whether land birds are found in Amsterdam Island; if so, they may help to account for the wonderful fact of the Tristan d'Acunha *Phylica* and *Spartina* being found also in it, though separated by 3,000 miles of ocean.

In conclusion, I have to state that no trace of the mountain flora of S. Africa has been found in any of the southern groups of islands.

* For the latest account of this group see Moseley in Journ. Linn. Soc. XIV., 377.

ENUMERATION OF THE PLANTS HITHERTO COLLECTED IN KERGUELEN ISLAND BY THE "ANTARCTIC," "CHALLENGER," AND "BRITISH TRANSIT OF VENUS" EXPEDITIONS.

I.—*Flowering Plants, Ferns, Lycopodiaceæ, and Characeæ.*
By J. D. HOOKER, P.R.S.

1. **Ranunculus crassipes,** *Hook. f. Fl. Antarct.* 224, t. 81.

Christmas Harbour, Observatory and Swain's Bay, Royal Sound (a form with petioles 5–7 inches long).

I have nothing to add to what I have said of this species in the Antarctic Flora, beyond that I can hardly doubt its being a derivative form of the Fuegian *R. biternatus,* Sm., with which it agrees in habit and its thick-walled beaked carpels, but differs chiefly in its robustness and simple leaves. *R. biternatus* has been found by Moseley in Marion Island, where it presents every character of the American plant.

2. **Ranunculus trullifolius,** *Hook. f. Fl. Antarct.* 226, t. 82 A.

In streamlets and lakes, Royal Sound, Swain's Bay, Betsy Cove; *Moseley, Eaton, Kidder.* (Fuegia and the Falklands).

Glaberrimus, caulibus prostratis radicantibus. *Folia* longe crasse petiolata, obovato-oblonga trulliformia v. fere orbicularia, apice obtuse 3–5-dentata v. lobata, carnosula, nervis obscuris; auriculis petiolaribus membranaco-dilatatis. *Flores* ad nodos solitarii, brevissime pedicellati. *Sepala* 3, orbicularia, concava, membranacea. *Petala* 3, sepalis æquilonga, obovato-oblonga v. spathulata, 3-nervia, nervo medio medium versus fossa nectarifera instructo. *Stamina* pauca. *Carpella* numerosa; matura cuneiformia, compressa, dorso incrassata, stylo gracili subulato.

I described this species in the Flora Antarctica from very imperfect specimens gathered by myself in the Falklands in mid-winter, along with the very similar *R. hydrophilus,* Gaud., and from a careful examination of the remains of the only flower found, which resembled in petals, sepals, and stamens those of its neighbour, I supposed it to be closely allied to it. Good specimens gathered by Cunningham in the Straits of Magalhaens, and by Eaton in Kerguelen, prove that it belongs to another section of the genus, differing from *R. hydrophilus* in the usually trimerous perianth and the long style of the flattened ripe carpels. *R. trullifolius* is, in fact, referable to St. Hilaire's genus *Casalia* (now reduced to *Ranunculus*), and its nearest ally is *R. bonariensis,* Poiret (*R. Kunthii Trian.* and *Planch.*), which differs by its ovate crenate leaves, long-peduncled flowers, and absence of style in the ripe carpels. *R. hydrophilus,* again, is probably a form of *R. adscendens,* St. Hil. (*R. humilis,*

Collie, in Hook. Bot. Beech. Voy. p. 4, t. ii.), which has similar minute subglobose ripe carpels without a style.

R. monanthos, Philippi of Chili, and *R. hemignostus* Steud. of Peru, are probably forms of *R. trullifolius,* which, as our figure shows, is a very variable plant in foliage and structure.

The *Ranunculus,* sp. 3, not in flower, of Kidder (Bull. U. S. Nat. Mus. 3, 21), of which Gray says it can hardly be a form of *trullifolius,* no doubt is this, if, as I apprehend, the term *caudate* as applied to the leaves is a misprint for *cordate.*

PLATE I., Figs. 1–5.—Plants in different states; of *natural size ;* 6, 7, reduced leaves and stipules; 8, sepal; 9, petal; 10 and 11, stamen; 12, immature, and 13, mature carpels:—all *enlarged.*

3. **Ranunculus Moseleyi,** *Hook. f.*;

pusillus, glaberrimus, acaulis, foliis radicalibus, petiolo in laminam obovatam v. oblongam integerrimam dilatato, floribus solitariis pedunculatis minutis 3–4-meris, petalis lineari-obovatis obtusis eglandulosis, staminibus 4–7, carpellis 10–12 maturis oblique subglobosis in stylum brevem gracilem abrupte attenuatis.—Ranunculus an nov. sp.; *Oliver, in Journ. Linn. Soc.* XIV., 389.

In the lake at Christmas Harbour, *Moseley.*

A very diminutive species, resembling in size and habit *R. limoselloides,* Muell, of Australia, but differing in the carpels, &c. In the latter respect it more nearly approaches *R. crassipes,* from which it differs in all other respects. Its allies are, no doubt, to be found amongst the S. American water-loving species.

PLATE II., Fig. 1—1 and 2, plants of *natural size ;* 3, leaf; 4, flower; 5, sepal; 6, petal; 7, stamen; 8, immature; and 9, mature carpel:—all *enlarged.*

4. **Pringlea antiscorbutica,** *Br. MSS.*; *Fl. Antarct.* 238, t. 90, 91;

Kidder in Bull. U. S. Nat. Mus., No. 321; *Oliver in Journ. Linn. Soc.* XIV., 389; *Dyer in Proc. Linn. Soc.* 1874, xxxiv.; *Hook. f.* l. c.

Throughout the island.—(Marion, Crozets, and Heard Islands).

Sepala lineari-oblonga, obtusa, membranacea, pilosa. *Petala* 0 in exemplaribus perplurimis a nobis scrutatis, in paucis 1–4, unguiculata, apice roseo-tincta, inconspicua, caduca. *Stamina* 6, subæqualia, filamentis elongatis complanatis, 4 longioribus per paria sepalis anticis posticisque opposita; antheræ magnæ, lineari-oblongæ, virescentes; pollen sphericum. *Disci* glandulæ 0 v. valde inconspicuæ. *Ovarium* oblongum, hirsutum, 2-loculare, carpellis lateralibus; stylus brevis, validus, glaber, stigmate capitato obscure 2-lobo dense villoso.

In the Proceedings of the Linnæan Society 1874, p. xxxiv, I have indicated the evidence of *Pringlea* being a wind-fertilized member of a natural order most or all the species of which are insect-fertilized. These indications are the usual absence of petals and disk-glands, the exserted anthers and long-tufted papillæ of the stigma, to which is to be added the absence of winged insects in Kerguelen Island. In reference to the last statement, it is a curious fact that wingless flies abound in the island, and on this very plant. Moseley, Journ. Linn. Soc. xv., 54, in his notes on Kerguelen botany, mentions an apterous fly as big as a blow-fly, nestling at the base of the leaves of *Pringlea* and laying its eggs in the fluid which is caught there; every cabbage yielding ten or a dozen specimens. He adds that he did not observe whether it climbs to the inflorescence in sunny weather.

Mr. A. W. Bennett, Proc. Linn. Soc. 1874, xxxix., has described the pollen of *Pringlea* as differing from that of nearly all other Crucifers in being much smaller and perfectly spherical, instead of ellipsoid with three furrows. This he considers to be a striking confirmation of my suggestion that the plant is wind-fertilized, and which is further confirmed by the total absence of hairs on the style.

Moseley found one plant with 28 flower-stalks, three of the one season growth, the others appearing to belong to eight preceding seasons.

It is a remarkable fact that all attempts to grow this plant in England, Scotland, and Ireland have failed; the young plants, after attaining a height of a few inches and a good crown of leaves, have invariably succumbed to the combined effects of summer's heat, and the attacks of the common parasite fungus, *Cystopus candidus*, which infests the *Capsella Bursa-pastoris*. Some few, out of many hundreds, sown at different seasons and under very varied conditions, survived one winter, but perished in the following summer.

PLATE II., Fig. 3.—1, 2, 3, apetalous flowers; 4, monopetalous, and 5, tripetalous flowers; 6, petal; 7, ovary; 8, the same laid open; 9, ovule:—all *enlarged*.

5. **Colobanthus kerguelensis**, *Hook. f. Fl. Antarct.* 249, t. 92.

Christmas Harbour, Swain's Bay, &c. (Heard Island, *Moseley*.)

(*Stellaria media* L.)

Introduced by sealers.

(*Cerastium triviale*, Link.)

Introduced by sealers.

6. **Lyallia kerguelensis**, *Hook. f. Fl. Antarct.* 548, t. 122; *Kidder in Bull. U.S. Nat. Mus.*, No. 3., 22. *Oliver in Journ. Linn. Soc.* XIV. 390. *Dyer in Proc. Linn. Soc.* 1874, xxxiv.

Christmas Harbour and Royal Sound.

The flowers have been described from Kidder's specimens by Asa Gray, and from Moseley's by Oliver and Dyer, the descriptions agreeing well. The stamens, which appear to be almost constantly three and hypogynous, are stated by Oliver to be variable in position. Kidder retains it in *Portulaceæ*, but Bentham and I had long previously placed it in *Caryophylleæ* in the Genera Plantarum and next to *Pycnophyllum*, a position which the discovery of the flowers confirms. It has many of the characters of *Colobanthus*, especially the androecium.

PLATE II., Fig. 2.—1, plant, of *natural size*; 2, leaves; 3, flower and bract; 4, flower laid open; 5, stamen:—all *enlarged*.

7. Montia fontana, *L.*

Common in wet places. (Marion Island, *Moseley*, and widely distributed in the N. and S. temperate regions).

8. Acæna affinis, *Hook. f. Fl. Antarct.* 268, t. 96 B.

Common throughout the island. (Marion and the Crozet Islands).

Called Kerguelen's tea, and used as a febrifuge by whalers (Kidder).

Unlike the *Pringlea* and *Cotula*, this plant has grown and flowered at Kew from roots sent by Moseley.

9. Callitriche verna, *L.*; Subsp. *obtusangula.* C. obtusangula, *Le Gall.*; *Hegelm. Monog. Gatt. Callit.* 54. C. antarctica, *Engelm. ex Hegelm.* l. c.; *Kidder in Bull. U.S. Nat. Mus.*, No. 3, 23. C. verna, *Hook. f. Fl. Antarct.* 272.

Common in wet places. (Marion and Heard Islands, *Moseley*, and widely distributed in the N. and S. temperate regions).

From a drawing of the ripe fruit which I made when in Kerguelen in 1840, I have no hesitation in referring this to Subspecies *obtusangula*, as Hegelmeyer did from his examination of my dried specimens. The fruit lobes are nearly semi-circular, and each pair is united by about two thirds of their faces. The free portions are obtusely trigonous at the back. Two forms are common in Kerguelen, as elsewhere in the south temperate zone, one aquatic with long stem and proportionally large spathulate leaves, the other smaller, terrestrial, suberect, with obovate or oblong leaves; this flowers the most abundantly.

10. Tillæa moschata, *D.C.* Bulliarda moschata, *D'Urv.*

Abundant in moist places near the sea. (Marion Island, *Moseley*, Crozets, *Kidder*; widely spread in high southern latitudes).

11. Azorella Selago, *Hook. f. Fl. Antarct.* 284, t. 99.

Very abundant throughout the island. (Marion and Heard Islands, Moseley; Crozets, Kidder; Fuegia; Mac Quarrie Island.)

Kidder remarks that the flowers are greenish yellow, not pale pink as I found them to be in winter. Also, that the leaves have not the bristles on the faces of the lobes as figured in the Flora Antarctica. I find them on specimens from all localities.

Moseley observes, in reference to this plant at Marion Island, that the mounds it forms evidently retain and store up a considerable amount of sun's heat, and that this fact probably explains its peculiar mode and form of growth, and that of many otherwise widely different Antarctic plants. He found that a thermometer plunged into the heart of a hummock rose to 50°, when the temperature of the air was 45°.

12. Galium antarcticum, *Hook f. Fl. Antarct.* 303 bis.

Common, but not found at Christmas Harbour. (Crozets, *Kidder*; Fuegia and Falkland Islands.)

Kidder remarks that the flowers are distinctly pedicelled, and as often 4- as 3-merous, and even 5-merous ones occur. Eaton's specimens confirm this.

13. **Cotula** (LEPTINELLA) **plumosa,** *Hook. f. Fl. Antarct.* 26 and 308, t. 20.

On cliffs, especially near the sea, often forming immense luxuriant blue-green patches where the soil is enriched by the dung of birds and seals. (Crozets, *Kidder*; Lord Auckland, Campbell's, and Mac Quarrie Islands.)

Reputed by the whalers to be a prompt and effectual emetic. Through a typographical omission of the word *not* at p. 308 of the Antarctic Flora, this plant is stated to be found on the continent of America. The genus *Leptinella* is reduced to a *Cotula* in the Genera Plantarum. This plant, like the *Pringlea*, proved so impatient of heat in this country, that of innumerable seedlings raised at Kew to several inches high all perished.

14. **Limosella aquatica,** *L.*

Common in the freshwater lagoon at Christmas Harbour. (Fuegia and all temperate regions.)

A very small form, with the leaf-blade hardly broader than the petiole. *Stamens* included. *Ovary* globose; style rather long.

15. **Juncus scheuzerioides,** *Gaud.; Hook. f. Flor. Antarct.*, 79, 358.

Common in spongy places. (Fuegia, the Falkland, Lord Auckland, and Campbell's Islands.)

16. **Uncinia compacta,** *Br.; Boott in Hook. f. Fl. Tasman*, ii. 103, t. 153 B.

Royal Sound and Observatory Bay, *Moseley, Eaton*. (Mountains of Tasmania and New Zealand.)

17. **Deschampsia antarctica,** *Hook. Ic. Pl. t.* 150 (Aira); *Hook. f. Fl. Antarct.* 377, t. 133.

Common and ascending to considerable altitudes. (Fuegia, Falkland Islands, South Shetlands.)

A true *Deschampsia*, as that genus is now defined, by its 4-toothed flowering glume and free caryopsis, *Munro*.

18. **Agrostis magellanica,** *Lamk.; Hook. f. Fl. Antarct.* 373. A. antarctica, *ibid.* 373, t. 132. A multicaulis, *ibid.* 95.

Common throughout the island. (Marion and Heard Islands, *Moseley*; Chili, Fuegia, Falkland, and Campbell's Islands.)

Since the publication of this plant as *A. antarctica*, I have examined a specimen of Lamarck's *A. magellanica* named by Nees in Arnott's Herbarium, and find it to be identical. Further, Munro informs me that it is fairly described by Trinius in his "Agrostideæ," and by Kunth in his supplemental volume (p. 175) from a Lamarckian specimen; he adds that the Kerguelen specimens agree with these descriptions, except in the flowering glume being larger and much longer than the ovary. This glume is sometimes obtuse or rounded, at others deeply divided. The beard on the callus, which is very indistinct on the Kerguelen's plant, is conspicuous on some Fuegian ones.

19. **Poa Cookii,** *Hook. f.; Fl. Antarct.* 382, t. 139 (Festuca).

Forma 1.; foliis culmum superantibus, panicula elongata interrupta.

Forma 2.; foliis culmum superantibus v. æquantibus acuminatis pungentibus, panicula densa sub-cylindracea.

Forma 3.; foliis culmum æquantibus subacutis v. obtusis, panicula minore laxiore, spiculis paucifloris coloratis.

Abundant and ascending to a considerable height:—Forma 1. Christmas Harbour; Forma 3. Royal Sound, on a high hill, *Eaton*. (Marion and Heard Islands, *Moseley*).

This fine grass should, unquestionably, be referred to *Poa* (as now defined by the compressed flowering glume, &c.), along with its near congener *Dactylis cæspitosa** of Fuegia and the Falklands, from which it differs, amongst other characters, in never forming tussocks. It is scarcely specifically distinct from *P. foliosa*, Hook. f. Handbook of N. Z. Flora 338 (Festuca foliosa, *Fl. Antarct.* i. 99, t. 55; *Fl. Nov. Zeald.* i. 308); and this, again, from the Fuegian *Poa lanigera*, Nees (Festuca fuegiana, *Fl. Antarct.* 380). The flowering glumes are often obscurely, or not at all toothed. The spikelets are 3–5-flowered and $\frac{1}{4}$–$\frac{1}{3}$ in. long (not eight lines as misprinted for three lines in the Antarctic Flora). A. Gray remarks of Kidder's specimens that they seem to be male only.

Poa pratensis, L.

Introduced by sealers.

Poa annua, L.

Introduced by sealers.

20. **Festuca erecta,** *D'Urv.*

Common and ascending to a considerable elevation. (Fuegia and the Falkland Islands.)

Often forming tussocks; panicles green or purplish.

21. **Festuca kerguelensis,** *Hook. f.* Triodia kerguelensis, *Fl. Antarct.* 379, t. 138 (*Poa*).

Common and ascending to 2,000 feet.

Spikelets sometimes 1-flowered. A very variable grass in stature, evidently allied to *F. erecta*, and more nearly still to *F. scoparia* (Fl. Antarct. 98; Fl. Nov. Zeald. i. 308), of which possibly it is a dwarf form, as suggested in the Handbook of the New Zealand Flora, p. 341. The naked base of the flowering glume, however, will always distinguish all the specimens I have examined.

Filices.

1. **Cystopteris fragilis,** *Bernh.*

Crevices of rocks near the hill-tops, Royal Sound, *Kidder*, *Eaton*. (Fuegia, Falklands, and N. and S. temperate regions generally.)

---

* The name *Poa cæspitosa* being occupied by Forster, though it is doubtful to what species it applies, I propose that of *flabellata* for the Tussock grass, which is the *Festuca flabellata*, Lamk.

2. **Lomaria alpina,** *Spreng.*

Common, often forming large beds, but not found at Christmas Harbour. (Marion Island, *Moseley;* Crozets, *Kidder;* all the colder S. temperate regions.)

3. **Polypodium** (GRAMMITIS) **australe,** *Mett.*

Crevices of rocks, Observatory Bay, *Kidder, Eaton.* (Marion Island, Moseley; Fuegia, and all the colder S. temperate regions.)

4. **Polypodium vulgare,** L. *var.* Eatoni, *Baker,* venis pellucidis.

Crevices of rocks by running streams, Observatory Bay, *Kidder, Eaton.* (Marion Island, Moseley; S. Africa; Sandwich Islands, and N. temperate hemisphere.)

This pellucid-nerved variety only occurs elsewhere in the Sandwich Islands.

Lycopodiaceæ.

5. **Lycopodum clavatum,** *L., var.* magellanicum; *Hook. f. Fl. Antarct.,* 113. L. magellanicum, *Swartz.*

Not uncommon throughout the island, but not met with at Christmas Harbour. (Var. *magellanicum,* Marion Island, Moseley; Fuegia, and all the colder S. temperate regions. The typical *L. clavatum* inhabits all northern cold damp climates.

6. **Lycopodium Selago,** *L.* var. Saururus, *Hook. f. Fl. Antarct.* 394. L. Saururus, *Lamk.*

Not uncommon throughout the island. (Var. *Saururus,* Marion Island, *Moseley;* Tristan d'Acunha, St. Helena, Bourbon, Peru. The typical form inhabits all damp cold climates.)

Characeæ.

7. **Nitella antarctica,** *Braun.* N. Hookeri, *Reinsch in Journ. Linn. Soc.* xv. 219. Chara flexilis, *Linn;* Fl. Antarct. 395.

In the Lake at Christmas Harbour; and in that next but one to the Observatory, in Observatory Bay, *Eaton.*

II.—*Musci*.

By William Mitten, A.L.S.

The first investigation of the mosses of Kerguelen was made by Dr. J. D. Hooker during the voyage of the "Erebus" and "Terror" in the winter of 1840.

From the collections made by him there were described 31 species and varieties, which were arranged as 25 species in 11 genera. Of the whole number six species were considered to be new and undescribed, and the remainder to have been found in other regions. The most remarkable species contained in this collection are the *Schistidium marginatum, Weissia stricta,* and *W. tortifolia.*

During the visit of the Challenger, there were collected by Mr. Moseley, in the summer of 1874, 28 species, of which number 20 were additional to those discovered by Dr. Hooker. Sufficient materials were obtained to establish the presence of eight more genera, all previously known to occur in austral lands, four of the species appearing to be new. Twenty-eight species were obtained by Dr. Kidder of the American Transit Expedition, of which number 12 were additions to the Flora, two being described as new. Following the above come the collections made by the Rev. A. E. Eaton, pending the observations of the transit of Venus, which include 38 species, of which 17 were additional to the Flora of Kerguelen Island, three being undescribed, and by this collection three genera were also added; thus raising the whole number of the species of mosses inhabiting the Island to 74. This, considering how much has been added by each collector to those which were previously known, is probably a low estimate of the entire moss flora.

No genera peculiar to Kerguelen are observable in the collections, unless a species here referred to *Blindia* and the *Schistidium marginatum* (here placed in *Streptopogon*) should be so considered. The remaining genera are universal in boreal as well as austral regions, with the exception of the three species of *Dicranum*, all which belong to extra-European sections of that genus. Twenty-three of the Kerguelen mosses are considered identical with species found in the north of Europe and America, of these *Bryum alpinum* and *Brachythecium salebrosum* had not before been identified in the southern hemisphere.

A few distinct and well-marked species have been gathered in Kerguelen Island which are also found at great elevations on the Andes of Quito and of New Grenada. Of these *Mielechhoferia campylocarpa* and *Psilopilum trichodon* are conspicuous instances; they probably inhabit the whole Andine chain. *Bartramia appressa, Brachythecium paradoxum,* and *Tortula Princeps* are found also in New Zealand and Tasmania; but with the exception of *Dicranum kerguelense* there is no species which points to any connexion with the mosses of South Africa.

1. **Ditrichium australe**, *Mitt.* l. c. (Cynontodium). Lophiodon strictus, *Hook. f. et Wils. Fl. Antarct.* 130, t. LIX., Fig. 1.

In dense fulvous tufts, with old capsules, *Moseley*. (Lord Auckland's and Campbell's Islands.)

In all the specimens referred to this species the dry young foliage is fulvous, the older brown or black; the terminal leaves are frequently longitudinally twisted, otherwise their direction is the same as when wet; the lower portion of the leaf is in outline of an elliptic oblong figure, from which the nerve is continued in a straight line, and is rather suddenly carried out so as to appear without a margin of leaf; a transverse section shows it to be concave above and convex beneath; the apex is abrupt, rounded, and nearly flat, so as to appear as if dilated, and, as stated in the Flora Antarctica, the species is distinguished from most of its allies by this particular. The substance of the base of the leaf is composed of elongated cells which, although shorter towards the top of the dilated portion, are not dense, so that the entire expansion is of a pellucid fulvous colour, the nerve being everywhere smooth, with a few small teeth at its apex.*

2. **Ditrichium Hookeri**, *C. Muller Syn.* I., 450 (Leptotrichum).

Royal Sound, with old capsules and young setæ, *Eaton*.

3. **Ditrichium conicum**, *Mont. in Ann. Sc. Nat. Ser.* 3, iv. 100. (Aschistodon.)

Near Vulcan Cave, barren, *Eaton*.

The imbrication of the leaves at the apices of the stems, when dry, so as to form an erect or curved point, renders this species not difficult to recognise in a barren state.

1. **Asiothecium vaginatum**, *Hook. Musc. Exot.* t. 141 (Dicranum).

---

* In the Journal of the Linnean Society, Sept. 1859, there was confused with the *Leptotrichum australe*, therein mentioned, the following apparently distinct species,—*D. punctulatum*, Mitt. ; dioicum ? dense cæspitosum, dichotome ramosum, folia inter se remotiuscula a basi erecta amplexante oblonga cellulis inferioribus elongatis superioribus abbreviatis rotundatis obscuriusculis veluti punctatis, subito in subulam patentem inferne canaliculatam apice angustam planiusculam denticulatam minutissime scabridam sublævem cellulis punctulatis areolatam producta, perichætii alia basi latiora et longiora parte subulato patentiora, theca in pedunculo breviusculo rubro parva ovali-cylindracea erecta leptoderma fulvo-fusca. Flos masculus in ramis terminalis, ovatus, e basibus foliorum dilatatis apice retusis vaginantibus involucratus. Distichium capillaceum, Fl. N. Zealand, II., 73.

Hab.—New Zealand, *Dr. Lyall*. Great Barrier Island, *Hutton and Kirk*. Fagus Forests, *Hopkins, Dr. Haast*.

In size colour and general appearance very similar to *D. australe*, having also the same, but narrower, flattened apices to its leaves; in the recurvation of the subulate portion from the top of the erect base it resembles *D. capillaceum*, and for this species Dr. Lyall's barren specimens were mistaken, although the leaves are not distichous, but so disposed that each fifth leaf occupies the same vertical position on the stem as the first counted from; the outline of the dilated base is not oval-elliptic as in *D. australe*, but oblong obtuse. The fruit in an old state is present on Dr. Haast's specimens; accompanying these fertile stems were many conspicuous male flowers, which do not appear to arise from the lower parts of fertile stems, but seem to be really distinct male plants.

Swain's Bay, *Eaton.*

Small barren stems, but not different from specimens from the Bogotian and Quitenian Andes.

1. **Blindia gracillima**, Mitt. Dioica, laxe cæspitosa. Caulis elongatus, gracillimus, inferne nudus, superne foliis remotiusculis laxe obtectus. Folia anguste lanceoloto-subulata, pagina folii e cellulis angustis elongatis parietibus pellucidis usque ad $\frac{2}{3}$ nervi apice vix denticulati longitudinis anguste continuata; cellulis alaribus in auriculam parvam dispositis rubris; folia perichætialia erecta, basi obovata, convoluta, sensim subulato-attenuata, nervo longius excurrente. Theca in seta brevi flexuosa arcuata pendula, subresupinata, globosa; operculo oblique rostrato; peristomii dentibus rubris latis teneris integris vel rarius pertusis intus lævibus extus parietibus transversalibus prominentibus appendiculatis; annulo nullo; calyptra parva, viridis, nigrescens. B. curviseta, *Mitt. in Linn. Soc. Journ.* XV., 193.

Royal Sound, in lakes, with young and nearly ripe fruit, *Eaton.*

Stems 2–4 inches long, forming loose tufts, the upper portions red, the lower black, denuded of leaves, and forming a loose entangled mass. Leaves at the apices of the stems fulvous and shining, the lower all blackened; in their direction the upper leaves are but little changed when wet or dry; they are $1\frac{1}{2}$–2 lines long; the areolation consists of elongate cells separated by pellucid walls; at the angles of the base of the leaf the alary cells are distinct and red. The nerve becomes indistinguishable at four-fifths of the whole length of the leaf, and is thence continued, and ends without forming a pungent point; leaves of the perichætium longer, and their dilated bases about twice the width of the cauline leaves. Seta 2–$2\frac{1}{2}$ lines, straight in its lower half, thence to the capsule twisted and variously curved. Capsule erect when dry, when wet with a swan's-neck-like curve, and so bent as to become horizontal; when mature spherical without any neck where it is affixed to the seta; colour reddish brown; substance thin but firm. Operculum always obliquely beaked, at length of the same colour as the capsule. Peristome perfectly formed; teeth red, broad at the base, thence with an even outline narrowed to their points, with the exception of a rare perforation there is no trace of their being composed of a double row of cells; at the base of the teeth the transverse divisions are close together, but above this they are much wider, and on turning the tooth on edge it is seen that each dissepiment of the articulations is prominent on the outer side, but not on the inner. Spores small, round. Calyptra coriaceous, brownish-green, deeply cleft, with a spreading base.

Tab. III., Fig. 1, plant of natural size; 2, cauline leaf; 3, perichætium with capsule; 4, portion of peristome; all *magnified.*

2. **Blindia microcarpa**, *Mitt. in Journ. Linn. Soc.*, XV., 65. Monoica, pulvinatim cæspitosa. Caulis dichotomus, fastigiatim ramosus. Folia patentia, stricta, plus minus falcata curvatave, dimidio inferiore lanceolato superiore carinato

anguste attenuato, integerrima, nervo angusto percursa, cellulis elongatis alaribus in auriculam parvam fuscam dispositis areolata; perichætialia brevia, parva, ovata, convoluta, in acumen subulatum producta. Theca in pedunculo gracili foliis caulinis dimidio breviore erecta, parva, ovalis; operculo subulato obliquo demum ore dilatato cyathiformi fusca; peristomii dentibus teneris; calyptra parva, dimidiata. Flos masculus foliis propriis perichætialibus similibus inclusus.

Kerguelen Island, *Moseley*.

This is the species mentioned in the Flora Antarctica, p. 128, under *Weissia contecta*, as being present in the Hookerian Herbarium, its habitat unknown.

In compact, but not coherent tufts. Stems fastigiately branched, about an inch high. Foliage shining, but little altered in direction wet or dry. The minute capsule is scarcely conspicuous amongst the leaves. Leaves at the tops of the stems yellowish green, below brown, erect or slightly falcate, about $2\frac{1}{2}$ lines long, composed of elongate cells with pellucid walls; nerve pale brown and with the pagina gradually attenuated into a very narrow flat entire point; alary cells at the angles of the base distinct, brown, forming sub-quadrate masses. Leaves of the perichætium $\frac{1}{4}$–$\frac{1}{2}$ as long as those of the stem, and quite concealed amongst them. Seta about 1 line long, straight, pale brown. Capsule as it reaches maturity appearing to pass from oval to nearly globular; after the fall of the operculum by the dilatation of its mouth it becomes cyathiform, with no distinct neck. Operculum with a very oblique subulate beak which is longer than the capsule. Peristome-teeth very thin, broad at base, narrowed upwards into entire points; transverse articulations remote. Calyptra small, coriaceous, brownish, scarcely reaching the base of the operculum. Male inflorescence in a small bud below the base of the perichætium.

Tab. III., Fig. ii.: 1, plant of natural size; 2, cauline leaf; 3, perichætium with comal leaf, capsule, and male flower; 4, old capsule; 5, portion of peristome; all *magnified*.

3. **Blindia contecta**, *Hook. f. & Wils. Flor. Antarct.* 404 t. 58, f. 3. (Weissia).

Christmas Harbour, on rocks, barren, *Hooker*.

In this species the perichætium is composed of enlarged leaves as in *Stylostegium*, Schimp., but the capsule which is immersed has a peristome.*

---

* These three species afford some considerations respecting their mode of fructification. The genus *Blindia*, Bruch et Schimp., at first included only the European *B. acuta*, with the "perichætium vaginans distinctum," the perichætial leaves being as large as the cauline and dilated below. To this was added by C. Müller (in the Synopsis) *B. cæspiticia*, which had been made into the genus *Stylostegium* in the Bryologia Europea, differing from *Blindia* in its gymnostomous capsule immersed in enlarged but not vaginant perichætial leaves, in these particulars analogous to some species of *Grimmia* of the section *Schistidium*, in which *B. cæspiticia* had itself sometimes been placed. The distinction between *Blindia* and *Stylostegium* is reduced by the presence of a peristome in *B. contecta* (which may be said to be a *Stylostegium* with a peristome) by the immersed capsule in *Stylostegium*, and the exserted one in *Blindia*. In *B.*

1. **Dicranum** (Isocarpus, *Mitt.*) **tortifolium,** *Hook. f. et Wils. Fl. Antarct.* 404, *t.* 152, *f.* 5 (Weissia).

Hab., Christmas Harbour, on gravelly banks, *Hooker.* Under the shoot of a waterfall near Vulcan Cove, with old capsules and young setæ, *Eaton.*

In compact tufts 1–1½ inch high. Foliage very green above, below becoming brown. Old capsules black and shining; young calyptras orange brown.

2. **Dicranum** (Isocarpus) **strictum,** *Hook. f. et Wils. Fl. Antarct.,* 404, *t.* 152, *f.* 4 (Weissia).

Christmas Harbour, on rocks near the sea, *Hooker.*

This has been described as dioicous, but the male flower is terminal on a branch arising some distance below the perichætium. The peristome has rather broad thin teeth; in the solitary example which could be examined, the teeth appeared to be partly adherent in pairs, the median line is obsolete. This species is closely related to *D. tortifolium.*

3. **Dicranum** (Hemicampylus, *Mitt.*) **robustum,** *Hook. f. et Wils. Fl. Antarct.* 406, *t.* 152, *f.* 8, *var.* lucidum; D. pungens, *var.* lucidum, *Hook. f. et Wils. l. c.*

Hab. Christmas Harbour, *Hooker, Moseley.*

Known only in a barren state.

3. **Dicranum** (Hemicampylus) **kerguelense,** *C. Muller, Syn. i.* 370. D. Boryanum, *Schwaegr;* Hook f et Wils Fl Antarct. 406. D. dichotomum, *Desv.* Prodr. 51 (Cecalyphum).

Christmas Harbour, *Hooker.* On an elevated moor, Royal Sound, *Eaton.*

microcarpa the perichætium is formed of leaves reduced in size like those which usually include the antheridia, and thus another modification of the perichætium is produced, all other particulars being as in *Blindia* proper. Thus, by the difference in the leaves of the perichætium, the species are separable into several groups:—

Stylostegium, B. & S.;—theca in perichætio e foliis caulinis ampliatis immersa.

Blindia, B. & S.;—theca e perichætio e foliis basi vaginantibus caulinorum magnitudinis exserta.

Homogastrium;—theca e perichætio microphyllo exserta.

The differences in the leaves of the perichætium between *Stylostegium* and *Blindia* are analogous to those which exist between the *Grimmiæ* of the sections *Schistidium* and *Grimmia;* between *Hedwigidium* and *Braunia;* between some *Bartramiæ* of the section *Leucomela* and *Eubartramia;* and also between the *Schlotheimiæ* of the sections *Stegotheca* and *Euschlotheimia.* States of the perichætium analogous to that observable in *B. microcarpa* occur chiefly in mosses which produce their fruit from the side of the stem, as *Anœctangium,* and in some species of *Fissidens.* Amongst the Neckeroid mosses perichætia may be observed in otherwise closely resembling species which are analogous to all three of the states here left in *Blindia.* Much time and many words might be saved in the description of mosses in which the perichætium is an important character, if some term at once conveying the essential part of the above information were employed, thus:—

Chanogastriati;—perichætium e foliis elongatis ampliatisque hians=*Stylostegium, Schistidium* (Grimmieæ), *Hedwigia, Hedwigidium, Cryphæa, Neckera.*

Heterogastriati;—perichætium e foliis elongatis inferne convolutis clausum=*Blindia, Dicranum, Hypnum,* &c.

Homogastriati;—perichætium e foliis abbreviatis iis perigonii similibus formatum=*Blindia microcarpa, Anœctangium, Pyrrhobryum,* &c.

In extensive tufts, with stems from 3-4 inches high, and fulvous green foliage, becoming when older, brown.

So far as can be seen from the small specimen in the Hookerian Herbarium of *Cecalyphum dichotomum*, it appears to be the same as the Kerguelen moss, as it was considered by Mr. Wilson. The chief distinction ascribed to *D. kerguelense* is to have the nerve vanishing towards the narrow flat point, and not as in *D. dichotomum* to have the nerve continued into the point and dentate on the back.

1. **Campylopus cavifolius**, *Mitt. Musc. Austr. Amer.* 87.

Kerguelen Island, in dense tufts, barren, *Moseley*. By some error this was enumerated in the Linn. Soc. Journal as *C. appressifolius*.

1. **Ceratodon purpureus**, *Linn. Sp. Pl.* 1575 (Mnium).

Hab.—Royal Sound and near Swain's Bay, in a dark purple almost blackened state, all barren, *Eaton*. (Heard Island, *Moseley*.)

This moss appears to be as common throughout the southern regions as it is in the northern. The southern states have generally a more robust appearance, but when *C. brasiliensis*, Hampe, from Brazil, *C. crassinervis*, Lorentz, from Valdivia, *C. capensis*, Schimp., from the Cape of Good Hope, and *C. convolutus*, Reichardt, from New Zealand, are compared side by side, the conclusion seems irresistible, that they are all forms of one species.

1. **Grimmia** (SCHISTIDIUM) **apocarpa**, *Linn. Sp. Pl.* 1579 (Bryum).

Christmas Harbour, *Hooker*. Cat Island, Royal Sound, *Eaton*.

A very small state; all the specimens unlike European, but not appearing to be really different.

2. **Grimmia** (SCHISTIDIUM) **falcata**, *Hook. f. et Wils. Fl. Antarct.* 401, t. 151, f. 8.

Christmas Harbour, on rocks and stones near a waterfall, *Hooker*.

This is either an aquatic species or an aquatic form of a species of which the form corresponding to rupestral states of *G. apocarpa* is unknown.

3. **Grimmia insularis**, *Mitt. in Journ. Linn. Soc.* XV., 73.

Heard Island, *Moseley*.

4. **Grimmia** (EUGRIMMIA) **Kidderi**, *James in Bull. U. S. Nat. Mus.*, 3, p. 25.

Near Swain's Bay, *Eaton*.

In small dense black cushions. Stems 3-4 lines high, with a few branches near the base, made up of repeated innovations from the base of the male flower, consisting of closely set widely ovate leaves, without diaphanous points, including a few antheridia. Leaves very small, canaliculate, margins erect, terminated by a short, nearly smooth hyaline point.

This ambiguous moss may be conjectured to represent a species near to the European *G. montana*.

5. **Grimmia** (DRYPTODON) **chlorocarpa**, *Brid., Mitt. in Hook. f. Handb. New Zealand Fl.*, II., 426 (sub Rhacomitrium crispulum).

Kerguelen Island, *Moseley*. Hill N.W. of Mount Crozier, barren, *Eaton*.

Very closely related to *G. Symphyodon* and *G. emersa*, C. Müller, and also to *D. crispulus*, Hook. f. et Wils.; all are possibly forms of one species.

6. **Grimmia** (DRYPTODON) **crispulus**, *Hook. f. et Wils. Flor. Antarct.* 124, et 402, *t.* 57, *f.* 9.

Christmas Harbour, in gravelly beds of rivulets, *Hooker*.

7. **Grimmia** (RHACOMITRIUM) **lanuginosa**, *Dill.; Brid.* i. 215.

Hab.—Kerguelen Island, *Moseley*; Royal Sound and near Vulcan Cove, barren, *Eaton*.

All the specimens are less robust than those collected by Dr. Hooker in Hermite Island; from the whitening of the tips of the leaves they are very hoary.

Many of the specimens brought from southern regions which appear to differ in only slight particulars from northern states have been described as distinct; of these, *Rhacomitrium firmum* De Notaris, from Chili, is a fulvous brown moss, *R. Geronticum*, C. Müller (Hedwigia, 1870), is possibly the same. *R. senile*, Schimp. (Lechler, 1089, from Magellan), with leaf points crisped and hoary, *R. incanum*, C. Müller (Hedwigia, 1870), from Cape of Good Hope, is, if specimens from the top of Table Mountain belong to it, scarcely in any particular different from Arctic examples.

8. **Grimmia** (RHACOMITRIUM) **protensum**, *A. Braun; Hook. f. et Wils. Flor. Antarct.* 402.

Christmas Island, barren, *Hooker*.

9. [G. FRONDOSA, *James in Bull. U. S. Nat. Mus.* 3, 25, is another Kerguelen Island species, found by Kidder.]

1. **Orthotrichum crassifolium**, *Hook. f. et Wils. Fl. Antarct.*, *p.* 125, *tab.* lvii. *f.* 8.

Christmas Harbour, common, *Hooker, Moseley*; Royal Sound, *Eaton*.

The specimens from Kerguelen have the points of the perichætial leaves reaching to three-fourths of the length of the capsule, which is thus only emergent, and in this respect they agree with some of the specimens gathered in Hermite Island by Dr. Hooker. No importance can be attached to this particular character, as in Dr. Hooker's specimens from Lord Auckland's Islands, emergent and exserted capsules may be seen on the same stems.

The capsules are either smooth or with a few folds regularly placed on one side, the remainder being smooth, and are more urceolate than any of the specimens collected by Dr. Hooker.

The inflorescence consists, as usual in the genus, of a male flower near the base of the perichætium in all the specimens.

2. **Orthotrichum atratum**, *Mitt. in Linn. Soc. Journ., XV., p.* 66. Monoicum. Caulis humilis, cæspitosus. Folia patentia, sicca incurva, laxe contorta, lanceolata, apice lata obtusiuscule acuta, nervo sub summo apice evanescente, cellulis

fere ubique parvis rotundatis obscuris; perichætialia majora. Theca in pedunculo longitudine perichætii subæquali ovalis, lævis, sicca infra os contracta, inferne collo crasso; operculo convexo, rostro angusto; peristomii dentibus 16, vel plus minus cohærentibus 8. Calyptra nigro-fusca, calva, ad medium usque thecæ descendens, nitida.

Kerguelen Island, *Moseley*.

Stems not more than half an inch high. Leaves a line long; a few of the youngest greenish, the rest all black, coriaceous. Capsule pale straw-coloured, somewhat fleshy, smooth when deoperculate, very slightly contracted just below the mouth at the base, when dry shortly plicate.

In all its parts larger than *O. crassifolium*, with leaves twice as wide, and without the horny appearance; it is, however, more nearly allied to that species than to any other, and approaches in some respects the *O. anomalum*, Hedw., which ascends far towards the Polar regions.

3. **Orthotrichum rupestre**, *Schleich.; Brid.* i. 279.

Royal Sound, with fruit nearly mature, *Eaton*.

The specimen is in good state, and appears to agree in all respects with the European, except that no internal peristome has been found; it does not correspond so well with either of the very closely allied species, *O. Sturmii* or *O. cupulatum*, which agree in being destitute of cilia.

1. **Zygodon Brownii**, *Schwaegr. t.* 317 *b*.

Kerguelen Island, *Moseley*.

The minute scrap rather establishes the fact that a species of the genus inhabits Kerguelen Island than provides materials for identifying with certainty that to which it is here referred.

Tortula (Syntrichia) **Princeps**, *De Notaris*; Barbula Mülleri, *Bruch et Schimp., Bryol. Europ. t.* 28. T. Fuegiana, *Mitt., Journ. of Linn. Soc.*, Sept. 1859. *Musc. Austr. Amer.* 174. Barbula, S. magellanica, *C. Müller* in *Bot. Zeit.* 1862, 349; B. antarctica, *Hampe;* Tortula antarctica, T. cuspidata, *et* T. rubella. *Hook. f. et Wils. Fl. Tasmanica, pl.* clxxii., *f.* 8, 9, 10.

Royal Sound, with abundant mature capsules; Observatory Bay, with older fruit, *Eaton*.

The first examination of the Kerguelen specimens yielded no male inflorescence, they were therefore considered to be *T. fuegiana*, with which in size, colour, and appearance they appeared to be identical, this being supposed to be a dioicous species, as no male flowers were observed in Lechler's Magellan specimens No. 1088, from Cabo Negro. The same specimens were again described by C. Müller as dioicous, under the name of *Barbula S. magellanica*. In seeking for the male flowers amongst Mr. Eaton's abundant specimens, it was, after the examination of many stems, ascertained that although no antheridia were present in the fertile flowers, a small proportion of the stems had a male flower without archegonia, either terminal on a short branch, or lateral from the growth of innovations. Finally it was discovered that there might be present on the same stem, flowers containing antheridia accompanied by others containing archegonia, and above both these another flower in which both organs were intermixed. Thus, with specimens in small quantity to examine, the inflorescence might be described as monoicous dioicous or synoicous, as might chance to happen to the investigator.

The European *T. Princeps* was at first correctly described by De Notaris as polygamous in the Bryologia Europea, where it is figured as *Barbula Mülleri*. It is there described as hermaphrodite, with a remark in a subsequent note that it occasionally produced flowers containing archegonia only. In Schimper's Synopsis and in the Bryologia Britannica it is simply stated to be synoicous. An examination of De Notaris's original specimen shows synoicous fertile flowers with innovations of the stem terminated by flowers with archegonia alone; in this particular coinciding with British specimens.

The distribution of this species appears to be very wide, and it would seem to be the preponderating if not the only species of the genus in southern regions. From N.W. America it extends to Mexico, Chili, and the Straits of Magellan; in Africa it is found at the Cape of Good Hope, and may be identical with the *Barbula mollis*, Schimp., of the Abyssinian Mountains; it occurs in N.W. India; it inhabits also New Zealand, Tasmania, and Australia, from whence several species have been described as dioicous, viz., *Barbula Latroheana*, C. Müller (Bot. Zeit. 1864, 358), *B. Preissiania* (ejusd. Synops. I. 642), *B. panduræfolia* (ejusd. et Hampe, Linnæa 1853, 493). No specimen, however, amongst those sent by Baron F. von Mueller to the Kew Herbarium has been examined without finding its inflorescence monoicous or synoicous. There is also *Tortula S. pusilla*, J. Angstr. from Magellan, described as dioicous? and *Barbula Lechleri*, C. Müller (Bot. Zeit. 1859, 229), as monoicous. All these species or supposed species may be well distinguishable, but if the certainty of the condition of their inflorescence is removed from their descriptions, the remainder becomes applicable to *T. Princeps*, in which the outline of the leaves even on the same stems is, as in the European *T. ruralis*, subject to a great amount of variation.

2. **Tortula** (Barbula) **serrulata**, *Hook. et Grev. in Brewst. Edinb. Journ.* i. 291, t. 12.

Kerguelen Island; a few small barren stems with other mosses, *Moseley*.

3. **Tortula** (Barbula) **erubescens**, *Mitt. in Hook. f. Handbook of New Zeald. Flora*, ii. 421 (Didymodon).

Kerguelen Island; a few fragments, *Moseley*.

Very closely related to the *T. rubella* so widely distributed in northern regions, differing chiefly in the longer operculum and larger size of the whole plant.

1. **Streptopogon australis**, *Mitt. in Linn. Soc. Journ.* xv. 66. Humilis. Folia inferiora patentia, spathulato-ligulata, obtusiuscule acuta, nervo in apice de-

sinente, margine apicem versus denticulata; superiora duplo latiora, a basi erectiore sensim recurva, patentia, apice cum nervo in acumen longitudine variabile sensim educto, margine superne serrulata.

Royal Sound; a single stem, *Eaton*. Two small stems amongst other mosses without precise locality, *Moseley*.

The small quantity found of this moss would be insufficient to give any idea of what might be supposed to be the usual appearance of the species were it not evidently a close congener to a very ambiguous moss found on thatch in the south of Britain, and which has been known first as a supposed gemmiferous variety of *Leptodontium flexifolium* (Sm.), and since as *Didymodon gemmascens*, Mitt. MSS. From this the Kerguelen species differs in the form of its lower leaves. In the British moss all the leaves are acuminate and tipped with a globular mass of individually obovate green gemmæ of a loose cellular substance, and gemmæ of the same form are present on the points of some of the upper leaves of *S. australis*.

Both species appear to be small, the British one is seldom more than half an inch in height; the entire plant, excepting a few rootlets, and the rarely present archegonia, which are red, is of a yellowish green. In the dry state it affords nothing to attract observation, but when wet, every leaf being terminated by its mass of gemmæ, it is unlike any other European moss, excepting the more robust *Orthotrichum phyllanthum* (Brid.). It comes nearer to some species of *Streptopogon*; the areolation of the leaves of *Calymperes* or of *Syrrhopodon* are widely different. The genus *Streptopogon* founded on *S. erythrodontus* (Tayl.), with the additional species discovered in the Quitenian Andes by Dr. Spruce, and those from the Bogotian Andes by Lindig and Weir, contains a number of species all seeming to have a tufted Orthotrichoid habit. They differ among themselves considerably, some of the Andean species having the leaf with a callous margin which is wanting in others, and the capsule immersed or shortly exserted from perichætial leaves which are not very different from the cauline. *S. mnioides*, Schw. t. 310 (*Barbula*), however, has the perichætium leaves much elongated, and different from those of the stem, simulating those of *Holomitrium*, and on this account should stand apart from the other species, thus—

STREPTOPOGON, *Wils.* Theca in perichætio e foliis caulinis subsimilibus immersa, emergens, vel breviter exserta. Calyptra breviter multifida.

CALYPTOPOGON, *Mitt.* Theca in perichætio e foliis elongatis a caulinis difformibus exserta. Calyptra profunde plurifida.

The first group contains all the species of which the fruit is known, and which correspond to the typical *S. erythrodontus*, together with probably some others which are known only in a barren state, including the two ambiguous species *S. australis* and *S. gemmascens*.

The second group consists of *S. mnioides* alone.

2. **Streptopogon? marginatus**;—Schistidium marginatum, *Hook. f. and Wils. Flor. Antarct.*, 399, *t.* 151. f. vi.

Christmas Harbour, forming large patches on wet rocks, *Hooker*.

This, which appears destitute of peristome, is in other respects more nearly related to *Streptopogon* than to any other genus, and if included in it would occupy a position analogous to that of *Stylostegium cæspiticium* and *S. contectum* before mentioned under *Blindia*.

1. **Entosthodon laxus,** *Hook. f. et Wils. Fl. Antarct.,* 399, t. 151, f. 5. (Physcomitrium).

Christmas Harbour, *Hooker*. Royal Sound, barren, and Swain's Bay, with nearly mature capsules, *Eaton*.

Traces of an internal peristome are present within the external teeth.

1. **Bartramia** (Philonotis) **appressa,** *Hook. f. et Wils. Fl. New Zeald.* ii. 89, t. 86, f. 5.

Royal Sound, barren; Observatory Bay, with a few nearly ripened capsules; and hill N.W. of Mount Crozier, a tall barren slender state, *Eaton*.

2. **Bartramia** (Philonotis) **australis,** *Mitt. in Hook. Handb. New Zeald. Flor.,* 448.

Swain's Bay and Royal Sound, all barren, *Eaton*.

The few small stems growing among other mosses appear to belong to this species.

3. **Bartramia** (Breutelia) **pendula,** *Hook. Musc. Exot.* t. 21.

Kerguelen Island, *Moseley*. Royal Sound, hill N.W. of Mt. Crozier; near Vulcan Cove, with abundant immature fruit, *Eaton*.

4. **Bartramia** (Eubartramia) **patens,** *Brid. Sp. Musc.* iii. 82.

Kerguelen Island, *Moseley*. Royal Sound, with old fruit; and hill N.W. of Mt. Crozier, *Eaton*.

5. **Bartramia** (Eubartramia) **robusta,** *Hook. f. et Wils. Fl. Antarct.* t. 59.

Kerguelen Island, *Moseley*. Royal Sound, with old capsules and young setæ rising, very fine tall specimens, and Swain's Bay, *Eaton*. (Heard Island, *Moseley*.)

[B. flavicans, *Mitt.*, is enumerated by James as amongst the U. S. collections, collected at the rear of the American Transit House.]

1. **Bryum** (Webera) **nutans,** *Schreb.*; *Hedw. Musc. Frond.* i. t. 4.

Near Vulcan Cove; hill N.W. of Mt. Crozier, a small state with unripe fruit growing amongst *Psilopilum trichodon*, *Eaton*.

2. **Bryum** (Webera) **elongatum,** *Dicks.*

Swain's Bay, a single stem with ripe capsule, *Eaton*.

3. **Bryum** (Webera) **crudum,** *Hedw. Musc. Frond.* i. t. 88 (Mnium).

Kerguelen Island, *Moseley*. Swain's Bay, with fruit just mature, *Eaton*.

4. **Bryum** (Webera) **albicans,** *Wahlenb.*

Christmas Harbour, *Hooker, Moseley*. Near Vulcan Cove, *Eaton*.

Specimens all barren.

5. **Bryum** (Eccremothecium) **pendulum,** *Hornsch.*

Royal Sound; and Cat Island, Royal Sound, *Eaton*.

The inflorescence, which is usually synoicous in capsuliferous flowers, is sometimes accompanied by unisexual flowers upon the same stem.

6. **Bryum** (ECCREMOTHECIUM) **Eatoni,** *Mitt. in Journ. Linn. Soc.,* xv., p. 195.
Synoicum. Caulis humilis, gracilis, innovationibus infra comalibus paucis ramosus. Folia erecto-patentia, inferiora minora, superiora elliptico-lanceolata, nervo in acumen tenue læve vel denticulis paucis asperum excurrente, margine limbo tenui e seriebus cellularum elongatarum 4–5 composito anguste reflexo integerrima, cellulis angustis limitibus teneris areolata; folia comalia longiora, basi subauriculato-dilatata, angulis rotundatis laxis areolatis. Seta elongata, recta, apice anguste curvata. Theca pendula, sporangio ovato collo subæquilongo; operculo depresse conico acuminulato; peristomio parvo, dentibus pallidis subsubulatis, apice punctulatis, processibus apice punctulatis ciliisque in unum angustissimum conflatis in membrana usque ad dentium longitudinis $\frac{1}{3}$ exserta impositis, annulo triplici circumdato.

Swain's Bay and Royal Sound, with fruit ripened, *Eaton.*

The very narrow leaves retain the same position in both the wet and dry state, they are narrower than observed in any form of *B. pendulum.*

Tab. III. f. iv.; 1, natural size; 2, cauline leaf; 3, leaf from perichætium; 4, capsule; 5, portion of peristome; all *magnified.*

7. **Bryum** (ECCREMOTHECIUM) **bimum,** *Schreb.; Bryol. Europ.* t. 21.
Christmas Harbour, *Hooker.* Near Swain's Bay; and Royal Sound, with ripe fruit. *Eaton.*

The specimens vary in size, the stems in some being nearly three inches high, the lower leaves are all blackened.

8. **Bryum** (ECCREMOTHECIUM) **alpinum,** *Linn.*
Royal Sound, with shining red foliage; and Swain's Bay, all barren, *Eaton.*

The red-leaved specimens are exactly similar to those states of this species which are found in sub-alpine regions in Europe; those states which are found in the plains have never the lustrous appearance which adorns this handsome moss.

The small specimen from Swain's Bay was mistaken for *B. lævigatum,* Hook. f. et Wils. (also a Kerguelen species), to which in colour it has a great resemblance, and the similarity was increased by the points of the upper leaves being broad and obtuse; the lower leaves are, however, of the usual form.

9. **Bryum** (ECCREMOTHECIUM) **argenteum,** *Linn.*
On sea cliffs near Observatory, barren, *Eaton.*

A small silvery state with the leaf points not produced.

10. **Bryum** (ECCREMOTHECIUM) **kerguelense,** *Mitt. in Journ. Linn. Soc.* xv. 67.
Monoicum, cæspitosum. Caulis brevis, ramosus. Folia erecto-patentia, imbricata, inferiora rameaque ovali-lanceolata, acuta, carinato-concava, nervo rubro percursa, margine integerrimo, cellulis angustioribus in seriebus duabus limbum subindistinctum formantibus; reliquis suboblongis; comalia longiora latioraque; perichætialia interna minora. Theca in pedunculo breviusculo rubro superne flexuoso

curvato horizontalis, tenui-membranacea, nitida; sporangio ovali collo recto æqui-longo sensim angustato; ore satis parvo coarctato; operculo convexo apice brevissime acuto; peristomii dentibus pallidis interni fragmentis externo usque ad medium adhærentibus.

Kerguelen Island, *Moseley*.

Stems including the numerous branches about 3 lines high, and with the foliage about half a line wide. Leaves appressed when dry, a few at the apices of the branches green, the lower all dark brown. Seta 3 lines long. Capsule about $1\frac{1}{2}$ lines long, ochraceous, almost shining. The male flowers are terminal on branches arising below the perichætium.

This small species appears to be nearly allied to *B. demissum*, Hook., but its capsule is symmetrical, and the peristome is different.

Tab. III. fig. iii.; 1, plants nat. size; 2, entire plant; 3, cauline leaf; 4, perichætial and comal leaves; 5, portion of peristome; all *magnified*.

11. **Bryum lævigatum**, var. β., *Hook. f. and Wils. Flor. Antarct.*, 415, t. 154, f. 3.

Christmas Harbour, barren, *Hooker*.

12. **Bryum Wahlenbergii**, *Schwæg*.

Christmas Harbour, common, *Hooker*.

[B. WARNEUM, Bland.; GAYANUM, Mont.; TORQUESCENS, Br. and Sch.; and PALLESCENS, Schwæg., are all enumerated by James as found by Kidder (Bull. U. S. Nat. Mus. 3, 26.]

1. **Mielichhoferia campylocarpa**, *Hook. et Arn. in Hook. Icon. Pl.*, t. 136 (Weissia).

Kerguelen Island, *Moseley*. Near Swain's Bay, with unripe fruit, *Eaton*.

First described from the Andes, where it was gathered by Jameson; it was afterwards found in Mexico, and may be one of those species extending throughout the Andine chain. *M. basillaris*, Bruch et Schimp., from the Abyssinian mountains, with entirely the same stature and appearance, differs in some particulars of the peristomial teeth, and in the nerve of the leaf vanishing below the point.

Plagiothecium antarcticum, *Mitt. in Journ. Linn. Soc.* xv. 71. Monoicum, cæspitosum, ramis ascendentibus. Folia compressa, subfalcata, nitida; caulina ovata, acuminata, integerrima, enervia; ramea ovato-lanceolata, tenuiter acuminata, subenervia; omnia basi subcordata, cellulis angustis elongatis areolata; perichætialia convoluta, late ovata, breviter acuminata. Theca in pedunculo elongato rubro ovalis, inæqualis, suberecta inclinatave; operculo breviter conico; peristomio interno ciliis in unum coalitis inter processus carinatos dentium longitudinis impositis in membranam usque ad dentium dimidiam longitudinem exsertam insidentibus.

Royal Sound, with mature and old fruit, *Eaton*. Marion Island, *Moseley*.

Stems forming extensive soft patches, with shining foliage about half a line wide. Seta about half an inch long, when dry twisted. Capsule obovate, the neck collapsing plicate, and so curved that the whole capsule is inclined; mouth large, pale peristome prominent. The male flower, as is frequent in this genus, forms one of a cluster of small bud-like flowers at the base of the perichætium.

Closely resembles the European *P. nitidulum, Wahl.*, scarcely differing except in the base of its leaves. This is the species which is mentioned in Hooker's Handbook of the New Zealand Flora, ii. 476, as Hypnum pulchellum Dicks? from the Canterbury Alps.

Tab. III. Fig. v.; 1, plant nat. size; 2, cauline leaf; 3, perichætium with male flower at base; 4, capsule; 5, portion of peristome—all *magnified*.

[P. DONIANUM, *Sm.*, is enumerated by James as having been collected by Kidder in the U. S. Transit Expedition.]

1. **Acrocladium politum,** *Hook. f. et Wils. Fl. Antarct.* ii., t. 154, f. 2 (Hypnum). *Mitt.* l. c.

Hab.—Christmas Harbour, slender, tufted state, *Hooker*. Royal Sound, small and barren, *Eaton*.

This moss resembles some species of *Plagiothecium*, but seems to differ in habit, its branches with conduplicate bifarious leaves having so close a resemblance to those of *Phyllogonium elegans*, Hook. f. et Wils., that it is frequently mistaken for that plant. In the review of the genus *Orthorhynchum*, Reich. by C. Müller (Linnæa Band, 36, p. 28), one of the species to be referred to this genus, the *O. Hampeanum*, C. Muller, sent from Australia Felix by Baron F. von Mueller, must, from the description, be *Acrocladium politum*, of which specimens have been seen from Von Mueller, not however exactly corresponding in locality.

1. **Stereodon cupressiformis,** *Linn.* (Hypnum).

Base of sea cliff, Royal Sound, barren, *Eaton*.

The small specimens obtained exhibit this variable species in that form which in Europe is found on the roofs of buildings or on the ground; they are very unlike *S. chrysogaster*, C. Muller, so common in New Zealand.

1. **Amblystegium uncinatum,** *Hedw.*

Christmas Harbour, *Hooker*. Near Vulcan Cove, a tall robust form with nearly mature fruit; and Royal Sound, a similar state, but barren, *Eaton*.

2. **Amblystegium fluitans,** *Dill.*

West side of Swain's Bay, barren, *Eaton*.

A large state, with all but the terminal leaves of a brown colour.

3. **Amblystegium riparium,** *Linn.*

In the lake at Christmas Harbour, *Hooker*.

Specimen in a very imperfect state. Also found by the U. S. Transit Expedition growing with *Ranunculus crassipes*.

4. **Amblystegium kerguelense,** *Mitt.* Dioicum? Caulis decumbens, ramis confertis ascendentibus pinnatim ramosis. Folia caulina laxe imbricata, stricta vel curvata, ovato-lanceolata, subulato-acuminata, integerrima, nervo basi lato sensim angustato et in acumen evanido percursa; cellulis parvis oblongis limitibusque pellucidis ad angulos paucis rectangulis latioribus areolata; folia ramea

erecto-patentia, angustiora, nervo crassiore.   Hypnum filicinum var. et H. serpens, var. *Flor. Antarct.*, p. 419 et 418.

Christmas Harbour, *Hooker*.   Near Swain's Bay, barren, *Eaton*.

The single patch of this moss gathered by Mr. Eaton exhibits the species as very closely resembling *A. filicinum*, Linn., when it has not assumed a pinnate form; it is larger than *A. serpens*.   The foliage is fulvous, neither wet nor dry is it altered in appearance.

5. **Amblystegium decussatum**, *Hook. f. et Wils. Fl. New Zeald.* ii. t. 90, f. 2. (Hypnum.)

Royal Sound, a slender straggling state, with irregular branches and an upright form, amongst *Bryum pendulum*; near Swain's Bay, an upright state more robust and more branched; near Vulcan Cove, a still larger state, with stems three inches high; all barren, *Eaton*.

All the specimens referred to this species have but little external resemblance to the complete state found fertile in New Zealand, but they agree very closely in the areolation of their leaves, and it is probable they are only slender forms similar to those produced by *A. filicinum*.

1. **Sciaromium conspissatum**, *Hook. f. et Wils. Fl. Antarct.* 419, t. 155, f. 3. (Hypnum).

Christmas Harbour; *Hooker, Moseley*.   A short barren state.

All the Kerguelen specimens are smaller than those from the Falkland Islands.

1. **Brachythecium subpilosum**, *Hook. f. et Wils. Fl. Antarct.* 418, t. 154, f. 4.   (Hypnum).

Kerguelen Island, *Moseley*.

More robust than the original specimens from Cape Horn, and in this respect nearer to the *Hypnum rutabulum*, var. 5, Fl. Antarct., from the Falkland Islands, which has since been named *H. subplicatum*, Hampe.   If, however, the species may be supposed to vary as much in aspect as the European *B. rutabulum*, these slightly larger forms may be fairly considered within the limits of probable variation.   Intermediate between the Hermite Island specimens and those from Kerguelen are some barren mosses from Otago, New Zealand, and some others collected in the Australian Alps by Von Mueller, to which it is probable the description of Dr. Hampe's *Hypnum austro-alpinum* may apply, as he says that the seta is thick and rough, and the capsule short, which are the most prominent characters appertaining to *B. subpilosum*.

2. **Brachythecium salebrosum**, *Hoffm.* (Hypnum).   Hypnum rutabulum, var. 4, *Hook. f. et Wils. Fl. Antarct.* 417.

Christmas Harbour, *Hooker*.   Hill N.W. of Mount Crozier, a fine silky state in large tufts, with stems 2–3 inches long; Swain's Bay, in boggy ground on the west side, a smaller state, all barren, *Eaton*.

This species is distinguished from *B. rutabulum* by the form of the leaves on the principal stems, which are not so dilated at their base, the outline being more nearly ovate and not deltoid. Specimens collected by Dr. Lyall in the Arctic regions at Beechy Island, correspond very nearly with the Kerguelen moss.

3. **Brachythecium paradoxum,** *Hook. f. et Wils. Fl. Antarct.* 449, t. 155, f. 2. (Hypnum).

Royal Sound, and Swain's Bay, with mature fruit, *Eaton.*

This species, which is found also in New Zealand and Fuegia, varies in size; the Kerguelen specimens are smaller than those from New Zealand; its affinity is with the European *B. velutinum* (Linn.), which is sometimes seen with falcate leaves, and then presents an appearance very different from its more usual state.

1. **Psilopilum trichodon,** *Hook. et Wils. in Hook. Lond. Journ. Bot.* vi. 289. (Polytrichum).

Hill N.W. of Mount Crozier, with narrow capsules, *Eaton.*

Originally described from the Andes of New Grenada, where it was found near the snow by Purdie; it was afterwards gathered by Jameson in a similar situation in the Andes of Quito.

Pogonatum alpinum, *Dill.*

Swain's Bay, with unripe fruit, *Eaton.*

This species occurs also in Australia, and has been described as *P. pseudo-alpinum* (C. Müller, Bot. Zeit. 1855, 750), but it is admitted that the southern specimens differ scarcely if at all from those of the boreal regions.

[CATHARINA COMPRESSA, *C. Müll.*; Polytr. compressum, *Hook. f. et Wils.*, is enumerated amongst the United States Expedition collections.]

1. **Andræa acuminata,** *Mitt.* A. acutifolia, var. γ, *Hook. f. et Wils. Fl. Antarct.* 396.

Christmas Harbour, *Hooker.* Kerguelen Island, with a few mature capsules, *Moseley;* Royal Sound, without fruit, *Eaton.*

In the outline of its leaves this species resembles *A. marginata,* Hook. fil. et Wils. Fl. Antarct. 396, t. 151, f. 1., but the areolation of their upper portion is different, the cells being about $\frac{1}{3000}$ of an inch long by $\frac{1}{5000}$ wide, those in the corresponding portion of the leaves of *A. marginata* being about $\frac{1}{5000}$ wide, and $\frac{1}{800}$ long.

2. **Andræa squarrosa,** *Mitt. Musc. Austr. Amer.* 629. A. alpina var. 1, *Hook. f. et Wils. Flor. Antarct.* 395.

Christmas Harbour, *Hooker.*

This species has the perichætial leaves in the Kerguelen specimens of the same form as in those collected by Prof. Jameson in the Andes of Quito.

[A. MARGINATA, *Hook. f. et Wils. Flor. Antarct.* 396, t. 151, fig. 1., has been found in Kerguelen Island by Kidder.]

III. *Hepaticæ*.

By William Mitten, A.L.S.

Nine species of *Jungermannia* and one *Marchantia*, were gathered by Dr. Hooker. These were arranged in 5 genera, and 5 of the species were described as new, the remainder being similar to species found elsewhere; none of the species were especially remarkable. Mr. Moseley collected at the time of the "Challenger's" stay 12 species, 7 of which were different from those obtained by Dr. Hooker, and 6 genera were also added to the flora. Fourteen species were found by Mr. Eaton; of these 8 species and 2 genera were additional to those previously known, bringing the whole number of the *Hepaticæ* up to 25.

The *Hepaticæ* of Kerguelen are allied most nearly to those of the Auckland and Campbell's Islands, and of Fuegia.

Noteroclada porphyrorhiza, *Leioscyphus pallens*, and *Teinnoma quadripartita*, are found also in Fuegia. *Jungermannia colorata*, and *Symphyogyna podophylla*, are found at the Cape of Good Hope. The former is, however, very widely distributed in austral regions. As with the mosses, it is remarkable how many additions were made to the flora by the small number of specimens obtained by each collector.

1. **Plagiochila heterodonta,** *Hook. f. et Tayl. Fl. Antarct.* 428, t. 157, f. 2.

Christmas Harbour, on moist rocks, *Hooker*. Royal Sound, barren, *Eaton*.

The specimens closely resemble those gathered by Dr. Hooker; it appears to be always a small species.

2. **Plagiochila minutula,** *Hook. f. et Tayl. Flor. Antarct.* 427, t. 157, f. 1.

Christmas Harbour, on rocks and the ground, *Hooker*.

1. **Leioscyphus turgescens,** *Hook. f. et Tayl. Fl. Antarct.* 150, t. 64, f. 2.

Hab. Royal Sound, amongst *Ditrichum Hookeri*, *Eaton*. (Lord Auckland's group).

2. **Leioscyphus pallens,** *Mitt. in Journ. Linn. Soc.* xv., 68. Caulis procumbens ascendensque, parce ramosus. Folia sursum secunda, conniventia, imbricata, orbiculata, caviuscula, integerrima, cellulis rotundis parietibus crassiusculis areolata. Amphigastria erecto-patentia, lanceolata, profunde bifida, laciniis elongatis subulatis. Folia involucralia majora, conformia; amphigastrio parvo quadrifido laciniis dentatis integerrimisve. Perianthium obovatum, ore truncato integerrimo.

Royal Sound, associated with *L. turgescens*, barren, *Eaton*.

Stems from an inch to an inch and a half long, seldom branched, with the leaves $\frac{1}{2}$ line wide. Leaves pale olive-green, becoming in age brown, rather firm, not collapsing when dry, composed of rounded cells which at first contain small round granules that disappear in the older leaves. Stipules ½ line long, the one immediately under the perianth is small and easily overlooked. Perianth compressed. No capsuliferous stems have been seen.

It appears that in this species, and in some others of the same genus, the compressed truncate perianth is the result of the small size of the involucral stipule, which in the coalescence of the leaves of which the perianth is theoretically formed, is too small to affect its form, the reverse of which is so evident in the perianth of *Lophocolea*.

Tab. III., Fig. vi., plant *nat. size*; 2. leaf detached; 3. stipule from the stem; 4. perianth as seen laterally with involucral leaves; 5. stipule next the perianth; all *magnified*.

1. **Lophocolea pallidovirens,** Hook. f. et Tayl. Fl. Antarct. 439, t. 159, t. 9.

Kerguelen Island, *Moseley*. Near Vulcan Cove, barren, *Eaton*. (Fuegia).

2. **Lophocolea Novæ Zealandiæ,** Lehm. et Lindenb. (Jungermannia).

Royal Sound, fragments amongst *Ditrichum Hookeri*. Hill N.W. of Mount Crozier, with young perianths, *Eaton*. (New Zealand and Lord Auckland's group).

3. **Lophocolea humifusa,** Hook. f. et. Tayl. Fl. Antarct. 436, t. 159, f. v.

Christmas Harbour, *Hooker*; near Observatory Bay, barren, *Eaton*.

The specimens are pale yellowish green, and seem not different from *L. bidentata*, with which it agrees in perianth.

1. **Teinnoma quadripartita,** Hook. Musc. Exot. 117 (Jungermannia).

Kerguelen Island, a few small fragments, *Moseley*. Gathered also amongst *Dicrana* at Christmas Harbour by *Hooker*.

1. **Jungermannia cylindriformis,** Mitt. in Journ. Linn. Soc. xv. Exilis. Caulis procumbens, ascendens, subsimplex, vix radicans. Folia subalterna, antice incurva, oblongo-ovalia, obtusa, sinu parvo obtuso obtuse bidentata, dentibus sæpe conniventer incurvis; involucralia minora, acute bidentata vel caulinis conformia. Perianthium elongatum, cylindraceum, obtusum, apice plicatum.

Royal Sound, in very small quantity with perianths amongst *Ditrichum Hookeri*, and hill N.W. of Mount Crozier, with *Scapania clandestina*, *Eaton*.

Stems about 2 lines long. Leaves ⅙ line long, brownish green. Perianth 1 line long, of the same colour as the leaves. This minute plant is nearly related to *J. inflata*, Huds., having the same cylindrical perianth, and involucral leaves not much different from those of the stem, which are the characters of the genus *Gymnocolea*, Dumort, which comprises besides the European *J. inflata*, and the *J. turbinata*, Raddi.

Tab. III., Fig. vii.; 1, plant *nat. size*; 2 and 3, perianth and involucral leaves, dorsal and lateral view; 4, cauline leaf, expanded; all *magnified*.

2. **Jungermannia leucorhiza,** *Mitt. in Journ. Linn. Soc.* xv. 68. Caulis procumbens, radicellis pallidis. Folia laxe inserta, quadrata subrotundave, sinu acuto obtusove bilobata, interdum lobo altero minore; lobis acutis obtusisve, incurvis; cellulis rotundatis et ovali-hexagonis areolata.

Kerguelen Island, in very small quantity amongst mosses, barren, *Moseley.*

Stems less than 1 inch long, with the leaves ½ line wide. Leaves green, tinged with brown.

Incomplete specimens of a species not before noticed in the Antarctic regions, but which appears to be near to the European *J. ventricosa,* Dicks, and to some states of *J. barbata.*

3. **Jungermannia colorata,** *Lehm. et Lindenb.*

Christmas Harbour, abundant on the hills, *Hooker, Moseley* (with perianths).

1. **Solenostoma humilis,** *Hook. f. et Tayl. Fl. Antarct.* ii. 434, t. 158, f. 6. (Jungermannia); J. inundata, *Flor. Nov. Zealand.* 128, t. 93, f. 3.

Hab. Christmas Harbour, barren, *Hooker.* A few fragments with one perianth, *Moseley.*

Both *S. humilis* and *J. inundata* were originally described as stipulate species, no amphigastria have, however, been since found on the specimens. It is probable that the figure of the supposed stipule of *J. humilis,* may have been drawn from a fragment of *Leioscyphus turgescens.*

Scapania, Lind (ex parte). Perianthium terminale, læve, a tergo ventreque compressum, ante capsulæ emissionem apice decurvum, herbaceo-membranaceum, ore truncato. Involucri folia 2, libera, caulinis conformia.—Plantæ terricolæ. Rami erecti ascendentesve, simplices vel furcati. Folia fere ubique æqualia, bifaria, equitantia, profunde biloba, laciniis subæqualibus apicibus rotundatis vel plus minus bifidis, textura e cellulis parvis. Amphigastria nulla.

This description is that of the Synopsis Hepaticarum, with slight modification, it applies to *S. densifolia, vertebralis,* and *chloroleuca,* all so intimately related that the possibility of their being forms of one species may be conjectured. These differ from the chiefly European species which were included in the original idea of *Scapania,* and which are now by right of priority assigned to *Martinellia,* Gray, in having leaves not keeled in the space between the equal lobes, a peculiarity which gives the plants a different aspect. The perianth known from a single example on *S. vertebralis,* is like that found in *Martinellia,* but is narrowed upwards, truncate, the mouth bent over and denticulate.

1. **Scapania densifolia,** *Hook. Musc. Exot.* 36 (Jungermannia).

Kerguelen Island, *Moseley.*

The specimens agree with those gathered by Menzies, and are of the same brown colour. The distinction between *S. densifolia* and its congeners may be thus stated:—
S. densifolia, Hook., lobis foliorum apice integris rarius emarginatis.—*S. vertebralis,* Tayl., lobis apice exsectis.—*S. chloroleuca,* Hook. f. et Tayl., lobis apice bifidis.

2. Scapania clandestina, *Mont. Bot. Crypt. Astrolabe*, t. 16, f. 4. Balantiopsis incrassata, *Mitt. in Journ. Soc. Linn.* xv. 197.

Hill N.W. of Mt. Crozier, in very small quantity with *J. cylindriformis, Eaton.*

The stems of this small plant are about ½ inch high, and with the leaves ½ line wide. Leaves firm, with small round cells; lobes unequal and differing in their direction, the dorsal patent, the ventral nearly twice as large and divergent. In the Kerguelen specimens the space between the lobes is keeled and curved, and both the lobes are denticulate, except the superior edge of the ventral lobe which is only denticulate towards the apex, and like that of the dorsal lobe is terminated by two larger teeth (hence bidentate, with a small rounded sinus). In this particular they nearly resemble the leaves of *Balantiopsis diplophylla* and *B. erinacea*, Tayl. (Scapania), but differ in their dense areolation. No authentic specimen has been seen of *S. clandestina*, Mont., but the figure quoted agrees except in the arcuation of the carina. A single stem picked from a tuft of *Aneura* from New Zealand has the lobes more nearly equal, the carina straight, very much longer, and all the marginal teeth more spiniform; it is probable as suspected in the Synopsis Hepaticarum, that the plant in a complete state would be different from the imperfect specimens yet seen. This species departs from *S. densifolia* and its allies in the leaves being carinate, and thus corresponds to *Martinellia*; it has, however, the apices of its leaves bedentate, which give it a different look from any of the species referred to that genus.

1. Cesia atrocapilla, *Hook. f. et Tayl. Fl. Antarct.* 423.

Foul haven, on clay banks, *Hooker;* in small blackish patches closely interwoven, *Moseley.*

From the examination of some branches of the specimens collected by Dr. Hooker it appears that fertile shoots would have their upper leaves nearly or quite entire and nearly orbicular in form.

1. Lembidium ventrosum, *Mitt. in Journ. Soc. Linn.* xv., 69. Caulis humilis, late compacteque cæspitosus, ascendens vel erectus, arcuatus, crassus, ramosus, innovationibus flagelliformibus ex amphigastriorum angulis emittens. Folia inferiora remota, superiora majora, insertione fere verticalia, patentia, apicem caulis versus imbricata, rotundata, profunde concava, apice rotundata sinuve subindistincto subretusa, cellulis parvis parietibus angustis areolata. Amphigastria parva, cauli appressa, subtriangulari-ovata, apice subemarginata. Perianthium in ramo superne valde incrassato, foliis amphigastrioque involucralibus convolutis ovatis apice breviter bitri-denticulatis, trigonum, ovatum, apice plicatum, ore laciniis conniventibus denticulatis obtusum.

Hill N.W. of Mt. Crozier, in dense tufts on the earth, with capsules just rising, *Eaton.*

In extensive brownish olive-green patches. Stems about 4 lines high, with the leaves scarcely ½ line wide, closely congested and cohering with very slender hyaline rootlets. Perianth large for the size of the plant, arising from the apex of a thickened branch; apex obtuse before the egress of the rather large spherical capsule, but afterwards sub-truncate. Spores minute, round, smooth, brown, accompanied by fusiform moniliate fibres.

L. nutans, Hook. f. et Tayl. Fl. Nov. Zeald. 160, t. 65, f. 8, is a larger species, and appeared by itself different from any genus that has been described, whereas *L. ventrosum* resembles the *Jungermannia Francisci* of Hooker's Brit. Jungermanniæ, t. 49, a species which also produces thickly fleshy stolons, is irregular in the emargination of its leaves, has the same kind of stipules as well as perianth, and is therefore a species of *Lembidium*. How this genus or group of species may be distinguished from the *Cephalozia* of Dumortier must remain for examination.

Tab. III., Fig. viii.; 1, plant of nat. size; 2, portion of stem with leaves and stipules; 3, perianth and involucral leaves on lateral branch;—all *magnified*.

1. **Herpocladium fissum**, *Mitt. in Journ. Linn. Soc. xv., 69*. Caulis perpusillus, firmus, crassiusculus. Folia alterna, patentia, ovata, obtusa, apice incurva, sinu parvo acuto breviter acute bidentata, concava, basi utroque caulem ad medium diametrum tegentia, margine dorsali interdum flexura sinuata rarius unidentata, cellulis densis obscuris areolata. Amphigastria foliis similia, patula divaricataue, apice obtusa, integra.

Kerguelen Island, *Moseley*.

Stems 3-4 lines long, with the leaves ¼ line wide. The entire plant almost black.

Tab. III., Fig. ix.; 1, plant of nat. size; 2, portion of stem, with leaves and stipule from the dorsal side; 3, lateral view of leaf and spreading stipule; 4, leaf detached and expanded;—all *magnified*.

1. **Tylimanthus viridis**, *Mitt. in Journ. Linn. Soc. xv., 197*. Humilis. Caulis erectus ascendensve, apice decurvus, subsimplex. Folia distiche expansa, oblongo-quadrata, apice oblique sinu lato subtruncato-bilobata; lobis obtusis, dorsali majore apicem versus interdum subdentatis, cellulis parvis rotundatis limitibus pellucidis areolata.

Royal Sound, and hill N.W. of Mount Crozier, all barren, *Eaton*.

Stems ½ inch high, green, with the leaves scarcely 2 lines wide. Leaves green, frequently convex from the recurvation of the margin. This nearly resembles *T. tenellus*, Tayl. (*Gymnanthe*) from Tasmania, but it seems to be a smaller species.

Tab. III., Fig. x.; 1, plant of *nat. size*; 2, portion of stem with leaves *enlarged*.

1. **Marsupidium excisum**, *Mitt. in Journ. Linn. Soc. XV., 69*. Caulis primarius repens, exinde ascendens, pallidus. Folia inferiora minora, deinde superiore minora, omnia oblongo-quadrata, concava, sinu obtuso bidentata, integerrima subcrenatave, lobis latis acutis incurvis, cellulis protuberantibus papulosa.

Royal Sound, with *Acrocladium politum* and *Pogonatum alpinum*, barren, *Eaton*.

Primary stems of the same colour as the leaves, fleshy, obscure, creeping; from these arise erect or ascending simple or branched shoots, which are arcuate, their points attenuated and decurved. The leaves where largest are about ½ line long, and when flattened of the same width, of a pale obscure olive-green; bases not decurrent; insertion variable but generally oblique; margins entire, or obtusely subcrenate; areolation of hexagonal or rounded cells with thin walls, enclosing a few green granules and projecting on both surfaces, but most on the external, as hyaline papulæ. Papulæ of the same kind are also present on the younger stems, but less prominent.

No kind of inflorescence has been seen on this species, and its location here is conjectured from its having the same habit as *M. Knightii*, from New Zealand.

Tab. III., Fig. xi.; 1, plant of *nat. size;* 2, part of stem with leaves; 3, cells from middle of leaf; both *magnified*.

1. **Fossombronia australis,** *Mitt. in Journ. Soc. Linn.* XV., 73. Caulis cæspitosus, prostratus vel ascendens, arcuatus, radicellis purpureis. Folia subquadrata, angulata, margine flexuosa, antice incurva. Perianthium turbinatum, margine flexuosum, angulatum, semina rotunda limbo hyalino lævia.

Kerguelen Island, *Moseley.* Royal Sound, and near Vulcan Cove, with young capsules, *Eaton.* (Heard Island, *Moseley*).

Some of the specimens are very large, with arcuate stems more than an inch long, producing many purple rootlets. The leaves are 1 line long by about 1½ wide, green, with pellucid cells.

2. **Fossombronia pusilla,** *Linn.*

Christmas Harbour, amongst moss, *Hooker.*

1. **Noteroclada porphyrorhiza,** Nees; N. confluens, *Fl. Antarct.* 446, t. 161, f. 7.

Christmas Harbour, on moist banks, *Hooker.*

1. **Symphogyna podophylla,** *Thunb.* (Jungermannia); *Gottsche, Lindenb. et Nees Syn. Hepat.* 481.

Near Vulcan Cove; Royal Sound; hill N.W. of Mount Crozier; all without fructification, *Eaton.*

1. **Aneura multifida,** *Linn.* (Jungermannia).

West side of Swain's Bay, on boggy ground, and near Vulcan Cove, all barren, *Eaton.*

2. **Aneura pinguis,** *Linn.* (Jungermannia).

West side of Swain's Bay, small and barren, *Eaton.*

1. **Marchantia polymorpha,** Linn.

Christmas Harbour; *Hooker, Moseley;* Royal Sound and Swain's Bay, *Eaton.* All the specimens produce scyphi, but are otherwise barren.

IV.—*Lichens.*

By the Rev. J. M. Crombie, F.L.S.

The first record that we can find of the Lichen-flora of this remote island, is contained in a preliminary account of the Antarctic Lichens collected by Dr. J. D. Hooker* during the voyage of the "Erebus" and "Terror," which was published by him and Dr. Thomas Taylor in the "London Journal of Botany" (1844), vol. iii. pp. 634–658. The Kerguelen Island lichens there enumerated amount in number to 17 species, named by Dr. Taylor; but at least one half of the names attributed to them are misapplied, and therefore must be excluded, owing chiefly to the determinations having been attempted in the absence of such microscopical analysis of the specimens as is now found to be essential for their discrimination. The number was subsequently raised to 27 species and varieties, when the list was revised by the Rev. Churchill Babington for publication in Dr. Hooker's "Flora Antarctica" (1847), vol. ii. pp. 519–542. A considerable proportion of the names in this later list must however be rejected for the same reason as those erased from the previous one. Unfortunately authentic examples of several of Dr. Taylor's critical species are wanting in the Kew Herbarium; † and his collection (now in the Herbarium of the Boston Society of Natural History), according to Professor Edw. Tuckerman, contains very little that is illustrative of his Kerguelen Island determinations. I have lately published a further revision of the Kerguelen Island Lichens collected by Dr. Hooker, in the "Journal of Botany" for April 1877, wherein the number of the species is reckoned to be 18 or 19 besides 2 named forms.

Mr. Moseley of the Challenger Expedition gathered in this island upwards of 13 species and 1 named form. (*Vide* Crombie in Journ. Linn. Soc. Bot. 1877.)

Dr. Kidder of the American Transit of Venus Expedition collected in the vicinage of Molloy Point 13 or 14 species and 1 named form. These with others from the Taylor collection are specified by Prof. Ed. Tuckerman in Bulletin U.S. Nat. Mus. No. 3 (1876), and are noticed by me in the "Journal of Botany" for April 1877.

The collection made by Mr. Eaton between the end of October 1874 and the end of February 1875, in the district immediately to the westward of Dr. Kidder's

---

* One (or more) species of Lichens was obtained in Kerguelen Island in 1776 by Mr. Anderson, the Surgeon and Naturalist who accompanied Captain Cook.—A. E. E.

† Dr. Taylor died shortly after the publication of his first rough determination of the Antarctic Lichens, and it was impossible to recover from the heap of his unarranged materials, which were in a confused state, all of the specimens which should have been returned. I strongly suspect, from the state of his notes sent to me from time to time, that he did not attend sufficiently to localities, and that some of the specimens in the Herbarium labelled as from Kerguelen Island did not come from that island.—J. D. H.

station, comprises 50 or 51 species and 9 named forms. Of these about 30 were described as new species referred to known genera, in the Journal of Botany for November 1875 and January 1876, and again with fuller diagnoses with which Dr. Nylander (who most kindly assisted me in their determination) favoured me in the Journal of the Linnæan Society (Botany) for July 1876. Though several of the new species bear a superficial resemblance to some of our northern lichens, yet on analysis they are found to be quite distinct, and for the most part are peculiar to Kerguelen Island.

The results obtained by the German Transit of Venus and Surveying Expedition at Betsy Cove, are not yet published.

The total number of species obtained from the island is 61, and 10 varieties. Traces of a few other species exist in the various collections, consisting either of sterile thalli or undeveloped apothecia which are necessarily indeterminable.

1. **Lichina antarctica,** *Cromb. in Journ. Bot.* v. 21 (1876); *et in Journ. Linn. Soc.* xv. 181.

Observatory Bay, on dry rocks near the sea, *Eaton*.

1. **Amphidium molybdophlacum,** *Nyl. in Journ. Bot.* iv. 333 (1875) (errone molybdophœum); *Cromb. in Journ. Linn. Soc.* xv. 181, *et in Journ. Bot.* vi. 103, 106. Lecanora melanaspis, *Bab. in Flor. Antarct.* 536 (excl. Syn. L. dichroa) Pannaria glaucella, *Tuckerm. in Bull. U. S. Nat. Mus.* iii. 28.

On earth and stones, Christmas Harbour, *Hooker;* on stones in wet places, Swain's Harbour, *Eaton;* Molloy Point, *Kidder*.

1. **Stereocaulon cymosum,** *Cromb. in Journ. Linn. Soc.* xv. 182, *et in Journ. Bot.* vi. 103. S. corallinum, *Hook. f. et Tayl. Flor. Antarct.* 528.

On rocks, altitude 6–1200 feet, Christmas Harbour, *Hooker, Moseley;* top of a hill on west side of Carpenter's Cove, barren, *Smith Dorien (Eaton)*.

1. **Cladonia fimbriata,** *Hoffm.; Cromb. in Journ. Linn. Soc.* xv. 182. C. pyxidata, *Linn.; Tuckerm. in Bull. Torr. Bot. Club,* October 1875, *et Bull. U. S. Nat. Mus.* 3, 29.

Dry slopes, Swain's and Observatory Bays, *Eaton;* Molloy Point, *Kidder*.

Var. costata, Flk. Observatory Bay, *Molloy*.

2. **Cladonia cornuta,** *Linn.; Cromb. in Journ. Linn. Soc.* 1877.

Kerguelen Island, *Moseley*.

3. **Cladonia acuminata,** *Ach.; Cromb. in Journ. Linn. Soc.* xv. 182. C. phyllophora, *Tayl.*

Christmas Harbour, *Hooker;* observatory Bay, common, but sparingly fertile, *Eaton*.

1. **Neuopogon melaxanthus,** *Ach.* (Usnea); *Cromb. in Journ. Linn. Soc.* xv. 182; *et in Journ. Bot.* vi. 103, 106 (1877). Usnea sulphurea, *Müll.; Tuckerm. in Bull. Torr. Bot. Club,* 1875, *et in Bull. U. S. Nat. Mus.* 3, 27. Ramalina scopulorum ε, *Hook. f. et Tayl. Flor. Antarct.* 522.

Rocks and boulders, Christmas Harbour, *Anderson, Hooker*; on the upper slopes and tops of hills, *Eaton*.

VAR. SOREDIIFER, *Cromb. l. c.* Common at all altitudes, *Eaton*.

VAR. CILIATA, *Nyl. Crombie, l. c.* Observatory Bay, *Eaton*.

2. **N. Taylori**, *Hook. f.; Flor. Antarct.* 521, t. 195, *Fig.* i. (Usnea); *Cromb. in Journ. Linn. Soc.* xv. 183, *et in Journ. Bot.* vi. 103 (1877).

Rocks ascending to 1,200 feet; Christmas Harbour, *Hooker, Moseley*; Swain's Bay and Carpenter's Cove (but not near Observatory Bay), *Eaton*; Molloy Point, *Kidder*.

1. **Parmelia stygioides**, *Nyl.; Cromb. in Journ. Bot.* iv. 333 (1875), *et in Journ. Linn. Soc.* xv. 183.

Dry rocks and stony slopes, Swain's Bay, *Eaton*.

1. **Peltigera rufescens**, var. spuria, *DC.; Cromb. in Journ. Linn. Soc.* xv. 183; *Journ. Bot.* vi. 103. Peltidea venosa, *Hook. f. et Tayl. Flor. Antarct.* 525.

On wet moss, &c., Christmas Harbour, *Hooker*; Swain's Bay, *Eaton*.

2. **Peltigera polydactyla**, forma HYMENINA, *Ach.; Cromb. in Journ. Linn. Soc.* xv. 183. P. polydactyla, *Flor. Antarct.* 524. P. horizontalis, *Ach.; Flor. Antarct. l.c.* 525.

Amongst moss, &c., Christmas Harbour, *Hooker*; Observatory Bay, *Eaton*.

1. **Pannaria dichroa**, *Hook. f. et Tayl. in Lond. Journ. Bot.* iii. 643; *Cromb. in Journ. Bot.* vi. 100 (1877), *et in Journ. Linn. Soc.* xvi. P. Taylori, *Tuckerm. in Bull. Torr. Bot. Club*, October 1875, *et in Bull. U. S. Nat. Mus.* 3, 28. P. placodicopsis, *Nyl. in Journ. Bot.* iv. 334 (1875); *Cromb. in Journ. Linn. Soc.* xv. 183. Lecanora melanaspis, *Ach.; Hook. f. et Tayl. Flor. Antarct.* 536.

On rocks, Christmas Harbour, *Hooker, Moseley*; Observatory Bay, sparingly, *Eaton*; Molloy Point, *Kidder*.

2. **Pannaria obscurior**, *Nyl.; Cromb. in Journ. Bot.* iv. 334 (1875), *et in Journ. Linn. Soc.* xv. 183.

On decayed moss, Observatory Bay, *Eaton*.

1. **Psoroma hirsutulum**, *Nyl.; Cromb. in Journ. Bot.* iv. 333 (1875), *et in Journ. Linn. Soc.* xv. 184.

On moss and dead stems of *Acæna*, very local, Observatory Bay, *Eaton*.

1. **Lecanora** (PLACOPSIS) **gelida**, *Linn.; Flor. Antarct.* ii. 535; *Tuckerm. in Bull. Torr. Bot. Club*, Oct. 1875, *et in Bull. U. S. Nat. Mus.* 3, 29; *Cromb. in Journ. Linn. Soc.* xv. 184, *et in Journ. Bot.* vi. 104, 106 (1877).

On stones, Christmas Harbour, *Hooker*; Observatory Bay, *Eaton*; Molloy Point, *Kidder*.

Var. VITELLINA, *Bab. in Flor. Antarct. l. c.; Cromb. l. c.* Christmas Harbour, *Hooker, Moseley*.

Var. LATERITIA, *Nyl.; Cromb. l. c.* Placodium bicolor, *Tuckerm. l. c.* Christmas Harbour, *Hooker, Moseley*. Swain's Bay and Royal Sound, *Eaton*.

2. **Lecanora** (PLACOPSIS) **macropthalma**, *Hook. f. et Tayl. in Lond. Journ. Bot.* iii. 660 (Urceolaria); *Tuckerm. in Bull. Torr. Bot Club, Oct.* 1875, *et in Bull. U. S. Nat. Mus.* 3, 29; *Cromb. in Journ. Linn. Soc.* xv. 185 and xvi., *et in Journ. Bot.* vi. 104.

On stones in moist places, *Hooker, Eaton,* &c.

Lecanora (PLACODIUM) **elegans**, *Ach.; Cromb. in Journ. Linn. Soc.* xv. 184, *et in Journ. Bot.* vi. 104 (1877); *Tuckerm. in Bull. U. S. Nat. Mus.* 3, 28. L. murorum, var. β, *Flor. Antarct.* 535.

On rocks and stones, Christmas Harbour, *Hooker;* Observatory and Swain's Bays, *Eaton.*

Var. LUCENS, *Nyl.; Cromb.* l. c. On dead stems of *Acæna* and *Pringlea*, Observatory Bay, *Eaton.*

4. **Lecanora subunicolor**, *Nyl.; Cromb. in Journ. Bot.* v. 19 (1876), *et in Journ. Linn. Soc.* xv. 184.

On rocks, Royal Sound, very sparingly, *Eaton.*

5. **Lecanora vitellinella**, *Nyl.; Cromb. in Journ. Bot.* iv. 334 (1875), *et* vi. 104 (1877), *et in Journ. Linn. Soc.* xv. 184. L. candelaria, *Ach.; Flor Antarct.* 537.

Maritime rocks, Christmas Harbour, *Hooker;* Observatory and Swain's Bays, *Eaton.*

6. **Lecanora cyphelliformis**, *Cromb. in Journ. Linn. Soc.* xvi.

Christmas Harbour, *Moseley.*

7. **Lecanora diphyella**, *Nyl.; Cromb. in Journ. Bot.* v. 21, *et in Journ. Linn. Soc.* xv. 184.

On rocks at low elevations, Observatory Bay, *Eaton.*

8. **Lecanora atro-cæsia**, *Nyl.; Cromb. in Journ. Bot.* iv. 334 (1875), *et* vi. 104 (1877), *et in Journ. Linn. Soc.* xv. 185. L. confluens, *Hook. f. et Tayl. in Lond. Journ. Bot.* iii. 636. L. albo-cœrulescans, *Ach.? Bab. in Flor. Antarct.* 538.

Rocks at Christmas Harbour, *Hooker;* Observatory Bay and stony slopes at Volage Bay, plentiful, *Eaton.*

9. **Lecanora brocella**, *Nyl.; Cromb. in Journ. Bot.* v. 21 (1876), *et* vi. 104 (1877), *et in Journ. Linn. Soc.* xv. 185.

On dead moss, &c., Christmas Harbour, *Hooker;* Observatory Bay, *Eaton.*

10. **Lecanora umbrina**, *Ach.; Cromb. in Journ. Linn. Soc.* xv. 185.

On dead plants, Observatory Bay, *Eaton.*

11. **Lecanora kergueliensis**, *Nyl.; Cromb. in Journ. Bot.* vi. 106 (1877). Urceolina kergueliensis, *Tuckerm. in Bull. U. S. Nat. Mus.* 3, 29.

Rocks at Molloy Point, *Kidder.*

12. **Lecanora sublutescens**, *Nyl.; Cromb. in Journ. Bot.* v. 21 (1876), *et in Journ. Linn. Soc.* xv. 186.

On a shaded sea cliff near Observatory Bay, colouring the rock, *Eaton*.

[L. CITRINA, *Ach.*, L. ERYTHROCARPA, *Fr.*, and L. HAGENI, *Ach.*, enumerated in the Flora Antarctica, 536, from very imperfect materials, are too doubtful to be enumerated.]

1. **Pertusaria perrimosa,** *Nyl.; Cromb. in Journ. Linn. Soc.* xv. 186; *Journ. Bot.* vi. 104. P. communis, *DC.; Flor. Antarct.* 540. ? Lecanora tartarea, *Ach.; Flor. Antarct.* 536.

On rocks, Christmas Harbour, *Hooker*; Observatory and Swain's Bays, *Eaton*.

2. **Pertusaria subferruginosa,** *Cromb. in Journ. Linn. Soc.* xv. 186.

On rocks, Observatory Bay, *Eaton*.

3. **Pertusaria cineraria,** *Nyl.; Cromb. in Journ. Linn. Soc.* xv. 186.

On rocks, Volage and Swain's Bays, *Eaton*.

1. **Lecidea variatula,** *Nyl.; Cromb. in Journ. Linn. Soc.* xv. 186.

On dead stems of *Acæna*, Observatory Bay, *Eaton*.

2. **Lecidea inundata,** *Fr.; Cromb. in Journ. Bot.* vi. 106 (1877). Biatora rubella, *Ehrh.; Tuckerm. in Bull. U. S. Nat. Mus.* 3, 29.

Molloy Point, *Kidder*.

3. **Lecidea assimilata,** *Nyl.; Cromb. in Journ. Linn. Soc.* xv. 187, *et in Journ. Bot.* vi. 104. L. aromatica *Ach. in parte Flor. Antarct.* 538.

Christmas Harbour, *Hooker*; Observatory Bay in turf, *Eaton*.

4. **Lecidea aromatica,** *Ach.; Nyl. in Journ. Bot.* vi. 104 (1877). *Flor. Antarct.* 538.

Christmas Harbour, *Hooker*.

5. **Lecidea enteroleuca,** *Fries?; Tuckerm. in Bull. Torr. Bot. Club,* Oct. 1875, *et in Bull. U. S. Nat. Mus.* 3, 30. *Cromb. in Bot. Journ.* vi. 106 (1877).

On dead grasses, Molloy Points, *Kidder*.

6. **Lecidea assentiens,** *Nyl.; Cromb. in Journ. Bot.* iv. 334, *et* vi. 105 (1877), *et in Journ. Linn. Soc.* xv. 187. L. contigua *var.* hydrophila, *Bab. in Flor. Antarct.* 538.

On rocks, Christmas Harbour, *Hooker*; Observatory Bay, *Eaton*.

7. **Lecidea intersita,** *Nyl.; Cromb. in Journ. Linn. Soc.* xv. 187.

On rocks, Observatory Bay, very sparingly; also ? one mile N.W. of Mount Crozier, *Eaton*.

8. **Lecidea phæostoma,** *Nyl.; Cromb. in Journ. Bot.* iv. 334 (1875), *et in Journ. Linn. Soc.* xv. 187.

On stones and bare soil, Observatory Bay, sparingly, *Eaton*.

9. **Lecidea amylacea,** *Ach.; Cromb. in Journ. Linn. Soc.* xv. 188 (1876), *et Journ. Bot.* vi. 104. L. spilota *Ach.; Bab. in Flor. Antarct.* 538. L. rivulosa, *Tayl. in Linn. Journ. Bot.* iii. 636.

On rocks and stones, Christmas Harbour, *Hooker*; Volage and Swain's Bays, *Eaton*.

10. **Lecidea subassentiens,** *Nyl.; Cromb. in Journ. Bot.* v. 21 (1876), *et in Journ. Linn. Soc.* xv. 188.

On rocks, Observatory Bay, very sparingly, *Eaton*.

11. **Lecidea perusta,** *Nyl.; Cromb. in Journ. Bot.* iv. 334 (1875), *et* vi. 105, 106 (1877), *et in Journ. Linn. Soc.* xv. 188. L. fusco-atra, *Ach.; Flor. Antarct.* 539; *Tuckerm. in Bull. Torr. Bot. Club, et in Bull. U. S. Nat. Mus.*, 3, 30.

On rocks, Christmas Harbour, *Hooker*; Observatory Bay, *Eaton*; Molloy Point, *Kidder*.

12. **Lecidea asbolodes,** *Nyl.; Cromb. in Journ. Bot.* v. 21 (1876); *et in Journ. Linn. Soc.* xv. 188.

On rocks, Observatory Bay, *Eaton*.

13. **Lecidea lygomma,** *Nyl.; Cromb. in Journ. Bot.* iv. 334 (1875), *et in Journ. Linn. Soc.* xv. 189.

On rocks, Observatory Bay, *Eaton*.

14. **Lecidea subcontinua,** *Nyl.; Cromb. in Journ. Linn. Soc.* xv. 189, *et in Journ. Bot.* vi. 104, 106 (1877). Urceolaria endochlora, *in parte Flor. Antarct.* 537.

On rocks and stones, Christmas Harbour, *Hooker*; Swain's Bay, *Eaton*.

Var. FERREA, *Nyl.; Cromb. l. c.* Swain's Bay, *Eaton*.

15. **Lecidea Eatoni,** *Cromb. in Journ. Bot.* iv. 334, 1875, *et in Journ. Linn. Soc.* xv. 189.

On rocks and boulders, Observatory, Volage, and Swain's Bays, *Eaton*.

16. **Lecidea homalotera,** *Nyl.* mss.; *Cromb in Journ. Bot.* vi. 105 (1877). Urceolaria endochlora, *Hook. f. et Tayl. in part. Flor. Antarct.* 537.

On rocks, Christmas Harbour, *Hooker*.

17. **Lecidea disjungenda,** *Cromb. in Journ. Bot.* vi. 105. Urceolaria endochlora, *Hook. f. et Tayl. in part. Flor. Antarct.* 537.

On rocks, Christmas Harbour, *Hooker*.

18. **Lecidea subplana,** *Nyl.; Cromb. in Journ. Bot.* iv. 334 (1875), *et in Journ. Linn. Soc.* xv. 189.

On boulders, sparingly, Observatory and Swain's Bays, *Eaton*.

19. **Lecidea stephanodes,** *Strn.; Cromb. in Journ. Linn. Soc.* 1877.

Kerguelen Island, *Moseley*.

20. **Lecidea Dicksonii,** *Ach.; Cromb. in Journ. Linn. Soc.* xv. 190. L. sincerula, *Nyl.; Cromb. in Journ. Bot.* v. 22 (1876).

On rocks, Royal Sound, Observatory, Volage, and Swain's Bays, *Eaton*.

21. **Lecidea tristiuscula,** *Nyl.; Cromb. in Journ. Linn. Soc.* xv. 190.

On stems, Volage Bay, sparingly, *Eaton*.

22. **Lecidea superjecta,** *Nyl.; Cromb. in Journ. Linn. Soc.* 1877.

Kerguelen Island, *Moseley*.

23. **Lecidea myriocarpa,** *DC.; Cromb. in Journ. Linn. Soc.* xv. 190, *et in Journ. Bot.* vi. 106 (1877). ? Buellia parasema, *Tuckerm. in Bull. Torr. Bot. Club, et in Bull. U. S. Nat. Mus.* 3, 30.

On rocks, Swain's Bay, sparingly, *Eaton*. Molloy Point, *Kidder*.

Var. ERUMPENS, *Crombie, l. c.* On dead *Acæna* stems, Observatory Bay (a single specimen), *Eaton*.

24. **Lecidea subplicata,** *Nyl.; Cromb. in Journ. Bot.* iv. 334 (1875), *et in Journ. Linn. Soc.* xv. 190.

On rocks, Observatory and Swain's Bays, *Eaton*.

25. **Lecidea cerebrinella,** *Nyl.; Cromb. in Journ. Bot.* v. 22 (1876), *et in Journ. Linn. Soc.* xv. 191.

On rocks, Observatory Bay, *Eaton*.

26. **Lecidea stellulata,** *Tayl. in Flor. Hibern.* 118; *Flor. Antarct.* 539; *Tuckerm. in Bull. Torr. Bot. Club, et in Bull. U. S. Nat. Mus.* 30; *Cromb. in Journ. Bot.* vi. 105 (1877).

On rocks, Christmas Harbour, *Hooker*; near Molloy Point, *Kidder*.

27. **Lecidea geographica,** *Linn.; Cromb. in Journ. Linn. Soc.* xv. 191, *et in Journ. Bot.* vi. 105 (1877). L. geographica, *var.* urceolata, *Schærer*; *Bab. in Flor. Antarct.* 539. Buellia geographica, *Tuckerm. in Bull. Torr. Bot. Club, et in Bull. U. S. Nat. Mus.* 3, 30.

On rocks frequent, *Hooker, Moseley, Eaton, Kidder*.

1. **Verrucaria tessellatula,** *Nyl.; Cromb. in Journ. Bot.* iv. 335 (1875), *et in Journ. Linn. Soc.* xv. 191.

On rocks and stones, Volage and Swain's Bays, and (where overflown by the tides) at Observatory Bay, *Eaton*.

2. **Verrucaria obfuscata,** *Nyl.; Cromb. in Journ. Bot.* v. 22 (1876), *et in Journ. Linn. Soc.* xv. 191.

On stones, hill N.W. of Mount Crozier, *Eaton*.

3. **Verrucaria æthiobola,** *Ach.; Cromb. in Journ. Linn. Soc.* xv. 193.

On rocks, Observatory Bay, *Eaton*.

4. **Verrucaria chlorotica,** *Ach.; Tuckerm. in Bull. Torr. Bot. Club, et in Bull. U. S. Nat. Mus.* 3, 30. (Sagedia); *Cromb. in Journ. Bot.* vi. 106 (1877).

Molloy Point, *Kidder*.

5. **Verrucaria prævalesens,** *Nyl.; Cromb. in Journ. Bot.* xv. 192.

Rocks at Observatory Bay, and ? hill N.W. of Mount Crozier (*Eaton*).

6. **Verrucaria kerguelina,** *Nyl.; Cromb. in Journ. Bot.* v. 22 (1876), *et in Journ. Linn. Soc.* xv. 192.

On rocks, Observatory Bay, sparingly, *Eaton*.

7. **Verrucaria insueta,** *Nyl.; Cromb. in Journ. Linn. Soc.* xv. 192.

On rocks and stones, Volage Bay, *Eaton*.

8. **Verrucaria congestula,** *Strn.; Cromb. in Journ. Linn. Soc.* 1877.

Kerguelen Island, *Moseley*.

[ISIDIUM OCULATUM, I. LUTESCENS, and LEPRARIA FLAVA, all enumerated in the Flora Antarctica as doubtful, are imperfect states of Lichens.]

V.—*Marine Algæ (exclusive of the Diatomaceæ).*

By G. DICKIE, A.M., M.D., F.L.S., Professor of Botany in the University of Aberdeen.

THE total number of marine species of *Algæ* known to be indigenous to Kerguelen Island (excluding *Diatomaceæ*) is 71. The collections upon which this estimate is based are those made respectively by—

Dr. Hooker (Antarctic Expedition) in the winter of 1840 (May—July), chiefly at Christmas Harbour, comprising 39 species;

Mr. Moseley (Challenger Expedition) in the summer of 1874 (January and February), chiefly at Christmas Harbour and the eastern coast as far as Betsy Cove, comprising 37 species;

Dr. Kidder (American Transit of Venus Expedition) in the spring and first part of the summer of 1874-5 (Sept. to Jan.) near Molloy Point, towards the entrance of Royal Sound, comprising 22 species;

And the Rev. A. E. Eaton (English Transit of Venus Expedition) in the spring and summer of 1874-5 (Oct. 11—Feb. 27), in the interior of Royal Sound (Observatory Bay) and in Swain's Bay, comprising 53 species.

The botanical results of the German Transit of Venus and Surveying Expedition, which was stationed for about two months at Betsy Cove, are not yet made known.

Mr. Eaton was at Observatory Bay during October, November, most of December, and the whole of February, during which time he made frequent use of the grapple. In Swain's Bay he collected *Algæ* on nine occasions between the 15th and the 30th of January inclusive. Of the 53 species in his collection 44 were obtained in Swain's Bay, and only 32 at Observatory Bay: 24 species (probably 27 or 28, *vide infra*) are common to both of the areas, 21 occurred (to Mr. Eaton) only in Swain's Bay, and 8 (from which 3 or 4 should be deducted, and added to the species common to both) were collected only in Observatory Bay. The preponderance of the Swain's Bay gatherings may partly be accounted for by the distance of Observatory Bay from the open sea. For Mr. Eaton noticed that in some very retired parts of Swain's Bay the components of the Alga flora and their state of growth were very similar to those prevailing at Observatory Bay. In advancing from the more sheltered to more open waters he observed considerable regularity maintained in the rate of change proceeding in the composition of the Alga flora; so that it was possible, while collecting in one place, to conjecture beforehand with tolerable accuracy the number of additional species that would be found in other positions more exposed to the slight swell that enters the bay from the outer sea. And he was of opinion that if it had been possible to have visited the coast external to the bay, 10 or 12 species would most likely have been added to the 53 in his collection. Judging from the number of species apparently indigenous to unsheltered situations which go to form the 18 that are not represented in his collection, this conjecture may have been not far from the mark.

But the advantages of situation afforded in Swain's Bay for the growth of various Algæ absent from the almost waveless shores of Observation Bay would have availed nothing, had it not been for the liberality and kindness of Captain Fairfax, R.N., in command of H.M.S. "Volage." Having invited Mr. Eaton to be his guest for three weeks, he conveyed him in his gig to almost every part of the bay that was accessible by boat in Kerguelen Island weather, and surrendered his cabins without reservation to the reception of buckets and specimens of all descriptions, excluding only seals and cetacea accommodated elsewhere.

The local distribution of the species round the coast may be ascertained roughly from a comparison of the constituents of the collections above mentioned. Of the 71 species, 14 are common to all of the collections, and 8 common to three out of the four, making together 22 species, which may be regarded as plants generally distributed round the island; 14 are common to Mr. Eaton's collection and one of the other three, and 1 species to Dr. Hooker's and Dr. Kidder's,—together making 15 local plants, mostly of frequent occurrence; 5 are common to Dr. Hooker's collection and Mr. Moseley's (gathered in Christmas Harbour), and 29 are in one of the collections only, making 34 scarce or rare species. Of the 29, there are in Dr. Hooker's collection 7 species, in Mr. Moseley's 4, in Dr. Kidder's 1, and in Mr. Eaton's 17.

As to their general geographical range, 20, or rather more than a quarter of them, are found in various parts of the shores of Europe, and some are cosmopolitan. The following 8, so far as is known, are peculiar to the island:—*Desmarestia chordalis, Sphacelaria corymbosa* and *S. affinis, Melobesia kerguelena, Nitophyllum fusco-rubrum, Epymenia variolosa, Ptilota Eatoni,* and *Callithamnion simile.*

The following are the numbers of the species after their respective families:—

Fucaceæ, 2.
Sporochnaceæ, 4.
Laminariaceæ, 2.
Dictyotaceæ, 2.
Chordariaceæ, 3.
Ectocarpaceæ, 3.
Rhodomelaceæ, 4.
Laurenciaceæ, 2.
Corallinaceæ, 3.

Sphærococcoideæ, 8.
Gelidiaceæ, 1.
Rhodymeniaceæ, 4.
Cryptonemiaceæ, 11.
Ceramiaceæ, 7.
Siphonaceæ, 3.
Ulvaceæ, 5.
Confervaceæ, 7.

Of these 16 belong to the Olive, 40 to the Red, and 15 to the Green Series.

There are also included in the present paper, for convenience, 4 freshwater species:—*Bostrychia vaga, Vaucheria Dillwynii, Ulva cristata,* and *Prasiola fluviatilis.*

1. **D'Urvillea utilis,** *Bory; Flor. Antarct.* 454; *Dickie in Journ. Linn. Soc.* xv. 43, 198; *Farlow in Bull. U. S. Nat. Mus.* 3, 30.

On exposed rocks at and below half-tide level, not in very sheltered situations; abundant.—In the Southern ocean, between lat. 45° and 55° S., reaching to lat. 65° S. in the meridian of New Zealand (Hooker).

2. **D'Urvillea Harveyi,** *Hook. f. Flor. Antarct.* 456, t. clxv., clxvi.; *Dickie in Journ. Linn. Soc.* xv. 44, 198; *Farlow in Bull. U. S. Nat. Mus.* 3, 30.

In positions still more open than *D. utilis.* (Cape Horn and the Falklands.)

1. **Desmarestia Rossii,** *Hook. f. and Harv., Flor. Antarct.* 467, t. clxxii., clxxiii.; *Dickie in Journ. Linn. Soc.* xv. 44, 198.

Swain's Bay on rocks in 3 fathoms, at the end of an island about 2 miles within the entrance of the bay, exposed to a slight swell from the open sea; local and not common, *Eaton*. (Fuegia, Falkland Islands, Heard Island, *Moseley*.)

2. **Desmarestia chordalis,** *Hook. f. and Harv., Flor. Antarct.* 467; *Dickie in Journ. Linn. Soc.* xv. 44, 198.

Swain's Bay, in 3 fathoms, with the preceding; very local, *Eaton*. Christmas Harbour, *Hooker, Moseley*. (Kerguelen Island only.)

A very graceful species. The fronds, upwards of 4 feet in length, are arranged in a manner similar to those of a fern, and cause the plant, as seen *in situ* from a boat, to bear a general resemblance in contour to such species as *Aspidium filix-mas.*

3. **Desmarestia aculeata,** *Lyngb.* Var. MEDIA, *Grev.; Hook. f. and Harv., Flor. Antarct.* 466; *Dickie in Journ. Linn. Soc.* xv. 44, 198.

Between tide-marks, Swain's Bay, *Eaton*. Cockburn Island (*Hooker*); and in North temperate and Arctic seas.)

4. **Desmarestia viridis,** *Lamx.;* D. VIRIDIS (and var. β. DISTANS), *Hook., Flor. Antarct.* ii. 466; *Dickie in Journ. Linn. Soc.* xv. 44, 198; *Farlow in Bull. U. S. Nat. Mus.* 3, 30.

Christmas Harbour, *Hooker, Moseley*. In 2 fathoms, Royal Sound and Swain's Bay; common, *Eaton*. (Marion Island; the Falklands; Cape Horn; American coast from New York northwards; Unalaschka; Hunde Island; W. coast of Europe.)

1. **Macrocystis pyrifera,** *Ag.; Flor. Antarct.* 461, t. clxix., clxx.; *Dickie in Journ. Linn. Soc.* xv. 44, 198; *Farlow in Bull. U. S. Nat. Mus.* 3, 30.

Abundant along rocky portions of the coast. (Antarctic Sea, from lat. 40° to 60° S.; New Zealand; Indian Ocean; Marion Island; Chili; California.)

1. **Lessonia fuscescens,** *Bory.; Flor. Antarct.* 457, t. clxvii., clxviii. A., and clxxi. D.; *Dickie in Journ. Linn. Soc.* xv. 44.

In exposed situations; Christmas Harbour, rare, *Hooker* and *Moseley*. (Chili, Fuegia, Falkland Islands, Cockburn Island, Auckland and Campbell Islands.)

1. **Asperococcus sinuosus,** *Roth; Flor. Antarct.* ii. 468; *Dickie in Journ. Linn. Soc.* xv. 198.

Crevices of rocks between tide-marks, Observatory Bay (two very small specimens), *Eaton*. (Widely distributed from the latitude of Spain to the Falklands; Florida; California; Japan.)

1. **Dictyosiphon fasciculatus,** *Hook. f. and Harv. Flor. Antarct.* 178, 467, t. lxix. 1.

Christmas Harbour, *Hooker*. (Falkland and Auckland Islands.)

1. **Adenocystis Lessonii,** *Hook. f. and Harv. Flor. Antarct.* i. 179, 468, t. lxix. 2 (details); *Dickie in Journ. Linn. Soc.* 1876, xv. 44, 198; *Farlow in Bull. U. S. Nat. Mus.* 3, 30.

Between tide-marks, abundant; scarcer but more finely grown in shallow estuaries (there, occasionally, as much as 5 or 6 inches long); Christmas Harbour, Royal Sound, Swain's Bay, &c. (Cape Horn; Falklands; Cockburn Island; Auckland and Campbell Islands.)

1. **Scytosiphon lomentarium,** *Grev.; Flor. Antarct.* 468 (*Chorda*).

Christmas Harbour, *Hooker*. (Falkland and Auckland Islands, Pacific Ocean to Japan and S. America; the Atlantic from the Faröe Islands to Cadiz; Mediterranean.)

1. **Elachista flaccida,** *Aresch.; Dickie in Journ. Linn. Soc.* xv. 199.

On *Rhodymenia palmata* in very shallow water along the shore in Observatory Bay, *Eaton*. (Atlantic coasts of France and Britain; Baffin's Bay.)

1. **Ectocarpus geminatus,** *Hook. f. and Harv. Flor. Antarct.* 469; *Dickie in Journ. Linn. Soc.* xv. 44, 199.

Plentiful on *Desmarestia* at Christmas Harbour, *Moseley*. Very slender solitary young plants on Annelid tubes, at 5 fathoms; stronger and more bushy, with trichosporangia only (but these abundant), on *Mytilus* at 1 fathom, and in tide pools, Observatory Bay; frequent in Swain's Bay among *Cladophora flexuosa* in pools and shallow water, *Eaton*. (Falklands and Cape Horn.)

1. **Sphacellaria corymbosa,** *Dickie in Journ. of Bot.* v. 50 (1876), *et in Journ. Linn. Soc.* xv. 199. ? S. funicularis, *Mont.; Flor. Antarct.* 469; *Farlow in Bull. U. S. Nat. Mus.*, 3, 30; fronde estuposa, filis cæspitosis, ramis inferne paucis dichotomis superne subpinnatim decompositis, ramulis alternis corymbosis.

On shells of *Mytilus* and on Annelid tubes; Swain's and Observatory Bays, *Eaton*.

The specimens are 2 to 3 inches long, cæspitose, but without fruit. Dr. Hooker's plant obtained at the Falklands, and Dr. Kidder's from the vicinage of Molloy Point in Royal Sound, are probably the same as the preceding.

2. **Sphacellaria affinis,** *Dickie in Journ. of Bot.* v. 50 (1876), *et in Journ.*

Linn. Soc. xv. 199; filis dense cæspitosis erectis parce dichotomis, articulis diametro subæqualibus vel paulo longioribus, trichosporangiis solitariis obovatis breviter pedicellatis.

On shells of *Mytilus* in rather open situations; Swain's Bay, *Eaton*.

The specimens are about ½ inch in height, and are similar in habit to the British *S. radicans*.

1. **Rhodomela Hookeriana,** *Ag.; Rhodomela Gaimardi Hook. f. and Harv. Flor. Antarct.* 481, t. clxxxiv. (*non Gaud.*); *Farlow in Bull. U. S. Nat. Mus.* 1876, 3, 30; *Dickie in Journ. Linn. Soc.* xv. 199.

Swain's and Observatory Bays, frequent, *Eaton*; near Molloy Point, one specimen, *Kidder*. (Falklands and Fuegia.)

1. **Polysiphonia abscissa,** *Hook. f. and Harv. Flor. Antarct.* 480, t. clxxxiii. 2; *Dickie in Journ. Linn. Soc.* xv. 199.

Forma microcarpa; P. microcarpa, *Hook. f. and Harv.* 479, t. clxxxii. 3; *Harv. Ner. Aust.* 42.

On roots of *Macrocystis*, and on tubes of Annelides, in 1 to 5 fathoms, Observatory Bay, *Eaton*. (New Zealand; Tasmania; Fuegia).

There are two forms of this species, one of them smaller and more rigid than the other. The *P. microcarpa* of the Flora Antarctica represents one of them, *P. abscissa* the other.

2. **Polysiphonia anisogona,** *Hook. f. and Harv. Flor. Antarct.* 478, t. clxxxii. 2; *Dickie in Journ. Linn. Soc.* xv. 44.

Kerguelen Island, *Moseley*. (Falklands and Fuegia, *Hooker*.)

1. **Dasya Berkeleyi,** *Mont.; Dickie in Journ. Linn. Soc.* xv. 44, et 199 (var. β *Davisii*); *Farlow in Bull. U. S. Nat. Mus.* 3, 30. Polysiphonia punicea, *Hook. f. and Harv. Flor. Antarct.* i. 182; (Heterosiphonia) Berkeleyi? var. β *Davisii, idem,* 480.

Swain's Bay, on the seaward sides of islands, *Eaton*; Royal Sound, *Kidder*. (Auckland Islands, Marion Island, Falklands, Fuegia, Chilöe.)

Mr. Eaton's specimens belong to the var. β *Davisii*, having a habit and colour different from those of the typical plant. There are examples in different stages; but in all the ramuli are heterosiphonous.

1. **Bostrychia vaga,** *Hook. f. and Harv. Flor. Antarct.* 484, pl. clxxxvi. i. (Stictosiphonia).

Christmas Harbour, on rocks and stones above high water mark, and in damp places a considerable distance from the sea, abundant, *Hooker*.

1. **Delisea pulchra,** *Mont.; Flor. Antarct.* 484; *Dickie in Journ. Linn. Soc.* xv. 45.

Christmas Harbour, *Hooker, Moseley*. (Heard Island, S. Tasmania, W. and E. Australia.)

1. **Ptilonia magellanica,** *Mont.; Dickie in Journ. Linn. Soc.* xv. 45, 200;

Farlow in Bull. U. S. Nat. Mus. 1876, 3, 30. Plocamium ? magellanicum, *Hook. f. and Harv. Flor. Antarct.* 474. Thamnophora magellanica, *Mont.*

Christmas Harbour, *Hooker, Moseley*. In tideways and on parts of islands open to a slight swell from the outer sea, not in sheltered waters; Swain's Bay, *Eaton*; Royal Sound, *Kidder*. (Falklands and Fuegia.)

1. **Melobesia antarctica,** *Ag.; Dickie in Journ. Linn. Soc.* xv. 200. M. verrucata, *Lamx; Dickie l. c.* 45. M. verrucata, var. antarctica, *Hook. and Harv. Flor. Antarct.* 482.

On *Ballia, &c.,* Swain's Bay, *Eaton;* Christmas Harbour, *Hooker*. (Fuegia, Falklands, Tasmania, New Zealand, Auckland, Antarctic Seas.)

2. **Melobesia lichenoides,** *Ell. and Sol.; Dickie in Journ. Linn. Soc.* xv. 200.

Swain's Bay, common, *Eaton*. The only example preserved was grappled in about 2 fathoms in a tideway between two islands, incrusting two sponges (*Microciona atrosanguinea,* Bk., and *Halichondria incrustans,* Jtn.; both British species.) It is normal in habit, texture, and in the character of the keramidia, and is very luxuriant, measuring about 3 inches square. (St. Paul's Island, Norfolk Island, the Mediterranean, France, Britain, Baffin's Bay.)

3. **Melobesia kerguelena,** *Dickie in Journ. Bot.* v. 51, 1876, *et in Journ. Linn. Soc.* xv. 200; simplex, 2½ poll. diam., dura, crassa, tantum in medio subtus adhærens, subconvexa, circumscriptione orbiculari, margine lævi parce undulata, keramidiis conspicuis numerosis plerumque in seriebus concentrice dispositis.

Swain's Bay in 2-3 fathoms, with the preceding, *Eaton*.

Mr. Eaton has an impression that this grows upon *Ballia* or *Ptilota,* but I should rather be disposed to suspect that it was attached to rocks. The description was taken from an almost complete specimen; there are fragments of others whose contour is less regular, probably through interference of external objects. All of them are in colour of a very pale buff or dull yellowish hue, varied with pale red tints.

1. **Delesseria Lyallii,** *Hook. f. and Harv. Flor. Antarct.* 471, t. clxxvi; *Dickie in Journ. Linn. Soc.* xv. 45, 200; *Farlow in Bull. U. S. Nat. Mus.* 1876, 3, 30.

Christmas Harbour, *Hooker* (but not seen attached), *Moseley*. Observatory Bay (ill grown), and Swain's Bay (well grown), abundant, *Eaton*. (Marion Island, Falklands.)

Dr. Hooker obtaining only wrecked specimens at Christmas Harbour, was led to suppose that this species was a resident of the exposed coast. It is however common in the very sheltered waters of Observatory Bay, though it certainly does not thrive there; and it is abundant and luxuriant round the islands in Swain's Bay, in 3-5 fathoms.

2. **Delesseria Davisii,** *Hook. f and Harv., Flor. Antarct.* 470, t. clxxv.; *Dickie in Journ. Linn. Soc.* xv. 45, 200.

Swain's Bay; normal on *Mytilus* in sheltered places at or just below low-water mark; varying in more open situations, *Eaton*. (Falklands and Fuegia, *Hooker*.)

3. **Delesseria quercifolia,** *Bory.; Flor. Antarct.* 471; *Dickie in Journ. Linn. Soc.* xv. 200.

Swain's Bay, frequent, *Eaton*. (Falklands and Fuegia, *Hooker*.)

4. **Delesseria crassinervia,** *Mont.; Flor. Antarct.* 471; *Dickie in Journ. Linn. Soc.* xv. 200.

Swain's Bay, near the surface of the water; Observatory Bay, only one ill-grown example, *Eaton*. (Fuegia, Falkland, Auckland and Campbell Islands.)

1. **Nitophyllum fusco-rubrum,** *Hook. f. and Harv., Flor. Antarct.* 472; *Dickie in Journ. Linn. Soc.* xv. 45, 200; *Farlow in Bull. U. S. Nat. Mus.* 1876, 3, 30.

Christmas Harbour, *Hooker*. Abundant in open water in Swain's Bay, not found in sheltered places, *Eaton*. There is a variety in the collection with coccidia. Mouth of Royal Sound, *Kidder*.

2. ?**Nitophyllum multinerve,** *Hook. f. and Harv., Flor. Antarct.* 473; *Dickie in Journ. Linn. Soc.* xv. 45.

Christmas Harbour (one specimen referred to this species with doubt, *Moseley*). (Falklands and Fuegia, *Hooker*.)

3. **Nitophyllum lividum,** *Hook. f. and Harv., Flor. Antarct.* 472, t. clxxix.; *Farlow in Bull. U. S. Nat. Mus.* 1876, 3, 30; *Dickie in Journ. Linn. Soc.* xv. 201.

In very sheltered water (one example only) at 6 to 10 fathoms, Swain's Bay, *Eaton*. Royal Sound, *Kidder*. (Falklands, *Hooker*.)

4. **Nitophyllum laciniatum,** *Hook. f. and Harv., in London Journ. of Bot.* iv. 256; *Dickie in Journ. Linn. Soc.* xv. 201. N. Bonnemaisoni *var.* laciniatum, *Hook. f. and Harv. l. c.* 474.

Swain's Bay, frequent, in 3 to 5 fathoms, *Eaton*. (Falklands and Fuegia, *Hooker*.)

1. **Chætangium variolosum,** *Mont.; Dickie in Journ. Linn. Soc.* xv. 45, 201. Notogenia variolosa, *Mont.; Flor. Antarct.* 487.

Christmas Harbour, very abundant, *Hooker, Moseley*. Observatory and Swain's Bays, abundant between tide marks. (Fuegia, Falklands, Auckland Islands.)

1. **Plocamium Hookeri,** *Harv. in Flor. Antarct.* 474; *Dickie in Journ. Linn. Soc.* xv. 45, 201.

Christmas Harbour, *Hooker, Moseley*. Swain's Bay, local, at 2 to 3 fathoms, in situations open to a slight swell from the outer sea, *Eaton*. (Heard Islands, *Moseley*.)

1. **Rhodophyllis capensis,** *Kütz.; Dickie in Journ. Linn. Soc.* xv. 201.

Swain's and Observatory Bays, sparingly, on tubes of Annelides, in 3–5 fathoms, *Eaton.* (Table and Simon's Bays.)

The few specimens collected by Mr. Eaton are dwarf and very narrow. They all have the structure of the genus, and must, I think, be referred to the above species.

1. **Rhodymenia palmata,** *Linn.; Flor. Antarct.* 475; *Farlow in Bull. U. S. Nat. Mus.* 1876, 3, 30; *Dickie in Journ. Linn. Soc.* xv. 201.

Swain's Bay and Royal Sound, common in tide pools and shallow water; very luxuriant specimens. Also a dwarf form of olivaceous complexion, growing between tide marks, dry at low water, in Swain's Bay, *Eaton.* (Falkland Islands; Fuegia; Unalaschka; Greenland; Newfoundland; Scandinavian, British, and French coasts.)

2. **Rhodymenia corallina,** *Grev.; Flor. Antarct.* 475; *Farlow in Bull. U. S. Nat. Mus.* 1876, 3, 30.

On roots of *Macrocystis*, Christmas Harbour, rare, *Hooker*; Royal Sound, *Kidder*.

1. **Phyllophora cuneifolia,** *Hook. f. and Harv., Flor. Antarct.* 486; *Dickie in Journ. Linn. Soc.* xv. 201. P. Brodiæi, *Turn.? Flor. Antarct. l. c.; Dickie l. c.*

Christmas Harbour, rare, *Hooker.* Swain's Bay, rare, in very sheltered water, at 5–10 fathoms, *Eaton.* (Falkland Islands.)

Professor Agardh (loc. cit.) considers with the authors of this species that it is probably a form of *P. Brodiæi*.

1. **Ahnfeltia plicata,** *Huds.; Dickie in Journ. Linn. Soc.* xv. 46, 201. Gigartina plicata, *Hook. f. and Harv., Flor. Antarct.* 487.

Local, between tide marks near Observatory Bay, *Eaton*; Christmas Harbour, abundant, *Hooker.* (Falkland Islands; temperate and colder seas in the northern hemisphere.)

1. **Callophyllis variegata,** *Bory.; Dickie in Journ. Linn. Soc.* xv. 46, 201; *Farlow in Bull. U. S. Nat. Mus.* 1876, 3, 31. Rhodymenia variegata (*in part*), *Hook. f. and Harv. Flor. Antarct.* 475.

Christmas Harbour, *Hooker*. Swain's and Observatory Bays, in sheltered situations, *Eaton*. Royal Sound, *Kidder*. (Auckland Islnnds; New Guinea; Falklands; Fuegia; Chili; Peru; California.)

Mr. Eaton's collection comprises different forms of this very variable species:— from Observatory Bay, on *Mytilus* in sheltered water, a variety with small marginal kalidia; from Swain's Bay var. β *atro-sanguinea*, also a narrow variety (?), torn at the apex and proliferous; and in addition var. γ on roots of *Macrocystis* in very sheltered water.

2. **Callophyllis dichotoma,** *Kutz.* Rhodomenia dichotoma, *Hook. f. and Harv., Flor. Antarct.* 186, t. lxxii. 1.

Swain's Bay, one specimen only, *Eaton.* (Marion Island, *Moseley;* Campbell Island, *Hooker.*)

The specimen obtained at Kerguelen Island by Mr. Eaton has the structure and kalidia of *Callophyllis.* The last are not marginal, and therefore it is not a form of *C. variegata.* [This species was not mentioned in Dr. Dickie's MS., nor in his list in the Linnæan Society's Journal; but the name and remarks were noted by him on the sheet containing the specimen in the collection, A. E. E.]

3. **Callophyllis tenera,** *J. Ag.; Dickie in Journ. Linn. Soc.* xv. 202.

Local in very sheltered water, Swain's Bay, *Eaton.* (South Shetlands.)

1. **Kallymenia dentata,** *Suhr.* (Halymenia), *Dickie in Journ. Linn. Soc.* xv. 46 (*vars. α and γ*), 202.

Swain's Bay; and (at 1 fathom, of inferior growth) Observatory Bay, *Eaton.* (Cape of Good Hope.)

1. **Gigartina Radula,** *Esp.; Dickie in Journ. Linn. Soc.* xv. 46, 202; *Farlow in Bull. U. S. Nat. Mus.* 1876, 3, 30. Iridæa Radula, *Hook. f. and Harv., Flor. Antarct.* 485.

Christmas Harbour, *Hooker.* Swain's and Observatory Bays, abundant on rocks from low-water mark to 1 fathom or more, *Eaton.* (Cape of Good Hope; New Zealand; Auckland and Campbell Islands; California.)

The collection includes various forms of this species.

1. **Iridæa capensis,** *J. Ag.; Dickie in Journ. Linn. Soc.* xv. 46.

Kerguelen Island, *Moseley.* (Cape of Good Hope.)

2. **Iridæa laminarioides,** *Bory.; Dickie in Journ. Linn. Soc.* xv. 46.

Kerguelen Island, *Moseley.* (Auckland Islands, and the S.W. shores of Chili.)

Mr. Moseley's collection contains several specimens which belong, I think, to this species.

1. **Epymenia variolosa,** *Kütz.; Dickie in Journ. Linn. Soc.* xv. 45. Rhodymenia variolosa, *Hook. f. and Harv., Flor. Antarct.* 476, clxxx.; *Dickie l. c.; Farlow in Bull. U. S. Nat. Mus.* 3, 30.

Christmas Harbour, *Hooker.* Swain's Bay, local, *Eaton.* Royal Sound, *Kidder.*

1. **Halymenia latissima,** *Hook. f. and Harv., Flor. Antarct.* 189, t. lxxiii., 1, 2; *Dickie in Journ. Linn. Soc.* xv. 202.

Swain's and Observatory Bays; common on rocks at low-water mark, and on *Mytilus* at 1 fathom, *Eaton.* (Auckland and Campbell Islands, *Hooker.*)

1. **Ceramium rubrum,** *Ag.; Flor. Antarct.* 488; *Dickie in Journ. Linn. Soc.* xv. 46, 202. C. rubrum var. secundatum, *Lyngb.; Farlow in Bull. U. S. Nat. Mus.* 3, 31.

Christmas Harbour, very abundant, *Hooker.* Common in Swain's Bay, Royal Sound, and near Vulcan Cove, *Eaton.* (General in the colder seas of both hemispheres.)

2. Ceramium diaphanum, *J. Ag.; Flor. Antarct.* 488.

Christmas Harbour, abundant, *Hooker.* (Cape of Good Hope and Atlantic coasts of Europe.)

1. Ptilota Eatoni, *Dickie in Journ. of Bot.* v. 51, 1876; *et in Journ. Linn. Soc.* xv. 202.;

rachide filiformi 5-6-pollicari, pinnis oppositis inæqualibus, unâ majore alternâ minore, pinnulis pinnarum longiorum apices versus pectinatis, reliquiis bases harum versus, cæterisque omnibus subulatis ex serie articulorum magnorum subsimplici structis, sphærosporis ad apices pinnularum subsolitariis, favellis terminalibus, ramis involucri 4–5 pectinatis conniventibus.

Swain's Bay, in 2 to 5 fathoms, on the side and end of a promontory and of an island about two miles from the entrance of the bay, exposed to the tide and a slight swell from the outer sea; very local. Usually parasitic upon *Ballia*, sometimes attached to *Mytilus*; *Eaton.*

This species resembles *P. Harveyi* in the character of the involucre, but differs from it in general habit, and in the structure of the larger and smaller pinnules. It is also dissimilar in colour, being dull purple.

PLATE V., Fig. iii.:—1, portion of frond of *nat. size*; 2, portion of stem with young branch; 3, apex of fully grown branch; 4, ditto with sphærospores; 5, sphærospores; all much *enlarged.*

1. Ballia callitricha, *Ag.; Dickie in Journ. Linn. Soc.* xv. 46, 202; *Farlow in Bull. U. S. Nat. Mus.* 1876, 3, 31.

Ballia Brunonis *var.* β Hombroniana, *Hook. f. and Harv., Flor. Antarct.* 488.

On *Mytilus*, roots of *Macrocystis*, and Annelid tubes, from tide pools down to 6 fathoms; very common in Christmas Harbour, Swain's Bay, and Royal Sound. (Falklands; Marion Island; Australia; Tasmania; New Zealand; Auckland Islands.)

1. Callithamnion simile, *Hook. f. and Harv., Flor. Antarct.* 489; *Dickie in Journ. Linn. Soc.* xv. 202.

Christmas Harbour, rare, *Hooker.* On *Mytilus*, Annelid tubes, and roots of *Macrocystis*, in 1 to 5 fathoms, in Observatory and Swain's Bays; frequent, *Eaton.*

2. Callithamnion Ptilota, *Hook. f. and Harv., Flor. Antarct.* 489, t. clxxxix. 1; *Farlow. in Bull. U. S. Nat. Mus.* 3, 31.

Royal Sound, *Kidder.* (Crozets, *Hooker.*)

3. Callithamnion Rothii, *Lyngb.; Dickie in Journ. Linn. Soc.* xv. 203.

In tide pools and at the extreme verge of low water, on *Mytilus*, in Swain's and Observatory Bays, local, *Eaton.* (Atlantic shores from Greenland to Africa; N.E. shores of the United States.)

I can see no essential difference between Mr. Eaton's specimens and the plant from the northern hemisphere. They agree in habit, and in the arrangement of the tetraspores. The articulations are a little longer than those of British examples.

1. **Codium adhærens**, *Ag.; Farlow in Bull. U. S. Nat. Mus.* 1876, 3, 31; *Dickie in Journ. Linn. Soc.* xv. 203.

On rocks in about 2 fathoms in Observatory Bay; frequent, *Eaton*. (Europe; Cape of Good Hope; Mauritius; Ceylon; Australia; Friendly and Loo-choo Islands.)

2. **Codium tomentosum**, *Stackh.; Flor. Antarct.* 491; *Dickie in Journ. Linn. Soc.* xv., 46.

Christmas Harbour, *Hooker*. (Tongabu; Banda Islands; and the colder seas of both hemispheres).

1. **Bryopsis plumosa**, *Grev.; Flor. Antarct.* 492; *Dickie in Journ. Linn. Soc.* xv. 203.

Dwarf or very young specimens on Annelid tubes in 5 fathoms, Observatory Bay, scarce, *Eaton*. (Greenland; widely distributed throughout both the temperate zones, and even in some of the warmer seas).

1. **Vaucheria Dillwynii**, *Ag.; Flor. Antarct.* ii. 492.

On the ground amongst Penguin rookeries, Christmas Harbour, *Hooker*.

1. **Ulva latissima**, *Linn.; Flor. Antarct.* 499; *Dickie in Journ. Linn. Soc.* xv. 47, 203; *Farlow in Bull. U. S. Nat. Mus.*, 1876, 3, 31.

Christmas Harbour, very common, *Hooker*. Royal Sound; Swain's Bay, *Eaton*. (Widely distributed in both hemispheres.)

2. **Ulva** (?) **cristata**, *Hook. f. and Harv.; Flor. Antarct.* 499.

In moist clefts of rocks overhanging Christmas Harbour, growing with *Trypothallus* [*Palmodictyon*, Ktz.] *anastomosans*, *Hooker*.

1. **Porphyra laciniata**, *Ag.; Flor. Antarct.* 500; *Dickie in Journ. Linn. Soc.* xv. 46, 203.

Christmas Harbour, very abundant, *Hooker*. Common in shallow water, Observatory Bay. (Temperate and colder seas of both hemispheres.)

2. **Porphyra vulgaris**, *Ag.; Flor. Antarct.* ii. 500; *Dickie in Journ. Linn. Soc.* xv. 46.

Christmas Harbour, very abundant, *Hooker*. (Widely distributed in the northern and southern hemispheres.)

1. **Enteromorpha compressa**, *Link.; Flor. Antarct.* 500; *Dickie in Journ. Linn. Soc.* xv. 46, 203.

Very abundant on rocks and in tide-pools; Christmas Harbour, *Hooker*; Royal Sound and Swain's Bay, *Eaton*. (Cosmopolitan).

2. **Enteromorpha intestinalis**, *Link.; Flor. Antarct.* 500.

Christmas Harbour, *Hooker*. (Cosmopolitan.)

1. **Prasiola fluviatilis,** *Sommerfeldt, Supplem. Fl. Lapp.* 1826 (*teste Dickie in Arctic Manual,* 1876.) P. Sauteri, *Meneghini MS., Kütz. Sp. Alg.* 1849, p. 473; *Dickie in Journ. Linn. Soc.* xv. 203.

On wet rocks in the bed of a streamlet descending from a patch of snow, about 400 feet above the sea, on the pyramidal hill on the W. side of Swain's Bay. (European Alps to 9,300 ft.; Spitsbergen; streams of W. Greenland.)

1. **Cladophora rupestris,** *Linn.; Flor. Antarct.* 495; *Dickie in Journ. Linn. Soc.* xv. 47.

Christmas Harbour, on rocks, *Hooker.* (General between the arctic circle and the Mediterranean; only at Kerguelen Island in the southern hemisphere).

2. **Cladophora arcta,** *Ktz.; Flor. Antarct.* 495; *Dickie in Journ. Linn. Soc.* xv. 47, 203. C. Hookeriana, *Ktz. Sp. Alg.,* p. 418.

Very abundant on *Mytilus* at low-water mark, and in shallow water, Observatory Bay, *Eaton.* (Falkland Islands; Fuegia; German and N. Atlantic Oceans; Greenland.)

3. **Cladophora simpliciuscula,** *Hook. f. and Harv. Flor. Ant.* 496, t. cxcii. 4, 1–3; *Dickie in Journ. Linn. Soc.* xv. 203.

One very small specimen, probably very young, too imperfect to be identified with absolute certainty, but which I think may be referred to this species, was obtained on an Annelid tube in Observatory Bay at 5 fathoms, *Eaton.* (Falklands and Fuegia, *Hooker.*)

4. **Cladophora flexuosa,** *Griff; Dickie in Journ. Linn. Soc.* xv. 203.

In tide-pools and at 5 fathoms in Observatory and Swain's Bays, specimens from the deeper water are poor, *Eaton.* (Shores of Europe; Massachusetts Bay.)

1. **Rhiozoclonium riparium,** *Roth.;* Cladophora riparia, *Hook. f. and Harv. Flor. Antarct.* 495.

Christmas Harbour, *Hooker.* (Cumberland Sound; British coasts, &c.)

2. **Rhizoclonium ambiguum,** *Ktz.; Conferva ambigua, Hook. f. and Harv. Flor. Antarct.* 494, t. cxci. 1.

Christmas Harbour, in the sea, *Hooker.*

1. **Chætomorpha Linum,** *Ktz.;* Conferva Linum, *Roth.; Flor. Antarct.* 493.

Christmas Harbour, on rocks near high-water mark, *Hooker.*

VI.—*Fresh-water Algæ collected by the Rev. A. E. Eaton.*

Algæ aquæ dulcis Insulæ Kerguelensis,

auctore

PAULO FRIDERICO REINSCH.

(Cum notulis de distributione geographica a G. Dickie adjectis.)

Tanto ampliores notitiæ de plantis simplicissima structura ac simplicissimis organis in terris diversissimis crescentibus, quo evidentius factum agnoscitur notandum: has plantas iisdem legibus non subjectas ex quibus dependet plantarum in systemate superiorum diffusio in orbi. Specierum plantarum microscopicarum diffusio universa determinatus rationibus peculiaribus: diffusione aeris meatus in superficie terræ effecta, mobilitate levissima cellularum propagativarum earumque vi vitali diu permanente in statu ipso siccato; neque minus efficitur diffusio rationibus vitæ multo simplicioribus accommodatis ad diversissima terræ cæla.

Materia hujus enumerationis algarum Insulæ selecta a Cl. A. Eaton in expeditione transitus Veneris in hieme 1874–5 continet numeros quatuordecim. Speciminum inquirendi causa ab Herbario Regio Kewensi mihi traditorum sunt: ampullulæ tres cum algis aquæ dulcis in spiritu vini asservatis, decem folia cum algis siccatis et capsula parvula cum algis siccatis. Omnia quæ ampullula major continuerat: Specimina compluria muscorum aquatilium densissime algis variis (*Schizosiphon* Spec. nova et *Nostoc* Spec. var.) vestita cum singulis speciminibus *Nitellæ antarticæ* et *Vaucheriarum* cæspitulis parvulis intermixta: mihi dedit materiam pro maximam partem hujus enumerationis. In hac ampullula inventæ erant 81 species algarum aquæ dulcis ad 45 genera spectantes; harum algarum sunt novæ species 28, nova genera 3.

Omnes in hac enumeratione receptæ species sunt conjunctæ in præparatorum collectione integra, nunc in Herbario Regio Kewensi deposita.

Insulæ Kerguelensis Specierum algarum aquæ dulcis hucusque cognitarum numerus totus est 106, numerus generum 67.

Ab his sunt

Diatomophyceæ	21 species, 13 genera.
Phycochromophyceæ	33 species, 18 genera.
Chlorophyllophyceæ	50 species, 34 genera.
Melanophyceæ et Rhodophyceæ	2 species, 2 genera.

Omnes Familiæ Algarum aquæ dulcis, *Ulothrichaceis, Chroolepideis, Confervaceis, Sphæropleaceis* exceptis, inveniuntur in Insula Kerguelensi. A *Cladophoris, Chætophoris, Rhizocloniis* nulla species est observata. In ordinem systematis adducta Flora Algarum aquæ dulcis Insulæ hucusque cognita, est hæc.

Diatomophyceæ.

13 genera, 21 species.   (2 Spec. novæ, 3 Spec. incert.)

Phycochromophyceæ.

Chroococcaceæ.—5 genera, 6 species (ab his 1 novæ, 1 incerta).
Oscillariaceæ.—3 genera, 3 species (ab his 2 novæ, 1 forma nova).
Nostochaceæ.—3 genera, 10 species (ab his 7 novæ, 1 forma nova).
Rivulariaceæ.—3 genera, 5 species (ab his 2 novæ, 2 formæ novæ).
Scytonemaceæ.—3 genera, 4 species (ab his 1 nova, 2 formæ novæ).
Sirosiphoniaceæ.—1 genus, 5 species (ab his 4 novæ, 1 forma nova).

Chlorophyllophyceæ.

Palmellaceæ.—7 genera, 9 species (2 novæ formæ).
Protococcaceæ.—4 genera, 5 species (1 genus novum, 1 species nova).
Volvocineæ.—1 genus, 1 species.   (Spec. nova?)
Desmidieæ.—4 genera, 5 species (1 nova, 3 formæ novæ).
Zygnemeæ.—4 genera, 7 species (1 nova, 1 forma nova).
Vaucheriaceæ.—3 genera, 6 species (2 novæ).
Ulvaceæ.—1 genus, 1 species.
Oedogoniaceæ.—2 genera, 5 species (2 species incertæ).
Chætophoraceæ.—7 genera, 10 species.
    a. Chætophoreæ.—4 genera, 6 species (1 genus novum, 5 species novæ, 1 forma nova).
    b. Gongrosireæ.—3 genera, 4 species (1 nova).

Rhodophyceæ.

1 genus, 1 species nova.

Melanophyceæ.

1 genus novum, 1 species nova.

Diatomophyceæ.*

1. **Stauroneis goeppertiana**, *Bleisch; Rabenhorst Alg. Europ. Nr.* 182; specimina kerguelenensia accuratissime consentiunt in magnitudine ac forma cellularum cum speciminibus Silesiacis in Collect. Algar. Rabenhorst. communicatis. Areola transversalis in speciminibus Silesiacis plurimis paulo est angus-

---

* Materiam enumerationis Diatomacearum in ampullula majore reperi in singulis terrosis corpusculis duris radiculis *Nitellæ antarcticæ* partim adhærentibus partim in liquore fluitantibus.

tior,—Longit. 0,0224 mm. ($\frac{1}{94}'''$ Engl.) Latit. 0,0058 mm. ($\frac{1}{354}'''$ Engl.)—(Distrib. Silesia.—Considered by some authorities to be probably a form of *S. dilatata*, which is widely distributed in Europe, *G. Dickie*.)

2. **Stauroneis anceps,** *Ehrenberg;* forma linearis. Maxime consentiunt specimina cum speciminibus Europæis a Erlangen et a Falaise leg. Brébisson.—(Distrib. Europa tota, California, Cayenne.)

3. **Stauroneis Phœnicenteron,** *Ehrenberg*.—Longit. 0,0952 mm. ($\frac{1}{22}'''$ Engl.)—(Distrib. Europa frequens, America, Persia.)

1. **Aclinanthes exilis,** *Kützing*. Longit. 0,0224 mm. ($\frac{1}{94}'''$ Engl.) Latit. 0,0028 mm. ($\frac{1}{757}'''$ Engl.) In quoque latere in medio cellulæ (a fronte visæ) nodulus singulus; in plurimis speciminibus Europæis tantum in interiore latere. In magnitudine ac forma maxime consentiunt cum speciminibus e Jura Franconia, e Hungaria, et e Falaise (Gallia).

In *Vaucheriæ* cellulis nidulans.—(Distrib. in Europa vulgaris.)

1. **Larirella diaphana,** *Bleisch*. Longit. 0,1008 mm. ($\frac{1}{21}'''$ Engl.) Latit. 0,0418 mm. ($\frac{1}{50}'''$ Engl.) In speciminibus singulis.—(Distrib. Silesia; an forma *S. splendidæ* in Europa vulgaris, *G. Dickie*.)

1. **Campylodiscus,** *species nova; Reinsch. in. Journ. Linn. Soc.* xv. 205; magnus, elliptico-ovalis, utroque polo rotundato-obtuso, costis marginalibus radialibus crassis usque ad tertiam partem latitudinis superficieis se pertinentibus in quoque latere 22is—24is, areolas 21as—22as rectangulares includentibus, area media lævi; frustulæ a latere visæ simpliciter spiraliter curvatæ areolis 21is—22is rectangularibus instructæ.—Diam. longit. 0,132 mm. ($\frac{1}{16}'''$ Engl.)—Diam. transvers. 0,0666 mm. ($\frac{1}{31}'''$ Engl.)—Costæ in 0,02 mm. ($\frac{1}{110}'''$ Engl.) tres.

In speciminibus singulis inter *Schizosiphontis kerguelensis* cæspites.

A *Campylodiscis* frustulis oblongis *Campylod. Lurirella*, Ehrenberg, (Abh. Berlin. Acad. 1845, p. 362), mihi tantum ex diagnosi nota, proxima species videtur.

1. **Gomphonema Brebissonii,** *Kützing, Spec. Alg.* p. 66; *Ralfs Brit. Infus.* p. 887. Gomph. acuminatum var. *Smith. Brit. Diat.*—Longit. 0,0478 mm. ($\frac{1}{44}'''$ Engl.) Latit. (in summo) 0,0112 mm. ($\frac{1}{189}'''$ Engl.)—Cum speciminibus e locis diversis Germaniæ et Austriæ et e Falaise Gallia maxime consentiunt.

In *Vaucheriæ sessilis* et *sericeæ* filis.—(Distrib. in Europa vulgaris, an forma *G. dichotomi? G. Dickie*.)

1. **Amphiprora Spec. nova,** *Reinsch in Journ. Linn. Soc.* xv. 205; parva, rectangularis, subtilissime striata, medio parum constricta, utroque polo late truncato-rotundato, lineis intermediis duabus in medio cellulæ æqualiter extrorsum curvatis aream mediam cruciformem lævem circumcingentibus, nodulo singulo et in quoque latere cellulæ in decussi linearum incluso et in summo utriusque lineæ. Longit. 0,0333 mm. ($\frac{1}{63}'''$ Engl.) Latit. 0,0084 mm. ($\frac{1}{283}'''$ Engl.)

Amph. Pockorngana Grunow: dimensionibus duplo majoribus cellulis ovato-oblongis rotundatis, nodulo centrali, *Amph. minor* Gregory : dimensionibus majoribus cellulis elliptico-oblongis polis rotundatis, striis radiatis differunt.

1. **Navicula elliptica**, *Kützing*, var. COCCONEIDES ; *Rabenhorst, Algenfl. Europ.* I., p. 180, dimensionibus duplo minoribus. Longit. 0,0201—0,0224 mm. ($\frac{1}{100}$—$\frac{1}{94}$''' Engl.) Latit. 0,013—0,0168 mm. ($\frac{1}{157}$—$\frac{1}{126}$''' Engl.)

In opere novissimo de Diatomaceis, Atlas der Diatomaceenkunde, Heft II. tab. VII., fig. 55, *Navicula* est delineata (e Germania) quæ maxime consentit in forma, magnitudine, ac structura cellulæ cum plantula Kerguelensis.—(DISTRIB. in Europa frequens, Java, Nova Zelandia.)

2. **Navicula dicephala**, *Ehrenberg*. Longit. 0,0248 mm. ($\frac{1}{84}$''' Engl.)

Specimina ad formam pertinent summis capituliformibus distinctius disjunctis. —(DISTRIB. Europa).

3. **Navicula minutissima**, *Grunow*. E. minimis cellularis lineari-oblongis, nodulo medio et linea longitudinali distincta, indistincte transversaliter striatis.— Longit. 0,0112 mm. ($\frac{1}{189}$''' Engl.) Latit. 0,0028 mm. ($\frac{1}{757}$''' Engl.)

Hæc *Navicula* in speciminibus numerosis in massa ex Diatomaceis exstituta; non sine dubio *Navicula kerguelensis* ad hanc speciem est posito.—(DISTRIB. Europa orientalis.)

4. **Naviculæ Spec.** Cellulis lanceolatis apicibus capituliformibus porrectis, nodulo centrali et linea media indistincta, marginibus distincte striatis striis ad mediam non pertinentibus. Longit. 0,0278 mm. ($\frac{1}{75}$''' Engl.) Latit. 0,0056 mm. ($\frac{1}{378}$''' Engl.)

1. **Amphora gracilis**, *Ehrenberg*. Longit. 0,0357 mm. ($\frac{1}{58}$''' Engl.) Latit. 0,0196 mm. ($\frac{1}{110}$ Engl.) Cellulæ ovato-ovales apicibus truncatis, nodulis circa tertiam partem diametri transversalis a margine distantibus, areola media subtiliter longitudinaliter striato. Specimina a Falaise (Gallia) et e Germania consentiunt in magnitudine ac forma cellularum. *Amphoræ gracilis* in Schmidt Atlas d. Diatomaceenkunde, vii. fasc. tab. 26, fig. 101, cellulæ, quæ ad *Amphoram angustam* Ehrenberg pertinent, graciliores et apicibus attenuatis.—(DISTRIB. Europa, Mexico, et in Kurdistania fossilis.)

1. **Pinnularia viridula**, *Smith Brit. Diatom.* 57, tab. xviii. fig. 179 ; *Rabenh. Eur. Alg.* i. p. 214. Forma apicibus subito attenuatis, striis transversalibus lineam mediam attingentibus distinctissimis. Longit. 0,0357mm. ($\frac{1}{58}$''' Engl.) Latit. 0,0123mm. ($\frac{1}{168}$''' Engl.)

Plantulæ Falaise (A. de Brebisson, leg.) et plantulæ Erlangensis in magnitudine consentiunt, sed differunt polis sensim attenuatis striis subtilioribus. (DISTRIB. Europa, America.)

2. **Pinnularia viridis**, *Ehrenberg*. Longit. 0,0648mm. ($\frac{1}{32}$''' Engl.) Latit. 0,013mm. ($\frac{1}{157}$''' Engl.) Specimina paulo minora speciminibus formæ apud Erlangam communis. (DISTRIB. Europa, America, Persia.)

3. **Pinnulariæ** species; cellulis ovato-ellipticis, polis rotundatis, nodulo cen-

trali firmo, striis transversalibus distinctis lineam mediam attingentibus. Longit. 0,0168—0,0196mm. ($\frac{1}{126}$—$\frac{1}{110}$'' Engl.) Latit. 0,0084mm. ($\frac{1}{283}$''' Engl.)

4. **Pinnulariæ** species; cellulis late ovato-ellipticis, polis subito angustatis apicibus rotundatis, nodulo centrali firmo striis transversalibus distinctis lineam mediam attingentibus. Longit. 0,0224mm. ($\frac{1}{94}$''' Engl.) Latit. 0,0112mm. ($\frac{1}{189}$''' Engl.)

1. **Synedra Vaucheriæ**, *Kützing.*; forma apicibus obtusis. Longit. 0,0448mm. ($\frac{1}{47}$''' Engl.) Latit. 0,0028—0,0056mm. ($\frac{1}{757}$—$\frac{1}{378}$''' Engl.)

Individua breviter stipitata basi radiatim conjuncta in *Schizosiphonte kerguelensi*, et in *Vaucheriæ* cellulis. (DISTRIB. in Europa frequens.)

1. **Eunotia pectinalis**, *Dillwyn.* Longit. 0,106mm. ($\frac{1}{20}$''' Engl.). Latit. 0,0393mm. ($\frac{1}{54}$''' Engl.) (DISTRIB. in Europa vulgaris.)

1. **Denticula thermalis**, *Kützing.*, var. minor. Longit. 0,0168mm. ($\frac{1}{126}$''' Engl.) Latit. 0,0056mm. ($\frac{1}{378}$''' Engl.) Cellulæ in quoque latere 9is nodulis instructæ. (DISTRIB. Aquis calidis Galliæ, Hungariæ, et Italiæ.)

1. **Cymbella gastroides**, *Ehrenberg.* Longit. 0,0421mm. ($\frac{1}{50}$''' Engl.) Latit. 0,013mm. ($\frac{1}{157}$''' Engl.)

Specimina minora speciminibus e locis variis Germaniæ. (DISTRIB. Per totam Europam.)

PHYCOCHROMOPHYCEÆ.

1. **Chroococcus macrococcus**, *Rabenh., Alg. Fl. Eur.* i. p. 33. Protococcus macrococcus, *Kütz., Tab. Phyc.* i. tab. 2. Forma cytiodermate tenuiore, cytioplasmate grossius granuloso.

Formam typicam tantummodo observavi in familia singula tricellulari inter algas unicellulares *Hormosiphonti* adhærentes. Diam. cellular. indivis. 0,0478mm. ($\frac{1}{40}$''' Engl.)

Formam in familiis singulis bi- et tricellularibus in massis minoribus algarum variarum foliis muscorum adhærentibus observavi. Hæc forma pertinet ad formam *Chr. macrococci=Chroococcus aureus*, Kütz., Tab. Phyc. ii. tab. 2, *Chrooc. macrococc.* Rabenh., var. β.; cytioplasmatis cellularum colorem nunc pallide-flavum in statu vivente cellulæ fuisse aureo-luteum non dubito. (DISTRIB. Europa tota.)

1. **Microcystis olivacea**, *Kütz., Tab. Phyc.* i. tab. 9. Diam. cellular. 0,0041mm. ($\frac{1}{567}$''' Engl.) Diam. famil. 0,066—0,0896mm. ($\frac{1}{31}$—$\frac{1}{23}$''' Engl.)

Observavi tantum familias singulas inter alias algas unicellulares muscis adhærentes. In colore quoque obscure olivaceo cum specim. Europæis consentit. (DISTRIB. Germania.)

2. **Microcystis parasitica**, *Kütz., Tab. Phyc.* i. tab. 9, fig. 1.

In physeumatum structura, magnitudine ac colore cellularum cum speciminibus Europæis et cum icone Kützingiana maxime consentiens. Physeumata minora et majora, partim cohærentia in *Nitellæ antarcticæ* cellulis affixa.

Diam. cellular. 0,003mm. ($\frac{1}{748}'''$ Engl.)   Diam. physeumatum 0,0278—0,112mm. ($\frac{1}{75}$—$\frac{1}{19}'''$ Engl.) (DISTRIB. Europa.)

1. **Gloœthece involuta,** *Reinsch. in Journ. Linn. Soc.* xv. 206; thallo non limitato inter algas minores disperso; cellulis oblongo-cyclindricis utroque polo rotundatis, diametro transversali duplo longioribus, singulis aut geminis, tegumentis binis crassissimis distincte plurilamellosis circumvelatis, cytioplasmate pallide-ærugineo subtiliter granuloso, plerumque granulo singulo majore instructo. Longit. cellular. (c. indum. exter.) 0,0278—0,0333mm. ($\frac{1}{75}$—$\frac{1}{63}'''$ Engl.) Longit. cellular. (c. indum. inter.) 0,0248—0,0278mm. ($\frac{1}{84}$—$\frac{1}{75}'''$ Engl.) Longit. cellul. (sine indum.) 0,0112—0,013mm. ($\frac{1}{189}$—$\frac{1}{157}'''$ Engl.)

Inter algas minores muscis aquaticis insidentes. Hæc *Gloœthece* aliqua similitudine consentit cum *G. monococca*, Rabenh, Fl. Eur. i. p. 62=*Gloœcapsa monococca*, Kütz., Tab. Phyc. i. tab. 23, itidem reperta plerumque in statu uni- et bicellulari; quæ species nova attamen est diversa indumenti structura valde distincte lamellosa et cellulis longioribus a *Gloœth. monococca*, cujus integumentum semper est structura homogenea.

2. **Gloœcapsa magna,** *Kütz., Tab. Phyc.* i. tab. 22, fig. 7.

Cellulæ singulæ et geminæ sphæricæ colore intensive ærugineo-viridi densissime positæ, physeumata sphærica plerumque cohærentia indumento colorato velata formantes. Cellular. diam. 0,028—0,0041mm. ($\frac{1}{757}$—$\frac{1}{567}'''$ Engl.) Diam. physeumatis 0,033—0,05mm. ($\frac{1}{63}$—$\frac{1}{42}'''$ Engl.)

Inter *Scytonema castaneum*, Kütz., in massis parvulis *Hormosiphonti coriaceo* adhærentibus " prope Vulcan Cove." Non est mihi certissime, hanc plantulam pertinere ad *Gl. magnam* propter speciminum observatorum minimum numerum. (DISTRIB. Europa, Greenlandia.)

1. **Anacystis marginata,** *Meneghini.*

Familiæ singulæ quarum diameter 0,17mm. ($\frac{1}{6}'''$ Engl.), inter Algarum massas minores natantes. (DISTRIB. Europa.)

1. **Leptothrix hyalina,** *Reinsch in Journ. Linn. Soc.* xv. 206; aggregata, cæspitulos dispersos et radicantes muscis aquaticis affixas formans, trichomatibus hyalinis, vaginis distinctissimis crassis hyalinis, superne sæpissime vacuis et in summo apertis, cellulis tenuissimis diametro æqualibus, cytioplasmate punctulato. Diam. trichomat. 0,0028—0,0041mm. ($\frac{1}{757}$—$\frac{1}{567}'''$ Engl.) Cæspitulorum altitudo, 0,084—0,112mm. ($\frac{1}{25}$—$\frac{1}{19}'''$ Engl.)

In foliis muscorum.

Leptothrix radians, Kütz., Tab. Phyc. ii. tab. 59, proxima species distinguitur vaginis multo angustioribus cellulis crassioribus.

1. **Lyngbya major,** *Kütz., Tab. Phyc.* i. *tab.* 90, *fig.* 8 ; var. *kerguelenensis;* trichomatibus inter alias algas dispersis subrectis, cellulis intensive ærugineis subtiliter distincte granulatis, diametro 8plo—10plo brevioribus, vaginis amplis hyalinis (interdum fuscescentibus) distincte 8-12—lamellosis, cellulis interstialibus

nullis. Diam. trichomat. (c. vagin.), 0,0361—0,0448mm. ($\frac{1}{60}$—$\frac{1}{47}$''' Engl.) Vaginar. crassitudo, 0,0084—0,0112mm. ($\frac{1}{283}$—$\frac{1}{189}$''' Engl.) Diam. cellular. 0,0196—0,0224mm. ($\frac{1}{110}$—$\frac{1}{94}$ Engl.) Trichomatum longitudo 8—15mm.

Inter alias algas natantes et affixas in dispersis trichomatibus. (DISTRIB. *L. majoris* in Europa orientali.)

Hujus formæ cellulæ cellulis interstitialibus non interruptæ cylindrum continuum formant, diametro trichomatum apicem versus non decrescente, ultima cellula late rotundata, vaginæ in trichomatum summis utplurimum sunt apertæ et cellulis vacuæ. In fere omnibus trichomatibus a me visis *Microthamnii* novi elegantis plantulas observavi, quæ *Lyngbyæ* sunt affixæ radiculis contortis sæpe circum circa trichoma procurrentibus.

Speciminum formæ typicæ ex mari Adriatico trichomata paulo sunt crassiora, sæpissime occurrunt cellulæ interstitiales colore rubro-lutescente distinctæ ceteris cellulis trichomatis.

1. **Limnactis minutula**, *Kütz., Tab. Phyc. ii. tab.* 63, *fig.* 1; var. trichomatibus rectis sensim attenuatis margine crenulatis, cellulis distincte separatis diametro triplo-quadruplo brevioribus, cytioplasmate dense grossius granuloso, cellulis summis diametro usque quadruplo longioribus, hyalinis distinctis, vaginis hyalinis, cellulis perdurantibus sphæricis cellularum diametro æqualibus. Diam. trichomatum, 0,0056—0,0075mm. ($\frac{1}{378}$—$\frac{1}{290}$''' Engl.)

In *Schizosiphontis kerguelensis* trichomatibus in cæspitulis parvulis usque 0,28mm. ($\frac{1}{7}$''' Engl.) latis. DISTRIB. Gallia, Germania, Dania, Suecia, Britannia.

1. **Dasyactis Kunzeana**, *Kütz.* Diam. trichomat. 0,0056—0,0068mm. ($\frac{1}{378}$—$\frac{1}{320}$''' Engl.) Diam. cum vagin. 0,0112mm. ($\frac{1}{189}$''' Engl.)

In physeumatibus parvulis singulis dispersis in *Nitellæ antarcticæ* cellulis nidulantibus. (DISTRIB. Germania.)

1. **Mastigothrix articulata**, *Reinsch in Journ. Linn. Soc.* xv. 207; trichomatibus prolongatis subcylindricis basin versus paulo incrassatis distincte articulatis, articulis inferioribus indistinctioribus, superioribus loculamentis distincte disjunctis diametro subæqualibus, cytioplasmate granulis majoribus instructo, sporis perdurantibus obovalibus dimidio (et paulo minus) trichomatis latitudinis æquantibus. Diam. trichom. (in basi) 0,0168mm. ($\frac{1}{126}$''' Engl.) Diam. trichom. (in superiore parte) 0,0112mm. ($\frac{1}{189}$''' Engl.)

In singulis trichomatibus partim in superficie partim in strato summo physeumatis *Hormosiphontis leptosiphontis*, s. n., nidulantibus observatum.

Mastigothrichi fusco, Kützing, simillima in forma ac crassitudine trichomatum distinguitur: cellulis distincte articulatis cellulis perdurantibus minoribus a cellula infima sejunctis. Cellularum *M. fusci* cytioplasma subtiliter granulosum, cellula perdurans diametro cellularum æquante, basi lata (interdum intus excavata) cellulæ infimæ trichomatis arctissime adpressa, cytioplasmate homogeneo.

2. **Mastigothrix æruginea**, *Kütz.* Trichomata vix discernenda a speci-

minibus Europæis in thallo *Chætophorarum* et *Nostochidis* nidulantibus. In singulis trichomatibus inter *Tolypothrichis flaccidæ*, Kütz., cæspitulos nidulantibus. In singulis trichomatibus cellulæ inferiores breviores et indistinctius disjunctæ. (Distrib. Germania.)

3. **Mastigothrix minuta**, *Reinsch in Journ. Linn. Soc.* xv. 207; trichomatibus distincte articulatis apicibus rectis, articulis inferioribus dimidio latitudine brevioribus (et paulo magis), sporis perdurantibus obovalibus usque subsphæricis diametro dimidio trichomatis latitudinis æquante. Latit. trichomatum (in basi) 0,0084—0,0097mm. ($\frac{1}{283}$—$\frac{1}{236}$''' Engl.) Diam. sporæ perdur., 0,0041mm. ($\frac{1}{567}$''' Engl.)

Inter algas minores (*Leptothrix, Coleochæte*) in foliis muscorum aquatilium insidentes.

A *Mastig. æruginea*, Kütz., dimensionibus duplo magis minoribus distincta species. In trichomatibus singulis vaginæ infima pars paulo incrassata et lamellosa, sed cellula perdurans non inclusa a lamellis.

1. **Hydrocoleum Eatoni**, *Reinsch in Journ. Linn. Soc.* xv. 207; fasciculis liberis inter alias algas dispersis usque ad 18mms. longis in summis sensim attenuatis, trichomatibus olivaceo-viridibus (a latere visis), 8is—12is consociatis et leviter contortis subtilissime distincte articulatis, cellulis distinctis omnibus homogeneis, diametro quintuplo brevioribus, cytioplasmate dense punctulato, vaginis achrois membranaceis duris subtiliter lamellosis, trichomatum fasciculi latitudinis dimidio crassis. Diam. fasciculi (in medio parte) 0,056—0,086mm. ($\frac{1}{38}$—$\frac{1}{25}$''' Engl.); (in summis) 0,0224—0,0333mm. ($\frac{1}{94}$—$\frac{1}{63}$''' Engl.) Diam. trichomatum 0,0041—0,0056mm. ($\frac{1}{567}$—$\frac{1}{378}$''' Engl.) Vagin. crassitudo 0,0028mm. ($\frac{1}{757}$''' Engl.)

Inter muscos aquaticas et aliis *algis* (*Vaucheria, Schizosiphon*) immixtum.

Hoc *Hydrocoleum* consentit cum *H. helvetico*, Nägeli, in fasciculorum dispositione, sed differt dimensionibus fasciculorum quintuplo magis majoribus, trichomatum diverso colore et cellulis brevioribus.

Tab. IV. Fig. i.—1, fasciculi media pars ($\frac{360}{1}$);—2, fasciculi summa pars ($\frac{360}{1}$).

Nostoc hydrocoleoides, *Reinsch in Journ. Linn. Soc.* xv. 208; subtilissimum, physeumate in modo *Hydrocoleorum* teretiformi prolongato peridermate distincto hyalino cincto ex trichomatibus et rectis et paulo contortis (5is—10is) fasciculatim conjunctis formato, trichomatibus pallide ærugineis parallelis leviter contortis vaginulis hyalinis velatis, cellulis vegetativis post divisionem diametro paulo longioribus, cytioplasmate punctulato, cellulis perdurantibus ceteris paulo majoribus sphæricis in trichomatibus sparsis. Diam. trichomat. 0,0022—0,0028mm. ($\frac{1}{940}$—$\frac{1}{757}$''' Engl.) Diam. fasciculi (in media parte) 0,0112—0,224mm. ($\frac{1}{189}$—$\frac{1}{94}$''' Engl.)

Inter *Tolypothrix Nægelii*, Kütz., et in massa parvula *Diatomacearum* foliis muscorum aquatilium et *Nitellæ* adhærente.

Hæc plantula paradoxa secundum structuram et physeumatis et trichomatum

Nostochidis generis bonam speciem se ostendit. Trichomata integumento communi distinctissime clausa vix sunt discernenda a trichomatibus specierum singularum. Species unica *Nostochidis* generis hucusque cognita physeumate filamentosa, a ceteris speciebus physeumate sive plano sive sphærico sat distincta.

Tab. IV. Fig. iv.—1, physeumatis pars media ($\frac{720}{1}$);—2, physeumatis summa pars, trichoma singulum usque in apicem excurrens ($\frac{720}{1}$).

2. **Nostoc polysaccum**, *Reinsch in Journ. Linn. Soc.* xv. 208; physeumate coriaceo irregulariter sphærico et subreniformi colore subaureo-fusco magnitudine seminis sinapeos ad Pisi sativi, intus loculamentoso ac dissepimentis coloratis lamellosis et radialiter et transversaliter positis percurso, peridermate firmo coriaceo fuscescente, trichomatibus centralibus paulo flexuosis, cellulis sphæricis colore pallide olivaceo, cytiodermate distincte dupliciter striato, cellulis perdurantibus sphæricis ceteris cellulis paulo latioribus. Diam. cellular. 0,0041mm. ($\frac{1}{567}$''' Engl.) Diam. cellular. perdurant. 0,0056mm. ($\frac{1}{378}$''' Engl.) Diam. physeumatis, 2,5—3mm.

Forma (an status peculiaris evolutionis?). Physeumate ex trichomatibus brevioribus vaginis amplis hyalinis homogenis (in modo *Hormosiphontis*) inclusis laxissime cohærentibus exstituto.

Tab. IV. Fig. i.—1, Physeumatis sectionis transversalis pars usque ad peripheriam physeumatis se pertinens, vesiculæ trichomata includentes, radialiter dispositæ, parietes vesicularum subcoloratæ ($\frac{65}{1}$). 2, Formæ physeumatis pars peripheriæ, sectio transversalis; physeuma ex vesiculis numerosissimis, trichomatibus singulis inclusis formatum, trichomata breviora in modo *Hormosiphontis* indumento crasso subhyalino inclusa ($\frac{65}{1}$). 3, Specimina plantulæ (in spiritu vini asservatæ) magnitudine naturali.

3. **Nostoc polysporum**, *Reinsch. in Journ. Linn. Soc.* xv. 208; physeumate sphærico magnitudine pisi minoris, indumento crasso hyalino distinctissime plurilamelloso velato, trichomatibus laxius positis subcontortis pallide ærugineis, cellulis sphæricis arctissime conjunctis, post divisionem transverse ellipticis, cellulis perdurantibus numerosissimis sphæricis ceteris cellulis duplo majoribus cytiodermate crasso. Diam. cellular. 0,0028mm. ($\frac{1}{757}$''' Engl.) Diam. cellular. perdurantium 0,0041—0,0058mm. ($\frac{1}{567}$—$\frac{1}{380}$''' Engl.). Diam. physeumatis, 3—4 mm.

Inter alias algas fluitans (in paucis speciminibus observatum).

A persimilibus: *N. gymnosphæricum* et *N. cæruleum*, Kützing, Tab. Phycol. ii. tab. 3, fig. 3, 4, differt indumento plurilamelloso, cellulis perdurantibus numerosioribus.

4. **Nostoc** species, e minoribus, physeumate irregulariter polyedrico, textura cartilaginea, colore rubro-fusco, trichomatibus contortis, cellulis subsphæricis arctissime adjacentibus, cellulis perdurantibus sphæricis ceteris cellulis paululo majoribus cytiodermate crasso distincto. Diam. cellular. 0,003—0,0041mm. ($\frac{1}{740}$—$\frac{1}{567}$''' Engl.) Diam. physeumatis 1,8mm.

Inter *Zygnemam*. In a freshwater pool, Swain's Bay.

In textura, forma irregulari physeumatis minus in trichomatum forma *N. eduli* Berkeley persimile.

Specimen unicum observatum speciem accuratius constituendam mihi non permittit.

5. **Nostoc paludosum,** *Kütz., Tab. Phyc. ii. tab.* 1, *fig.* 2. Specimina singula observata insidentia plantulis *Bulbochæteis* foliis muscorum insidentibus et in trichomatum crassitudine et in cellularum forma maxime consentiunt cum speciminibus Germanicis et eum icone Kützingiana. Diam. cellular. trichomat. 0,0011—0,0018mm. ($\frac{1}{1890}$—$\frac{1}{1300}$''' Engl.)

Ab omnibus *Nostochidis* speciebus pagnitis species cellulis minimis. DISTRIB. In Europa vulgaris.

6. **Nostoc leptonema,** *Reinsch in Journ. Linn. Soc.* xv. 209; physeumatibus usque semini sinapeos æqualibus sphæricis paulo elasticis arctissime conjunctis cohærentibus, indumento exteriore subcrasso hyalino homogeneo, trichomatibus prolongatis multipliciter contortis laxius (in majoribus) et densius (in minoribus) intricatis, cellulis oblongis polis attenuatis laxe se adtingentibus; cellulis perdurantibus sphæricis usque subovalibus sparsis ceteris cellulis duplo paulo magis majoribus. Diam. cellular. 0,0015—0,0024mm. ($\frac{1}{950}$—$\frac{1}{810}$''' Engl.) Diam. cellular. perdur. 0,0056mm. ($\frac{1}{378}$''' Engl.) Diam. physeumatis 0,2—1,5mm.

In muscorum caulibus et foliis physeumatibus cohærentibus, partim corpora uvæformia formans.

A *Nostochidibus* physeumate sphærico *Nostoc aureum*, Kütz., Tab. Phyc. ii. tab. 1, fig. 4 (planta marina) proximum in magnitudine et textura physeumatis ac crassitudine trichomatum; hoc *Nostoc* differt trichomatibus brevissimis valde curvatis cellulis perdurantibus minoribus.

Inveniuntur interdum muscorum foliis insidentia corpora ex parenchymatice conjunctis physeumatibus varia magnitudine formata.

Forma: *Crystallophorum*. Physeumate corporibus crystallisatis subsphæricis inclusis ex crystallis (Ferri oxydati?) radialiter dispositis formatis. Diam. corpor. crystallisat. 0,0224—0,05mm. ($\frac{1}{94}$—$\frac{1}{42}$''' Engl.)

1. **Anabaina confervoides,** *Reinsch in Journ. Linn. Soc.* xv. 209; e subtilioribus stratum tenue formans, trichomatibus prolongatis rectissimis parallelis in muco communi nidulantibus, cellulis distinctissimis rectangularibus usque subquadraticis, spatiis interloculatis angustioribus distinctis sejunctis, diametro transversali paulo longioribus (usque duplo), cytioplasmate subtiliter granuloso colore pallide ærugineo; cellulis perdurantibus ellipticis ceteris cellulis paulo latioribus et longioribus. Diam. cellular. 0,0022—0,0028mm. ($\frac{1}{940}$—$\frac{1}{760}$''' Engl.)

In stratis tenuioribus inter alias Algas.

Hæc species peculiaris distinguitur ab omnibus hucusque cognitis speciebus cellulis angulosis (nec sphæricis nec ellipticis).

2. **Anabaina involuta,** *Reinsch in Journ. Linn. Soc.* xv. 299; libere natans, e tenuioribus, trichomatibus prolongatis multipliciter involutis, cellulis sphæricis et ellipticis (in statu indiviso) intermixtis, cytioplasmate subtiliter granuloso, cytiodermate extrorsum muco hyalino tenui velato, cellulis perdurantibus (sporis) sphæricis sparsis ceteris cellulis paululo latioribus, cytiodermate crasso distincto, cytioplasmate granuloso. Diam. cellular. 0,0024—0,0032mm. ($\frac{1}{945}$—$\frac{1}{710}$''' Engl.) Diam. cellular. perdurant. 0,0041mm. ($\frac{1}{567}$''' Engl.)

In trichomatibus singulis inter alias algas Phycochromaceas interjectis.

Ab *A. circinali* in trichomatum forma persimili differt cellulis quadruplo magis minoribus, cytioplasmate subtiliter granuloso.

1. **Hormosiphon leptosiphon,** *Reinsch in Journ. Linn. Soc.* xv. 210; globosum, magnitudine Pisi sativi, olivaceo-viride, physeumate subcartilagineo intus molli, indumento exteriore subtenace, trichomatibus prolongatis marginem physeumatis versus radialiter dispositis subcontortis pallide ærugineis, indumento hyalino homogeneo decolorato subtenui velatis, cellulis subsphæricis subarcte conjunctis, cellulis perdurantibus sphæricis ceteris cellulis duplo latioribus. Diam. cellular. 0,0028mm. ($\frac{1}{757}$''' Engl.) Diam. cellular. perdurant. 0,0056mm. ($\frac{1}{378}$ Engl.) Diam. physeum. 3,5—6mm.

In physeumatibus singulis inter *Schizosiphontis kerguelensis* cæspites.

Physeumatum observatorum dua procreant intus physeumata singula filialia ellipsoidea trichomatibus brevissimis subrectis et leviter contortis indumento crasso hyalino decolorato velatis densissime repleta, indumento communi distincto velata. Physeumatis externa pars plerumque ex trichomatibus physeumatum filialium trichomatibus simillimis formata. Superficies physeumatum est vestita plantulis variis egregie *Stigeoclonio subtili*, n. sp., singulis filis *Euactidis Künzeanæ* et *Tolypothrichis flaccidæ*.

Tab. IV. Fig. vii.—1, physeumata dua magnitudine naturali.—2, trichomatis singuli pars maxime aucta, α. cellula perdurans ($\frac{720}{1}$).—3, physeumatis sectionis transversalis pars exterior, cum physeumate filiali singulo trichomatibus brevissimis indumento crasso velatis dense repleto, α. indumentum exterius physeumatis.

2. **Hormosiphon coriaceus,** *Kütz., Tab. Phyc.* ii., *tab.* 14, *fig.* 1; var. KERGUELENSIS, *Reinsch in Journ. Linn. Soc.* xv. 211; physeumate irregulariter expanso subplano subcoriaceo colore obscure rubro-fusco, in sectione transversali ex stratis 5is—7is formato, peridermate (sectionis transversalis physeum.) lateris superioris crassiore lamelloso fusco, lateris inferioris peridermate tenuiore, trichomatibus vermiculiformibus multipliciter contortis, vaginis fuscis amplis distinctissimis plurilamellosis, cellulis sphæricis colore pallide ærugineo. Diam. cellular. 0,0041 mm. ($\frac{1}{567}$''' Engl.) Diam. trichom. (cum vaginis) 0,0224—0,0278 mm. ($\frac{1}{94}$—$\frac{1}{75}$''' Engl.) Physeumatis crassitudo 0,139—0,168 mm. ($\frac{1}{14}$—$\frac{1}{12}$''' Engl.)

Specimina majora in spiritu vini asservata ac in charta intenso.

Marshy ground near Vulcan Cove. (DISTRIB. Gallia, Germania, Italia.)

1. **Schizosiphon kerguelensis**, *Reinsch in Journ. Linn. Soc.* xv. 211; cæspitosus, cæspitulos confertos radialiter dispositos usque 6 mms altos in muscis aquaticis affixos formans, trichomatibus radiantibus e basi repetito dichotomoramosissimis in summis fastigiatis, pseudoramulis ultimis corymbosis fasciculatis apicibus paulatim subangustatis, vaginis pseudoramulorum ultimorum fuscis integerrimis celullarum diametro subæqualiter crassis, vaginis trichomatum inferioris partis crassioribus dense subtiliter lamellosis, cellulis omnibus æqualibus distinctis, diametro subæqualibus apicem pseudoramulorum versus non decrescentibus, cytioplasmate colore pallide olivaceo-viridi granulis majoribus distinctis dense repleto, cellulis perdurantibus singulis aut compluribus basilaribus subsphæricis diametro cellularum æqualibus. Diam. trichomat. (in diversis locis mensuratus) 0,0168—0,0333 mm. ($\frac{1}{126}$—$\frac{1}{63}$''' Engl.) Diam. pseudoramulorum ultimorum 0,013—0,0168 mm. ($\frac{1}{157}$—$\frac{1}{126}$''' Engl.)

Hab. in muscis aquaticis caules densissime peliculæ formiter inducens.

Hæc species elegantissima in cæspitulis muscis in caliculo vitreo inclusis copiosissima est reperta, in primis speciminis majoris caules densissime erant obtecti. Cellularum funiculus singulis locis haud raro et simpliciter et dupliciter contortus, quæ partes trichomatum paulo sunt incrassatæ; basin trichomatum versus cellularum funiculi plerumque sunt contorti; trichomatum infimæ partis vaginæ plurilamellosæ et trichomatum infima pars cuneiformiter angustata in filum singulum producta.

1. **Tolypothrix flaccida**, *Kütz. Tab. Phyc.* ii. *tab.* 32, *fig.* 2. Forma cellulis diametro transversali æqualibus et paulo longioribus. Diam. trichom. 0,0056—0,0084 mm. ($\frac{1}{378}$—$\frac{1}{283}$''' Engl.)

In cæspitulis parvulis in foliis muscorum aquaticorum insidens. (DISTRIB. *T. flaccidæ*, Britannia, Gallia, Germania, Helvetia.)

Hæc formæ sunt peculiares ut in forma typica, cellulæ perdurantes complures postpositæ, cellulæ complures funiculi trichomatum sæpissime interstitiis hyalinis sunt disjunctæ et trichomatum summa pars cellulis vacua.

2. **Tolypothrix Nægelii**, *Kütz.*

Hæc *Tolypothrix* a forma typica est distincta trichomatibus paulo tenuioribus, pseudoramulis crebrioribus, quæ sunt brevissimæ in singulis trichomatibus; summa pars cellulæ perdurantis singulæ in pseudoramulorum basi nonnunquam est truncata.

Inter *Schizosiphontis* cæspites et affixi et fluctuantes cæspituli. (DISTRIB. *T. Nægelii*, Helvetia.)

1. **Schizothrix hyalina**, *Kütz. Spec. Alg. Tab. Phyc.* ii. *tab.* 40, *fig.* 1.

Var. RAMOSISSIMA, *Reinsch in Journ. Linn. Soc.* xv. 211; trichomatibus *Schizosiphonti* insidentibus subtilissimis, funiculis et submoniliformibus et subcylindraceis pallide ærugineis, vaginulis amplis hyalinis cinctis; pseudoramulis numerosis erectis flagelliforme attenuatis. Diam. trichom. (cum vaginulis) 0,0022—0,0056 mm.

($\frac{1}{940}$—$\frac{1}{378}$''' Engl.) Diam. trichom. intern. 0,0011 mm. ($\frac{1}{1890}$''' Engl.) Altitudo plantulæ 0,8 mm. ($\frac{1}{4}$''' Engl.)

In *Schizosiphontis kerguelensis* trichomatibus in cæspitulis dispersis. (DISTRIB. *S. hyalinæ*, Montibus Europæ.)

Hanc formam peculiarem, verisimile speciem propriam, tantummodo in paucis sed bonis speciminibus observavi, quæ erant apta ad constituendum genus. Est similitudo maxima cum *Schizothr. hyalina* in trichomatum et vaginarum crassitudine et cellularum funiculi forma, quamquam incrementi modus et loci natalis est diversissimus.

1. **Sirosiphon vermicularis**, *Reinsch in Journ. Linn. Soc.* xv. 211; e minimis, cæspitulis parvulis trichomatibus subrectis summis attenuatis procumbentibus intertextis, plus minusve ramosis, ramulis alternantibus apicem versus sensim attenuatis ramulis summis diametri trichomatis primarii dimidio tenuioribus, trichomatum cellulis uniseriatis arctissime conjunctis, cytiodermate subcrasso firmo fuscescente, cytioplasmate subtilissime granuloso, ramulorum cellulis apicem ramulorum versus angulosis confervoideis, vaginis (trichomat. primarior.) tenuioribus (vix cellular. diametri octavam partem) simpliciter striatis cellulas arctissime includentibus; cellulis interstitialibus nullis. Diam. trichom. primar. 0,0112 mm. ($\frac{1}{189}$''' Engl.) Diam. ramulorum 0,0056 mm. ($\frac{1}{378}$ Engl.)

In cæspitulis singulis inter alios *Sirosiphontes Hormosiphonti coriaceo* prope Vulcan Cove adhærentes.

Ab omnibus *Sirosiphontibus* hucusque cognitis species minutissima. *Sirosiphon* in ramulorum cellulis diversis a cellulis trichomatis primarii. *Sirosiphonti sylvestri*, Itzigsohn. proximus sed sat distinctus trichomatibus tenuioribus cellulis cytiodermate tenuiore indistincte articulatis.

2. **Sirosiphon pulvinatus**, *Kütz.*; var. cellulis trichomatis primarii cytiodermate crassissimo colorato absque ordine biseriatis cellularum ramulorum uniseriatis aut absque ordine biseriatis. Diam. cellular. 0,0056—0,0068 mm. ($\frac{1}{378}$—$\frac{1}{330}$ Engl.) Diam. cellular. c. vagina 0,013 mm. ($\frac{1}{157}$''' Engl.) Trichomat. crassit. 0,0224—0,0306 mm. ($\frac{1}{94}$—$\frac{1}{69}$''' Engl.) Trichomata perpauca dispersa. Forsan propria species. (DISTRIB. *S. pulvinati*, Europa, Americ. boreal.)

Var.; trichomatibus irregulariter ramosis, ramulis apice obtusis numerosis subcontortis, cellulis omnibus æqualibus subovalibus, cytiodermate tenuiore hyalino decolorato, irregulariter biseriatis. Dimensionibus iisdem præced.

3. **Sirosiphon** species nova, *Reinsch in Journ. Linn. Soc.* xv. 212; e minoribus, trichomatibus singulis inter alias algas dispersis, irregulariter pinnatoramosis, ramis bilateralibus trichomati primario æqualiter formatis et æqualiter crassis, summis non attenuatis, cellulis subsphæricis spatiis hyalinis disjunctis, cytiodermate tenui homogeneo subhyalino, cytioplasmate subhomogeneo pallideærugineo, vaginis crassis hyalinis subhomogeneis decoloratis, cellulis interstiti-

alibus? Diam. cellular. 0,0041—0,0056 mm. ($\frac{1}{567}$—$\frac{1}{378}$''' Engl.) Trichomat. crassit. 0,0112—0,013 mm. ($\frac{1}{189}$—$\frac{1}{157}$''' Engl.)

In trichomatibus singulis inter alias *Sirosiphontes* et inter *Scytonemam castaneum* inter *Hormosiphontem coriaceum* (near Vulcan Cove).

S. velutinus et *S. hormoides* Kützing trichomatibus crassioribus fasciculato-ramosis et dichotome ramosis distincti. *S. panniformis*, Kütz., distinguitur ramis elongatis trichomate primario tenuioribus et cellulis interstitialibus.

4. **Sirosiphon kerguelensis**, *Reinsch. in Journ. Linn. Soc.* xv. 212; trichomatibus ramosissimis, trichomate primario procumbente ramis irregulariter ramosis ramulis ultimis apicem versus æqualiter latis, cellulis trichomatis primarii ac ramulorum ovalibus usque irregulariter sphæricis in seriem simplicem dispositis, intervallis hyalinis usque cellularum longitudini æquantibus disjunctis, articulis tubuliformibus angustissimis (lacunis tubuliformibus in muco vaginæ) conjunctis, cytioplasmate subhomogeneo dilute ærugineo, cytiodermate subtili decolorato (cellularum trichom. primarii crassiore fuscescente), cellulis summis ramulorum cohærentibus lyngbyaceis, vaginis crassis hyalinis decoloratis subhomogeneis (vaginis trichom. primarii sublamellosis aureis). Diam. trichom. primar. 0,0278—0,0333mm. ($\frac{1}{75}$—$\frac{1}{63}$ Engl.) Diam. ramulorum 0,0248 mm. ($\frac{1}{84}$''' Engl.) Diam. cellular. 0,013 mm. ($\frac{1}{157}$''' Engl.)

In trichomatibus singulis inter alias *Sirosiphontes*. Cum præcedente.

Hic *Sirosiphon* primo pro formam propriam *Sirosiphontis ocellati* habitus, cui est persimilis in trichomatis ramificatione et crassitudine, sed propter propriam de ceteris *Sirosiphontibus* discedentem structuram trichomatis propriam speciem se offert.

Tab. IV. Fig. vi.—1, trichomatis pars summa $\frac{180}{1}$;—2, trichomatis pars maxime aucta, $\frac{720}{1}$.

In singulis speciminibus observavi *Sirosiphontem* sequentem quem hujus *Sirosiphontis* varietatem puto. Trichomata ramosa ramis subintegris adscendentibus, cellulis ovalibus usque subsphæricis, intervallis hyalinis disjunctis. Articuli tubuliformes cellulas singulas conjungentes plurimum desunt.

5. **Sirosiphon Oliveri**, *Reinsch in Journ. Linn. Soc.* xv. 213; cæspitulis parvulis, trichomatibus adscendentibus prolongatis subramosis, ramulis singulis (et raro ramulis compluribus brevioribus approximatis) et leviter contortis, e serie simplice cellularum formatis, cellulis ovalibus diametro dimidio brevioribus (et paulo magis et minus), cytiodermate firmo crasso extrorsum fuscescente, cytioplasmate subhomogeneo obscure-ærugineo, vagina membranacea simplici subtenui, cellulis interstitialibus nullis. Diam. trichomatum (cum vaginis) 0,0196—0,0224 mm. ($\frac{1}{110}$—$\frac{1}{94}$''' Engl.)

In cæspitulis parvulis inter *Hormosiphontem coriaceum* cum cæspitulis *Scytonematis castanei* intermixtis; cum præced.

S. velutino, Kütz., et *S. hormoide*, Kütz., speciebus proximis in cellularum forma ac dispositione differt trichomatibus subintegris, vaginis tenuioribus.

Tab. IV. Fig. ii.—1, trichomatis summa pars ($\frac{180}{1}$);—2, trichomatis pars maxime aucta, vagina dupliciter striata, cellularum cytioderma dupliciter striatum, cellula singula longitudinaliter divisa ($\frac{720}{1}$).

6. **Sirosiphon secundatus,** *Kützing, Tab. Phycol.* ii. *tab.* 37, *fig.* 1; forma trichomate primario partim incrassato, ramis prolongatis apice incrassato; cellulis parvulis trichomatis primarii numerosis absque ordine dispositis, cellulis ramorum uni- aut irregulariter biseriatis cytiodermatibus crassis confluentibus. Diam. trichomatis primarii 0,0333—0,0393 mm. ($\frac{1}{63}$—$\frac{1}{54}$''' Engl.) Diam. ramorum 0,0224—0,0278 mm. ($\frac{1}{94}$—$\frac{1}{75}$''' Engl.) Diam. cellular. 0,0056 mm. ($\frac{1}{387}$''' Engl.)

In specimine singulo observato, inter alias *Sirosiphontes*. Cum præcedente. (DISTRIB. Europa.)

CHLOROPHYLLOPHYCEÆ.

1. **Glœocystis vesiculosa,** *Nægeli.* Cellulæ indivisæ usque ad 0,011 mm. ($\frac{1}{180}$''' Engl.) diam.; familiæ bicellulares 0,0058 mm. ($\frac{1}{378}$''' Engl.) diam.

Inter algas unicellulares adhærentes foliis muscorum. (DISTRIB. Germania, Helvetia.)

1. **Palmella mucosa,** *Kütz. Tab. Phyc.* i., tab. 16, fig. 7; cellul. diam. 0,0056—0,0112 mm. ($\frac{1}{378}$—$\frac{1}{189}$''' Engl.)

Inter alias algas unicellulares. (DISTRIB. Europa.)

Distinguitur a forma communi cellulis paulo minoribus et integumentis crassioribus distinctius limitatis.

1. **Pleurococcus vestitus,** *Reinsch, Algenfl. Frank.*, p. 56, tab. iii., fig. 4). Var. MINOR; cellulis sphæricis singulis aut binis aut quaternis et compluribus sphærice conjunctas familias formantibus, cytioplasmate dense subtiliter granuloso, cytiodermate crasso (interdum colorato) verruculis acutis dispersis instructo. Diam. cellular. 0,0112—0,013 mm. ($\frac{1}{189}$—$\frac{1}{157}$''' Engl.)

Inter alias algas unicellulares. (DISTRIB. *P. vestiti*, Germania).

2. **Pleurococcus angulosus,** *Corda.* Protoc. palustris Kütz. Tab. Phyc. i. tab. 9; forma, cellulis sphæricis in familias minores in modo *Merenchymatis* cohærentes collocatis. Cellular. diam. 0,0041—0,0056 mm. ($\frac{1}{567}$—$\frac{1}{378}$''' Engl.) Diam. familiar. 0,0224—0,0278 mm. ($\frac{1}{94}$—$\frac{1}{75}$''' Engl.)

Cum præcedente. (DISTRIB. *P. angulosi*, Europa.)

Scenedesmus acutus, *Meyen.*

In singulis speciminibus observatum inter *Zygnemæ* cæspites; in a freshwater pool on the W. of Swain's Bay. (DISTRIB. Europa.)

1. **Botryococcus Braunii,** *Kütz.*

Maxime consentit cum speciminibus Europæis e diversis locis. Inter cæspites *Schizosiphontis* et in massis parvulis algarum unicellularium muscis aquaticis adhærentibus. Inveniuntur familiæ et virides et fuscescentes. (DISTRIB. Germania, Helvetia.)

1. **Oocystis Nægelii,** *Al. Braun.* Longit. cellular. 0,0278—0,0306 mm. ($\frac{1}{75}$—$\frac{1}{60}$''' Engl.) Latit. cellular. 0,0168 mm. ($\frac{1}{126}$''' Engl.)

Magnitudine ac forma cellularum ac cytioplasmatis textura maxime consentiunt specim. cum speciminibus Germanicis. Indumentum familiarum bi- aut quadricellularium distincte dupliciter striatum.

In singulis familiis in massa parvula *Phycochromophycearum* unicellularium *Hormosiphonti* adhærente; marshy ground near Vulcan Cove. (DISTRIB. Germania).

1. **Dictyosphærium Ehrenbergii,** *Nægeli;* cellulis paulo majoribus. Diam. cellular. 0,0084—0,0112 mm. ($\frac{1}{280}$—$\frac{1}{189}$''' Engl.)

Inter algas varias muscorum foliis insidentes. (DISTRIB. Europa meridionalis.)

1. **Pediastrum ellipticum,** *Ralfs Brit. Desmid.;* var. ÆQUILOBUM; cœnobio elliptico continuo, cellulis disci regulariter 5-6-gonis, membrana hyalina achroa lævissima, cellulis periphericis leviter obtusangule-emarginatis, lobulis æqualibus cellulæ dimidio brevioribus apice truncatulis. Longit. maxima cœnobii 0,278—0,336 mm. ($\frac{1}{8}$—$\frac{1}{7}$''' Engl.) Diam. cellular. 0,0278—0,032 mm. ($\frac{1}{75}$—$\frac{1}{60}$''' Engl.)

In speciminibus duobus inter *Hormosiphontis* physeumata observatum. (DISTRIB. *P. elliptici,* Britannia.)

Asterosphœrium,[*] genus novum *Protococcacearum.* Cœnobium sphæricum, intus excavatum, libere natans, e cellulis angulosis parenchymatice arctissime conjunctis (sicut in *Pediastris*), extrorsum pyriforme ampliatis et subito angustatis formatum.

1. **Asterosphærium elegans,** *Reinsch in Journ. Linn. Soc.* xv. 213. Cœnobium sphæricum e cellulis 64 aut 128 formatum. Diam. cœnobii ex cellulis 128 formati 0,144 mm. ($\frac{1}{13}$''' Engl.)

Inter algas minores libere natans (in paucis speciminibus observatum.)

Hoc genus proxime se continuatur generibus *Protococcacearum* cœnobio ex cellulis parenchymatice conjunctis formato (*Hydrodiction, Pediastrum, Cœlastrum, Staurogenia*). Cœnobii dispositio fit in quoque hemisphæra secundum seriem: 1, 6, 11, 16, 21, 26 (seriem arithmeticam primæ ordinis cum numero differentiali=5). Quo dispositionis modo hoc genus discedit a *Pediastris,* generi proximo. *Pediastrorum* plurimum specierum dispositio cœnobii fit, in speciminibus regulariter formatis, secundum seriem: 1, 5, 10, 16 (seriem arithmeticam secundæ ordinis cum serie differentiali prima: 4, 5, 6, et numero differentiali=1).

In *Asterosphærii* cœnobiis legem dispositionis cellularum, ut fere fit, ad explicanda cœnobia pervenire in omnibus casibus, certissime puto; sicut per analogiam in *Protococcaceis* cœnobio pluricellulari sphærico (*Cœlastrum* et *Sorastrum*), quorum cœnobia abnormiter disposita rarissime observari possunt.

---

[*] ἀστήρ stella, σφαῖρα globus.

Omnium speciminum observatorum cellulæ erant vacuæ, veluti sæpe observamus in *Pediastris* majoribus.

Tab. IV., Fig. viii.—1, specimen integrum ex cellulis 128 exstitutum ($\frac{360}{1}$);—2, cœnobii marginis pars magis aucto ($\frac{720}{1}$).

1. **Glœocystis botryoides,** *Nægeli, Gatt. einzell. Alg.* Cellular. diam. 0,0022—0,004 mm. ($\frac{1}{940}$—$\frac{1}{567}$''' Engl.) Thallus gelatinosus, cellulis singulis et quaternatis, tegumentis crassis hyalinis distinctis.

In massis parvulis cum aliis algis *Phycochromaceis* inter *Hormosiphontem coriaceum* var., prope Vulcan Cove. (DISTRIB. Europa orientalis.)

2. **Glœococcus** species. Diam. cellular. 0,0041—0,0056 mm. ($\frac{1}{567}$—$\frac{1}{378}$''' Engl.) Cellulæ subsphæricæ in familiis 4- et 8-cellularibus consociatæ, cytioplasmate colore intensive viridi, locello hyalino decolorato singulo instructo.

In familiis singulis dispersis inter alias algas *Hormosiphonti coriaceo,* var. adhærentes.

1. **Polyedrium tetrætricum,** *Nægeli.* Cellulæ angulis acutiusculis (vix aculeolatis), marginibus lateralibus subrectis. Diam. cellular. 0,0224 mm. ($\frac{1}{94}$''' Engl.)

In speciminibus singulis inter alias algas unicellulares *Hormosiphonti coriaceo* var. adhærentes, prope Vulcan Cove. (DISTRIB. Europa australis.)

2. **Polyedrium minimi,** *Al. Braun, Alg. unicellul.,* p. 94, forma; cellulæ regulariter tetragonæ (quadraticæ) marginibus lateralibus omnibus æqualibus (vix leviter repandis), angulis obtuso rotundatis. Latit. 0,0066—0,0075 mm. ($\frac{1}{335}$—$\frac{1}{200}$''' Engl.)

In speciminibus singulis in massa parvula algarum *Zygogonio toruloso* var. adhærente. (DISTRIB. *P. minimi* Europa orientalis.)

Polyedrium Pynacidium, Reinsch, Algenflora von Franken. 1866, p. 80, tab. iii. a.—d., complures formas comprehendit. Specimen fig. *d.* delineatum est *P. minimum* Al. Braun, "lateribus alternis profundius emarginatis;" specimen fig. *a.* delineatum cum speciminibus Kerguelensibus exacte consentit; specimen fig. *b.* formam repræsentat marginibus æqualiter emarginatis.

1. **Chlamydococci** species, *Reinsch in Journ. Linn. Soc.* xv. 214; cellulis globosis vel ellipticis magnitudine paulo diversis, cytioplasmate et subhomogeneo et granuloso (granulis amylaceis dense repleto), in statu progressiore corpusculis sphæricis majoribus colore intensive luteo-purpurascente binis-quinternis instructo (cellulis filialibus, Zoogonidiis), cytiodermate hyalino crassissimo plurilamelloso (interdum unilateraliter incrassato). Cellular. diam. (ante divis.) 0,0278—0,0393 mm. ($\frac{1}{75}$—$\frac{1}{54}$''' Engl.) Diam. post divisionem 0,0196—0,0224 mm. ($\frac{1}{110}$—$\frac{1}{94}$''' Engl.)

Hab. in foliis musci aquatici.

Hujus plantulæ vera natura initio mihi erat aliquid dubia. In cellularum plurimum magnitudine cellulas filias non procreantium, cytioplasmatis colore, cytiodermatis structura valde consentiens cum *Chroococco aureo*, nihilominus inveniuntur cellulæ singulæ cytioplasmatis valde diversa structura a *Chroococcis*. Sed post observatis cellulis minoribus (Zoogonidiis) sine dubio in cohærentia organica cum cellulis majoribus *Chroococcoideis*, hujus plantulæ positio in systemate est constituta. Quæ cellulæ forma late pyriformi, polo subito angustato, cytiodermate tenui, cytioplasmate homogeneo colore intensive purpureo erant inventæ in consortio cellularum majorum cytioplasmate vacuarum. In singulis cellulis sunt inclusæ complures cellulæ filiales sphæricæ colore luteo-purpureo, aliis cellulis sunt corpuscula bina (interdum singulum, cellula super.). Nonnullarum cellularum cytioplasma densissime est repleta corpusculis amylaceis. *Chlamydococci* species duæ cognitæ differunt cytiodermate multo tenuiore non lamelloso. Cellulæ filiales duæ ($\frac{360}{1}$); cellula singula, cytiodermate unilateraliter incrassato, cytioplasmate cellulis filialibus (*gonidiis*) α. compluribus ($\frac{360}{1}$); cellula singula cytiodermate tenuiore, cytioplasmate corpusculis amylaceis densissime repleto ($\frac{360}{1}$). Cellula minor pyriformis (*Zoogonidium*).

1. **Cosmarium pseudo-nitidulum**, *Nordstedt, Bydr. till Kaenned. om sydl. Norges Desmid. Lund.* 1872, tom. ix., p. 46, tab. i., fig. 4); var. semicellularum semicircularium cytioderma in apice intus nodulo singulo incrassatum. Longit. 0,033mm. ($\frac{1}{63}$''' Engl.) Lat. 0,0248 mm. ($\frac{1}{92}$''' Engl.)

In speciminibus singulis in massis minoribus algarum variarum in muscis aquaticis adhærentibus.

2. **Cosmarium crenatum**, *Bréb.* var. kerguelense; cellula in ambitu late ovali, diametro longitudinali diametro transversali paulo longiore ($\frac{7}{6}$), semicellulis subsemicircularibus basi arctissime se adtingentibus incisura non disjunctis, margine undato exciso, gibberulis truncatulis 14is—15is instructo, superficie verruculis in seriebus radialibus dispositis verruculosa, areola media lævi, semicellulis e vertice visis ambitu ellipticis (in laterum medio leviter tumidis), isthmi latitudo $\frac{1}{5}$ diametri transversalis. Diam. transv. 0,033 mm. ($\frac{1}{63}$''' Engl.); diam. longit. 0,039 mm. ($\frac{1}{54}$''' Engl.) Isthmi latitudo 0,0067 mm. ($\frac{1}{360}$''' Engl.)

In specimine singulo observatum inter *Vaucheriæ* et *Schizothrichis* cæspites. (DISTRIB. Europa, America borealis, Greenlandia.)

Formis singulis *Cosm. pulcherrimi* Nordstedt (Symb. ad Flor. Brasil. Desmid. p. 175. tab. iii., fig. 24) simillimum in semicellularum ambitu et forma (*C. pulcherrim. β. boreale*, Nordst. Desmid. Spetsberg. et Beeren Eiland. p. 32, tab. vi. fig. 14), sed differt semicellulis e vertice visis in medio utrinque non productis, a fronte visis in medio lævibus.

1. **Staurastrum kerguelense**, *Reinsch in Journ. Linn. Soc.* xv. 214; semicellulis a latere late trapezicis angulis longe productis, margine terminali subrecto a vertice visis regulariter trigonis, marginibus lateralibus rectis angulis in cornulum rectum margine regulariter crenulatum longe productis, cornulis summis bispinosis, cytiodermate lævi seriebus tribus verrucularum marginibus semicellulæ

parallelis et in cornulis excurrentibus ornato, isthmi latitudine quinta pars cellulæ latitudinis. Latit. cellulæ 0,1038 mm. ($\frac{1}{20}'''$ Engl.); isthmi latitudo 0,0146 mm. ($\frac{1}{110}'''$ Engl.)

Observavi tantum specimina dua in massa algarum muscis aquaticis adhærente.

S. gracili, Ralfs. simile semicellularum forma, sed differt dimensionibus duplo magis majoribus, cornulis multo gracilioribus.

E Familia pulcherrima *Desmidiacearum* specierum numerosissima sunt reperta tantum *Cosmaria* dua hac *Staurastrum Palmogloeæ* species et *Euastrum binale var.*

1. **Euastrum binale,** *Turpin, var.* GIBBOSUM; semicellulis in sciagraphia trapezicis, margine terminali recto in medio levissime emarginato, angulis obtusis non productis, marginibus lateralibus gibberulis binis æqualibus rotundatis, superficie semicellulæ in quoque latere gibberulis binis æqualibus instructa, semicellulis a latere apice truncatis. Longit. 0,0306 mm. ($\frac{1}{69}'''$ Engl.) Latit. 0,0224 mm. ($\frac{1}{94}'''$ Engl.) Latit. margin. termin. 0,013 mm. ($\frac{1}{157}'''$ Engl.) Isthmi latitudo 0,0041 mm. ($\frac{1}{567}'''$ Engl.)

In singulis speciminibus inter algas unicellulares *Hormosiphonti* adhærentes. (DISTRIB. Europa, America borealis.)

A ceteris formis *Euastri binalis* hæc forma distinguitur superficie gibbosa semicellularum. *Euastrum binale var. dissimile*, Nordstedt (Desmid. Arctoæ, Konigl. Wetensk. Akad. Forhandl. Stockholm 1875, Nr. 6, p. 31, tab. viii. fig. 31), persimile in semicellularum sciagraphia, differt lobulis basalibus leviter repandis, angulis paulo productis, superficie non gibbosa.

1. **Palmogloeæ** species; cellulis ellipticis polis angustatis, diametro transversali dimidio diametri longitudinalis breviore, cytiodermate subcrasso, cytioplasmate granulis singulis majoribus instructo in massa gelatinosa irregulariter expansa nidulantibus. Longit. cellular. 0,0068—0,0084 mm. ($\frac{1}{340}$—$\frac{1}{283}'''$ Engl.) Latit. cellular. 0,0041 mm. ($\frac{1}{567}'''$ Engl.)

In massis parvulis *Hormsiphonti* adhærentibus.

Granulis amylaceis cytioplasmatis ad *Palmogloeas* spectans, a *P. macrococca* et *micrococca* distinguitur cellulis minoribus et polis angustatis.

1. **Vaucheria sessilis,** *Vaucher.* Maxime consentit cum speciminibus Europæis. Oosporæ maturæ membrana trilamellosa.

Filum unicum fructiferum observari potuerat in massa ex algis diversis composita. (DISTRIB. Europa, America borealis.)

2. **Vaucheria sericea,** *Lyngbye.* Filum singulum florescens observatum. Oogonium ad fecundationem aptum, antheridia bina horizontaliter flexa nondum aperta. In filo singulo observato oosporas maturas evolvente oosporæ in oogonio laxe inclusæ. (DISTRIB. Europa, America borealis.)

3. **Vaucheria pachyderma,** *Synon.* VAUCH. DILLWYNI, *Web. et Mohr. exp.*

Fila compluria oosporis maturis observata. Oosporarum membrana plurilamellosa duplo crassior membrana *V. sessilis*.

In cæspitulo parvulo ex filis intertextis *Vaucheriæ* specierum variarum composito inter *Nitellæ* specimina incluso, pauca observavi fila quæ pertinerent ad aliquam *Vaucheriam* ad Corniculatas spectantem. (DISTRIB. Europa frequens.)

De antheridiis et ogoniis nondum evolutis non potuerat discerni aliquid certi, nescioque, hæc fila pertinere ad *V. sericeam*, pachydermam an ad speciem propriam.

4. **Vaucheria geminata,** *De Candolle*. Fila compluria oogoniis immaturis sine dubio ad *V. geminatam* spectantia; thalli ramulus lateralis minutus flores evolvens paulo longior et gracilior ramulo speciminum Europæarum, cornulum jam in positione propria, summo—ad observatorem verso—minime lateraliter contorto. *V. hamatæ* ramulus lateralis dimidio brevior ac ramuli oogonia procreantes duplo longiores. Antheridium a basi curvatum in uno anfractu contortum. (DISTRIB. Europa, America borealis.)

Status evolutionis partium florum *Vaucheriarum* perfecte congrunt cum eodem statu evolutionis florum *Nitellæ antarcticæ*. Tempus anni, respondens statu analogo vitæ harum plantularum in nostris latitudinibus ver est (menses Aprilis, Maii, usque ad initium mensis Junii). Quarum plantularum phænomena vitalia normam dare ad dijudicandas ullæ regionis terræ rationes in respectu commutationum temporum quadripartitarum anni, verisimile videtur.

1. **Olpidium caudatum,** *Reinsch in Journ. Linn. Soc.* xv. 215; cellulis sphæricis sine radiculis substrato viventi insidentibus, in polo processu singulo spini formi cellulæ diametro subæquante postremo aperto instructis, cytiodermate distincto subcrasso, cytioplasmate dense granuloso.—Diam. cellular. 0,0112—0,013 mm. ($\frac{1}{189}$—$\frac{1}{157}$''' Engl.)

In *Schizosiphontis kerguelensis* trichomátibus.

O. ampullaceum (*Chytridium ampullaceum*, A. Braun, Ber. d. Berlin. Acad. 1855, p. 66; Rabenhorst, Fl. Eur. Alg. ii. p. 282) est distinctum ab hoc *Olpidio* dimensionibus duplo minoribus (0,0064 mm.; $\frac{1}{312}$''' Engl. diam.).

Tab. IV., Fig. vi.—1, *Schizosiphontis* trichomates pars cum plantula parasitica insidente, $\frac{360}{1}$;—2, cellula singula parasitica major amplificata, $\frac{720}{1}$.

1. **Chytridium pyriforme,** *Reinsch in Journ. Linn. Soc.* xv. 215; cellulis zoogonidiis nondum egressis operculose apertis ovato-pyriformibus, basi sensim angustata, in radiculum in substrato vivente radicantem prolongatis diametro transversali dimidio diametri longitudinalis angustiore, cytioplasmate dense subtiliter granuloso, cytiodermate distincto dupliciter striato, cellulis zoogonidiis egressis subcylindricis usque subcuneatis, operculo transversaliter a cellula se sejungente subhemisphærico apice rotundato (non acuminato), radiculo usque tertiam partem diametri longitudinalis cellulæ æquante, in medio plus minusve incrassato apiculo prolongato deorsum verso. Diam. transvers. cellulæ 0,013—

0,0168 mm. ($\frac{1}{157}$—$\frac{1}{126}$''' Engl.) Diam. longitud. cellulæ 0,0258—0,0278 mm. ($\frac{1}{80}$—$\frac{1}{75}$''' Engl.)

In *Vaucheriæ* cellulis.

A *Chytridiis* cognitis proximum in cellularum forma *Chytr. Olla*, A. Braun, (Verjung. p. 198. Ber. Berlin. Acad. 1855, p. 380; Rabenh. Fl. Eur. ii. p. 277), quod *Chytridium* distinguitur cellulis latioribus operculo obtuse umbilicato; *C. acuminatum* et *C. brevipes*, A. Braun, sunt distincta operculis acuminatis. In omnibus *Vaucheriæ* cellulis, quæ portaverunt plantulas parasiticas, sunt observatæ prolongationes utriculiformes dense positæ e *Vaucheriæ* cellula egressæ. Parasita invenitur plerumque in iisdem locis *Vaucheriæ* cellulæ infectæ ubi sunt evolutæ hæc prolongationes utriculiformes. In his locis abnormiter transmutatis apparent parietes intercalares quæ separant lumen transmutatum a cellulæ cetero lumine. Certissime adducta est transmutatio abnormis *Vaucheriæ* a plantulis parasiticis. Complures casus hucusque sunt observati, in quibus efferuntur transmutationes morphologicæ plantarum altiorum per plantulis parasiticis unicellularibus.* Singula fila *Vaucheriæ* abnormiter transmutatæ observavi, quibus desunt *Chytridii* cellulæ, sed in cytioplasmate *Vaucheriæ* sunt impositæ cellulæ sphæricæ magnitudine varia manifesto alienæ *Vaucheriæ* cellulæ. Utrum aliquam connexionem esse geneticam inter *Chytridii* cellulas pyriformes *Vaucheriæ* insidentes et cellulas entophyticas, an non, incertum est.

1. **Microthamnion cladophoroides**, *Reinsch in Journ. Linn. Soc.* xv. 216; e maximis, fruticulosum, filis solitariis erectis regulariter ramosis, radiculis singulis contortis in substrato (algis viventibus) insidentibus, ramulis erecto-patentibus attenuatis unilateraliter dispositis (in speciminibus minoribus) aut verticillatim dispositis (in speciminibus majoribus), cellulis fili primarii apicem versus paulo incrassatis diametro 4plo–6plo longioribus, cellulis ramulorum in basi paulo constrictis diametro 10plo–20plo longioribus, cytioplasmate omnium cellularum subhomogeneo, colore pallide luteo-olivaceo, granulis singulis dispersis instructo. Fili primarii cellularum latit. 0,0056 mm. ($\frac{1}{378}$''' Engl.) Ramulorum cellularum latit. 0,0028—0,0041 mm. ($\frac{1}{757}$—$\frac{1}{565}$''' Engl.) Plantulæ altit. 0,556 mm. ($\frac{1}{4}$''' Engl.)

In *Lyngbyæ majoris*, Kütz. forma trichomatibus et in *Chlorococci* spec. cellulis radiculis brevissimis affixum. Hæc plantula elegantissima *Cladophoris* singulis in habitu haud dissimiles, sed sat distincta a *Cladophoris* cytioplasmate subhomogeneo ac dimensionibus minimis, ad *Microthamnia* spectat quibuscum consentit in cytioplasmatis structura. Generis specierum trium hucusque cognitarum nulla aliqua similitudine consentit cum plantula Insulæ Kerguelensis.

1. **Stigeoclonium Hookeri**, *Reinsch in Journ. Linn. Soc.* xv. 216; læte viride, parasiticum, cæspitulos chætophoræformes basi radicante formans; filis ætate provectiore inferne nudæ et subintegræ superne ramosissimis, basi

---

* *Synchytrium Taraxaci* (De Barg. et Woron. Ber. d. Naturf. Gesellsch. Freiburg, iii. 2. tab. i. ii., fig. 1-7.

radiculis anastomosantibus instructis, ramis spicatis (plerumque) integerrimis approximatis stricte erectis, cellulis filorum primariorum hyalinis cytioplasmate contracto (in statu vegeto?), diametro transversali (inferiorum) duplo—triplo longioribus et æqualibus (superiorum), cellulis ramorum tumidis, omnibus in sporangia zoogonidia evolventia transmutatis diametro æqualibus et dimidio brevioribus. Diam. cellularum filorum primar. 0,0112—0,013 mm. ($\frac{1}{189}$—$\frac{1}{158}$''' Engl.) Diam. ramorum sporidiferorum 0,0084—0,0112 mm. ($\frac{1}{283}$—$\frac{1}{189}$''' Engl.) Altitudo plantulæ 1—1,5 mm.

In *Nitellæ* cellulis et in foliis muscorum.

Hoc *Stigeoclonium elegantissimum* cum *S. debili* et *uniformi*, Kütz. (Tab. Phyc. iii. tab. 3), aliqua similitudine consentit. Primum differt ramulis longioribus longius distantibus non fasciculato-racemosis; secundum, ramificatione verticilliformi caulis primarii. Hæc species tres cum *S. gracili*, Kütz., subspecies formant specieis unæ typicæ.

Tab. V., Fig. i.—1, *Nitellæ* pars cum cæspitulo *Stigeoclonii* insidente ($\frac{60}{1}$);— 2, fili singuli summa pars major aucta, omnes cellulæ zoogonidia procreantes ($\frac{360}{1}$).

2. **Stigeoclonium subtile**, *Reinsch in Journ. Linn. Soc.* xv. 217; minutissimum, parasiticum, ex filis sterilescentibus tenuioribus longioribus integerrimis erectis e filis procumbentibus dense intertextis crassioribus ortis formatum, cellulis ramulorum erectorum tenuioribus diametro 4plo–8plo longioribus, cellulis filorum procumbentium latioribus diametro subæqualibus, filis propagativis paulo crassioribus, cellulis zoogonidia procreantibus cellulis filorum sterilescentium multo brevioribus subquadraticis arctissime conjunctis. Diam. filiorum erectorum 0,0048—0,0056 mm. ($\frac{1}{567}$—$\frac{1}{378}$''' Engl.)

In muscorum foliis, in *Nitellæ* et *Vaucheriæ* cellulis, et in *Schizosiphontis kerguelensis* trichomatibus.

Hoc *Stigeoclonium* ramulis prolongatis tenuissimis flagelliformibus erectis ex ramulis crassioribus ortis aliqua similitudine consentit cum *S. setigero*, Kütz. (Tab. Phyc. iii. tab. 5), quod distinguitur cæspitulis multo majoribus fluctuantibus (usque tres lineas longis).

Cæspituli tantum fila propagativa procreantes haberi possunt pro Speciem propriam. In singulis speciminibus plantulæ inveniuntur et fila sterilescentia et fila propagativa. Ulteriora paulo crassiora sed breviora saepe inveniuntur ex uno ramulo evoluta cum filis sterilescentibus. Cæspitulos quoque singulos in *Hormosiphontis* sp. n. physeumate crescentes una cum *Choreoclonii procumbentis* gen. n. cæspitulis observavi; in his plantularum duarum infimæ partes adeo sunt inter se coalitæ ut plantulas duas valde diversas in cohæsione genetica putare possis.

Choreoclonium, genus novum.* Plantula parasitica ex filis ramosis procumbentibus densius aut laxius intricatis substrato dense adpressis interdum parenchymatice inter se conjunctis formata; cellulæ rectangulares usque quadraticæ.

* κορέω expando, κλόνις clunis.

Propagatio ?—Synon. Genus s. n. in Reinsch, Contribut., p. 76, tab. iv. (Chloroph.) descriptum et delineatum genus ad *Chætophoraceas* spectans, *Stigeoclonio proximum*.

1. Choreoclonium procumbens, Reinsch in Journ. Linn. Soc. xv. 217.

Cellular. diam. 0,0028—0,0041 mm. ($\frac{1}{757} - \frac{1}{567}$''' Engl.) Cellular. longit. 0,0112—0,0224 mm ($\frac{1}{189} - \frac{1}{94}$''' Engl.)

In foliis muscorum et in *Nitellæ* cellulis.

Hanc plantulam primo observavi anno 1872 in Germania in plantis aquaticis (*Hottonia, Utricularia*) crescentem, deinde in compluribus formis variis locis Germaniæ. In contributionibus meis formas varias in uno genere conjunctas sine nomine recepi; post plantulam Kergulenensem inventam nimirum dubitare possum in identitate plantularum e locis duobus remotissimis.

Tab. IV., Fig. ix.—1, folii musci aquatici pars cum plantula singula minore in nervo folii crescenti ($\frac{360}{1}$); —2, alteri folii pars cum plantula majore obtecta ($\frac{360}{1}$).

1. Draparnaldia subtilis, Reinsch in Journ. Linn. Soc. xv. 218;

filis ramisque primariis hyalinis, ramis e basi repetito dichotome ramosissimis, ramulis furcatis acutis plerumque in pilum hyalinum ex cellulis compluribus exstitutum longe productis, cellulis infimis fili primarii diametro æqualibus cytiodermate crasso lamelloso, cytioplasmate subhomageneo subtilissime granulato, cellulis superioribus diametro usque duplo longioribus, cellulis ramulorum diametro usque triplo longioribus, cytioplasmate dense granulose. Diam. fili primarii 0,0168—0,0232 mm. ($\frac{1}{126} - \frac{1}{84}$'''.) Diam. ramulorum 0,0056—0,0084 mm. ($\frac{1}{378} - \frac{1}{283}$''' Engl.) Plantulæ altitudo 1—2 mm.

In *Vaucheriæ* cellulis et in muscis aquaticis in plantulis dispersis radiculis numerosis radicantibus. Hæc plantula elegans tantummodo in speciminibus paucis observata differt a ceteris *Draparnaldiis* et magnitudine et loco natali.

2. Draparnaldia distans, Kütz., Tab. Phyc. iii., tab. 14, fig. 2;

forma tenuis, cellulis fili primarii duplo-quadruplo diametro transversali longioribus, ramis primariis perpaucis, ramulis sparsis crebrioribus brevioribus cum ramulis longioribus in ambitu lanceolatis perpaucis intermixtis, ramulis ultimis plerumque in pilum achroum cellulare attenuatis, cellulis ramulorum tumidis diametro subæqualibus. Diam. cellular. fili prim. 0,0278—0,056 mm. ($\frac{1}{75} - \frac{1}{38}$''' Engl.)

In speciminibus exsiccatis cum *Zygnemate* intermixtis. " In a freshwater pool on the W. of Swain's Bay." (DISTRIB. Europa.)

1. Proterderma viride, Kützing.

Familiæ singulæ in foliis musci aquatici laxius insidentes, in magnitudine cellularum ac forma (0,0084 mm.; $\frac{1}{283}$''' Engl. diam.) cum speciminibus Franconicis maxime consentiunt.

1. Zygogonium torulosi, Kütz., Tab. Phyc., tab. 14, fig. 1;

forma crassior. Cellulæ diametri transversalis dimidio brevioribus (ante divisionem usque æqualibus) cytiodermate interiore crassissimo plurilamelloso, cytiodermate exteriore subtoruloso. Diam. cellular. 0,0168—0,0196 mm. ($\frac{1}{126}$—$\frac{1}{110}$''' Engl.) Diam. filorum (c. indum.) 0,033—0,0393 mm. ($\frac{1}{63}$—$\frac{1}{54}$''' Engl.)

In cæspitulis inter *Hormosiphon coriaceum*, var. " In moist places near Vulcan Cove."—DISTRIB. *Z. torulosi* Europa orientalis.

In filis singulis observantur sicut in speciminibus Europæis cellulæ subsphæricæ laterales filis adhærentes indumento crassissimo velatæ. Quæ cellullæ—nullo modo cellulæ propagativæ—oriuntur in hoc *Zygogonio* et in *Z. anomalo* divisione longitudinali interdum incidente cellularum singularum fili.—Hæc forma a forma typica in Tab. Phycol. delineata cellulis angustioribus cytiodermate crassiore et filis crassioribus distinguitur. *Z. torulosum, Kütz.*, cum serie specierum: *Z. ericetorum, anomalum, delicatulum*, a Cl. Rabenhorst (Fl. Eur. Alg. ii., p. 254) in una specie contrahuntur, sed characteres constanter observatæ horum *Zygogoniorum* a speciebus Kützingianis discedere mihi non permiserunt.

2. **Zygogonium tenuissimum**, *Reinsch in Journ. Linn. Soc.* xv. 218; filis tenuissimis cellulis diametro duplo longioribus (et paulo minus) regulariter rectangularibus, cytiodermate subcrasso homogeneo hyalino, cytioplasmate contracto colore luteo-viridi granulis majoribus instructo. Diam. cellular. 0,0068—0,0084mm. ($\frac{1}{320}$—$\frac{1}{283}$''' Engl.)

In singulis filis inter *Scytonemam castaneum* dispersis, "near Vulcan Cove."—Differt a *Z. delicatulo* et *Z. salino* cellulis longioribus, a *Z. gracili* et *Z. Ralpsii* cellulis brevioribus, ab omnibus *Zygogoniis* autem filis multo tenuioribus.

1. **Spirogyra longata**, *Kütz., Tab. Phyc.* v. tab. 20, fig. 1; cellular. diam. 0,039—0,05mm. ($\frac{1}{54}$—$\frac{1}{42}$''' Engl.) Longitudo cellularum 5plum—7plum latitudinis.

In a freshwater pool W. of Swain's Bay (specim. exsiccat.).

Structura fasciæ spiralis latæ anfractibus 4is—5is maxime consentit cum speciminibus Europæis. Fila omnia incopulata sunt latiora (usque duplo) filis formæ communis Europeæ per totam Europam diffusæ.—(DISTRIB. Europa, America borealis.)

2. **Spirogyra** *Spec.*; Cellularum diam. 0,0278—0,0393mm. ($\frac{1}{75}$—$\frac{1}{54}$''' Engl.) Longitudo 4plum—5plum latitudinis. Fila omnia incopulata ad quandam speciem *Spirogyræ* spectantia, quæ pertinet ad *Spirogyras* cytiodermate in utroque polo cellulæ nec protenso nec replicato; sed structura fasciarum spiralium adeo est transmutata ut numerus fasciarum et forma certe non potest explicari.

Fila dispersa inter *Draparnaldiam distantem.* " In a freshwater pool on the W. of Swain's Bay" (specim. exsiccat.).

1. **Sirogonium sticticum**, *Kütz.* Cellular. diam. 0,045—0,050mm. ($\frac{1}{47}$—$\frac{1}{42}$''' Engl.) Fasciæ *chlorophyllaceæ* ternæ—quaternæ in quaque cellula nucleis ex substantia proteinicis cum jodinis agentia fuscescentibus majoribus (nunc decoloratis) instructæ. Qui nuclei sunt majores nucleis speciminum ex Germania.

In singulis filis incopulatis inter *Zygnemæ* cæspites. "In a freshwater pool W. of Swain's Bay."—(DISTRIB. Europa borealis et centralis.)

1. **Zygnema Vaucheri,** *Agardh;* Z. SUBTILE, *Kütz., Tab. Phyc.* v. tab. 16, fig. 1. Diam. cellular. 0,0168 mm. ($\frac{1}{120}$''' Engl.) Fila omnia incopulata.

"In a stream W. of Swain's Bay, 20. 1. 1875."—(DISTRIB. Europa.)

Speciminum in charta siccatorum cytioplasmatis structura distincte non perspicua, attamen tinctura intensive lutea in charta effusa post aqua conspersa *Zygnemam* certissime indicat.

2. **Zygnema affine,** *Kütz., Tab. Phyc.* v. tab. 16, fig. 3.—Diam. cellular. 0,0196—0,0224mm. ($\frac{1}{110}$—$\frac{1}{94}$''' Engl.) Longitudo duplum usque triplum latitudinis. Structura cytioplasmatis in singulis cellulis bene conspicua.

In filis dispersis singulis inter *Draparnaldiam distantem*. "In a freshwater pool W. of Swain's Bay."—(DISTRIB. Europa orientalis.)

1. **Bulbochæte** *Species.* Cellular. diam. med. 0,013mm. ($\frac{1}{157}$''' Engl.) Cellularum longitudo triplum latitudinis, cellula basalis oblongo-pyriformis pedicello pediformi breviore.

In foliis musci aquatici.

A *Bulbochæteis* genere species observavi tres, specimina omnia sine florescentia et fructificatione. Apud species singulas cellulæ filorum et dimensionibus et forma consentiunt; qua de causa difficillime possunt determinari specimina sterilescentia. Cellulæ basalis et cellularum forma ac magnitudo aliqua similitudine consentit cum *B. crenulata*, Pringsh.

2. **Bulbochæte** *Species.* Cellula basalis pyriformis basi attenuata. Diam. cellular. 0,013 mm. ($\frac{1}{157}$''' Engl.)—Forsan species propria.

Plantula singula in musci aquatici folio crescens.

3. **Bulbochæte** *Species.* Cellular. diam. 0,0168—0,0224mm. ($\frac{1}{126}$—$\frac{1}{94}$''' Engl.) Cellulæ paulo longiores quam latiores. Cellula basalis late pyriformis basi subito in pedicellum pediforme angustata; ramificatio repetito dichotome ramosa.

In foliis musci aquatici, in *Nitellæ* cellulis, et in *Schizosiphonte kerguelensi*.

Aliqua similitudine consentit cum specie nova descripta et delineata in contributionibus meis, p. 81, tab. xiv. (chloroph.), fig. 4.

Specimina fere omnia observata ostendunt incrementi intensi, ut observamus in speciminibus collectis in mensibus Maii et Junii in Europæ regionibus mediis. In evolutione setarum, quæ brevissimo tempore fit, cellulæ matricalis cytiodermatis partes per cellulam filialam se evolventem ad latus premuntur, et brevi tempore permanentes in hoc statu, post evolutis setis, dejiciuntur. Quæ lamellæ cytiodermatis in hoc tempore vitæ in plurimis plantulis sunt observandæ. Quod factum tempus certum in historia incrementi hujus plantulæ nobis indicat. Plantulæ Kerguelenenses in hieme collectæ repræsentant statum evolutiones ejusdem spatii in nostris regionibus mensibus Maii et Junii respondentem. Idem factum est observatum in speciminibus *Nitellæ antarcticæ*, quæ sunt in primo statu florescentiæ.

1. **Oedogonium delicatulum,** *Kützing.* Diam. cellular. 0,0056mm. ($\frac{1}{378}'''$ Engl.), longitudo 3plum—4plum latitudinis.

In singulis filis sterilescentibus in foliis musci aquatici, et in *Nitella* insidentibus observatum; cellulæ basalis forma filorum latitudo, cellularum longitudo cum speciminibus Europæis maxime consentit.—(DISTRIB. Europa, frequens.)

2. **Oedogonium** *Species.* Diam. cellular. 0,0112mm. ($\frac{1}{189}'''$ Engl.), longitudo 8plum—9plum latitudinis. Non possunt determinari specimina propter penuriam florescentiæ et fructificationis. Complures species *Pedogoniorum* maxime consentiunt in cellularum forma ac magnitudine, ex qua causa non potest dici de earum fila sterilia aliquid certi.

In filis singulis steriliscentibus in foliis musci aquatici.

1. **Coleochæte scutata,** *Brébisson;* Phyllactidium setigerum, *Kützing, Tab. Phyc.* iv., tab. 87, iii.

In foliis musci aquatici in familiis planis planitieformibus plus minusve regulariter circumscriptis forma. Familiæ complures in foliis angustioribus et in foliis quibus lamina deest, in massas irregulariter sphæricas aggregatæ.

Phyllactidium setigerum, Kütz., et *Coleochæte scutata,* Kütz., plantulæ synonymicæ sunt, prima haberi potest forma setigera *Coleochæteis scutatæ.* Speciminum Kerguelenensium setæ paulo sunt longiores et robustiores setis speciminum Europeorum. (DISTRIB. Europa, America borealis.)

2. **Coleochæte irregularis,** *Prings., Jahrb. f. wissensch. Bot.* 1860, ii., tab. i., fig. 6, tab. vi. fig. 3–9. Diam. cellular. 0,0112—0,0126mm. ($\frac{1}{189}-\frac{1}{126}'''$ Engl.)

In *Nitellæ* superficiem cellularum dense incrustans.—DISTRIB. Europa, orientalis.

Cum speciminibus Europæis maxime consentit plantula Kerguelenensis in magnitudine et forma cellularum. Specimina plantulæ quæ observavi e diversis locis Europæ (Galliæ et Germaniæ) consentiunt quoque in loco natali (in *Nitellæ cellulis,* plerumque in *Nitella syncarpa*), et in crescendi modo.

1. **Aphanochæte repens,** *Al. Braun.* Diam. cellular. 0,011mm. ($\frac{1}{189}'''$ Engl.)

In foliis musci aquatici. Sine *oogoniis,* fila in foliis sæpe inter Algas minores (*Leptothrix, Tolypothrix*) dispersa. Cum speciminibus Europæis in omnibus partibus consentit.—(DISTRIB. Europa, America borealis.)

Extra formam typicam filis substrato dense adpressis invenitur quoque forma peculiaris in speciminibus singulis dispersis forma. Filis in corpuscula sphærica et uvæformia accumulatis (pressione in filis singulis soluta).

1. **Gongrosira pachyderma,** *Reinsch in Journ. Linn. Soc.* xv. 218.
Aquatica, parasitica, pulvinulos ex cellulis absque ordine (rarius in modo parenchymatis) cohærentibus formans, cellulis irregulariter sphæricis usque ellipsoidicis (omnibus? fructiferis) cytioplasmate granuloso colore pallide luteo-viridi, cytiodermate crassissimo plurilamelloso (usque cellulæ diametri transversalis dimidio

æquali). Diam. cellularum (cytioderm. incl.) 0,039—0,051mm. ($\frac{1}{54}$—$\frac{1}{42}$''' Engl.) Cytiodermatis crassitudo 0,0041—0,0056mm. ($\frac{1}{567}$—$\frac{1}{378}$''' Engl.)

In foliis musci aquatici. *G. De Baryana*, Rabenh. (Fl. Europ. Alg. ii., p. 388; Alg. Eur. Nro. 223), proxima species (in lapidibus lignisque pr. Francofurtem ad Oderam, Itzigsohn et De Bary.), differt cellulis minoribus in filis subramosis dispositis, cytiodermate tenuiore.

MELANOPHYCEÆ ET RHODOPHYCEÆ.

Rhizocladia,* *Nov. Genus* (ad *Phæosporeas* Thuret spectans, *Pleurocladiæ* Al. Braun proximum).

Plantula ex strato procumbente ex filis ramosis substrato viventi dense adhærentibus formato et ex filis erectis ramosis fructiferis exstituta. Cellulæ filorum procumbentium primo rectangulares, ætate provectiore ovales usque lageniformes. Fila erecta singula aut bina ex cellulis filorum procumbentium orta, primo integra et ex cellulis æqualibus formata, demum subramosa et fructifera et ex cellulis inæqualibus formata. Trichosporangia in apice filorum erectorum ex 3is—5is cellulis quadraticis usque rectangularibus formata. Oosporangia?

1. **Rhizocladia repens,** *Reinsch in Journ. Linn. Soc.* xv. 220. Character idem generis. Longitudo cellul. explic. filor. repent. 0,0097—0,0112mm. ($\frac{1}{236}$—$\frac{1}{189}$''' Engl.) Latitudo cellular. filor. erect. juvenil. 0,0041mm. ($\frac{1}{567}$''' Engl.) Latit. trichosporang. 0,0058mm. ($\frac{1}{382}$''' Engl.) Plantulæ explicitæ altitudo 0,056—0,089mm. ($\frac{1}{38}$—$\frac{1}{23}$''' Engl.)

In foliis muscorum aquaticorum et in cellulis *Nitellæ*.

Hanc plantulam primo habui proximam algæ jam descriptæ et delineatæ in contributionibus meis (p. 76, tab. 14, Chlorophylloph.), quæ planta est constituta ex filis ramosis procumbentibus. Ad hanc plantam nunc ad *Chætophoraceas* positam pertinentes formas postea inveni et ex filis procumbentibus et ex ramulis brevissimis erectis exstitutas. Sed per compages peculiares ex cellulis compluribus brevioribus supra positis exstitutas sine dubio trichosporangiis *Melanophycearum* proximas hujus plantulæ positionem veram agnoscimus in systemate. Cellulæ singulæ dispersæ observantur in filis erectis singulis, ceteris cellulis crassiores et breviores (oosporangia?), quarum natura vera in statu vivente tantummodo agnosci potest. A genere *Pleurocladia* unico aquæ dulcis hucusque cognito *Melanophycearum* organis peculiaribus fœcundationis præditarum differt incrementi modo ac filorum structura.

Tab. V., Fig. ii.—1, plantulæ juvenilis filis sterilescentibus æqualiter altis particulus, cellulæ filorum repentium arctissimæ intricatæ ($\frac{720}{1}$); fig. 2, aliud specimen cæspituli expliciti filis erectis in statu vario evolutionis, filum singulum in apice trichosporangium (α) evolvens ($\frac{720}{1}$).

* ῥίζα radix, κλάδος ramus.

1. **Batrachospermum minutissimum,** *Reinsch in Journ. Linn. Soc.* xv. 220; e minimis, oculis inarmatis vix conspicuum, parasiticum, filis integerrimis erectis singulis aut perpaucis aggregatis, articulis inferioribus subcuneiformibus apice paulo incrassatis, cellulis corticalibus 4is—6is obtectis superioribus nudis rectangularibus, ramulis verticillorum integerrimis (rarius singulis ramulis instructis) æqualibus, apicibus paulo angustatis ex cellulis 5is—7is exstitutis apicem fili versus sensim decrescentibus, inferioribus articulorum longitudine subæquantibus, summis duplo-triplo longioribus, cellulis ramulorum rectangularibus usque subquadraticis, cytiodermate exteriore tenuissimo vix conspicuo, cytioplasmate subhomogeneo colore obscure olivaceo-viridi; fructificatio?—Diam. articulorum 0,0041—0,0056mm. ($\frac{1}{567}$—$\frac{1}{378}$''' Engl.) Diam. ramulorum 0,0041mm. ($\frac{1}{567}$''' Engl.) Plantulæ altitudo 0,37—0,45mm. ($\frac{1}{7}$—$\frac{1}{5}$''' Engl.)

In *Nitellæ* cellulis et in muscorum foliis, in filis singulis sparsis cum aliis algis (*Tolypothrix, Leptothrix*) intermixtis, rarius in cæspitulis parvulis.

B. tumidum, in *Chara vulgari* crescens (Reinsch, Contributiones, p. 69, tab. xliv., Rhodoph. fig. 1-5) a speciebus hucusque cognitis proxima species, sed valde diversa dimensionibus omnium partium multo majoribus (3-4 lineas longa), verticillorum ramulis numerosis repetito dichotome ramosis; in ramulorum cellularum forma aliqua similitudine consentit.

[The fresh-water species recorded by Reinsch are 106, to these may be added the following mentioned in the Antarctic Flora:—*Oscillatoria purpurea,* Hook. f. & Harv., *Calothrix olivacea,* H. f. & H., *Ulva cristata,* H. f. & H., *Mastodia tessellata,* H. f. & H., *Trypothallus anastomosans,* H. f. & H., *Nostoc commune,* Vaucher, and *N. microscopicum,* Carm., making a total of 113 species. This interesting Antarctic island, therefore, so far as explored, appears to be very rich in certain forms of fresh-water Algæ.—*G. Dickie.*]

VII.—*Fungi*.

By the Rev. M. J. Berkeley, M.A., F.L.S.

[The Fungi collected in Kerguelen Island amount to 9 or 10 (the tenth being still an undetermined form).*

Dr. Hooker obtained 2 species in the winter (May and June) 1840; Mr. Moseley 3 in addition to the same, during summer (December and January) 1873-4; Mr. Eaton, also in summer, 5 determinable species, and 1 that could not be identified (*see* footnote), besides the species found by Dr. Hooker.

Until a few days before Midsummer (*i. e.* Christmas) no Fungi were seen in the vicinage of the English Observatory Bay. The first to appear was the common mushroom, a single specimen of which was found on an island in the sound by some officers from H.M.S. "Volage." Later in the summer the other four species came up in a few places on the mainland. They were not by any means of frequent occurrence, and probably scarcely any of them would be found at the time of year corresponding with the date of Dr. Hooker's visit to the island.—*A. E. Eaton.*]

1. **Agaricus** (Galera) **kerguelensis**, *Berk. in Journ. Bot.* v. 51 (1876); *et in Journ. Linn. Soc.* xv. 22. Cæspitosus, fulvus, pileo e breviter campanulato convexo lævi carnuloso, margine tenui striatâ, stipite æquali apice pulverulento-granulato, lamellis distantibus ventricosis adnatis.

Amongst moss in a bog on the eastward portion of the base of a promontory E. of Vulcan Cove, January 1875, *Eaton*.

Cæspitose, attached by abundant mycelium. Pileus $\frac{1}{2}$ inch across; stem nearly 1 inch high, $\frac{1}{2}$ to $\frac{3}{4}$ line thick; principal gills about 12 in number, shortly but truly adnate, and not in the least decurrent.

It is far more fleshy than any variety of *A. hypnorum*, to which species no doubt it is closely allied; and while agreeing with *A. embolus* in possessing comparatively few gills, it differs from that species in the mode of their attachment.

2. **Agaricus** (Galera) **hypnorum**, *Batsch.; Berk. in Journ. Linn. Soc.* xv. 53.

Hab.—On *Azorella*. January 1874. *Moseley.*

Spores ·0004 inch long.

3. **Agaricus** (Naucoria) **furfuraceus**, *Pers.; Berk. in Journ. Linn. Soc.* xv. 221.

---

* This species is referred to by Mr. Eaton (in Proc. Roy. Soc. 1875, May. xxiii. 355) as "a peculiar "parasite on *Azorella*, which grows out of the rosettes" of the leaves "in the form of a clear jelly, which "becomes changed into a firm yellowish substance of indefinite form." It was common on the sides of hills in the neighbourhood of the observatory towards the end of December, and a series of examples was preserved in spirit, but they could not be worked out.

In the same bog as *A. kerguelensis*, and at the same time, *Eaton*.

4. **Agaricus** (Naucoria) **glebarum,** *Berk. in Flor. Antarct.* 447, t. clxii. fig. iii.; *et in Journ. Linn. Soc.* xv. 53.

On *Azorella*, January 1874, Kerguelen Island, *Hooker, Eaton*. (Marion Island, *Moseley*. On tufts of *Bolax*, Falkland Islands, *Hooker*.)

Spores ·0003 inch long.

5. **Agaricus** (Psalliota) **campestris,** *Linn.; Berk. in Journ. Linn. Soc.* xv. 221. A. (P.) arvensis, *Eaton in Proc. Roy. Soc.* xxiii. 355.

On an island near Observatory Bay, in Royal Sound, 16th December 1874. A solitary specimen, *Eaton*. (Almost cosmopolitan.)

1. **Coprinus atramentarius,** *Fries; Eaton in Proc. Roy. Soc.* xxiii. 355; *et in Journ. Linn. Soc.* xv. 222 (footnote).

Two or three specimens were found singly close to the margins of two of the lakes among the hills near Observatory Bay, in February 1875, *Eaton*.

2. **Coprinus tomentosus,** *Fries; Berk. in Journ. Linn. Soc.* xv. 53.

On dung, January 1874, *Moseley*.

1. **Peziza** (Sarcoscyphæ) **kerguelensis,** *Berk. in Flor. Antarct.* 451, t. clxiv. fig. iii.; *Cooke, Mycol.* fig. 134; *et in Journ. Linn. Soc.* xv. 53, 222.

Christmas Harbour, May and June, on bare boggy ground near the sea, growing amongst *Confervæ*, *Hooker*. Amongst dwarfed *Leptinella plumosa*, on wet ground close to the shore, growing in rings, rare. One ring on an island in Swain's Bay, January, and another on the mainland near Observatory Bay, February 1875, *Eaton*. Royal Sound and Betsy Cove, *Moseley*. (Hermite Island, Cape Horn, alt. 1,000 ft., *Hooker*.)

1. **Sphæria herbarum,** *Pers.*

On dead stems of *Pringlea*, *Eaton*.

ZOOLOGY.

SEALS AND CETACEANS. *By William Henry Flower, F.R.S.*

THREE species of seals and two of cetacea are known to be indigenous to Kerguelen Island. Single examples of two of the seals and one of the cetacea have been submitted to me, comprising—a skeleton of a female "sea leopard" *Ogmorhinus leptonyx* (Blainv.), one of a very young male "sea elephant" *Macrorhinus leoninus* (L.), and one of a full grown female cetacean of the genus *Globicephalus*. These are now in the British Museum.

PINNIPEDIA.

Ogmorhinus leptonyx (Blainv.).

Phoca leptonyx, Blainville, Journ. Phys. 1820, xci. 288.

Stenorhynchus leptonyx, F. Cuvier, Mem. du Mus. 1824, xi. 190, pl. xiii. 1 (not *Stenorhynchus*, Lamk. 1819, of the Crustacea; nor *Stenorhynchus*, Lat. 1823 of the Insecta).

Ogmorhinus leptonyx, Peters Monatsb. K. Akad. d. Wissenschaft. Berlin, 1875 Juni 10, &c.

"Sea Leopard."—The female example killed in Observatory Bay, Royal Sound, 13th October 1874, judging from the condition of the bones, was not quite full grown. The skeleton does not differ to any appreciable degree from another of corresponding age and sex in the Museum of the Royal College of Surgeons, received from Tasmania. It has 7 cervical, 14 thoracic, 6 lumbar, and 3 sacral vertebræ. The tail was left in the skin. Extreme length from muzzle to tip of tail 7 feet 8 inches.

Dist.—This species is extensively distributed in the southern hemisphere, having been met with in South Georgia, the Falklands, Kerguelen Island, South Australia, Tasmania, New Zealand, Campbell Island, &c.

Professor Peters has lately proposed the name *Ogmorhinus* for the genus to which this seal belongs, the name *Stenorhynchus*, by which it is generally known, having been preoccupied in zoology.

Macrorhinus leoninus, (L.)

Phoca leonina, Lin. Syst. Nat. ed. xii. 1766, i. 55.

Ph. elephantina, Molina, Saggio 1782, p. 260.

Ph. proboscidea, Perron, Voy. aux Ter. Austr. 1816, ii. 34.

Macrorhinus proboscideus, F. Cuv., Mem. du Mus. 1824, xi. 200, pl. xiii. (not *Macrorhinus* Lat., Fam. Nat. du Regne Anim. 1825, a genus of the *Coleoptera*).

Mirounga patagonica, J. E. Gray, Griffith's Animal Kingdom, 1827, v. 179.

Morunga elephantina, Gray, Cat. Osteol. Spec. Brit. Mus. 1847, p. 33; idem Cat. Seals and Whales, 1866, p. 38, fig. 13 (skull).

Cystophora proboscidea, Wyv. Thomson, in Good Words, 1874, November, p. 748 (Chal. Exped.).

C. leonina, J. W. Clark, in Nature, 1875, Sept. 2, p. 366 (German Tr. Exped.).

Macrorhinus leoninus, Kidder, Bulletin U. S. Nat. Mus. 1876, iii. 39 (Am. Tr. Exped.).

"Sea Lion and Lioness" Anson's Voy. 1748, p. 122 and figure.

"Sea Elephant" of authors; Eaton, Proc. Roy. Soc. 1875, xxiii. 502; "Platyrhine seals," idem, op. cit. p. 353.

Two young males casting their coats were killed with stones on the beach near Thumb Peak on the 8th December 1874; and several of each sex of corresponding age and size were shot with explosive bullets in Swain's Bay in the following month. These last were valueless as specimens.

Dist.—The sea elephant has long been known as an inhabitant of Kerguelen Island, and formerly was widely distributed along the coasts of the Antarctic and southern temperate seas.

Mr. Eaton in Proc. Roy. Soc. 1875, xxiii. 502, says:—"Some examples are "uniformly reddish brown; others are pale, blotched and spotted with darker grey. "They usually lie just above the beach, separately, in hollows among the *Acæna* and "*Azorella*, where they are sheltered from the wind. On being approached they "make no attempt to move away (possibly because there are no land animals, in-"digenous to the country, capable of molesting them, to cause them to acquire a "habit of flight) but raise up the fore part of their body, open the mouth wide, and "utter a peculiar slobbering cry."

There is much confusion as to the synonymy of this species. It is the *Phoca leonina* of the Systema Naturæ, ed. xii., founded upon the "sea lion and lioness" of Juan Fernandez, described and figured in Anson's Voyage, 1748, the *P. elephantina* of Molina, 1782, and the *P. proboscidea* of Perron, 1815, and of many later authors. *Leonina* therefore is the earliest specific appellation.

With regard to the generic name, after *Phoca* (which of course is inadmissible, having since the dismemberment of the group been restricted by common consent of naturalists to *P. vitulina* and its immediate allies), *Cystophora* Nilsson, 1820, has priority for those who hold that there is not sufficient difference between the southern Sea Elephant and the well-known Hooded Seal (*Cystophora cristata*, Fab., or *borealis*, Nils.) of the Arctic seas to separate them generically. But those who hold the contrary opinion (and the remarkable dissimilarity of the auditory ossicles inclines me to this view) should adopt *Macrorhinus* F. Cuv., 1824, which was afterwards needlessly superseded by *Morunga*, J. E. Gray.

It should be mentioned that *Macrorhinus* F. Cuv., is often (as in Agassiz, Nomenclator Zoologicus) quoted as if it had been first used in the Dict. de Sc. Nat. xxxix., 1826, and priority is accordingly misattributed to Latreille's use of the same name in the Fam. Nat. du Régne Animal, 1825, for a genus of the *Coleoptera*.

Otaria gazella, (Peters).

Arctophoca gazella, Peters, Monatsb. K. Akad. d. Wissensch. Berlin, 10 Juni 1875.

[The Fur Seal of Kerguelen Island does not resort to the sheltered waters visited by the American and English Expeditions. A single example was captured by the German Expedition at Betsy Cove, and according to Wyville Thomson (op. cit. p. 748) the Challenger Expedition seems to have obtained two in Fuller's Harbour.

It is due to Dr. Günther to state that, prior to my leaving England, he informed me of the occurrence at Kerguelen Island of an undetermined small species of seal besides the Sea Leopard and Sea Elephant, and showed me an unpublished drawing of its head. While the expedition was at Observatory Bay, I wrote to Captain Fuller of the "Roswell King," endeavouring to negotiate for a complete skeleton and skin of the Fur Seal, which he expected to meet with at Swain's Islands,—a small group off the N. E. coast of the main island. He returned in about a month afterwards without having killed or seen one in the course of his trip. The species was doubtless *A. gazella*, Peters.—*A. E. E.*]

CETACEA.

Balæna australis, Desmoulins.

[A pair of Right Whales were seen occasionally among the islands in Royal Sound, between the English Station at Observatory Bay, and the Prince of Wales' Foreland. I noticed portions of two old skeletons of this species stranded on the eastern shore of a promontory immediately to the eastward of Vulcan Cove, about five miles (an hour and a half's walk) from the head of Carpenter's Cove. They comprised the skulls, the lower jaws, the cervical and some of the other vertebræ, the scapulæ, and a few ribs. They were in tolerable condition, but it was impossible to convey even the scapulæ over so great an extent of bog and rocky hills to a boat at Carpenter's Cove.—*A. E. E.*]

Globicephalus melas, Traill.

? *Catodon svineval*, Lacepéde, Hist. Nat. de Cétacés, 1804, p. 216, pl. xiii.

Delphinus melas, Traill, Nicholson's Journ. 1809, xxii. 81–83.

D. globiceps, Cuv. Ann. du Mus. 1812, xix., pl. i. 2.

Globicephalus, Lesson, Compt. de Buffon, 1828, i.

Globiocephalus svineval, Gray, Zool. Ereb. & Ter., 1846, p. 32.

G. macrorhynchus, auctorum (not Gray, op. cit., p. 33); Van Bened. & Gerv., Osteogr. der Cét. pl. lii.; Hector Trans. N. Zeal. Instit. vii., pl. xxvi. 3.

The "Blackfish," "Cäaing Whale," or "Pilot Whale."

This was the animal mentioned in Mr. Eaton's First Report (Proc. Roy. Soc. 1875, xxiii. 501), as found by Mr. Midshipman Forrest, dead in shallow water in Swain's Bay. The example is an adult female, all the epiphyses being united to the bodies of the vertebræ. The skeleton was complete; but about five of the caudal vertebræ and a few of the terminal phalanges have been lost. There are 7 cervical vertebræ, of which five are coherent, 11 dorsal, and 35 lumbo-caudal. Maxillæ visible as a narrow strip along the whole of the outer border of the rostrum. Detention $\frac{10-10}{10-10}$ (teeth much worn). Extreme length of body in the flesh, 19 ft. 4 in.

Principal measurements of the skull:—

Total length	26 inches.
Length of rostrum from the tip to the centre of a line drawn across the anteorbital notches	13·5 ,,
Width of base of rostrum at the anteorbital notches	10·3 ,,
Width of rostrum in its middle	8·7 ,,
Greatest width of skull	19·0 ,,
Breadth of occipital condyles	7·1 ,,
Mandible, length of	20·6 ,,
,, length of tooth line of	6·2 ,,
,, height of, at coronoid process	6·1 ,,

I can detect no difference between this skeleton and that of the "Blackfish" from the seas round Tasmania, five examples of which, presented by Mr. W. L. Crowther, F.R.C.S., of Hobart Town, are in the Museum of the Royal College of Surgeons. Indeed I must go further, and add that I can find no tangible osteological grounds for separating either the Kerguelen or the Tasmanian specimens from the well known Caaing Whale, *Globicephalus melas*, Traill, of the N. Atlantic, the Grindval of the Faröese. When only one or two examples of each are compared together, it is not difficult to discover distinctions between them, because numerous individual variations occur in different parts of the skeleton; but when larger series of both Northern and Southern specimens are examined, these variations do not appear to resolve themselves into constant characters by which the one can be distinguished from the other. The reasons for giving to the specimens from distant seas a distinctive appellation have been mainly derived from the supposition that it was impossible that cetacea from widely remote localities could be specifically identical with one another; but this is an assumption too great to be made without proof, and in the present instance is especially inadmissible, considering that *Globicephali* almost, if not quite, identical have been observed by voyagers in all intermediate localities. If an assumption such as this could be allowed, it would strike at the root of all our knowledge of the geographical distribution of animals. It is, however, possible that future observers with still more ample opportunities may succeed better in discriminating between the two. Meanwhile, until some valid distinctive characters shall be pointed out, I think it will be safest and most in accordance with the true principles of zoological nomenclature to designate them all by the specific name *melas*, under which one member of the group was first clearly described by Traill.

The *Catodon svineval* of Lacepede (1804) is possibly founded on the present species; but the description is so vague and inaccurate that it can hardly be conceded priority over the excellent account and figure given by Traill (1809). Cuvier, unacquainted with Traill's memoir, described and figured the animal independently in 1812 under the name of *Delphinus globiceps*. This hybrid word was modified by Lesson in 1828 into the generic term *Globicephalus*. Dr. Gray in 1846 revived the specific term *svineval*, and definitely applied it to this species in his subsequent catalogues, altering Lesson's generic designation at the same time to *Globiocephalus*.

All the specimens of *Globicephalus* from the southern hemisphere are often catalogued in systematic works under the name of *G. macrorhynchus*, Gray. In the Osteographie des Cétacés, pl. lii., a skull from New Zealand, closely resembling the one from Kerguelen, is figured by Van Beneden and Gervais under this name; but nothing is apparent in the figure to distinguish it from the skulls of *G. melas* from the coasts of Brittany and Iceland which are figured with it, excepting characters appertaining to a somewhat inferior age. The letterpress of this section of their work, which it is hoped may clear up some of the difficulties existing at present in the history and synonymy of this group, has not yet appeared. A skull figured by Dr. Hector in the Transactions of the New Zealand Institute (vol. VII., pl. xvi., fig. 3) under the name in question also resembles that of *G. melas*. And M. Fischer (Journal de Zoologie, 1872, i. 273) has been unable to detect any distinction of specific value between a skeleton of a *Globicephalus* brought from the Cape of Good Hope, by M. Verreaux, and those of individuals from the northern seas. The present specimen from Kerguelen Island, as well as the five from Tasmania mentioned at the commencement of these remarks, agree in the form of the cranium and the number of their teeth $\left(\frac{10\text{---}10}{10\text{---}10}\right)$ far more closely with skulls of *Globicephali* from Iceland (as figured in Van Beneden & Gervais Osteogr. des Cet. pl. li.), the Faroë Islands (Mus. Roy. Col. Surgeons), the British Coast (British Museum; figured in Gray Cat. Seals and Whales, 1866, p. 316, fig. 62), and the Mediterranean (as figured in Van Beneden & Gervais Osteogr. des Cet. pl. li.), than they do with the type of Gray's *G. macrorhynchus*. In all of them a strip of the maxilla is distinctly visible along the whole of the outer border of the rostrum.

It follows therefore, judging from the osteological characters only (for the descriptions of the external appearance of the "blackfish" of the south seas which we possess at present are too vague to furnish good zoological characters) that *Globicephali* almost if not quite identical are found in the North Atlantic and the Mediterranean, as well as in the seas of the Cape of Good Hope, Kerguelen, Tasmania, and New Zealand..

[Several decayed skulls of *Globicephalus* were found here and there on the shores of retired inlets in Royal Sound, and a pair of these animals often frequented a rocky bay adjacent to Observatory Bay, sounding along the edge of the kelp, and coming quite close to the land.—*A. E. E.*]

Note.—*Globicephalus macrorhynchus*, Gray, Zool. Ereb. and Ter. 1846, p. 33, is a well marked second species of this genus, whose geographical distribution is not yet clearly ascertained. The species was based upon a skull in the Museum of the Royal College of Surgeons (Osteol. Cat. No. 2519) presented by F. D. Bennett, Esq., and thus characterised by Dr. Gray :—

"Nose of skull short and broad, rounded in front, nearly as broad in the middle
" as at the preorbital notch. Teeth sub-cylindrical $\frac{8}{8}$. Lower jaw rounded in
" front."

This skull, which is quite adult, besides being smaller than that of *G. melas*, has the premaxillary bones so wide in their anterior half that they extend from side to side over the whole of the upper surface of the rostrum, concealing the maxillæ completely. A precisely similar skull from Guadeloupe is figured by Van Beneden and Gervais in the Osteographia des Cétacés, pl. lii. 3, under the name of *G. intermedius*, and another, from the Atlantic coast of the United States, by Cope in the Proceedings of the Academy of Natural Sciences of Philadelphia, 1876, p. 129, as *G. brachypterus*.

This is certainly not the species to which the Kerguelen specimen can be referred.

BIRDS. *By R. Bowdler Sharpe, F.L.S., F.Z.S., &c., Senior Assistant in the Zoological Department, British Museum.*

(Plates VI.–VIII.)

I PROPOSE on the present occasion to give a complete account of the Avifauna of Kerguelen Island, founded on the collections in the British Museum (partly made by the Rev. Mr. Eaton, partly by the naturalists of the Antarctic expedition) and on the reports of the German and American expeditions, which have been recently published.* The American expedition obtained twenty-one species, the German naturalists recording twenty-three.†

With regard to the Antarctic Expedition, Sir Wyville Thomson thus writes:—
" This expedition had the extraordinary advantage of having Dr. Hooker attached
" to it as one of the assistant-surgeons, and the surgeons to the *Erebus* and
" *Terror*, Dr. M'Cormick and Mr. Robertson, and the assistant-surgeon of the
" *Terror*, Dr. Lyall, were all zealous naturalists, and co-operated heartily with
" Dr. Hooker in his work, so that every possible advantage was taken of the
" sixty-eight days of their stay in Christmas Harbour. Their visit was, however,
" in the depth of winter, and although the actual difference between the winter
" and summer temperature is not so great as might have been anticipated, the
" winter weather is so boisterous and unsettled, that on forty-five of the sixty-eight
" days it blew a gale, and on three days only neither snow nor rain fell." Under these circumstances, therefore, it is not a little creditable to the officers of this expedition that they managed to collect a series of *nearly* every species obtained by the more recent visitors to the island, while they procured several species which none of the latter have met with.

Before proceeding to a detailed account of the species, it may not be uninteresting to give a list of the birds noticed in Christmas Harbour by Captain Cook, when he visited the island one hundred years ago. We are able to give a tolerably correct idea of these from the paintings by Ellis preserved in the British Museum. Ellis was a draughtsman employed by Sir Joseph Banks, whom he accompanied on Cook's voyage. All these paintings are well executed, and the species are recognisable; each one has written on the back the locality, and there can be little

* Contributions to the Natural History of Kerguelen Island, made in connection with the American Transit of Venus Expedition, 1874–75; by J. H. Kidder, M.D. I. Ornithology. Edited by D. Elliott Coues, Washington, 1875. 8vo. pp. i–ix, 1–51. (Bulletin of the United States National Museum, No. 2.) The same. By J. H. Kidder, M.D. Part II. pp. 1–122 (Bull. U. S. Nat. Mus., part 3).

Uebersicht der auf der Expedition Sr. Maj. Schiff "Gazelle" gesammelten Vögel. Zusammengestellt von J. Cabanis und A. Reichenow, J. f. O., 1876, pp. 319–330.

† One species of *Prion* is deducted, as I consider that *P. ariel*, recorded by them, is nothing but the young of *P. turtur*. (Vide infrà.)

doubt but that either these drawings themselves or the actual specimens from which they were taken, formed the types of some of Latham's species. The following species are figured from the "Island of Desolation," as Cook called Kerguelen Island.

No. 39. Ossifraga gigantea.
No. 43. Prion desolatus.
No. 45. Eudyptes saltator * (fig. mala).
No. 46. Aptenodytes longirostris.
No. 54. Sterna vittata.
No. 59. Chionis minor.

In the Narrative of the "Wreck of the Favorite" † is given a very good list of the birds of Kerguelen Island, with figures of many of the species (pp. 173–199). Who was the author of this list I have not been able to discover, but that assistance was rendered by some one who had access to the collections of the Antarctic Expedition I have very little doubt, as nearly every bird procured by the latter is mentioned. Sir J. Hooker, to whose kindness I owe the loan of the work, does not remember the name of the compiler, but he informs me that the late Professor Henslow took a great interest in it.

Lastly, Sir J. Hooker has allowed me to incorporate in this paper certain notes extracted from his "Journal," an act of courtesy that I gratefully acknowledge; so that by references and quotations I hope to be able to present to ornithologists a tolerably correct idea of the Avifauna of Kerguelen Island, as at present determined.

The notes on the habits of the birds are contributed by the Rev. A. E. Eaton, whose initials are in each case appended. Many of the soft parts of the birds described are taken from the careful notes made by Dr. Kidder in his "Report," some few from Gould's works.

CHIONIDÆ.

Chionis minor.

Chionis minor, *Hartl. Rev. Zool.* 1841, *p.* 5; *id. op. cit.* 1842, *pl.* 2, *figs.* 2, 2*a*, 2*b*; *Lafr. t. c. p.* 402; *Gray, List Anseres, &c. B. M. p.* 52 (1844); *Gray & Mitch. Gen. B. iii., p.* 522, *pl.* 133 (1845); *Schl. Handl. Dierk. p.* 400, *pl. viii, fig.* 98 (1857); *id. de Dier. fig. to p.* 232; *Gray, Handl. B. iii., p.* 21, *No.* 10,056 (1871); *Kidder, Bull. U. S. Nat. Mus. No. ii., pp.* 1–4 (1875); *id. & Coues, op. cit. iii., p.* 7; *Reichen. J. f. O.* 1876, *p.* 89; *Cab. & Reichen. t. c. p.* 327, *pl.* 1, *fig* 2.

Chionarchus minor, *Kidder and Coues, Bull. U. S. Nat. Mus. iii., pp.* 85–116.

---

* This figure is not so good as most of them, and apparently represents two immature Penguins, which may be the young of *E. saltator*.

† Narrative of the wreck of the "Favorite" on the Island of Desolation; detailing the adventures, sufferings, and privations of John Nunn; an historical account of the Island, and its whale and seal fisheries. Edited by W. B. Clarke, M.D. 8vo. pp. i–xx, 1–236. London, 1850.

Ad. pure albus, primariorum recticumque scapis flavicanti-albis; rostro nigro; regione palpebrali et oculari pallide coccinea; caruncula lorali nigra; iride purpurascenti-nigra; tarso pedibusque sordide albis, carneo tinctis. Long. tot. 17·5, culmen 1·35, alæ 9·4, caudæ 5·4, tarsi 1·7. ♀ mari similis, sed caruncula lorali paullo minore. Long. tot. 16·0, alæ 8·9, caudæ 4·9, tarsi 1·6. *Juv.* similis adultis, sed rostro multo minore et remigibus pallide roseo apicatis.

In the American account of their Kerguelen collections, it is stated by Dr. Kidder (Bull. U. S. Nat. Mus. iii., p. 89) that he has only succeeded in finding a record of four specimens of this species in European collections. The writer has omitted to notice that in the British Museum " List of Anseres," &c. (p. 52), published by the late Mr. G. R. Gray in 1844, no less than seven specimens are mentioned, and I now give a complete list of the birds at present in the national collection. I would call attention to the fact that the bird from the Crozettes seems to have darker legs than the Kerguelen bird: whether this occurs only in the dried skin, or is to be found in the living Sheathbill, must be left for future visitors to these islands to determine.

a. b. ad. *c.* juv. st. Kerguelen Island. Lieut. Alexr. Smith.

d. ♀ ad. sk. Kerguelen Island. Antarctic Expedition.

f. g. ad. sk. Christmas Harbour, May 30, 1840. Antarctic Expedition.

h. ad. sk. Crozette Islands. Captain Armson.

Sir J. Hooker's journal contains the following note:—" The young birds have pink tips to their wings. When hard pressed it takes to the water (but this very seldom), and swims slowly."

[Where the coast is rocky and sheltered, Sheathbills are common. In some of the most favourable situations they were in flocks of a dozen, thirty, or more; but the birds which are breeding live in separate pairs. They frequent the shore to feed between the tide marks upon mussels (whose shells they break by pecking at them), isopod crustacea, *Enteromorpha*, and *Ulva*. Other kinds of seaweeds do not seem to be relished by them; one was observed to peck at a piece of *Delesseria*, but it did not eat it. They are also very assiduous in their attendance upon colonies of Shags and Crested Penguins, whose eggs they greedily devour. The sitting birds stretch out their necks and croak at the Sheathbills sauntering past their nests; but the marauders, keeping just out of reach of their bills, pay little regard to them, and proceed in a business-like manner to eat up the first eggs they may chance to find unguarded. It occasionally happens that while an old Shag is gesticulating violently at a *chionis* in front of her, his friend pecks from behind at the eggs which in the excitement of the moment are not covered completely by her. When she finds out what is taking place, she drives him away with a croak, and true to her sex affects to have won her point in the affray. Reseating herself upon the nest with great dignity of deportment, and gently replacing with her bill the broken eggs under her feathers, she resigns herself to the task of trying to hatch them. The Sheathbills withdraw. Some time after they have gone away, the broken eggs are inspected, and if there is only a small hole pecked in each of them, they are kept in the nest.

Their appearance and manner of caressing one another led the blue jackets to call Sheathbills "white pigeons." In their gait and flight they closely resemble Ptarmigan; and like these they utter their cry when starting on the wing, as well as during flight. After they have attained a fair amount of speed they sail along from time to time with outstretched wings. On alighting at their destination they often greet one another with a gentle chuckle while they nod their heads.

Shortly before they began to build it became evident that they were about to do so. Scarcely a hen could fly anywhere without being attended by two males. The holes that seemed available for nests underwent frequent and searching examination by them, and in a short while the nests began to be built. Each male then became extremely jealous of his mate; and if a neighbour happened to approach too near the nest, he was at once warned off by the legitimate owner. Mr. Midshipman Stuart of H.M.S. Volage saw some fighting, springing up into the air like partridges.

The nest is usually built in a hole between or behind rocks which constitute the extreme limit of the shore where the land plants meet it. Its position can generally be ascertained approximately at a glance by looking for the most conspicuous boulder on the beach frequented by the pair of birds; for Sheathbills usually alight upon a rock of that description just before they enter their hole. Another easy method of finding the nest, is to startle the cock bird whilst he is feeding along the shore alone, for he is apt to fly to his mate when alarmed. The holes selected for nests are generally rather spacious within; and pieces of building materials are frequently dropped by the birds outside the entrances. The nest is of very simple construction without a lining; it consists of a heap of dried seed-stalks of *Pringlea antiscorbutica,* or tufts of *Festuca erecta,* slightly hollowed out on the top. Occasionally old burrows of *Prion* and *Halobæna* are occupied by sheathbills: when this is done the nest is constructed, not in the terminal chamber of the Petrel's burrow, but a foot or two feet within the entrance of the tunnel, which is thus far enlarged for the purpose by the *Chionis.* One nest was in a burrow of this kind at the base of a large stone standing on a slope amongst *Acæna,* upwards of 10 yards from the shore.

The usual number of eggs in a nest is two or one; three is exceptional. Those in the same nest are not always alike in colour, nor in the style of their marking. The first eggs were found on the 23rd of December; the first brood of nestlings about the middle of January. The young are clothed with unicolorous slate grey down.

The "sheath" of the bill is immovable. In the chick it is concrete with the horny substance of the bill, and is only indicated by a faint line of demarcation. In the course of growth it becomes developed into a distinct lamina, embracing the base of the upper mandible.

The bird-louse parasitic on this species was not observed by us. The Patagonian chionis (*C. alba*), according to Giebel, is infested with an undescribed *Liotheum*. (*See* Zeitschr. f. d. gesammt. Naturwiss. Berlin, 1861, Heft. 8, 9, p. 311.)—A. E. E.]

ANATIDÆ.

Querquedula eatoni. (Plate VI.)

Querquedula *sp.*, *Gray, List Anseres, &c., Brit. Mus.*, p. 138 (1844).

Querquedula Kerguelensis, of the "Wreck of the 'Favorite,'" *p.* 186 (1850, *descr. nullâ*).

Querquedula eatoni, *Sharpe, Ibis*, 1875, *p.* 828; *Coues and Kidder, Bull. U. S., Nat. Mus.*, ii. *p.* 4; *iid., op. cit.* iii. *p.* 7; *Cab. and Reichen. J. f. O.* 1876, *p.* 329.

♂. supra brunneus, plumis plurimis griseo marginatis, rufescenti-fulvo maculatis aut fasciatis; scapularibus nigricantioribus; pileo paullo rufescentiore, plumis nigro medialiter striatis; facie laterali et gutture albicantibus, minute nigro striolatis, mento fulvescenti-albo; corpore reliquo subtus albicante, brunneo marmorato, plumis plerisque pectoralibus versus basin griseo brunneis aut medialiter brunneo striatis; hypochondriis brunneis, albido terminatis et rufescenti-fulvo transfasciatis; subcaudalibus rufescenti-fulvis, nigro adumbratis, longioribus nigricantibus fulvo terminatis; tectricibus alarum superioribus cinerascenti-brunneis, majoribus pallide badio terminatis, fasciam alarem formantibus; remigibus cinerascenti-brunneis, secundariis extus purpureo æreis albido terminatis, speculum alare æreum vix sub certa luce olivascente nitens exhibentibus; secundario proximo nigricante vel aspectu externo viridi nitente, medialiter cinerascente strigato, albo apicato; secundariis interioribus nigricantibus extus pallide brunnescentibus albo limbatis; rectricibus mediis nigricantibus, reliquis brunneis albo marginatis, nonnullis rufescenti-fulvo notatis; tectricibus subalaribus brunneis, inferioribus intimis et axillaribus albis brunneo maculatis, tectricibus majoribus cinerascentibus alæ inferiori concoloribus; rostro virescenti-plumbeo, culmine nigro; pedibus ochrascentibus vel sordide viridescentibus; iride purpurescenti-nigra. Long. tot. 15·5, alæ 8·5, caudæ 4·8, tarsi 1·2.

The above is a copy of the description published by me in the "Ibis;" and I then described the female as being similar to the male, but distinguished by the want of the alar speculum. According to Dr. Coues' observations, however, the female differs only in having the green speculum duller and tinged with brown: this was seen in specimens determined by dissection by Dr. Kidder. It is probably a young female that I diagnosed in the original article as follows:—

♀ mari similis sed speculo alari absente, secundariis albo terminatis; caudâ brunneâ, rufescenti-fulvo fasciatim marmoratâ.

This plain-coloured Teal is allied to *Q. gibberifrons* and *Q. creccoides*. From the former it is at once to be distinguished by the fawn-coloured bar on the wing and the bronzy speculum, the wing-bar being broadly white, and the speculum black in *Q. gibberifrons*.

Q. creccoides resembles *Q. eatoni* in having the fawn-coloured wing-bar; but then the speculum is black, and the greater part of the bill is yellow. *Q. eatoni* also has the axillaries whitish barred with brown, whereas they are quite white in the allied species; and, moreover, it has remains of rufous-buff bars on most of the feathers of the upper surface, the back being uniform in the other species.

a. ♀ ad. sk. Kerguelen Island. Rev. A. E. Eaton.
b. ad. c. d. ad. sk. Kerguelen Island. Antarctic Expedition.

[We found this bird very tame, and only after the officers from the ships had bagged upwards of 2,000 head within a radius of eight miles of the observatory did the survivors acquire habits of caution, and learn to restrain their curiosity. On the bogs they are conspicuous objects.

Until they commence to breed these Teal associate in small "springs," and not like mallards in "bunches." Every day when the tide is out they leave the bogs and hills to congregate upon the rocks and the mud of the estuaries left bare by the ebb. The scene presented at their favourite resorts at these times is an animated one. Hundreds of Gulls flocking from their resting places in the neighbouring cliffs are filling the air with their clamour, and leisurely walking about where the mussels lie thickest upon the mud-bank. The Teal in small parties are busily dabbling along the water's edge and round the stones in tide pools for the abundant isopod *Sphæroma gigas*. Here and there a Sheathbill rambles apart over the sea-weed covered rocks. Elegant in form and graceful in their flight the Terns flit lightly to and fro, hovering now and again and plunging into the kelp, or playfully engaged in clamorous pursuit of their successful companions, presently alight upon a boulder side by side twittering amicably. Here comes a bustling Cormorant hurrying past, turning his head from side to side occasionally as he looks about him in his flight, a contrast to the sooty Albatross which is sailing so silently across the sky.

On the return of the tide the Teal withdraw from the shore, some of them to the cliffs, others to the hill sides and to marshes bordering streams and lakes, where they strip blossom and seed from the Kerguelen cabbage (*Pringlea*), and sift the soft ground for food.

Their nests are built in the large majority of instances in crowded patches of *Pringlea* near the sea. Very few were found under solitary plants, and when they were the plant was sure to be growing close beside a lump of *Azorella*. Scarcely ever was a nest placed amidst *Acæna* only. All of them were thoroughly protected from the weather by the leaves, and well concealed. They were all neatly lined with a thick layer of down. The teal build apart from one another, and not in company, consequently I never found more than thirty nests in one day. There appears to be much irregularity in their time of breeding. The greater number of them had eggs in December, but whilst some had even hatched theirs so early as the day of the Transit (*i.e.* December 9), others had not begun to lay before the first week in February. The eggs are few in number. Whether this is due to the effects of inbreeding attendant upon the isolation of the species is open to conjecture. It may be occasioned by the coldness of the climate, as in the case of Arctic ducks. Usually there are only three, but occasionally there are four or five eggs in a nest. This last is the largest number of eggs observed together, and of ducklings in any one brood.

The louse of this Teal was not found. The British *Q. crecca* has a *Menopon*, a *Nirmus* and a *Lipeurus*.

In the South African Museum at Cape Town there are two undetermined specimens of *Q. eatoni* from the Crozettes.—*A. E. E.*]

LARIDÆ.

Larus dominicanus.

Gaviota Mayor, *Azara, Apunt.* ii., *p.* 338, *No.* 409.

Larus marinus, *pt. Vieill. N. Dict. d'Hist. Nat.* xxi., *p.* 507.

Larus dominicanus, *Licht. Verz. Doubl. p.* 82; *Neuwied. Beitr. Naturg. Brasil.* iv., *pt.* 2, *p.* 850; *Gray, List Anseres, &c., Brit. Mus. p.* 169; *id. Voy. Ereb. and Terror, Birds, p.* 18; *id. and Mitch. Gen. B. p.* 654, *pl.* 180; *Reichenb. Handb. Longipennes, pl.* xxvi., *fig.* 883; *Schl. Mus. P. B. Lari, p.* 12; *Blasius, J. f. O.* 1865, *p.* 378; *Layard, B. S. Afr. p.* 367; *Gray, Handl. B.* iii., *p.* 112; *Finsch, J. f. O.* 1872, *p.* 241; *Scl. and Salv. P. 3. S.* 1871, *p.* 576; *Gurney in Anderss. B. Dam. Ld. p.* 357; *Buller, B. N. Zeal. p.* 270, *pl.* 28, *fig.* 2; *Scl. and Salv. Nomencl. Av. Neotr. p.* 149; *Finsch, J. f. O.* 1874, p. 203; *Sharpe, Voy. Ereb. and Terror, App. p.* 32; *Coues and Kidder, Bull. U. S. Nat. Mus.* no. ii., *p.* 13; *iid. op. cit.* no. 3; *Cab. & Reichen. J. f. O.* 1876, *p.* 328.

Larus littoreus, *Forster, Descr. Anim. p.* 56.

Larus antipodus, *Gray, List Anseres, &c. p.* 169; *id. Ibis,* 1862, *p.* 245.

Dominicanus vociferus, *Bruch, J. f. O.* 1853, p. 100, 1855, p. 281.

Dominicanus pelagicus, *Bruch, J. f. O.* 1853, *p.* 100, *pl.* 2, *fig.* 3, 1855, p. 280; *Bp. Consp.* ii., *p.* 214; *id. C. R.* xlii., *p.* 770.

Dominicanus vetula, *Bruch, ll. cc. pl.* 2, *fig.* 4; *Bp. Consp.* ii. *p.* 214.

Dominicanus antipodus, *Bruch, ll. cc. pl.* 2, *fig.* 8; *Bp. Consp.* ii., *p.* 214.

Larus verreauxi, *Bp. Rev. Zool.* 1854, *p.* 7; *id. Naum.* 1854, *p.* 211; *Gray, Handl. B.* iii., *p.* 112.

Dominicanus fritzei, *Bruch, J. f. O.* 1855, *p.* 280; *Bp. Consp.* ii., *p.* 214; *id. C. R.* xcii., *p.* 770.

Dominicanus verreauxi, *Bruch, J. f. O.* 1855, p. 281.

Dominicanus azaræ, *Bp. Consp.* ii., p. 214.

Clupeilarus verreauxi, *Bp. Consp.* ii., *p.* 221, *id. C. R.* xlii., *p.* 770.

Clupeilarus antipodum, *Bp. C. R.* xlii., *p.* 770.

Larus vociferus, *Burm. Th. Bras.* iii., p. 448.

Larus antarcticus, *Ellman, Zool.* 1861, p. 7472.

Larus fuscus, *Ellman, Zool.* 1861, *p.* 7472; *Chapm. Trav. S. Afr. App. p.* 425.

Larus azaræ, *Pelz. Reis. Novara, Vög. p.* 151.

Larus vetula, antipodum, fritzei, pelagicus, *Gray, Handl. B.* iii., *p.* 112.

Ad. dorso toto et scapularibus schistaceo-nigris, his late albo terminatis; tectricibus alarum dorso concoloribus; remigibus nigris, albo terminatis, primariis quibusdam plus minusve subterminaliter albis; secundariis schistaceo-nigris, late albo terminatis; uropygio, supracaudalibus caudaque pure albis, pileo et collo undique cum corpore subtus toto, subcaudalibus, subalaribus et axillaribus pure albis; rostro flavo, mandibulæ gonydis angulo rubro; pedibus flavis, antice viridibus, iride flava. Long. tot. 23, culm 1·85, alæ 16·5, caudæ 7·5, tarsi 20.

Young birds are brown, just as in the case of its northern ally *L. marinus*. I can see no difference at all between adult specimens of this gull from all the localities mentioned below, whence the British Museum possesses examples. The description is taken from an adult bird from Kerguelen Island, presented by Lieut. J. B. Smith.

Besides specimens from the Falkland Islands (*Antarctic Expedition*), Straits of Magellan, East Patagonia (*Admiral Fitzroy*), Valparaiso (*Captain W. S. Brett*), Cape of Good Hope (*Sir A. Smith; E. L. Layard*), and New Zealand (*Sir G. Gray; Capt. Stokes*), the British Museum contains the following from Kerguelen Island:—

a. b. c. ad., *d. e.* juv. Kerguelen Island. Antarctic Expedition.

f. ad. Royal Sound. Lieut. J. B. Smith.

N.B.—In the list of birds appended to Nunn's narrative *Larus pacificus* is also included as an inhabitant of Kerguelen Island, as well as *L. dominicanus*. Not having seen an actual specimen from thence I have not included the latter species, though the description of the bird leaves little doubt of its identity.

[This Gull does not build in cliffs, but amongst the land plants or rocks which are immediately bordering upon the shore. On low promontories the nests are often placed within a few yards of each other in hollows amongst *Azorella*; but where the coast-line is tolerably even they are usually a considerable distance apart, and frequently adjacent to some rock or hump of *Azorella* to which the respective pairs of Gulls have been accustomed to resort. In the breeding season the sites of these solitary nests are roughly indicated by lonely birds sauntering between tide marks in their neighbourhood. As the females do not sit close when approached, but steal away unobtrusively whilst out of gunshot, this clue to the positions of their nests afforded by their mates is worth observing. The principal materials employed in the construction of the nest are dead seedstalks of *Pringlea* disposed like rushes in a Moorhen's nest, which serve as the basis, and dried tufts of *Festuca erecta*, which constitute the lining. The eggs are usually two or three in number, and there are two principal varieties in their coloration, the one paler than the other. The dark variety appears to be the commonest, as among upwards of 30 nests examined by me on the 8th of December, only three or four contained light-coloured eggs. In the same nest both varieties are sometimes met with. The young wander from the nest soon after they are hatched, and hide amongst the neighbouring plants when alarmed. Sometimes whilst down-clad they venture to swim into the kelp amidst the outcries of all the Gulls in the neighbourhood, who attend them in their progress, swooping down at them from time to time as if to seize them; but they are then liable to be blown out to sea by the prevailing wind, and to be killed by cramp. The nestlings' food call is very peculiar, and can be heard a long way off; it has some resemblance to the squeak of a hungry Long-eared Owl.

The Gulls which frequented Observatory Bay, bred for the most part upon islands in the Sound; there were no nests on the shores of the bay itself, and scarcely half-a-dozen on the mainland within several miles of it. But although they were widely dispersed, every day as soon as eight bells (noon) was struck on board the ships, they might be seen hurrying towards them from various directions until a large flock had assembled to await the emptying of the wash buckets after dinner, although not a bird might have been in sight a minute or two before.

The South African museum contains only Cape examples of this Gull.

The louse was not observed. *Larus* is frequented by species of *Docophorus*, *Nirmus*, *Colpocephalus*, *Menopon*, and *Trinoton.—A. E. E.*]

Stercorarius antarcticus. (Plate VII., figs. 1, 2.)

Lestris cataracctes (*nec L.*), Quoy et Gaim. *Voy. de l'Uranie, p.* 137, *pl.* 38; *Gould, B. Austr. pl.* 21; *Hutton, Ibis,* 1867, *p.* 185, 1872, *p.* 248; *Finsch, J. f. O.* 1872, *p.* 240, 1874, *p.* 203.

Lestris antarcticus, *Lesson, Traité, p.* 616; *Scl. P. Z. S.* 1860, *p.* 390; *Abbott, Ibis,* 1861, *p.* 165; *Scl. & Salv., P. Z. S.,* 1871, p. 579 (*pt.*); *iid. Nomencl. Av. Neotr.* p. 148; *Cab. & Reichen., J. f. O.* 1876, *p.* 328.

Stercorarius antarcticus, *Gray, List Anseres, Brit. Mus.* p. 167; *Bp. Consp.* ii. *p.* 207; *Pelz. Reis. Novara, Vög. p.* 150; *Giglioli, Faun. Vertebr. Oceano, p.* 61; *Buller, B. N. Zeal. p.* 267; *Saunders, P. Z. S.* 1876, *p.* 321.

Megalestris antarctica, *Gould, P. Z. S.* 1859, *p.* 98.

Lestris fuscus, *Ellman, Zool.* 1861, p. 7472.

Buphagus antarcticus, *Coues, Pr. Phil. Ac.* 1863, p. 127; *id. B. N. West,* p. 604.

Stercorarius catarractes (*nec L.*); *Schl. Mus. P. B. Lari, p.* 45; *Gould, Handb. B. Austr.* ii., *p.* 389; *Layard, B. S. Afr. p.* 366; *Sharpe, Voy. Ereb. & Terror, App. p.* 32.

Buphagus skua antarcticus, *Coues & Kidder, Bull. U. S. Nat. Mus. No.* 2, *p.* 9; *iid. op. cit., No.* 3, *p.* 9.

Ad. fuliginoso-brunneus, plumis dorsalibus et scapularibus albido terminaliter lavatis, colli postici plumis et quibusdam dorsalibus medialiter fulvo striatis; alis saturatius brunneis, anguste cineraceo marginatis; remigibus cinerascenti-brunneis, scapis albidis, primariis versus basin conspicue albis; dorso postico et uropygio cinerascenti-brunneis, supracaudalibus magis chocolatinis; cauda cinerascenti-brunnea, versus apicem nigricante, scapis albis: corpore subtus toto fuliginoso, plumis paucis sordide ferrugineo lavatis, hypochondriis quibusdam obscure albicante lavatis; subalaribus saturate brunneis, majoribus et remigibus subtus cinerascenti-brunneis, primariis intus ad basin conspicue albis; rostro nigro; pedibus ex virescente schistaceo-nigris; inde intense chalybeo. Long. tot. 23, culm, 2·8, alæ 16·3, caudæ 6·0, tarsi, 2·8.

The above description is taken from the only skin brought by Mr. Eaton, and is in nearly uniform brown plumage. Two other skins are in the Museum, from Christmas Harbour, where they were procured by the Antarctic Expedition. These two show more clearly the buff streakings on the neck, and have the feathers of the mantle mottled with yellowish buff or whitish centres. Mr. Howard Saunders, who has examined our series, has called attention to the remarkable colouring of the skins obtained by the Antarctic Expedition in the Southern Seas among the ice. They are very light coloured, and have quite a fringe of yellow round the neck, while the peculiar pale ashy brown colour of the under parts is very conspicuous. I agree with Mr. Saunders that it would not be wise to separate these Antarctic birds from the ordinary *S. antarcticus* without further information regarding them, for the Kerguelen birds show a certain approach to the golden neck-feathers; and moreover, I find on referring to the registers that one is said to be from the Auckland Islands. It may have been sent home in the same box as the Auckland collection, and registered by mistake with them. It still preserves its original ticket,

"Antarctic Gull, Male," but no locality is attached, and its companion bird, a female, is also without locality. These are the two birds catalogued by Mr. Gray in his List of Anseres (p. 168) as from the Antarctic Seas.

I further agree with Mr. Saunders that no one comparing the two species of Skuas would unite *S. antarcticus* with *S. catarractes*, the bills being so different, as will be seen in the figures in the plate.

Besides the undermentioned specimens from Kerguelen Island, the British Museum contains examples from South Africa (*Sir A. Smith*), Campbell Island (*Lieut. A. Smith*), Antarctic Seas, and the Pack Ice, Antarctic Ocean (*Antarctic Expedition*), New Zealand (*Sir G. Grey*), Norfolk Island (*F. M. Rayner*), and from Lat. 36° 8′ S., Long. 88° 55′ E. (*Sir G. Grey*).

a. ad. Royal Sound, Kerguelen Island; Rev. A. E. Eaton.
b. c. ad. Christmas Harbour, Kerguelen Island; Antarctic Expedition.

[Every marsh near Royal Sound used to have its pair of Skuas. Many were destroyed within a radius of four miles from the ships; and before the expedition sailed from the island it was impossible to walk far without coming across dead bodies of the poor creatures. The cause of this useless slaughter was the menacing aspect of the birds, who swooped with fierce impetuosity directly towards the face of any one approaching their domain, rising only just in time to clear his head, and uttering short despairing cries. They did not feign to be crippled quite so much as the Skuas in Spitzbergen, but preferred intimidation as a means of averting danger from their nest. When they thought they had succeeded in making the enemy retreat, they celebrated their triumph standing face to face upon the ground, with their wings extended vertically so as almost to meet above their back, whilst one or both loudly chaunted a pæan, consisting of a dozen notes or so delivered in the tones of a carrion crow. In October they also used to croak now and then during their flight; and this croak, which was discontinued in the breeding season, was very like the lower croak of a raven. Indeed, it was at first difficult to reassure oneself that they were not a species of *Corvus* as they circled in the air far off, and the Blue Jackets used to call them "Black Crows" for some time, but before long the designation "Molly-hawks" came to be applied to them. This change of name took place at the commencent of petrel digging. If Blue Petrels were turned loose in the daytime, they were almost invariably chased by Skuas, and killed on the wing before they had flown half a mile. Petrels of one sort or another seem to constitute the staple diet of these Skuas. They hunt for them in the evening when it is becoming dusk, flying rapidly along the hillsides close to the ground, like hawks, ready to pounce upon any that they may see emerging from the mouth of their burrows. Again in the early morning they are upon the wing to waylay Petrels returning late from the sea. Nor are they idle during the rest of the day. I have mentioned their fondness for eggs in my paper in the 'Proceedings.'

The nest is built amongst *Azorella*, where the ground is dry and slightly raised. It consists of a hollow scraped in the soil lined with dead tufts of *Festuca erecta*. The eggs are two in number, and do not vary much in colour. The statements in Capt. Hutton's paper on Birds inhabiting the Southern Ocean [Quart. Journ. Sc. (1866), vi. 77], that the Kerguelen Skua breeds on flats among grass two feet high, and lays three eggs, and that the young are dark brown spotted with white, do not accord with our observations. They young are dark brown, without any pale *spots* whatever. The ordinary food-call of the nestlings is rather plaintive and tremulous; they also quack like Mallards.

The old Skuas were much puzzled when they saw rabbits come out of Petrels' holes. They hovered for a long time over their heads, and at length used to stand beside the mouths of the burrows waiting for the young ones to creep forth, just as if they were watching for Petrels. It is doubtful whether they will succeed in ridding the island of these mischievous vermin, although the young birds reared by me readily fed upon rabbits procured with the sling.

Twenty or thirty adult Skuas used to assemble every afternoon upon a small sheltered lake near Swain's Harbour, where they washed and basked.

Under the name of *S. catarrhactes*, two examples of this Skua from the Crozettes are exhibited in the South African Museum.

Louse not found. *Stercorarius* serves as host to species of *Docophorus, Nirmus, Lipeurus,* and *Colpocephalum* (Giebel). *A. E. E.*]

Sterna virgata.

Sterna arctica, (pt.) *Gray, List Anseres Brit. Mus. p.* 178 (1844).

Sterna virgata, *Cab. J. f. O.* 1875, *p.* 449; *Saunders, P. Z. S.* 1876, *p.* 646; *Cab. & Reichen. J. f. O.* 1876, *p.* 328.

Sterna vittata, *Coues, in Kidder's Report, Bull. U. S. Nat. Mus.* ii. *p.* 17 (1875, nec. Gm.); *Coues & Kidder, op. cit.* iii, *p.* 11 (1876).

Ad. supra fumoso-cinereus, tectricibus alarum dorso concoloribus; remigibus fumoso-cinereis intus albicanti-cinereis, scapis pallidissime brunneis, primario primo extus nigricante, scapa pure alba, secundariis albo terminatis; uropygio et supracaudalibus albis vix cinereo lavatis; rectricibus cinereis, intus albis; pileo summo nuchaque nigris; facie laterali pallide cinerea; striga lata superciliari a basi narium ducta, alba; corpore reliquo subtus fumoso-cinereo, subalaribus et subcaudalibus concoloribus, iis imis vix albicantibus; rostro et pedibus corallinis; iride indigotico-nigra. Long. tot. 12, culm. 1·1, alee 10·3, caudæ 3·1, rectrice extima 5·4, tarsi 0·65.

Juv. supra nigricans, plumis ad basin cinereis, subterminaliter nigris, ochrascenti-brunneo marginatis et transfasciatis; uropygio et supracaudalibus albicanti-cinereis, ochrascenti-brunneo lavatis; remigibus secundariis et rectricibus eodem modo quam dorsum terminaliter marmoratis; regione parotica cinerea postia nigra; corpore subtus ubrascenti-brunneo, nigricanti-brunneo transversim marmorato, plumis ad basin albis, subalaribus purè albis.

This seems to be quite a distinct species. It is closely allied to *Sterna antarctica* of New Zealand, and in coloration is almost exactly similar. The latter bird, however, according to Dr. Buller (B. N. Zeal. p. 238), has the bill and feet " bright yellow," whereas these parts in *S. vittata* are coral-red.

a. imm. sk. Kerguelen Island. Antarctic Expedition.

b. ad., c. juv. Royal Sound, Kerguelen Island. Rev. A. E. Eaton.

[Crustacea and the young of *Notothenia* inhabiting the kelp afford subsistence to the Terns in Kerguelen Island. Skimming along the shore in pairs and threes, and over the belt of *Macrocystis*, these with their graceful flight and elegant form enliven the prevailing desolation of the coast scenery. In general their eggs are laid on barren stony ground one or two hundred yards from the beach, and the brows of terraces of the seaside hills are more frequently selected for the nesting places than any other situations. There is rarely (if ever) more than one egg in a nest, and this is deposited in a slight hollow amongst the stones, and generally upon the bare soil. The birds breed irregularly throughout the early summer months. Fresh eggs were taken by the officers of the " Volage " about the third week in November; a nestling recently hatched was found on the 8th of December, and on the 23rd of the same month a young Tern fully fledged was shot while on the wing. Its beak and feet were black. Another young one able to fly was kept for some time on board the " Supply." It became quite tame, and ate boiled rice and soaked bread readily. After we had been a few days at sea it died, probably through want of water. The adult examples in the collection were secured in the absence of a gun by a couple of stones thrown at them consecutively from a sling while they were passing overhead. A third specimen was caught by hand during our last ten minutes on the island. It was sitting upon a rock near the pier with the eye on my side of its head blinded, so I crept up quietly and laid hold of it.

The two specimens marked as *Sterna meridionalis* from the Crozettes in the South African Museum are identical with the Kerguelen Tern.

The louse of this species was not obtained. *Sterna* is infested with species of *Nirmus*, *Docophorus*, *Colpocephalus*, and *Analges*, according to Giebel. A. E. E.]

Sterna vittata.

Wreathed Tern, *Lath. Gen. Syn.* iii., *pt.* 2, *p.* 359.

Sterna vittata, *Gm. S. N.* i., *p.* 609 (*ex Lath.*); *Lath. Ind. Orn.* ii., *p.* 807; *Pelz. Reis. Novara, vög. p.* 152; *Saunders, P. Z. S.*, 1876, *p.* 647.

Ad. pileo nuchaque nigerrimis; corpore reliquo suprà clarè cinereo; tectricibus alarum dorso concoloribus; remigibus cinereis, intùs albis, primariis extùs nigricanti-cinereis, rachidibus albis; secundariis et scapularibus conspicuè albo terminatis; uropygio et supracaudalibus purè albis; caudâ maximè furcatâ, purè albâ; facie laterali cinereâ, fasciâ distinctâ a narium basi, infrà oculos usque ad nucham lateralem ductâ, et maculâ infraoculari, albis; corpore reliquo subtùs saturatè cinereo, gulâ magis albicante; tibiis, crisso, subalaribus et subcaudalibus, purè albis. Long. tot. 15·3, culm. 1·4, alæ 10·0, caudæ 3·0, rectric. ext. 7·0, tarsi 0·7.

Juv. similis avi juveni *S. virgatæ*, et eodem modo nigro et ochrascenti-brunneo varius.

The view of Herr Von Pelzeln, and Mr. Howard Saunders, who refer this Tern to the *Sterna vittata* of Gmelin, is doubtless correct. The latter gentleman observes :— " Gmelin's description, founded on Latham, fairly suits this species, " although I am inclined to doubt the correctness of the locality assigned, viz., " Christmas Island, especially as there is no mention in Cook's Voyages of any Tern " being found there, except the Sooty Tern, of which there is a full description." Many of the birds described by Latham were taken from specimens or drawings in the collection of Sir J. Banks. A list of the few birds of Kerguelen Island as noticed on Cook's voyage has been drawn up at the commencement of the present paper, and I believe that No. 54, representing the Tern from the " Island of Desolution," formed the type of *Sterna vittata*, and that the locality Christmas *Island* is a mistake for Christmas *Harbour*, which was the place where Cook's expedition landed on Kerguelen Island. The picture represents *S. vittata*, and not *S. virgata*, and that the species occurs there is proved by the under-mentioned specimen collected in the island by the Antarctic Expedition.

a. ♂ ad., *b.* juv. S. Paul's Island, Jan. 1853 (*J. Macgillivray*). Capt. Stanley.
c. ad. Kerguelen Island. Antarctic Expedition.

PROCELLARIIDÆ.
Pelecanoides urinatrix.

Tee-tee, *Forster, Icon. Ined. No.* 88.

Diving Petrel, *Forster, Voyage* i. *p.* 189; *Lath. Gen. Syn.* iii. *pt.* 2, *p.* 413; *id. Gen. Hist.* x. *p.* 194.

Procellaria urinatrix, *Gm. S. N.* i. *S.* 560 (*ex Lath.*).

Pelecanoides urinatrix, *Lacép. Mem. de l'Inst.* 1800, *p.* 517; *Gray, List Anseres Brit. Mus. p.* 158; *id. Gen. B.* iii., *p.* 646; *id. Voy. Ereb. & Terror, Birds, p.* 17; *id. Ibis,* 1862, *p.* 243; *Coues, Pr. Philad. Acad.* 1866, *p.* 190; *Gray, Handl. B.* iii., *p.* 102, *No.* 10825; *Sharpe, Voy. Ereb. & Terror, App. p.* 33; *Buller, B. N. Zeal. p.* 313; *Coues & Kidder, Bull. N. S. Nat. Mus.* ii., *p.* 36; *iid. op. cit.* iii., *p.* 17.

Halodroma urinatrix, *Illiger, Prodr. Syst. Mamm. &c. p.* 274; *Reichenb. Handb. Longipennes, pl.* ix., *figs.* 762, 763; *Bp. Consp.* ii., *p.* 206; *Schl. Mus. P. B. Procell. p.* 37; *Gould, Handb. B. Austr.* ii., *p.* 483; *Finsch. J. f. O.* 1872, *p.* 256, 1874, *p.* 210; *Cab. & Reichen. J. f. O.,* 1876, *p.* 328.

Procellaria tridactyla, *Forster, Descr. Anim. p.* 149; *Ellman, Zool.* 1861, *p.* 7473.

Puffinuria urinatrix, *Gould, B. Austr* vii., *pl.* 60.

Pelecanoides berardi, *Quoy et Gaim. Voy. Uranie,* p. 135, *pl.* 37; *Temm. Pl. Col.* 517; *Gray, List Anseres Brit. Mus.* p. 158; *id. Gen. B.* iii., p. 646; *Gould, P. Z. S.* 1859, p. 98; *Scl. P. Z. S.* 1860, p. 390; *Coues, Pr. Philad. Acad.* 1866, p. 190; *Gray, Handl. B.* iii., p. 102, No. 10827; *Buller, B. N. Zeal.* p. 314.

Halodroma berardi, *Reichenb. Handb. Longipennes,* pl. ix., *fig.* 764; *Bp. Consp.* ii., p. 206; *Schl. Mus. P. B. Procell.,* p. 38; *Finsch., J. f. O.* 1874, p. 210; *Scl. & Salv. Nomencl. Av. Neotr.* p. 149.

♀ *Ad.* suprà niger, plumis basaliter cineraceis; scapularibus magis cinereis albo terminatis; alis dorso concoloribus, primariis intùs brunnescentibus, secundariis magis cineraceis angustè albo terminatis; rectricibus nigricantibus, intùs cineraceis angustè albido terminatis; loris et regione oculari nigricantibus; regione paroticâ obscurè cinereâ, plumis vix albo terminaliter marginatis; genis et corpore subtus toto albis, genis, gutture imo, colli et pectoris lateribus cineraceis albo terminatis; hypochondriis cineraceo lavatis; subalaribus albis, majoribus cineraceo lavatis; remigibus infrà cinerascenti-brunneis; rostro nigro, ad basin mandibulæ cœruleo; iride cinerascente; pedibus cœruleis.

The specimen described above is one of Mr. Eaton's. Considerable variation exists with regard to the amount of grey mottling on the throat and sides of the body and neck, but I cannot determine whether this arises from difference of sex or age. Mr. Gould has figured the adult bird with a pure white breast, and the young one with the under parts shaded with grey, but neither of them have the banded appearance which is apparent in most of the specimens in the British Museum. The young birds in the collection, some of which have remains of down still adhering to them, are quite as pure white on the breast as the adults.

The principal differences between *Pelecanoides urinatrix* and its allies *P. berardi* and *P. garnoti,* have been stated by authors to consist in size and the colour of the feet. Thus Professor Schlegel (*l. c.*), who is followed by Dr. Coues in his well-known paper (Pr. Philad. Acad. 1866, p. 190), separates them on characters which may be tabulated as follows:—

 a. feet blackish; bill longer and more slender; size larger . . *garnoti.*
 b. feet bluish: bill robust; size smaller . . . . . *urinatrix.*
 c. feet pale with black membranes; size small · . . *berardi.*

The colour of the feet, therefore, plays an important part in the differentiation of the species, so that it will be of interest to note the colouring of these organs, as given in various works. Thus for *P. urinatrix* Dr. Kidder gives the "tarsus and foot lavender blue," while Mr. Gould writes:—"Tarsi and toes beautiful light blue; webs transparent bluish white, tinged with brown." Forster gives the "feet blue, the soles and webs black," while Dr. Buller has them as follows:—"Legs and feet cobalt, tinged with green, the webs bluish white." According to the last-named observer, in *P. berardi* the legs and feet are "yellowish with dark webs."

It is, therefore, no wonder, with the discrepancy in colour mentioned in the works of the above-named authors, that Dr. Coues, after studying the series brought home by Dr. Kidder from Kerguelen Land, should observe as follows:—"As very "strongly intimated in my paper, satisfactory diagnoses of the three currently "reported species of the genus are wanting. Nor is my faith in their distinctness "increased on finding that these specimens, which from the locality undoubtedly "represent the original *P. urinatrix*, are fully up to the dimensions of the supposed "larger *P. garnoti* from the west coast of South America. Observed variation in "the colour of the feet, which is one point that has been relied upon, lessens the "probability of distinctness, especially as the ascribed coloration does not coincide "in every case with the dimensions. The size and proportions of the examples "examined, as carefully measured in the flesh by Dr. Kidder, warrant me in "adducing the *P. garnoti* of Lesson as a synonym of *P. urinatrix*, to which I "still refrain, however, from adding the *P. berardi* of Quoy and Gaimard." (Report; Bull. N. S. Mus. N. H. *l. c.*)

Although my measurements of skins do not agree exactly with those of Dr. Kidder made from the recently-killed bird, I add the dimensions of the series in the British Museum, arranged according to the size of the bill, irrespective of locality.

	Total Length.	Culmen.	Wing.	Tail.	Tarsus.
a. Auckland Isles (*Antarctic Exp.*)	7·4	0·55	4·15	1·6	0·6
b. Kerguelen Island (*A. E. Eaton*)	7·9	0·6	4·7	1·75	1·0
c. Straits of Magellan	7·5	0·6	4·45	1·65	0·95
d. ♀ Kerguelen Island (*A. E. Eaton*)	7·6	0·65	4·55	1·75	0·95
e. Christmas Harbour (*Antarctic Exp.*)	9·3	0·65	4·95	1·65	1·0
f. Auckland Islands (*Smith*)	7·6	0·65	4·85	1·8	1·05
g. „ „ „	8·0	0·65	4·7	1·7	1·05
h. Straits of Magellan	7·5	0·65	4·7	1·7	1·05
i. New Zealand (*Hector*)	9·0	0·7	5·1	1·75	1·05
k. Kerguelen Island (*Antarctic Exp.*)	8·0	0·7	4·55	1·6	1·05
l. „ „ „	8·0	0·7	4·75	1·65	0·95
m. New Zealand (*Sir George Grey*)	7·8	0·7	4·6	1·65	0·95
n. Chili (*Brydges*)	8·0	0·8	5·2	1·7	1·25
o. „ „	7·7	0·8	5·4	1·8	1·25
p. Valparaiso (*Brett*)	—	0·85	5·45	1·7	1·3

Although the general length of the skin in these birds goes for very little, the bill is a reliable character, and from the above table of measurements it will be seen that even in the same localities the size of the bill and feet is not constant. My conclusions differ from those of Dr. Coues, insomuch that I consider that *P. berardi* is nothing but the young bird of *P. urinatrix*, and that *P. garnoti* on the contrary must be held to be distinct on account of its very much larger size; at all events the examples from Western South America indicate a distinct race.

The British Museum possesses specimens from the Auckland Islands (*Lieut. A. Smith, Antarctic Expedition*), from New Zealand (*Dr. Hector, Sir G. Grey*), and from the Straits of Magellan, as well as the following from Kerguelen Island.

a. b. ♂. ♀. ad.   Observatory Bay, Kerguelen Island.   Rev. A. E. Eaton.
c. ad.   Christmas Harbour, July 20, 1840.   Antarctic Expedition.
d. ad.   Kerguelen Island.   Antarctic Expedition.

[This bird, as Prof. Wyville Thomson well observes, has a close general likeness to *Mergulus alle*. Both of them have a hurried flight; both of them, while flying, dive into the sea without any interruption in the action of their wings, and also emerge from beneath the surface flying, and they both of them swim with the tail rather deep in the water. But this resemblance does not extend to other particulars of their habits. The Rotche, when breeding, usually flies and fishes in small flocks of six or a dozen birds, and builds in communities of considerable size, which are excessively noisy. Diving Petrels, on the other hand, are more domestic in their mode of living, fishing and flying for the most part in pairs or alone, and building sporadically.

They had begun to pair when we reached Kerguelen Island. The first egg was found on the 31st of October. Their burrows are about as small in diameter as the holes of Bank Martins (*Cotyle riparia*) or Kingfishers (*Alcedo ispida*). They are made in dry banks and slopes where the ground is easily penetrable, and terminate in an enlarged chamber on whose floor the egg is deposited. There is no specially constructed nest. Some of the burrows are branched, but the branches are without terminal enlargements, and do not appear to be put to any use by the birds. Before the egg is laid, both of the parents may be found in the nest-chamber, and may often be heard moaning in the daytime: but when the females begin to sit, their call is seldom heard, excepting at night, when the male in his flight to and from the hole, and his mate on her nest, make a considerable noise. There seems to be a difference of a semi-tone between the moans of the two sexes. The call resembles the syllable "oo" pronounced with the mouth closed while a slurred chromatic ascent is being made from E. to C. in the tenor. This kind of Petrel has much difficulty in taking flight from ground which is comparatively level; it it only by running against the wind, or by starting from a lump of *Azorella*, that the birds are able to rise upon the wing if they happen to alight upon a flat. During my walks on calm nights I used frequently to hear them fluttering along the ground in the dark, and (if I had a lantern) easily caught them by uncovering the light and turning it on to them. They sometimes lay still in my hand without trying to escape; but when they flew off from it, they did so in a manner which showed that they were not at all crippled. They flew to light on board H.M.S. "Supply" on dark nights in October, when there was snow upon the deck.

There is a pair of *P. urinatrix* from the Crozettes in the S. African Museum.

One of the birds dug out near Observatory Bay was inconveniently crowded with a species of *Pulex* and *Nirmus setonis*, sp. n. *A. E. E.*]

Daption capensis.

White and black-spotted Petrel, *Edwards, N. H. Birds, pl.* 90.
Le Petrel tacheté vulgairement Damier, *Briss. Orn.* vi., *p.* 146.
Procellaria capensis, *Linn. S. N.* i., *p.* 213; *Hahn & Kuster, Vög. Asien, Lief.* xix., *taf.* 6; *Gray, List Anseres, &c., Brit. Mus. p.* 164; *Schl. Mus. P. B. Procell. p.* 14; *Pelz. Reis. Novara, Vög. p.* 146; *Layard, B. S. Afr. p.* 361.
Daption capensis, *Steph. Gen. Zool.* xiii., *p.* 241; *Gould, B. Austr.* vii., *pl.* 53; *Bp. Consp.* ii., *p.* 188; *Reichenb. Handb. Longipennes, taf.* xii., *figs.* 337, 338; *Lawr. N. Am. B. p.* 828; *Gould, Handb. B. Austr.* ii., *p.* 469; *Coues, Pr. Philad. Acad.* 1866, *p.* 162; *Degl. and Gerbe, Orn. Eur.* ii., *p.* 469; *Giglioli, Faun. Vertebr. nell' Oceano, p.* 46; *Cab. Von der Decken Reis.* iii., *p.* 62; *Finsch & Hartl. Vög. Ostafr. p.* 816; *Finsch, J. f. O.* 1872, *p.* 256, 1874, *p.* 208; *Buller, B. N. Zeal. p.* 299; *Scl. and Salv. Nomencl. Av. Neotr, p.* 149; *Sharpe, Voy. Erebus and Terror, Birds, App. p.* 33; *Kidder, Bull. U. S. Nat. Mus. p.* 39, *note; Cab. & Reichen. J. f. O.* 1876, *p.* 329.
Procellaria punctata, *Ellman, Zool.* 1861, *p* 7473
Fulmarus capensis, *Gray, Handl. B.* iii., *p.* 107.
Calopetes capensis, *Sundev. Av. Meth. Tent. p.* 142.

Ad. Schistaceo-niger, albo variegatus, plumis ad basin cineraceis subterminaliter albis, dorsi plumis ad apicem triangulariter schistaceo-nigris; scapularibus dorso concoloribus et eodem modo notatis; tectricibus alarum minimis nigricanti-brunneis; reliquis brunneis ad basin albis et angustè albo extùs limbatis, majoribus intimis purè albis, quibusdam brunneo terminatis; tectricibus primariorum et remigibus brunneis, intùs albis brunneo terminatis; caudâ albâ latè brunneo terminatâ; facie laterali pileo concolori; fasciâ parvâ suboculari albâ; gulâ summâ brunneâ, plumis ad basin celatè albis; gutture et colli lateribus albis, plumis brunneo terminatis; corpore reliquo subtùs purè albo, axillaribus, tectricibus subalaribus majoribus et subcaudalibus albis brunneo terminatis; subalaribus omnibus marginem alarum formantibus nigricanti-brunneis; rostro nigricanti-brunneo; pedibus saturatè brunneis. Long. tot. 15·5, culm. 1·35, alæ 10·5, caudæ 4·0, tarsi 1·7.

There seems to be very little variation in the plumage of these birds, but the throat is more uniform dusky in some individuals, while in others it is mottled with white; the amount of white mottling on the mantle also varies somewhat, the more uniformly coloured birds being apparently the younger ones.

Mr. Eaton did not bring home a specimen of the Cape Petrel from Kerguelen Island, but he noticed it in the vicinity, as did also Dr. Kidder (l. c.) It is included in the list of Drs. Cabanis and Reichenow, as having been obtained at Kerguelen, and John Nunn's narrative speaks of it as occurring there, being called by him and his companions the "Spotted Eaglet," or "Spotted Night-hawk." He says it was extremely abundant near the island. As will be seen below, the British Museum possesses two specimens captured in the vicinity by the Antarctic Expedition.

That the Cape Pigeon breeds on Kerguelen Island is proved by the following note in Sir J. Hooker's Journal: "It builds in sheltered ledges of cliffs about 50 or 100 " feet above the level of the sea. I found two on a nest, but quite mature. Its " note is a short hoarse croak."

In the British Museum are specimens from the Cape Seas (*Capt. Harry, E. M. Langworthy*), Lat. 34° 37′ S., Long. 22° 29′ (*D. Blewitt*), Western Australia (*Sir G. Grey*), New Zealand (*Dr. Hector*), off Valparaiso (*Capt. W. S. Brett*), as well as the following.

a. b. ad. Off Kerguelen Island. Antarctic Expedition.

[Cape Pigeons were plentiful on 9th of October near Bligh's Cap, and on the following day there were many of them off the coast in the neighbourhood of Mt. Campbell. A few followed the ships into Observatory Bay, but they did not stay there. I do not know whether they ever breed in the island. In December and the first week in January there were many of them still about the entrance of Royal Sound and Swain's Bay; but when we passed by at the end of the latter month none were visible in either of these places, nor did we see any in March between Kerguelen Island and the cape. It is possible that their nests might be found in the cliffs of the Prince of Wales Foreland and the islands in the mouth of Swain's Bay, or in those of the more exposed parts of the coast farther west; but there is no evidence of their breeding there. *A. E. E.*]

Majaquens Æquinoctialis.

The Great Black Petrel, *Edwards, Nat. Hist. B.* ii., *pl.* 89.

Le Puffin du Cap de Bonne Espérance, *Briss. Orn.* vi., *p.* 137.

Procellaria æquinoctialis, *Linn. S. N.* i., *p.* 213; *Burm. Th. Bras.* iii., *p.* 345; *Schl. Mus. P. B. Procell. p.* 19; *Layard, B. S. Afr. p.* 360; *Finsch & Hartl. Vög. Ostafr. p.* 817; *Cab. & Reichen. J. f. O.* 1876, *p.* 329.

Procellaria æquinoxialis, *Vieill. N. Dict. d'Hist. Nat.* xxv., *p.* 422.

Puffinus æquinoctialis, *Steph. Gen. Zool.* xiii., *pt.* 1, *p.* 229; *Gray, List Anseres, &c., Brit. Mus. p.* 160; *id. Gen. B.* iii., *p.* 647; *Reichenb. Handb. Longipennes, pl.* xii., *figs.* 340, 341; *Pelz. Reis. Novara, p.* 142.

Procellaria nigra, *Forster, Descr. Anim. p.* 26.

Priofinus æquinoctialis, *Jacq. & Pucher. Voy. Pole Sud.* iii., *p.* 146.

Majaquens æquinoctialis, *Reichenb. Handb. Longipennes, pl.* xii., *figs.* 340, 341; *Bp. Consp.* ii., *p.* 200; *id. C. R.* xlii., *p.* 768; *Coues, Pr. Philad. Acad.* 1864, *p.* 118; *Giglioli, Faun. Vertebr. Oceano, p.* 35; *Scl. & Salv. Nomencl. Av.*

Neotr. p. 149; *Coues and Kidder, Bull. U. S. Nat. Mus.* ii., p. 25; *iid. op. cit.* iii., p. 13.

Fulmarus æquinoctialis; *Gray, Handl. B.* iii., p. 108, no. 10915.

♀ *Ad.* suprà fuliginoso-niger, plumis dorsalibus vix brunnescente marginaliter lavatis; alis nigricanti-brunneis, tectricibus alarum brunneo lavatis; remigibus nigricanti-brunneis, intùs chocolatinis, scapis primariorum pallidè et conspicuè brunneis; uropygio et supracaudalibus cinereo-nigricantibus; caudâ cinereo-nigrâ, rectricibus medianis nigricanti-brunneis; facie et collo lateralibus fuliginoso-nigris, fasciâ latâ e genis medianis per mentum ductâ, albâ; corpore reliquo subtus fuliginoso-nigro, plumis omnibus brunneo lavatis; subalaribus brunneis, cinerascente lavatis; rostro virescenti-albo, suturibus nigris; iride nigra. Long. tot. 19·5, culmen 2·55, alæ 15·0, caudæ 5·5, tarsi 2·4.

The above description is taken from the only skin brought back by Mr. Eaton. The white mark is well developed on the chin and the fore part of the cheeks, occupying all the latter, excepting a brown patch at the base of the lower mandible. The way in which the white facial markings vary is shown in the following examples:—

1. ? Pacific Ocean. Sir Joseph Banks (spec. *a* of Gray's List of Anseres, 1844, p. 160). The white mark is confined to a small spot on the chin. The locality is doubtless erroneous.

2. ♂ Simon's Bay, Cape of Good Hope. The white mark occupies the entire chin and upper throat, and encroaches slightly on the centre of the cheeks.

3. Cape of Good Hope. The white marking is developed on the chin as in the preceding bird, occupying on one side the anterior portion of the cheeks, but totally absent on the other cheek, which is sooty brown. This is spec. *e* of Gray's list, considered by him to be a variety or a young bird.

Considering that the Australian *M. conspicillatus* only differs in the extent of the white markings on the head, it might be supposed that it was not really distinct from *M. æquinoctialis*, in which these white markings form such a varying peculiarity, the extent of white being, as shown above, sometimes not the same on both sides of the face. The Australian bird, however, has a white band across the crown, and a second one extending from the cheeks round the side of the head below the ear-coverts, as well as a white patch on the chin; and in addition to this the nostrils and sides of the mandibles are yellowish horn-colour, according to Mr. Gould. In *M. æquinoctialis* none of this yellow colouring in the bill is seen.

Mr. Layard, who has had opportunities of observing both species alive, considers them distinct, as does also Dr. Coues, who has devoted great time and attention to the Petrels. The following specimens are in the British Museum:—

a. ad. Royal Sound, Kerguelen Island. Rev. A. E. Eaton.
b. ad. Pacific Ocean? Sir Joseph Banks.
c. ad. st. Cape of Good Hope. M. Verreaux.

[In Kerguelen Island a hole similar to a deserted rabbit's earth, excavated in wet ground, with water standing (in early summer) an inch or two inches deep within the entrance, especially if it is in a slope near the sea, may be regarded as the burrow most likely to be that of a White-chinned Petrel. If it is occupied by the birds there will probably be some green shoots of *Acæna* clipped off from plants near its mouth dropped by them in the water. During the season while the birds are pairing before their egg is laid they make an extraordinary cackle in the nest-chamber; the sound of approaching footsteps, or a thump upon the ground some distance away from the nest, and even a shout at the mouth of the burrow, will cause them to commence it in the day time. During the night this call is uttered by the female sitting on her nest or in the entrance of the tunnel; and she can be heard at the distance of a quarter of a mile when there is a calm. Much trouble may be saved in digging out the nest by sounding with the spade along the course of the burrow until the situation of the nest-chamber is ascertained. This is spherical and tolerably large. Being in most instances near the surface of the ground, care must be taken in the removal of its roof, or the bird's back may be broken by the spade while she is sitting upon her egg. As soon as the chamber is laid open it is well to catch the hen by her beak and drag her out of the hole while she is still dazzled by the light, giving her no time to use her claws. On being released she usually makes no attempt to fly, unless she is purposely chased down the hill; but after waddling away a few yards she returns to her burrow (or to where its entrance used to be before it was dug into and choked with clods), and begins at once to dig her way into the tunnel through the obstructions with which it has been blocked up. She takes little notice of bystanders so long as they remain still, passing leisurely by them or even over their feet if they happen to be in her way. The nest is built of mud and pieces of plants arranged in the form of an inverted saucer three or four inches high, slightly hollowed out at the top. A space is left between its base and the sides of the nest-chamber. Some of the birds had no white patch under their chin; when it was present it varied in extent in different examples. In most instances it formed a small triangular blotch occupying the apical portion of the angle enclosed by the lower mandible; but in a few cases the white was limited to one or two feathers only. In none of the Kerguelen specimens did the patch extend to the forehead, as it does in the birds from Australia.

The S. African Museum contains the following examples of this species:—1 (young, in down) from the Crozettes; 2 from Table Bay with the white chin patch very large; and 1 from Australia with the white on each side of the face below the eyes, and a white band across the forehead, in addition to the chin patch.—*A. E. E.*]

Puffinus kuhli.

Le Puffin, *Buff. Pl. Enl.* x., *pl.* 962.

Procellaria puffinus, *Temm. Man. d'Orn.* ii., *p.* 805.

Nectris cinerea, *Kuhl, Beitr. Zool. p.* 148, *pl.* 11, *fig.* 12 (*nec Gm.*); *Linderm. Vög. Griechenl. p.* 170.

Puffinus cinereus, *Cuv. Règne Anim.* 1829; *Less. Traité, p.* 613; *Savi, Orn. Tosc.* iii., *p.* 38; *Schl. Rev. Crit. p.* cxxxii.; *Webb & Berth. Orn. Canar. p.* 43; *Bolle, J. f. O.* 1855, *p.* 177, 1857, *p.* 344; *Vernon Harcourt, Ann. N. H.* (2) xv., *p.* 438; *Loche, Expl. Sci. Alger. Ois.* ii., *p.* 174; *Degl. et Gerbe, Orn. Eur.* ii., *p.* 376; *Elwes & Buckley, Ibis,* 1870, *p.* 336; *Gigl. Faun. Vertebr. nell' Oceano, p.* 47; *Doderl. Avif. Sicil. p.* 228; *Fritsch, Vög. Eur. tab.* 58, *fig.* 1; *Godman, Ibis,* 1872, *p.* 223; *Bree, B. Eur.* v., *p.* 86 (1876).

Puffinus kuhli, *Boie, Isis,* 1836, *p.* 258; *Bp. Consp.* ii., *p.* 202; *Pelz. Reis. Novara, Vög. p.* 142; *Gray, Handl. B.* iii., *p.* 102; *Salvad. Ucc. Ital. p.* 298; *Shelley, B. Egypt, p.* 357; *Heugl. Orn. N. O. Afr.* iv., *p.* 1367; *Scl. & Salv. Nomencl. Av. Neotr. p.* 149; *Irby, B. Gibr. p.* 217.

Nectris macrorhyncha, *Heugl. Syst. Uebers. p.* 68, *no.* 711.

Nectris gama, *Hartl. Orn. Madag. p.* 84 (*nec Bp.*).

Procellaria cinerea, *Schl. Mus. P. B. Procell. p.* 24.

Ad. suprà brunneus, pileo concolori, plumis dorsalibus et scapularibus cinerascenti-brunneo marginatis; scapularibus longioribus nigricantibus brunneo marginatis; tectricibus alarum nigricanti-brunneis, majoribus pallidioribus brunneis vix cineraceo lavatis et angustissimè albido limbatis; remigibus nigricanti-brunneis, intùs chocolatinis; supracaudalibus cineraceo-brunneis, ad apicem albicantibus brunneo vermiculatis; rectricibus centralibus saturatè brunneis, reliquis cinerascenti-brunneis versùs apicem saturatiùs brunneis; facie laterali pileo concolori, vix saturatiore; genis et colli lateribus brunneis obscurè albido irroratis; corpore reliquo subtùs albo, pectoris lateribus obscurè cineraceo lavatis; subalaribus et axillaribus albis, margine alari saturatè brunneo; tectricibus majoribus externis versùs apicem brunneo lavatis; rostro flavo, versùs apicem nigro; pedibus flavis; iride nigricante. Long. tot. 20·5, culm. 2·15, alæ 13·4, caudæ 6·6, tarsi, 2·05.

♀ mari similis.

Juv. similis adultis sed sordidior, subtùs sordidiùs albus; rostro nigricante; pedibus cœrulescentibus.

On comparing the two specimens from Kerguelen Island, collected by the Antarctic Expedition, with Mediterranean examples of *P. kuhli* in the British Museum, I am unable to find any real specific distinctions between them, the Northern birds being perhaps a shade larger, and having rather a stouter bill.

a, b. ad. Kerguelen Island. Antarctic Expedition.

It is probably the foregoing species which is alluded to in Nunn's "Narrative" under the name of *Puffinus major*. In the above-mentioned work occurs the following sentence: "As there appears to be some doubts whether this species is " an inhabitant of Desolation, some elucidation is required; it is considered to have " an extremely wide range." The description given agrees best with *P. major*, which is found in the Cape Seas. *Adamastor cinereus* is stated by Captain Hutton to go to Kerguelen Island (Ibis, 1865, p. 286) to breed. No specimens have been obtained by the recent expeditions, but so good an observer as Captain Hutton would not include the species without good reason; it may, therefore, be looked for.

Thalassoica tenuirostris.

Procellaria glacialis, *Forster, Descr. Anim. p.* 25. (*nec L.*)

Procellaria tenuirostris, *Audub. Orn. Biogr.* v., *p.* 333; *id. B. N. Amer.* vii., *p.* 210; *Cass. U. S. Expl. Exp. p.* 409; *Lawr. in Baird's B. N. Amer. p.* 826; *Elliot, B. N. Amer. Intr. cum fig.* (head).

Procellaria glacialoides, *Smith, Illustr. Zool. S. Afr. pl.* 51; *Gray, List Anseres Brit. Mus. p.* 162; *id. Gen. B.* iii., *p.* 648; *Gould, B. Austr.* vii, *pl.* 48; *Reichenb. Handb. Longipennes, pl.* xiii., *fig.* 789; *Cass. U. S. Expl. Exp., Birds, p.* 409; *Pelz. Reis. Novara, p.* 146; *Layard, B. S. Afr. p.* 361; *Buller, B. N. Zeal. p.* 301; *Sharpe, Voy. Ereb. and Terr. App. p.* 33.

Priocella garnoti, *Hombr. & Jacq. Voy. Pole Sud, pl.* 32, *fig.* 43.

Thalassoica glacialoides, *Reichenb. Handb. Longipennes, pl.* xiii., *fig.* 789 (*pess.*); *Bp. Consp.* ii., *p.* 192; *Gould, Handb. B. Austr.* ii., *p.* 467; *Coues, Proc. Philad. Acad.* 1866, *p.* 31; *Giglioli, Faun. Vertebr. Oceano, p.* 47.

Thalassoica glacialoides *var. polaris et tenuirostis, Bp. Consp.* ii., *p.* 192.

Thalassoica polaris, *Bp. C. R.* xlii., *p.* 768.

Procellaria smithi, *Schl. Mus. P. B. Procell. p.* 22; *Finsch, J. f. O.* 1872, *p.* 255, 1874, *p.* 174.

Ad. suprà clare argentescenti-canus, plumis celatè albicantibus; pileo antico et facie laterali albicantibus, regione paroticâ vix cano lavatâ; colli lateribus clare canis; tectricibus alarum et scapularibus dorso concoloribus; remigibus cinerascenti-brunneis, intùs versùs basin albicantibus, secundariis extùs cinereo lavatis; caudâ omninò canâ; corpore subtùs toto purissimè albo, crisso laterali cano lavato; subalaribus et axillaribus purissimè albis, margine alari cano; rostro carnescente, versùs apicem carnescenti-corneo, ad apicem nigro; pedibus cinereis, tarso pallidè coccineo lavato. Long. tot. 16·5, culmen 1·8, alæ 12·8, caudæ 5·5, tarsi 2·15.

The description is taken from a specimen procured at Kerguelen Island by Lieut. Alexander Smith, who accompanied the Antarctic Expedition, and on his return presented it to the British Museum (Cf. Gray, Cat. Anseres, p. 162). None of the recent expeditions seem to have obtained it, but John Nunn's Narrative states that this species "was used, when young, by our party as food, and our supplies were "obtained by digging the young birds from the burrows in the sand or tussock- "banks on the lee or S.E. side of the island. The bird was known and eaten by us "under the name of the White Night-hawk." The Antarctic Expedition also procured a specimen of this species at Kerguelen Island. Great difference exists in the size and thickness of the bill and legs.

Specimens are in the national collection from South Africa (Type of *P. glacialoides*: *Sir A. Smith*); Antarctic Seas (*Antarctic Expedition*); Louis Philippe Island; Lat. 44° S., Long. 110½ W. (*J. Macgillivray*); Straits of Magellan; and off Valparaiso (*Capt. W. S. Brett*); as well as the following from the present locality:—

a. ad. Kerguelen Island; Lieut. A. Smith.
b. ad.   „   „   Antarctic Expedition.

Sir J. Hooker's Journal contains the following important note:—"A smaller "bird than *D. capensis*, about the size of a fowl, of a dusky brown colour, with a "white bar across the wings. It is not uncommon." This surely must have been *Thalassœca antarctica*, figured in the "Voyage of the Erebus and Terror."

Œstrelata brevirostris.

Procellaria grisea, *Kuhl. Beitr. Zool.* 1820, *p.* 144, *fig.* 9 (*nec Gm.*); *Schl. Mus. P. B. Procell. p.* 9.

Procellaria brevirostris, *Less. Man. d'Orn.* ii., *p.* 611; *Gray, List Anseres, etc., Brit. Mus. p.* 163; *id. Gen. B.* iii., *p.* 648.

Œstrelata grisea, *Coues, Pr. Philad. Acad.* 1866, *p.* 148 (*nec Gm.*).

Fulmarus griseus, *Gray, Handl. B.* iii., *p.* 107.

Œstrelata kidderi, *Coues, Bull. U. S. Nat. Mus.* ii., *p.* 28; *Kidder, l. c. p.* 15.

Œstrelata brevirostris, *Salvin in Rowley's Orn. Misc. p.* 235.

♀ *Ad.* saturatè cinereus, plumis clariore cinereo marginatis ad basin multo pallidioribus; pileo laterali concolori cinereo, regione oculari saturatiore, fronte lorisque vix pallidioribus; corpore reliquo subtùs cinereo, plumis basaliter albicantibus, pectoris lateribus clarè cinereo lavatis; tectricibus alarum superioribus et subalaribus cinerascenti-brunneis, cinereo lavatis; tectricibus primariorum et remigibus nigricanti-cinereis, intùs pallidioribus; infrà pallidè cinereis; caudâ sordidè cinereâ; rostro nigro; tarso pedibusque fuscis, unguibus nigris. Long. tot. 12·8, culm. 1·1, alæ 10·2, caudæ 4·0, tarsi 1·35.

This species may almost be said to have been rediscovered by the recent expeditions to Kerguelen Island, as it had not been satisfactorily identified by naturalists for some years, and was apparently unknown to Dr. Coues when he wrote his well-known papers on the *Procellariidæ*. In his account of the American collection of birds from Kerguelen, he named this species after Dr. Kidder, who accompanied the American expedition as naturalist; but Mr. Salvin shortly after showed that it was the same as the bird described by Lesson in 1828. In justice to the late Mr. George Robert Gray it must be noted that he correctly identified the specimens brought by the Antarctic Expedition (*Cf. List Anseres, p.* 163). This useful little list appears to have been overlooked by Dr. Coues.

Afterwards in the "Handlist" (p. 107), Gray made the mistake of referring *Œ. brevirostris* as a synonym to *Pterodroma macroptera*; and the Kerguelen birds he referred with a query to *Fulmarus griseus* (*Kuhl.*). In this arrangement he was following Dr. Coues' identifications of 1866.

With regard to the name *unicolor* of Gould, quoted by Gray as a synonym and noticed by Coues, I cannot find that it was ever published. A specimen with this name attached, in Mr. Gould's handwriting, is in the Museum, and is doubtless the authority for Mr. Gray's quotation.

a. ad. Tristan d'Acunha. Capt. Carmichael, R.N.*

b. c. ad. Kerguelen Island. Lieut. Alex. Smith.

d. e. f. ad. Kerguelen Island. Antarctic Expedition.

g. a. d. Christmas Harbour, Kerguelen Island, Feb. 1840. Antarctic Expedition.

♀. ad. Royal Sound, Kerguelen Island. Rev. A. E. Eaton.

[This Petrel is less common than *Œ. lessoni* about Observatory Bay. I am unacquainted with its call, the birds not even screaming when dug out, although they bit and scratched the hand. They burrow into clayey soil near lakes and upland marshes. The burrow is rather smaller in diameter than that of *M. æquinoctialis*, but in all other respects is very similar to it. During the early portion of the breeding season the floor of the tunnel leading to the nest chamber is flooded with water an inch or two inches deep; and any one who saw it then for the first time in his life, with water trickling in a little streamlet out of its mouth, or standing stagnant within it, would readily suppose it to be an old hole abandoned long ago. Towards the beginning of autumn (February), however, the ground becoming dry, the water disappears. The nest is composed of damp and decayed vegetable matter, comprising sprigs of *Acæna* and *Azorella*, tufts of *Festuca erecta*, &c. It is two or three inches in height, and slightly concave.

The first nest was taken on the 8th of November. The embryo was tolerably advanced in growth. In January, in a branch of a *Majaqueus* burrow, was a nestling which seemed to be the young of *Œ. brevirostris*; the *Majaqueus* egg was in the chamber of the main burrow, to which there was only one entrance.—*A. E. E.*]

* Printed in Gray's List of Anseres "Tristan de Chusan," which we believe to be a misprint for Tristan d'Acunha.

Œstrelata lessoni.

Procellaria lessoni, *Garnot, Ann. Sci. Nat.* vii., *p.* 54, *pl.* 4; *Less. Traite, p.* 611; *Gray, List Anseres Brit. Mus. p.* 163; *id. Gen. B.* iii., *p.* 648; *Gould, B. Austr.* vii., *pl.* 49; *Reichenb. Handb. Longipennes, pl.* xx., *fig.* 339; *Finsch, J. f. O.*, 1872, *p.* 255, 1874, *p.* 207; *Buller, B. N. Zeal. p.* 303, *pl.* 29, *fig.* 3; *Sharpe, Voy. Ereb. and Terror, Birds, App. p.* 33; *Cab. & Reichen. J. f. O.* 1876, *p.* 329.

Procellaria leucocephala, *Forster, Descr. Anim. p.* 206; *Gould, Ann. N. H.* xiii., *p.* 363; *Pelz. Reis. Novara, Vög. p.* 145.

Œstrelata leucocephala, *Bp. Consp.* ii., *p.* 189; *Gould, Handb. B. Austr.* ii., *p.* 451.

Rhantistes lessoni, *Bp. C. R.* xlii., *p.* 768.

Œstrelata lessoni, *Cass. Pr. Philad. Acad.* 1862, *p.* 327; *Coues, op. cit.* 1866, *p.* 142; *Giglioli, Faun. Vertebr. Oceano, p.* 40; *Coues and Kidder, Bull. U. S. Nat. Mus.* ii., *p.* 27; *iid. op. cit.* iii., *p.* 14.

♂ *Ad.* suprà clarè canus, plumis puriùs cinereo marginatis, basaliter albidis, scapis linealiter nigris; pileo antico albido, posticè cinereo, plumis ad basin cinerascentibus: loris cinerascentibus; regione oculari nigricante, et regione paroticâ nigricanti-schistaceo lavato; facie laterali reliquâ albâ; colli et pectoris lateribus clare cinereis albido tanquam irroratis; corpore reliquo subtùs purè albo, lateribus corporis et hypochondriis angustè nigro striolatis; subalaribus cinerascenti-brunneis, angustè cinereo terminatis; axillaribus cinerascentibus, basaliter et apicaliter albidis; alis suprà nigricanti-brunneis, tectricibus medianis et majoribus cinereis albo marginatis; remigibus sordidè cinerascenti-nigris, intùs cinereo-brunneis; scapularibus ad basin cinereis, ad apicem nigris; uropygio saturatè brunneo, cinereo lavato; supracaudalibus canis, albido terminatis; caudâ albâ, plumis cano irroratis, rectricibus duabus mediis canis; rostro nigro; pedibus carneis, digitis suprà nigris; membrano ad apicem nigro; iride saturatè brunnea. Long. tot. 15·5, culm. 1·5, alæ 12·4, caudæ 5·5, tarsi 1·7.

♀ *Ad.* mari similis, seb pileo puriore albo, et scapis plumarum dorsalium et hypochondriarum minùs distinctè indicatis. Long. tot. 16·5, alæ 12·3, caudæ 5·2, tarsi 1·7.

The descriptions are taken from a pair of birds collected by Mr. Eaton in Kerguelen Island. Compared with some other specimens from Australian seas, they do not show any differences, unless it be that the black on the feet is not so extended. Dr. Kidder describes the latter as follows:—Tarsus and foot flesh-pink, " black along upper surfaces of digits, and on the web near the claw." This agrees with the markings exhibited by the Kerguelen Island birds, but in a South Australian specimen quite the terminal half of the webs are black, and the birds collected by the " Rattlesnake " also show this peculiarity. In the plumage a certain amount of variation is shown in the vermiculations of grey on the head, and in the amount of grey on the tail, the latter being almost entirely white in some with grey mottlings, while in others it is almost uniform grey, mottled only on the outer web.

For a description of the young birds, Dr. Kidder's paper (*l. c.*) must be consulted. This species, which was obtained by all the recent expeditions, does not appear to have been collected in Kerguelen Island during the Antarctic Expedition, nor is it mentioned in Nunn's Narrative. In the National collection are examples from New Zealand (*Sir George Grey*); Southern Seas (*Antarctic Expedition*); South Indian Ocean, Lat. 40¾° S., Long. 125½° E., Jan. 14, 1847 (*J. Macgillivray*); South Pacific Ocean, Lat. 44° S., Long. 110½° W. (*J. Macgillivray*); Lat. 36° 39 S.; Long. 10° 3' S. (*Sir George Grey*), and the following :—

a. b. ♂. ♀. ad. Royal Sound, Kerguelen Island. Rev. A. E. Eaton.

[In Captain Hutton's paper before referred to, allusion is made to an undetermined species of Petrels to which the euphonious sobriquet *Procellaria diabolica* has been applied. It was said to be a bird inhabiting Desolation Island, which flew about by night uttering unearthly shrieks. There are good reasons for supposing the sprite to be Lesson's Petrel. It is difficult to describe the cry of this bird. For a long time there was no finding out which of the Petrels gave utterance during its flight to its weird sounds. Whenever its cry was heard, I went out with a lantern to endeavour to get a sight of the bird, but without success. At last, near Thumb Peak, we dug up some large birds whose outcries, when caught by the beak, plainly identified them beyond all question with *Œ. lessoni*.

The burrow of this Petrel can be recognised externally by its being about as large as an ordinary rabbit's hole and dry, and by its entrance being generally sparsely bestrewn with green shoots of *Acæna*. It is usually excavated in *Azorella*, the tunnel is short, the large terminal chamber contains no special nest, and when the hand is cautiously introduced to feel after the egg, it is promptly and severely bitten by the old bird. It is therefore well to take the precaution of dragging her forth from the interior before an attempt is made to secure the egg. Her removal can be easily effected. While she is stooping forward at the entrance of the nest-chamber looking out in readiness to bite, a piece of stick is presented to her, which she seizes instantly, and whilst it is being shaken to make her hold it fast, her beak is suddenly grasped with the hand, and she is drawn up by it out of the burrow shrieking loudly. Care must meanwhile be taken to prevent her from thrusting her claws into the hand. Nests were found from the extreme confines of the sea shore to an altitude of about 300 feet above the mean level. They were common amongst *Azorella* at the foot of the cliffs near Thumb Peak, and on the summit of the lower terraces; also on some of the hills near the Swain's Harbour Transit Station. There were also some nests near the principal station on a slope by a freshwater lake on the landward side of a hill.

There is a specimen of Lesson's Petrel in the S. African Museum, captured in Lat. 32° 46′ S., Long. 59° 13′ E.—*A. E. E.*]

Œstrelata mollis.

Procellaria mollis, *Gould, Ann. N. H.* 1844, *p.* 363; *id. B. Austr.* vii., *pl.* 50; *Cass. U. S. Expl. Exp.* 1858, *p.* 410; *Schl. Mus. P. B. Procell. p.* 11; *Pelz. Reis. Novara, Vög. p.* 146; *Finsch, J. f. O.* 1874, *p.* 255, 1872, *p.* 207; *Cab. & Reichen. J. f. O.* 1876, *p.* 329.

Procellaria inexpectata, *Forster, Descr. An. p.* 204.

Cookilaria mollis, *Bp. Consp.* ii., *p.* 190.

Rhantistes mollis, *Bp. C. R.* xlii., *p.* 768.

Œstrelata mollis, *Coues, Pr. Philad. Acad.* 1866, *p.* 150; *Gould. Handb. B. Austr.* ii., *p.* 453; *Giglioli, Faun. Vertebr. Oceano, p.* 42.

Ad. cinerascenti-brunneus, plumis clariore cinereo marginatis; plumis frontalibus albido marginatis; alis saturatiùs brunneis, tectricibus majoribus vix cinereo lavatis; supracaudalibus caudâque cineraceis; loris albis; regione oculari et parotica anticâ cinereo-nigris; facie laterali et corpore subtùs toto purè albis, colli lateribus et hypochondriis cineraceo irroratis, pectore laterali concolori cineraceo; subalaribus brunneis, albido marginatis, majoribus interioribus pallidius cinerascenti-brunneis; remigibus infrà saturatè brunneis, secundariis intùs pallidioribus; rostro nigro; tarso, digitis ad basin, et membrano interdigitali dimidio basali carnescenti-albis, pedibus aliter nigris. Long. tot. 12·5, culm. 1·2, alæ 10·1, caudæ 5·85, tarsi 1·45.

For description of the young bird and variations in the plumage, Dr. Coues' article in the Philadelphia 'Proceedings' must be consulted.

As will be seen by Mr. Eaton's note below, he believes that he saw this species, which was procured in Kerguelen Island by the German expedition.

The British Museum contains specimens of this Petrel from South Australia (*Sir George Grey*); and the South Atlantic, Lat. 36° 50′ S., Long. 27° 50′ W.; Lat. 34° 43′ S., Long. 40° W. (*J. Macgillivray*).

[Off Cape Sandwich and the neighbouring low land, and out at sea during the first few days sail from Kerguelen Island, I noticed a Petrel very like *Œ. lessoni*, but differing from that species in having a dark coloured tail and back. This may have been *Œ. brevirostris;* but when I was looking through the collection in the S. African Museum on my return to the Cape, I was led to believe the species I had seen to be *Œ. mollis*, Gould, which is represented in the collection mentioned by a specimen taken in Lat. 31° 26′ S., Long. 30° 26′ E., exhibited as *Procellaria mollis*, Gould.

The lice of *Œ. mollis*, according to Bulow, are *Trabeculus schilingii* and *Colpocephalum furcatum* of that author.—*A. E. E.*]

Procellaria nereis.

Thalassidroma nereis, *Gould, P. Z. S.* 1840, *p.* 178; *id. B. Austr.* vii., *pl.* 64; *Bennett, Gath. Nat. p.* 240; *Abbott, Ibis,* 1861, *p.* 164; *Gray, Ibis,* 1862, *p.* 245; *Buller, B. N. Zeal. p.* 322; *Finsch, J. f. O.* 1872, *p.* 257, 1874, *p.* 213; *Sharpe, Voy. Ereb. & Terror, Birds, App. p.* 34.

Procellaria nereis, *Gray, Gen. B.* iii., *p.* 648; *Bp. Consp.* ii., *p.* 196; *Coues, Pr. Philad. Acad.* 1864, *p.* 81; *Gould, Handb. B. Austr.* ii., *p.* 476; *Giglioli, Faun. Vertebr. Oceano, p.* 36; *Gray, Handl. B.* iii., *p.* 104, *no.* 10852; *Scl. and Salv. Nomencl. Av. Neotr. p.* 148; *Coues & Kidder, Bull. U. S. Nat. Mus.* ii., *p.* 31; *iid. op. cit.* iii., *p.* 16; *Cab. & Reichen. J. f. O.* 1876, *p.* 329.

Oceanitis nereis, *Cab. J. f. O.* 1875, *p.* 449.

♀ *Ad.* suprà sordidè cinerea, uropygio et supracaudalibus clariùs cinereis, his subbasaliter albis; caudâ quoque clarè cinereâ, nigro terminatâ, rectricibus extimis nigricantibus; pileo et collo undique et collo postico cinerascenti-fumosis, interscapulio clariore cinereo lavato; scapularibus dorso concoloribus, longioribus nigricantibus; tectricibus alarum cinerascenti-fumosis, marginalibus angustè cinereo lavatis, medianis clariùs cinereis angustè albo terminatis; remigibus sordidè cinereis, primariis extùs ad basin nigricantibus, secundariis etiam magis nigricantibus; gutture, præpectore, et pectoris summi lateribus fumosis; corpore reliquo subtùs purè albo, hypochondriis imis et subcaudalibus cinereo lavatis; subalaribus et axillaribus purè albis, minoribus et marginalibus fumoso-brunneis, his imis albo terminatis; remigibus infrà cinerascentibus, intùs versus basin albidis; rostro et pedibus nigris; iride nigrâ. Long. tot. 7, culm. 0·55, alæ 5·0, caudæ 2·8, tarsi 1·2.

Compared with an Australian example in the collection, the Kerguelen skin has rather a larger bill, and is more dusky brown on the head and throat, and has the brown colour extending lower down on to the fore-neck; these characters are, however, exhibited in another Australian example, so that there seems to be only one species. The Sea-Nymph Petrel was not previously known to inhabit Kerguelen Island, but specimens were collected by the English, American and German expeditions.

The British Museum contains specimens from New Zealand (*Sir George Grey*); off the eastern coast of New South Wales (*J. Macgillivray*); Lat. 43° S., Long. 140° E. (*J. Macgillivray*), and the following :—

a. ♀ ad. Royal Sound, Kerguelen Island. Rev. A. E. Eaton.

[Late in the night of the 6th of November the faint cry of a strange bird in the distance roused me from sleep. Calling for a dark lantern, I proceeded with George Wilson, the sapper on watch, to search for the Petrel, guided by its call, which was uttered at intervals until we were quite close to it. The light being now turned on in the right direction, the bird was discovered sitting upon the open ground within a yard or so of us, and it was so dazzled that it made no attempt to escape, being caught by hand. It proved to be a female of *P. nereis*. Its call was very similar to the crake-like cry of *P. oceanica*. Dr. Kidder* said that this Petrel was common at Molloy Point. As no other example was either seen or heard by me besides that mentioned above, I suspect that this species rarely ventures so far up the Sound as Observatory Bay.

From the specimen of *P. nereis* whose casual capture is mentioned above, six examples of its parasite, *Lipeurus clypeatus*, Giebel, were obtained.—A. E. E.]

Oceanitis tropica.

Procellaria grallaria, *Licht. Verz. Doubl. p.* 83.
Procellaria oceanica, *Bp. Zool. Journ.* iii., *p.* 89 (*nec Kuhl.*).
Thalassidroma tropica, *Gould, Ann. N. H.* xiii., p. 366.
Thalassidroma melanogastra, *Gould, Ann. N. H.* xiii., *p.* 367; *id. B. Austr.* vii., *pl.* 62; *Layard, B. S. Afr. p.* 358; *Finsch, J. f. O.* 1872, *p.* 257, 1874, *p.* 212; *Buller, B. N. Zeal. p.* 319; *Hutton, Ibis,* 1874, *p.* 42; *Buller, l. c. p.* 121; *Sharpe, Voy. Ereb. and Terr. Birds, App. p.* 34; *Cab. & Reichen. J. f. O.* 1876, *p.* 329.
Fregetta tropica, *Bp. Consp.* ii., *p.* 197; *id. C. R.* xlii., *p.* 769; *Coues, Pr. Philad. Acad.* 1864, *p.* 87.
Fregetta melanogastra, *Bp. C. R.* xlii., *p.* 769; *Coues, Pr. Philad. Acad.* 1864, *p.* 87; *Gould, Handb. B. Austr.* ii., *p.* 479; *Giglioli, Faun. Vert. Ocean., p.* 38.
Procellaria melanogaster, *Schl. Mus. P. B. Procell. p.* 6; *Gray, l. c. p.* 105.

♀ *Ad.* Fumoso-nigricans, uropygio nigerrimo, plumis imis albo terminatis; supracaudalibus conspicuè albis, fasciam latam exhibentibus; tectricibus alarum cinerascenti-brunneis, majoribus pallidioribus vix albido terminatis; tectricibus primariorum et alâ spuriâ saturatius fumoso-nigricantibus; remigibus nigris vix versùs apicem cinereo lavatis, intùs pallidioribus, fumoso brunneis; secundariis clariùs cineraceo lavatis et anguste albo limbatis; caudâ nigrâ, vix cinereo lavatâ, ad basin extremam albâ; gutture toto, faciei et colli lateribus, præpectore et pectore summo fumoso-nigricantibus, gulæ plumis basaliter albis; corpore reliquo subtùs purè albo, medialiter fumoso-nigricante, fasciam latam per pectus et abdominem ductam exhibentibus; subcaudalibus albis longissimis basaliter albis, ad apicem dimidialiter fumoso-nigricantibus; subalaribus et axillaribus purè albis, his versùs basin fumosis, tectricibus marginalibus brunneis, majoribus pallidè cinerascenti-brunneis anguste

---

* Since the preceding paragraph was written Dr. Kidder's Report has been sent me from the Smithsonian Institute. At p. 32 he writes of this species as follows:—" The first specimens were taken on the " 28th and 29th of October, being dug out by the dogs from small burrows under clumps of *Azorella*. A " pair captured on the latter date were found under a tussock not two yards above high-water mark, on the " beach, under a high cliff. No eggs were found at that date. Eggs were first found, December 12, under " the overhanging margins of clumps of grass and 'Kerguelen tea' (*Acæna ascendens*), in a bit of swampy " lowland near the sea. Strange to say, I have only found the male with the egg. In this locality there " were no burrows, the overhanging herbage seeming to afford sufficient protection to the nests."

albo terminatis; rostro et pedibus nigris; iride nigrâ. Long. tot. 7·6, culmen 0·6, alæ 9·4, caudæ 3·15, tarsi 1·55.

♂ *Ad.* similis fœminæ adultæ. Long. tot. 7·2 culm 0·6, alæ 6·8, caudæ 3·0 tarsi 1·6.

The differences between *O. melanogastra* and *O. tropica* are extremely slight, consisting in the white throat and the greater amount of black on the abdomen and centre of the body in the latter bird. I believe it possible that *O. leucogastra* is also only a stage of plumage of the same species, the four specimens in the Museum being apparently immature, with narrow whitish edgings to the feathers of the upper surface. Whether this is the case I am, however, unable to prove at present. The following are the series of measurements of the specimens in the British Museum.

			Total Length.	Culmen.	Wing.	Tail.	Tarsus.
a.	♂	O. melanogastra, Kerguelen Island	7·2	0·6	6·8	3·0	1·6
b.	♀	" "	7·6	0·6	6·4	3·15	1·55
c.	♂	O. tropica, Lat. 6° 33′ N., Long. 18° 6′ W.	7·0	0·6	6·7	3·0	1·65
d.	♀	" Lat. 12′ S., Long. 30½° W.	7·0	0·6	6·55	3·1	1·6
e.	♂	O. leucogastra	7·0	0·6	6·55	3·3	1·45
f.		" S. Australia	6·8	0·6	6·1	2·95	1·6
g.		"	7·1	0·6	5·95	3·1	1·6
h.		" Lat. 37½° S., Long. 42° E.	7·6	0·6	6·3	3·0	1·5

[Occasionally late in the evening and during the night a piercingly shrill piping note repeated singly at intervals of four or six seconds used to be heard on the hills about Observatory Bay. Generally the sound changed its direction, showing that the bird which uttered it was flying. This call might be imitated on a piccolo fife in the key of G or F. In its complete form it consists of a series of single notes separated by pauses of four seconds or more, followed by a jerky succession of notes in the same tone.

One night the sound was traced to a crevice in a cliff beneath an immovable rock. The place was marked by a pile of stones, and visited early the next morning. While efforts were being made to move the rock the bird within the recess became alarmed, and uttered a cry somewhat like that of a kestril hawk in its tone, but not nearly so loud. On another night the sound was followed up to a hill. Every now and then the bird ceased piping, but it recommenced whenever the call was imitated with the lips. Its nook was therefore easily discovered; it was in a terrace on the hillside under a piece of rock. The stone was pulled away, the nesting place laid open, and two birds in it disclosed, of which one escaped. The female was caught, and she proved to be an *O. melanogastra*. A third pair was caught in a slope of broken rocks near the top of a hill, a few nights later, in a similar way. Their nesting place had been used before, as there were fragments of an old egg-shell in the hollow that they had prepared for laying in. After this I went for three weeks to Swain's Bay. On returning to Observatory Bay only one bird was heard on only one night. No eggs were found by anybody.

In the South African Museum are two examples of this petrel—one from the Southern Ocean, and another from the South Atlantic.—*A. E. E.*]

Oceanitis oceanica.

Procellaria pelagica, *Wilson, Am. Orn.* vi., *p.* 90, *pl.* 69, *fig.* 6 (*nec L.*).

Procellaria oceanica, *Kuhl. Beitr. Zool. p.* 136, *tab.* x., *fig.* 1; *Gray, Gen. B.* iii., *p.* 648; *Schl. Mus. P. B. Procell. p.* 6.

Procellaria wilsoni, *Bp. Journ. Acad. Philad.* iii., *pt.* 2, *p.* 231, *pl.* ix., *fig.* 2; *Yarr. Br. B.* iii., *p.* 517; *Fritsch, Vög. Eur. tab.* 61, *fig.* 3.

Thalassidroma wilsoni, *Audub. B. Amer. pl.* ccclx.; *id. B. Amer.* 8vo, viii., *p.* 106, *pl.* 460; *Gould, B. Austr.* vii., *pl.* 65; *Macgill. Br. B.* v., *p.* 456; *Burm. Th. Bras.* iii., *p.* 446; *Cass. U. S. Expl. Exp. p.* 402.

Thalassidroma oceanica, *Schinz. Europ. Faun. p.* 397, *pl.* 1; *Gray, List Anseres, Brit. Mus. p.* 161; *id. Gen. B.* iii., *p.* 648; *id. List Br. B. p.* 225; *Pelz. Reis. Novara, Vög. p.* 145; *Degl. & Gerbe, Orn. Europ.* ii., *p.* 386.

Oceanitis wilsoni, *Keys. & Blas. Wirb. Eur. p.* 238; *Bp. C. R.* xlii., *p.* 769; *Salvad. Ucc. Ital. p.* 301; *Gigl. Faun. Vertebr. nell' Oceano, p.* 38.

Oceanitis oceanica, *Bp. C. R.* xlii., *p.* 769; *Gould, Handb. B. Austr.* ii., *p.* 478; *Salvad. Cat. Ucc. Sard. p.* 132; *Giglioli, Faun. Vertebr. Oceano, p.* 37; *Coues and Kidder, Bull. U. S. Nat. Mus.* ii., *p.* 30; *iid. op. cit.* iii., *p.* 16.

♂ *Ad.* fuliginoso-niger, pileo undique aliquot cinerascente, regione auriculari magis nigricante; tectricibus alarum fumoso-nigricantibus, majoribus versùs apicem pallidè brunneis, plagam formantibus; tectricibus primariorum remigibusque nigris, intùs brunnescentibus, secundariis quoque pallidioribus extùs brunnescentibus; plumis uropygialibus imis nigris albo terminatis; supracaudalibus purè albis; caudâ nigrâ, rectricibus ad basin albis, externarum albedine magis extensâ; corpore subtùs fumoso-brunneo, lateraliter saturatiore, subcaudalibus brunneis ad basin albis; crissi lateribus conspicuè albis, quibusdam fumoso lavatis; subalaribus fumoso-brunneis, intimis vix pallidioribus; rostro nigro; pedibus nigris, membranis flavis; iride nigra. Long. tot. 6, culm. 0·45, alæ 5·35, caudæ 2·45, tarsi, 1·35.

The above is a description of a Kerguelen Island skin, and after a comparison of our series I am unable to find any grounds for separating a northern and a southern species. Some examples are blacker and some are grayer, as is the case with other Petrels.

The British Museum specimens are from the following localities:—Yarmouth; Atlantic Ocean (*Rev. W. Hennah*); South Africa (*Sir A Smith*); South Australia (*Sir G. Grey*); Lat. $36\frac{3}{4}°$ N., Long. $12\frac{1}{4}°$ W. (*J. Macgillivray*); and from the ice off Louis Philippe Island (*Antarctic Expedition*). To these is now added the following:—

a. ad. Royal Sound, Kerguelen Island.   Rev. A. E. Eaton.

[From the 10th of October, when we passed Cape Sandwich, until the middle or third week of November, we completely lost sight of the Storm Petrels. About the period last mentioned, however, they began to frequent Observatory Bay in large numbers. Their first appearance in it took place during a strong breeze which lasted several days. When this was succeeded by more moderate weather, we saw little of them in the day-time; but towards evening they used to fly over the water like Swallows, and some of them might be observed flying near the ground far away into the country, following the course of the valleys, or playing round the inland cliffs. We tracked them along the lower hill-sides and the margins of lakes over rocks and bogs; but our efforts to learn what became of them were unattended with success. Probably at that time they were not preparing to breed, and the birds were merely going overland from the bay to other inlets of the sea. At length when we went to Thumb Peak their mode of nesting was discovered. Carefully watching, with Lieut. Goodridge, R.N., the birds flying to and fro about the rocks, we observed that they occasionally disappeared into crevices amongst piles of loose stones, and crept under loose masses of rock. Having meanwhile ascertained their call, we were able by listening attentively to detect the exact positions of several of these hidden birds. They were easily caught when the stones were rolled aside; but they were in couples, merely preparing for laying, and therefore we did not find any eggs. On our way back to Observatory Bay after the Transit we called at the American Station, and were informed by Dr. Kidder that he had observed this Petrel on the shore near Molloy Point. The sea-shore in the neighbourhood of Observatory Bay is of a different character (for the most part) from that which is adjacent to the American Station, and, being less favourable than it, was seldom resorted to for nesting by the Petrels. But the country in general about our bay afforded them unlimited accommodation. For, provided that they can find a slope of shattered rocks with suitable chinks and crevices, or dry spaces under stones or large boulders sheltered from draughts, whether they be near the Sound or on the sides and summits of high hills, they readily appropriate them. The egg is laid upon the bare ground within the recess selected by the birds, either in a chance depression formed by contiguous stones or in a shallow circular hollow excavated in the earth by the parent. Having found numbers of their nesting-places I will describe my method of searching for them. Whenever there was a calm night I used to walk with a darkened bull's-eye lantern towards some rocky hillside, such as the Petrels would be likely to frequent. It was best to shut off the light and keep it concealed, using it only in dangerous places where falls would be attended with injury, and progress in the dark was hardly possible, lest the birds seeing it should be silenced. On arriving at the ground selected it was probable that Storm Petrels would be heard in various directions, some on the wing, others on their nests, sounding their call at intervals of from two to five minutes. Those on nests could be distinguished from others flying by their cries proceeding from fixed positions. Having settled which of the birds should be searched after, a cautious advance had to be made in her direction, two or three steps at a time, when she was in full cry. As soon as she ceased an abrupt halt was imperative, and a pause of some minutes might ensue before she recommenced her cry and permitted another slight advance to be effected. In the course of this gradual approach the position of the bird might be ascertained approximately; but it had to be determined precisely, and to learn exactly where she was she had to be stalked in the dark noiselessly. No gleam could be permitted to escape from the lantern. Loose stones and falls over rocks,—to avoid them it was sometimes necessary to dispense with slippers, and feel one's way in stockings only, for should the Petrel be alarmed once with the noise or the light, she would probably remain silent a considerable time. Now and then it would happen that upon the boulder beneath which she was sitting being almost attained the bird would cease calling. When this occurred, and many minutes elapsed without her cry being resumed, it was advisable to make a detour and approach the rock from the opposite side, as her silence might be attributed to her seeing a person advancing towards her, and she would probably recommence her call so soon as he was out of sight. If she did not, a small pebble thrown amongst the rocks would usually elicit some sounds from her, as she would most likely conclude that the noise was being made by her mate returning to the nest. When the stone beneath which the bird was domiciled was gained at last, redoubled care had to be exercised. By stooping down and listening very attentively her position could be accurately ascertained. Then the lantern was suddenly turned upon her before she had time to creep out of sight, and her egg could be secured with the hand, or with a spoon tied on to a stick. Sometimes I worked without a lantern, and marked the positions of the nests with piles of stones so that they might be revisited by day. Several eggs were obtained in February from nests which had been thus marked early in the previous month. The first egg taken by us was found by a retriever on the 22nd of January, on an island in Swain's Bay. Captain Fairfax sent me a nestling a day or two before we sailed for the Cape. Two of the eggs were laid in unusual situations. One of them was found by a man under a *Pringlea* plant; but this may have been an egg of *Procellaria nereis*. The other was deposited just above the tide-mark in a cavity of a rock rather open to the air and light. I had found the bird there one night, had taken her up into my hand, and had gently replaced her in the hollow, nearly a month before the egg was laid.

The young bird in the egg has the tarso-metatarsal joint short.

In the S. African Museum there is a specimen of *P. oceanica* from the S.E. coast of Africa, another from the S. coast of Africa, and two from Table Bay. *A. E. E.*]

Prion vittatus.

Blue Petrel, *Forster, Voyage,* i., *pp.* 91, 153; *id. Drawings, No.* 87. Petrel bleu, *Buffon, H. N. Ois.* ix., *p.* 316. Broad-billed Petrel, *Lath. Gen. Syn.* iii., *pt.* 2, *p.* 414; *id. Gen. Hist.* x., *p.* 194.

Procellaria vittata, *Gm. S. N.* i., *p.* 560; *Kuhl. Beitr. Zool. pl.* xi., *fig.* 13; *Forster, Descr. Anim. p.* 21; *Schl. Mus. P. B. Procell. p.* 16; *id. & Poll. Faun. Madag., p.* 144.

Procellaria forsteri, *Lath. Ind. Orn.* ii., *p.* 827.

Prion vittatus, *Lacép. Mem. de l'Inst.* 1800, *p.* 514; *Gray, Gen. B.* iii., *p.* 649, *pl.* 178, *fig.* 1; *id. List Anseres, &c., Brit. Mus. p.* 165; *id. Voy. Ereb. & Terror, Birds, p.* 18; *Gould, Ann. N. H.,* xiii., *p.* 366; *id. B. Austr.* vii., *pl.* 55; *Bp. Consp.* ii., *p.* 192; *Reichenb. Handb. Longipennes, pl.* x., *figs.* 771, 772; *Gray, Ibis,* 1862, *p.* 247; *Pelz. Reis. Novara, Vög. p.* 147; *Gould, Handb. B. Austr.* ii., *p.* 474; *Coues, Pr. Philad. Acad.* 1866, *p.* 169; *Giglioli, Faun. Vertebr. Oceano, p.* 44; *Gray, Handl. B.* iii., *p.* 108; *Buller, B. N. Zeal., p.* 312; *Finsch, J. f. O.* 1872, *p.* 256, 1874, *p.* 211; *Sharpe, Voy. Ereb. & Terror, Birds, App. p.* 33.

Pachyptila vittata, *Illiger, Prodr. Syst. Mamm., &c., p.* 274 : *Temm. Pl. Col.* 528; *Burm. Th. Bras.* iii., *p.* 444; *Cab. & Reichen., J. f. O.* 1876, *p.* 328.

Procellaria latirostris, *Bonn. et Vieill. Enc. Méth.* i., *p.* 81.

Pachyptila forsteri, *Steph. Gen. Zool.* xiii., *p.* 251; *Less. Traité, p.* 613; *Jard. & Selby, Ill. Orn.* i., *pl.* 47; *Swains. Classif. B.* ii., *p.* 374.

Pachyptila banksii, *Smith, Ill. Zool. S. Afr. pl.* 55.

Prion banksii, *Gould, Ann. N.H.,* xiii., *p.* 366; *Gray, List Anseres Brit. Mus.* p. 165; *id. Gen. B.* iii., *p.* 649; *Bp. Consp.* ii., *p.* 193; *Gray, Ibis,* 1862, *p.* 247; *Kirk, Ibis,* 1864, *p.* 338; *Layard, B. S. Afr., p.* 362; *id. Ibis,* 1867, *p.* 460; *Giglioli, Faun. Vertebr. Oceano, p.* 44; *Gray, Handl. B.* iii., *p.* 108; *Finsch & Hartl. Vög. Ostafr., p.* 815; *Finsch, J. f. O.* 1872, *p.* 256, 1874, *p.* 211; *Buller, B. N. Zeal., p.* 311; *Sharpe, Voy. Ereb. & Terror, Birds, App. p.* 34.

Procellaria banksi, *Schl. Mus. P. B. Procell. p.* 17; *id. & Poll. Faun. Madag.* p. 145.

Prion magnirostris, *Gray, Handl. B.* iii., *p.* 108.

Prion australis, *Potts, Ibis,* 1873, *p.* 85.

♂ *Ad.* clarè cinereus, pileo obscuriore; vittâ uropygiali indistinctè nigricante; tectricibus alarum brunneis vix cinereo lavatis; medianis et majoribus clarè cinereis; alâ spuriâ et tectricibus primariorum brunneis vix cinereo limbatis; remigibus cinereis, intùs albis, primariis longis extùs nigricantibus, secundariis intimis subterminaliter nigricantibus, albo terminatis; caudâ clarè cinereâ, versus apicem nigricante, rectrice extimâ omninò cinereâ; regione lorali cinereâ albo variegatâ; vittâ latâ superciliari albâ; macula ante-oculari nigricante; regione paroticâ sordidè cinereâ, plumis versùs basin albis; genis et corpore subtùs albis, colli et pectoris lateribus clarè cinereis; hypochondriis et subcaudalibus cinereo lavatis, his longioribus apicaliter nigricantibus; subalaribus et axillaribus albis, his vix cinereo lavatis; rostro pedibusque cœruleis; iride saturatè brunneâ. Long. tot. 13·0; culm. 1·4; alæ, 8·3; cauda, 4·25; tarsi, 1·25; digit. med. c. u., 1·65.

♀. mari similis, sed rostro angustiore et laminibus minus conspicuis.

Although Mr. Eaton did not bring back a skin of this species, its occurrence in Kerguelen Island was certified by the discovery of a head in the stomach of a Giant Petrel. The American naturalists did not notice it, but the German expedition is stated to have brought back this species of Prion from the island.

Prion vittatus is a bird easily recognisable by the *shape* of its bill; in colour it exactly resembles *P. desolatus*, and a few words on the subject of these birds may not be out of place here, seeing that my conclusions are of a very different nature from those of recent writers on the broad-billed Petrels. To take first the paper by Dr. Coues in the "Proceedings" of the Philadelphia Academy for 1866 (*p.* 162), we find that he separates *Prion* (type *P. vittatus*) as a distinct genus from *Pseudoprion* (type *P. turtur*), keeping in the latter genus four species, *P. banksii, P. turtur, P. ariel, P. brevirostris*. The chief differences between *Prion* and *Pseudoprion* are to be found in the shape and laminations of the bill; and indeed the latter form almost the only specific characters for the distinguishing of the above-named four species.

Mr. G. R Gray, in the "Handlist," follows the arrangement of Dr. Coues, keeping his genera as sub-genera, but he adds to *P. vittata* a second species, *P. magnirostris*, Gould (ubi?); and to *P. banksii* and its allies he adds *P. desolata* of Gmelin and of Kuhl, concerning which species see below.

The chief points to be noted are the reference of *P. brevirostris* of Gould (1855) to *P. ariel* of Gould (1844), in which I agree with Mr. Gray, and the reference of *P. rossii* of Gray to *P. banksii*, in which I do not agree with him, for an examination of the types shows me that if *P. ariel* were a species, it is also *P. rossii* of Gray. But *P. ariel* is *not* a species, according to my studies of the genus, and is only *P. turtur*, when not full grown. Again, I consider *P. banksii* to be also no species, but to be a stage only of *P. vittatus*, the laminations in the bill being developed with age, and not being specific characters.

The solution of this question has been much simplified for me by the examination of three birds presented to the British Museum by Sir George Grey, of which the following are illustrations (*Plate* VII., *Figs.* 2–5). They were sent as exemplifying the old male, old female, and young male of *P. vittatus*.

There are no differences in colour or markings. Then, again, any number of intermediate links are to be found in a series of specimens, and as a variation in the extent of the laminations accompanies a difference in size of bill, it follows that the birds must either be regarded as stages of one species, or must be divided into several species, which has been the plan adopted by many recent writers. In the plate are figures of the largest and smallest bills in a series of *P. vittatus* (*Figs.* 3-7).

On comparing these figures with those of *P. desolatus* given on the same plate (*Figs.* 8-10), it will be seen that though the *size* varies, the *shape* is constant, the bill being bowed out from the base and gently incurved towards the tip. The variations in the dimensions of the wing and middle toe are shown in the accompanying list of specimens in the British Museum:—

 a. ♀. Cape Seas (*A. Smith*; type of *P. banksii*). Wing, 7·55; middle toe, 1·45.

 b. ♀. Coast of Australia (*Sir G. Grey*). Wing, 7·2; middle toe, 1·45.

 c. ♂. Ad. Australian Seas (*Sir G. Grey*). Wing, 7·6; middle toe, 1·5.

 d. ♀. Ad. Australian Seas (*Sir G. Grey*). Wing, 7·15; middle toe, 1·4.

 e. ♂ juv. Lat. 35° 1′ S.; long. 6° 15′ E. (*Fd.*) Wing, 7·6; middle toe, 1·45.

 f. ♂. Eastern entrance to Bass' Straits (*Macgillivray*). Wing, 7·4; middle toe, 1·5.

 g. New Zealand (*Sir G. Grey*). Wing, 7·2; middle toe, 1·35.

 h. New Zealand (*Sir G. Grey*). Wing, 6·8; middle toe, 1·35.

 i. New Zealand (*Dr. Hector*). Wing, 6·8; middle toe, 1·6.

 k. ♂. Ad. Pitt's Isl., Chatham Isl. (*W. L. Travers*). Wing, 7·85; middle toe, 1·65.

 l. Auckland Islands (Antarctic Expedition). Wings, 6·65; middle toe, 1·45.

 m. n. ♂. South Seas (Antarctic Expedition; types of *P. rossii*).

 ♀. Skeleton. Menado; *Dr. A. B. Meyer.*

The soft parts in the description are copied from an original label attached to Mr. Travers' specimen from Pitt's Island. Mr. Gould gives them as follows:—
" Bill light blue, deepening into black on the sides of the nostrils and at the tip, " with a black line along the sides of the under mandible; irides very dark brown; " feet beautiful light blue." The plate represents the bill with a yellow " nail " at the tip, an important feature when considered along with the occasional appearance of a nail in specimens of *P. desolatus* (*vide infrà*). Dr. Buller, in describing the bill, does not notice this yellow nail, and further information is desirable on this point.

Prion desolatus.

Brown-banded Petrel, *Lath. Gen.* iii., *pt.* 2, *p.* 409; *id. Gen. Hist.* x., *p.* 187.

Procellaria desolata, *Gm. S. N.* i., *p.* 562; *Kuhl, Beitr. Zool. pl.* xi., *fig.* 7; *Gray, Gen. B.* iii., *p.* 648.

Daption desolatum, *Shaw, Gen. Zool.* xiii., *p.* 244.

Procellaria turtur, *Kuhl, Beitr. Zool. p.* 143, *pl.* xi., *fig.* 8; *Smith, Ill. Zool. S. Afr. pl.* 54; *Gray, List Anseres, &c., Brit. Mus. p.* 165; *id. Gen. B.*, iii., *p.* 648; *Schl. Mus. P. B. Procell. p.* 17; *Layard, B. S. Afr. p.* 361.

Prion turtur, *Gould, Ann. N. H.* xiii., 1844, p. 366; *id. B. Austr.* vii., *pl.* 54; *Bp. Consp.* ii., p. 193; *Reichenb. Handb. Longipennes, pl.* x., *figs.* 774, 775; *Coues, Pr. Philad. Acad.* 1866, p. 166; *Gould, Handb. B. Austr.* ii., p. 472; *Pelz. Reis. Novara, Vög.* p. 147; *Giglioli, Faun. Vertebr. Oceano,* p. 45; *Gray, Handl. B.* iii., p. 108; *Buller, B. N. Zeal.* p. 309; *Finsch, J. f. O.* 1872, p. 256, 1874, p. 311; *Sharpe, Voy. Erebus & Terror, Birds, App.* p. 34, *pl.* 29.

Prion ariel, *Gould, Ann. N. H.* xiii., p. 366; *id. Intr. B. Austr.* p. 117; *Bp. Consp.* ii., p. 194; *Gray, Ibis,* 1862, p. 247; *Giglioli, Faun. Vertebr. Oceano,* p. 45; *Gould, Handb. B. Austr.* ii., p. 473; *Gray, Handl. B.* iii., p. 108.

Prion rossii, *Gray, List Anseres, &c., Brit. Mus.* p. 165; *Bp. Consp.* ii., p. 195.

Prion brevirostris, *Gould, P. Z. S.* 1855, p. 88, *pl.* 93.

Halobæna typica, *Bp. Consp.* ii., p. 194.

Procellaria ariel, *Schl. Mus. P. B. Procell.* p. 19.

Æstrelata desolata (pt.), *Coues, Pr. Philad. Acad.* 1866, p. 155.

Pseudoprion banksii, turtur, ariel, brevirostris, *Coues, t. c. p.* 166.

Prion desolata, *Gray, Handl. B.* iii., p. 108.

Pseudoprion desolatus, *Coues, Bull. U. S. Nat. Mus.* ii., p. 32.

Pachyptila ariel et turtur, *Cab. J. f. O.* 1875, p. 449; *id. & Reichen. J. f. O.* 1876, p. 328.

Ad. clarè cinereus, pileo vix saturatiore, vittâ uropygiali nigricante indistinctâ; tectricibus alarum maximis brunneis, paullò cinereo lavatis, reliquis clarè cinereis; alâ spuriâ et tectricibus primariorum cinerascenti nigris vix cinereo lavatis; remigibus cinereis, intùs albis, primariis longis extùs nigricantibus, secundariis intimis subterminaliter nigricantibus, albo terminatis; caudâ clarè cinereâ, versùs apicem nigricante, fronte canescente, plumis albido terminatis; loris et vitta latâ superciliari albis; facie laterali albâ, regione paroticâ superiore cinereâ; plagâ anteoculari fuscescente, albo variegatâ; corpore subtùs toto albo, colli et pectoris summi lateribus clarè cinereis; hypochondriis et subcaudalibus paullò cinereo lavatis; subalaribus et axillaribus albis; rostro coeruleo; pedibus lilacino-coeruleis, unguibus nigris ad basin lilacinis aut albis; iride cyanescenti-cinereo. Long. tot. 10·5, culm. 1·1, alæ, 6·95, caudæ 3·7, tarsi, 1·25, digit. med. c. u. 1·5.

The name of *P. desolatus* was originally conferred by Gmelin on Latham's "Brown-banded Petrel." The latter is stated to have been in the collection of Sir Joseph Banks, from the "Island of Desolation." The description is apparently taken from a dried specimen, as the colours assigned to the soft parts show:—"The "bill is black with the tip yellowish, the legs brown, webs yellow, claws black." These are the colours which dried skins exhibit, but they are not found in any species of Prion when alive. It is, therefore, most probable that Latham's description was taken from an actual skin, as in 1824 he reproduces it almost verbatim in his "General History of Birds," with the habitat "Island of Desolation; Sir Joseph "Banks." In many instances Latham appears to have drawn up his descriptions from the paintings and drawings made by Forster, Parkinson, and Ellis, for Sir J. Banks, but, when he has done this, he generally states the fact in his latest work. We may therefore conclude that the type of his "Brown-banded Petrel" existed as a skin in Sir J. Banks' collection, probably as late as 1824, but it is not now in the British Museum. Ellis' drawing is unmistakeable, and was perhaps taken from the actual type-specimen when in the flesh; it was from access to this drawing that the late Mr. Gray was enabled to make out that *P. desolatus* was a *Prion*, and the collection of specimens by the recent expeditions has led to the re-discovery of an interesting species. On looking over the series of *Prion* skins in the Museum, I have found a specimen from Christmas Harbour, Kerguelen Island, collected during the Antarctic Expedition, and I have carefully compared it with the other birds in the Museum collection. It is of the species usually called by naturalists *Prion turtur*, and is, I believe, the true *P. turtur* of Kuhl. At first I was inclined to consider this a different species, as it has such a distinct yellow nail at the end of the bill, but whether this is a specific character or not, I cannot at present say for certain. It is very plain in some individuals, and in others very indistinct, so that it may be merely the fading of the bill after death. In the allied species, *P. vittatus*, Mr. Gould describes the bill as entirely blue, but then he *figures* it with a yellow nail to the bill: this character, therefore, should be looked after by any one who may have the opportunity. On mature consideration, I believe that *Prion turtur* is only the male of *P. desolatus*, and is distinguished by its larger bill. This organ alone defines it from *P. vittatus*, which has a differently *shaped* bill, more bowed sideways from the base to the tip. *P. ariel*, *P. rossii*, and *P. brevirostris*, are only young birds apparently with the bills not fully developed. The following is a list of the specimens of *P. desolatus* in the national collection.

a. Madeira. J. Gould, Esq.; type of *P. brevirostris*; wing 6·55, middle toe 1·4.

b. Cape Seas. Sir A. Smith; fig. *l. c.* as *P. turtur*; wing 7·3, middle toe 1·5.

c. Royal Sound, Kerguelen Island. A. E. Eaton; wing 6·95, middle toe 1·5.

d. Christmas Harbour. Antarctic Expedition; wing 7·15, middle toe 1·6.

e. Australian Seas. Wing 7·1, middle toe 1·4.

f. Cook's Straits, New Zealand. Dr. Lyall; wing 6·9, middle toe 1·5.

g. ♂. Indian Ocean, Lat. 40¾° S., Long. 123½° E. (*J. Macgillivray*). Capt. Stanley; wing 6·6, middle toe 1·5.

h. Indian Ocean (as above). Capt. Stanley; wing 7·3, middle toe 1·5.

The figures (*Pl.* vii., *pp.* 8–10) illustrate the variation in size of bill in this species. They show the constancy of the shape, notwithstanding a slight variation in size. On comparing these figures with those of *P. vittatus*, it will be seen that the outline of the bill in the present species is different from that of the last-named bird. When seen from above, the sides of the bill are nearly *straight*, and this form is constant, even when the size varies conspicuously.

[The burrows both of this species and of *H. cærulea* resemble rats' holes. They are usually made in *Azorella* or amongst *Acæna* growing upon dry rocky slopes or stony ground; but a few of the birds took possession of some of the burrows out of which *H. cærulea* had been evicted, deepening them to adapt them to their own requirements. A well marked track leading to the mouth of the hole is worn by the birds running down the slope to gain impetus for their start on taking flight, which path they also use in returning to the nest if they chance to fall short of the entrance. Sometimes the burrows are branched, and have two or three entrances; occasionally their sidings are *culs de sac*, and only abandoned "leads." The egg is laid upon loose debris of *Azorella*, &c., or on the bare ground constituting the floor of the terminal chamber, as is that of *H. cærulea*.*

When we disembarked in Observatory Bay, *P. desolatus* was pairing. Eggs obtained on the 29th of November were fresh. Most of the nestlings had flown before we left the Island.

It has already been stated by Dr. Kidder, and also in the Proc. Roy. Soc. 1875, that during the breeding season, the various species of burrowing Petrels are found in their holes in pairs until the egg is laid; after that, until the young is hatched, only one bird at a time remains in the hole by day, the other returning with food at intervals during the night, and that when the nestling issues from the egg, the parents leave it by itself the whole of the day, and visit it only in the night. The Storm Petrels have the same habits.

The call uttered by this species in its flight may be denoted thus : u–u, u–u, u–u, and so on. Now and then, it also (as I suppose) uses another call, which is repeated only at distant intervals. It consists of three short notes slurred, and the intermediate note is three tones higher than the other two. At a distance, this has a resemblance to the mew of a cat.

The multitudes of Blue Petrels which breed in Kerguelen Island are hardly conceivable. Every dry hillside and knoll in the neighbourhood of Royal Sound was populous with them to a remarkable extent. During the day the birds were silent, excepting when a noise happened to disturb them and cause them to coo. But on calm nights at the end of October and beginning of November, their mingled cries produce a low continued murmur like the sound of distant street traffic in a large town, in which the calls of only the nearer birds can be distinguished, and the rustling of their wings as they fly by is almost incessant. Father Sidgreaves thought it would have been worth while to ascertain how many of them on an average crossed the disk of the moon viewed through a telescope in the course of five minutes; but more important work prevented this being done.

---

* According to Dr. Kidder (Bull. U. S. Nat. Mus. 1875, p. 36) the egg is covered with the debris. I am inclined to suspect that his finding it so was either dependent upon the style of the implement used by him in digging the birds out, or was attributable to the vivacity of his dogs when they drew the Petrel; for I employed a sharp shovel-headed spade, and no dogs, and usually saw the egg when I stooped down to look into the chamber.

They left the land before the Expedition sailed, so that towards the end of February scarcely a bird could be found anywhere in a burrow, and rarely could even one be heard at night, perhaps only one in ten or twenty minutes. They had withdrawn from the sheltered sound to the more open sea. When we were steaming past the Prince of Wales' Foreland *en route* for the Cape, large flocks of Blue Petrels were in close attendance upon dense shoals of fish between us and the shore, which were playing here and there at the surface of the sea, beating the water into foam.

This Petrel burrows rapidly, loosening the soil with its beak, and shovelling the earth backwards with its feet like a domestic fowl engaged in dusting itself, whilst its wings are held just a little apart.

Several examples of this species are in the S. African Museum under the names of *Prior banksii* and *P. turtur*, which have been captured at Green Point, near Table Bay. The louse obtained from *P. desolatus* is *Lipeurus clypeatus*, Giebel (one example only.)—*E. A. E.*]

Halobæna cœrulea.

Blue Petrel, *Forster, Voyage,* i., *p.* 91; *Lath. Gen. Syn.* iii., *pt.* 2, *p.* 415.

Procellaria cœrulea, *Gm. S. N.* i., *p.* 560; *Bonn. et Vieill. Enc. Méth.* i., *p.* 80; *Gray, List Anseres, &c., Brit. Mus. p.* 165; *id. Gen. B.* iii., *p.* 648 : *id. Ibis,* 1862, *p.* 247; *Layard, B. S. Afr. p.* 361; *Buller, B. N. Z., p.* 306; *Finsch, J. f. O.,* 1872, *p.* 255, 1874, *p.* 208; *Sharpe, Voy. Ereb. & Terror, Birds, p.* 33.

Pachyptila cœrulea, *Illiger, Prodr. Syst. Mamm., &c., p.* 275.

Procellaria similis, *Forster, Drawings,* No. 86; *id. Descr. Anim., p.* 59.

Procellaria forsteri, *Smith, Ill. Zool. S. Afr., pl.* 411.

Halobæna cœrulea, *Bp. Consp.* ii., *p.* 193; *id. C. R.* xlii., *p.* 768; *Gould, Handb. B. Austr.* ii., p. 457; *Coues, Pr. Philad. Acad.* 1866, *p.* 163; *id. & Kidder, Bull. U. S. Nat. Mus.* ii., *p.* 34; *iid. op. cit.* iii., *p.* 17.

Halobæna typica, *Bp. C. R.* xlii., *p.* 768.

Zaprium cœruleum, *Coues, Bull. U. S. Nat. Mus.* ii., *p.* 34.

Ad. suprà saturatè cinereus, pileo distinctè saturatiore; fronte albicante; tectricibus alarum cinereis, minimis nigricanti-brunneis, majoribus albo terminatis; tectricibus primariorum nigricantibus; remigibus cinereis, intùs albis, primariis extùs nigricantibus, secundariis intimis albo terminatis, subterminaliter nigricantibus, scapularibus secundariisque concoloribus et eodem modo coloratis; caudâ quadratâ, cinereâ, conspicuè albo terminatâ; facie laterali et regione supraoculari albis, regione paroticâ summâ cinereâ; corpore subtùs toto albo, pectoris et colli lateribus et hypochondriis imis cinereo lavatis; subalaribus et subcaudalibus albis; rostro nigro; pedibus nigris; iride saturatè brunneâ vel nigrâ. Long. tot. 10, culm. 1·1, alæ 8·5, caudæ 4·0, tarsi 1·25.

The description is taken from Sir A. Smith's specimen from the Cape Seas, figured by him as *P. forsteri*. The single specimen brought by Mr. Eaton is not adult, as it has still a few remains of down attached to the sides of the neck. It is much darker than the one described, but I am unable to judge whether this is a sign of age, or arises from the exposure of Smith's specimen to the light, which may have bleached it. The British Museum now possesses the following skins:

a. ad. Cape Seas. Sir A. Smith. Type of *P. forsteri*.
b. jun. Royal Sound, Kerguelen Island. Rev. A. E. Eaton.

[The resemblance between this Petrel and the *Prion desolatus* extends even to their coo. Their calls underground are so much alike, that on hearing one it is difficult to say to which of the two species the bird cooing should be referred without digging it up for inspection; and their tone is very similar in sound to the cooing of some foreign doves. But their calls during flight are very different from one another.

The comparative immunity of this species from the ravages of the men was due partly to its commencing to lay eggs later than the former, and partly to its nests being less easy of access than those of that Petrel. For *H. cærulea* is in the habit of burrowing into *Azorella* growing upon dry soft loam where no obstacles impede its progress; its eggs are therefore obtainable without much trouble. It had only just begun to lay when we first landed. So long as its eggs continued to be fresh, the liberty men dug out as many as they could, cruelly destroying the old birds, which they flung away in heaps; but when most of the eggs became uneatable through incubation, they abandoned petrel digging. About this time *P. desolatum*, burrowing in *Azorella* and *Acæna* where the ground was rocky, commenced laying, and thus naturally escaped their notice.

The eggs of *H. cærulea* were fresh and profusely plentiful so early as the 23rd of October. A nestling almost fully fledged was killed on the 9th of February.

Some of the old birds while they were dying cast up the contents of their crop, which were green like *ulva*.

Louse not observed; but there is a *Lipeurus* from this bird in the Halle Museum according to Prof. Giebel.—*A. E. E.*]

Ossifraga gigantea.

Giant Petrel, *Lath. Gen. Syn.* vi., *p.* 396, *pl.* 100.

Procellaria gigantea, *Gm. S. N.* i., *p.* 563; *Gray, List Anseres, &c., Brit. Mus. p.* 162; *id. Voy. Erebus & Terr., Birds, p.* 17; *id. Gen. B.* iii., *p.* 648; *Gould, B. Austr.* vii., *pl.* 45; *Reichenb. Handb. Longipennes, pl.* xii., *fig.* 332; *Lawr. B. N. Amer., p.* 825; *Cass. U. S. Expl. Exp. Orn., p.* 407; *Schl. Mus. P. B. Procell., p.* 18; *Pelz. Reis. Novara, Vög., p.* 144; *Layard, B. S. Afr. p.* 360; *Finsch, J. f. O.* 1872, *p.* 255, 1874, *p*, 206.

Procellaria ossifraga, *Forster, Descr. Anim., p.* 343.

Ossifraga gigantea, *Reichenb, Syst. Av. Tubinares, pl.* 20, *fig.* 332; *Bp. Consp.* ii., *p.* 186; *id. C. R.* xlii., *p.* 768; *Gould, Handb. B. Austr.* ii., *p.* 443; *Coues, Pr. Philad. Acad.* 1866, *p.* 32; *Giglioli, Faun. Vertebr. Oceano, p.* 48; *Gurney in Anderss., B. Dam. Ld. p.* 354; *Buller, B. N. Zeal., p.* 297; *Sharpe, Voy. Ereb. & Terror, Birds, App. p.* 33; *Coues & Kidder, Bull. U. S. Nat. Mus.* ii., *p.* 23; *iid. op. cit.* iii., *p.* 13; *Cab. & Reichen, J. f. O.*, 1876, *p.* 329.

Suprà brunneus, plumis omnibus marginaliter cineraceis; tectricibus alarum dorso concoloribus; remigibus saturatè brunneis, intùs chocolatinis; caudâ saturatè brunneâ; subtùs cinerascens, plumis brunneis cineraceo marginatis; subalaribus brunneis, imis cinerascentibus; remigibus infrà pallidioribus brunneis, propè rachidem quasi albicantibus; rostro perlato, vix carnescente, vel flavicanti-albo; pedibus sordidè nigris; iride saturatè brunneâ. Long. tot. 33, culmen 5·1, alæ 21·5, caudæ 7·7, tarsi 4·0.

The above is a description of a Kerguelen skin in the ordinary brown plumage, but whether this is the adult stage it is difficult to say. Many writers have drawn attention to the variation in plumage in the "Nelly," and the general opinion is that the white plumage represents an albinism, but in the "Wreck of the Favorite" (p. 187), the young bird is described as "grey, darker on the back," while the adult bird is said to be white. This view is borne out by the fact that the white-plumaged birds have generally some remains of brown feathers about their bodies. Dr. Kidder describes the nestlings as follows:—"The down of the young bird is entirely grey "in colour; the head is partly naked, and the bill, tarsi, and feet are coloured "nearly as in the adult, but somewhat paler. The first fully-formed feathers are "similar to the adult plumage." At the same time Dr. Kidder seems not to have noticed any white specimens, and that these are rare near the Cape is proved by the fact that Mr. Layard (*l. c.*) states that the species is common in Table Bay throughout the year, and that "a white variety is common up the west coast towards "Walwisch Bay."

In the British Museum are specimens from Wellington, New Zealand (*Dr. Hector*), and the South Pacific Ocean, as well as one from the present locality.

a. ad. Kerguelen Island. Antarctic Expedition.

[The breeding places of the Giant Petrel in Royal Sound, which had long been an enigma to us, were discovered on Long Island by a shooting party from the "Volage" on the 23rd of December. The nests (according to the statements of the officers) were of a similar make to those of Albatrosses, and contained half-grown nestlings. They were constructed above ground amongst *Azorella*, about 200 yards from the sea, not very far apart from each other. There were two groups of them on the S.W. side of the island, each consisting of about 30 nests, which were situated on the upper parts of very gradual slopes. One of the nestlings was brought off to the ship. It was about as large as a Cochin fowl. Whenever anybody walked past, it ejected oil from its mouth to the distance of a yard, after the manner of Petrels; on this account it was summarily set upon and despatched. Its down was very dense and thick, and formed a regular jacket beneath which the young feathers were well developed. Dr. Garrod, of the Zoological Society, on dissecting it, found portions of two Prion's skulls in its crop. The eggs are probably laid as early in the season as those of the King Penguin.

Giant Petrels are well known as scavengers. When the *Globiocephalus* was exposed upon the beach in Swain's Bay, a couple of dozen of these birds used to take possession of the carcass during the absence of the flensing party. It was amusing to watch them assembling to feed upon it. Those which happened to be flying about in its immediate neighbourhood began to circle round it as soon as the men retired; and after passing and repassing it a few times they settled one by one upon the water, swam to the shore, and waddled quickly towards the krang. Others attracted to the spot by their movements might now be seen hurrying up from various directions to partake of the banquet. The largest of the birds meanwhile had taken her stand upon the krang with outstretched wings. There she stood tugging at loose ends of sinews, and with difficulty tearing off with her beak morsels of the tough flesh. If any of the others ventured to approach too near before she had allayed her hunger, she ran open-mouthed at them with wings half spread, and drove them off with loud croaks. Seldom did any dare to withstand her attack; if he did she allowed him to get what he could from off the tail of the carcase where he could make very little impression upon it, whilst she returned to the more fleshy portion of the trunk. When the stronger birds grew tired of eating, the others were permitted to feed; and this they did greedily, quarrelling from time to time amongst themselves. A few yards away upon the slopes of *Azorella*, small groups of Skuas were standing waiting impatiently for an opportunity of gorging themselves, but not daring to associate with the Giants. When the boat was rowed towards them, the Petrels alarmed waddled off in haste to the water, and swam away at full speed, looking like prototypes of the roc. We chased them, and they tried to rise from the water, running with their great feet splashing along the surface and flapping heavily with their wings (feet keeping stroke with wings), making quite as much noise as Swans starting to fly. Where it was perfectly calm they could hardly take flight; but where the breeze was blowing they easily rose into the air by running to windward. We drove some on shore up a hill; it was a most exciting chase as we gained rapidly upon them; but becoming fatigued with their climb they turned round and rushed past us down the slope with an impetus that sufficed to start them on the wing.

The Giant Petrel is troubled with an undescribed *Lipeurus* (Giebel), and with *Docophorus coloratus*, Rudow, neither of which were found by me.

There is an example of *O. gigantea* in the South African Museum from the Cape Seas.—*A. E. E.*]

Diomedea exulans.

The Albatross, *Edwards, Nat. Hist. B.* ii. *pl.* 88; L'Albatros, *Briss. Orn.* vi., p. 126.

Diomedea exulans, *Linn. S. N.* i., p. 214; *Gm. S. N.* i., p. 506; *Vieill. Gal. Ois.* p. 234, *pl.* 293; *Gray, Voy. Erebus and Terror, Birds*, p. 18; *id. Gen. B.* iii., p. 650; *Gould, B. Austr.* vii., *pl.* 38; *Bp. Consp.* ii., p. 184; *id. C. R.* xlii., p. 768; *Lawr. in. N. Amer. B.* p. 821; *Cass. U. S. Expl. Exp.* p. 397; *Schl. Mus. P. B. Procell.* p. 333; *Pelz. Reis Novara, Vög.* p. 147; *Gould, Handb. B. Austr.* ii., p. 427; *Coues, Pr. Philad. Acad.* 1866, p. 175; *Degl. et Eerbe, Orn. Eur.* ii., p. 366; *Giglioli, Faun. Vertebr. Oceano*, p. 49; *Gray, Handl. B.* iii., p. 109; *Anderss. B. Dam. Ld.* p. 355; *Finsch, J. f. O.* 1872, p. 254, 1874, p. 206; *Buller, B. N. Zeal.* p. 289; *Scl. und Salv. Nomencl. Av. Neotr.* p. 148; *Sharpe, Voy. Ereb. and Terror, Birds, App.* p. 32; *Bree, B. Eur.* 1876, v., p. 90; *Kidder & Coues, Bull. U. S. Nat. Mus.* iii., p. 11, *Cab. & Reichen. J. f. O.* 1876, p. 328.

L'Albatros du Cap de Bonne Esperance, *Buff. Pl. Enl.* 237.

Chocolate Albatros, *Lath. Gen. Syn.* iii., *pt.* 1, p. 308.

Diomedea spadicea, *Gm. S. N.* i., p. 595.

Diomedea albatrus, *Pall. Zoogr.* ii., p. 308; *Forster, Descr. Anim.* p. 27.

Diomedea adusta, *Tschudi, J. f. O.* 1856, p. 157.

Ad. purè albus, dorsi plumis, scapularibus et corpore subtùs plus minusve cinereo fasciatim irroratis; tectricibus alarum nigricanti-brunneis, basaliter albis; remigibus nigricantibus, secundariis et scapularibus albis, ad apicem cinerascenti-brunneis; caudâ albâ, cinereo versùs apicem marmoratâ, pogonio externo terminaliter nigricante; corpore subtùs toto cum subalaribus et subcaudalibus purè albis; rostro albido, vix coccineo tincto, versùs apicem flavicante; palpebris pallidè viridibus; pedibus albis, coccineo tinctis; iride saturate albâ. Long. tot. 38, culmen 7·6, alæ 25·5, caudæ 10·0, tarsi 4·9.

Juv. brunneus, alis caudâque saturatioribus; pileo dorsoque dilute brunneis, basaliter albis et pallidius marginatis; facie laterali, regione paroticâ, gulâque purè albis; subtùs pallidè brunneus, subcaudalibus saturatiùs brunneis.

Considerable variation is seen in the plumages of the Albatross from youth to maturity, the young birds being brown, and gradually becoming whiter and whiter with age.

According to Sir J. Hooker's "Journal" none of this species were observed at Christmas Harbour, but above Cape François "the nests were huddled together, as " many as 50 or 60 of them, and were built on the grassy slopes above the precipice " 700 or 800 ft. above the sea. A good deal of straw and stubble was mixed with " them, or rather plastered up with the clay to give it consistency. Their height was " about 1½ ft., and their breadth much the same. From a distance they looked " like so many Cheshire cheeses."

The Museum contains specimens from the Cape Seas (*Capt. Harry, E. M. Langworthy*, &c.); South Australia (*Sir George Grey*); New Zealand; and one from Kerguelen.

a. *Ad.* Kerguelen Island. Antarctic Expedition.

[The Great Albatross bred on the flat ground near Shoal Water Bay. The birds captured on their nests were destroyed by liberty men from the U.S.S. Monongahela for the sake of their wing-bones and feet (just as the Sooty Albatrosses were by our men), much to the regret of Dr. Kidder and the American Astronomers. Capt. Fuller of the whaling fleet said that a few pairs of the birds build near Sprightly Bay. Their occurrence in the neighbourhood of Mount Campbell was reported by H.M.S. "Challenger."

The vicinage of the Prince of Wales Foreland would have been worth a visit; but there was no means of getting there from Observatory Bay by boat. The prevalence of sudden and violent squalls makes boat navigation in the open Sound extremely dangerous. An American boat's crew sailing from Three Island Harbour was once detained nearly a fortnight at the Foreland by strong winds blowing out of the Sound; and the men were almost starved before their schooner could rescue them.

An adult Wandering Albatross can breathe without much difficulty with a weight of about 130 lbs. upon its back. When specimens had to be killed we employed large men to sit down upon them, holding their beaks to prevent the birds from biting. The pink stains on the sides of the neck mentioned in letters from the "Challenger" and in Dr. Kidder's report were well marked in our adult examples.

Two new species of lice were obtained from this Albatross, namely, *Docophorus dentatus* and *Nirmus angulicollis*, Giebel. Previously *Lipeurus thoracicus*, Rudow, was known to occur upon it.—*A. E. E.*]

Diomedea melanophrys.

Diomedea melanophrys, *Temm. Pl. Col.* 456; *Gray, Gen. B.* iii., *p.* 650; *Gould, B. Austr.* vii., *pl.* 43; *Reichenb. Handb. Longipennes, pl.* xvi., *figs.* 797, 798; *Bp. Consp.* ii., *p.* 185; *Schl. Mus. P. B. Procell., p.* 33; *Gould, Handb. B. Austr.* ii., *p.* 438; *Pelz. Reis. Novara, Vög. p.* 148; *Coues, Pr. Philad. Acad.* 1866, *p.* 181; *Gray, Handl. B.* iii., *p.* 109; *Finsch, J. f. O.*, 1872, *p.* 254, 1874, *p.* 206; *Anderss. B. Damara Land, p.* 355; *Buller, B. N. Zeal. p.* 292; *Scl. & Salv. Nomencl. Av. Neotr. p.* 148; *Sharpe, Voy. Ereb. & Terror, Birds, App. p.* 32.

Thalassarche melanophrys, *Bp. C. R.*, xlii., *p.* 768; *Giglioli, Faun. Vertebr. Oceano, p.* 57.

Ad. suprà brunneus, dorso paullo cinerascente, scapularibus dorso concoloribus, imis nigricantibus; alis brunneis, tectricibus majoribus versùs basin cinereo lavatis; remigibus saturatè brunneis, intùs cinerascenti-brunneis, versùs basin albis; uropygio et supracaudalibus purè albis; caudâ cinerascente, versùs apicem brunnescentiore, ad basin albidâ; pileo et collo undique albidis, hoc vix brunneo lavato;

plumis supraocularibus saturatè brunneis; corpore subtùs toto purè albo; subalaribus et axillaribus fumoso-brunneis, majoribus versùs apicem cinerascenti-albis; remigibus infrà pallide fumoso brunneis, intùs basaliter albis; rostro fulvescenti-albo, basaliter nigro marginato; pedibus flavicanti-albis, pallide cœruleis; iride pallide brunneâ, saturatius marmoratâ. Long. tot. 28, culmen, 5·2, alæ 20·5, caudæ 8·0, tarsi 3·3.

[Not observed by us. The Challenger Expedition believed that they saw this Albatross on the eastern side of the island, but it was not seen by Dr. Kidder or by the German expedition. Its louse is *Lipeurus prox.*—A. E. E.]

The specimens in the British Museum are two in number.

a. ad. Cape Seas. Purchased.

b. ♂ jun. Off the entrance to Bass' Straits, July 11, 1847 (*J. Macgillivray*). Capt. Stanley.

Diomedea culminata.

Diomedea culminata, *Gould, P. Z. S.* 1843, p. 107; *id. Ann. N. H.* xiii., p. 361; *id. B. Austr.* vii., pl. 41; *Gray, Gen. B.* iii., p. 650, pl. 179; *Bp. Consp.* ii., p. 185; *id. C. R.* xlii., p. 768; *Schl. Mus. P. B. Procell.*, p. 35; *Coues, Pr. Philad. Acad.* 1866, p. 183; *Gould, Handb. B. Austr.* ii., p. 437; *Gray, Handl. B.* iii., p. 109; *Cab. in Von der Decken's Reis.* iii., p. 52; *Finsch, J. f. O.* 1872, p. 254, 1874, p. 206; *Buller, B. N. Zeal.* p. 295; *Sharpe, Voy. Ereb. & Terror, Birds, App.* p. 32; *Cab. & Reichen. J. f. O.* 1876, p. 328.

Diomedea chlororhynchus, *Audub. B. Amer.* 8vo. viii., p. 79.

Thalassarche culminata, *Giglioli, Faun. Vertebr. Oceano*, p. 59.

Ad. pileo colloque undique pallidè cinereis; plumis ante-ocularibus et regione oculari saturatè cinereo lavatis; fasciâ infra-oculari albâ; dorso cinerascenti-brunneo, scapularibus alisque nigricanti-brunneis; remigibus nigricantibus, intùs chocolatinis, scapis flavicantibus; uropygio et supracaudalibus purè albis; caudâ saturatè cinerascenti-brunneâ, scapis flavicanti-albidis; corpore subtùs toto purè albo, facie et colli lateribus vix cinereo lavatis; pectoris lateribus intimis celatè et axillaribus intimis saturatè brunneis, his longioribus et subalaribus purè albis, subalaribus alæ margini proximis brunneis; rostro nigro, culmine corneo, mandibulâ imâ et gonyde aurantiacâ. Long. tot. 30, culmen, 4·7, alæ 20, caudæ 8·0, tarsi 2·8.

Juv. similis adulto, sed saturatior, pileo colloque saturatiùs cinereis.

This species occurs in the German list as obtained at Kerguelen Island, a skeleton of the bird having been preserved. Dr. Kidder in the American Report also states that it was common along the coast, and was occasionally seen in Royal Sound.

The following specimens form the series at present in the British Museum.

a. ad. Lat. 36½° S., Long. 95½° E. Capt. Stanley. June 5, 1847 (*J. Macgillivray*).

b. ad. Australian seas. J. Gould, Esq.

c. juv. Off Van Diemen's Land, August 3, 1839.

d. juv. Off Veragua. Mr. Brydges.

Diomedea fuliginosa.

Albatross with a white eyebrow, *Cook's Voyage,* i., *p.* 38.

Sooty Albatross, *Lath. Gen. Syn.* iii., *pt.* i., *p.* 309.

Diomedea fuliginosa, *Gm. S. N.* i., *p.* 595; *Gray, Gen. B.* iii., *p.* 650; *Bp. Consp.* ii., *p.* 186; *Lawr. B. N. Amer. p.* 823; *Schl. Mus. P. B. Procell., p.* 35; *Pelz. Reis. Novara, Vög. p.* 149; *Gray, Handl. B.* iii., *p.* 109; *Finsch, J. f. O.* 1872, *p.* 254, 1874, *p.* 206; *Buller, B. N. Zeal. p.* 296; *Sharpe, Voy. Ereb. & Terror, Birds, App. p.* 33; *Cab. & Reichen. J. f. O.* 1876, *p.* 328.

Diomedea spadicea, *Less. Man. d'Orn.* ii., *p.* 391.

Diomedea fusca, *Audub. B. Amer. pl.* ccccliv.; *id. op. cit.* 8vo. viii., *p.* 83, *pl.* 454.

Diomedea palpebrata, *Forster, Descr. Anim. p.* 55.

Phœbetria fuliginosa, *Bp. C. R.* xlii., *p.* 768; *Gould, Handb. B. Austr.* ii., *p.* 441; *Giglioli, Faun. Vertebr. Oceano, p.* 60; *Coues, Pr. Philad. Acad.* 1866, *p.* 186; *id. and Kidder, Bull. U. S. Nat. Mus.* ii., *p.* 21; *iid. op. cit.* iii., *p.* 12.

Ad. suprà fuliginoso-brunneus, plumis dorsalibus sæpè pallidiùs marginatis; loris, facie laterali, mentoque saturatiùs brunneis; annulo supra- et post-oculari argentescenti-albo; alis fuliginoso-brunneis, remigibus saturatiùs brunneis, scapis albicantibus, secundariis basaliter cinerascentibus; caudâ saturatè brunneâ, rectricum scapis conspicuè albis; corpore subtùs cinerascenti-brunneo, plumis obscurè fulvescenti-brunneo marginatis; subalaribus et remigibus infrà chocolatino-brunneis; rostro nigro; pedibus pallidè carneis; iride purpurascenti-griseo. Long. tot. 27, culm, 4·7, alæ 20·5, caudæ 10·5, tarsi, 3·0.

Besides three examples from South Australia (*Sir George Grey*), and Lat. 38° S., Long. 30° E. (*J. Macgillivray*), the British Museum has the following:—

a. ad. Royal Sound, Kerguelen Island. Capt. Inglis, R.N.

[The Sooty Albatross is common in Royal Sound. The hills near the sea on the mainland and islands present occasionally places suitable for its nidification. As a rule the nests are built in the most sheltered situations that can be found at the foot of the precipitous terraces of volcanic rock which are so characteristic of the neighbourhood. Here and there recesses hollowed out at the base of these terraces and cliffs are thoroughly protected by the overhanging rock from wind and rain. In dry nooks of this nature *D. fuliginosa* constructs its nests of pieces of adjacent plants (especially *Festuca erecta*) disposed in the form of a low truncated cone hollowed out at the top. The nests appear to be used many years in succession, as the original materials of several that were examined seemed to have been reduced by age to vegetable mould. These old fabrics are relined with fresh dry grass when the birds return at the commencement of the breeding season. The position of her nest is liable to be betrayed to persons walking within sight of the female when she is sitting, for every now and then while she is observing their movements she will utter her cry, and thus reveal her situation. If anyone goes near her she assumes a rather formidable attitude, and ruffling up the feathers of the neck snaps fiercely and loudly with her beak at the intruder, the noise resembling that made by a large dog in catching flies. But notwithstanding her menacing gestures the egg can be secured (if it be desired) without displacing her from the nest. A pocket handkerchief presented to her with the left hand, or a hat placed gently upon her head, will completely engross her attention while the egg is being abstracted from beneath her with the right; and she will afterwards remain in the nest complacently watching her visitor's retreat. Nearly a dozen of their nests were taken by the English Expedition. From one found near Thumb Peak a female was brought by Staff Commander Inglis, of H.M.S. "Supply," from whom the skin preserved was obtained. On the 23rd of October a female was killed, while she was sitting with the male in her nest, by Lieut. Dowding, R.M.I., and Mr. Edwardes, Assist. Surgeon of H.M.S. "Volage." They found that she would have laid her egg in the night, its shell being spotted already. Another female subsequently laid in the same nest. The Rev. J. B. Budds found a nestling about a week before we sailed from Royal Sound; a newly born kid was sitting upon it. On our way to the Cape we saw a Sooty Albatross the pale band on whose neck was of a dirty white instead of the usual ash colour.

A single specimen of a *Lipeurus* was found upon a Sooty Albatross in Royal Sound (probably a moulted skin).—*A. E. E.*]

PELECANIDÆ.

Phalacrocorax verrucosus.

Graculus carunculatus, *Schl. Mus. P. B. Pelecani*, p. 20 (*spec. b.*).

Halieus (Hypoleucus) verrucosus, *Cab. J. f. O.* 1875, p. 450, pl. 1, fig. 1.

Graculus carunculatus, *Coues & Kidder, Bull. U. S. Nat. Mus.* ii., p. 7; *iid. op. cit.* iii., p. 8.

Halieus verrucosus, *Cab. & Reichen. J. f. O.* 1876, p. 329.

Ad. suprà purpurascenti-niger, pileo paullò cristato; interscapulio vix viridescente; scapularibus et tectricibus alarum distinctè viridescentibus; remigibus brunneis, extùs viridescente lavatis; caudâ nigrâ; facie laterali et genis anticis purpurascenti-nigris; genis posticis, colli lateribus, et corpore subtùs toto, purè albis, hypochondriis imis tibiisque purpurascenti-nigris; subcaudalibus albis, longioribus purpurascenti-nigris; pectore laterali celatim viridescenti-nigro; subalaribus sordidè viridescenti-nigris, his imis et remigibus infrà brunneis; rostro nigro; carunculis rostri basalis laetè flavis; pedibus flavis; iride cyaneâ. Long. tot. 25, culm. 2·25, alae 11·2, caudae 5·0, tarsi 2·4.

Juv. purpurascenti-brunneus, dorso obscurè viridescente; subtùs brunneus, plumis basaliter albidis, faciem striolatam exhibentibus; genis posticis gulâque purè albis.

Professor Schlegel, who has a Kerguelen specimen in the Leiden Museum, considers it to be specifically the same as the Falkland Island Cormorant, "très "reconnaissable à une large raie blanche s'étendant sur les plumes de l'aile couv- "rant l'avant bras." Dr. Coues, after examining the specimens brought by Dr. Kidder from Kerguelen, writes: "I have no hesitation in identifying this species "as above, although the single adult specimen collected does not show the white "trans-alar fascia spoken of by authors." He is, however, evidently influenced by Schlegel's determination. On the other hand, Dr. Cabanis considers that the Kerguelen bird is different on account of its smaller size, especially of the feet and bill, and from the want of the white band in the wing. The other differences mentioned by Dr. Cabanis do not seem to me to be of any great importance, but the Kerguelen skins in the Museum certainly do exhibit the differences enumerated by him. I consider that the material at my disposal is too limited to decide the question, and I therefore follow Dr. Cabanis in his determinations.

a. ad.; *b.* juv. Kerguelen Island. Antarctic Expedition. [C.]

[The habits of this Cormorant are so similar to those of the common British species, that it is needless to describe them. Mention has already been made elsewhere (Proc. Roy. Soc. 1875) of their remarkable tameness. Another strongly marked trait in their disposition is inquisitiveness. Not only will they direct their flight towards a man walking by the shore, in order to have a good look at him; but if he chances to be standing still, the Shags will not unfrequently alight close to him to stare at the stranger. Sometimes they are attracted by noise, not however quite to so great an extent as the Spitsbergen Guillemots. On the day of our first landing in Observatory Bay, whilst standing amongst a flock of Cormorants which were basking on a point, I fired at a Teal. The Shags fled precipitately at the sound of the report; but three minutes had hardly elapsed before five and twenty of the birds were standing round me in a circle almost within reach of the gun, mute with astonishment, looking at me first with one eye and then with the other.

There is much virtue in a mere name. Our men called these Cormorants "Shags," and would not touch them. Some of our American friends (not the astronomical party) having designated them "Shag-ducks," shot a few dozens of them with rifles, to eat them. They could have killed as many as they pleased, for in the intervals of fishing the Cormorants rest in the cliffs, and do not readily take flight unless thoroughly alarmed.

The Shags in Observatory Bay were commencing to build on the 16th of October. The first eggs were found by us about the middle of November.

The British Cormorant (*P. carbo.*) has a *Docophorus*, but the louse of the Kerguelen species was not discovered, though I procured a carpet bag full of the nestlings for my Skuas. Dr. Kidder found "a tick of prodigious size" upon some young birds.—*A. E. E.*]

Tachypetes aquila.

The Man-of-War Bird, *Edwards, Gleanings,* vi., *p.* 209, *pl.* 309.

La Frégate, *Briss. Orn.* vi., *p.* 506, *pl.* xliii., *fig.* 2 A.

Pelecanus aquilus, *Linn. S. N.* i., *p.* 216.

Le Grand Frégate de Cayenne, *Buff. Pl. Enl.* vii., *pl.* 961.

Frigate-Pelican, White-headed Frigate-Pelican, and Palmerston Frigate-Pelican, *Lath. Gen. Syn.* iii., *pt.* 2, *pp.* 587, 591, 593.

Pelecanus leucocephalus and Pelecanus palmerstoni, *Gm. S. N.* i., *p.* 572.

Fregata aquila, *Illiger, Prodr. Syst. Mamm. & Av. p.* 279; *Gray, List Anseres Brit. Mus. p.* 190; *Gosse, B. Jamaica, p.* 477; *Reich. Handb. Steganopodes, pl.* xxxi., *figs.* 372; *G. C. Taylor, Ibis,* 1859, *p.* 150; *Schl. Mus. P. B. Pelec. p.* 2; *Buller, B. N. Zeal. p.* 339; *Gigl. Faun. Vertebr. nell' Oceano, p.* 63; *Sharpe, Voy. Erebus & Terr., Birds, App. p.* 35.

Tachypetes aquila, *Vieill. N. Dict. d'Hist. Nat.* xii., *p.* 143; *id. Gal. Ois.* ii., *p.* 187, *pl.* cclxxiv.; *Kittl. Kupf. Vög. p.* 15, *taf.* xx., *fig.* 1; *Less. Traité, p.* 606; *Audub. B. Amer. pl.* ccccxxi.; *id. B. Amer.* 8vo. vii., *p.* 169, *pl.* 421; *Bp. Consp.* ii., *p.* 167; *Lawr. B. N. Amer. p.* 873; *Hartl. Orn. W. Afr. p.* 260; *Burm. Th. Bras.* iii., *p.* 549; *Cass. U. S. Expl. Exp. p.* 358; *Newton, Ibis,* 1859, *p.* 369; *Blasias & Baldam. in Naum. Vög. Deutschl.* xiii., *p.* 287; *Gould, Handb. B. Austr.* ii., *p.* 499; *Finsch & Hartl. Faun. Central-Polyn. p.* 265; *Finsch, J. f. O.* 1872, *p.* 260, 1874, *p.* 216; *Cab. & Reichen. J. f. O.* 1876, *p.* 329.

Tachypetes leucocephalus, *Kittl. Kupf. Vög. p.* 15, *taf.* xx., *fig.* 2.

Attagen aquila, *Gray, Gen. B.* iii., *p.* 669; *Gould, B. Austr. Intr. p.* c; *Jerd. B. Ind.* iii., *p.* 853.

Fregata leucocephala, *Reichenb. Handb. Steganopodes, pl.* xxxi., *fig.* 373.

Tachypetes palmerstoni, *Cass. U. S. Expl. Exp. Birds, p.* 359.

Tachypetes minor, *Hartl. Orn. Madag. p.* 87.

Ad. Niger, dorso toto æneo viridi et purpureo nitente, plumis lanceolatis; pileo saturatiore, saturatè viridi; alis caudâque nigris vix bronzino lavatis; gulâ nudâ rubrâ; corpore reliquo subtùs nigro, pectore medio purpureo, lateraliter æneo-viridi; subalaribus nitidè nigris; rostro cinerascente, versùs apicem nigro; pedibus carneo-brunneis; iride nigra. Long. tot. 34, culm. 4·0, alæ 23·0, caudæ 16·0, tarsi circa 2·85.

Juv. brunneus, purpureo et viridi nitens; tectricibus alarum pallidè brunneis albicante marginatis; pileo nigro, collo postico brunneo, plumis pallidius marginatis; gulâ nudâ minùs extensâ, gutture et facie laterali saturatè brunneis; pectore albo, abdomine brunneo, subcaudalibus et subalaribus nigris; alis caudâque nigris.

The German expedition brought back a head of this species obtained at Kerguelen Island; it has not been met with by any other of the visitors. I have not assured myself that *T. minor* is really a distinct species, for although some of the birds are evidently less bulky in their proportions, on measuring them, the supposed differences of size are found to be very slightly pronounced. As the Museum series is not sufficient to enable me to determine the question for certain, I have refrained from giving a list of the specimens.

IMPENNES.

Aptenodytes longirostris.

Patagonian Penguin (*pt.*), *Penn. Phil. Trans.* lviii., *p.* 91, *pl.* 9; *Lath. Gen. Syn.* vi., *p.* 563.

Aptenodyta patachonica (*pt.*), *Gm. S. N.* i., *p.* 556.

Le Manchot de la Nouvelle Guinée, *Sonn. Voy. N. Guin. p.* 180, *pl.* 113.

Le Manchot des Isles Malouines, *Buff. Pl. Enl.* x., *Pl.* 975.

Apterodyta longirostris, *Scop. Del. Faun. et Flor. Insubr.* ii., *p.* 91, *no.* 69.

Pinguinaria patachonica (*nec Forster*), *Shaw in Miller's Cimel. Phys. pl.* 45; *id. Nat. Misc.* xi., *pl.* 409.

Hairy and Woolly Pinguin, *Lath. Gen. Hist.* x., *p.* 392.

Aptenodytes pennantii, *Gray, Ann. N. H.* xiii., *p.* 315; *id. Gen. B.* iii., *p.* 642; *Reichenb. Handb. Pygopodes, pl.* 1, *figs.* 1, 2; *Bp. C. R.* xlii., *p.* 775; *Gould, P. Z. S.* 1859, *p.* 98; *Scl. P. Z. S.* 1860, *p.* 390; *Hyatt, Cat. Orn Coll. Boston Soc. N. H.* i., *p.* 11; *Cab. & Reichen. J. f. O.* 1876, *p.* 330.

Spheniscus pennantii, *Schl. Mus. P. B. Urinatores, p.* 3; *id. Dirent.* p. 268.

Aptenodytes longirostris, *Coues, Pr. Philad. Acad.* 1872, *p.* 193; *Sharpe, Voy. Erebus and Terror, Birds, App. p.* 37, *pl.* 32; *Coues and Kidder, Bull. U. S. Nat. Mus.* ii., *p.* 39; *iid. op. cit.* iii., *p.* 18.

Suprà cinerascens, plumis omnibus cinereo apicatis, supracaudalibus majus distincte terminatis; alis cinereis, pennis remigialibus seriatim cinereo terminatis, margine alari summo nigricante; caudâ rigidâ nigrâ; pileo summo usque ad nucham nigro; facie laterali gulâque totâ nigris; plagâ latâ aurantiacâ à regione paroticâ posticâ per collum lateralem angustante et gulam nigram marginante; colli lateribus cinereis dorso concoloribus, anticè latè nigro marginatis: jugulo medio aurantiaco; corpore reliquo subtùs sericeo-albo, pectoris lateribus dorso concoloribus; alâ subtùs albâ, versùs basin et apicem nigricante, margine alari latè nigricante; rostro nigro, versùs basin mandibulæ carneo; pedibus nigris; iride lætè brunneâ. Long. tot. 34, culm. 3·4, alæ 11·5, cauda 4·4.

The Kerguelen Island specimens in the British Museum are of larger bulk than those from the Falkland Islands.

a. ad. Kerguelen Island. Antarctic Expedition.

b. ad. Swains' Bay, Kerguelen Island. Rev. A. E. Eaton.

c, d. ad. Falkland Islands. Antarctic Expedition.

e. juv.   ,,   ,,

[The King Penguin does not breed in Royal Sound nor in Swain's Bay. The master of the schooner " Roswell King," Capt. Fuller, stated that it breeds in very few places upon the island; that a large community occupy a position on the hills west of Mount Ross near Table Bay (Kerguelen Island), and that there are others near Cape Sandwich. They were seen from the " Volage " in December at this last-named locality. The eggs are laid about the beginning or middle of October.

In December and January small parties of these Penguins come into sheltered inlets to moult. We used to find them in Swain's Bay, Carpenter's Cove, and in a bay near Vulcan Cove. Usually they were standing amongst the herbage within a few yards of the shore; occasionally they were between the tide marks. The officers of H.M.S. " Volage," who had more opportunities than I had of seeing these birds, were of opinion that they remained on shore without food until the moulting was completed; because if the Penguins while the change of plumage was progressing came from the sea every day, their breasts would in all probability be denuded of feathers, for they are then so easily detached from the skin that they could hardly fail to be stripped off in the efforts of landing; whereas their breasts were well clothed with old loose feathers until the new plumage was matured. There are so few land animals in Kerguelen Island that the unwonted sight of people walking never failed to attract the notice of the King Penguins. Standing at their ease in their sheltered hollows they uttered as it were derisive cries from time to time while the strangers laboured through the *Azorella*. Seldom did they take the trouble to stir when anyone approached them, but remaining in a group, some standing still, others lying down, they quietly awaited the progress of events. Their unconciousness of danger was singularly shown by the following incident. One day while grappling for *Algæ* in Swain's Bay I came with one of the men upon six Kings in a group. Seeing that some of them had finished moulting and were well coloured, we walked up to them, seized the two finest by their necks, and sat down upon their backs. The others stayed beside us unconcerned at the fate of their companions, though they were beating the ground beneath us with their wings and gasping for breath within a yard or so of them. " What shall be will be :" so they made themselves comfortable, and they were not molested. Meanwhile my bird was becoming moribund; and happening to look at its eyes I noticed that the colour of the iris was a very dull hazel, and that the pupil in contracting assumed a quadrangular form. The eyes of the survivors of the party presented the same peculiarity, which appears to be a characteristic of the species.

There are four examples of this Penguin from the Crozettes in the South African Museum, 2 young and 2 adults.

The louse of the King Penguin is *Goniodes brevipes*, n. sp., of which I found only one example.—*A. E. E.*]

Pygoscelis tæniata.

Le Manchot Papou, *Sonnerat, Voy. N. Guin.* p. 181, *pl.* cxv.

Aptenodytes papua, *Forster, N. Comm. Götting.* iii., *p.* 140, *pl.* 3; *Gm. S. N.* i., *p.* 556; *Vieill. Gal. Ois.* ii., *p.* 246, *pl.* 299 (*var.*).

Papuan Penguin, *Lath. Gen. Syn.* iii., *pt.* 2, *p.* 565.

Apterodyta papuæ, *Scop. Del. Faun. et Flor. Insubr.* ii., *p.* 91, *No.* 71.

Chrysocoma papua, *Steph. Gen. Zool.* xiii., *p.* 59.

Pygoscelis papua, *Gray, List Anseres, &c., B. M.*, *p.* 153; *id. Voy. Ereb. & Terror, Birds, pl.* 25; *Reichenb. Handb. Pygopodes, pl.* ii., *fig.* 738; *Hyatt, Cat. Orn. Coll. Boston Soc. N. H.* i., *p.* 13; *Cab. & Reichen., J. f. O.* 1876, *p.* 330.

Eudyptes papua, *Gray, Gen. B.* iii., *p.* 641; *Cass. U. S. Expl. Exp. p.* 264; *Gould, P. Z. S.* 1859, *p.* 98; *Abbott, Ibis,* 1860, *p.* 336.

Aptenodytes tæniata, *Peale, U. S. Expl. Exp. p.* 264.

Pygoscelis wagleri, *Sclater, P. Z. S.* 1860, *p.* 390.

Spheniscus papua, *Schlegel, Mus. P. B. Urinatores, p.* 5.

Pygoscelis tæniata, *Coues, Pr. Acad. N. Sci. Philad.* 1872, *p.* 195; *Scl. & Salv. Nomencl. Av. Neotr., p.* 151; *Sharpe, Voy. Ereb. & Terror, Birds, App. p.* 38; *Coues & Kidder, Bull. U. S. Nat. Mus.* ii., *p.* 41; *iid. op. cit.* iii., *p.* 18.

Ad. suprà nigricans vix cinereo lavatus; alis magis cinereis, margine alari conspicuâ et remigum apicibus fasciam terminalem latam formantibus albis; supracaudalibus rigidis nigricantibus cinereo lavatis; rectricibus nigris, marginaliter brunnescentibus; fasciâ latâ verticali albâ ab utroque oculo per verticem ductâ; facie laterali et gutture cinerascentibus, gutturis plumis albido variis; corpore reliquo subtùs sericeo-albo; alâ inferiore albâ, remigibus extimis apicaliter cinereis plagam conspicuam exhibentibus; pectore subalari et plagâ alterâ ad ortum alæ positâ cinereis; rostro lætè aurantiaco, culmine nigro; pedibus aurantiacis; iride læte brunneâ. Long. tot. 31, culm. 2·4, alæ 8·5, caudæ 5·5.

a. ad. Kerguelen Island. Antarctic Expedition.

b. ad. Kerguelen Island. Antarctic Expedition.

c. juv. Falkland Islands. Sir W. Burnett and Admiral Fitzroy.

[The Johnnie (as the whalers call this bird) is common in Royal Sound. It builds in communities, some of only a dozen, others from 70 to 150 families. A more populous colony upon the mainland was visited by six officers from the ships, who estimated the number of nests in it to amount to 2,000 or more. These larger communities are approached from the sea by regular paths, conspicuous at a distance, like well-worn sheep tracks, which lead straight up the hill from the water. Their formation is due to the Penguins being very particular about where they land and enter the sea. A small party of the birds occupied a position upon the neck of a low promontory within an hour's work of Observatory Bay. Their nests were nearest to the farther side of the isthmus; but when they were approached the male birds used to run to the water, not by the shortest route where it was deep close to the rocks, but by the longest to a place where the shore was shelving. It was amusing to see them start off in a troop as fast as their legs could carry them, holding out their wings and tumbling headlong over stones in their way, because as they ran they would keep looking back instead of before them, and to hear their outcries. Panic and consternation seemed to possess them all; but the females (possibly because they could not keep up with their mates) seldom went far from their nests; and, if the intruder stood still, soon returned and settled down again upon their eggs. Not many weeks had passed before a change was effected in their conduct. The young were hatched, and now the mothers anxiously endeavoured to persuade them to follow the example of their fathers and run away to sea. But the nestlings preferred to stay in their nests; they did not mind if the stranger did stroke them; although their anxious mothers ran at him with open mouths whenever he dared to do so. Only a few of the older chicks could be prevailed upon to stir; and they after waddling a few yards became satisfied with their performance and turned to go home again. The mothers, who had straggled to a greater distance, began to return too. It was now that the more tardy youngsters began to experience the ills of life. Every Penguin that had reached its place before them aimed blows at them as they passed by towards their own abodes. One of the little birds certainly did seem to deserve correction. It saw its neighbour's nest empty and sat down in it. The old female Johnnie, the rightful occupier, presently returned in company with her own chick, to whom, having put her head well into his mouth, she began to administer refreshment after his run. Seeing them so pleasantly engaged, the small vagrant, thoughtlessly presuming upon her generosity, went nearer and presented himself to be fed also, as if he had a right to her attention and care. She looked at him while he stood gaping before her with drooping wings, unable for the moment to credit what she saw. But suddenly the truth flashed upon her, and provoked by his consummate audacity she gave vent to her indignation, pecked his tongue as hard as she could, chased him out of the nest, darting blows at his back, and croaked ominously after him as he fled precipitately beyond the range of her beak, leaving trophies of down upon the scene of his unfortunate adventure. The whole of this community of Penguins was subsequently boiled down into "hare soup" for the officers of H.M.S. "Volage;" and very nice they found it.

The nests were composed of dried leaf-stalks and seed-stems of *Pringlea*, together with such other suitable material as happened to be at hand. There were two eggs in every nest, and one of them was invariably larger than the other. Most likely the birds hatched from the larger eggs are of the opposite sex to those which are produced from the smaller. Whether the big or the little egg is the first to be laid was not ascertained.

As is the case with many other kinds of birds, Johnnies are very regular in their habits. Every afternoon at nearly the same time they repair to the shore when they have done fishing, landing in small parties at their accustomed places at the heads of shallow inlets. On issuing from the water they dispose themselves to rest, seldom proceeding beyond the verge of the shore. Those which are inclined to sleep put their heads behind their flippers; the others stand amongst them with the neck shortened so as to bring the head down close to the body with the beak slanting upwards and forwards, somewhat in the manner of a very young thrush during repose. Their eyes present a rather tearful appearance, and resemble bits of dull black glass set in their heads,—perhaps the nictitating membrane may be kept drawn over them. At frequent intervals a kind of watery fluid is ejected from their mouth by a shake of the head.

I was led to suspect that these Penguins are liable to be attacked by seals,* for in places not much frequented by man, if they once effect a landing they do not readily return to the water on being alarmed, but run away from the sea up hill as fast as they can go. After they have gone some distance they turn round and look back, while they take breath; but as soon as they are rested sufficiently they willingly resume the ascent. It is not until they have been driven so far as to become thoroughly tired that they refuse to proceed further; but when this stage has been reached it is useless to urge them to advance without a pause. As they face about, the sight of the boot ready to push them over is greeted with deprecating sighs, and should these be disregarded, and they be sent over upon their backs, as soon as they regain their feet they rush at their driver, launch their bill at his knees, beat their wings furiously against his calves and shins, and make a dash on all fours down the hill at full speed to regain the sea.

When they became accustomed to being chased by men, the Penguins acquired the habit of betaking themselves to the water at the first alarm.

A small party of these birds used persistently to land in Observatory Bay every evening at the very time when the men erecting our huts were returning to the ship after their work. Such of the Johnnies as managed to escape being caught one day were sure to reappear the following evening just at the critical time, dragging themselves out of the water to afford sport to the men. By the time that the huts were completed the survivors were reduced in number to a couple of birds; and there can be little doubt that these would have followed their late companions into the soup-kettle had the putting up of the Observatory occupied one more day.

* In the Arctic Regions Loons are occasionally caught on the water by Walrus.

The cry of the Johnnie distantly resembles the short bark of the fox.

In the South African Museum there is an adult specimen from the Crozettes.

Its louse is unknown. A full-grown bird sent me by Captain Fairfax was infected with a tick.—*A. E. E.*]

Eudyptes chrysolophus. (Plate VIII., fig. 2.)

Catarractes chrysolophus, *Brandt, Bull. Acad. Sci. St. Petersb.* ii., *p.* 314 (*nec auct. recent.*).

Eudyptes chrysocome (nec Forster), *Abbott, Ibis,* 1860, *p.* 337; *Scl. P. Z. S.* 1860, *p.* 390; *Cab. & Reichen. J. f. O.* 1876, *p.* 330.

Eudyptes diadematus, *Gould, P. Z. S.* 1860, *p.* 419; *Schl. Mus. P. B. Urinatores, p.* 8; *Coues, Pr. Philad. Acad.* 1872, *p.* 206; *id. & Kidder, Bull. U. S. Nat. Mus.* ii., *p.* 47; *iid. op. cit.* iii. *p.* 20.

Eudyptes catarractes, *Gray, Handl. B.* iii., *p.* 98.

Ad. suprà nigricans cinereo lavatus, alis cinereo nigricantibus, margine alari summâ vix albicante, margine remigiali medialiter albâ; caudâ rigidâ, dorso concolori; facie laterali gulâque dorso concoloribus; pilei plumis nitidis nigris elongatis cristam formantibus, frontis plumis basaliter aurantiacis; fasciâ superciliari cristali a loris suprà oculum per latera capitis ductâ; corpore reliquo subtus purè albo, pectoris lateribus dorso concolori; alâ subtùs albâ, margine alari nigricante, plagâ nigricante etiam propè ortum alæ et ad apicem remigialem positâ.

Long. tot. 24, culm. 2·35, alæ 7·15, caudæ 4·8.

Juv. similis adulto, sed minor et fasciâ superciliari sulphureâ nec aurantiacâ distinguendus.

I never commenced the study of a bird under greater disadvantages than in the present instance. Dr. Coues and Drs. Cabanis and Reichenow both record a species of Penguin from Kerguelen Island under the names of *E. diadematus* and *E. chrysocome* respectively, and I have no specimen before me wherewith to test these identifications. This is the more unlucky as I do not agree entirely with the identifications of recent authors, and a good series of the Kerguelen Penguins would have been of great service. The latest writers on these birds have been Professor Schlegel, Prof. Hyatt, and Dr. Coues, and they all agree in recognising four species, where at the most I can only find three distinct ones, as follows:—

   a. superciliary streak golden yellow, commencing above the eyes, not from the base of the bill; forehead golden yellow at the base of the feathers, black at the tips . . . . *chrysolophus.*

   b. superciliary streak sulphur yellow commencing at the base of the bill.

a. forehead crested, no yellow bases to the feathers, the sulphur-coloured eyebrow produced backwards, longer than the black plumes of the head . . . . *chrysocome.**

b. forehead crested, no yellow bases to the feathers; the sulphur-coloured eyebrow very long and drooping and coterminous with an inner black crest . . . . *saltator.*

Dr. Coues in his usual painstaking manner has worked out these Yellow-crested Penguins from the material at his disposal in America, and after examining them carefully and describing the differences in plumage in detail he observes :—" Although I " am able to distinguish the three currently accredited species, in the few specimens " examined, yet the distinctions are not of a very satisfactory nature, and I strongly " suspect that when specimens enough shall have been compared, the supposed " specific characters will melt insensibly into each other, so that at most only varietal " distinction can be reasonably asserted. Indeed I am not sure that differences of age " or season or special conditions of plumage may not be the sole basis of the supposed " species." These remarks apply to *E. catarractes, E. chrysocome,* and *E. chrysolopha* of his paper. As regards the first of these, " at once distinguished by the shortness " of its tail . . . . also known by its inferior size, &c.," I am convinced that these are only signs of immaturity, and Von Pelzeln's plate in the ' Novara ' Voyage opens our eyes to the way in which these birds progress from the nestling to the adult. Size alone appears to me to be of no value as a character, and it is curious to see how some of these Penguins (*E. chrysolophus,* i.e. *E. diadematus,* for instance) differ in their bulk, though apparently in full, richly-crested, adult plumage. With

---

* The synonymy of the true *E. chrysocome* appears to be as follows :—

Eudyptes chrysocome. (Plate VIII., fig. 3.)

Aptenodytes chrysocome, *Forster, N. Comm. Götting.* iii., *p.* 135, *pl.* 1; *id. Descr. Anim. p.* 99.
Pinguinaria cirrhata, *Shaw in Miller Cimel. Phys. pl.* xlix.
Pinguinaria cristata, *Shaw, Nat. Misc. pl.* 437.
Eudyptes chrysocome, *Gould, B. Austr.* vii., *pl.* 83.
Eudyptes pachyrhynchus, *Gray Voy. Erebus and Terror, p.* 17; *id. Gen. B.* iii., *p.* 641; *Finsch, J. f. O.* 1872, *p.* 261, 1874, *p.* 217; *Gray, Handl. B.* iii., *p.* 98.
Chrysocoma catarractes, *Bp. C. R.* xlii., *p.* 775; *Gould, Handb. B. Austr.* ii., *p.* 517.
Chrysocoma pachyrhynchus, *Bp. C. R.* xlii., *p.* 775.
Spheniscus chrysocome, *Schl. Mus. P. B. Urinatores, p.* 6.
Eudyptes catarractes, *Giglioli, Faun. Vertebr. Oceano, p.* 28.
Eudyptes chrysolopha, *Gray, Handl. B.* iii. *p.* 98.
Eudyptes chrysocomus, *Buller, B. N. Zeal., p.* 345, *pl.* 33, *fig.* 1.

Hab—South Australia, Tasmania, and New Zealand. (N.B. If the species also occur in the Falklands, then the following synonyms must be added :—Eudyptes nigrivestis, *Gould, P. Z. S.* 1860, *p.* 418; E. nigriventris (lapsu), *Gray, Handl. B.* iii., *p.* 98.

Spec. in Mus. Brit.

a. ad. New Zealand. Type of *E. pachyrhynchus.*
b. ad. „ Dr. Lyall.

this exception the conclusions of the present paper are much the same as those of Prof. Schlegel, Dr. Coues, and Prof. Hyatt, as regards the number of species recognisable. The last-named gentleman, in addition to certain characters, such as the shape of the black on the throat, &c., brings forward the number of the scutellæ on the first joint of the toes, a distinction which I have not found to hold good, and the writer himself does not seem to have much confidence in it, as in one species he says, "there *may* be three or four scutellæ on each toe."

A few words as to the nomenclature of these Penguins. I must protest against the introduction of *E. catarractes* into the genus, a name founded on an old plate of Edwards and quite irrecognisable, as will be seen by any one consulting the following references which apply to it:—

The Penguin, *Edwards, N. H. Birds*, i., *p.* 49, *pl.* 49 (no locality given, no eyebrow mentioned or depicted; clearly an immature bird and a bad figure).

Le Gorfou, *Briss. Orn.* vi., *p.* 102 (bird not seen by him, and the description evidently derived from Edwards).

Phaeton demersus, *Linn. S. N.* i., *p.* 219 (*ex Briss. nec Pelecanus demersus, L.*).

Red-footed Pinguin, *Lath. Gen. Syn.* iii., *pt.* 2, *p.* 570.

Aptenodyta catarractes, *Gm. S. N.* i., *p.* 558.

Chrysocoma catarractes, *Steph. Gen. Zool.* xiii., *p.* 61.

Then as regards the name *chrysolophus* of Brandt, this appears to me to have been misapplied by all the recent writers on Penguins, and I cannot resist the impression that the original description has not been consulted. I therefore transcribe it:—

"Catarractes chrysolophus, Brandt. Cristâ in mediâ fronte incipiens, maxima ex parte e pennis vitellinis compositâ; color niger in gulâ triangularis; tectricum caudæ superiorum mediæ albido-flavicantes.

"Catarractes chrysocome, Forster. Crista intùs nigra extrinsecus sulphurea angustè in rostri basi incipiens posticè dependens; color nigra in gula trinicatus; tectrices caudæ superiores omnes dorso concolores."

As no locality is given for these two birds, we have nothing but the bare descriptions to go upon, and there seems to my mind no doubt that *C. chrysolophus* of Brandt is *C. diadematus* of Gould, the latter name becoming of course a synonym. At the same time Brandt's *C. chrysocome* is not the true *chrysocome* of Forster, but is the bird called by authors *Eudyptes chrysolophus* (Brandt), but which turns out not to be the true *chrysolophus*. Apparently we have here as complete a tangle as can be found in the annals of ornithology, and that is saying much!

Although no specimens were brought by Mr. Eaton, Dr. Coues identifies the species without hesitation as *E. diadematus*, Gould.

Spec. in Mus. Brit.

a. b. ad. [Antarctic Expedition]. The Admiralty.

c. juv. Falkland Islands. Antarctic Expedition.

d. ♂ ad. Falkland Islands. Oct. 1842. Antarctic Expedition.

e. ad. Berkeley Sound, Falkland Islands. Antarctic Expedition.

[This species was not met with by the English transit party on Kerguelen Island. The sealers brought some of its crests to our ships, and spoke of it as the "Macaroni." From what the officers of the "Volage" told me, I was led to understand that the sealers said that the bird was not found anywhere near the southern end of Kerguelen Island. The crests obtained were brought from Herd's Island. Sir Wyville Thomson in one of his letters to "Good Words" (op. cit. 1874, Nov., p. 314) states that the "Macaroni" occurs at Christmas Harbour in small numbers in company with the Rock-hopper; and that on the outer cliffs beyond the mouth of the harbour there are some strong Penguin rookeries consisting almost exclusively of the "Macaroni."—*A. E. E.*]

Eudyptes saltator. (Plate VIII., fig. 1.)

Le Manchot hupé de Sibérie, *Buff. Pl. Enl.* 984.

Chrysocoma saltator, *Steph. Gen. Zool.* xiii. p. 58, *pl.* 8.

Eudyptes chrysolophus (*nec Brandt*), *Gray, Gen. B.* iii., p. 641; *Abbott, Ibis,* 1860, p. 338; *Scl. P. Z. S.* 1860, p. 390; *Schl. Mus. P. B. Urinatores,* p. 7; *Gray, Handl. B.* iii., p. 98; *Scl. & Salv. Nomencl. Av. Neotr.,* p. 151; *Coues, Bull. U. S. Nat. Mus.* ii., p. 15; *Kidder, op. cit.* III., p. 10; *Cab. & Reichen. J. f. O.* 1870, p. 330.

Catarractes chrysolopha, *Reichenb, Handb. Pygopodes, pl.* 1*a. figs.* 12–14 (*pt.*).

Chrysocoma chrysolopha, *Bp. C. R.* xlii., p. 775.

Eudyptes chrysocome, *Pelz. Reis. Novara, Vög.* p. 140, *pl.* 5.

Ad. suprà sordidè cinereus, pilei plumis rigidis, elongatis, cristam frontalem exhibentibus, verticis lateralis plumis quoque elongatis, cum fasciâ latâ superciliari cristam duplicem formantibus; facie laterali cum colli lateribus gulâque totâ brunnescenti-cinereis: corpore reliquo purè albo; pectore laterali, hypochondriis, imis, et tibiis posticè cinereis; alâ suprâ saturatè cinereâ, margine alari summâ vix albidâ, secundariis etiam albo terminatis; caudâ rigidâ dorso concolori; alâ subtùs albâ, ad basin et juxtà marginem alarum summarum cinereâ; remigibus primariis versùs apicem cinereo-nigricantibus; rostro aurantiaco; pedibus albicantibus; iride coccineâ. Long. tot. 23, culmen 2·0, alæ 7·0, caudæ 3·5.

Juv. similis adulto, sed cristâ absente et gutture albido brunneo mixto distinguendus.

I have fully discussed my reasons for changing the name of *chrysolophus* for this species under the heading of the previous bird, and it seems curious that Stephens' name of *saltator* has not been applied to it before, as the plate, though apparently derived from Buffon's illustration in the "Planches Coloriées," leaves no doubt as to the species represented.

The following specimens constitute the series in the British Museum:—

a. ad. [Cape of Good Hope]. Sir A. Smith.
b. ♂ ad. Tristan d'Acunha. J. Macgillivray, Esq.
c. ♂ juv. „ „
d. min. Bounty Island. Purchased.
e.f. min. Falkland Islands. The Admiralty.
g. ad. Kerguelen Island. Capt. Fairfax, R.N.

[On some parts of the coast where the interstices of fallen rocks piled up at the base of cliffs afford them suitable shelter, the *Eudyptes* abound. Their colonies in Royal Sound were smaller than those in Swain's Bay. The most populous of their communities visited by us were situated on the shores of the promontory to the eastward of Vulcan Cove. There were there some thousands of the birds, and they were very noisy. Their cry is a kind of guttural cackle somewhat like the syllables " Gurougha, gurougha, gurougha," pronounced rapidly. The designation "Rock-hoppers" applied to them by the whalers is extremely appropriate; for although they occasionally walk a few paces at a time over a plane surface of rock, with the confined gait of competitors walking in a sack race, their ordinary mode of progression is a series of bounds executed with much apparent ease and with an elasticity of motion such as is exhibited by Kangaroos. Standing amongst them silent (but most certainly not in silence) it was interesting to watch their proceedings. Those birds which had eggs far advanced in incubation remained in their nests, scarcely noticing the hand which stroked their backs unless they saw it move. Others stooping low peeped out of the neighbouring crevices beneath the boulders, or jumping a few yards away stood upon the rocks to gaze at their visitor, leaving their fresh eggs or empty nests unguarded. Birds more confident than these then began to come near, Sheathbills for the eggs, carefully avoiding the Penguins who croaked at them as they sauntered past them; and other Penguins who were ready to risk an approach for the sake of choice materials for nests so unexpectedly left at their disposal. Two or three hard pecks with its bill, and the *Chionis* is happy over its new laid Penguin's egg; two or three journeys to and fro, and the thieving Rock-hopper has carried off to his own nest the choicest portions of his timid neighbour's. The despoiled birds soon see what is going on at their homes, and come bounding back to the rescue. Satisfied with the dispersion of their depredators, this most hopeful of mothers seats herself upon her broken eggs demurely, while her neighbour with an air of resignation betakes himself once more to the task of collecting dried *Pringlea* seed-stalks and other rubbish for the repair of his ruined nest. Sometimes the eggs are laid upon the bare rock without anything else to rest upon. There is the same sexual difference in size between the two eggs in a nest of this *Eudyptes* as there is in those of the *Pygosceles*. The crest on each side of the head is separated into two divergent drooping plumes one above the

x

other, excepting in swimming, when the bird rising to the surface has the crest closely flattened down upon the sides of the head, where it forms a yellow streak. The eyes are reddish orange, with circular black pupils.

In the S. African Museum this species is represented from the Crozettes by two specimens (and perhaps a third) under the names *Eudyptes nigrivestis* and *E. chrysocoma*.*—A. E. E.]

* In the same collection is another species from the Crozettes represented by one specimen marked *Eudyptes chrysolophus*, which was not found by us in Kerguelen. A second specimen under the same name *E.* "*chrysolophus*, var." from the same islands appeared to be an albino of the same species.

Explanation of the Plates.
(Plate VI.)

Querquedula eatoni, Sharpe. Figure of an adult specimen procured by the Antarctic Expedition.

(Plate VII.)

Fig. 1. *Stercorarius antarcticus*. Head of a specimen collected by Mr. Eaton in Kerguelen Island.

Fig. 2. *Stercorarius catarractes*. Head of a British specimen.

Fig. 3. *Prion vittatus*. Bill of an adult female from South Australia. Presented by Sir George Grey.

Fig. 4. *P. vittatus*. Bill of a young male from the same collection as the foregoing.

Fig. 5. *P. vittatus*. Bill of adult male from the same collection.

Fig. 6. *P. vittatus*. Bill of a very old male from the Chatham Islands.

Fig. 7. Bill of the type of *P. banksii*, from the Cape of Good Hope; probably of the same age and sex as fig. 4.

Fig. 8. *Prion desolatus*. Bill of the type specimen (probably a young male) of *P. brevirostris*, Gould, from Madeira.

Fig. 9. *P. desolatus*. Bill of a male specimen from Christmas Harbour, Kerguelen Island, with a very distinct yellow "nail" to the upper mandible; probably an old male bird.

Fig. 10. *P. desolatus*. Bill of a female bird brought by Mr. Eaton from Royal Sound, Kerguelen Island.

(Plate VIII.)

Fig. 1. *Eudyptes saltator*. Head of an adult bird from Tristan d'Acunha.

Fig 2. *Eudyptes chrysolophus*. Head of an adult bird, procured by the Antarctic Expedition.

Fig. 3. *Eudyptes chrysocome*. Head of an adult bird from New Zealand. (Type of *E. pachyrhynchus*, Gray.)

Eggs.—*By Howard Saunders, F.L.S., F.Z.S.*

Chionis minor, *Hartl.*
(Lesser Sheathbill.)

The general character of 19 eggs is a dirty white ground, splashed and blotched with brown. At the first glance there is a startling superficial resemblance in coloration, and sometimes in shape, to a very common dark form of the egg of the Razorbill (*Alca torda*); other specimens are in shape and markings like boldly blotched examples of eggs of the *Œdicnemus* group of Plovers. On shining them to the light, the eggs show a green membranous lining. Unfortunately the egg of the other species *Chionis alba* (*Gm.*), of the Falkland Islands, is not known, for Capt. Abbott, who wrote an account of the birds of the Falklands (Ibis 18), did not obtain it, and the statements made to him by the sealers as to its egg being white, must be received with doubt. The first eggs were obtained on 23rd Dec., and in stating that none were found by Mr. Eaton until 10th January, the American Naturalist, Dr. Kidder, must have been labouring under an error. The complement of eggs seems to be one or two, and rarely three.

The average dimensions of the egg are 2·2 in. × 1·5 in.

Querquedula eatoni, *Sharpe*.

Thirty eggs of this species present remarkable variation, the general hue being of a pale green or greenish buff. Laying appears to commence early in December. The average dimensions are 2 in. × 1·4 in.

Larus dominicanus, *Licht*.

Thirty eggs of this bird present the usual characters of the eggs of the larger species of Gulls, being of various shades of olive-green, ranging to stone-colour, and occasionally brown, spotted and blotched with darker shades of brown and streaks of black. Eggs of this species in my collection from the Crozettes, and from the Falkland Islands present precisely similar characters. The first egg was obtained on 14th October; the complement is three, as is usual with Gulls. Average dimensions, 2·85 in. × 1·9 in.

Stercorarius antarcticus, *Lesson*.

In five eggs of this species, three taken about the middle of November are of a pale olive-green indistinctly blotched with brown, and two others, obtained on 8th December, apparently from the same nest, as that is the usual complement, are of a brownish buff ground with rather bolder markings. Although slightly larger than the eggs of *Stercorarius catarractes*, the representative species of the Northern Hemisphere, there is no noticeable difference in character. Length 3 in., breadth 2·1 in.

Sterna virgata, *Cab*.

Eight eggs of this species are of an olive colour blotched with black, the marks tending to form a zone. They do not differ much from eggs of typical *Sterna*, but there is a tendency to greenish, and an absence of the white in the ground colour, which approximates them to eggs of *S. antarctica* of New Zealand, the near ally of this species, the eggs of which, so far as I may venture to judge from only two specimens, average somewhat less.

It appears to lay but one egg, on the terraces on the sea-side hills, commencing towards the end of November. The dimensions are 1·75 in. × 1·2 in.

Pelecanoides urinatrix, *Gm*.

Ten eggs are all pure white, except where peat-stained, nearly equal at each end, or but very slightly pointed. Dimensions, 1·5 in. × 1·1 in.

Majaqueus æquinoctialis, *Linn*.

Twelve eggs of a pure white colour (except where the granulations of the shell are filled up with the yellowish dirt from the burrows where they are deposited) have a more repulsively musky smell than any other eggs of the group. Length 3·2 in., breadth 2·1 in.

Æstrelata brevirostris, *Less*.

Two eggs of this species, badly cracked owing to their fragile texture, are of a pure dull white, nearly ovoidal in shape, and measure 2·15 in. × 1·7 in.

Æstrelata lessoni, *Garnst*.

Eleven eggs of this species, of the characteristic dead white colour, have somewhat less of the musky smell than those of most of the burrowing Petrels. Length 2·75 in., breadth 1·85 in. Slightly pointed at smaller end.

Procellaria oceanica, *Kuhl*.

Nine eggs average 1·3 in. × ·9 in., and are of a dull white colour, with minute purple-red spots which generally form a zone, usually, although not invariably, at the larger end; they are however at times distributed sparingly over the whole surface. There is not much difference between the shape of either end.

Prion desolatus.

Six eggs of the same character and texture as the preceding species, and distinguished by their greater length in proportion to breadth, the average of the former being 1·8 in; nearly as much as in *Halobæna cærulea*, whilst the breadth is only 1·3 in.

Halobæna cœrulea, *Gm*.

There are ten specimens in the collection, of the rough granulated texture, blunt ends torn, and dead-white colour, characteristic of the eggs of this family. There is no sign of a zone of rust-coloured markings, but there is the usual musky smell about the shell. They vary in dimensions a good deal, and as from the delicate nature of the shell, several specimens are total wrecks, it is not easy to take an exact average, but I am inclined to think that one of the above eggs belongs to the former species. 1·9 in. in length, and 1·5 in. in breadth, is tolerably close. The single egg is laid as early as 23rd October, onwards.

Diomedea fuliginosa, *Gm*.

There are two examples of this egg; white, spotted towards the larger end, so as to form a zone, with minute reddish marks. The egg is long in proportion to its breadth, being 4 in. × 2·65 in.

Phalacrocorax verrucosus, *Cab*.

The six eggs are of the usual pale greenish blue, with the chalky incrustations characteristic of those of the Cormorant family. Average length 2·3 in., breadth 1·45 in.

Aptenodytes longirostris, *Scop*.

(Eggs noticed in MS., but none appear to be in collection.)

Pygoscelis tæniata, *Peale*.

A solitary egg of this species is of a pale blue thickly coated with calcareous matter. It measures 2·5 in. × 2 in.

Eudyptes saltator.

Two eggs ascribed to this species differ considerably in size, the larger measuring 2·7 in. × 2 in.; the other 2·4 in. × 1·65 in. The colour is very pale blue with a white calcareous coating irregularly disposed over the surface. In shape they are somewhat pointed at one end.

Fishes.—*By Dr. A. Günther, F.R.S.*

The species collected are four in number, three of which were previously known to occur on the shore of the Island, viz.:—*Harpagifer bispinis, Chœnichthys rhinoceratus*, and *Notothenia coriiceps*. The fourth is a Ray, apparently undescribed, which may be characterised thus:—

Raja eatonii.

Allied to *R. smithii*. Snout of moderate length, the anterior margins meeting at a right angle; the width of the inter-orbital space is two-sevenths of the distance of the eye from the end of the snout. The anterior profile, from the snout to the angle of the pectoral fin, is slightly emarginate, the outer pectoral angle being rounded. The greater part of the upper surface of the body is smooth, minute spines are distributed between the eyes, and in a narrow stripe along the margins of the body; a broad band of minute spines along the median line of the back and the upper surface of the tail; a single larger recurved spine in the middle of the back; a series of nine or ten rather small spines placed at a considerable distance from each other along the median line of the tail, no spines on the side of the tail. Lower parts smooth. Upper lip fringed on the side; teeth pointed, conical, in about 30 series in the upper jaw. *Male* with a patch of claw-like spines on each pectoral fin. Brownish black above, with indistinct round whitish spots; whitish below, with some irregular brownish-black spots; lower part of the tail brownish-black.

A single adult male was obtained in Royal Sound. It is 26½ inches long, the tail measuring 14 inches; its greatest width is 18 inches.

A specimen of the female sex was obtained during the visit of the "Challenger" to the island. It differs very little from the male, but, of course, the claw-like spines on the pectoral are absent. The lower part of the body is entirely white.

A second species of Ray (*Raja murrayi*, Gthr.) was found in Kerguelen's Land also by the naturalists of the "Challenger." This will be described in connexion with the collection made during that expedition.

MOLLUSCA.—*By Edgar A. Smith, F.Z.S., Senior Assistant in the Zoological Department, British Museum.*

(Plate IX.)

Of the thirty-three species treated of in the following paper, and representing twenty-five genera, twenty-five were obtained by the Rev. A. E. Eaton, the remaining seven having been collected during the Antarctic Expedition under the command of Sir James Ross. Of this number, 18 are new to science, and nearly all are mentioned for the first time as inhabiting this locality.

Some of the species are of very great interest, especially the new genus *Neobuccinum*, the *Struthiolaria*, and the new genus *Eatoniella* among the *Gastropods*, and of the bivalves the *Saxicava*, *Lissarca*, and the magnificent *Solenella*, by far the largest known species of that genus.

The Malacological fauna resembles generally that of the Falkland Islands and South Patagonia. More than half of the genera and seven or eight of the species found at Kerguelen Island are known to occur at those localities, and further research will probably discover a still greater number of genera and species to be common to these two, longitudinally, so widely separated localities. With respect to their latitudes the difference is unimportant, since they both range between 49° and 54° S. lat. As the Cape of Good Hope, Tasmania, and South West Australia, are the nearest points of mainland, it might be expected that some resemblance to the fauna of those countries might be observable. However, it is not so, as far as our present knowledge extends. Many of the shells from Kerguelen Island have the generally unattractive appearance as regards coloration which so frequently obtains in species found in cold climates. Indeed, some of them seem to be southern representatives of boreal types. The *Neobuccinum, Trophon, Saxicava, Kellia, Yoldia, Radula,* and *Doris,* are remarkable instances of similarity to northern forms.

The following table shows the great affinity which exists in the fauna of South Patagonia and Kerguelen Island. Only those species are quoted from Patagonia which are identical with or nearly allied to Kerguelen forms, and a blank signifies that the genus has not yet been recorded from there.

Genera.	Patagonian Species.	Kerguelen Species.
Neobuccinum		N. eatoni, *Sm.*
Trophon	T. philippianus, *Dunker*	T. albolabratus, *Sm.*
Struthiolaria		S. mirabilis, *Sm.*
Purpura		P. striata, *Martyn*, (American Exp.)
Admete (—?)		A. (—?) limnææformis, *Sm.*
Littorina		L. setosa, *Sm.*
Hydrobia		H. pumilio, *Sm.*
,,		H. caliginosa, *Gould.*
Eatoniella		E. kerguelenensis, *Sm.*
,,		E. caliginosa, *Sm.*
,,		E. subrufescens, *Sm.*
Skenea		S. subcanaliculata, *Sm.*
Rissoa		R. kergueleni, *Sm.*
Scissurella		S. supraplicata, *Sm.*
Trochus (Photinula)	T. (P.) expansus, *Sowb.*	T. (P.) expansus, *Sowb.*
Patella	P. ænea, *Martyn*	P. kerguelenensis, *Sm.*
,,	P. fuegiensis, *Reeve*	P. fuegiensis, *Rve.*
,,	P. mytilina, *Gmelin*	P. mytilina, *Gmel.*
Siphonaria	S. magellanica, *Philippi*	S. redimiculum, *Rve.*
Hemiarthrum		H. setulosum, *Cpr.* (Amer. Exp.)
Doris		D. tuberculata, *Cuv.*
Helix		H. hookeri, *Reeve.*
Saxicava	S. antarctica, *Philippi*	S. bisulcata, *Sm.*
Kellia	K. miliaris, *Philippi*	K. consanguinea, *Sm.*
Lepton		L. parasiticum, *Dall.*
Arca (Lissarca)		S. rubro-fusca, *Sm.*
Yoldia	Y. woodwardi, *Hanley*	Y. subæquilateralis, *Sm.*
Solenella	S. magellanica, *Smith*	S. gigantea, *Sm.*
Mytilus	M. magellanicus, *Chemnitz*	M. magellanicus, *Chem.*
,,	M. edulis, *L.*	M. edulis, *L.*
Modiolarca	M. trapezina, *Lamarck*	M. trapezina, *Lamarck.*
,,	M. exilis, *H. and A. Adams*	M. exilis, *H. and A. Ad.*
,,	M. pusilla, *Gould*	M. minuta, *Dall.*
,,	M. pusilla, *Gould*	M. pusilla, *Gld.*
Radula	R. pygmæa, *Philippi*	R. pygmæa, *Phil.*
Waldheimia	W. dilatata, *Lamarck*	W. dilatata, *Lamarck.*

CEPHALOPODA.

[In the Bull. U. S. Nat. Mus. 1876, No. iii., p. 42, Dr. Kidder records finding a species of Octopus too much mutilated for identification.]

GASTROPODA.

Neobuccinum, *gen. nov.*

Testa bucciniformis; canalis latus, brevis; *operculum* ovatiusculum, *uni-spirale* (*nucleo vix terminali*), ad marginem prope nucleum leviter sinuatum, concentrice lineis incrementi curvatis striatum.

This genus is allied to *Buccinopsis*, but differs from it somewhat in the dentition of the animal, and with regard to the operculum.

In *Buccinopsis* the nucleus of the operculum is terminal, but in the present genus it is situated on the inner side about one-tenth of the entire length from the extremity, and just at this point the outline is interrupted by a slight sinus. It consists of one whorl, which gradually increases by concentric layers well defined by the lines of growth, and the inferior surface is somewhat thickened along the outer edge, that is, that opposite the nucleus.

Neobuccinum eatoni.

(Plate IX., fig. 1.)

Buccinopsis Eatoni, Smith, Annals and Mag. Nat. Hist. July 1, 1875, xvi., p. 68.

Testa elongato-ovata, turrita, tenuis, laevis, pallide livido-fuscescens, haud nitens; anfractus 6? (apice fracto), reliqui 4 perconvexi, lente crescentes, laeves, lineis incrementi flexuosis insculpti, sutura profunda, fere canaliculata sejuncti; apertura ovata, longitudinis totius circiter ⅘ aequans; columella laevis, polita, in medio leviter arcuata, basin versus obliqua; canalis latissimus, perbrevis, vix recurvus; labrum simplex, tenue.

Operculum ovatum, concentrice plicato-striatum, nucleo laterali, vix terminali.

Long. 56 mill., diam. 27; apertura long. 27 mill., diam. 14.

Animal (in spirit) uniformly buff colour; foot broad in front, and somewhat truncated, narrowed posteriorly; head of moderate size, furnished with two rather short tentacles not adjacent at their base; eyes situated on prominences on the outer side of the tentacles towards their bases; proboscis very long; siphonal expansion of the mantle thick, of medium length.

Lingual ribbon very long; teeth in three series, central (rachidian) tooth tricuspidate, prongs straight, nearly equal in size, the central one a little the longest; lateral teeth (uncini) tricuspidate also, prongs hooked, outer one the largest, the inner rather smaller, the median very much smaller still and close to the latter.

Hab.—Obtained in shrimp pots and by dredging in 3-7 fathoms at Royal Sound and Swain's Bay.

This remarkable species is chiefly characterized by its smooth convex whorls which are destitute of all sculpture and ornamentation, with the exception of the lines of growth. The suture is particularly deep, almost channelled. Around the short cauda of the body-whorl, from a little below the middle of the columella, runs a carination (as in certain species of *Bullia*), which joins the basal channel near the labrum.

Since publishing the description of the shell of this species, I have examined the animal, and find that its dentition, which bears a close resemblance to that of *Neptunea dilatata* (Troschel, Gebiss der Schnecken, ii., 1868), does not agree exactly with that of *Buccinopsis*. The tongue of the latter is described by Alder as having "a single plain and slightly curved tooth on each side, and a very thin non-"denticulated plate in the centre."

This difference in the odontophore, and the dissimilarity of the opercula, are, I think, sufficient to entitle the present species to generic rank.

Trophon albolabratus.

(Plate IX., fig. 2.)

Trophon albolabratus, Smith, Annals & Mag. Nat. Hist. 1875, xvi., p. 68.

Testa ovato-fusiformis, turrita, alba; anfractus 6, primi duo (nucleus) læves, cæteri convexi, liris spiralibus (in anfr. superioribus 4–5, in ultimo circiter 13) æqualibus subæquidistantibus cincti, et lamellis foliaceis numerosis subconfertis et prominentibus instructi; apertura superne ovalis, infra in canalem prolongata, intus saturate fusca, longitudinis testæ circiter $\frac{3}{5}$ æquans; labrum intus sub-late albo marginatum, leviter expansum; columella in medio parum arcuata, basi obliqua, callo inferne crassiusculo, superne tenui labroque juncto induta, cæruleo-alba, margine interno fusca; regio umbilici leviter rimata; canalis angustus, obliquus, paululum recurvus, modice elongatus.

Operculum flavo-corneum.

Long. 40 mill., diam. 18; apertura long. 24 mill., diam. 11.

Animal (in spirit) uniformly pale buff; foot rather truncate in front and somewhat acuminated posteriorly; head small; tentacles adjacent at their base, not very long (contracted?), much thicker below than above the eyes, which are situated on prominences towards their base.

Teeth of lingual ribbon in three series; laterals consisting of a single spine-like tooth longer than the others; rachidian teeth in three rows, small, spine-like, the two outer rather smaller than the central one.

Hab.—On rocks just below low-tide mark in Swain's Bay and Royal Sound; frequent.

The nearest ally of this species appears to be *T. philippianus* of Dunker, which is found in the Straits of Magellan, at Cape Horn, and the Falkland Islands. From this species it differs in having the whorls rounded above, and not flattened or excavated, the penultimate is larger and more elevated, the body-whorl is more inflated below the middle and not produced into such an elongated cauda, and the canal is shorter and the aperture rather larger, the longitudinal lamellæ are more prominent and not nearly so numerous.

Struthiolaria mirabilis.

(Plate IX., fig. 3.)

Struthiolaria mirabilis, Smith, Annals & Mag. Nat. Hist., July 1st, 1875, p. 67.

Struthiolaria costulata, Martens Bericht Gesellsch. Naturforsch. Freunde, Berlin (July 24th, 1875, teste Martens in litt.), p. 66.

Testa ovata, tenuis, imperforata, leviter turrita, alba, epidermide tenuissima fugaci olivaceo-alba amicta; anfractus $6\frac{1}{2}$, convexiusculi, superne anguste planulati, lente accrescentes, longitudinaliter oblique arcuatimque crebre plicati (plicis inferne ad suturam vix attingentibus); liris spiralibus prominentibus supra plicas undulatis (in anfr. superioribus 7-8, in ultimo circiter 22, illis infra medium simplicibus) succincti; apertura longitudinis totius circiter $\frac{4}{7}$ æquans; columella arcuata.

Operculum corneum unguiculatum, subtus costis duabus a nucleo unguiformi divergentibus munitum, superne in medio longitudinaliter unisulcatum, concentrice striatum.

Long. 42 mill., diam. 22.

Proboscis conical, rather compressed and annulated; *tentacles* two, short, tapering, situated at the base of the proboscis; *eyes* small, placed on very slight prominences very near the base of the tentacles; *foot* small, somewhat truncate in front and acuminate posteriorly; operculigerous lobe nearly terminal, with the operculum placed transversely, the nucleus to the left when viewing the lobe, foot downwards. Teeth — ? The radula I believe has been described by Shacke from specimens obtained by the German expedition.

Hab.—Dredged at a depth of 3-7 fathoms at Swain's Bay.

But a single specimen of this very remarkable shell was obtained by Mr. Eaton. This unfortunately has the labrum so much broken away, that it is impossible to describe the form of the aperture and the nature of the basal channel. However, the animal and operculum agree in all respects externally with *Struthiolaria*; and although the shell has more the general aspect of *Buccinum*, there can be no doubt of its true location. The species which compose this genus are strong thick shells; this, on the contrary, is particularly fragile, and clothed with a very thin deciduous epidermis.

The name given to this species by Von Martens was first announced at a meeting of the "Gesellschaft naturforschender Freunde zu Berlin" on the 15th of June 1875. Dr. Martens informs me in a letter dated December 22nd, 1876, that the report of this meeting, in which a brief description is given of the species (p. 66), was published on July 24th, 1875, which was unusually late, as generally the report ought to be issued before the next meeting, which was July the 20th. At any rate the name *mirabilis* would have priority as it was published on July 1st, and the manuscript was in the printer's hands almost a month before that date.

[Purpura striata, *Martyn*.

Dall, Bull. U.S. Nat. Mus. iii., p. 43.

Buccinum striatum, *Martyn*.

Two specimens probably of this species were obtained by Dr. Kidder.

Hab.—Kerguelen Island (Kidder); New Zealand (Martyn).]

Admete (— ?) limnææformis, sp. nov.

(Plate IX., fig. 4.)

Testa tenuissima, subdiaphana, lævis, parum nitens, vitreo-alba; anfractus 3, primi duo parvi, convexiuusculi, ultimus amplissimus, convexus; sutura simplex, aliquanto profunda; apertura ovata, superne paululum acuminata, longitudine testæ totius circiter $\frac{3}{5}$ æquans; columella leviter obliqua, vix arcuata, inferne oblique et subabrupte truncata, superne callo tenuissimo, super anfractum expanso, labro juncta; labrum simplex, tenue.

Long. $2\frac{1}{2}$ mill. Diam. $1\frac{1}{2}$.

Hab.—On Sea-weed and Polyzoa in Swain's Bay at a depth of 4–5 fathoms.

This species has the appearance of a minute *Limnæa* with the columella truncated a little below the middle. No sculpture is visible under a simple lens, but by the aid of the microscope very fine lines of increment are discernible. Although this is a very curious form for an *Admete*, I am not acquainted with any other genus which it more resembles. The character of the truncation of the columella is similar, and the absence of an operculum is also congeneric.

From the fewness of the whorls and the thinness of the shell it may be conjectured to be but the young of some larger species. This may be the case, but at present I am unable to identify it as the fry of any genus with which I am acquainted.

Littorina setosa.

(Plate IX., fig. 6.)

Littorina setosa, Smith, Annals & Mag. Nat. Hist. July 1875, xvi., p. 69.

Testa imperforata, ovato-turrita, tenuis, pallide rosea, circa medium anfractuum linea spirali rufa cincta, epidermide fugaci villosa vel setosa olivacea induta; anfractus 6, convexi, superne aliquanto tabulati, sutura profundiuscula discreti, ubique spiraliter et oblique minute punctato-striati; apertura subquadrato-circularis, longitudinis totius $\frac{1}{2}$ æquans; columella perparum arcuata, ad basin leviter patula; labrum simplex. Operculum paucispirale, ovatum, superne acuminatum, tenuissimum, flavocorneum.

Long. 14 mill., diam. $8\frac{1}{2}$; apertura long. 7 mill., diam. 5.

Animal (in spirit) small, pinkish yellow, the tentacles and top of the proboscis stained with purplish-black; tentacles short and thick; eyes at the tips of slender peduncles which arise from the outer bases of the tentacles; foot oblong, rounded in front and rather more so behind.

Hab.—Swain's Bay in 3–4 fathoms, at the extremity of the promontory farthest from Royal Sound and nearest to Mt. Ross, two or three miles within the entrance of the bay.

The epidermis of the shell of this species is very deciduous and is minutely hairy, the hairs being disposed in oblique series corresponding with the lines of growth.

Hydrobia caliginosa.

(Plate IX., fig. 8.)

Littorina caliginosa, Gould, Wilkes' Explor. Exped. 1852, p. 198, Atlas fig. 240.

Testa ovato-conica, angustissime vel vix rimata, tenuis, fusca, labrum versus pallidior; anfractus 5 convexi, læves, striis incrementi obliquis tenuiter insculpti, sutura distincta, leviter obliqua, sejuncti; anfr. ultimus amplus; apertura ovatiuscula, longitudinis totius circiter $\frac{5}{7}$ æquans; labrum tenue ad basin parum patulum; columella obliqua, minime (nisi prope basin) arcuata, incrassata, calloque tenui labro juncta; rima angustissima, interdum columella omnino obtecta. Long. $4\frac{1}{2}$ mill. Diam. $2\frac{1}{2}$.

Operculum corneum, tenuissimum, flavescens, paucispirale, nucleo paululum a margine inferiori remoto, subtus marginibus anfractuum externis carinatis.

Hab.—Kerguelen's Island. Terra del Fuego (*Gould*).

The brown colour of this species is produced probably by the dried remains of the animal within the shell, and could these be removed the whole of the structure would have the same horny appearance as the labrum presents. Nearly all specimens have the two or three uppermost whorls more or less eroded. The aperture oblique, somewhat ovate in form, being rather broader inferiorly than above, and faintly patulate on the basal margin. The columella is slightly thickened and reflexed, at times wholly concealing the extremely narrow umbilical fissure. The operculum is very thin, consists of about two and a half very rapidly increasing volutions, is finely striated by the incremental lines, and the nucleus is situated at about one-third the entire length from the lower end and on the underside, the outer margin of the whorls next the suture is raised or keeled.

Hydrobia pumilio, *sp. nov.*

(Plate IX., fig. 7.)

Testa depresso-globosa, minuta, anguste rimata, purpureo-fusca; anfractus 2, rapidissime accrescentes, lineis incrementi tenuissime striati, primus convexiusculus, vix super ultimum elatus, ultimus maximus, perconvexus; sutura simplex, distincta; apertura subcircularis, amplissima, longitudine tota testæ paulo minor; peristoma vix continuum, sed marginibus callo tenuissimo conjunctis, pallidum, rimam versus leviter incrassatum; operculum corneum flavescens, unispirale, lente crescens, nucleo fere centrale. Diam. max. 1 mill. Diam. min. $\frac{3}{4}$; axis 1.

Hab.—Swain's Bay.

Although this species is so minute, the shells do not appear to be very young. In one example the last whorl slightly descends near the peristome, which cannot strictly be called continuous, the two extremities being united by a very thin callous deposit on the body whorl. The nuclear portion of the single whorl which constitutes the operculum appears to be convex.

This species differs from normal *Hydrobiæ*, whose spire is generally rather elevated, in having the apex but very slightly exserted above the last whorl, and consequently the contour of the spire is unusually convex.

Eatoniella, *Dall*.

Eatonia, Smith, Ann. and Mag. Nat. Hist. 1875, xvi. p. 70 (name preoccupied).
Eatoniella, Dall, Bulletin U. S. Nat. Mus. 1876, iii. p. 42.

Testa formæ rissoideæ; apertura subcircularis; peristoma simplex, continuum, margine labrali haud incrassatum. Operculum ovatum, pauci- vel unispirale, nucleo subterminali a latere columellari paululum remoto, subtus ossiculo prominenti a nucleo exsurgente et marginem columellarem versus directo munita.

There are two genera which have affinity to the present one—*Jeffreysia* and *Rissoina*. With the former it agrees in the form and character of the aperture, but differs in having the nucleus of the operculum not lateral, but situated within the margin and towards the lower end. It agrees in this respect with *Rissoina* (see Adams, "Genera of Recent Mollusca," vol. iii. pl. 35. f. 1, *a* & *b*), but is distinguished from that genus by the absence of the slight basal channel of the aperture and the lack of any incrassation to the labrum.

The operculum of *Jeffreysia* is composed of concentric layers (as in *Purpura*), commencing from a nucleus situated on the margin of the inner or columellar side; and the ossicle or rib proceeds "from the nucleus in the direction of the *outer* margin" (Jeffreys, "Brit. Conch." iv. p. 58; in the figure, *l. c.* pl. 1. f. 3, it is apparently the reverse).

In *Eatoniella* the operculum is spiral, consisting of one or more whorls, the nucleus is situated within the margin and about one fourth the entire length from the lower end, and the ossicle is directed towards the *inner* margin.

The name originally given by me to this genus having been preoccupied, has been modified to that in use, by Mr. Dall.

Eatoniella kerguelenensis.

(Plate IX., fig. 10.)

Eatonia kerguelenensis, Smith, Annals and Mag. Nat. Hist. 1875, xvi. July, p. 70; *Eatoniella kerguelenensis*, Dall, Bulletin U. S. Nat. Mus. No. iii. p. 42.

Testa ovato-conica, tenuis, olivaceo-nigrescens, labrum versus pallidior semipellucida, vix rimata; anfractus 6, convexi, læves, parum nitidi, lineis incrementi striati, sutura simplici sejuncti; apertura fere circularis, longitudinis totius $\frac{5}{12}$ æquans; peristoma simplex, continuum, ad regionem umbilicalem leviter incrassatum et vix reflexum. Long. 3 mill., diam. $1\frac{2}{3}$.

Operculum ovatum, intus concavum, nucleo posteriore non tamen terminali, crassiusculum, super marginem externum lira incrassatum, unispirale, supra incrementi lineis valde striatum, infra ossiculo elongato a nucleo exsurgente munitum.

Hab.—On a sponge (*Tethya antarctica*), Royal Sound, in 40 fathoms.

This species was found in company with *Rissoa Kergueleni*. It is of a very different form, the spire being conical, the last whorl shorter and a trifle broader; and it also differs in colour. In general aspect it very much resembles several species of *Hydrobia*; but the operculum at once distinguishes it.

Eatoniella caliginosa.
(Plate IX., fig. 9.)

Eatonia caliginosa, Smith, Annals and Mag. N. H. 1875, xvi. July, p. 71.
Eatoniella caliginosa, Dall, Bulletin U. S. Nat. Mus., No. iii. 43.

Testa ovata, modice tenuis, nigra, vix rimata; anfractus $4\frac{1}{2}$, convexi, læves, vix nitidi, sutura simplici discreti, lineis incrementi obsolete striati: apertura fere circularis, superne paululum acuminata, longitudinis totius $\frac{1}{2}$ fere æquans; peristoma continuum, levissime incrassatum, ad regionem umbilicalem albidum, aliquanto reflexum, et basin versus parum effusum. Long. 2 mill., diam. 1.

Operculum ei *E. kerguelenensis* fere simile.

Hab.—Swain's Bay. Found with the preceding species.

This minute shell, with a simple style of sculpture, is of a very black olive-colour, with a nearly circular aperture, the peritreme of which is black outwardly, and whitish in the columellar region.

Eatoniella subrufescens.
(Plate IX., fig. 11.)

Eatonia subrufescens, Smith, Annals and Mag. N. H. 1875, xvi. July, p. 71.

Testa ovata, leviter conica, tenuis, semidiaphana, vix rimata, subrufescens, labrum versus albida; anfractus $4\frac{1}{2}$, lente accrescentes, convexi, sutura subprofunda divisi, læves nisi striis incrementi tenuiter sculpti; apertura subcircularis, longitudinis testæ $\frac{1}{3}$ paulo superans; peristoma continuum, ad marginem columellarem leviter incrassatum et reflexum, rimam umbilicalem indistinctam effingens.

Operculum ei *E. kerguelenensis* subsimile, ossiculo tamen fortissimo munitum. Long. $1\frac{1}{2}$ mill., diam. $\frac{2}{3}$.

Hab.—Royal Sound, on a sponge (*Tethya*) at a depth of 7 fathoms.

The reddish colour of the upper whorls is attributable to the dried remains of the inhabitant.

Skenea subcanaliculata.
(Plate IX., fig. 15.)

Skenea subcanaliculata, Smith, Annals and Mag. Nat. Hist. 1875, xvi., p. 71.

Testa minuta, orbiculata, depressa, tenuis, subdiaphana, albida, late profundeque umbilicata; spira minime elevata; anfractus $3\frac{1}{2}$, sublente accrescentes, perconvexi, ad suturam valde incurvati, fere canaliculati, læves nisi striis incrementi levissime insculpti; apertura subcircularis, leviter obliqua; peristoma continuum, simplex.

Operculum subcirculare, paucispirale, nucleo subcentrali.

Diam. max. $1\frac{1}{3}$ mill., diam. min. 1, alt. $\frac{1}{2}$.

Hab.—Kerguelen Island. Royal Sound, on *Tethya* with *Eatoniella*, in 7 faths.

Some specimens are of a faint reddish colour in the upper whorls; but this may be from the dried animal within. The whorls are very much incurved at the suture, so much so that almost a channel is produced.

Rissoa kergueleni.

(Plate IX., fig. 12.)

Rissoa kergueleni, Smith, Annals and Mag. Nat. Hist. xvi. 1875, p. 69.

Testa ovata, semipellucida, vitrea vel lactea, ad apicem pallide rubescens, tenuis, imperforata; anfractus 5, convexi, politi, sutura angustissime marginata divisi; apex obtusus; apertura ovata, superne acuminata, longitudinis totius $\frac{5}{12}$ adæquans; peristoma continuum, leviter incrassatum et expansum.

Operculum paucispirale, corneum, simplex.

Long. 3 mill., diam. $1\frac{1}{2}$.

Hab.—Swain's Bay, on a sponge (*Tethya*) at 7 fathoms depth.

This pretty species is of a glassy texture, sometimes streaked longitudinally with opaque white. The whorls are divided by a narrowly margined suture, below which there is a faint depression; the first two form an obtuse apex, and the penultimate is large.

Scissurella supraplicata.

(Plate IX., fig. 5, 5a.)

Scissurella supraplicata, Smith, Annals and Mag. N. H. 1875, xvi., p. 72.

Testa heliciformis, spira brevi, anguste perforata, tenuis, semipellucida, alba, epidermide caduca crassiuscula pallide olivacea amicta; anfractus 3, primus —— ? (abruptus), secundus convexiusculus, superne aliquanto planulatus et radiatim arcuate plicatus, ultimus magnus, paululum supra medium carina duplici tenui (cum scissura continua) succinctus, supra carinam radiatim arcuate plicatus, infra eam lineis incrementi striatus; apertura maxima, irregulariter circularis, ad marginem basalem levissime expansa; peristoma continuum, scissura profunda angusta.

Operculum corneum, —— ?

Diam. max. $1\frac{1}{3}$ mill., diam. min. 1, alt. 1.

Hab.—Swain's Bay.

The deep narrow slit in the lip is situated between the two threadlike keels, as is the case in several other species. The operculum is too far within the aperture to allow of examination.

Trochus (Photinula) expansus.

Margarita expansa, Sowerby, Conchol. Illus. f. 16 and 17.

Photina expansa, H. and A. Adams, Proc. Zool. Soc. 1851, p. 191.

Photinula expansa, H. and A. Adams, Genera Rec. Shells, i., p. 428.

Trochus (Margarita) hillii, Forbes, Proc. Zool. Soc. 1850, p. 272, plate 11, fig. 10.

The shells of this species are generally of a pale pinkish colour, varying to reddish brown, at times banded with faint reddish lines, and some specimens which have the dried remains of the animal in them are stained with a greenish tint. The whorls are five in number, convex, and the last is slightly depressed close to the suture. They are devoid of sculpture, excepting the fine striations of growth. The aperture is large, oblique, and roundly subquadrate; the columella is very oblique, straightish in the middle, and in the umbilical region it is excavated rather deeply. The interior of the aperture is beautifully iridescent.

Animal with the pedal disk somewhat rounded in front, narrowed and acuminated posteriorly; head provided with two subulate (apparently not very elongate) tentacles; eyes placed at the tips of two short peduncles on the outside of the tentacles, between which, on the frontal region, are two compressed lobes; the buccal regions produced into largish compressed lobes; the edge of the mantle, on the right side, furnished with four distant elongate cirri, and close to the neck with a large flattened lobe; on the left side are four corresponding cirri, but, in place of the lobe, three small tentacular filaments. Odontophore six mill. in length; dentition similar in arrangement to that of *Trochus (Gibbula) cineraria*, as figured in Gray's Guide to Mollusca, 1857, p. 152, consisting of 11 rachidian teeth, the central one of which is largest, while the others on either side gradually diminish in size as they approach it, and very numerous slender contiguous lateral teeth (uncini).

Hab.—Royal Sound and Swain's Bay, common; dredged in 3–5 faths.

The above description shows that the animal of *Photinula* is truly Trochoid. The circumstance of the lappet on the right side towards the neck being replaced on the left by three small filaments in this species is only a specific character, for in *Phot. cærulescens* the lappets on each side are similar. All the specimens of this species (*P. expansa*) in the British Museum are from the Falkland Islands, and this appears to be the first record of its existence at Kerguelen Island. The nearest ally is *P. violacea*, King, with which *Margarita magellanica*, Hombron and Jacquinot, is synonymous. Indeed the relationship is so close that it becomes questionable whether there is any distinction other than variation would afford.

Patella (Patinella) kerguelenensis, *sp. nov.*
(Plate IX., figs. 13, 13a.)

Patella ferruginea, Sowerby MS. in Mus. Cuming, Reeve Con. Icon., viii., sp. 40. (Nec *P. ferruginea*, Gm., nec *P. ferruginea*, Sowerby, Gen. of Shells, fig. 4.)

Patinella magellanica, Dall, American Journ. Conch. 1871, vi., p. 273; Bulletin U.S. Nat. Mus. 1876, iii., p. 43 (nec *P. magellanica*, Gmel.).

Testa ovalis, antice paululum angustata, convexe satis elevata, apice prominenti beneque ante verso præsertim in exemplis junioribus, late radiatim costata, costis parum prominentibus, et sæpe aliis minoribus interjectis, lineis incrementi concentricis undulatis eleganter crebreque insculpta; extus cœruleo-cinerea, costis sæpissime saturatioribus, et apicem versus in exemplis detritis ferruginea; intus jucunde æneo-fuscescens, plerumque marginem versus pallidior (hac parum undulata); cicatrix muscularis perspicua.

Long. (exempl. max.) 82 mill., lat. 70, alt. 45.

The above description is based upon the examination of about twenty specimens.

Animal (in spirit) with the sole of the foot greenish-ash colour, inky-black on the sides, encircled around the middle by a bluntly serrated frill which is interrupted in front of the head; branchiæ pale buff-colour; margin of the mantle blackish, furnished with alternately small and smaller cirri along the edge, the smaller black; head and tentacles black; lips pale buff.

Lingual ribbon very long and narrow; teeth not hooked, in diverging pairs; median pair two-pronged, the inner prong much the larger, and resembling in form a flat spear-head; lateral pairs alternate with the median pairs, four-pronged, the second prong from the centre and the outermost considerably larger than the other two, which are of nearly equal size, the innermost however rather the larger.

Hab.—Swain's Bay usually on rocks about 1 fathom below the surface; but one specimen was also obtained at the extreme verge of low water. Dead shells were occasionally found on some of the islands scattered over the recreation grounds of Cormorants and Gulls (Eaton); also Royal Sound (Kidder).

The figure in the Con. Icon. of this species (as *P. ferruginea*) is very good as regards form, but does not show the prominent apex, and only represents a small specimen. This name adopted by Reeve being preoccupied by Gmelin cannot be retained, and consequently I have substituted for it *kerguelenensis*, as in all probability the species is exclusively indigenous to Kerguelen Island. The *P. ferruginea*, Gmelin, founded on a figure in Martini's Conchylien Cabinet, I., pl. viii., fig. 66, is well refigured by Reeve (Con. Icon. viii., f. 14a-b) under the name of *P. costosoplicata*, which is the first adjective in Martini's brief description. Gmelin's name must therefore be retained for the same species.

The *P. ferruginea*, var. (no author mentioned) figured by Sowerby in his Genera of Recent and Fossil Shells, fig. 4, is one of the endless varieties of *P. ænea*, Martyn. From this species *P. kerguelenensis* differs in having the shell less prominently costated, differently coloured, and in the apex being very prominent and much curved over so as to give it a capuliform appearance, a character constant in all specimens, young and old, elevated or depressed; it also differs in the coloration of the animal. Now in some depressed varieties of *P. ænea* the apex is somewhat curved over, but it is not prominent and has not the appearance of an umbo, as exhibited by the apex of *P. kerguelenensis*.

In the determining of the present species, having had occasion to examine the Patagonia shell, I will give the results of my investigation. The synonymy is as follows:—

Patella (Patinella) ænea.

Patella ænea, Martyn, Universal Conch. (1784), I., f. 17; Reeve, Con. Icon. viii., f. 9a–b.)

var. = *Patella magellanica*, Gmelin, Syst. Nat. 1789, p. 3703; Martini, Con. Cab. i., f. 40a–b; Reeve, Con. Icon. viii., f. 19a–b.

Patinella magellanica, Dall, Annals & Mag. Nat. Hist. 1871, vii., p. 289.

var. = *Patella deaurita*, Gmelin, S. N., p. 3719; Chemnitz, Con. Cab. x., f. 1616.

var. = *Patella cymbularia*, Delessert (non Lamarck), Recueil Coq. Lamk. pl. xxiii., f. 8a–c.

var.= *Patella delesserti*, Philippi, Abbild. & Beschreib. Conch. iii., p. 9, pl. 1, f. 5a–b;— ? Dall, Bull. U.S. Mus. 1876, iii., p. 44.

var. = *Patella varicosa*, Reeve, l. c., f. 21a–c.; = *Patella atramentosa*, Reeve, f. 41a–b.; = *Patella venosa*, Reeve, f. 18a–c.; = *Patella chiloensis*, Reeve, f. 98a–b.

On comparison of all the figures above quoted the conclusion at which I have arrived, namely, that they represent but varieties of this Protean species, may be somewhat startling, but the large series (about 180 specimens) before me fully justifies the result.

The apex is found in all positions from the very centre almost to the margin of the shell; every degree of coarseness or fineness of the costation exists, some ribs being very rugose and others only slightly so or quite smooth, some broad and others much finer. The typical form is oval, slightly narrowed in front and subdepressed; in these respects agree *ærea* proper, and the varieties *varicosa*, *deaurata*, *ferruginea* (Sow. non Gmel.), and *delesserti*. The varieties *magellanica*, *atramentosa*, *venosa*, and *chiloensis* are more roundly oval and usually more elevated. The form figured by Delessert as *P. cymbularia*, of which there are several specimens in the British Museum, is remarkable for its white colour and the strongly contrasting dark coppery brown scutum, which is well defined by the muscular scar. In *P. varicosa* the ribs are described by Reeve as "nearly obsolete," in *venosa* as "more or less obsolete with age," in *chiloensis* as "worn, nearly obsolete." Now their obsoleteness is occasioned merely by erosion, for the types of these species are all in an extremely worn condition, and it is evident by carefully examining the margins of the shells where they are least abraded, that the ribs have existed as in normal specimens. The animals (in spirit) of different varieties do not offer any distinctive characters. The freshest are of a buff colour, with the sole of the foot olive, and the chief tentacular cirri (those which usually mark the number of principal ribs on the shell) on the edge of the mantle are black. Tentacles buff, with a black spot on the upper surface near the tips. A frill similar to that of *kerguelenensis* encircles the foot.

Patella (Patinella) fuegiensis.

(Plate IX., figs. 14, 14a.)

Patella fuegiensis, Reeve, Conchol. Iconica, viii., sp. 78.

The description given by Reeve is very good, but he does not lay sufficient stress upon the beautiful raised concentric ridges. He calls them striæ, which term scarcely gives the idea of thread-like lirations such as these. They are very closely packed and undulate very prettily on and between the numerous radiating ribs.

The figure, except in outline and the position of the apex, gives but a poor idea of this beautifully sculptured Patella. It represents the number of ribs at about forty, whereas there are usually about sixty. The specimens from Kerguelen's Island are a trifle narrower and much more depressed than examples from the Falkland Islands; in fact, it is only near the apex that they are at all raised, and towards the margin they are up-turned, so that the dorsal surface is concave, and this form of the shell certainly prevents the animal from entirely concealing itself when adhering to a flat surface. But this peculiarity of form only exists in adult specimens, for several small ones are like ordinary species in this respect. The radiating ribs are almost obsolete in the flat examples, but the undulating concentric lirations, which are more prominent and farther apart than in the type form of the species, define their position; in young shells they are more pronounced. Colour generally uniformly purplish slate, with the apical region ferrugineous; interior similarly tinted, but rather more deeply. One shell has a white border. They are all very thin and fragile, and the edge is very liable to break off in a line with the concentric raised lines of growth.

The animal has the sides and sole of the foot greenish-grey, the edge of the mantle and gills pale buff, the tentacular filaments on the margin of the mantle blackish except at their tips, tentacles short and thick, pale buff, with a black spot above.

The frill-like expansion of the mantle, similar to that of *P. ænea* and *P. kerguelenensis*, is a little above its edge, is bluntly serrated, and interrupted beneath the head.

Teeth of the lingual ribbon slightly hooked, in pairs, scarcely diverging; the central pairs two-pronged, the inner prong much the larger, spear-head shaped; the lateral pairs alternating with the central ones are four-pronged, the innermost prong smallest, the next two subequal, and the outside one situated nearly at right angles to the rest of the tooth, about the same size or a trifle larger.

Hab.—Royal Sound and Swain's Bay, everywhere very common on the submerged fronds of long floating kelp (*Macrocystis*) bordering the shore.

Patella (Nacella) mytilina.

Patella mytilina, Gmelin, Syst. Nat. p. 3698.
Patella conchacea, Gmelin. l. c. p. 3708.
Nacella mytiloides, Schumacher, Syst. Vers Testac. p. 179.
Patella cymbularia, Lamk. Anim. S. Vert. ed. 1, vol. vi., p. 335; ed. 2, vol. vii., p. 541; Philippi, Abbild. & Beschr. iii., p. 1, f. 2.
Patella cymbium, Phil. Archiv. Naturgesch. 1845, p. 60; Abbild. p. 7.
Nacella mytilina, Gmel., Dall, Annals & Mag. Nat. Hist. 1871, vii., p. 289.
Nacella cymbularia, Lamk., Adams Genera Rec. Mol. i., p. 467.
Var. = *Patella vitrea*, Phil. l. c. p. 9, pl. i., f. 4 a, b.
Var. = *Patella hyalina*, Phil. l. c. p. 8., pl. i., f. 3 a, b.
(Not. *Patella cymbularia*, Delessert, Recueil. Coq. Lamk. pl. 23, f. 8 a–c, which is one of the many varieties of *Patella ænea*, Martyn.)

On carefully examining a large series of specimens (about 40) of this species, I can arrive at no other conclusion than that the two forms described by Philippi, above quoted, are but variations of Gmelin's shell.

The principal difference is in the position of the apex. In a long series every position is found; in some shells it is quite marginal or nearly so (*P. hyalina*); in others it is slightly more remote from the margin, and again still more so, until it takes that position which it occupies in the typical *mytilina*. The character of the sculpture varies to no material extent. All examples are more or less radiately costated, sometimes conspicuously, and at other times almost obsoletely, the crenulation or undulation of the margin varying in coarseness with the strength or feebleness of this costation. The concentric lines of growth are of the same general nature in every specimen, and all possess minute scratch-like striation (visible only under a lens) radiating irregularly from the apex. The form of the marginal outline is considerably altered in those specimens whose apex is quite marginal, being considerably narrowed at the apical end and somewhat acuminated.

The coloration varies in shells of similar form, some being of a general greyish tint, varied at intervals with darker concentric rings and often a few radiating palish stripes on the ribs. Others are uniformly yellowish-brown, others pale luteous broadly striped with black, and finally, others are of a uniform pale horny colour; but all have the apex cupreous.

The six Kerguelen specimens obtained by Mr. Eaton are uniform in shape, oval, but rather acuminate towards the apex, which is only slightly removed from the margin, rather depressed, more so than Magellan specimens. They are of a brownish-red colour for the most part, and gradually blend into olive towards the margin, coppery within. Thus it will be seen that they only differ from *P. mytilina* proper, in form somewhat and coloration.

Animal similar to that of *P. kerguelenensis*.

Hab.—On young and short *Macrocystis* at a depth of two fathoms in Swain's Bay, at the end of the same promontory as *Littorina setosa*.

From the descriptions of the four preceding species of *Patellidæ* it will be observed that the animals offer no particular differences exteriorly except in coloration. In all " *a scalloped frill, interrupted only in front* " encircles the sides of the foot, and this peculiarity induced Dall to impose upon those species possessing it the subgeneric name of *Patinella*.

As *P. mytilina* possesses a shell differing somewhat in form and texture, perhaps it may be well to retain the subgeneric title *Nacella* proposed by Schumacher for this species.

With this group Messrs. Adams place *P. pellucida*, Linn., but since this animal has the branchial cordon interrupted in front of the head, it becomes necessary to locate it in another section, which Leach has styled *Patina*.

Siphonaria redimiculum.

Siphonaria redimiculum, Reeve, Conchol. Iconica, ix. sp. 24, 1856.
Siphonaria magellanica, Philippi? Malakozool. Blatt. 1857, vol. iii. p. 165.
Siphonaria lateralis, Couthony MS.? in Coll. Cuming.
Siphonaria tristensis, Dall, Bulletin U. S. Nat. Mus. 1876, iii. p. 145 (not *S. tristensis*, Leach).

It is somewhat questionable whether the young examples of *S. magellanica*, which are described by Philippi as being hooked at the apex, are not slight varieties of Reeve's species. The former is said to have the siphonal angle very prominent and the inner margin white articulated with black, and in these respects it differs from *redimiculum*, the siphonal angle of which is not conspicuously prominent (a character very often exceedingly variable in shells belonging to the same species) and the interior is very dark brown or purplish black, a little paler at the margin, which is at times articulated with white.

The shells from Kerguelen Island consist of two varieties, or perhaps two stages of growth. The first and most abundant form agrees with certain specimens collected in the Straits of Magellan by Dr. R. O. Cunningham (presented to the Museum by the Admiralty), and also with others in the Cumingian collection labelled *S. lateralis* of Couthouy, of which I find no description, from the Falkland Islands. This variety is considerably different in form from the typical *redimiculum*, being much flatter, with the apex less cap-shaped and not so terminal, and the costation rather more rugose; but the coloration is the same. The second variety may be said to be quite normal, but as they are smaller shells than those which belong to the other variety, it is very probable that they are but the young form of this species, which is borne out by the statement of Philippi that the young shells of his *S. magellanica* have the apex *aduncum, sæpe margini incumbentem*, which applies exactly to this variety. *S. Tristensis*, Leach (not *S. Tristensis*, Reeve, Con. Icon.

sp. 23, which is *S. Lessoni* of Blainville) and *S. Macgillivrayi*, Reeve, perhaps only a local variety of the former, are the closest allies of this species.

The animal is of a dark slate colour indistinctly dotted with a darker tint; the sole of the foot and beneath the head greenish buff; edge of the mantle very pale bluish, at times spotted with slate colour. The foot oblong, rounded posteriorly and somewhat truncated in front; the head large, flattened beneath, and arched in front. The radula is very short and broad, and presents toward the mouth of the animal a somewhat semi-lunate surface, which is slightly keeled longitudinally down the centre. It is armed with numerous close transverse series of minute teeth very slightly divergent from the central tooth. They gradually diminish in size towards the margin, are but slightly uncinate, and the central tooth is similar to the rest.

[Hemiarthrum setulosum.

Hemiarthrum setulosum, Dall, Bulletin U. S. Nat. Mus. 1876, iii., p. 14.

Hab.—Kerguelen Island, on stones at low water (Dr. Kidder).]

Doris tuberculata.

Doris tuberculata, Cuvier, Ann. Du Mus. v. p. 469, pl. 74, f. 21; Alder and Hancock Brit. Nad. Moll. fam. 1, pl. 3; Jeffreys Brit. Conch. v. p. 83, pl. 3, f. 4.

? Doris (sp. undetermined), Kidder, Bulletin U. S. Nat. Mus. 1876, iii. p. 48.

A Nudibranch brought from Kerguelen Island by the Antarctic Expedition has been identified as a variety of this common European species, by Mr. P. S. Abraham, who has recently been studying the species of this genus in the national collection. He says that it possesses no characters of specific distinction from *D. tuberculata*, and differs from it only in a few slight and unimportant particulars attributable to mere variation.

The unequal tubercles, flattened at their tips, are more numerous and crowded at the extremities and along the sides of the mantle than in the middle of it.

The undetermined *Doris* found by Dr. Kidder in tide pools at low-water in Royal Sound will very likely prove to be the same species.

Helix (Patula) hookeri.

Helix hookeri, Reeve, Con. Icon. vii., pl. 208, f. 1474. *Helix (Patula) hookeri*, Pfeiffer, Malaco. Blätt. ii., p. 126; Monog. Helic. iv., p. 87; vol. v., p. 152.

Helix (Hyalina) hookeri, Dall, Bulletin U. S. Nat. Mus. No. iii., 45.

The animal of this species (in spirit) with a narrow foot, rather narrower posteriorly than in front. The sole of it a pale livid olive, and the sides dark slate-colour. The mantle above the head pale livid, dotted with dark slate spots.

Mr. Eaton says:—" During life the animal (viewed through a lens) is black " reticulated with grey; tentacles either black above and dark grey beneath longi-

" tudinally, or dark grey throughout; foot bordered above by a ribbon-like stripe
" which is composed of long oblong tessellations, whose interstices are grey, which is
" separated by a thin pale irregular line from the more finely reticulated upper
" portion of the sides and back; the interspaces of the reticulation of these last are
" slightly raised and black, and cause the surface to be somewhat granulated.
" Some of the lines of growth in the shell are occasionally straw colour."

Hab.—Common in the neighbourhood of Royal Sound, especially on the terraces of basalt.

CONCHIFERA.

Saxicava bisulcata, *sp. nov.*
(Plate IX., fig. 21.)

Testa transverse subrhomboidalis, mediocriter crassa, valde inæquilateralis, tumida, alba, epidermide tenuissima lutescente partim induta, sulcis duobus minime profundis ab umbonibus usque ad medium marginis ventralis, ibique productis dentes duos effingentibus arata, lineis incrementi concentricis subtenuibus insculpta; umbones parvi, contigui, fere terminales, incurvati; margo ventralis minime arcuatus, medium versus bidentatus; latus anticum extra umbones vix prominens, posticum a margine dorsali decliviter arcuatum; ligamentum distinctum, flavescens.

Diam. transversa 6 mill. Alt. $3\frac{1}{3}$. Crass. $3\frac{1}{3}$.

Hab.—Kerguelen's Island.

This very curious species of *Saxicava* is the only one with which I am acquainted having dentate ventral margins to the valves. These dentitions are the prolongations of two shallow sulci which radiate from near the umbones, not quite down the centre of the valves but a trifle posteriorly, to their edge. When the shell is closed the ventral margin has an undulating aspect, which is produced by the interlocking of these little tooth-like projections. Besides these two sulcations in some specimens there are very faint traces of others, or raised ridges radiating down the posterior dorsal slope. Many of these shells have several porcellaneous tubercles which adhere to the inner surface of the valves, and all of them have a single perforation at their base.

Kellia consanguinea *sp. nov.*
(Plate IX., fig. 20.)

Lasea rubra, Dall, (non Montagu), Bulletin U. S. Nat. Mus. 1876, No. iii., p. 45.

Testa transversa, valde inæquilateralis, paululum oblique ovalis, postice leviter angustior, tumida, haud perfragilis, flavescens, prope cardines et latera purpureo-roseo tincta, intus dilute rosea, ad umbones plerumque erosa, epidermide mediocriter crassa, flavescente induta, concentrice tenuiter striata, hic illic fortius concentrice plicata; umbones aliquanto prominentes, contigui, ad circiter longitudinis $\frac{1}{3}$ a latere antici siti; linea cardinalis arcuata, purpureo-rosea; valva dextra dente conico infra apicem umbonis (qui in fossa parva in valva sinistra accommodat) munita, et utrinque dente laterali prominenti (qui in sulco profundo in valva sinistra inter dentem similarem marginemque dorsalem accommodat) instructa; valvarum margines generaliter ubique arcuati.

Diam. transversa 3 mill. Diam. longitud. 2⅓. Crass. 2.

Hab. Royal Sound; abundant under stones between tide-marks at Observatory Bay.

At a first glance this species might easily be mistaken for the European *Lasæa rubra*, to which it has a very great resemblance. It is, however, of a rather stronger structure, the epidermis is thicker, the form too transversely rather more elongate, the umbones always, in the seven examples at hand, considerably eroded, and there is not the faintest trace under a powerful microscope of that minute (apparently punctate) radiating striation which is observable in *L. rubra*. The dentition is a little different also. The *Kellia miliaris* described by Philippi in Wiegmann's Archiv für Naturgeschichte, 1845, p. 51, from the Straits of Magellan, is another closely allied species but differently coloured, it has no epidermis, and the umbones are rather more prominent.

Lepton parasiticum.

(Plate IX., fig. 22.)

Dall., Bulletin U.S. Nat. Mus. 1876, iii., p. 45.

Hab.—Parasitic on a Sea-urchin (*Hemiaster cavernosus*), living in the deep ambulacra, and also on the surface of the test.

This species is remarkable for its parasitic nature and being viviparous. On opening an adult specimen I found it to be filled with about a dozen very small ones.

Lissarca, *subgen. nov.*

Testa æquivalvis, subrhomboidalis, valde inæquilateralis (umbonibus fere terminalibus), concentrice striata; linea cardinalis utrinque paucidentata, in medio lævis; margines valvarum intus dentati.

This sub-genus of *Arca* is distinguished from *Barbatia* of Gray, which is the nearest allied group, in having the shell concentrically (and not radiately) striated, with the umbones nearly terminal, and consequently it is very much more inæquilateral than is usually the case in *Barbatia*.

Arca (Lissarca), rubro-fusca, *sp. nov.*

(Plate IX., fig. 17.)

Pectunculus miliaris? *Phil.* Wiegman's Archiv. Naturgesch. 1845, p. 56.

Testa valde inæquilateralis (umbonibus fere terminalibus), ventricosa, irregulariter subrhomboidalis, antice oblique aliquanto truncata, postice latior, arcuata, mediocriter crassa, ubique rubro-fusca epidermide luteo-olivacea tenuiterque concentrice laminata amicta; area dorsalis angustissima, linearis; umbones magni, mediocriter prominentes, fere contigui; linea cardinalis in medio rectiuscula, lævis, utrinque leviter arcuata, dentibus albidis, obliquis tribus vel quatuor (posticis quam anticis longe majoribus) munita; margo ventralis parum arcuatus, latus versus anticum levissime sinuatus; valvarum margines (præter prope sinum levissimum et ad medium lateris postici) intus fortiter denticulati.

Diam. transversa 4 mill. Alt. $2\frac{2}{3}$. Crass. $2\frac{1}{2}$.

Hab.—Kerguelen Island (Antarctic Exped. and Transit Exped.).

It might be thought that this remarkable little species has been described from young shells on account of their smallness, however their comparative solidity, the strong teeth on the hinge and on the margin, give them the appearance of being adult. These marginal dentations are interrupted at the posterior extremity of the valves, and also towards the anterior end of the ventral margin, where it is faintly sinuated, and where also the little byssus protrudes.

The hinge line is moderately straight, and between it and the dorsal line, which is slightly arcuate, there is an extremely narrow linear area. The umbones are rather tumid and project slightly above the dorsal line. From the umbo to the posterior end a little above the middle of the valves there is the faintest depression. The posterior muscular scar is sub-pear-shaped, largish and well defined, the anterior is small and indistinct, and the pallial line is simple and continuous. The hinge-ligament is central, and so small as to be scarcely traceable. Its teeth are strongly developed, especially the three or four posterior ones, which are conspicuously stronger than the anterior and more oblique.

I have quoted Philippi's *Pectunculus miliaris* with a note of interrogation, as I cannot reconcile certain peculiarities in the present species with his description. At all events they are congeneric, and therefore *P. miliaris* must be placed in *Lissarca*, which is distinguished from *Pectunculus* by its transverse trapezoidal form and the subterminal position of the umbones, which in the latter genus are almost central; the teeth are fewer and the valves not radiately striated or ribbed.

L. rubro-fusca apparently differs from *L. miliaris* somewhat in form, the umbones not being very acute, the margins of the valves are crenulated only in certain places, and not in others, the number of teeth on the hinge-line is smaller, and a dorsal area exists although it is extremely narrow.

These are, it is true, but small distinctions, and had Philippi's description been more copious, possibly these shells might have been referred to his species without doubt. The habitat of his species is the Straits of Magellan, and this is favourable to the identity of the two species; however, until an opportunity is offered for the comparison of authentic examples, it seems to me that it will be the safest course to apply a distinctive name to the Kerguelen Island form.

Two specimens collected by Mr. Eaton differ from those obtained by the Antarctic Expedition many years ago in being rather shorter, and wider posteriorly, in having the umbones less terminal, and the ventral margin of the valves being without denticulations; the last characteristic may be due to immature age.

Yoldia subæquilateralis.

(Plate IX., fig. 18.)

Yoldia subæquilateralis, Smith Annals & Mag. N. H. 1875, xvi., p. 73.

Testa ovalis, postice acuminata, subæquilateralis, postice paululum brevior, convexiuscula, epidermide olivacea vel flavo-olivacea induta, concentrice rugose striata, utrinque umbonibus ad marginem subventralem striis paucis subgranosis radiantibus insculpta, utrinque leviter hians, intus cæruleo-alba; margo dorsalis utrinque multum declivis, antice levissime convexo-arcuatus; postice fere rectus; margo ventralis ubique arcuatus; latus anticum late rotundatum, posticum subacuminate productum; fovea ligamentalis parva triangularis; dentes cardinales utrinque 11; sinus pallii latissimus parum profundus.

Lat. 34 mill., long. 23, crass. 9.

The animal resembles that of *Solenella gigantea* in all respects excepting that the edge of the foot is bluntly serrated or scalloped.

Hab.—Swain's Bay. Dredged in 7–10 fathoms in very sheltered water.

I know but one species which approaches the present one somewhat closely, namely *Y. eightsii* of Couthouy. From this, however, it is well distinguished by its different form. By reference to Jay's figure upon which *Y. eightsii* is founded (for no description is given; Cat. Shells, 1839, ed. 3, pl. i. f. 12 & 13), it will be perceived that a very inequilateral shell is there represented, with a much *excavated* posterior dorsal slope; on the contrary, *Y. subæquilateralis* is almost equilateral, with a *straight* posterior dorsal acclivity.

Solenella gigantea.

(Plate IX., fig. 19.)

Solenella gigantea, Smith Annals and Mag. N. H. 1875, July, p. 72.

Testa elongato-ovalis, postice subrhomboidalis, parum inæquilateralis, postice longior, aliquanto ventricosa, marginem versus posticum compressiuscula, epidermide nitidissima (vel fusco- vel flavo-olivacea) induta, lineis incrementi concentricis (interdum prominentibus) ornata, et striis paucis tenuissimis et confertis ab umbonibus usque ad medium lateris antici radiata, intus alba, iridescens; margo dorsalis utrinque leviter declivis, ventralis vix arcuatus; extremitas lateris antici brevioris paululum medium supra leviter acuminato-rotundata; postica superme subrostrata, inmedio leviter sinuata; dentes cardinales postice circiter 32, antice 11; impressio pallii perprofunde sinuata.

Lat. 62 mill., long. 32, crass. 19.

Animal furnished with a large foot of an oval form beneath, acuminated at both ends, deeply cleft down the centre, and deeply striated across, surrounded by a nearly even margin; siphons small, united, retractile, very unequal in size; the upper or exhalant one very slender, the lower considerably larger; gills small, terminating at the side of the body; mouth very small; palpi very large, lamellately wrinkled within, the terminal appendages towards the siphons with undulating margins; the margin of mantle treble-edged, simple, only fringed near the siphonal extremity.

Hab.—Royal Sound, both at Observatory Bay and near the eastern shore of Swain's Harbour, on mud in about 10 fathoms.

This magnificent species is by far the largest yet described of this genus, and is at once known from the other three species by its different form. The posterior end pouts in the same manner as in the North American *Yoldia thraciæformis*. The epidermis in young and half-grown specimens is of a bright yellowish olive colour; but in the adult shell it becomes of a dark olive-brown, and is much eroded in the umbonal region; it is slightly reflexed within the margin of the valves, and is held between the two outer edges of the mantle. The few radiating contiguous striations towards the anterior end furnish another very distinctive character.

Mytilus magellanicus.

M. magellanicus, Chemn. Con. Cab. viii., pl. 83, f. 742; Knorr, Vergnügen iv., pl. xxx., f. 3; Reeve, Con. Icon. x., pl. 6, f. 22; Dall, Bulletin N. S. Nat. Mus., No. iii., p. 47.

M. bidens (Linn.? part.) et auctorum, vide Born. Mus. Vindobon. p. 128; Gmelin, p. 3354; Dillwyn, Descript. Cat. p. 313.

Hab.—Royal Sound, on roots of kelp and on rocks at a depth of 3 fathoms, obtained by means of a grapple.

This common Magellan species varies in colour very considerably. The largest specimens are generally of the form and dark purplish-black colour, as represented by Reeve's figure; others are of a more purplish-slate tint, clothed with a rich olive-brown epidermis, and again others are altogether bright yellow. In the British Museum there are several examples of this species with the locality New Zealand attached to them. I cannot trace any character whereby they can be separated specifically, and the habitat is certainly correct, as the shells were received from a reliable authority.

Mr. Eaton says that " the specimens brought up by the kelp often had extremely thick shells, and occasionally measured upwards of five or six inches in length; the thin yellow examples appeared to be young, and those which were olive-brown to be less aged than the purple-black shells, as a rule. The animals were in great request among the blue jackets (when the expedition first landed), who used to haul up the kelp and collect them by buckets-full; but early in November they went out of season and became uneatable."

Mytilus edulis.

Mytilus edulis, Linn. Syst. Nat. ed. 12, p. 1157.

Mytilus canaliculus, Dall (non Hanley), Bull. U. S. Nat. Mus. 1876, 3, p. 41; (non *M. canaliculatus*, Martyn).

Probable varieties are *M. ungulatus* (Linn. part), Lamarck; *M. chilensis*, Hupé; *M. chilensis* (Philippi), Reeve; *M. obesus*, Dunker; *M. trossulus*, Gould.

Hab.—Abundant on rocks between and a little below tide-marks at Kerguelen's Island.

After a careful consideration of this species, I cannot arrive at any other conclusion but that the Kerguelen shells undoubtedly are specifically the same as the common edible mussel (*M. edulis*). No definite distinction can be traced in the shells (unfortunately only eleven in number) collected at Kerguelen, from specimen from the Dutch coast bought in the London market. The form of the shell (always more or less variable), colour of the exterior and interior, the hinge with the few irregular teeth, muscular scars, and the punctures in the interior towards the ventral margins, are precisely alike in both local forms. Mr. W. H. Dall, who has given an account of the Mollusca obtained by the American Transit party (*see* Bulletin U. S. Nat. Mus. p. 48) remarks that "the shell of this species closely " resembles some varieties of *Mytilus edulis*, but the soft parts are quite different. " The foot is large and quite flat beneath. The viscera and branchiæ are white, " the foot and mantle edge streaked with dark brown."

I have closely examined the soft parts of four Kerguelen specimens, and contrary to Mr. Dall's assertion that they are quite different, I find them to be exactly the same as in European specimens, excepting that the foot is smaller if anything, and not "larger," but this may possibly be due to contraction, since the foot in some Dutch examples is larger than in others of similar size. In coloration not the slightest difference is discoverable, both forms having the foot and the mantle-margin more or less brown.

Several other species, for instance, *M. chilensis*, Hupé, described in the zoological portion of Gay's History of Chile, *M. chilensis* of Philippi, published in Reeve's Conchologia Iconica, and *obesus*, Dunker, in the same work, apparently do not offer any appreciable specific characters whereby they may be separated from *M. edulis*, and I do not feel convinced that the large *M. ungulatus* is anything more than a gigantic form of this species. And again Jeffreys, speaking of *M. trossulus*, Gould, says that it " probably differs in no other respect than being called a 'representative'

" species." Thus it would appear that these so-called species probably are but locality-species, but until an opportunity occurs of studying the Mytilidæ *en masse*, it would be hazardous to affirm so definitely. None of the specimens from Kerguelen Island exceed two inches and a half in length.

Modiolarca trapezina.

Modiola trapesina, Lamk. Anim. S. Vert. ed. 2, vol. vii., p. 24.
Modiola trapezina, Küster, Con. Cab. viii., heft. 3, pl. 6, f. 16 & 17.
Modiolarca trapezina, Gray, Synopsis Brit. Mus. 1840, p. 151; Proc. Zool. Soc. 1847, p. 199; Adams, Gen. Rec. Moll. iii., pl. 122, f. 1 and 1a.
Phaseolicama trapezina, Hupé, Gay's Hist. Chile, Malacologia, pl. 8, f. 9.
Phaseolicama magellanica, Rouss., Voy. au Pol Sud, Moll., p. 116, pl. 26, f. 2a–d.
Gaimardia trapesina, Gould, Atlas United States Explor. Exped. pl. 41, f. 568.
Hab.—Kerguelen Island.

This species is also found at the Falkland Islands. Messrs. H. and A. Adams describe " the hinge with two small oblique teeth in the right valve, which receive " two corresponding ones on the left." All the specimens which I have examined, be they adult or young, have but a single tooth in each valve, or more strictly speaking a single rounded tubercle situated just below the apex of the umbo, and sometimes the faintest indication of a second.

The ligament is very slender, only just visible exteriorly, and placed in a narrow elongate groove posterior to the umbones. The shells present various grades of colour, some being as Lamarck states, " luteo-fulva," others gradually passing into dark purplish-red.

Modiolarca exilis.
(Plate IX., fig. 24.)

Modiolarca exilis, H. & A. Adams, Proc. Zool. Soc. 1863, p. 435.

Testa parva, inæquilateralis, irregulariter transverse ovata, antice breviter rostrata, mediocriter convexa, sub epidermide tenui fugaci flavo-olivacea (plerumque partim detrita), fusco-purpurea, concentrice tenuiter striata, umbones prominentes, incurvati, contigui, propius ad latus anticum siti; margo dorsalis antice subrecte declivis, postice arcuate declivis; latus anticum angustum paululum rotundate rostratum, posticum late arcuatum; margo ventralis leviter arcuatus antice levissime sinuatus; dentes 2 sub apicem utræque valvæ; ligamentum fere omnino internum, in sulco elongato postico situm.

Diam. transversa 5 mill. Alt. fere 4. Crass. 2½.

Hab.—Kerguelen Island and " Falkland Islands," Mus. Cuming.

The only specimen of this species from Kerguelen Island agrees precisely with others from the Falkland Islands, of which there is a large series in the Cumingian Collection, and also with the types in the collection of Mr. Henry Adams.

In every example the thin yellowish-olive epidermis is worn off from a large portion of the surface of the valves, and is retained only near the ventral margins and on the sides.

Modiolarca minuta.

(Plate IX., fig. 23.)

Kidderia minuta, Dall, Bulletin U. S. Nat. Mus. 1876, No. 3, p. 46.

There are three specimens of this species obtained by the early Antarctic Expedition in the Museum. At first they appeared to me to differ so slightly from *M. pusilla* of Gould from Terra del Fuego that I had labelled them as a variety of that species; but Mr. Dall, who possibly has a larger series at hand, has pointed out certain differences, which although slight may be sufficient to separate the two forms. However, I cannot retain the genus *Kidderia* as described by him. He says that "it differs from *Modiolarca* in its single anterior muscular scar, the " presence of strong *nymphæ* for the sub-internal ligament, and in the full deve- " lopment of the cardinal teeth." Of the three examples at hand I have opened two, and in both distinctly observe that there are two anterior muscular scars as in *Modiolarca*. The second, the upper one, is extremely difficult of observation, because it is situated deep within the shell and under the apex of the umbo. In *M. pusilla* two scars are also present.

The second character referred to as distinguishing *Kidderia* from *Modiolarca*, namely, "the presence of strong *nymphæ*," is merely one of degree, and equally untenable. In the present species they are less strongly developed than in *pusilla*, and rather more so than in the type of the genus (*M. trapezina*.)

The third distinction is likewise one of degree of development, the teeth of the Patagonian species being rather stronger than those of the Kerguelen shell, and both more developed than those of the common *trapezina*, in which they are represented by a very small tubercular tooth in each valve, and sometimes a faint trace of a second one.

Radula (Limatula) pygmæa.

(Plate X., fig. 16.)

Lima pygmæa, Philippi, Archiv. f. Naturgesch., 1845, p. 56.
Limatula falklandica, A. Adams, Proc. Zool. Soc., 1863, p. 509.
Hab.—Swain's Bay.

There can be no doubt, I think, that the above two names have been applied to the same species. Both authors describe their shells as ovate, equilateral, and the costation obsolete on the sides. In the Museum there is a specimen from the Strait of Magellan, the locality cited by Philippi, rather larger than his example, which was evidently a young shell, but answering to his description in every respect, and agreeing, excepting in size, perfectly with the type of *Falklandica*, which measures

15 mill. long and 11 in breadth and 10 in thickness. Nothing need be added to Adams's excellent description excepting the number of the ribs, which average about 24 in the largest specimens.

BRACHIOPODA.

Waldheimia dilatata.

Terebratula dilatata, Lamarck, Anim. sans Vert., ed. 2, vol. vii., p. 330; Sowerby, Thesaurus Conch., i., p. 352, Pl. lxx., figs. 48, 49.

Terebratula gaudichaudi, Blainville, Dict. Sci. Nat., 1828.

Waldheimia dilatata, Gray, Cat. Brachiopoda Brit. Mus., p. 59.

Terebratula (Waldheimia) dilatata, Reeve, Con. Icon., xiii., Pl. II., fig. 2.

Hab.—Observatory Bay. On rocks at 4 fathoms, obtained with the aid of a grapple out of a cleft in the rocks.

Reeve questions the correctness of the habitat attributed to this species by Gray. But considering how many species of animals found at Kerguelen Island are also indigenous to Patagonian seas, there can be little doubt that Gray was quite correct in this instance.

EXPLANATION OF PLATE IX.

Fig. 1. Neobuccinum Eatoni, *a*. operculum.
2. Trophon albolabratus.
3. Struthiolaria mirabilis, *a*. operculum.
4. Admete limnææformis.
5, 5*a*. Scissurella supraplicata.
6. Littorina setosa.
7. Hydrobia pumilio.
8. ,, caliginosa, and operculum.
9. Eatoniella caliginosa.
10. ,, kerguelenensis, and operculum.
11. ,, subrufescens, and operculum.
12. Rissoa kergueleni.

Fig. 13, 13*a*. Patella (Patinella) kerguelenensis.
14, 14*a*. ,, ,, fuegiensis.
15. Skenea subcanaliculata.
16. Radula pygmæa.
17. Arca (Lissarca) rubro-fusca.
18. Yoldia subæquilateralis.
19. Solenella gigantea.
20. Kellia consanguinea.
21. Saxicava bisulcata.
22. Lepton parasiticum.
23. Modiolarca minuta.
24. ,, exilis.

POLYZOA.—By G. Busk, F.R.S.

(Plate X.)

Owing to the absence* of any published accounts of the Polyzoa collected at Kerguelen Island by the Challenger, the American Transit of Venus, and the German Surveying and Transit of Venus Expeditions in 1874–75, the subjoined list treats exclusively of Mr. Eaton's collection. The 26 or 27 species comprised in it are all of them inhabitants of the littoral or Laminarian zone, and were obtained with the grapple in Swain's Bay and Observatory Bay. Of the whole number 17 or 18 belong to the suborder *Cheilostomata*, 9 to the *Cyclostomata*. No representative of the *Ctenostomata* was collected.

The collection affords nine or ten forms previously undescribed; the remainder belong to a fauna which ranges from the southern extremity of S. America (which may be regarded as its "centre") to New Zealand in a westerly direction, one or two species extending even farther, to Australia and the Cape of Good Hope. It is observable that no Arctic form has been brought from Kerguelen Island, although some have been met with further south, two instances of the occurrence of the Arctic *Hornera lichenoides* obtained during the voyage of H.M.SS. "Erebus" and "Terror" having been communicated to me by Sir J. Hooker. Mr. Eaton suspects their absence may be attributed to the shallowness of the areas searched by him, the greatest depth being not more than 10 fathoms.

CHEILOSTOMATA.

SALICORNARIIDÆ.

Salicornaria malvinensis.

Busk, Brit. Mus. Cat. Polyzoa, part i., p. 18, Pl. lxiii. 1, 2, and lxv. (bis), 1.

Hab. and Dist.—Swain's Bay. Also East Falklands and S. Patagonia (Darwin).

Onchopora.

Ann. and Mag. Nat. Hist. 1876, xvii. 116.

The genus *Onchopora*, as originally constituted, embraced *Tubicellaria* of D'Orbigny, but in the place cited above I have proposed to restrict it to those forms which have no tubular prolongation of the mouth. They certainly constitute a very distinct type.

Onchopora sinclairii. (Plate X., figs. 1, 2.)

Busk, Quart. Journ. Micr. Soc. v. 172, Pl. xv. 1–3.

Hab. and Dist.—Swain's Bay. Also New Zealand (Sinclair).

The Kerguelen Island specimens have afforded me an opportunity of giving better figures of this species than the earlier drawings.

* This account was drawn up in 1876.

CELLULARIIDÆ.

Cellularia cirrata.

Cellularia cirrata, Ellis and Solander, Zooph. 29, tab. iv. D.

Menipea cirrata, Lamx., Exp. Méth. p. 7, pl. iv., fig. D.D 1; Bk. Brit. Mus. Cat. Poly. i., p. 21, pl. xx., 1, 2.

Hab.—Swain's Bay. Also S. Africa, Krauss.

Menipea fuegensis.*

Busk, Cat. Polyz. i., p. 21, pl. xix.

Hab. and Dist.—Swain's Bay. Also Falkland Islands (Hooker); Tierra del Fuego (Darwin).

Menipea patagonica.

Busk, Cat. Polyz. i., p. 22, pl. xxiii., 1, xxv., xxvi., 1, 2.

Hab. and Dist.—Swain's Bay. Also Falklands (Hooker); Port Desire, Patagonia, at low water (Darwin).

CABEREIDÆ.

Caberea boryi.

Caberea boryi, Aud. Savig. Descript. del Egypt. Explic. tab. xiii., 4; Bk. Brit. Mus. Cat. Polyz. 1., p. 38, pl. xvi., 4, 5.

Caberea zelanica, idem, pl. xxxviii.

Hab. and Dist.—Swain's Bay. Also New Zealand, Cumberland Id., Australia (Hooker); Algoa Bay, Hastings (Tumanowicz); East Falklands, Patagonia (Darwin); coast of Devon (Miss Cutler); Jersey (Alder).

FLUSTRIDÆ.

Carbasea ovoidea.

Busk, Cat. Polyz. i., p. 52, pl. xlix., 5-7.

Hab. and Dist.—Swain's Bay. Also S. Patagonia (Darwin).

Diachoris magellanica.

Busk, Cat. Polyz. i., p. 54, pl. lxvii.

Hab. and Dist.—Swain's Bay. Also New Zealand (Dr. Lyall); Straits of Magellan (Darwin); Adriatic (Heller; sub nomine *D. buskii*).

Diachoris inermis.

Busk, Cat. Polyz. i., p. 54, and ii., pl. lxxii.

Hab.—Swain's Bay. Also New Zealand (Dr. Lyall); Straits of Magellan (Darwin).

---

* Subsequent consideration of this form, which occurs abundantly in the "Challenger" collection, induces me to refer it to *C. aculeata*, D'Orb.

Diachoris costata. (Pl. x., figs. 4–6.)

Busk, Ann. & Mag. of Nat. Hist. 1876, xvii., p. 116.

Cells elongate-oval, posterior surface glistening; aperture covered in by numerous (9–12) acute, sometimes furcate costæ, which arch over and interdigitate in the middle line; 4–6 strong oral spines; a pedunculate, reclinate avicularium on one or (more usually) both sides near the upper part of the cell.

Hab.—Swain's Bay. Also Falklands (Darwin).

The cells in this very distinct form have some resemblance to those of *Beania australis* (Bk. Brit. Mus. Cat. Polyz. i., 32, pl. xvi., 1–3); but in the genus *Beania* the cells are more or less erect, and are attached to a connecting tube in a linear series, and there are no avicularia.

In *D. hirtissima*, Heller, the aperture is covered over in front by numerous marginal arched costæ or spines, and usually there are also two strong oral spines at the summit of the cell; but there are no *avicularia*, and the back of the cell is set with numerous forked spines or setæ.

MEMBRANIPORIDÆ.

Membranipora galeata.

Busk, Cat. Polyz. i., pl. lxv., 5, ii., p. 62,

Hab.—Swain's Bay. Also E. Falklands in 4–10 faths. on *Laminaria* (Darwin).

Membranipora spinosa. (Pl. x., fig. 3.)

Flustra spinosa, Quoy & Gaim. Voy. de l'Astrolabe.

Hab. and Dist.—Swain's Bay.

This species is apparently the same as that described by Quoy and Gaimard (or probably by Lamouroux) in the place referred to; but I have thought it well to give an original figure.

Lepralia, *Johnston*.

§ *Armatæ*.

Lepralia galeata.

Busk, Cat. Polyz. ii., p. 66, pl. xciv., 1, 2.

Hab. and Dist.—Swain's Bay. Also Falklands, Fuegia, on Fucus and shell (Darwin).

Lepralia margaritifera.

Lamouroux, in Quoy & Gaimard, Voy. de l'Uranie, pl. xcii., 7, 8; Bk. Cat. Polyz., ii., 72, pl. ci., 5, 6.

Hab. and Dist.—Swain's Bay. Tierra del Fuego (Darwin).

In the description of this species given in the British Museum Catalogue (loc. cit.) no mention is made of the small avicularium with a semicircular mandible, placed (as in *L. verrucosa*, a very closely allied form) on the upper side of the umbo. It should be remarked also that in the Kerguelen Island specimens the cells are much larger than those in examples from Tierra del Fuego, and that the avicularia (which in these last form a prominent feature) are very few and scattered. The Kerguelen Island form is in fact altogether a more robust variety than the other.

Lepralia ciliata.

Eschara ciliata, var. β, Pallas, Elench. 38.

Lepralia ciliata, Johnst. Brit. Zooph. 279, pl. xxxiv., 6; Bk. Cat. Polyz. ii., 73, pl. lxxiv., 1, 2, & lxxvii., 3, 4, 5.

Hab. and Dist.—Swain's Bay. Also S. Coast of England; Belfast Bay (Will. Thomp.); Adriatic (Heller); Mediterranean; America (Pallas); Beaufort Dyke 110–115 faths. (Capt. Beechy).

I have been unable to find more than one very small specimen of this species in Mr. Eaton's collection; but the characters are well marked.

§§ *Inarmatæ*.

Lepralia eatoni. (Plate X., figs. 7, 8.)

Busk, Ann. & Mag. of Nat. Hist. 4th Ser. 1876, xvii., p. 117.

Cells broadly oval, distinct; mouth semicircular, lower lip straight, notched in the middle; 4 to 8 erect oral spines. Surface of cells in the interior of the zoarium smooth, entire, or obscurely pitted round the border, sometimes umbonate; in the marginal cells there is a row of distinct pores round the border: ovicell prominent, subglobose, with faint radiating lines in front, and a row of small pores round the base.

Hab.—Swain's Bay.

Lepralia hyalina. (Pl. X., fig. 9 normal; fig. 10 var. ζ; fig. 11 var. η.)

Linn. Syst. Nat. ed. xii., 1286; Bk. Brit. Mus. Cat. Poly. ii., p. 84, pl. lxxxii., 1, 2, 3; xc.; xcv., 3, 4, 5; ci., 1, 2.

Lepralia hyalina usually forms very regular circular patches (most commonly upon seaweeds) composed of hyaline, elongated, cylindrical or barrel-shaped cells, having the mouth circular and of variable size, sinuated or notched in the lower margin, the peristome thin and usually quite unarmed (though in some instances there are two short blunt marginal projections on the upper edge,—processes rather than spines—as in the var. *cornuta*, Bk. Brit. Mus. Cat. Poly. ii., 84, and in M. D'Orbigny's *E. chilina*). But with these general characters occur several forms, apparently varieties of one species, which differ very widely from one another in the extremes. Some of them indeed might (perhaps more properly) be regarded as distinct species; but on the whole, when the intermediate forms are taken into consideration, it seems allowable to treat them as varieties.

In addition to such varieties, then, of this protean species, as are given in the British Museum Catalogue, Kerguelen Island affords what appear to be three others. These, together with one of those referred to in the Catalogue, which is in Mr. Eaton's collection, make four forms in the present series doubtfully ascribed to *L. hyalina*; viz. :—

Var. δ, *discreta*, Bk. Cat. Polyz. ii., p. 85; pl. ci., 3, 4.

Hab. & Dist.—Swain's Bay. Also, Falkland Islands, 4–10 faths.; Fuegia (Darwin); California (Greville).

Var. ε, *conferta*, Bk. Ann. & Mag. of Nat. Hist. 4th ser. 1876; xvii. p. 117.

Characterised by the crowded and compressed growth of the cells and ovicells in the central portion of the patch, which gives the zoarium the aspect of a cellepore, and by the wide patulous mouths of the cells, and more especially of the marginal cells.

Hab.—Royal Sound, or Swain's Bay.

Var. ζ *bougainvillei*, Bk., Ann. & Mag. of Nat. Hist. 4th ser. 1876, xvii. 117. (Pl. X., fig. 10.)

This appears to be identical with the form figured by M. D'Orbigny, and therefore I have retained his specific name.

Hab.—Swain's Bay, or Royal Sound.

Var. η. *muricata*, Bk., l. c. (Pl. X., fig. 11.)

Characterised by the smaller size of the cells, and by their surface like that of the ovicells being thickly studded with short spines.

CYCLOSTOMATA.

CRISIIDÆ.

Crisia edwardsiana.

Crisidia edwardsiana, D'Orbigny Voy. d. l'Amér. Merid., Polyp. 7, pl. i., 4–8.
Crisia edwardsiana, Bk. Brit. Mus. Cat. Poly. iii., p. 5, pl. ii., 5, 8.

Hab.—Swain's Bay. Also, New Zealand (Sinclair); Australia (M'Gillivray); Tierra del Fuego (Darwin); Patagonia (D'Orb.).

Crisia kerguelensis. (Pl. X., figs. 17, 18.)

Busk, Ann. & Mag. of Nat. Hist. 4th ser. 1876; xvii., 117.

Zoœcia 3 to 5 in each internode, branches arising from the second or third, elongated, curved abruptly forwards, mouth slightly expanded, peristome thin membranous. *Oœcia* pyriform, somewhat compressed and subacuminate at the top; opening behind, curved, tubular. Growth lax, straggling, irregular.

Hab.—Swain's Bay.

It has much of the habit and general aspect of *Crisidia geniculata*, but differs from it in the number of cells in each internode, the very sparse punctulation of the surface, and the peculiar form of the oœcia.

IDMONEIDÆ.

Idmonea marionensis. (Pl. X., figs. 15, 16) (young state).

Busk, Brit. Cat. Poly. iii., p. 13, pl. vii. 78 (young state); pl. xiii. 3-5.

Hab.—Swain's Bay. Also, Auckland (and Orakei Bay, fossil ?), New Zealand (Stoliczka); Marion Island, 80 faths. (Hooker); Gulf of Florida, Bahia (fossil; Smitt).

Numerous instances of the initial growth of this species, as well as of *Pustulopora delicatula*, *Tubulipora*, &c., affixed to the surface of seaweed, occur in the collection. Two such specimens of the *Idmonea* are here figured (Plate X., figs. 15, 16). The zoœcium springs in the usual way from a small hemispherical vesicle, rising at once in a tubular form, and soon sending forth a lateral bud or secondary tube, and so on. In no stage does it resemble a *Tubulipora*, not being adnate except at the extremity, or merely by a few lateral struts from the first cell or two.

Pustulopora delicatula.

Busk, Cat. Poly. iii. 20, pl. vi., B. 3.

Hab. and Dist.—Swain's Bay. Also Australia, 15 fathoms (Voyage of Rattlesnake); Madeira? (J. G. Jeffreys).

TUBULIPORIDÆ.

Tubulipora organizans. (Plate X., figs. 20-25.)

Tubulipora organizans, D'Orbigny, Voy. d. l'Amér. Mérid. p. 19, pl. ix., 1-3.

Hab.—Swain's and Observatory Bays; abundant on *Macrocystis*. Also Falkland Islands (D'Orb.).

M. D'Orbigny's figure appears to represent the mode of growth of this form; and as the species is extremely abundant on the kelp at Kerguelen Island, I have little hesitation in applying to it the appellation given by him to the Falkland Islands form. The manner of growth in narrow, ligulate, dichotomously dividing branches that hardly expand at all at the extremity and which are composed of short irregular series of tubes diverging on either side from the median line, is not unlike the growth of *T. serpens*; and I am not certain whether *T. organizans* might not well be regarded as a variety of that species. Upon the whole, however, sufficient diversity is apparent between the two to justify their specific distinction. One marked character which I consider important, is the manner in which the growth of the zoarium commences. This proceeds as usual from a semiglobose vesicle; but in the form considered by me to be *T. organizans* this vesicle or *bulla* is supported by numerous short processes, which form a kind of denticulate border round its attached base. From this *bulla* arises a single tube whose mouth expands and gives rise from its interior to a second tube which buds forth laterally in the usual way,—and so on.

Tubulipora stellata. (Plate X., fig. 26.)

Busk, Ann. & Mag. of Nat. Hist. 4th. ser. 1876, xvii. 118.

Zoarium irregularly stellate. Zoœcia diverging from the centre in all directions.

Hab.—On *Macrocystis*; Swain's Bay.

DISCOPORELLIDÆ.
Discoporella infundibuliformis. (Plate X., fig. 19.)

Busk, Ann. & Mag. of Nat. Hist. 4th ser. 1876, xvii. 118.

Zoarium stipitate, infundibuliform. Zoœcia arising from the interior of the funnel; mouth expanded, with 5–6 acute teeth.

Hab.—Swain's Bay.

Not more than two or three specimens of this very peculiar form have been met with, and these may eventually prove to be only the initial stage or young growth of the species.

Discoporella fimbriata.

Busk, Cat. Poly. iii. 32, pl. xxvii.

Hab. and Dist.—Swain's Bay. Also Tasmania? (Mrs. Smith); Tierra del Fuego, Cape Horn at 40 faths., Chiloë at 96 faths., Chonos Archipelago at 13 faths. (Darwin).

Discoporella canaliculata. (Plate X., figs. 12–14.)

Busk, Ann. & Mag. of Nat. Hist. 4th ser. 1876, xvii. 118.

Zoarium circular, bordered, slightly convex; tubes very irregularly uniserial, with a raised canalicular fillet on one side; interspaces cancellous.

Hab.—Swain's Bay.

[A considerable proportion of the specimens stated to have been obtained at Swain's Bay must have come really from Observatory Bay.—*A. E. E.*]

CRUSTACEA.—*By Edward J. Miers, F.L.S., F.Z.S., Zoological Department, British Museum.*

(Plate XI.)

The species of Crustacea hitherto obtained at Kerguelen Island are so few in number that they cannot be supposed to represent in an adequate manner this department of the Fauna.

The Antarctic Expedition under Capt. Sir James Ross visited the island in the winter (May and June 1840), and of the species of Crustacea in the British Museum, apparently brought back by officers of that expedition from " Kerguelen Land," two are new to science.

A brief summary of the results obtained by the Challenger Expedition by dredging in the neighbourhood of Kerguelen Island, was given by the late Dr. R. von Willemöes-Suhm in a letter to Professor C. Th. E. von Siebold, published in the Zeitschrift f. Wissenschaft. Zool. xxiv., 1874. From it we learn that upon the island itself nothing was found except a small Brachyurous Decapod (probably the *Halicarcinus planatus*). The inhabitants of the 1st zone (not deeper than 40 faths.) comprised several species of *Serolis, Sphæroma, Arcturus,* some *Gammaridæ,* several species of *Caprella,* and some *Pycnogonida*. Richer and more interesting results were obtained in the 2nd zone (40 to 120 faths.), where *Tanais* and *Praniza,* very remarkable *Amphipoda, Mysidæ,* and a *Nebalia* were discovered.

The German Surveying and Transit of Venus Expedition dredged off the open coast, and collected, for the most part, along the rather exposed shores in the neighbourhood of Betsy Cove. They entered also Vulcan Cove and Foundry Branch, but whether they searched at all for Crustacea in these sheltered inlets is uncertain.

The collectors who accompanied respectively the English and American Transit of Venus Expeditions, were advisedly unprovided for dredging in deep water, and consequently their operations were confined to the beach, the laminarian zone, and depths external to the latter, not exceeding 10 fathoms. Mr. Eaton obtained 10 species, of which 7 were new to science. Dr. Kidder with more limited opportunities for work, obtaining 7 species, added 3 to the fauna (2 of them were new to science), and thus raised the total to 15.

No species of terrestrial *Isopoda,* of the family *Oniscidæ,* have been discovered on this island.

Among the 15 indigenous species, several are characteristic of the Antarctic region, which in its widest sense embraces Tierra del Fuego, the Falklands, and the lands and islands of the Antarctic Ocean. *Halicarcinus planatus* and *Sphæroma gigas* are known to inhabit the seas of Patagonia and New Zealand, and are especially abundant in the former area; *Cassidina emarginata* is indigenous also to the Falklands and Patagonia; *Jæra pubescens* occurs at the Falklands, and *Serolis latifrons* has been obtained from New Zealand and the Aucklands.

In addition to the Kerguelen Island species, I have included *Serolis septem-carinata* from the Crozets in the present paper.

DECAPODA.

PINNOTHERIDÆ.

Halicarcinus planatus.

Cancer planatus, Fab. Ent. Syst. 1793, ii., 446.

Leucosia planata, idem, Ent. Syst. Suppl. 1798, 350.

Hymenosoma leachii, Guer.-Mén., Icon. Rég. Anim. iii. Crust. 10, pl. x., 1; id., Voy. Coq. 1828, ii. 22.

Halicarcinus planatus, White, Ann. & Mag. of Nat. Hist. 1846, xviii. 178, pl. ii., 1; idem, List Crust. Brit. Mus. 1847, p. 33; Dana, U. S. Explor. Exped. 1852, xiii., Crust. part i., 385, pl. xxiv., 7; M. Edwards, Ann. Sc. Nat. 1853, ser. 3, xx. 223; Heller, Reise der Österr. Freg. Novara, 1865, Crust. p. 66; Miers, Cat. N. Zeal. Crust. 1876, p. 49; Smith, Bull. U. S. Nat. Mus. 1876, iii. 57.

Hymenosoma tridentatum, Jacq. & Lucas, Voy. del' Astrolabe, 1853, Zool. iii. Crust. p. 60, pl. v. 27.

Hab.—Kerguelen Island, very common everywhere on the *Macrocystis* (Eaton); on rocky beaches and at 5 fathoms (Kidder). Also Tierra del Fuego; the Falklands, abundant; New Zealand; the Auckland Islands.

It is to be noted that in White's figure of this curious little flat crab, the three frontal teeth are not made sufficiently prominent: they rise from under the raised marginal line bordering the front of the carapace.

The largest specimen in the series collected by Mr. Eaton is a little over 13 mm. in length.

ISOPODA.

ASELLIDÆ.

[Jaera pubescens.

Dana, U. S. Explor. Exped. 1853, xiv., Crust. part ii., 744, pl. xlix., 9; Smith, Bulletin U. S. Nat. Mus. 1876, iii. 63.

Hab.—In company with *Sphæroma gigas* on rocky beaches. Kerguelen Island (Kidder); also Tierra del Fuego (Dana).]

ÆGIDÆ.

Æga semicarinata.

(Plate XI. fig. 1.)

Miers, Ann. & Mag. of Nat. Hist. 1875, xvi. 115.

Corpus punctatum, elongato-ovatum; coxis oblique bilineatis; ultimo segmentorum post-abdominis postice truncato levissiméque emarginato, carinâ longitudinali lævi ad marginem posticam haud attingente; penultimo exterioribus ramorum appendicium subovatis, interioribus truncato-triangularibus, haud usque ad apicem segmenti sequentis attingentibus.

Body elongate-ovate, moderately convex, punctate; the punctulations very sparse or wanting on the anterior halves of the segments of the pereion, but more numerous and coarser upon those of the pleon, especially the last of them. Eyes large. Each segment of the pereion is traversed by a faint impressed transverse line in or about the middle; the sixth segment is slightly the longest. First five segments of the pleon subequal and very short; the sixth (the last) about as long as three-fourths of its width at the base, narrowed from its proximal to its truncate and slightly emarginate distal extremity, and with a slight impression near the base on each side of a rather indistinct, smooth, longitudinal median carina, which falls short of the posterior margin. Upper antennæ with the first two joints greatly dilated, the third joint very slender; the flagellum with about 11 joints hardly reaching as far as the anterior margin of the first segment of the pereion. Lower antennæ with the first five joints dilated, the first three of them very short, the fourth and fifth longer, and subequal to one another; the flagellum with about 21 joints, reaching just over the posterior margin of the first segment of the pereion. Pereiopoda with coxæ rather acute and produced behind, traversed by two oblique raised lines; the meros-joints of the four posterior pairs rather slender, the posterior margins entire, slightly carinated. Rami of the appendages of the penultimate segment of the pleon subequal, entire, ciliated, not reaching so far as the extremity of the segment; the outer ramus sub-oval; the inner triangular, broad and truncate at the end.

Length of the largest female 58 mm.

Hab.—Kerguelen Island.

Of this fine species one adult female, and four smaller examples, are in the collection of the British Museum.

Æ. semicarinata resembles *Æ. serripes*, M. Edwards (Hist. Nat. Crust. 1840, iii. 241) in having the caudal segment truncate; but is distinguished from it by the posterior margins of the thighs being entire, as well as by the appendages of the penultimate segment of the pleon not reaching to the posterior margin of the terminal segment.

Plate XI., fig. 1. *Æ. semicarinata* (nat. size); *a*, cephalon (enlarged); *b*, maxillipes; *c*, 1st pereiopus; *d*, 4th pereiopus.

SPHÆROMIDÆ.

Sphæroma gigas.

Leach, Dict. Sc. Nat. 1818, xii. 346; M. Edw. Hist. Nat. Crust. 1840, iii. 205; Dana, U. S. Explor. Exped. 1853, xiv., Crust. ii. 775, pl. lii. 1; Miers, Cat. N. Zeal. Crust. 1876, p. 110; Smith, Bull. U. S. Nat. Mus. 1876, iii. 63.

Hab.—Royal Sound (Am. & Engl. Tr. Ven. Exp.). Also "New Holland" (M. Ed.); Bay of Islands, N. Zealand (Dana); and the Auckland Islands (Brit. Mus.).

Mr. Eaton only collected a single small specimen of this species, but states that it is very common under stones on the shore, and in the kelp. A large series is in the collection of the British Museum, and many specimens are stated by Mr. Smith to have been obtained by the officers of the American Expedition. It is worthy of remark, that all of the specimens from Kerguelen Island collected by the American naturalists, as well as those in the British Museum, are the true *S. gigas* of Leach; and that there is no example among them of the variety named by White *lanceolata*, which occurs, as well as the normal form of the species, at the Falklands, where it appears to be as common as the typical *S. gigas* alone is at Kerguelen Island. The variety mentioned is characterised by the acutely lanceolate rami of the appendages of the penultimate segment of the pleon. I was at first disposed to suspect that this difference between the two forms was merely one of sex; but I have since observed adult males of both of them.

Dynamene eatoni.

(Plate XI., fig. 2.)

Miers, Ann. & Mag. of Nat. Hist. 1875, xvi. 73.

Corpus late ellipticum, lateribus subparallelis, convexum; ultimo segmentorum post-abdominis convexo, lateribus subrectis convergentibus, postice emarginato, excisurâ fere eâdem longitudine ac latitudine, rotundatâ; penultimo ramis appendicium subæqualibus, integris, ovalibus, usque ad excisuram segmenti sequenti vix attingentibus.

Body broadly elliptical, convex, smooth, naked, the sides of the pereion almost parallel. Cephalon transverse, closely encased within the first segment of the pereion, bordered in front with a thin raised line. Eyes very small. Segments of the pereion short, of equal length above; the first three with the posterior edge nearly straight, the next four bent slightly backwards at the sides. Segments of the pleon (the last excepted) coalescent, with the lines of union marked by incised lines at the sides; last segment convex, with the lateral margins almost straight, and with a rounded emargination, which is about as wide as deep, at the distal extremity. Upper antennæ with the basal two joints dilated, the first about twice as long as the second; the third joint very slender, and about as long as the second; the flagellum, with about 14 joints, reaches as far as the posterior margin of the first segment of the pereion. Lower antennæ with the first four joints slightly dilated, the first two short, the next two longer, the fourth usually the longest; the flagellum, composed of about 24 joints, reaches to the posterior margin of the third segment of the pereion. Pereiopoda slender, almost naked;

dactyli with two claws. Rami of the lateral appendages of the pleon subequal, oval, entire, not ciliated, reaching almost to the terminal notch of the following segment. Colour reddish or greyish brown, with darker spots.

Length of the largest (♂) 17; of the smallest 3 mm.

Hab.—Swain's Bay and Observatory Bay; common.

Mr. Eaton collected a good series of examples of different ages and sizes.

In contour *D. eatoni* somewhat resembles *D. dumerili*, Aud.,* which has been recorded from the coast of Natal; but in that species the terminal notch is much deeper, and the rami of the penultimate segment are much shorter.

Plate XI., fig. 2, *D. eatoni* (enlarged); *a*, cephalon (enlarged); *b*, mandible; *c*, maxillipes; *d*, 4th pereiopus.

Cassidina emarginata.

Cassidina emarginata Guérin-Ménev., Icon. Règne Anim., Texte Crust. p. 31; Cuningham, Trans. Lin. Soc. 1871, xxvii., part iv., 499, pl. lix. 4.

Cassidina latistylis, Dana, U.S. Explor. Exped. 1852, xiv.; Crust. part ii. 784, pl. lii. 12.

Hab.—Royal Sound and Swain's Bay. Also W. coast of Patagonia and the Straits of Magellan (Cuningh.), plentiful; Falkland Islands (Guér. Mén.).

The form of this species changes considerably with the advance of age. In young examples (those seen by me are nearly all of them females) the body is more convex, and is proportionally narrower than it is in the others; resembling the form figured as *C. latistylis* by Dana. In older specimens (males) from Kerguelen Island, the body is more depressed and is much broader than in the others; resembling Guérin-Méneville's description and Cuningham's figure. There are specimens of this latter form in the British Museum from the Straits of Magellan, and it is probable that the two may be identified as conditions of the same species.

ANISOPODA.

SEROLIDÆ.

Serolis latifrons.

Serolis latifrons (White, List Crust. Brit. Mus. 1847, p. 106:—name only, without description); Miers, Ann. & Mag. of Nat. Hist., 1875, xvi. 74; id. Cat. N. Zeal. Crust. 1876, p. 117, pl. iii. 7; Smith, Bull. U. S. Nat. Mus. 1876, iii. 63.

Corpus culminatum itaque convexum, acute ovatum; ultimo segmentorum post

---

* **Dynamene dumerili.**

Sphæroma dumerilii, Aud. in Sav. Desc. Ég. 1809, Explic. tab. Crust. i. 95, pl. xii. iv.; Krauss Südafr. Crust. 1843, p. 65.

S. savignyi, M. Edwards, Hist. Nat. Crust., 1840, iii. 208.

abdominis subtriangulari, postice emarginato, supra tricarinato, carinâ in medio unâ e basi ad apicem altâ, rectâ, alterâque utrinque minus expressâ prius ad marginem anticum et adjacente atque parallelâ, deinde juxta latus retrocurvatâ, non tamen ad marginem attingente: penultimo ramis appendicium acuminatis, imparibus, exteriore brevissimo.

Body roof-shaped, with a series of impressed lines and punctulations near the posterior margins of the segments. Segments of the pereion sinuated, but not (as in some of the species in this genus) much prolonged posteriorly. Terminal segment of the pleon large, sub-triangular, with the apex semicircularly emarginate; an elevated keel extends directly from the middle of the base to the terminal notch, and on each side of it one less prominent runs outwards close to and parallel with the base of the segment, and towards the lateral margin is curved backwards so as to terminate eventually not far from the same. Upper antennæ short; the joints of the peduncle dilated, the last of them less so than the preceding. Lower antennæ with the last joint of the peduncle as stout as but about twice as long as the penultimate joint. First pair of gnathopoda with the carpus very short, slightly prolonged and acute at its distal extremity; propodus broadly ovate, articulated with the carpus at the middle of its lower edge, which in front of the joining is armed with a close-set series of short spines; dactylus acute, arcuate. Second pair of gnathopoda (in the male) slender, the propodus not dilated, the dactylus when retracted fitting into a cavity in the lower surface of the propodus. Pereiopoda slender. Rami of the appendages of the penultimate segment of the pleon narrowly acuminate, the outer ramus very small and not half as long as the other. Colour brown, with irregular paler blotches.

Length of the largest example 30 mm.

Hab.—Royal Sound; common about rocks in shallow water, *e.g.* in a cove full of reefs adjacent to Observatory Bay, and along the rocky beach near the other English station at Swain's Haulover. Dr. Kidder also obtained a specimen near Molloy Point; and it was observed seemingly by Sir J. Hooker at Christmas Harbour (*fide* Eaton). Also Rendezvous Cove, Aucklands (Brit. Mus.).

The present species as well as *S. septem-carinata* are clearly distinguished from all that are enumerated by Professor Grube in his Monograph of the genus* by the form and direction of the ridges upon the last segment of the pleon.

[Serolis bromleyana.

Will.-Suhm,, Zeitsch. f. Wiss. Zool. 1874, xxiv., App. p. 19.

Hab.—In 1975 faths., S. of Kerguelen Island (Ch. Exped.).

This species is very large, and has the segments of the pereion produced at the sides into very long spines.]

* Beiträge zur Kenntnis der Gattung Serolis, in Archiv. für Naturgesch. Berlin, 1875, xli. 208-234, pls. v. & vi.

[Serolis septem-carinata.
(Plate XI., fig. 3.)

Serolis quadricarinata, White, MS. List Crust. Brit. Mus. 1847, p. 106 :—name only, no description :—*nomen ineptum*.

Serolis septem-carinata, Miers, Ann. & Mag. of Nat. Hist. 1875, xvi., 116.

Corpus depressum, rugosum; ultimo segmentorum post-abdominis lateribus sinuatis, apice lente emarginato, dorso in longitudinem recte septem-carinatum; penultimo ramis appendicium parvis lamelliformibus subacutis, horum interiore paulo majore.

Body depressed, and (especially at the sides) rugose. Segments of the pereion with the postero-lateral angles prolonged backwards and acute; the corresponding angles of the penultimate segment of the pleon similarly produced, so that their apices lie almost in a straight line with the distal extremity of the ultimate segment. The last segment of the pleon very slightly concave at the sides, with a shallow apical emargination; dorsum traversed longitudinally by seven carinæ; of these one in the middle extends to the terminal emargination but is indistinct, while the carina next to it and next but one are somewhat thickened posteriorly, and terminate before meeting the lateral margins. Upper antennæ small and very slender; peduncle with the first two joints short and dilated, the third joint longer and slender, the fourth very small; flagellum hardly reaching to the postero-lateral angle of the first segment of the pereion. Lower antennæ with the terminal and subterminal joints of the peduncle long and subequal to one another (flagella imperfect). Gnathopoda almost similar to those of *S. latifrons*. Pereiopoda slender, the claws small. Rami of the appendages of the penultimate segment of the pleon small, lamelliform, subacute; the inner ramus rather the larger.

Length of the largest example 13 mm.

Hab.—The Crozets. Three specimens obtained by Lieut. A. Smith, R.N., are in the British Museum.

The catalogue name applied to this species by White, was probably given to it on account of the prominence of four of the ridges (the nearest two on each side of the median ridge) of the terminal segment.

Plate XI., fig. 3. *S. septem-carinata* (enlarged 2 × 2); *a*, appendage of the penultimate segment of the pleon.]

AMPHIPODA.

ORCHESTIIDÆ.

[Hyale villosa.

Hyale villosa, Smith, Bull. U. S. Nat. Mus. 1876, iii. 58.

Hab.—Kerguelen Island, on rocky beaches (Kidder).

Mr. Smith described this species from a single somewhat mutilated male example, nearly 10 mm. long.]

[Lysianassa kidderi.

Lysianassa kidderi, Smith, Bull. U. S. Nat. Mus. 1876, iii. 59.

Long. 3 to 4 mm.

Hab.—Kerguelen Island, on rocky beaches, with *Hyale villosa* (Kidder).

According to Mr. Smith (loc. cit.), "all the specimens received are apparently "immature, and the males evidently, and very likely the females also, have not "attained the adult characters." He further observes that "the characters "assigned to the genus *Lysianassa* (as restricted by Boeck) would require con- "siderable modification to admit our species."]

Anonyx kergueleni.
(Plate XI., fig. 4.)

Lysianassa kergueleni, Miers, Ann. & Mag. of Nat. Hist. 1875, xvi. 74.

Cæca, corpore lævi; angulis capitis antero-lateralibus acutis, prorsus productis; iis tertii segmentorum post-abdominis postero-lateralibus in lobis angustis et acutis reflexis, ad apices sursum leviter curvatis; segmento terminali lamelliforme bipartito.

Smooth; eyes invisible; antero-lateral angle of cephalon acute, produced below and beyond the base of the upper antenna. Third segment of the pleon with the postero-lateral angles prolonged as narrow lobes, which are more or less curved upwards at the tips, and which at their base are nearly at right angles with the posterior margin of the segment. Upper antennæ subpyriform; their first joint large and stout, the next two short; flagellum composed of about 14 joints, the first of which is longer than the next; accessory appendage 5-jointed, with a slender hair at its tip. Lower antennæ slender, rather longer than the upper; the last two joints of the peduncle longer than the one preceding them; flagellum composed of about 21 joints. The mandibles have the slender palpus inserted on a level with the strong molar tubercle. First pair of maxillæ with the inner lobe slender, ovate, and armed at its apex with two setæ; the outer lobe strong, truncate, armed at the apex with three or four spines; the last two joints of the palpus lamelliform, ovate, finely denticulated along the apical margin. Second pair of maxillæ with the lobes rather narrowly ovate, ciliated at the apex; the outer lobe the larger. The maxillipedes have the inner lobe long, reaching nearly to the extremity of the antepenultimate joint of the palpus; the outer lobe extends almost to the apex of the penultimate joint of the same, and is minutely denticulated at its rounded apex and along its inner edge; terminal joint of palpus unguiform. First pair of gnathopoda with the carpus about as long as the propodus; the propodus more than twice as long as wide, with the sides parallel up to the ciliated and obliquely truncate distal extremity; dactylus acute, reversible. Second pair of gnathopoda longer than the first, slender and weak; propodus shorter than the slender carpus, and with dense long hair at its distal end; dactylus obsolete. Coxæ of the second pair of pereiopoda deeply emarginate behind, and with the postero-lateral angle shortly produced backwards at the apex into a broad obtuse lobe. Third, fourth, and fifth pairs of pereiopoda with the basa broadly oblong, and their straight posterior margins very minutely serrated. All of the pleopoda are biramose; the rami slender, acutely lanceolate. Telson lamelliform, longer than broad, slightly narrowed towards its apex, cleft almost to its base; the lobes mucronate.

Length of the largest example about 15 mm.

Hab.—Royal Sound. Common.

In the form of the antero-lateral angles of the cephalon, and of the postero-lateral angles of the third segment of the pleon, this species to some extent resembles *(1) *Hippomedon holbölli*, Kröyer, as described by Böeck, as well as (2) *H. abysii*, Goës, and (3) *Anonyx pumilus*, Lilljeborg,—all from the Northern Seas. But it differs from these species in having the inner lobes of the maxillipedes proportionately much longer; and in this respect it approaches more nearly to the type of structure exhibited in *Orchomene*, Böeck. The eyes also, which are well marked in the species just referred to, are not visible in any of the specimens of *A. kergueleni*.

On account of the subcheliform character of the first pair of the gnathopoda, and the divided telson, I refer this species to the genus *Anonyx*, as defined by Mr. C. Spence Bate, instead of retaining it in *Lysianassa*, where I placed it at first. I cannot refer it with certainty to any one of the numerous genera recently established by Boeck in his systematic arrangement of the Scandinavian and Arctic *Amphipoda*;† I believe, indeed, that it will be found necessary to introduce important modifications of the systematic arrangement and generic characters proposed by this author into any general revision of this difficult order, which may hereafter be undertaken, based upon the comparison of species from foreign as well as the European and Arctic seas.

Plate XI., fig. 4, *A. kergueleni* (enlarged, 2×2); *a*, end of pleon (side view); *b*, telson; *c*, 1st maxilla; *d*, 2nd maxilla; *e*, maxillipes; *f*, 1st gnathopus; *g*, 2nd gnathopus.

Atylus australis.

(Plate XI., fig. 5.)

Paramæra australis, Miers, Ann. & Mag. of Nat. Hist. 1875, xvi. 75.

Atylus australis, id. op. cit., p. 117; Smith, Bull. U. S. Nat. Mus. 1876, iii. 61.

---

* (1.) *Hippomedon holbölli*, Kröy., Böeck Forhandl. Vidensk. Selsk. 1871, p. 102.=*Anonyx denticulatus*, S. Bate Cat. Amphipod. Crust. Brit. Mus. 1862, p. 74, pl. xii. 2.

(2.) *Hippomedon abysii*, Goës, v. Böeck, op. cit., p. 103.

(3.) *Anonyx pumilus*, Lilljeborg, v. Böeck, op. cit., p. 110.

† De Skandinaviske og Arktiske Amphipoder, 1872–76. 4to. Christiania.

Corpus læve, carinis spinisque dorsalibus carentibus; oculis subreniformibus; antennis subæqualibus exappendiculatis; pedum primo et secundo subparibus, manu lateribus subparallelis apiceque oblique truncato, dactylo brevi; tertio segmentorum post-abdominis margine posticâ subrectâ, angulisque postero-lateralibus rotundatis; ultimo segmentorum longo, bipartito, lobis ad apices emarginatis.

Body smooth, without dorsal carinæ. Cephalon with a small subtriangular median lobe, and broad obtuse slightly prominent lateral lobes; eyes sub-reniform, black. First, three segments of the pleon with the inferior margins rounded, forming a distinct angle with the posterior margin in the second segment, and appearing to be minutely serrulate, owing to a series of small submarginal spines. Antennæ about half as long as the animal, sub-equal, slender, without an accessory appendage; the upper pair with the first two joints of the peduncle each about as long as the cephalon, and the third joint short; the flagellum with its joints increasing in length but diminishing in thickness towards the extremity; lower antennæ with the first three joints short, the fourth and the fifth joints longer; the flagellum as in the upper antennæ. Mandibles spinose at the apex; palpus inserted on a level with the strong molar tubercle, and triarticulate, the second joint much the stoutest, furnished like the third joint with long cilia towards its distal extremity. Palpi of the first pair of maxillæ 2-jointed; the second joint ciliated at the apex and more than twice as long as the first. Lobes of the second pair of maxillæ oval, ciliated at the extremities; the outer lobe rather the larger. Maxillipedes with the palpi 5-jointed, and ciliated, their third joints the largest, and their apical joints unguiform; inner lobe ciliated and at the apex spinose, reaching as far as the distal end of the second joint of the palpus, outer lobe denticulated on the inner edge, reaching to the apex of the antepenultimate joint of the palpus. Gnathopoda with the carpus shorter than the propodus, narrow at the base, enlarging distally; propodus with subparallel sides, even margins, and obliquely truncate distal extremity; upon which the acute, slender, and slightly arcuate dactylus can be closed. Third, fourth, and fifth pairs of pereiopoda with small transverse coxæ; the basa longer, with the inferior margins rounded. Pleopoda with acute, slender, subequal rami. Telson lamelliform, reaching beyond the peduncles of the last pair of the pleopoda, cleft nearly to its base; the lobes slightly emarginate at the apices. Antennæ, gnathopoda, pereiopoda, and the rami of the pleopoda, fringed with short hairs.

Length 17 mm.

Hab.—Swain's Bay and Observatory Bay. Four specimens were obtained; two of them adult.

The present species resembles *Atylus fissicauda*,* Dana, from Valparaiso, in having reniform eyes, and the lobes of the telson emarginate; but it is distinguished from it by the greater length of the telson, by the lobes of the same being somewhat narrowed towards the apices, and by their emarginations being very small and placed a little on one side. Also the gnathopoda are subequal; the fourth segment of the pleon is slightly produced over the fifth, and its posterior margin is straight; and the postero-lateral angles of all of the segments of the pleon are rounded, not acute as in *A. fissicauda*.

It is probable that a separate genus will eventually have to be formed for the reception of the two species just mentioned, and *A. austrinus*,† Spence Bate. They differ from the normal species of the genus *Atylus*, as restricted by Böeck, in being destitute of dorsal carinations, and in some other particulars. For *A. australis* I originally founded a new genus *Paramœra*, allied to *Melita* in having the inner rami of the posterior pair of pleopoda short or rudimentary, but differing from it in the absence of an accessory appendage to the upper antennæ. A subsequent examination of a series of younger examples showed, however, that my original types had sustained injury, the rami in question having been broken off and lost; and that in reality the inner rami are as well developed as the outer in *A. australis*. Yet though the genus *Paramœra* is unavailable for *A. australis*, it will hold good for the reception of *Melita tenuicornis*,‡ Dana ♀, and *Gammarus Fresnelii*,§ Audouin, mentioned at the time of its publication as apparently included in it; unless, as is probable, there be some error in the figures and descriptions published of these species.

The specimens obtained by Dr. Kidder at Kerguelen Island, and doubtfully referred to *A. australis* by Prof. Smith, differ from the typical form principally in possessing a minute accessory appendage to the upper antennæ, which does not exist in any of the specimens examined by me.

Plate XI., fig. 5. *A. australis* (enlarged); *a*, end of pleon (side view); *b*, telson; *c*, mandible; *d*, 2nd maxilla; *e*, maxillipes; *f*, 1st gnathopod; *g*, 2nd gnathopod.

Podocerus ornatus.
(Plate XI., fig. 6.)

Miers, Ann. & Mag. of Nat. Hist. 1875, xvi. 75.

Corpus læve; tribus prioribus segmentorum post-abdominis postice utrinque emarginatis, et angulis postero-lateralibus rotundatis; ultimo eorundem simplici

---

* Atylus fissicauda.
Iphimedia fissicauda, Dana, U. S. Explor. Exped. 1853, xiv. 929, pl. lxiii. 4.
† *Atylus austrinus*, S. Bate, Cat. Amphip. Crust. Brit. Mus. 1862, p. 137, pl. xxvi. 4.
‡ *Melita tenuicornis*, Dana, in U. S. Expl. Exped. 1853, xiv. Crust. i. 963, pl. lxvi. 5 g–m.
§ *Gammarus Fresnelii*, Aud. in Sav. Descr. Égypte, 1809, Crust. texte i. 93, pl. xi. 3.

conico; pedum anticis parvis, carpo haud manu minori, secundis magnis, carpo parvo, manu validâ ovatâ integrâ.

Cephalon small, its anterior margin forming almost a right angle with the inferior margin in front of the small round black eyes. Posterior margins of the first three segments of the pleon notched above their rounded and not prominent postero-lateral angles. Upper and lower antennæ subequal, very robust; peduncles with the last two joints subequal, much longer than the one before them, densely fringed beneath with long flexible hair; the upper with a small accessory appendage. Palpus of the mandible ciliated, very stout. Inner lobe of maxillipes very short, reaching to the apex of the antepenultimate joint of the palpus; outer lobe of the same reaching beyond the distal extremity of the penultimate joint, and denticulated along the inner margin; the joints of the palpus are ciliated, and the penultimate is the longest joint. First pair of gnathopoda small and weak; merus, carpus, and propodus somewhat dilated beneath, with long hairs on their inner margins; dactylus arcuate, acute. Second pair of gnathopoda large and well developed; merus very small, carpus inserted toward the middle of the inferior margin of the propodus, propodus large, ovate, entire beneath, dactyl strong and arcuate. Last three pairs of pereiopoda with the propodus shortly spinose, and the dactylus arcuate, acute, and reflexible. Pleopoda biramose, the last pair the shortest; each ramus has a series of short spines. Telson small, conical, simple, ciliated at the apex. Colour pale, varied with numerous small black spots.

Length 13 mm.

Hab.—Swain's Bay. Only two examples; females with ova, in a mutilated condition.

This species is distinguished by the long hairs of the antennæ, the form of the second pair of the gnathopoda, whose propodus is not dentate, the shape of the segments of the pleon, &c.

Plate XI., fig. 6. *P. ornatus* (much enlarged); *a*, end of pleon (side view); *b*, maxillipes; *c*, 1st gnathopus; *d*, 2d gnathopus.

PYCNOGONIDA.

NYMPHONIDÆ.

Nymphon antarcticum.*
(Plate XI., fig. 7.)

Nymphon gracilipes, Miers, Ann. & Mag. of Nat. Hist. 1875, xvi. 76.

Corpus gracillimum, pilis brevissimis sparsis; capite longitudine collo æquali, tuberculo oculigero obtuse subconico; prolationibus segmentorum thoracis laterali-

---

* It is necessary to alter the designation of this species; the name *gracilipes*, which I applied to it in 1875, having been adopted in the same year by Dr. Heller, for a species collected by the recent Austrian Expedition to the North Pole. I have no means of ascertaining to which of the two species priority of publication, in the year referred to, belongs.

bus longe distantibus; pedibus secundo articulorum tertio duplo longiori, septimo octavi æquilongo.

Very slender, sparsely clothed with very short hairs, which become more crowded towards the extremities of the legs. Head and neck subequal in length, and together about as long as the rest of the body; the former cylindrical, stout; the latter somewhat constricted in the middle. Oculigerous tubercle prominent, situated in front of the foremost pair of legs. Body terminated behind by a short obtuse cylindrical process. First (mandibular) pair of appendages triarticulate; the first joint long and slender, the second and third hairy, forming a complete chela; dactylus or third joint slender, arcuate. Second pair of appendages slender; the first joint very short, the second the longest, the last three joints hairy. Third (ovigerous) pair of appendages 11-jointed; the first joint very short, the second scarcely longer, the third and fourth again longer, the fifth the longest, the sixth to the tenth gradually decreasing in size, the eleventh minute. Legs very long and slender; the first joint and the third very short, the second joint rather longer, the fourth, fifth, and sixth very long, the seventh and eighth (first and second tarsal) subequal, straight; claws two, one very small.

Length 13 mm.

Hab.—Observatory Bay. On roots of *Macrocystis* grappled in 5–7 fathoms; a single example.

This species is allied to *N. grossipes*, O. Fab., as described by Kroyer, from the northern seas, but differs somewhat in the length of the neck, and in the proportions of the joints of the legs and appendages.*

Plate XI., fig. 7. *N. antarcticum* (natural size); *a*, body (enlarged); *b*, mandible; *c*, tarsus.

Nymphon brevicaudatum.

(Plate XI., fig. 8.)

Miers, Ann. & Mag. of Nat. Hist. 1875, xvi. 117.

Corpus robustum pilosum; capite sessili crasso, tuberculo oculigero elato obtuse cylindrico; prolationibus lateralibus thoracis parum distantibus, pedibus secundo articulorum tertio parum longior, septimo octavi æquilongo.

Rather stout, hairy. Head sessile, thick, subcylindrical; oculigerous tubercle high, slender, subcylindrical. Body short; the lateral leg-bearing lobes not remote from one another as in the preceding species, nor yet so closely contiguous as in *Tanystylum*, terminated behind by a short subcylindrical process. First pair of appendages (mandibles) well developed; the first two joints rather long, the chelæ slender. Second pair of appendages with the first joint very short, the second the

---

* *Pycnogonum grossipes*, O. Fab. Faun. Grœnl. 1780, 229.

Nymphon grossipes, Kröyer, Naturh. Tidsskr. 1844–45. ii. R. i. 108; id. Voy. en Scand. Crust. tab. xxxvi. 1.

longest, the third long, the fourth and fifth shorter, subequal to each other. Third (ovigerous) pair of appendages 11-jointed; the first three joints very short, the fourth and the fifth subequal in length and the longest, the sixth rather shorter, the next four very short, the eleventh minute, unguiform. Legs with the first three joints very short, the second very slightly the longest of them, fourth, fifth, and sixth long, seventh and eighth (first and second tarsal) subequal to each other, straight and slender; claws two, one very small.

Length about 7 mm.

Hab.—Kerguelen Island (Antarctic Expedition). Several specimens, mostly females bearing ova.

This species is allied to the boreal *N. brevitarse*,* Kröyer; but it is distinguished by its more robust form, its long and slender oculigerous tubercle, its longer tarsal joints, &c.

Plate XI., fig. 8. *N. brevicaudatum* (natural size); *a*, body (enlarged); *b*, appendage of second pair; *c*, ovigerous appendage; *d*, tarsus.

ACHELIIDÆ.

Tanystylum, gen. nov.

Caput sessile, crassum. Appendicium primæ 1-articulatæ, non cheliformes; secundæ 5-articulatæ (?); tertiæ 10-articulatæ. Abdomen postice processu longo styliformi desinitum.

Head sessile, thick. First (mandibular) pair of palpiform appendages inarticulate, simple, not cheliform; second pair five-jointed (?); third (ovigerous) pair ten-jointed. Abdomen terminated by a long styliform process.

The family to which this genus belongs was characterised by Dr. Semper † in 1874 as distinguished by the possession of simple mandibular appendages from all others of the *Pycnogonida*. It occupies a position intermediate between the *Nymphonidæ* (which have the mandibular palpi fully developed, triarticulate and cheliform) and the restricted *Pycnogonidæ* (which are completely destitute of mandibular palpi.

Tanystylum differs from all the other genera of the *Acheliidæ* in having the mandibles reduced to a single joint, and in the slender styliform termination of the abdomen.

Tanystylum styligerum.

(Plate XI., fig. 9.)

Nymphon styligerum, Miers, Ann. & Mag. of Nat. Hist. 1875, xvi. 76.

Corpus robustum, hirtum; capite sessili; prolationibus segmentorum thoracis

---

* *Nymphon brevitarse*, Kröyer, Naturh. Tidsskr. 1844-45, ii. R. i. 115; Voy. en Scand., &c. Crust. tab. xxxvi. 4.

† *Acheliidæ*, Semper in Verh. Phys.-Med. Gesellsch. Wurzburg, 1874, vii. 274.

lateralibus contiguis; articulorum pedis secundo et tertio longitudine subæqualibus, septimo brevissimo, octavo curvato.

Rather stout. Head very thick and somewhat barrel-shaped, widest nearly in the middle. Body with the lobes of the leg-bearing segments in close contact with one another, and so constituting a broad mass or plastrum. First (mandibular) pair of palpi-form appendages uni-articulate (*i.e.* of a single piece); second pair apparently 5-jointed, with the first joint and the third very short, the second, fourth, and fifth longer; third (ovigerous) pair 10-jointed, the first three joints very short, the fourth and fifth longer, the next four very short, the tenth minute unguiform. Legs with the first three joints very short, the next three longer, the seventh joint (first tarsal) very short, the eighth (second tarsal) longer and curved; claws two, unequal.

Length of body about 3 mm.

Hab.—Observatory Bay; on roots of *Macrocystis* grappled in 5–7 faths. (two specimens only.)

The hairiness of the palpi makes it extremely difficult to ascertain positively the number and proportionate lengths of their component joints. The styliform termination of the body appears to arise from the dorsal surface of the abdomen and from between the bases of the last pair of the legs.

Plate XI., fig. 9, *T. styligerum* (nat. size): *a*, body (enlarged); *b*, oculigerous tubercle; *c*, ovigerous appendage; *d*, leg.

ENTOMOSTRACA.—By *George Stewardson Brady*, *M.D.*, *F.L.S.*, *C.M.Z.S.*

(Plate XII.)
[OSTRACODA.

All of the *Ostracoda* collected by me in Kerguelen Island were lost by a breakage.—A. E. E.]

COPEPODA.

The Entomostraca submitted to me were taken in the following localities :—One surface-net gathering, in lat. 35° 9′ S., long. 45° 30′ E.; another gathering from a freshwater lake, and a third from a pool above high-water mark, both in Kerguelen Island. The oceanic species were *Calanus finmarchicus* and a *Sapphirina*, either identical with or very closely allied to *S. danæ*, Lubbock; those from Kerguelen Island were a freshwater species, apparently new, described by me briefly in Ann. and Mag. Nat. Hist., Sept. 1875, under the name *Centropages brevicaudatus*; and a species from brackish water, *Harpacticus fulvus*. No species have yet been recorded by the other Expeditions.

Harpacticus fulvus, *Fischer*.

Hab.—Royal Sound, Kerguelen Island; abundant in pools above high-water mark. It was first noticed by Mr. Eaton at the American station (Molloy Point), in pools by the landing place; but the specimens preserved came from Observatory Bay.

The occurrence of this species in Kerguelen Island is particularly interesting from the fact that it is found all over the European shores in precisely similar situations, that is to say in brackish pools, at or above high-water mark, which are liable to become warm through exposure to the sun's rays. These are in no respect distinguishable from European specimens.

Centropages brevicaudatus, *Brady*.

Centropages brevicaudatus, Brady, Ann. & Mag. Nat. Hist., ser. 4, Sept. 1875, vol. xvi. 162; Eaton, op. cit., vol. xvii. p. 264.

Long. ♀ $\frac{1}{10}$ inch. *Male* not observed. *Female* robust, rostrum short and blunt, last segment of cephalothorax produced on each side into an acutely angular oblong ala-form process. Upper antenna as long as the first two segments of the cephalothorax, slightly tapering towards the extremity, beset with short setæ, 25-jointed; joints subequal, somewhat increasing in length as far as the penultimate joint. Lower foot-jaw rather short, of moderate strength, armed at the apex with two slender falcate claws. The other mouth organs present no distinctive peculiarities. Swimming-feet with both of their branches 3-jointed, the inner branch short; first pair much smaller than the next three, with the terminal spines of the outer branch smooth, and of moderate length and strength; fifth pair with the second joint of the outer branch produced internally into a strong denticulated spine, and the marginal setæ of both of the branches extremely short. Abdomen short, composed of two segments exclusive of the caudal segments; the vulva forms a large rounded protuberance on the first segment; the caudal setæ are short, subequal, half as long as the abdomen, and plumose.

Hab.—This species is very plentiful in freshwater lakes in the neighbourhood of Observatory Bay. It swims slowly, and with an even motion; and hence the females have the appearance of small brown seeds borne along with the water. Their abdomen is somewhat jecinoreus in colour, paler than the cephalothorax. By some accident they were at first stated to be oceanic (Ann. & Mag. Nat. Hist., Sept. 1875), but the tube containing the specimens bore a record of their true habitat, and this statement was afterwards corrected (op. cit. March 1876). No males were preserved.

Plate XII., figs. 11—19, *C. brevicaudatus* ♀ (enlarged): 11, female (side view); 12, upper antennæ; 13, mandible and palp; 14, maxilla; 15, upper foot-jaw; 16, lower foot-jaw; 17, a swimming-foot of the 1st pair; 18, one of the 5th pair; 19, abdomen.

Calanus finmarchicus, *Gunner*.

Cetochilus septentrionalis, Goodsir, Edinb. New Phil. Journ. 35, p. 339, Pl. vi., figs. 1-11; Baird, Nat. Hist. Brit. Entom., p. 235, T. xxix., figs. 1 *a–g*.

Hab. and Dist.—Two specimens were taken in the surface-net in lat. 35° 9′ S., long. 45° 30′ E. It is found also in the British, European, and Greenland seas.

Sapphirina danæ, *Lubbock*.

Trans. Ent. Soc. London, N. S. 1856, vol. iv., part ii., p. 23, pl. xii., 9–11.

Dimensions ♂ $\frac{15}{100}$ inch long, $\frac{7}{100}$ inch broad; ♀ $\frac{14}{100}$ inch long, $\frac{4}{100}$ inch broad.

Male.—Outline from above sub-elliptical, rather widest in front; first cephalothoracic segment considerably broader than long, equal in length to the four following conjoined. Superior antennæ short, stout, moderately setose, 5-jointed, with the second joint about as long as the four others united; inferior antennæ somewhat longer, 4-jointed, geniculated between the second and the third joints; first and second joints by far the longest, third very short, fourth long, terminated by a small claw and a bristle about as long as the claw. Mandible (?) slender, with a single apical tooth and two stout plumose lateral filaments; no palpi. Upper foot-jaw short and stout, 3-jointed, ending in a simple claw; lower foot-jaw larger, stout, 3-jointed, bearing a large curved apical claw which is fully as long as the three preceding joints taken together. Swimming-feet in four pairs, all nearly alike, 2-branched, the branches 3-jointed and subequal in length to each other; marginal spines of the outer branch sharp and slender, the other setæ of moderate length and very distinctly plumose. Caudal laminæ ovate, about twice as long as wide, armed with 4 very short setæ, of which two are apical, one is on the middle of the outer margin, and the fourth is intermediate between this and the other two; at the extremity of the inner margin is a minute spine.

In the median line, behind the bases of the first pair of antennæ, is a very distinct nervous mass in the form of a ganglionic ring elongated backwards, which emits numerous diverging lateral filaments, and separates posteriorly into two large nerve trunks. From the front of this ganglion are given off two short processes, supporting at their extremities two bulb-like "conspicilla" or lenses, slightly in advance of which are two nebulous spots which seem to be of the nature of "ocelli."

Behind the bifurcation of the cephalo-thoracic ganglion lies a glandular organ of considerable size, covering the upper portion of the alimentary canal; and from it, or from its immediate neighbourhood, two curved tubes are prolonged backwards to the sixth body-segment, where they converge towards the intestinal tube and terminate in cœcal expansions; these are the "vasa deferentia."

The alimentary canal is a straight funnel-shaped tube, which extends along the median line to the apex of the abdomen.

Female.—Very different in shape from the male; the abdomen being abruptly narrower than the cephalothorax. Cephalothorax of 4 segments, the first of which is as long as the three remaining together. Abdomen 6-jointed (exclusive of caudal segment), only about half as wide as the cephalothorax, from which it is separated very distinctly. The first segment is constricted anteriorly, and gives attachment to the last (fifth) pair of feet; these feet are rudimentary, and consist of a single joint armed with two terminal setæ. The other abdominal segments are about twice as broad as long, and are subequal to one another. The caudal segments are rather wider in proportion than those of the male.

Hab.—Taken by the surface-net in Lat. 35° 9′ S., Long. 45° 30′ E., March 1875 (Eaton); Lat. 27° 30′ N., Long. 20° W. (Lubbock). A very similar species, perhaps the same, was obtained by Mr. Eaton on the 25th of September 1874 on his outward voyage, within 60 miles W. of the Crozets; but none of this earlier gathering could be preserved.

Mr. Eaton says of this species:—

"The large oval flattened Entomostraca in the water reflected from their surface a blue light which changed to opal in certain positions, and were conspicuous like phosphorescent spots from the deck of the vessel. The colour was entirely due to reflection, their actual substance being colourless and diaphanous. They swam back downwards, and occurred mostly in narrow bands of brownish water, some of which from 50 to 200 yards in width extended for miles as far as we could see on both sides of us. The water derived its tint from jelly-like granulated oval flattened disks easily injured by removal from the sea, which resembled in appearance spawn of Mollusca; but having nothing better than a Coddington lens at hand to examine it with, I am unable to vouch that it was spawn. There were also various kinds of small jelly-fish in some of the bands,—here and there a miniature *Velella* of ultra-marine blue,—most of them the produce of Sertularians; and the brilliant Entomostraca entangled in their trailing tentacles could be distinguished from among the others by the constancy with which their position relative to one another was maintained. As a rule these cœlenterata did not contribute largely to the formation of the bands, although they were abundant; but now and then bands were crossed by us which consisted of little besides these Acalephæ. When it is considered that the individual jelly-fish are colourless, with the exception of their proboscis (when there is one), eye-specks, and ovaries,—so transparent indeed that often eye-specks are the only parts of them visible in a white basin—the difficulty of conceiving the innumerable myriads of them in a strip of water miles in length coloured brown by them alone, may be imagined. The *Sapphirinæ* are less plentiful in these bands than they are in those which are due to the 'spawn.' The smaller Entomostraca were mostly of a beautiful ultramarine blue."

The small form here referred to is, I presume, the female, as the gathering sent to me contains no other species except *Calanus finmarchicus*, which, so far as I know, never exhibits any bright colouring. These Sapphirinæ even after their preservation in spirits with glycerine present a very vivid opalescence.

Plate XII., figs. 1–10, *S. danæ* (enlarged); 1, female (from above); 2, male (from beneath); 2*a*, upper antennæ; *b*, cerebral ganglion; *c*, lenses; *d* and *d*, nerve filaments; *e*, testis; *f* and *f*, vasa deferentia; *g*, alimentary canal; 3, upper antenna ♂; 4, lower ditto; 5, mandible (?); 6, upper foot-jaw; 7, lower ditto; 8, a swimming-foot of the first pair; 9, cerebral ganglion with lenses; 10, caudal lamina ♀.

ARACHNIDA. *By Rev. O. P.-Cambridge, M.A., C.M.Z.S.*

(Plate XIII.)

The few examples of Arachnida found during the Transit of Venus Expedition to Kerguelen's Land, were all apparently new to science. One, indeed, at present seems to me incapable of inclusion in any previously recognised order of Arachnids. The whole collection consisted of but five species; one of Araneidea, and three of Acaridea, the fifth, being that upon which I have founded a new genus, family, and order. At first sight this delicate little Arachnid gave me the idea of being a Chelifer deprived of its forcipated palpi; but a subsequent examination with a stronger lens showed me that it possessed palpi of an entirely different character from those of the pseudo-Scorpiones; and a final scrutiny under a still higher power led to the detection of the eyes: in the number and position of these there is displayed a remarkable similarity to the Solpugidea, while there are not wanting some general indications of affinity to the Araneidea. Its small size and general appearance when alive would probably induce one to refer it to the Acaridea; but the structure of the mouth-parts, the distinct cephalothorax and abdomen, and especially the character of the eyes, seem to preclude this allocation. It is possible, however, that when the Acaridea shall be more thoroughly worked out by some future arachnologist, the present anomalous little creature may become the type of a suborder, or perhaps of only a family of that order. Meanwhile in forming a distinct order for its reception, I desire to obtain the free criticism and opinion of arachnologists * more conversant than myself with some obscure groups of Acaridea, as to its true systematic position.

The following descriptions are reprinted from the Proc. Zool. Soc. Feb. 1876, pp. 259-265; and the plate is almost the same as in op. cit. pl. xix., the difference consisting chiefly in the addition of the figures of *Tullbergia*, and the greater enlargement of some of the others.

ACARIDEA.

ACARIDES.

Torynophora.

Body oval; a slightly indented transverse line towards the fore part on the upperside appears to mark the junction of the cephalothorax and abdomen. *Mouth-*

---

* Dr. T. Thorell inclines to the opinion that this Arachnid is an *Acarid*.

parts almost soldered together, leaving only the short palpi and the extremities of the falces traceable. *Legs* 8, slender, in 4 pairs (1–2 and 3–4 on each side), 5-jointed, and terminating with two somewhat S-curved claws springing from a small supernumerary or heel-joint. *Eyes* four, in two pairs, one pair on either side of the caput. *Falces* armed on the underside with serrated opposed edges. *Palpi* short, strong, 4-jointed, with a single strong curved jaw-like claw springing from its base on the upperside.

Torynophora serrata. (Plate XIII., fig. 2.)

Length about · 5 mm.

This minute Acarid is of an oval form, tolerably convex above, and of a uniform pale luteous colour. From the fore part of the cephalothorax four pointed processes project, each one terminating with a very small joint, from which springs a curious clavate or spoon-shaped bristle or tag; a few with a somewhat similar tag are dispersed thinly over the upper surface of the body, which is closely wrinkled, the wrinkles taking different, but regular, directions on the different parts of the body.

The *eyes* are very minute, in two pairs, one on either side of the caput; those of each pair are near together but not contiguous.

The *legs* are 5-jointed, slender, and not very long; they are armed with fine spines, bristles, and hairs, and terminate with two tarsal S-shaped claws, springing from a small terminal joint, and furnished beneath with some slender prominent clavate hairs. The legs are in pairs, the first and second, and third and fourth legs on each side having their basal joints respectively contiguous to each other, as in the genus *Trombidium*, and articulated to the fore half of the lower surface of the body.

The *palpi* are short strong, 4-jointed, and to the upperside of the base of the digital joint is articulated a strong curved claw.

The *maxillæ, labium,* and *falces* coalesce and form a kind of suctorial apparatus, towards the fore part of which on the underside are two opposed curved saw-edged processes.

Several examples of this curious Acarid, found under stones, were contained in in the Rev. A. E. Eaton's Kerguelen's Land collection. Being so very minute and delicate, they had suffered considerably by being preserved in strong spirit.

Fig. 2. *a*, upperside, highly magnified; *b*, underside without the legs; *c*, profile; *d*, leg of first pair; *e*, extremity of tarsus of ditto; *f*, palpi; *g*, extremity of one of the cephalic projections, highly magnified; *h*, one of the clavate hairs on abdomen, highly magnified; *k*, mouth-parts on underside, highly magnified; *o*, natural length.

Acarus.

Acarus neglectus, sp. n.

Adult female. Length one third of a line.

The body is of an oval form, broadest towards the fore part, whence it narrows quickly to the extremity of the caput.

The *Cephalothorax* is very short and coalesces with the abdomen, being scarcely traceable by a fine suture. The whole is tolerably convex above, and of a dull yellowish, whitey-brown colour, furnished with a few fine bristly hairs on the upper side, and two longish divergent tapering plumose hairs projecting horizontally from the hinder part of the abdomen.

Eyes, none.

The *palpi* and *falces* are short, but of equal length. The former are 5-jointed, the basal joint being the longest, and the latter have a strong denticulate forcipated claw.

The *legs* are short, articulated to the under side of the fore half of the body, and do not differ much in length, their relative length being apparently 4, 3, 2, 1, or 4, 3, 1, 2. They are strong, tapering, and 6-jointed, the basal joints being, apparently, soldered to the under surface of the body, and the terminal or tarsal joint is undivided; their colour is a dull yellowish brown, and they are furnished with some strong spines beneath the two last joints, as well as with a few hairs of different lengths; the longest of the latter being one or two near the extremity of each of the tarsal joints, which last terminate with a strongish sickle-shaped claw.

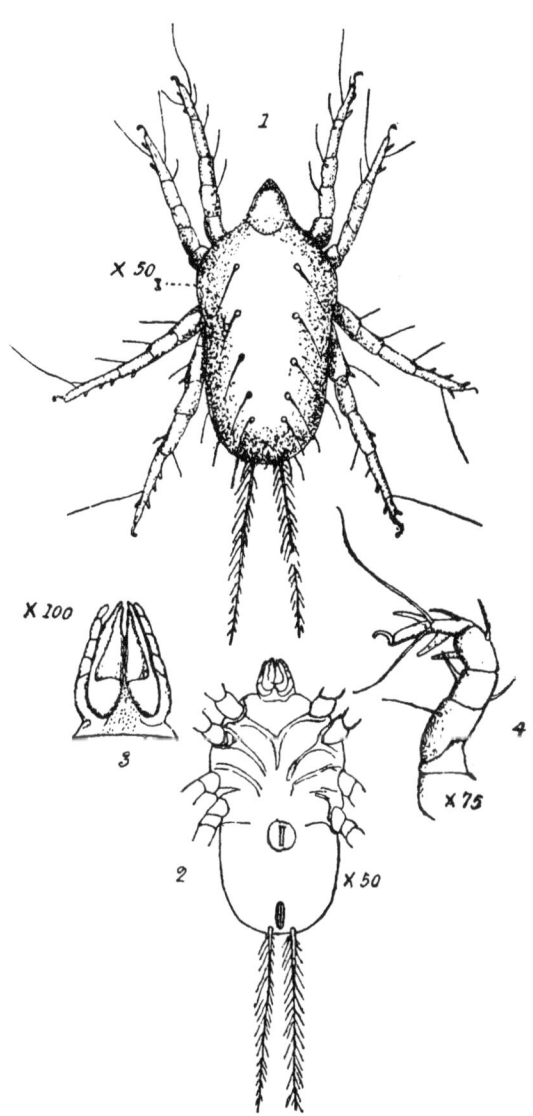

Fig. 1, from above; 2, from beneath; 3, falces and palpi beneath; 4, left leg of the second pair.

The genital aperture is placed just behind the basal joints of the fourth pair of legs, and consists of an elongate brown slip on a circular pale yellowish-brown patch. The basal orifice is situated just beneath the hinder extremity of the abdomen, and has the appearance of a simple brown longitudinal line.

The examples from which this description has been made were contained amongst a mass of fibrous web-like substance received from Mr. Eaton with the other Kerguelen Island Arachnids, but were overlooked at the time when those were described and figured in the Zoological Society's Proceedings for February 1876, pp. 259–265.

BDELLIDES.

Scirus. *C. Koch?*

Scirus pallidus. (Plate XIII., fig. 3.)

Length about 1 mm.

As far as I could ascertain from the single example contained in the Kerguelen Island collection, this small Acarid is an undescribed species of the genus *Scirus*. Its colour is a dull yellowish white, and there are a few obscure blackish markings in two parallel longitudinal lines along the upperside of the abdomen. The body and legs are furnished with a few longish pale semidiaphanous hairs. The *eyes* are in two pairs, those of each pair contiguous, and in the position indicated by the two small oval markings in figure 2 *b*. The only example received was injured by the action of the spirit in which it had been preserved, so that the exact details of its structure could not be satisfactorily observed; in the general appearance, however, of the beak-like mouth-parts there seemed to be but little difference from the genus *Bdella* and others nearly allied.

Hab.—Under stones near the lakes at Observatory Bay, in rather dry places.

Fig. 3. *a*, upperside; *b*, upperside of caput and mouth-parts, more enlarged, showing the position of the eyes; *c*, natural length.

IXODIDES.

Hyalomma, *C. Koch.*

Hyalomma puta. (Plate XIII., fig. 4.)

Length about 1·25 mm.

Body oval. *Cephalothorax* yellowish brown, strongly tinged with red on either side of its fore part and on the fore part of the caput.

Abdomen dark yellow-brown, and (as well as the cephalothorax) thinly clothed with short pale hairs; the hinder part of the abdomen is of a pale dull yellowish hue, and its margin is indented with four small elongate notches. On each side of the underpart, just behind the basal joints of the 4th pair of legs, is a roundish patch, whose surface appears to be covered with minute points, which may possibly be the plates of spiracular organs.

The *legs* are moderately long and tolerably strong, 7-jointed, the last or tarsal joint being very small; they are of a pale yellowish colour, marked underneath with patches of a bright shining orange red, and furnished with a few short hairs; each tarsus terminates with two curved diaphanous claws springing from a small claw-joint; and beneath them is an oval sucker-like pad.

The *palpi* are five-jointed (including the basal joints or maxillæ); these latter are of a reddish colour and soldered to the labium; the colour of the palpi is similar to that of the legs; the terminal (or digital) joint is short and small; the next to it (or radial) is large and tumid, the other two joints short. The length of the palpi slightly exceeds that of the falces.

The *falces* are porrected in the form of a beak, and are armed beneath with numerous sharp hooks or teeth directed backwards.

Several examples of this Acarid were found by the Rev. A. E. Eaton on a Penguin (*Pygosceles tæniatus*) in Kerguelen's Land; and he also found some on a reef of rocks between tide-marks, lurking in crevices.

Fig. 4. *a*, upperside; *b*, underside without legs; *c*, caput, palpi, and falces, more enlarged; *d*, natural length.

Pœcilophysidea.

External characters of the Order:—*Cephalothorax* and *abdomen* covered with a continuous epidermis of a rather slight texture, unsegmented, and united to each other throughout their whole breadth, the point of junction being clearly indicated by a traverse line or suture. *Palpi* filiform, and terminating with a single minute claw. *Legs* eight in number, their basal joints closely grouped together on the sternal surface of the cephalothorax, the tarsi terminate with two claws, between which is a slender pectinated style. *Falces* didactylous. *Maxillæ* coalescing at their base. *Labium* (properly so called) wanting. *Eyes* two.

Pœcilophysides.

In its *general appearance* this curious little Arachnid seems to be a compound of the Spiders, Solpugids, Chelifers, and Acari. On its upperside it reminds one strongly of the Solpugidea, both in the massive falces, and its two eyes on a small tubercle at the fore extremity of the caput; its underside bears a strong resemblance in the maxillæ and palpi to the Araneidea; its profile resembles that of the pseudo-Scorpiones; while in its small size, continuous, delicate epidermis, and closely approximated thorax and abdomen, it shows a strong likeness to the Acaridea.

Pœcilophysis.

Eyes two, closely grouped on a small tubercle at the fore extremity of the caput, just between and above the base of the falces. *Falces* massive, as long as the cephalothorax, two-clawed, the upper claw fixed, the lower movable, and both curved. *Maxillæ* large, coalescing at their base, and produced at their inner extremity into a strong angularly pointed projection. *Labium* none, and *sternum* none, properly so called, the basal joints of the legs being articulated to the inferior surface of the cephalothorax. *Abdomen* longer than the cephalothorax. A small elongate oval aperture towards the hinder part of the underside is probably the genital opening, while a still more minute orifice beneath its extremity is probably the anal aperture.

Legs moderately long and tolerably strong; they are 6-jointed, furnished with long bristles, and terminating in two S-curved claws, beneath which is a longish, slender, slightly upturned style, plumose or finely pectinated along its underside.

Palpi 4-jointed, similar to the legs in armature; the terminal (or digital) joint ends with a small hooked claw; and the bristles or hairs on it are long and plumose.

Pœcilophysis kerguelenensis. (Plate XIII., fig. 5.)

Adult female. Length about ·75 mm.

The *cephalothorax* is of a somewhat quadrate form, narrower before than at its junction with the abdomen; it is moderately convex above, and has a few long pale hairs or slender bristles directed forwards on its upperside; its profile line is nearly level; and the colour of the cephalothorax and abdomen is pale yellow, the legs and other parts being of a whitish hue.

The *eyes* are small, but close together near the hinder part of a small roundish tubercle or eminence, at the middle of the fore extremity of the caput.

The *legs* are 6-jointed, rather long, tolerably strong, not greatly differing in length, their relative length being 1, 4, 2, 3; they are furnished with long pale bristles; and the tarsi, which are undivided and with two claws, are curved somewhat in the form of an S; beneath them is a largish bristle or style, pectinated or plumose on its inferior side. The joints do not differ greatly in length, the first two or basal ones being the longest, and the rest nearly equal.

The *palpi* are similar, in their general armature, to the legs. The digital joint is longer than the radial, and of an ovoid form; its hairs are plumose, and the single terminal claw is sharply hooked and minute.

The *falces* are as long as the cephalothorax, very massive at the base and didactyle, the lower claw being movable and opposed to the upper one; both claws are curved, but project in the same straight line and in the same plane as the cephalothorax, which the falces equal in length.

The *maxillæ* are long, their inner extremities considerably produced into an angularly pointed form, and extending close beneath the falces, to about two-thirds of their length.

The *abdomen*, looked at in profile, is higher and more convex than the cephalothorax, and about double its length; its fore extremity is conterminous in its breadth with the cephalothorax, but is constricted laterally near the middle, the hinder extremity being rounded and obtuse; its upper surface is furnished with a few long pale hairs or bristles.

Several examples of this minute but most interesting Arachnid were found under stones in Kerguelen's Land by Mr. Eaton. Unfortunately, from their extreme delicacy and small size, they had suffered severely from the action of the strong spirit in which they were immersed.

Fig. 4. *a*, upperside; *b*, profile without legs or palpi; *c*, extremity of tarsus of leg of second pair; *d*, digital joint of palpus; *e*, portion of falces and maxillæ in profile; *f*, natural length.

ARANEIDEA.

AGELENIDES.

Myro.

Cephalothorax oval, roundly truncated before, and moderately constricted on its lateral margins at the caput. Upper convexity moderate; profile-line slopes very gradually in a slightly curved line from the hinder slope to the ocular region; clypeus unusually narrow, almost obsolete. Spinners short, those of the inferior pair longest and strongest. *Eyes* eight, unequal in size and forming a rather large and somewhat oval area, enclosed by two longitudinal curved rows of three eyes each; the curves directed from each other; within this area, and towards its fore part, are two minute eyes near together in a transverse line. *Legs* not greatly different in relative length, which is 4, 1, 2, 3. Each tarsus terminates with three claws. *Maxillæ* large, curved towards the labium, much and roundly protuberant on the outer sides towards their extremity, which is rather obliquely truncated; the palpi issue from unusually near their lower extremities. *Labium* rather more than half the length of the maxillæ, very difficult to be seen clearly; but its form is apparently oblong, slightly rounded at the apex.

Myro kerguelenensis. (Plate XIII., fig. 6.)

Adult male. Length nearly 4·7 mm.

The *cephalothorax* is of a yellow brown colour, the margins surrounded with a fine black line; the normal grooves and indentations are well marked, and suffused with dusky black, giving the thorax somewhat the appearance of radiating markings; the ocular region is furnished with some bristly black hairs; and some longer and finer ones are distributed along the central longitudinal line to the hinder slope.

The *eyes* are unequal in size, and form a largish hexagonal area on the fore part of the caput close to its fore margin; they may be described either as in two longitudinal curved rows of three eyes each, with two minute ones in a transverse line towards the fore extremity of the enclosed area, or as four pairs, a hinder pair, two fore lateral pairs, and a fore central pair; those of the hinder pair are separated by an interval of an eye's diameter, and each is also divided by an equal interval from the hinder eye of the lateral pair on its side; the eyes of each lateral pair are separated by a slight interval, not exceeding half the diameter of one of the fore central eyes; the hind laterals are the largest of the eight; and the interval between the fore laterals is equal to about 1½ diameter; the interval between the eyes of the fore central pair is equal to a diameter, the distance of each from the fore lateral on its side being rather greater; and that of each from the hind lateral is equal to the diameter of the latter; the interval between the fore lateral eyes and the fore margin of the caput is very slight indeed, being less than the diameter of one of the central eyes.

The *legs* are moderately long and strong; their relative length does not differ greatly, being 4, 1, 2, 3. They are of a brownish yellow colour, faintly and imperfectly annulated with dusky brown; the annulations are scarcely perceptible in some examples. They are furnished with spines, bristles, and hairs, some of the latter being at right angles to the different joints; each tarsus terminates with three curved claws, of which the inferior is much the smallest, and sharply bent downwards.

The *palpi* are similar in colour to the legs; furnished with hairs and a few bristles, and of moderate length and strength. The cubital joint is short and bent; the radial is slightly longer and spreads out gradually on its outer side into a somewhat irregularly shaped, prominent, but not very large apophysis at its fore extremity; the digital joint is long and rather narrow, similar in form to that of some species of *Tegenaria*; the palpal organs are simple, and have a long curved filiform spine connected with them, and by which they are very nearly encircled; this spine issues from the base of the palpal organs, and curving first round their inner margin terminates on the opposite side.

The *falces* are long, strong, and vertical, prominent near their base in front, and similar in colour to the cephalothorax.

The *maxillæ* are similar in colour to the falces; their form has been already described above.

The *labium* has also been described. Its colour is dark blackish brown.

The *sternum* is heart-shaped, similar in colour to the legs, and furnished with hairs.

The *abdomen* is oval, moderately convex above, and does not project greatly over the base of the cephalothorax. It is clothed with hairs, some on the fore part of the upperside being prominent and of a bristly nature. Its colour is brownish yellow mixed with brown and black, a tolerably distinct pattern being visible on the upperside; that on the fore half consists of two longitudinal curved rows, each of three irregular yellowish spots, followed (on the hinder half) by a series of alternate yellowish and black angular bars or chevrons, the angles directed forwards; or the hinder half may be described as of a yellowish colour marked with a series of four or six black angular bars, which diminish in size towards the spinners; the underside of the abdomen is of a uniform blackish brown colour with an indistinct pale longitudinal line on either side.

The spinners of the inferior pair are strong but short, though longer than those of the superior pair.

The *female* resembles the male in general colour and markings; the sides of the abdomen, however, on the hinder half are paler than those of the male, and are marked with two differently sized oblique irregular blackish markings; the genital aperture is simple, but of a distinctive form.

Eight males and one female were received from the Rev. A. E. Eaton, by whom they were found on Kerguelen's Land under stones, and running also at times on the ground.

This spider (as above observed), the only one found on the island, is of great interest, being unmistakeably allied to *Tegenaria* and *Agelena*, though quite distinct from both.

Fig. 5. *a*, upperside; *b*, fore part of caput and falces, from the front, showing the position of the eyes; *c*, profile; *d*, maxillæ and labium; *e, g*, left palpus in two positions; *h*, genital aperture (♀); *f*, natural length.

OBSERVATIONS on the INSECTS collected in KERGUELEN ISLAND.—*By the Rev. A. E. Eaton, M.A.*

There are doubtless some other species of insects indigenous to Kerguelen Island besides those which are enumerated in the following papers. For although the climate is bleak and the vegetation poor, the amount of ground suitable for the habitation of terrestrial articulata is rather extensive, and in point of humidity and altitude presents much diversity of character. On the shores of the intricate inlets of the sea and of the islets adjacent to them, ample space is available for the maintenance of littoral species. The low flat ground at the eastern base of the Mt. Crozier range, many square miles in extent, though saturated with the drainage from the neighbouring mountains, might be expected to yield good entomological results, if it should ever be visited by a collector. The slopes about Thumb Peak, unusually sunny for Kerguelen Island, could hardly fail to be productive of at least *Coleoptera*. It is true that none were captured there, but that was because they were not sought for, the eggs of a new species of bird having to be collected during the short stay that was made in that neighbourhood. But probably the most favourable situations of all would be the valleys and low hills intervening between Mt. Ross, the inland snowfields and glaciers, Vulcan Cove, and Royal Sound.

Along the shore amongst the tide-pools, undisturbed by waves from the open sea, *Halyritus* of the *Chironomidæ* may be readily observed, while *Hyalomma* lurks in the crevices. More numerous are the denizens of tide-refuse on the upper part of the beach. There the hairy little *Apetenus* can be seen creeping leisurely over the boulders, and *Amalopteryx*, curiously winged, which also abounds among the nests of the Cormorants and Rock-hoppers. Beneath the stones, too, *Anatalanta* is plentiful, though it is not restricted to this zone like the *Phytosus*, but is distributed over the neighbouring slopes and valleys. Beyond the limits of the shore the forms of life become more varied. The sluggish *Calycopteryx* awaits inspection on the *Pringlea* leaves, ready to drop into their axils so soon as it shall be approached too nearly. There roam the spiders *Myro* over stones and *Azorella*; the moth *Embryonopsis*, clambering the tufts of withered grass, leaps lightly off like a splinter chipped away by the boot, and feigns to be dead; while, if the day be calm and fine, the minute *Limnophyes* may be seen upon the wing. *Ectemnorhinidæ* are often to be met with under stones, especially upon the hill tops and on banks and ridges. Several species of *Acaridea* resort to similar shelter on the slopes and low ground; some are gregarious on open surfaces of rock, while others live among the herbage. They are often associated with *Collembola*, which also frequent wet moss in the marshes.

It will be seen from the descriptions and figures that many of the insects are remarkable for peculiarities of structure. The predominance among them of forms incapable of flying constitutes one of the most salient features of this portion of the fauna. A few of the *Diptera* are in addition noticeable on account of the exceptional poverty of their hair.

Where species have been alluded to without description, in literature relating to the island, the references to such allusions are given in a distinct paragraph supplementary to the citations of earlier descriptions.

COLEOPTERA.—*By C. O. Waterhouse, Senior Assistant in the Zoological Department, British Museum.*

(Plate XIV.)

The *Coleoptera* in Mr. Eaton's collection represent 7 species, of which only one was previously known. Of these 6 belong to the *Rhynchophora* and 1 to the *Brachelytra*.

The *Rhynchophora* are very peculiar, and possess an assemblage of characters not to be met with in any other species of this tribe. Consequently their systematic position is very doubtful. The species originally described under the name *Ectemnorhinus viridis* (G. R. Waterh.) as an ally of *Phyllobius* is placed near *Rhinomacer* by Lacordaire in his work on the Genera of the Coleoptera, and on first inspection was referred by me to the *Brachyderidinæ*. I am still of opinion that it accords better with either *Phyllobius* or *Brachyderes* than it does with *Rhinomacer*; but though I have examined it and the other allied species very closely, I am unable even now to assign to them a definite place in the series of genera.

The *Brachelytron* is a *Phytosus*, which differs only as a species from the normal British representatives of the same genus.

All of Mr. Eaton's species are apterous; but Dr. Kidder, of the American Transit of Venus Expedition, states that a large brilliantly coloured specimen of a beetle flew to a light in his tent one night at Molloy Point. (Kidder, Bulletin U. S. Nat. Mus., 1876, iii. 49.)

BRACHELYTRA.

ALEOCHARIDÆ.

Phytosus atriceps.

(Plate XIV., fig. 15.)

C. Waterhouse, Ent. Mo. Mag., 1875, xii. p. 55.

Rufo-testaceus, breviter pubescens; capite abdominisque in quatuor segmentis basalibus nigrescentibus.

Long. 3·2 mm.

General form that of *P. nigriventris*, but a little broader. Reddish-yellow, scarcely shining. Antennæ scarcely longer than the head and thorax together, slightly thickened towards the apex, the three basal joints elongate, the first a little longer than the second, the third shorter than the preceding, elongate-obconic, the fourth subquadrate, the following joints blackish, the fifth to tenth short, the four penultimate transverse, the last ovate. Head rounded, thickly and finely punctulate.

Thorax a little broader than the head, scarcely broader than long, depressed, very finely coriaceous, slightly narrowed towards the base, all the angles rounded. Elytra scarcely narrower than the thorax, narrowed towards their base, one-third broader than long, very finely coriaceous, shoulders oblique. Abdomen less dull, blackish, the apex reddish, the sides nearly parallel.

The claws in this species are distinctly longer and more slender than in *Ph. nigriventris*, but the tarsi present no other difference.

Hab.—Under stones near high-water mark, Observatory Bay. Only two specimens seen, one of which has been lost. January.

Sir C. Wyville Thomson in "Good Words" for November 1874, p. 750, enumerates 3 species of the *Curculionidæ* and 1 "small *Staphylinus*" (? *Phytosus*) found on the island by the Challenger Expedition.

PHILHYDRIDA.

[Ochthebius, sp. ?

Dr. Kidder, in Bulletin U. S. Nat. Mus., 1876, iii. 49, records the finding of a very few specimens of a small black *Ochthebius* (sp. undescribed) on rocks near the sea and about the roots of wet tufts of grass. It was not observed by Mr. Eaton.]

RHYNCHOPHORA.
ECTEMNORHINIDÆ.
Canonopsis.

C. Waterh., Ent. Mo. Mag., 1875, xii. 55.

Rostrum short, thick, truncate at the apex; maxillæ visible; antennal scrobes small, deep, and somewhat reniform, open in front. Antennæ placed very near the apex of the rostrum, long, but not very slender, scape just reaching the front margin of the thorax, slightly enlarged at the apex; funiculus with the 1st and 2nd joints elongate, the 1st rather longer than the 2nd, 3rd joint short (about two-thirds the length of the 2nd), the 4th, 5th, and 6th joints globular, the 7th a little broader than the 6th, the club cone-shaped. Eyes round, moderately prominent. Thorax as long as broad, gently constricted in front and behind the middle, truncate in front and behind. Scutellum very small and triangular. Elytra united at the suture, not broader than the thorax at their base, convex, gradually enlarging to the middle, and then again narrowed to the apex. Wings absent. Legs moderately long, femora strongly clavate; anterior tibiæ nearly straight; apex of the posterior tibiæ truncate, hollowed, the margins ciliated. Tarsi spongy below, the 1st joint as long as the two following together, the 4th joint a little longer; the claws curved, separated. Metasternum very short. Intercoxal projection of the abdomen wide, nearly straight in front and at the sides; 3rd and 4th segments equally short, much shorter than the 2nd. Pygidium slightly visible from above.

Canonopsis sericeus.

(Plate XIV., fig. 9.)

C. Waterh., op. cit., p. 55.

Subfusiformis, convexus, flavo-griseo-sericeus; antennis piceis, clavâ nigrâ; fronte foveâ magnâ impressâ; thorace sub æque latitudine ac longitudine, supra in longitudinem canaliculato, canaliculâ ante medium expansâ et tuberculo obsoleto utrinque instructâ; elytris elongato-ovalibus, apud bases thorace haud latioribus attamen $3\frac{1}{2}$ longioribus, punctato-striatis, insterstitiis vix convexis, quorum secundo vittis brevibus velutinis nigris duabus ornato; femoribus apud bases piceis.

Long. 12; elytr. lat. 4 mm.

Hab.—Cat Island, Three Island Harbour, Royal Sound. Common under stones, 11th October.

Ectemnorhinus.

Ectemnorhinus, G. R. Waterh., Trans. Ent. Soc. London, series 2, ii. p. 184.

Agonelytra, C. O. Waterh., Ent. Mo. Mag., 1875, xii. 55.

M. Lacordaire (Gen. des Coleopt., vi., p. 562) says of this genus, "Corps oblong, ailé." The type specimen which Lacordaire had for examination is now in the British Museum collection, and is undoubtedly without wings.

Ectemnorhinus viridis.

(Plate XIV., fig. 10.)

Ectemnorhinus viridis, G. R. Waterh., loc. cit.

Agonelytra longipennis, C. Waterh., Ent. Mo. Mag., 1875, xii. p. 56.

Elongatus, nigro-piceus, cinereo-sericeus; capite antice angustato, fronte foveolâ leviter impressâ, rostro supra bicarinato, antennis piceis, clavâ nigrescenti; thorace leviter convexo, sub æque longitudine ac latitudine, ante et postea paulo angustato, lateribus in medio leviter rotundatis, in dorso carinâ longitudinali nitidâ; elytris apud bases thorace $\frac{1}{3}$ latioribus, et fere quadruplo longioribus, paulo in medio ampliatis, leviter punctato-striatis, interstitiis planis ad apices rotundatis humeris obtusis; corpore subter pedibusque piceis nitidis, femoribus supra tarsisque nigrescentibus.

Long. 7; lat. 2·5 mm.

Antennæ moderately long and stout, 1st and 2nd joints of the funiculus slightly elongate, subequal, the 3rd joint shorter, the 4th to 7th becoming gradually shorter and slightly transverse. The silky pubescence which covers the head, thorax, and elytra is fine, and not very close on the former. The elytra strongly embrace the abdomen, and are broadest about the middle, truncate at the base; the shoulders angular, but obtuse. The tibiæ are very slightly flexuous.

Var.—Silky pubescence green, appearing golden in some lights.

Hab.—Common in the neighbourhood of Royal Sound.

Ectemnorhinus angusticollis.

(Plate XIV., fig. 11.)

Agonelytra angusticollis, C. Waterh., op. cit., p. 56.

Ec. viridi affinis, distinguendus tamen a tertiis ad septimos articulos funiculorum antennarum transversis, a thorace supra haud carinato, elytrisque postice latioribus.

Long. 7 mm.

This species is very closely allied to the preceding; but the antennæ are shorter, owing to the 3rd to 7th joints of the funiculus being transverse, the 6th and 7th very strongly so. There is no distinct carina on the thorax, and the elytra are broadest behind the middle, very broadly rounded at their apices. The silky pubescence is yellowish in the female, with indications of two paler, broad, abbreviated bands across the suture, one next the scutellum, the second about the middle.

The male is narrower, darker in colour, and has the bands narrower, shorter, and less distinct.

Hab.—Near Royal Sound.

Ectemnorhinus gracilipes.

(Plate XIV., fig. 12.)

Agonelytra gracilipes, C. Waterh., op. cit., p. 56.

Niger, griseo-pubescens; capite supra fere plano; thorace capite latiori, in medio parum ampliato, paulo in latitudinem minus quam longitudinem; elytris ad bases thorace $\frac{2}{5}$ latioribus, gradatim usque ad medium ampliatis, postice angustatis, convexis, supra depressiusculis, leviter punctato-striatis, interstitiis planiusculis; antennis piceis, clavâ nigrescenti; pedibus gracilibus, longis, tarsis articulo tertio bene dilatato.

Long. 3·5—4·7; lat. 1·25—1·8 mm.

The elytra are gently rounded at the base; the shoulders are distinct, but very blunt.

Hab.—Near Royal Sound.

Ectemnorhinus brevis.

(Plate XIV., fig. 13.)

Agonelytra brevis, C. Waterh., op. cit., p. 57.

Niger, ænescens, viridi-griseo parce pubescens; rostro in longitudinem supra leviter impresso; thorace in longitudinem haud minus quam latitudinem, convexo, ante et postea augustato, lateribus bene rotundatis; elytris apud bases thorace $\frac{2}{5}$ latioribus, $2\frac{1}{2}$ eodem longioribus, medium tenus gradatim ampliatis, apicem versus angustatis, convexis, sat fortiter punctato-striatis, interstitiis planiusculis; antennis piceis, clavâ magnâ nigrescenti; pedibus nigrescentibus, femoribus ad bases tibiisque intus piceis.

Long. 3·2—4·7; lat. 1·7—2 mm.

The pubescence is short and thick (more like narrow scales), and is so arranged as to give the elytra a slightly mottled appearance, a spot on each side of the suture being particularly distinguishable.

Hab.—Near Royal Sound.

Ectemnorhinus eatoni.
(Plate XIV., fig. 14.)

C. Waterh., Ent. Mo. Mag., 1876, xiii. p. 51.

Pyriformis, convexiusculus, niger, parce et brevissime viridi-griseo-pubescens; capite rostroque in longitudinem fortiter impressis, hôc brevi, haud apicem versus angustato; antennis nigris; thorace paulo capite latiori, vix in longitudinem minus quam latitudinem, antice posticeque angustato, subtilissime coriaceo, subopaco, in medio dorso longitudinaliter carinato, post utrinque obsolete tuberoso; elytris ad bases thorace paulo latioribus, post bene ampliatis, supra depressiusculis, distincte striatis, striis lateralibus obsolete punctatis, interstitiis planis, transversim subtilissime strigosis; pedibus longis, femoribus ad bases piceis, et in medio inflatis, tibiis intermediis paulo curvatis.

Long. 5·5; lat. 2·05 mm.

This species is most nearly allied to *E. brevis*, but is much less short, and has the shoulders of the elytra effaced. The thorax is distinctly carinate above; the forehead and rostrum are deeply and broadly impressed, the antennæ are rather slender; the apex of each elytron is broadly and bluntly rounded, and rather expanded on the outer side; the femora are somewhat strongly inflated in the middle, the tarsi are rather narrower than in the other species of the genus, and the claw joint is very long.

A single example; probably a female.

Hab.—Near Royal Sound.

LEPIDOPTERA.—By the Rev. A. E. Eaton.

Larvæ of at least two, perhaps of three, species of Lepidoptera were obtained in the vicinage of Observatory Bay. One sort was occasionally brought off to the ships by the men in gatherings of *Pringlea* in October and the early part of November. It would probably develop into a moth about as large as an *Agrotis* of medium size, and was conjectured to belong to the *Noctuina* (Etn., Proc. Roy. Soc. 1875, xxiii. 354). Its affinities, however, may be very different, as it is likely to have been the larva of the insects referred to by Dr. Kidder as "lepidopterous " insects of moderate size, with very imperfect and abbreviated wings, active in " their movements, and . . . with . . . antennæ . . . long and thread-like " . . . [and labial palpi] "pectinate, and curling backward over the top of the head." These were obtained by him " on the evening of December 18 . . . . from the " roots of grass" (Kidder, Bulletin U. S. Nat. Mus. 1876, iii., 50). If the larvæ above mentioned and these moths are the same insect, the condition of the palpi of the imago seems to indicate that it is related to the *Gelechiidæ*. And this supposition is quite consistent with the form of some other larvæ (believed to be younger examples of the species obtained in October) which were found commonly in moss in December and February near Observatory Bay. They could be identified on the island, in situ, by the following brief description.

Larva Gelechiidiform with 16 legs, grey, paler beneath and along the spiracles; head and dorsum of first segment pale corneous, the usual raised dots of the other segments dark grey, shining, each bearing,—some a short, others a long,—testaceous hair. Length 13 mm. and upwards. Common in wet places on the hill sides near the sea, making galleries in moss.

All of these larvæ died before our arrival at the Cape. (Etn., Proc. Roy. Soc. 1875, xxiii. 504).

In Sir J. Hooker's MS. Journal mention is made of a moth with rudimentary wings clothed with mouse-coloured hair found at Christmas Harbour, which may have been Dr. Kidder's insect.

The other larvæ observed belonged to the moth described below.

Dr. Kidder reports " a single flying tineid moth was observed soon after our " landing, but supposed to be a clothes-moth from our own boxes." This could not have been the *Embryonopsis* because that is unable to fly; and as no other instance is on record of a moth being seen on the wing in Kerguelen Island, there is good reason for concurring in the opinion that it was an introduced species. *Endrosis fenestrella* had been transported to Cape Town in some boxes belonging to the English expedition a few weeks earlier.

Embryonopsis, *Eaton*, 1875.

Imago. Labial palpi long, recurved, smooth; the first joint very short, the second of moderate length, the third longer than the other two together, tapering to an acute point. Antennæ simple, filiform, the basal joint slightly larger than any of the others; in the ♂ as long as the abdomen, in the ♀ rather shorter. Anterior wings subcorneous, acuminately ovate, convex, extending almost to the apex of the abdomen in the ♂, and only as far as its middle in the ♀; longitudinal nervures 5, simple, the subcosta extending nearly to the middle of the costa, the next two nervures subequal and just falling short of the apex of the wing, the fifth terminating about as far from the fourth, as the fourth from the third; no transverse veinlets; the posterior wings extremely minute, not reaching even to the base of the abdomen; posterior femora thickened; genital appendices of ♂ abnormal, very broad, comprising two lateral pairs arising from the pleuro-ventral region; ♀ the ovipostor 2-jointed, extensile.

Larva Gelechiidiform, with 16 legs.

The smoothness of the head, and the conformation of the tri-articulate labial palpi, together with the form of the larva, attest the relationship of *Embryonopsis* to the *Gelechiidæ*. The exceptional shortness of the wings, and the presence in the male of an outer pair of lateral appendages ensheathing an inner pair, and also the broadness of these last, readily distinguish it from other genera in that family.

Embryonopsis halticella. (Pl. XIV., fig. 8.)

Eaton, Ent. Mo. Mag. 1875, xii. 61. See also E. Doubleday, Ann. Ent. Soc. Fr. 1848, Bull., p. lxiii; Thomson, Good Words, 1874, November, p. 750; Eaton, Proc. Roy. Soc. 1875, xxiii. 354; Moseley, Journ. Lin. Soc. Botany, 1876, xv. 54.

Imago ♂ ♀, sooty black varied with ochraceous; antennæ black, their basal joint, the vertex, and the palpi, sprinkled with ochraceous; forewings with a wide longitudinal ochraceous streak through the middle; hind wings pale; legs somewhat ochraceous; abdomen ochraceous at the sides; in the ♂ the last dorsal segment is triangular and is bifid at the apex, the last ventral segment is semi-elliptical; the outer pair of lateral appendages parabolic, papyraceous, about half as long as the inner, externally clothed with scales, internally glabrous and smooth; inner pair of lateral appendages corneous, very closely invested by the outer pair, and very broad, subcircular, externally flatly convex, lutescent, polished and glabrous, traversed with a few fine raised lines, and slightly incrassated towards the margin, furnished at the extreme base with a series of very long appressed hairs which spread fanwise over them; internally they are very abundantly furnished with an ochraceous tomentum disposed in slender toothlike bundles; lower penis cover (?) linear lanceolate, longitudinally concave above, narrowed suddenly to a slender point whose extreme apex is slightly turned upwards, piceous; penis sheaths strong, slender, tapering, finger-like, slightly arcuate, connivent towards their tips, and strongly bearded beneath with rigid obliquely set testaceous bristles; upper penis cover scaphoid, carinate, obtuse, testaceous. Long. corp. 5—5·5 mm.

Larva.—Dull pale yellowish, with the paler dorsal line bordered by a pale brownish lilac stripe on each side extending from the first to the penultimate segment. Head dark piceous; first segment with a transverse dorsal fuscous blotch which posteriorly is incised by the dorsal line; spiracles, the usual dots and their hairs, two dorsal spots on the penultimate segment, the dorsum of the last segment and a spot at the base of each of its prolegs on the outer side, black. Frons pale green. Long. corp. " 5 lines," probably a note-book error for 5 mm.

Hab.—Common in the vicinage of Royal Sound. The larva was found on the 17th of November (the first day the imago was noticed), feeding upon and residing within the sheathing leaves of the young shoots of *Festuca cookii*. Several specimens were obtained in patches of that grass close to the shore in Observatory Bay. It was afterwards detected in other places, and in shoots of *F. erecta*. The moth also occurs on the slopes of the inland hills. It is unusual for larvæ belonging to this family to burrow within grass shoots.

The homologies attributed to the elements of the genital armature above, are different from those which were given in the original description. Subsequent examination of the genitalia in recent representatives of most of the principal groups of Lepidoptera has led to a modification of the views entertained in 1875 concerning the nature of some of the parts. What was then taken to be the termination of the anus is now believed to be in no way connected with that outlet, and is considered to be a portion of the accessories of the genital apparatus; and similarly that which was specified as the penis in the earlier diagnosis, is now described as the lower penis cover. Many persons would doubtless be disposed to question the validity of this last correction, and maintain that the earlier view was the right one; but so far as has been ascertained the penis in the Lepidoptera is in most, if not in all cases, an erectile, membraneous introvertible tube, which is invisible in dried examples. This tube, which can be protruded by means of pressure in the living insect or one freshly killed, is closed at the apex by a sphincter, which is relaxed spasmodically at intervals to discharge a coloured fluid supposed to be *meconium*, until a doubt led to its being examined under a high power, when it was discovered to be in all cases crowded with spermatozoa. The lower penis cover projects immediately below this membraneous tube, and in the present insect is longitudinally concave above, as if with a view to its accommodation; and there is some reason for suspecting that it enters the vagina simultaneously with the penis. The upper penis cover is used in the introversion of the penis after it has been distended, thrusting it inwards as it were with the pecks of a bird's beak by its up and down motion.

Plate XIV., fig. 8, *Embryonopsis halticella* (from above). a, head (from the side); b, anterior wing; c, appendages of ♂ (from the side); c^1 and c^{1*} inner lateral appendages (from without and from within); c^2 genitalia ♂ (from the side, the lateral appendages being removed); p, upper penis cover; q, penis sheaths; r, lower penis cover (?).

DIPTERA.—*By G. H. Verrall.*

(Plate XIV.)

On comparing the Kerguelen Island Diptera with their nearest foreign allies, some are found to have retained the habits of the families to which they belong, and not to have departed in any way from their normal condition. These are they which frequent the denser portions of the herbage, and are efficiently sheltered by it even when the weather is at its worst, *e.g.*, the *Sciara* and the *Limnophyes*. A larger number, also remaining true to the habits of their kindred, are less protected from the cold than the forms before mentioned, being denizens of rocks and crevices; these have small rudimentary wings (one is apterous) and a fair amount of hair and bristles (the apterous species being the worst off in these respects). Another is likewise constant to the habits of its nearest relations; and in consequence it is more exposed to the rigours and vicissitudes of the climate than any of the other species, frequenting as it does the leaves of the Kerguelen Island cabbage (*Pringlea*); it is almost apterous, and its hair and bristles are of the most rudimentary description.

The Diptera of the Falkland Islands have their wings fully developed.

MUSCIDÆ.

§ Calyptera.

ANTHOMYINÆ.

Homalomyia, Bouché.

Homalomyia canicularis, Linné.

Musca canicularis, Lin. F. S. 1761, No. 1841.
Homalomyia canicularis, Schiner, F. Aust. Dipt. 1862, i. 654, &c.

Hab. and Dist.—In the house at Observatory Bay; introduced. This species is attendant upon man in almost every part of the world.

§§ Acalyptera.

MICROPEZINÆ, Verral.

Calycopteryx, Eaton, 1875.

Frons broad, slightly concave, produced at the base of the antennæ more than the width of the eye; bristles short, only two near the eye margin (as in Calobata), rather distant and outwardly ascending, two others near the hinder angle of the eye, of which the inner is the largest and ascends inwards, while the outer is very short and ascends outwards, also with a very short one on each side of the middle of the occipital margin projecting backwards, and with one prorect decumbent on each side of the foremost ocellus. Antennæ with short minute hair, which along the back and apex of the second joint is sparsely intermixed with some coarser and more rigid; third joint scarcely longer than the second, its outer side subcircular, its inner side somewhat reniform, deeply umbilicate at the insertion of the funiculus, the arista glabrous, arising just beyond the summit of the dorsal half of the third joint (viewing the joint as a circle, it would be just within the second superior quadrant), and so distant from the base; in the dried insect the third joint through shrinking becomes slightly transverse, and the arista seems nearer to the tip; the basal joints of the arista are distinct but short. Proboscis stout; palpi subcylindrical, mandibles lanceolate, of moderate size. Thorax completely concrete with the scutellum; the transverse suture broad and ill-defined; disc minutely scabrous, with three short bristles transversely disposed on each side behind, apparently representing the long bristles that usually precede the scutellum, and with a solitary very short bristle on each side, a little in advance of the wings. Wings represented by a minute vesicle adnate to a smaller tubercle (tegula?); halteres extremely small, clavate. Legs without long bristles of any kind or spurs, excepting 3 or 4 at the apices of the coxæ, clothed with short dense appressed hairs, which become rather longer and somewhat spreading at the tips of the femora and tibiæ; hind tarsi long and slender, with the basal joint as long as the next three. Abdomen sessile, 6-jointed, exclusive of the genital segments; ovipositor considerably extended, jointed; penis when not in use folded down upon the underside of the genital segment between the rather small lateral appendices, the same segment likewise doubled forwards to fit into an excavated protuberance at the ventral apex of the antepenultimate segment.

The larvæ feed on decaying vegetable matter.

In the structure of the antennæ, the form of the frons, the sparsity and relative positions of the bristles of the same, the round bare eyes, the absence of vibrissæ, the condition of the proboscis and the palpi, the great length of the legs and the absence of the preapical spine of the posterior tibia, the length and slimness of the fusiform abdomen, and in the construction of the genital apparatus, the affinity between *Calycopteryx* and *Micropeza* is clearly displayed.

Calycopteryx moseleyi.

(Plate XIV., fig. 1 *a–e*.)

Eaton, Ent. Mo. Mag. 1875, xii. 59.

See also Thomson, Good Words, 1874, Nov., p. 750; Eaton, Proc. Roy. Soc. 1875, xxiii. 354; Moseley, Journ. Lin. Soc. Botany, 1876, xv. 54; Kidder, Bulletin U. S. Nat. Mus. 1876, iii., 51 and 52.

Imago. Atro-corvinus, clothed with a dense appressed very minute sub-olivaceous pubescence; the dorsum, notum, and frons with very minute scattered setulæ, which are very limited in number on the frons, but numerous on the other parts, causing them to appear microscopically scabrous. Frons, oral parts, and the adjacent portion of the orbits of the eyes, coxæ, and venter luteous or orange yellow; wings and halteres testaceous; tarsi somewhat jecinoreus with stiff and moderate spreading testaceous hairs, anterior tibiæ with short testaceous hairs on their posterior surface. The ovipositor becomes flattened in drying; the ♂ lateral appendices are rounded and small; the penis for some distance is tubular and then suddenly expands into an open cup-like termination, whose edges are deeply, widely, and unevenly sinuated.

Long. corp. ♂ 8–9, ♀ 8–10·5 mm.

Hab.—Kerguelen Island, and (according to Mr. Moseley) Heard Island. The imago is very plentiful creeping over the leaves of *Pringlea antiscorbutica*; and its maggots abound in the decaying leaves accumulated at the base of the stem of the same plant. When these flies are alarmed they fold up their legs and drop down into the axils of the leaves, remaining motionless for some time before they again timidly unbend their legs and struggle to stand upon their feet. In snowy weather they appear to leave the head of the cabbage and retire to the rubbish on the ground round the base of the stem.

Mr. Moseley's statement that the insect deposits its eggs in the fluid which is caught at the bases of the leaves does not correspond with Mr. Eaton's observations.

Dr. Kidder mentioning "small scale-like bodies which Mr. Eaton supposed to " represent the balancers ('halteribus brevibus et parvis')" in this insect, adds " Baron Osten Sacken, however, finds that these scales are really representatives " of the wings."

Mr. Eaton suspects that the Baron failed to distinguish the halteres. Judging from the context, it is probable that the specimens examined by him were in fluid; and if this were the case, his oversight may be easily accounted for. Yet even in fluid they can be demonstrated without difficulty if the light be judiciously manipulated: in dried or living examples they are, however, seen to the best advantage.

The question raised by Mr. Moseley as to whether *Pringlea* flowers are frequented by *Calycopteryx* in sunshiny weather, may be answered in the negative. Mr. Eaton visited the plants at various times and at various hours of the day and night with a view to ascertain whether insects visited the inflorescence. He occasionally found *C. moseleyi* on the lower bracts of the spike, but the flowers appeared to be quite unattractive to the flies as well as to all other insects. Sir J. Hooker had invited his attention to the subject of insects visiting the flowers of plants indigenous to the island, irrespective of entomological considerations.

(Plate XIV., fig. 1.) *Calycopteryx moseleyi* (from side); a & a^1, antennæ, from within and from without; b, proboscis; x, mandible; c, thorax and 1st segment of the abdomen, from above; d, wing and halter; e, appendices ♂.

EPHYDRINÆ.

Amalopteryx, *Eaton*, 1875.

Head transverse, subquadrate, frons inclined, rather tumescent about the ocelli, bristles long, 2 ascending separately near the inner orbit, 2 suberect slightly divergent near the posterior orbit, and 2 prorect ascending and divergent by the foremost ocellus; epistoma very steep, below very prominent, widely convex and strongly bristled, above keeled in the middle with a concavity on each side; genæ large, bristled below; back of the head wide, descending below the eyes; antennæ short, second joint with spinulose bristles towards the apex; third joint circular compressed; arista very shortly pubescent; proboscis stout, the second joint of the stipes somewhat callous and pubescent beneath; palpi very short, subcylindrical; lamellæ deficient altogether. Thorax convex, well bristled; scutellum semi-elliptical, the four bristles long; wings almost as long as the abdomen, linear, with five simple longitudinal nervures, the subcosta becomes confluent with the costa at the transverse fold, the radial and cubital are united to the discoidal near its base, the radial also joins the costa just before the apex of the wing, the cubital joins the apical margin immediately behind the apex and is thickened towards its termination, the discoidal is confluent with the margin near the middle and thickens it near their junction, it is united to the apex of the thickened anal vein by the second transverse vein; halteres of moderate size, not very long, clavate; legs of moderate length, strong, without spines, the posterior femur somewhat thickened. Abdomen 5-jointed, ovate oblong; the apex in ♂ slightly obtuse, genitalia concealed, in ♀ produced into a short stout tooth-like ovipositor directed downwards and forwards.

This genus resembles *Hecamede* in the depth of the back of the head, its round naked eyes, the carination of the upper part of the epistoma, the five-jointed abdomen, and the spinose bristles of the second joint of the antenna.

Amalopteryx maritima.

(Plate XIV., fig. 2.)

Eaton, Ent. Mo. Mag. 1875, xii. 58.

See also Thomson, Good Words, 1874 (Nov.), p. 750; Eaton, Proc. Roy. Soc. 1875, xxiii. 355; Kidder, Bulletin U. S. Nat. Mus. 1876, iii. 52.

Fuligineous, with legs and setæ deep black; body completely invested with an extremely minute very closely appressed fuligineous pubescence, and with numerous short fine appressed black hairs, which are very sparse on the frons; epistoma somewhat cærulescenti-griseus; antennæ dull black; wings pale nigrescent with piceous nervures, slightly spinulose along the costal and apical margins; pulvilli whitish.

Long. corp. ♂ 3, ♀ 4·5; al. ♀ 3 mm.

Hab.—Royal Sound; common near Thumb Peak and Observatory Bay, among grass and stones along the upper limits of the sea-shore, and also in cliffs where Cormorants and Rock-hopper Penguins build. It can jump but not fly.

Dr. Kidder remarks of these flies that "they do not appear to jump in any " definite direction, but spring into the air, buzzing the small winglets with great " activity, and seem to trust to chance for a spot on which to alight, tumbling over " and over in the air. I never observed them jumping when undisturbed." This gives an accurate idea of their performance.

It may be mentioned that the only known species of the nearest allied British genus *Hecamede* has somewhat similar habits, being found on fresh marine rejectamenta, and but seldom attempting to fly. *Amalopteryx* is apparently less strictly confined to the shore, being commonest among the grass bordering the beach and among the birds' nests in the cliffs.

(Plate XIV., fig. 2.) *Amalopteryx maritima* (from the side); *a*, antenna; *b*, proboscis (distended); *c*, wing; *d*, posterior leg.

Apetenus, Eaton, 1875.

Head roundly subquadrate, slightly transverse; eyes oval convex; epistoma carinate in the middle, somewhat retiring, about as long as wide, naked; upper lip prominent, mouth opening of moderate size; genæ deep and wide, concave below the eye, their peristomal edge rather prominent, slightly everted and strongly bristled, the vibrissæ are short, but one of the bristles below the eye is patent and rather long; back of the head deeper than the width of the eye, convex, bristly; occiput with 2 erect little bristles in the middle, and with 4 or 5 longer bristles on each side just behind the eye; frons convex, rather broad, with long bristles, slightly tumid about the ocelli, the inner orbits rather broad with 3 almost equidistant bristles, of which the foremost two ascend upwards and backwards, and the hind one ascends outwards, there are also 2 bristles widely and transversely divergent near the hinder angle of the eye, and 2 prorect beside the foremost ocellus; antennæ short, the second joint spineless but with a long erect dorsal bristle in the middle, the third joint subrotund, compressed, the arista arising near the base, bare, the basal joints distinct; proboscis short, stout; labial lobes short, broad and ciliated, stipes with a small tuft of bristles beneath; palpi short, clavate; mandibles squamiform, very minute. Thorax truncate behind, slightly convex in front; setæ long; scutellum small, semi-elliptical, with 4 long setæ; wings minute squamiform; halteres small; legs simple, densely pubescent, with setæ; ungues slightly curved, rather short; pulvilli distinct. Abdomen 5-jointed, hirsute, the first joint very long; the ♂ genitalia and the ♀ first joint of the ovipositor exposed.

The larvæ feed upon tide refuse and *Enteromorpha*.

This genus apparently has relationship to *Pelina* and *Parydra*; but its affinities are not quite satisfactorily determined.

Apetenus litoralis.

(Plate XIV., fig. 3.)

Eaton, Ent. Mo. Mag., 1875, xii. 58.

See also Eaton, Proc. Roy. Soc., 1875, xxiii., p. 354.

Imago. Black, with very black bristles, and with a dense microscopically minute cinerescent pubescence: mouth pale, eyes piceous; wings nigrescent, oblong, slightly emarginate near the apex, pubescent, setulose along the costa and at the apex, halteres pale testaceous or, like the pulvilli, whitish; legs hairy, the tibiæ externally with dense spreading setulæ, and at the apex internally often with some testaceous pubescence. Abdomen with a pale cinereous spiracular line; beneath pale with a longitudinal median black stripe, which is divided into two spots at the second segment, and in the ♀ is continued as a black line along the proximal and the apical joints of the ovipositor; eggs pale ochraceous; the appendices of the male comprise a pair of arcuate finger-like connivent processes slightly pectinate towards the apex beneath, and a much shorter exterior pair of broadly triangular or ovate convex incurved flaps which enclose the others, there are besides a pair of very short linear and obtuse superior appendices.

Long. corp. ♂ ♀ 4·5—5 mm.

Hab.—Royal Sound and Swain's Bay, common amongst shore refuse. The pale grey larvæ were found amongst *Enteromorpha*.

(Plate XIV., fig. 3.) *Apetenus litoralis* (from the side). *a*, antenna; *b*, proboscis; *x*, mandible; *c*, wing; *d*, posterior leg; *e*, genitalia ♂.

BORBORINÆ.

Anatalanta, *Eaton*, 1875.

Head somewhat rounded; the epistoma retreating rapidly and deeply below the antennæ curves out again to the produced mouth border, its limits are distinctly circumscribed by a fine line; upper lip slightly projecting, the mouth opening very large; peristoma rather bristly with the large vibrissa inserted at some little distance from the margin, genæ broader than the eyes are deep; back of the head inflated, more than half as wide as deep; eyes prominent, small, round; frons broad, with 2 very short little bristles near one another on the inner orbit ascending outwards, 2 longer widely divergent ascending as usual near the hinder angle of the eye, and 2 short prorect and slightly divergent near the foremost ocellus; antennæ short, rather distant at their insertion, second joint with two dorsal setæ near the apex, third joint roundly reniform, somewhat deeper than long, arista bare, very long, with distinct and slender basal joints, inserted nearly midway between the proximal and distal edges of the joint; proboscis short and stout, the stipes very thick, hairy beneath, palpi slender clavate with rather compressed apices, mandibles small acute. Mesothorax truncate before and behind, with scarcely any slope from the disc anteriorly, and without a trace of the suture; wings utterly wanting; one long bristle at the side about the middle just behind where the suture should be, one just above the position that would be occupied by the base of the wing, one behind it nearer the scutellum, and one on the disk just before the scutellum, on each side. Scutellum transverse, with the hinder angles well defined, and with 4 long almost equidistant setæ. Halteres totally absent. Legs slender and rather long without bristles excepting on the coxæ and about the apex of the intermediate and posterior tibiæ; hind tarsus with the first joint dilated, more than half as long as the second joint, which is also slightly dilated and is almost as long as the remaining three taken together; the thickened joints are densely pubescent beneath; abdomen 6-jointed, flattened, in ♂ oval, in ♀ broadly ovate and obtuse, the basal joint as long as the next two together, the bristles all very minute; genitalia concealed.

Larva carnivorous, and probably also capable of thriving upon putrescent vegetable substances.

This genus comes very near *Borborus*, and especially to the subgenus *Apterina* of Macquart, in the condition of the hind tarsus and other salient characteristics. It differs from it in the form of the palpi, in having no trace of either the wings or the halteres, and in the thorax being conspicuously narrow and small in comparison with the head and abdomen.

Anatalanta aptera.

(Plate XIV., fig. 4.)

Eaton, Ent. Mo. Mag. 1875, xii. 59.

See also Eaton, Proc. Roy. Soc. 1875, xxiii. 354; Moseley, Journ. Lin. Soc. Botany, 1876, xv. 54 (apterous fly as large as a house fly); Kidder, Bulletin U. S. Nat. Mus. 1876, iii. 52.

Imago closely invested with a black dense microscopically minute down; eyes and legs piceous, femora nigrescent above; mesothorax, legs, and abdomen with rather minute, crowded, more or less appressed, very black coarse hairs, which give to the dorsum of the mesothorax a peculiar falsely rugose appearance; they are less obvious on the frons and more sparsely scattered; the thickened joints of the posterior tarsi have a dense yellowish pubescence beneath, and the intermediate tibiæ have one or two spine-like setæ near their apex.

The larvæ are found in dead birds.

Hab.—Kerguelen Island, generally distributed near the sea. This species was common under stones along the landward border of the shores of Observatory Bay, and also on dead birds in many places.

(Plate XIV., fig. 4.) *Anatalanta aptera* (♂ from the side), 4' (♀ from above). *a*, *a*¹, antennæ from within and from without; *b*, proboscis; *x*, mandibles; *c*, legs, *c*¹, anterior; *c*², intermediate; *c*³, posterior leg.

NEMOCERA.
MYCETOPHILIDÆ.
SCIARINÆ.
Sciara.

Sp. —— ?

A single female *Sciara* probably indigenous to the island, was taken on a window at "Flamsteed House," Observatory Bay, on the 4th of January 1875. It would be absurd to describe it.

CECIDOMYIDÆ.
LESTREMINÆ.
Limnophyes, *Eaton*, 1875.

Imago; head small, ovately triangular; eyes roundly oval, hardly reniform; ocelli absent; antennæ divergently prorect, filiform, 6-jointed, with sparse verticils of spreading hairs, the basal joint very stout, the second much smaller than the first, but yet slightly thicker than the remaining joints, which are of even width, the apical joint as long as the preceding two together; mouth short, the margin hairy, palpi 4-jointed. Thorax robust, above arched anteriorly and produced like a hood over the head, its contour viewed from above is somewhat ovate, and it has about four longitudinal rows of short fine sparse hairs ascending upwards and inwards; scutellum moderately large, prominent, semicircular or roundly subquadrate. Wings oblong, suddenly constricted at the base, rather straight along the costa, the apex almost parabolic, margins ciliated; 4 longitudinal veins (Nos. 1, 3, 4, and 5); No. 1 very short, becoming obsolescent in the marginal area; No. 3 extending beyond the middle of the costa; No. 4 deeply forked, united by a crossveinlet to No. 3 just beyond the point of furcation, its upper branch like No. 3 accompanied by a slight crease in the membrane; No. 5 rather deeply forked, the furcation acute, similarly accompanied by a crease which follows its lower branch; this last vein is succeeded by one or two longitudinal folds simulating additional nervures. Halteres large. Legs slender, with fine short hairs; tibiæ almost scabrous, with a minute spine at the apex interiorly; the first tarsal joint much longer than the next. Abdomen slender, 8-jointed, with a few fine hairs above; ovipositor formed of two very short lamellæ.

Larva not observed.

Allied to *Campylomyza*, but differing in the neuration of the wings, and in the deficiency of ocelli.

In the original diagnosis it was stated that the number of joints in the palpi and abdomen were respectively five and seven. It appears to be more correct to regard them as being four and eight jointed. The antennal joints are likely to vary in number with the sex, and to be more numerous in the male than in the female.

Limnophyes pusillus.

(Plate XIV., fig. 5.)

Eaton, Ent. Mo. Mag. 1875, xii. 60.

See also Eaton, Proc. Roy. Soc. 1875, xxiii. 354; Moseley, Journ. Lin. Soc. Botany, 1876, xv. 54 ("winged gnat"); Kidder, Bulletin U. S. Nat. Mus., 1876, iii. 52 ("small gnat").

Imago.—Head and thorax lutescent, eyes black, antennæ griseous with the basal joint pale; mesothorax with a large lanceolate black dorsal spot in the middle, anteriorly ochreous at the sides above, the mesosternum nigrescent, and the wings almost imperceptibly cinerescent; legs griseous with the coxæ whitish. Abdomen dull virescenti-griseous, with the last three segments nigrescent beneath.

Long. ♀ 1 mm.

Hab.—Royal Sound, abundant. Especially plentiful on moss in boggy places, and in windows. It flies freely in calm sunny weather.

(Plate XIV., fig. 5.) *Limnophyes pusillus* (from the side.)

a, antenna; *b*, palpus and mouth (compressed); *c*, head and thorax (from the side).

CHIRONOMIDÆ.

Halirytus, *Eaton*, 1875.

Imago ♀.—Head suborbicular, palpi very short, 2-jointed; antennæ divergent 6-jointed, the basal joint very large, nearly orbicular, the next four much smaller, submoniliform, the apical joint oval, about as long as the preceding two together, the basal joint has one rather short, and a few still shorter bristles near its middle, and the apical joint has a short bristle on one side, and a finer hair on the other side near its base, and some extremely minute pubescence, which is hardly discernible even under the microscope; genæ each with one minute bristle below the eye; epistoma scutiform; eyes suborbicular, protuberant, close to their upper orbit behind are three short bristles, the hinder two of which are near together; ocelli absent. Mesonotum somewhat cucullate, being strongly arched in front and projecting forwards above the head; scutellum semi-elliptical, prominent, with a transverse line of minute erect bristles; metanotum very transverse, exceedingly short; the spiracles on each side of the mesothorax are very prominent; wings rudimentary, somewhat narrowly obovate, reaching to the apex of the first abdominal segment; halteres small, clavate and slender; legs very long, the posterior tibiæ not thickened nor spurred, the proximal joints of tarsi very long, ungues and pulvilli very small. Abdomen with 7 dorsal and 6 ventral segments (exclusive of the base supporting the valves of the ovipositor), subcylindrical; ovipositor pointed obliquely downwards, composed of a stout basal joint terminated by a pair of acute short lanceolate lamellæ enclosing a smaller pair of spicules. Male unknown.

The larva probably feeds on *Enteromorpha*.

This genus is akin to *Corynoneura*, from which it is separated by its 2-jointed palpi, the comparative nakedness of its antennæ, its entire eyes, the spurless tibiæ of which the hind pair is not thickened, and perhaps the number of abdominal segments. If the portion reckoned above as the base of the ovipositor be regarded as a segment, then there is no difference between these genera in that last particular. All the known species of Corynoneura are extremely minute. In the original diagnosis the number of the segments was said to be five; they were enumerated from below, and the proximal segment was taken to be metathoracic.

Halirytus amphibius.
(Plate XIV., fig. 6.)

Eaton, Ent. Mo. Mag. 1875, xii. 60.

See also Eaton, Proc. Roy. Soc. 1875, xxiii. 354. ("Another seems to be a degraded member of the Tipulidæ," as limited in Westw. Introd.) Possibly also Mr. Moseley's "apterous gnat (*Culex*)" belongs here; Moseley, Journ. Lin. Soc. Botany, 1876, xv. 54.

Imago. Black above, virescenti-griseous beneath and at the sides of the thorax; head virescenti-griseous, with eyes and labrum black, antennæ pale cinereous; wings and halteres opaque, whitish; legs virescenti-griseous, with scattered minute black hairs; abdominal segments above, each with a fine transverse black line at the base, whose extremities are produced obliquely backwards and downwards for a short distance on each side, the tips of the segments narrowly whitish; beneath the tips of the segments are pale, the remaining dark portion of each segment is enclosed by a black line, and stippled with pale dots (at the insertion of minute hairs), and in the middle of their bases some segments have two diverging black lines, others have black stripes; the base of the ovipositor is black, its ventral portion is scutiform with the apiculus bifid, and anteriorly is punctulated like the previous ventral segments; the laminæ of the ovipositor are testaceous.

Long. corp. ♀ 4—5 mm.

Hab.—Royal Sound and Swain's Bay, at the verge of the tide, creeping over *Enteromorpha* and Mussels exposed by the recess of the sea, and walking upon the surface of puddles and tide-pools. Near Observatory Bay the fly was common upon some small isolated rocks which were always submerged at high water. The adults in that locality must spend a large portion of their lives under water, and hence the species was named *H. amphibius*. All of them were females; none were actually seen beneath the surface. Probably whenever the water has retired sufficiently from the top of the rocks, all the flies hurry up from below to take an airing.

It is rather likely that the males have fully developed wings, and are able to fly.

Plate XIV., fig. 6, *Halirytus amphibius* (from the side): *a*, antennæ; *b*, legs— b^1, anterior; b^2, intermediate; b^3, posterior.

NEUROPTERA.—*By the Rev. A. E. Eaton.*

[PSOCIDÆ.

Rhyopsocus eclipticus, *Hagen.*

Bulletin U. S. Nat. Mus., 1876, iii. 52—57.

The true Neuroptera of Kerguelen Island are as plentiful as the snakes of Iceland; and it is doubtful whether there is any representative of this order indigenous to it, unless the *Mallophaga*, which *must* be placed somewhere, are reckoned as members of it. For the species of the *Psocidæ* cited above, described from a single example taken at Molloy Point, and mounted on glass in balsam as a microscopic object, is of uncertain nationality, and may have accompanied the American Transit of Venus Expedition from Washington. Dr. Kidder (loc. cit.), recording its apprehension " on October 17, within doors," remarks, " Shortly before " its capture some instrument-boxes, brought from Washington, and containing a " quantity of packing straw, had been unpacked in the same room; a circumstance " rendering the habitat of the insect very doubtful at the time."]

COLLEMBOLA.—*By Sir John Lubbock, F.R.S.*

The Collembola collected in Kerguelen Island comprised three forms.

One was a species of *Isotoma*, apparently, but the examples were not in a condition to be determinable with certainty.

One was a *Smynthurus*. In this genus the forms of the feet and of the saltatory apparatus afford good specific characters. There being only one specimen in the collection, this was sent to a professional mounter of microscopic preparations. Unfortunately it was put up with the legs bent under the body so much as to prevent the feet being examined; and in attempting to rearrange the specimen so as to show them, he destroyed it. Under any circumstances, however, it would have been unsatisfactory to describe a new species of *Smynthurus* from a single specimen, and without knowing the true colour. The only example was obtained under a stone near a lake not far from the chief English Observatory, in January.

The third species constitutes a new genus, which I have dedicated to M. Tullberg, who has done so much to extend our knowledge of the group to which it belongs.

LIPIURIDÆ.

Tullbergia.

Lubbock, Ann. & Mag. Nat. Hist., 4th ser., 1876, xviii. 324.

Corpus elongatum. Antennæ non clavatæ, 4-articulatæ. Organa post-antennalia transversa. Unguiculi inferiores nulli. Spinæ anales magni.

The *Lipiuridæ* differ from all other Collembola, except the *Anouridæ*, in the absence of the remarkable saltatory organ which is so characteristic of the *Poduridæ*. From the *Anouridæ* they are at once distinguishable by the mandibulate mouth.

The present genus is characterised by the cylindrical antennæ, the uni-unguiculate feet, and the large anal spines.

Tullbergia antarctica. (Plate XIII., fig. 1, *a—c*.)

Lubbock, loc. cit.

White (colourless in spirits). Skin granular, and with scattered hairs. Antennæ 4-articulate, non-clavate. Ocelli none, or not apparent. Post-antennal organ transverse, placed directly behind the antennæ; it has numerous oval tubercles. Feet with only one claw, and no tenent hairs. Anal spines large and strong; their apex oblique and outwardly prolonged into a somewhat slender triangular point, not acuminate. Length 3 mm.

Hab.—Common in wet moss on hillsides and low ground in the neighbourhood of Observatory Bay, Royal Sound.

Plate XIII., fig. 1, *T. antarctica*, seen from above (magnified): *a*, antenna; *b*, foot; *c*, anal spines.

MALLOPHAGA.—*By C. Giebel.**

(Plate XIV.)

The Mallophaga entrusted to me for examination were collected from *Pelecanoides, Prion, Halobæna, Diomedea,* and *Aptenodytes*, and comprise species of the genera *Docophorus, Nirmus, Goniodes,* and *Lipeurus*. Only one, and that the commonest, obtained from *Procellaria nereis, Prion desolatus,* and perhaps also from *Diomedea exulans*, was previously known, having been recorded from *Halobæna cærulea*. The four additional species described as new are separated by very marked characters from all others hitherto known in their respective genera. Among them *Goniodes brevipes* from *Aptenodytes longirostris* is particularly interesting, because no *Philoptera* at all were known from this bird, and this genus has been observed on Natatores only once before, being especially parasitic upon Gallinaceæ. *Pelecanoides* also had previously furnished no Mallophaga, and the species of *Nirmus* procured from it is very distinct. The *Docophorus dentatus* from *Diomedea exulans* is remarkable on account of its relationship to a species observed on a *Vultur*, from which it is distinguished by very precise differentia.

Docophorus dentatus.

(Plate XIV., fig. 16.)

Giebel, Ann. & Mag. of Nat. Hist. 1876, May, xvii. 388.

Brevis, latus, capite rotundato truncato-trigono, marginibus lateralibus multisetosis, posticis gracillime bidentatis; antennis setaceis; signatura frontali feminis triangulari in lineam mediam occipitalem exeunte. Thorace brevi, lato; metathoracis hexagonis angulis posticis dentiformibus; pedibus brevibus, tibiis multispinosis. Abdomine orbiculari, maculis marginalibus intus rotundatis, ventralibus partitis.

Mas obscurior, marginibus profunde crenatis, fasciis in medio divisis.

Long. corp. 3·—3·75 mm., capit. 1·25 mm., thorac. 0.75 mm., abd. 1·75 mm.

This species is distinguished from all other known *Docophori* by its abundant and strong lateral setæ, the two backward processes of the hind margin of the temporal border, and by the excised apex of the retro-duced metanotum. It is short, broad and plump, like *D. brevicollis* from *Vultur monarchus;* but differs from that species in the peculiar formation and the characteristic markings of the body.

Head short, broad, the fore-part shorter than the hinder; clypeus broad, almost truncate in front, beset with short scattered hairs; trabeculæ very long, acute,

* Translated from the German by the Rev. A. E. Eaton and Mr. R. M'Lachlan.

extending backwards over the antennal sinus; tempora outwardly expanded, broad, convex, with long setæ at the lateral margins, posteriorly with a concolorous, irregular, tooth-like process prolonged backwards on either side. Antennæ inserted in a deep sinus before the middle of the head, filiform; the basal joint the strongest and somewhat shorter than the second joint which is the longest, the others gradually tapering, the third joint shorter than the second and longer than the fourth which is subequal to the fifth. The brown frontal marking forms an equilateral triangle, extending beyond the middle of the head, and emits a fine pale line from its apex to the middle of the occipital border, and another from each of its obtuse lateral angles to the frenal margin. Thoracic segments transverse, bordered with brown, divided longitudinally by a fine median line: prothorax narrowed posteriorly, the lateral angle bearing one seta and slightly projecting shortly before the sub-convex hinder margin; meta-thorax broad, transversely 6-sided, lateral angles rounded and beset with several long setæ, hinder angles acute. Legs very short and stout, the femora and tibiæ with scattered hairs, the latter fringed with shorter stiff bristles along the inner edge, and with two long strong spines at the apex; ungues long. Abdomen sub-orbicular, the sides only very slightly crenate, beset with the usual setæ, and with short scattered hairs on the disk above and beneath: segments above marked with brown freckles which coalesce at the lateral borders, and are produced inwards into slightly narrower stripes of even width whose obtuse ends surround a narrow middle space; from the second segment up to the antepenultimate these markings are intersected by a pale band running parallel with the side of the abdomen: on the ventral surface the inner row of the spots is more conspicuous than the outer, and under a high magnifying power each of the spots composing it is seen to be subdivided into three contiguous with one another. In the smaller and paler female the apical segment is sharply excised.

In the male (one specimen), which is dark brown, and larger than the female, the frontal marking is much shorter and broader than that of the other sex, is not triangular, and terminates in a median tooth-like point exactly between the antennæ. The abdomen above up to the penultimate segment has the brown bands interrupted by a pale median line only, instead of by a band: beneath, this pale median space is present only in the anterior part, the bands being continuous in the hinder segments, and the brown lateral margin is sharply separated from the bands. The sharply produced posterior angles of the apical segment are serrate.

Hab.—On *Diomedea exulans*, among the breast feathers. Five examples collected in March 1875.

Although similar sexual differences have not yet been observed in any other species of *Docophorus*, yet on considering the agreement in other respects between the female examples and this darker and larger male, I dare not separate these two forms.

From *Diomedea* Nitzsch mentions only *Docophorus thoracicus*. His types being no longer at hand a comparison is not possible; but since he states that that species agrees with *Lipeurus taurus* in size and colour, it must be distinct from ours. *Doc. brevicollis* (Giebel, Insecta epizoa tab. x. 7) from *Vultur monarchus* is allied to *Doc. dentatus* in general habit and markings; but a close comparison reveals very considerable differences in the form, and also in the lesser details of the ornamentation.

Nirmus angulicollis.

(Plate XIV., fig. 17.)

Giebel, Ann. & Mag. of Nat. Hist. 1876, xvii. 388.

Oblongus, fulvus, fusco-pictus; capite semi-elliptico, antice brevi-rotundato, antennis ante medium insertis; prothoracis angulis anticis acute exstantibus, metathoracis coarctati angulis obtusis; abdomine angusto marginibus crenatis, maculis rectangulatis ventralibus bipartitis.

Long. corp. 3·25 mm., capit. 0·50 mm., thorac. 0·75, abd. 2 mm.

A species distinguished from all others similar to it by the configuration of the head, and still more by the peculiar form of the two thoracic segments.

Head as broad in front as behind, anteriorly rounded abruptly and furnished with 8–10 marginal bristles on each side. Foremost angle of the deep antennal sinus not at all prominent; the hinder angle, on the contrary, has a very considerable eye knob. Temporal margins parallel with one another, studded with minute and distant hairs; occipital margin slightly emarginate. The antennæ reach about as far as the occipital margin; the basal joint stout, the second the longest, the third equal to the fourth and the shortest, the fifth thicker and somewhat longer; all have long, fine, minute, distant hairs, and the apical joint is terminated with a tuft of bristles. Head pale brown, with dark brown spots before and after the antennal groove, and with two such spots, triangular, at the occipital border; tempora rather darker than the middle of the head. Thorax brown, with a pale longitudinal median line: prothorax as wide as the occiput, somewhat transverse; the neck contracted in front in the sinus of the occipital margin; its acute anterior lateral angles extend sideways as widely as the rounded temporal angles; in advance of the rounded posterior angles the sides of the prothorax appear somewhat narrowed, and they are destitute of marginal setæ. The much longer metathorax is slightly narrowed before the middle, and has rounded angles destitute of bristles. Legs slender, brown, with pale apices to the joints, and with fine little hairs; femora and tibiæ of almost equal length and strength, the last with several strong apical spines; ungues strongly curved. The abdomen attains its greatest width at the 5th and 6th segments; at the 7th it again becomes somewhat narrower; the last two segments are greatly abbreviated and suddenly narrowed: the hinder angles of the segments, though acute, project only a little, and hence the sides of the abdomen are merely slightly crenate; their setæ usually become more numerous towards the apex, but are not very long; the 8th segment has on each side a marginal process, and as well as the 9th has numerous setæ at the hinder margin. Above, the foremost seven segments are marked with transverse quadrangular brown spots separated from one another only by the pale median line and the paler joinings of the segments, so that the segment in front and behind has a very dark lateral spot. Beneath, these quadrangular spots are separated into an inner paler row and an outer row of dark spots marked with the stigmata, by means of a pale longitudinal line parallel with the side at the inner half of the spiracular row.

Hab.—On *Diomedea exulans*, with *Docophorus dentatus*; three examples.

The broad rounding of the short fore-head, the acute tooth-like four angles and the lateral marginal excision of the very broad prothorax, as well as the conspicuous contraction of the metathorax before its middle, prevent this new species from being confounded with any of the very numerous known species of this genus. In general habits and marking, it may be placed next to *Nirmus fenestratus* (Giebel, Insekt. Epizoa, 1875, tab. vi. 7) of the Cuckoo.

Nirmus setosus.

(Plate XIV., fig. 18.)

Giebel, Ann. & Mag. of Nat. Hist. 1876, xvii. 388.

Flavus, fusco pictus; capite obtuse trigono-cordato, temporibus late rotundatis, multisetosis; prothorace lato, metathorace trapezoidali angulis lateralibus obtusis, setis multis atque longis instructis; abdomine oblongo marginibus obtuse crenatis, segmentisque fusco-vittatis.

Long. corp. 2 mm., capit. 0·20 mm., thorac. 0·20 mm., abd. 1·20 mm.

This elongated pale yellow species with darker edging has its most striking characters in the copious bristling of the tempora and of the thorax.

Head somewhat longer than its width behind, narrowed rather suddenly in front of the antennæ to about $\frac{1}{3}$ of its greatest breadth, the clypeus border strongly convex, terminated by two setæ on each side, behind which stands on both sides a long marginal seta in the midst of the labrum. Anterior angle of the antennal sinus prominent and acute, furnished at the base with a small spine. The hind-head has broad rounded tempora, each of which carries six long setæ, which project in part over the metanotum. Antennæ with the basal joint long and stout, the second somewhat shorter and more slender, the third equal to the fourth and shorter still, the fifth again rather longer and with a tuft of bristles at the end. Prothorax transverse, widest behind the middle, bearing four long setæ behind the obtusely rounded lateral angles. Metathorax longer but not broader, with a tuft of yet longer bristles at its even blunter posterior lateral angles. Legs slender; the femora rather shorter and stouter than the tibiæ, both of them with minute scattered hairs. Abdomen narrow, slender, widened only a little posteriorly, and narrowed again so as to terminate obtusely: the first of the posterior lateral angles of the segments has one minute spine, the second two, and the following have three and four long setæ of unequal length; the broadly trilateral apical segment has a short seta, and in some examples bears a triangular plate covering the anal aperture. Above and beneath, the segments have small scattered hairs.

Body pale yellow (in immature specimens whitish, with narrow brown edgings), the labrum and antennal sinus, as well as two slender triangular occipital spots which are prolonged as temporal lines convergently to the dark eye-spots, brown. Thorax margined at the sides with brown. The anterior abdominal segments have acute marginal brown spots about their middle, and these are themselves connected by means of bands on each side, which, though narrowed in the middle, are still continuous; in the posterior segments these markings become paler, and they are wanting in the apical segment; the spots encircling the stigmata become dark brown at the lips of these orifices.

Hab.—On *Pelecanoides urinatrix*, in the white feathers of the breast and belly. Six examples, two of them immature, captured on the 14th October 1874 at Observatory Bay, Kerguelen Island.

Among the known species, *Nirmus fusco-marginatus* is nearly related to the present insect; but it is distinguished from the Kerguelen animal by its more decidedly narrowed fore-head, and by its wanting the abdominal bands. *Nirmus depressus* on *Phalacrocorax brasiliensis* differs in the widened form of the thoracic segments, and in its oval, very differently marked abdomen. The similar species inhabiting the Gulls and Terns are readily separable by their white colour and black decoration. The long strong bristles of the temples and of the angles of the thoracic segments also afford to the eye marked peculiarities, from which the name is taken.

Goniodes brevipes.

(Plate XIV., fig. 19.)

Giebel, Ann. & Mag. of Nat. Hist., 1876, xvii. 389.

Capite thoraceque flavis, fusco-marginatis; abdomine albido, maculis marginalibus fuscis oblique fusiformibus: capite antice parabolico temporibus dilatatis, angulatis, postice in dentem prolongatis; antennis brevibus; prothorace transverse oblongo; metathoracis latioris lateribus angulatis margine postico valde convexo; pedibus brevissimis; abdomine late ovali, marginibus lente crenatis, segmentis setigeris, ultimo lato emarginato.

Long. corp. ♀ 1·50 mm.

Parabolic margin of the fore-head evenly beset with only fine short hairs; the obtuse anterior angles of the outwardly and posteriorly greatly expanded tempora have three minute spines apiece; the tooth-like posterior temporal angles extend backwards as far as the middle of the prothorax, and from each of them a strong bristle is prolonged further over it; the occipital margin, excavated between these acute angles, is, in the middle, again rather convex. Antennæ sunk in a shallow lateral sinus in the middle of the head, very short, hardly reaching as far as the anterior temporal angles; the short, stout, and obtusely conical basal joint with short scattered hairs, the next three joints obviously shorter and successively decreasing in length; the obliquely truncate extremity of the apical joint has a little tuft of bristles. Prothorax twice as long as wide, the sides slightly convex, the hind margin very convex, destitute of marginal setæ: metathorax shorter but conspicuously broader, its obtuse lateral angles very prominent, each of them with three long setæ and a minute spine in front of these, its very convex, almost angular, posterior margin encroaches deeply upon the abdomen. Legs short and weak, with minute scattered hairs; tibiæ cylindrical with a brown terminal annulus; tarsal joints short, annular, with two spines on the inner margin; ungues short, conical. Abdomen compressed, sides sub-parallel, slightly narrowed only at the posterior segments so as to terminate very obtusely: the first segment with 1, the following segments with 2, the posterior segments with 3 or 4 strong marginal setæ, the apical segment with two pairs of them on each side of the median incisure: above, along the middle of the back are dense decumbent setæ, which are most numerous on the fourth segment, but afterwards becoming gradually fewer are altogether absent on the last two segments; beneath, along the middle of the segments the decumbent setæ are less crowded.

Head yellow edged with brown, which colour extends to the anterior temporal angles: on the occipital margin are two narrowly triangular dark brown marginal spots from whose apices the line bounding the region of the vertex is prolonged. In the thorax the brown edging becomes paler especially about the middle. Abdomen yellowish white; the segments have on both sides a pale brown marginal spot pointed inwards and outwards and marked in its middle by the pale stigma; the yellow apical segment has no marginal spot: beneath, the stigmata are very conspicuous through their dark edging.

Hab.—On *Aptenodytes longirostris* among the neck feathers. One specimen obtained in January in Swain's Bay.

G. mammillatus, from *Pelecanus ruficollis*, differs from the present species in its almost quadrate head, its obtuse temporal angles, its much longer antennæ, and its stout femora and long tibiæ. *G. heterocerus* is nearer to it in general habit, and so is *G. chelicornis*; but these are separated clearly by their short fore-head, their shorter posterior temporal angles, their longer antennæ and legs, and by their altogether different setæ and abdominal markings.

Lipeurus clypeatus.

(Plate XIV., fig. 20.)

Giebel, Insecta Epizoa, 236; idem, Ann. & Mag. of Nat. Hist., 1876, xvii. 389.

Oblongus fulvus fusco-pictus; clypeo excisuris lateralibus definito; antennis gracilibus; prothorace trapezoidali, metathorace longiore; abdomine anguste oblongo, marginibus profunde crenatis, nigro fuscis, feminæ fasciis fuscis.

Long. corp. 2·5 mm., capit. 0·2 mm., thorac. 0·35 mm., abd. 1·75 mm.

♀ Head elongate, narrowed gradually in advance of the antennæ to about $\frac{1}{3}$ of its breadth, ending with the very convex fore margin of the clypeus, which itself is sharply limited on each side by an acutely defined marginal notch; in this notch stand 2 bristles of equal length, and beneath in front of the mandibles is 1 on each side, while posteriorly at the winged margin are 3—5 bristles half as long. The parallel temporal margins, wanting setæ, posteriorly turn inwards at right angles at the slightly concave occipital margin, and each curve has one long marginal bristle. Antennæ inserted after the middle of the head, each in a deep sinus whose anterior angle is acute; their first joint equal in length to the second, the third and fourth considerably shorter, the fifth longer than the penultimate and terminated with tufts of bristles. Prothorax somewhat transverse, and like the metathorax trapezoidal with the sides almost straight and without bristles, only the metathorax has four unequal bristles at the angles. Anterior legs as usual the shortest and stoutest, the other two pairs more slender, the hindermost reaching backwards as far as the middle of the abdomen: coxæ long; trochanters sharply defined; femora slender, longer than the tibiæ, which like them have but few hairs; ungues slender, strongly curved. Abdomen long and narrow, very slightly widened in the middle: segments somewhat transverse, all of equal length from the first, their apical margin slightly convex, their posterior lateral angles obtusely rounded but projecting so as to give a crenate outline to the side; the said angles of the first segment with one seta, the following with three very long unequal setæ, the last two segments with short and minute bristles: above and beneath, the surface has only very scattered bristles.

♀ Head pale yellow, laterally bordered with brown; the clypeus projecting in front, clear and transparent; ocelli black; antennæ without markings. Thorax edged with darker at the sides; legs without markings. Abdomen pale brown bordered with black brown, the edging becoming paler at the last two segments.

♂ Antennæ longer than in the female. Abdomen whitish, with blackish brown marginal spots and pale stigmata.

Hab.—♀ Five examples were captured in the feathers of the neck and breast of *Procellaria nereis* on the 6th Nov. 1874 at Observatory Bay. Another example was previously taken there on *Prion desolatus*, 14th Oct.

♂ One example from the neck feathers of *Diomedea exulans*, in March 1875. As only one specimen lies before me, I am unable to decide whether the differences above described which it presents are indicative of specific distinction from the females with which I have associated it, or are merely sexual peculiarities.

L. clypeatus, first described from several examples from *Halobæna cærulea* in the collection of the Museum of Halle, cannot be confounded with any other known species occurring on the Storm Petrels, owing to the peculiar form of the clypeus, the short antennæ, the trapezoidal segments of the thorax, and the abdominal markings.

Lipeurus —— sp ?

A white skin of a *Lipeurus* 2 mm. long, from the head of *Diomedea fuliginosa*, was obtained in Dec. 1874; but no systematic position can be assigned to it. It is true that Rudow gave a diagnosis of a species from the same bird, under the name *Lipeurus meridionalis* (See Giebel, Insecta Epizoa, 255); but that was stated to have an octagonal metathorax, and a broad pale-dull-yellowish abdomen, which peculiarities, to say nothing of the markings, are not shared by our insect; besides the head of that species is compact, whereas in ours it is elongate.

MARINE ANNELIDA.—*By W. C. McIntosh, M.D., F.R.S.*

(Plate XV.)

The collection of Marine Annelida made during the stay of the English Transit of Venus Expedition comprises seven species, representing five families, one of which, however, is Nemertean. Six appear to be new. Like the Polyzoa and Cœlenterata they were procured by a grapple in the Laminarian region, from depths of 10 fathoms and under. The Rev. A. E. Eaton states that the shore where he was stationed was somewhat unfavourable for collecting between tide-marks, as it consisted for the most part of ledges of rock without loose boulders, or of a coarse and barren shingle. The mean temperature of the water between tide-marks was 36° F. Mr. Eaton found the same paucity of Annelida in the littoral region at Spitzbergen.

The American Transit of Venus Expedition obtained 4 species.

The tubicolar forms and Polynoidæ occurred on the roots of *Macrocystis*, and some of the young Nereids in the usual silken tubes on the fronds of *Delesseria*. None of the Annelids were found under stones, excepting the Earthworm described by Professor Lankester.

POLYNOIDÆ.

Hermadion longicirratus.

(Plate XV., figs. 1—4.)

Hermadion longicirratus, Kbg. Fregatten Eugen. Resa, &c., p. 22, taf. vi. 33.

This form seems to be identical with Kinberg's species from York Bay, Straits of Magellan, though the scales and bristles differ slightly from the published figures—the former being densely covered with minute spinulose papillæ (fig. 1), and the latter (fig. 2) showing dorsally a less expanded distal region, with a close series of oblique rows of spines (fig. 3). The tip in some is slightly dilated. The ventral bristles, again, have the curve of the terminal hook pronounced, while the spinous region is rather narrow and short (fig. 4). All the bristles are of a deep brownish yellow hue. The antennæ, tentacular cirri, and dorsal cirri have a filiform tip attached to a bulbous region, the latter and the rest of the cirrus beneath being furnished with small clavate papillæ. Much more minute clavate papillæ occur on the palpi. The brownish scales generally have a few whitish touches: the first is circular, the succeeding reniform, and the posterior elongated from before backwards. It is a large and broad form, one specimen being about $2\frac{1}{4}$ inches long.

Hab.—Swain's Bay and Royal Sound, Kerguelen Island (*Eaton*); York Bay, Straits of Magellan (*Kinberg*).

Eupolynoë mollis.

(Plate XV., figs. 5—9.)

M'Intosh, Ann. & Mag. of Nat. Hist. 4th ser., 1876, xvii., p. 319.

This species superficially resembles *Alentia gelatinosa*, Sars, though a close examination shows many points of difference, and leaves a general impression that the form is intermediate in character between the latter and such types as *Harmothoë imbricata*, L.

The head is proportionally larger, and does not exhibit the nuchal process so characteristic of *A. gelatinosa*; while instead of the closely approximated pair of large eyes on each side, the lateral pairs are widely separated, a large one occupying the anterior prominence and a small one being situated at the posterior border. Moreover they nearly constitute a square, whereas in *A. gelatinosa* they lie in the processes of a V. The tentacle is absent; but its basal segment is very large in comparison with the antennæ and tentacular cirri. In *A. gelatinosa* they do not differ much.

The scales appear to be fifteen on each side, and they are nearly as soft as those of *A. gelatinosa*, which they further resemble (though smaller) in shape and smoothness. With regard to the latter, however, a high power shows that there is a limited area, near the outer and anterior border, covered with distinct papillæ which are low and truncate (fig. 5). The dorsal cirrus has a very slight enlargement below the tapering tip (as in *A. gelatinosa*); but, in addition, it has a few minute clavate papillæ. The latter also occur on the ventral cirri.

The feet are as distinctly marked as in *Alentia*; but there is a much greater disproportion between the dorsal and ventral bristles, both of which are pale. The dorsal fascicle consists of a short series of somewhat translucent bristles with distinct spinous rows (almost as well marked as in *Evarne*), and gently tapering to a smooth portion at the tip (fig. 6) the fine longitudinal lines being somewhat wavy. The long ventral bristles, again, consist of two groups, more evidently separated than in *Alentia* or *Eupolynoë anticostiensis*. The superior tuft arises behind the spine, and is composed dorsally of slender bristles (fig. 7) with very elongated and delicately tapered spinous regions, ending in minutely bifid tips like those in *Eupolynoë anticostiensis*.[*] A gradual change ensues toward the lower bristles (of this tuft), which have a stouter shaft, a shorter spinous region, and a strong hook with a secondary process at the tip (fig. 8). The bristles of the next series have still stronger shafts, shorter spinous regions; and the hook at the tip increases in size, while the secondary process diminishes (fig. 9). Inferiorly, again, there is a tendency to repeat the elongated spinous region and slender forked tip of the upper series.

---

[*] Ann. & Mag. Nat. Hist. ser. 4, vol. xiii. p. 265, pl. x. f. 3.

There are nine papillæ on the dorsal border of the extruded proboscis, and as many on the ventral surface. A filiform cirrus occurs under each inferior maxilla.

Hab.—Royal Sound.

NEREIDÆ.

Nereis eatoni.

(Pl. XV., figs. 10.—12.)

M'Intosh, Ann. & Mag. of Nat. Hist. 4th ser., 1876, xvii., p. 320.

This species somewhat resembles *Nereis dumerilii*, Aud. & Ed. The head has four large eyes, the anterior pair being somewhat ovoid and by far the larger. When turned backward the long tentacular cirri reach to the fourteenth segment. The maxillæ have about eight distinct teeth behind the point. The paragnathi form, near each maxilla, five long rows and four shorter; and there are besides several interrupted transverse rows between the former on the ventral surface. All are composed of denticulate horny processes of microscopic size. The anterior feet have blunt processes; their cirri are shorter; and the bristles have on the whole shorter tips than in *N. dumerilii*, ranging from those with long tips (fig. 10) to those with short terminal processes (fig. 11). The articulating end of the shaft in the latter organs has also a somewhat wider pit for the terminal process. At the twenty-fifth foot (fig. 12) the superior lingula is rather larger than in *N. dumerilii*, and the outline of the other processes also differs. Towards the posterior extremity (*e.g.* the sixtieth foot), again the superior lingula forms a very prominent elongated process, which is much thicker and less pointed than in the British form; and it also differs from *N. polyodonta*, Schmarda, in this respect.

Hab.—Royal Sound.

[Nereis antarctica.

Nereis antarctica, Verrill, Bulletin U. S. Nat. Mus., 1876, May, iii. 64.

Hab.—Royal Sound, on the beach (Kidder).]

TEREBELLIDÆ.

Amphritrite kerguelenensis.

(Pl. XV., fig. 13.)

M'Intosh, Ann. & Mag. of Nat. Hist. 4th ser., 1876, xvii., p. 321.

A large form with seventeen setigerous tubercles. The cephalic region shows four lobes, viz. the ventral anterior lobe, a large process in front and beneath the first branchia, a fan-shaped lobe, and finally a large fold running from the root of the last branchia downwards. The long branchiæ spring from three short trunks on each side. There is a prominent papilla below each setigerous tubercle in the first six segments, and in addition a similar process below the second branchia. The ventral scutes appear to be twelve. The hooks (fig. 13) somewhat resemble those of *A. affinis*, Mgrn., but differ in the anterior curvature. The colour of one specimen was purplish brown.

This species forms a heavy tube of fine mud, lined by a thin chitinous secretion; and, from the flattening of the ventral surface, it would appear to lie on the bottom.

Hab.—Royal Sound.

Neottis antarctica.
(Pl. XV., figs. 14, 15.)

M'Intosh Ann. & Mag. of Nat. Hist. 4th ser., 1876, April, xvii., p. 321.

A very large member of the family, differing from *Thelepus* in having three groups of branchiæ on each side, and from *Grymæa* by the fact that the bristle-tufts commence on the third segment, and also by the structure of the hooks. The cephalic lobe is furnished with numerous ocular specks. The bristles resemble those of *Thelepus*, as also do the hooks, which are borne on a thin lateral lamella marked by a band of dark pigment. A single process only appears in profile (fig. 14) above the large tooth of the hook, though two are very evident in oblique views (fig. 15). The brownish body is peculiarly streaked posteriorly by pale transverse lines.

The animal constructs a large chitinous tube of a dark brownish colour, on which Polyzoa, Zoophytes, and Algæ flourish.

Hab.—Royal Sound, very common.

[Neottis spectabilis.

Verrill, Bulletin U. S. Nat. Mus. 1876, May, iii., 66.

Hab.—Royal Sound, in 12 fathoms (Kidder).]

SERPULIDÆ.
Serpula, sp.
(Plate XV., fig. 16.)

The softened specimen resembles *S. vermicularis*, L., in external appearance; but the operculum is absent. The branchiæ appear to be about forty in number on each side. The anterior hooks (fig. 16) are larger than in *S. vermicularis*, and form a triangle of quite a different shape. The uncini along the edge of the organ are seven or eight in number, the inferior, as usual, surpassing the rest in size. The posterior hooks present the same structure, and are accompanied by the brush-shaped bristles as in *S. vermicularis*.

The tube resembles that of the latter, even to the double funnels so often seen in front.

The absence of the operculum prevents further definition. The undeveloped left opercular process resembles that in *S. vermicularis*, though it is somewhat longer.

Hab.—Swain's Bay.

NEMERTINEA.

LINEIDÆ.

Lineus corrugatus.

(Plate XV., fig. 17.)

M'Intosh, Ann. & Mag. of Nat. Hist. 4th ser., 1876, xvii., p. 322.

Body (in spirit) flattened, rather abruptly pointed anteriorly, and more gradually posteriorly. The œsophageal region is marked externally by a series of prominent and somewhat regular rugæ, which sweep from the mouth dorsally and ventrally; so that the dorsal view recalls that observed in *Arion ater*.

Colour dark olive throughout, with the exception of a white band, which crosses the anterior border of the snout, and passes backward to the posterior third of the lateral fissure, where it bends dorsally and terminates.

The special characters are the very large mouth, with the prominent rugæ, which show that the animal probably possesses unusual powers of œsophageal protrusion—a supposition borne out by the great development of the external circular muscular fibres (fig. 17 cm), the dorsal longitudinal coat, and the other fibres of the organ. The internal glandular lining j is also very firm. The outer layers of the proboscis correspond with the type in the Lineidæ; but the internal longitudinal layer e, observed in an imperfect condition in *Micrura fusca*,[*] is largely developed.

Hab.—Swain's Bay.

Explanation of Plate XV.

Fig. 1. Portion of the scale of *Hermadion longicirratus*, Kbg., the inferior edge being slightly turned so as to show the papillæ in profile. × 210 diam.

Fig. 2. Dorsal bristle of the foregoing form. × about 20 diam.

Fig. 3. Tip of another dorsal bristle. × 90 diam.

Fig. 4. Ventral bristle from the middle of the fascicle. × 97 diam.

Fig. 5. Portion of the outer and anterior border of the scale of *Eupolynoë mollis*. On the inferior margin the papillæ are seen in profile. × 210 diam.

Fig. 6. Tip of a dorsal bristle of the same. × 210 diam.

Fig. 7. Tip of a slender bifid bristle from the superior ventral series. × 350 diam.

Fig. 8. Tip of one of the lower bristles from the same tuft. × 350 diam.

[*] Brit. Annelida, Ray Soc., Pt. i. p. 103, Pl. 20, fig. 4.

Fig. 9. One of the smaller bristles from the middle of the next ventral series. × 210 diam.

Fig. 10. Bristle with a long tip, from the ventral series of *Nereis eatoni*. × 350 diam.

Fig. 11. Bristle with a short tip from the ventral series of the same.

Fig. 12. The twenty-fifth foot of the foregoing form. × about 12 diam.

Fig. 13. Anterior hook of *Amphitrite kerguelenensis*. × 350 diam.

Fig. 14. Anterior hook of *Neottis antarctica*. × 350 diam.

Fig. 15. Tips of two of the former seen obliquely. × 350 diam.

Fig. 16. Anterior hook of *Serpula* ——? × 350 diam.

Fig. 17. Vertical transverse section of the ventral body-wall in *Lineus corrugatus*, showing the thick circular muscular layer (cm) enveloping the œsophageal region; d'', pigmentary layer divided (as in *Lineus marinus*) by a definite black band (2); 3, curious translucent stratum cut into somewhat regular spaces; e, external longitudinal muscular layer of the body-wall; e', circular muscular coat; e'', inner (longitudinal) muscular layer; j, firm glandular lining of the œsophagus; v, vascular meshes around the œsophageal region. × 55 diam.

Fig. 18. Transverse section of the proboscis of the same: a, external coat; b, great longitudinal muscular layer; c, belt of circular muscular fibres; d, basement layer; e, internal longitudinal muscular layer, specially developed in this form; f, glandular lining of the organ thrown into various folds; g, lozenge-shaped portion of longitudinal fibres formed by the crossing of two bands from the circular muscular coat; g', separate segment at the other pole of the circle. The two latter are somewhat indistinct. × 55 diam.

Terrestrial Annelida.—*By E. Ray Lankester, M.A., F.R.S.*

The Rev. A. E. Eaton, on his return from the Expedition of the Transit observers to Kerguelen's Land, placed in my hands for description two small earthworms obtained by him in the island, and preserved in strong spirit. The specimens were small and immature, not exceeding $1\frac{1}{2}$ inch in length; but by cutting transverse sections of one, and slitting the other up the median dorsal line, staining with carmine, and mounting in Canada balsam, I have succeeded in making out the affinities of the species.

The study of the various species of Earthworms (Lumbricidæ proper) has only recently been attempted with due attention to anatomical detail. Their excessively complicated generative glands, ducts, and pouches present the greatest diversity of arrangement, so as to enable us to establish a series of strongly marked genera, which, while differing in the arrangement of these parts, yet present but slight differences in external form, or in the arrangement of their setæ. Professor Edmond Perrier, availing himself of the very fine collection of exotic Lumbricidæ in the Jardin des Plantes, has been the pioneer in this branch of investigation, and in his memoir "Recherches pour servir a l'histoire des Lombriciens terrestres," published in the "Nouvelles Archives du Muséum d'Histoire Naturelle, 1872," he has established a series of genera on the only possible characters in modern zoology—namely those derived from thorough anatomical examination. M. Perrier has studied earthworms from North and South Africa, from the East Indies, from the West Indies, from North and South America, and a number of scattered islands, and has rendered it evident that he has tapped a rich storehouse of zoological facts of first-rate importance. Presenting, as they do, a considerable number of genera, and occurring as they do almost universally on the earth's surface where there is vegetable soil—being moreover absolutely destitute either of means of transport or of power to resist deleterious agents whilst being passively transported (earthworms and probably their eggs are rapidly killed by sea-water), the *Lumbricidæ* promise to yield, when fully investigated, a mass of information bearing upon the problems of the causes of geographical distribution and the connections of continents and islands in past epochs—more decisive and indisputable in its character than that presented by any similar small group of the animal kingdom. The essential feature of their organisation which gives to the *Lumbricidæ* so interesting and important a position, is the possession of a most sensitive generative apparatus—*sensitive*, that is to say, in the sense of responding by innumerable modifications of its highly-developed male ducts, prostatic glands, seminal reservoirs, penial setæ, copulatory pouches, and other accessory glands, to those slight differences of environment which whilst thus affecting the genitalia so as to create generic distinctions, have yet left the external form and character unaffected.

The two small specimens from Kerguelen's Land are the first Earthworms of special interest which I have received, though for some time, through the kindness of Sir J. Hooker, Earthworms, found in the Wardian cases sent from abroad to the Royal Gardens at Kew, have been forwarded to me for examination. I may take this opportunity of saying that persons who may wish to preserve specimens of exotic earthworms for examination in this country should either send them home alive, which is easy and the most satisfactory to the student, or should kill them with chloroform, by which means they are prevented from shrinking, and then place them first for twenty-four hours in weak spirit, and afterwards in the strongest which can be procured.

The Earthworms brought from Kerguelen by Mr. Eaton are small specimens of a species of *Acanthodrilus*. The genus *Acanthodrilus* is established by Edmond Perrier for the reception of three species, two of which come from New Caledonia (*A. obtusus* and *A. ungulatus*), whilst the third (*A. verticillatus*) is an inhabitant of Madagascar. The addition of Kerguelen's Land to the distribution already indicated by Perrier for *Acanthodrilus*, is a matter of some consequence, though until our collections of *Lumbricidæ* are more exhaustive than at present, it would be very rash to discount the conclusions to which we shall be ultimately led.

I propose now to give the characters of the genus *Acanthodrilus* as indicated by Perrier in his classical work, and then to point out the distinctive characters of the Kerguelen species.

Characters of the genus ACANTHODRILUS, *with notes on the new species.*

The *Lumbricidæ* are divided by Perrier into three sections according as the male generative apertures are in front of, within, or behind the clitellum. The genus *Lumbricus* alone is *Præclitellian*, the genera *Anteus*, *Titanus*, *Rhinodrilus*, *Urochæta*, and *Geogenia* are *Intraclitellian*, whilst *Pontodrilus*, *Eudrilus*, *Moniligaster*, *Acanthodrilus*, *Digaster*, *Perionyx*, and *Perichæta* are *Postclitellian*.*

The genus *Acanthodrilus* is especially characterised amongst the *Postclitellian Lumbricidæ* by the possession of two pairs of male generative orifices which are placed in the 17th and 19th, 18th and 20th, or 19th and 21st segments.† These orifices are so placed as to give exit each to a bundle of greatly elongated and specially modified "penial" or "genital" setæ. The term "penial" proposed by Perrier is more appropriate than that which I had previously used in describing

* It is impossible to determine the true value and position of the genera of Lumbricidæ established by Kinberg, since he has not furnished the necessary anatomical details.

† The cephalic lobe and the buccal ring form the *first* segment.

similarly modified setæ in Chætogaster and Nais (see my paper " On distinct larval and sexual forms in the Oligochæta." Ann. & Mag. Nat. Hist. 1870). The existence of these penial setæ is what has suggested the name of the genus, since they appear to be unique amongst the *Lumbricidæ*, though we find similar setæ in the *Naididæ*, and in *Lumbricus* an enlargement and elongation of the setæ in several of the segments connected with the reproductive organs, though not a well marked specialization of form is noticeable. The setæ which are thus modified in *Acanthodrilus* are those which correspond to the two ventrally placed pairs (one on each side the median line) of a segment of *Lumbricus*, the dorsally placed pairs being unmodified. In the new *Acanthodrilus* the penial setæ are in two bundles of four each, or eight altogether to each male genital pore. They are notched near the anterior extremity as in Perrier's *A. verticillatus* (see fig. 6.)

Perrier gives as a character of the genus that the locomotor setæ are arranged as in *Lumbricus* in four series, each group of bristles containing as in *Lumbricus* two functional setæ. This character must be amended, since in the new *Acanthodrilus* of Kerguelen's Land the setæ are arranged, not in *four* series of bundles or groups, each containing *two* setæ, but in *eight* series, each seta standing alone, and widely separate from its fellows of neighbouring series (fig. 4). Thus on each segment we can distinguish, on each side of the median antero-posterior vertical plane, a medio-ventral seta and a latero-ventral seta, a latero-dorsal seta, and a medio-dorsal seta. It becomes quite clear that the double ventral series in *Lumbricus* and the other species of *Acanthodrilus*, is formed by the approximation of two single series such as we see in the medio-ventral and latero-ventral series of the *Acanthodrilus* of Kerguelen's Land, since in certain segments of this species, namely, the 16th, 17th, 18th, and 19th, the two separate ventral series of single setæ approach one another, and form a double ventral series (see fig. 2), exactly comparable to the arrangement which obtains throughout the series both dorsal and ventral in *Lumbricus* and most other *Lumbricidæ*. *Acanthodrilus* is stated by Edmond Perrier to possess the median dorsal pores leading from the body-cavity to the exterior, which are wanting in some genera.

A full description of the genitalia of *Acanthodrilus* is still a desideratum. The exact position of the testes and ovaries is not known, nor do my very young specimens from Kerguelen enable me to supply the required information. Opening close by the side of the penial bristles, and with its orifice covered by a flap of integument (fig. 7), is a tube (one on each side in each of the two penial segments) which runs horizontally, and expands into a short, undulated, thick walled cæcum. These four cæca have been observed by Perrier in the various species of *Acanthodrilus* studied by him, and appear to be prostatic glands connected each with a distinct vas deferens, which place the four male genital orifices in continuity with the testes situated about the 11th segment. The four cæca (see fig. 7 *pr.*) are well developed in the specimen from Kerguelen's Land, but I was unable to find the testes or vasa deferentia in these small specimens.

The cingulum noticed by Perrier in his *Acanthodrilus ungulatus* on the 14th, 15th, 16th, 17th segments, was not developed in my specimens.

The copulatory pouches, which in the species described by Perrier are placed to the number of two pairs in the 8th and 9th or 8th and 10th segments, have a similar position in the new species, namely, on the line of the latero-ventral series of setæ between the 7th and 8th and the 8th and 9th segments.

The cephalic protuberance, prostomium, or upper lip (fig. 3 *pr*) is worthy of note from its peculiar setting in the buccal ring; a similar form of prostomium is described by Perrier in the *Acanthrodrilus verticillatus* of Madagascar.

Distinctive features of the ACANTHRODRILUS of Kerguelen's Land.

Male orifices and penial setæ placed in the 17th and 19th segments; orifices of the copulatory pouches between the 7th and 8th, and 8th and 9th segments. Setæ arranged, not in four double, but in eight single rows, viz.: right and left medio-ventral, latero-ventral, latero-dorsal, and medio-dorsal. The latero-ventral and medio-ventral rows converge in the 16th, 17th, 18th, and 19th segments. The penial setæ are formed by eight setigerous sacs, a latero-ventral and a medio-ventral to each of the four male genital pores. The prostomial lobe is short, and sunk within the buccal ring.

The genus and species are briefly characterised thus :—

Acanthrodrilus, *Edm. Perrier*.

Lumbricidæ post-clitelliani, poris genitalibus masculis quattuor, duobus in seg. 17, 18, vel 19; duobus in seg. 19, 20, vel 21, prope setas ventrales positis. Setæ ventrales, poros genitales juxta, valde elongatæ et numerosæ, *peniales* dicuntur. Bursæ copulatrices utrinque duæ in seg. 7, 8, vel in segmentis vicinalibus.

Acanthodrilus kerguelenensis, sp. n.

A. poris genitalibus masculis in seg. 17 et seg. 19 positis; bursis copulatricibus inter seg. 7 et 8, et inter seg. 8 et 9. Setæ locomotores in lineis 8 ordinatæ, utrinque medio-ventrales, latero-ventrales, latero-dorsales et medio-dorsales. Lobus prostomialis brevis, rotundus, annulo buccali immersus.

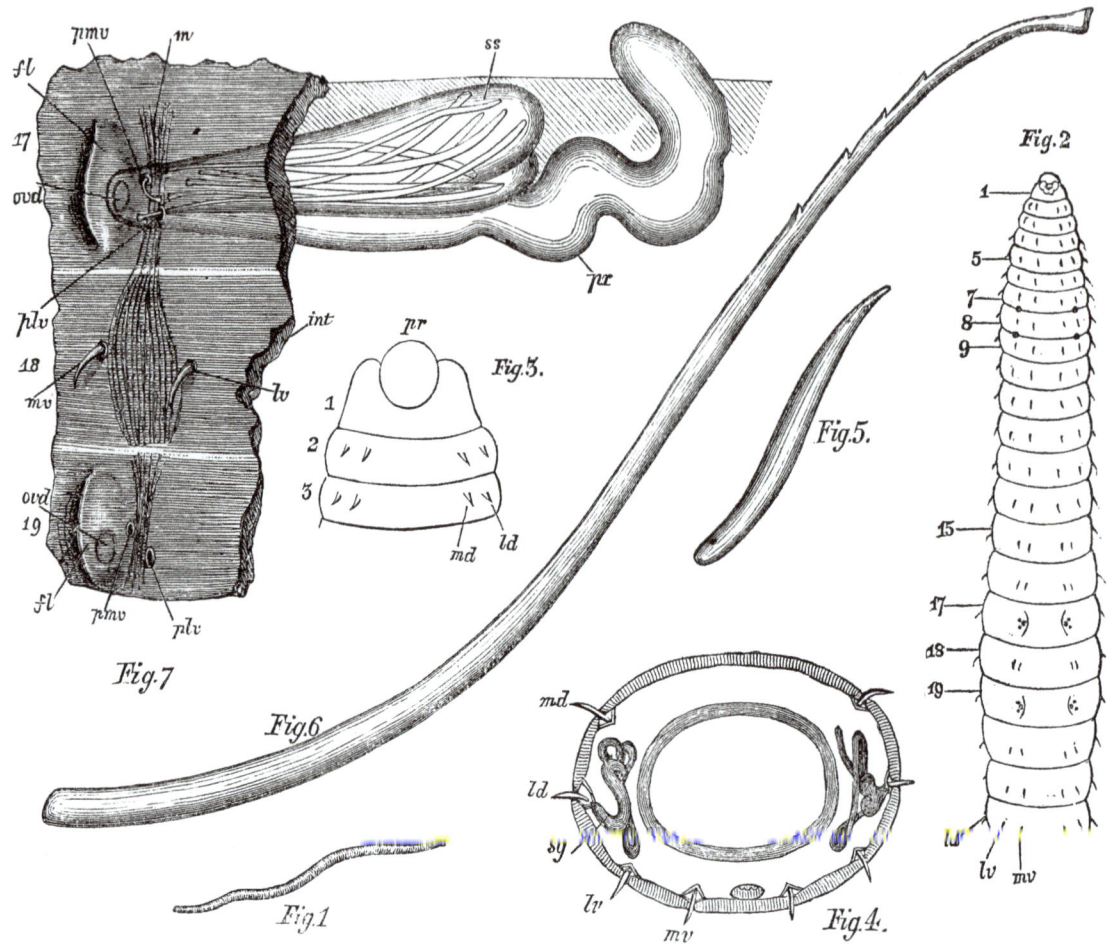

Explanation of the Woodcut.

Fig. 1. Young specimen of *Acanthodrilus kerguelenensis*, natural size.

Fig. 2. The anterior twenty-one segments of the same magnified and seen from the ventral aspect, to show the position of the setæ and the genital orifices.

ld, latero-dorsal series of setæ.

lv, latero-ventral series of setæ.

mv, medio-ventral series of setæ.

The medio-dorsal series are not visible.

Between segments 7, 8, and 8, 9, are seen the orifices of the copulatory pouches.

In segment 17, and segment 19 are seen the two pairs of male genital pores, which are provided with the penial setæ.

Fig. 3. Dorsal view of the three anterior segments to show the form of the prostomial lobe *pr*. *md*, medio-dorsal setæ. *ld*, latero-dorsal setæ.

Fig. 4. Diagrammatic section through the thirtieth segment to show the position of the eight series of setæ, and of the openings of the segmental organs.

md, medio-dorsal setæ.

ld, latero-dorsal setæ.

lv, latero-ventral setæ.

mv, medio-ventral setæ.

sg, segmental organs.

Fig. 5. Ordinary locomotor setæ.

Fig. 6. One of the penial setæ ("genital setæ" *mihi* of *Chætogaster* and *Nais*), drawn to the same scale as fig. 5.

Fig. 7. The male genital orifices and setigerous pores of the left side (diagrammatic).

int, cut edge of the integument.

fl, flap of the integument which overlies the male genital orifice *ovd*.

ovd, orifice of the vas deferens.

pmv, penial seta and orifice of the medio-ventral series.

plv, penial seta and orifice of the latero-ventral series.

ss, penial setæ in the setigerous sac.

pr, prostatic cæcum of Perrier.

mv, medio-ventral locomotor setæ.

lv, latero-ventral locomotor setæ.

m longitudinal muscular band separating the orifices of the medio-ventral and latero-ventral setigerous sacs.

The setæ, setigerous sacs, and prostatic cæcum of the 19th segment are not represented. They would be identical with those of the 17th.

ECHINODERMATA.—*By Edgar A. Smith, F.Z.S., Senior Assistant Zoological Department, British Museum.*

(Plates XVI. and XVII.)

Opportunity was taken some pages back of exhibiting the relations of the Molluscan-fauna of Kerguelen Island to that of the Falklands and Patagonia; and it was pointed out that representatives of boreal types entered into its composition. Materials for similar comparisons between the Echinodermata indigenous to the same regions scarcely exist, but such as there are, make it apparent that what obtains in the Mollusca holds good also in the Echinodermata with respect to geographical distribution.

The sources of information relating to the animals of this class inhabiting the vicinage of Kerguelen Island, are at present limited to a few relics of the collections made there by the Antarctic Expedition in 1840; the collection made in 1874 by Dr. Kidder of the American Transit of Venus Expedition, part of the Challenger's collections, and Mr. Eaton's specimens. The remaining portion of the *Echinodermata* obtained by the Challenger Expedition is still in course of investigation; and the collections of the German Transit and Surveying Expedition are likewise not yet worked out. This will account for the fewness of the species at present known from the island, which altogether amount in number to 15. Two of them were obtained by the Antarctic Expedition, one by the Challenger, four others by Dr. Kidder; besides five out of the preceding seven, eight others were obtained at Observatory or Swain's Bay. The species not found by Mr. Eaton are *Asterias rupicola*, which was common on the beach near Dr. Kidder's station, and *Cidaris nutrix*: those which he did get were procured almost exclusively from the Laminarian zone. This will suffice to show how far dependence can be placed upon the following summary of the geographical relations of the little known fauna.

The species hitherto recorded from Kerguelen Island comprise 1 Holothurian, 2 Echini, 9 Asteriidæ, and 3 Ophiuridæ,—in all 15 species referable to 12 genera.

From Patagonia are known 1 Holothurian, 1 Echinoid, 3 or 4 Asteriidæ, and 5 Ophiuridæ,—in all, 10 or 11 species comprised in as many genera. Of these the Holothurian, the Echinoid, one of the Asteriidæ, and one of the Ophiuridæ,—that is 4 species, are indigenous also to the coast of Kerguelen Island.

A similarity to certain boreal terms is exhibited by some of the species. Thus *Porania antarctica* strangely resembles *P. pulvillus* of the northern seas of Europe; the *Pedicellaster* represents another septentrional genus; *Pteraster affinis* imitates closely *Pt. militaris* of boreal waters. The genera *Ophioglypha* and *Ophiacantha* are almost cosmopolitan in distribution; yet the Kerguelen Island representative of the former, *O. hexactis*, in colour and in *tout ensemble* approaches *O. Sarsii* of the Greenland coast.

The new form *Leptoptychaster* has been found only at Kerguelen Island.

HOLOTHURIIDÆ.
Pentactella lævigata.

Verrill, Bull. U. S. Nat. Mus. 1876, iii., p. 68.

Hab.—"Kerguelen Island, twelve fathoms" (Kidder). Very common on the laminæ of *Macrocystis* in Royal Sound and Swain's Bay (Eaton). Also Patagonia (Antarctic Exped.).

The colour of this species during life is dull whitish; the body is somewhat opaque, but the distended tentacles are semi-transparent white, tinged very faintly at the tips of their ultimate ramifications with pale rose colour. Owing to their whiteness, specimens can be seen on the kelp, sometimes at a considerable depth below the surface, appearing to be much larger than their real size through the refraction of the light.

ECHINIDÆ.
[Cidaris nutrix.

Wyville Thomson, Journ. Linn. Soc. 1876, vol. xiii., p. 62, fig. 4.

Hab.—Balfour Bay, Royal Sound, 50–70 fathoms. This most interesting species forms part of the collection of the Challenger.]

Hemiaster cavernosus.

Tripylus cavernosus, Philippi, Wiegm. Archiv. f. Naturgesch. 1845, p. 345, pl. xi., f. 2.

Brissopsis cavernosa, Agassiz, Ann. Sci. Nat. 1847, viii., p. 5.

Hemiaster cavernosus, Id., Revision of the Echini, 1873, part iii., p. 587, pl. xxi^c., fig. 1, 2.

Tripylus australis, Philippi, op. supra cit., 1845, p. 347, pl. xi., fig. 3.

Brissopsis australis, Agassiz, Con. R. Ann. Sci. Nat. viii., p. 5.

Faorina australis, Gray, Ann. & Mag. of Nat. Hist. 1851, vii., p. 132.

Hemiaster australis, Agassiz, *l. c.* 1873, part iii., p. 586, pl. xxi^c., fig. 3.

Faorina antarctica, Gray, op. supra. cit., 1851, vii. p. 132; idem, Cat. Rec. Echin. 1855, p. 57.

Hemiaster cordatus, Verrill, Bulletin U. S. Nat. Mus. 1876, iii., p. 69.

Hemiaster sp. Wyville Thomson, Journ. Linn. Soc. 1876, xiii., p. 67, fig. 6.

The spines when first observed were greenish olive in colour, but are gradually acquiring in spirit a brownish olive hue.

Hab.—Swain's Bay and Royal Sound, on mud, in 5–10 fathoms. Extremely common; Accessible Bay, 20–50 fathoms (*Wyville Thomson*); also Patagonia (*Philippi*).

A comparison of the types of *Faorina antarctica* with the *Hemiastri* collected by Mr. Eaton at Kerguelen Island, shows that the species are identical with one another. Dr. Gray in 1855 says, in reference to this species and the two forms described by Philippi, cited above,—" Perhaps these three are only one species." Mr. Verrill, in a footnote at op. cit. p. 71, states that M. A. Agassiz is of the same opinion, viz., that the Kerguelen species is identical with Philippi's. After a careful study of Philippi's figures and descriptions, Dr. Gray's types, and Mr. Eaton's specimens, I thoroughly concur in believing them to be all of one species (the form with deep ambulacral furrows the female; that with shallower furrows the male).

If it should hereafter be discovered that the Patagonian species is, on the contrary, distinct from the Kerguelen Island species, the name proposed by Gray (*antarctica*) will still have to be retained.

ASTERIIDÆ.

Asterias meridionalis.

(Plate XVI., fig. 1.)

Perrier, Revision d. Stell. 1875, p. 76; idem, Ann. & Mag. of Nat. Hist. 1876. xvii., p. 36.

The specimens from Kerguelen Island exhibit, in certain particulars, inconstant variations from the type of this species, and also from one another. The following remarks. however, are not confined to these aberrations and differences.

Body 6-rayed; rays rather stout, gradually tapering, convex above, rather more than twice as long as the disk is wide. Ambulacral spines irregularly biserial, but not constantly so, rather stouter than in the type; in one example they are at irregular intervals uni-serial, and towards the inter-radial angles of the mouth they are also in single series. Next to the ambulacral spines succeeds a series of tentacular pores; and then the lower or ventral margin of the ray bears a double (rarely triple) series of short stout spines. Above these extends a naked band with small groups of papulæ in it; and above this naked band, limiting the dorsal area, is a row of isolated spines similar to the spines in the other series. The dorsal spines of the disk and rays are numerous, irregularly scattered, short, blunt, scabrous to the touch, and striated; and among them are interspersed numerous small groups of papulæ: one example has the spines of the upper and lateral surfaces conformable to those of the type; but another example bears spines which are acute and shortly conical, particularly so on the disk and thickest portion of the arms; and in this specimen they are also disposed more regularly than is usual, tending to arrange themselves in longitudinal series down the arms. These spines are about 2 mm. long in the disk of the example last mentioned, but become shorter and less regular in arrangement towards the extremities of the rays. The madreporic plate is placed in an angle of the rays, and about midway between the centre of the disk and the margin. Colour reddish brown above, pale buff beneath.

Dimensions.—Length of arm, 3 inches; width of arm at the base, 7 lines; width of disk nearly 1½ inches.

Hab.—On roots of *Macrocystis* in Observatory Bay. Also obtained by the Antarctic Expedition.

M. Perrier in his earlier description likens this species to *A. cuninghami*, to which he accidentally ascribed 6 rays instead of 5. This might have been misleading, had he not subsequently in the Annals and Magazine of Natural History (loc. cit.) corrected the mistake.

Asterias perrieri.

(Plate XVI., fig. 2, 2a–b.)

Smith, Ann. & Mag. Nat. Hist. 1876, xvii., p. 106.

Discus sex-radiatus, modice amplus, in latitudine ad spatium maximum inter radiorum oppositorum apices circiter $\frac{2}{7}$ æqualis; radii cylindraceo-attenuati inferne anguste planati. Sulci ambulacrales haud latissimi; spinæ ambulacrales subgraciles, subcylindricæ, non clavatæ, in serie unicâ positæ, in exempl. maxim. circiter 3 mm. longæ; spinæ ventrales in serie duplici spinis ambulacralibus adjacentes, binatæ (nisi rarissime ternatæ), divergentes, spinarum intima subacuminata, major intermediâ, sed minor et crassior ambulacrali. Dorsum lateraque spinis brevissimis minutis subconicis diverse aspersis, papulis innumeralibus interpositis. Tessella madreporiformis parviuscula, inter centrum marginemque disci intermedia. Color saturate fusco-rufus.

Disci diam. 45; radiorum longit. 150; ad bases crass. 19 mm.

(*Young.*) Six-rayed; the rays very short, nearly as broad as long, with only two rows of ambulacral tentacles bordered by a single series of spines; the latero-dorsal margin with a single row of large spines, the dorsal area with a similar longitudinal series down the middle of the ray.

Hab.—Common on the roots of *Macrocystis*, and also taken in shrimp pots, at Observatory Bay. The largest specimen has a cluster of some hundreds of young ones clinging to its ventral disk.

A. rugispina of Stimpson is allied to this species, with which I have had much pleasure in associating the name of M. Edmond Perrier of the Jardin des Plantes, who has recently identified many species of the Asteriidæ in our national collection.

Pedicellaster scaber.

(Plate XVI., fig. 3.)

Smith, Ann. & Mag. Nat. Hist. 1876, xvii., p. 107.

Discus quinque radiatus, in latitudine ad radii longitudinem circiter semiæqualis; radii subcylindrici, sensim attenuati, modice acuti. Spinæ ambulacrales graciles æquilongæ, in seriebus tribus positæ, duplo longiores dorsalibus; anguli oris interradiales spinis parvis ad apices binis. Spinæ dorsales disci radiorumque brevissimæ, obtusæ, scabræ, distantes, in substructuram quasi clathratam diverse dispositæ, eæ prope ambulacra paulo longiores cæteris; maculæ reticuli nudæ. Anus subcentralis; tessella madreporiformis subrotunda, prope marginem in angulo interradiali sita.

Disci diam. 9, crass. 6; radiorum longit. 18, ad bases crass $5\frac{1}{2}$ mm.

Hab.—On roots of *Macrocystis*, Observatory Bay.

This species appears to agree very fairly with Sar's description of his genus *Pedicellaster*, excepting that the ambulacral furrows cannot be said to be "broad," and that the ambulacral spines are in three rows instead of two. But these differences are more specific than generic; and therefore I think that this may, notwithstanding them, be properly considered a second species of that northern genus.

Echinaster spinulifer.

(Plate XVI., fig. 4.)

Othilia spinulifera, Smith, Ann. & Mag. Nat. Hist. 1876, xvii., p. 107.

Discus quinque radiatus, ad radii longitudinem in latitudine semiæqualis, modice crassus, superne leviter rotundatus; radii cylindraceo attenuati, breves, spinis scabris brevissimis numerosis in substructuram quasi clathratam diverse dispositis, intervallis clathrorum nudis; prope spinas ambulacrales iisque parallele spatium lineare fere nudum, radii basin versus sensim latius, serie solum unicâ spinarum minutarum (longe minorum etiam spinis ambulacralibus) munitum adjacet. Spinæ ambulacrales divergentes, transverse super tessellas singulas quaternis dispositæ, duæ interiores paulo longiores; spina gracillima parva recta est intimæ harum apud basin. Anguli oris interradiales spinis parvis duabus terminantur. Anus subcentralis; tessella madreporiformis in angulo interradiali submarginalis.

Disci diam. 7; crass. 6; radiorum longit. 14, diam. apud basin $4\frac{1}{2}$ mm.

Hab.—Observatory Bay.

This curious little species is remarkable for the shortness of the rays, and for the minute slender spine within the ambulacral groove at the base of the innermost spine being straight, instead of being curved or hooked as the homologous spine is in the normal species of the genus. The spines on the dorsal and lateral surfaces display no regularity in their arrangement; but parallel with the ambulacral spines in the almost naked narrow space adjacent to them, the series of very small spines is disposed definitely, one spine on each plate, and above this the spines are placed two or three on a plate.

Pteraster affinis.

(Plate XVI., fig. 5.)

Smith (sp. nov. ?) Ann. & Mag. Nat. Hist. 1876, xvii., p. 108.

Discus magnus, quinque radiatus, subtus planus, supra convexus, modice crassus, ad radii longitudinem in latitudine æqualis. Radii breves, e basibus latis statim ad apices angustati, ibique ita recurvati ut sulcos ambulacrales exponant; subtus utrinque extra sulcos membranâ tenui spinis gracilibus circiter 30 apicibus vix ultra marginem membranæ projicientibus induti; tessellæ inter-ambulacrales spinas graciles quatuor in membranâ tenuissimâ fere ad apices producta gerentes, intimâ harum paulo cæteris breviori; anguli oris interradiales spinis similibus octonis, pariter membranâ conjunctis, duabus extimis longe brevissimis, duabusque intimis longissimis, spinis atque supra illas crassis duabus sibi parallelis, leviter in medio concavis et apices versus acuminatis, longissimæque spinarum 30 lateralium in longitudine æqualibus. Dorsum lateraque supra projecturis spiniferis minimis, foraminibus minutis haud numerosis interpositis; illæ spinis scabris in membranâ connexivâ pæne apicium tenus conditis diverse a quaternis usque ad denas munitis; foramen centrale modice amplum, fimbriâ spinarum brevium membranâ conjunctarum circumdatum. Color (in spiritu vini) sordidus, pallide ochraceus.

Exempl. max. disci diam. 15, crass. 7 ; radii longit. 17, diam. ad basin 8 mm.

Exempl. minor. disci diam. 10, crass. 5 ; radii. longit. 9 mm.

Hab.—On roots of *Macrocystis*, Observatory Bay.

This species approaches very closely to *P. Danæ* of Verrill,[*] which is supposed to have been found at Rio Janeiro. It appears, however, to possess longer arms (though the smaller specimen, it will be noticed, is considerably shorter in the rays than the larger); the spines of the dorsal fascicles are everywhere uniform and scabrous; there are only eight spines at each of the inter-radial angles of the mouth, and the two larger spines above them are not very long, but are stout. In these respects chiefly it differs from Mr. Verrill's species.

Porania antarctica.

(Plate XVII., fig. 1.)

Smith, Ann. & Mag. Nat. Hist. 1876, xvii., p. 108.

Discus quinque radiatus, radiis sub-breviter conicis apices versus acuminatis, modice crassus, ad radii longitudinem in latitudine æqualis, subtus planus, supra convexus, ubique indutus cutem crassam carnosam, infra transverse inter margines sulcosque ambulacrales lineariter sulcatam, superne lævem, nisi apud medium disci et etiam super radios spinis tubercularibus parvis sparsim ornatam; margines laterales inferne spinis brevibus compressis ad apices latis et truncatis fimbriatæ, spinâ unicâ super tessellam singulam sulcis linearibus jam supra dictis

---

[*] Described in Proc. Boston Soc. Nat. Hist. 1869, vii., pp. 386, 387

definitam; spinæ ambulacrales biseriatæ, exteriores duplo longiores multoque robustiores interioribus, latæ, ad apices abrupte truncatæ neque angustatæ, extus leviter sulcatæ ita ut duplices esse videantur; tessella madreporiformis rotunde ovalis, paulo ad centrum propior quam marginem; anus centralis, papillis brevissimis spiniformibus circiter 12 circumjectis. Color carneus vel sanguineus.

Diam. max. 90, minim. 48 mm.

Hab.—Dredged in a retired inlet of Swain's Bay, in about 10 faths. of water, and outside the belt of *Macrocystis*.

The furrows on the outside of the exterior ambulacral spines are formed chiefly by the skin which clothes them. The minute tubercles on the back exhibit no regularity in their arrangement; there are about a dozen of them on the middle of the disk and a few on the short conical arms.

This species may be distinguished from the rather closely related northern *Porania pulvillus*, Müller, by differences in the ambulacral spines, and in the number and character of the marginal spines. *P. pulvillus* has 3 or 4 spines on each of the marginal plates, which are much smaller than the single spines in *P. antarctica*.

Asteropsis is the name usually adopted by authors for the genus to which Müller's species belongs. The genus thus designated was founded by Müller and Troschel for the reception of *Asterias carinifera*, Lamk. (see Wiegm. Archiv. 1840, p. 323); und in their System der Asteriden, 1842, pp. 62–4, they include in the same genus *A. ctenacantha*, *A. pulvillus*, and other species in addition to the one just mentioned.

Porania is the appellation given to Müller's *A. pulvillus* by Gray, who described it as *Porania gibbosa*, Leach, in the Ann. & Mag. of Nat Hist., 1840, December, p. 288.

It has appeared advisable, to me, to restrict *Asteropsis*, (Mül. & Trosch. (of which *Gymnasteria*, Gray, is a synonyme) to *A. carinifera*, Lamk., *inermis*, Gray, &c., and to retain *Porania*, Gray, for *P. pulvillus* and *antarctica*.

Pentagonaster meridionalis.
(Plate XVI.., fig. 6, 6a.)

Astrogonium meridionale, Smith, Ann. & Mag. Nat. Hist., 1876, xvii., p. 109.

Discus quinque radiatus, ad radii longitudinem in latitudine circiter $\frac{4}{5}$ æqualis, depressus, subtus supraque leviter convexus. Radii modice longi, apud bases lati, apices versus sub-repente attenuati, spinis ambulacralibus in seriebus quatuor dispositis, interioribus compressis apices versus dilatatis tandemque truncatis, exterioribus teretis ad apices rotundatis in longitudine subparibus. Anguli oris interradiales singuli spinâ unicâ conicâ robustâ exstante super angulum inter-ambulacralem (similiter ac in *Pterastro*) et infra hanc spinis parvis 6—8 in serie reclinantibus. Discus radiique subtus fasciculis spinosis in seriebus ex sulcis ambulacralibus ad margines laterales excurrentibus, fasciculis ipsis ex spinis parvis cylindricis subacutis, multo quam spinis ambulacralibus brevioribus, compositis: latera angusta, fasciculis parvis spinosis quadratis (circiter 20 super radios singulos) confertim in seriebus angustis duabus margini appositis, iis in serie inferiore cæteris super superficiem ventralem, iisque in serie superiore cæteris super superficiem dorsalem consimilibus. Fasciculi dorsales numerosi, breviter pedunculati et fere contigui, singuli ex spinis brevissimis ad apices obtuse rotundatis 10—20 constructis; interstitia fasciculorum omnia nuda, pedicellariis magnis multis munita. Radii ad apices supra tuberculo magno, unico; tessella madreporiformis rotunda, elevata, fere in medio inter centrum marginemque, posita; anus subcentralis.

Disci diam. 24; crass. 10; radiorum long. 29 mm.

Hab.—Dredged in Observatory Bay in 5-10 faths. on mud.

This species apparently belongs to the section of the genus *Astrogonium* defined by Gray for the reception of a form from Port Essington (Proc. Zool. Soc. 1847, p. 79). For it differs from Dr. Gray's short divisional description in no respects excepting that the ventral surface is not *entirely* covered with granules (or spines), and that the disk is not *flat*. Probably this last distinction is due to Dr. Gray's example having been dried, whereby it very likely had become shrunken and more depressed than it would have been had it been preserved in fluid like the Kerguelen Island example. This species, *A. paxillosum*, Gray, is now placed in the genus *Pentagonaster*, by M. Perrier.

Leptoptychaster.

Leptychaster, Smith, Annals & Mag. Nat. Hist. 1876, xvii., p. 110.

Derivation (Gr.) λεπτός *'narrow*, πτύξ *a plate*, and ἀστήρ *star*; in reference to the narrowness of the ventral plates.

Discus quinque-radiatus, depressus; radii modice longi; superficies dorsalis fasciculis spinularum minutarum pedunculatis confertim obsita; radii serie unicâ laterali tessellarum tenuium transversarum lamelliformium usque ad ambulacra vix productarum muniti, seriequo alterâ fasciculorum spinarum minutarum (fasciculis unicis cum tessellis ordinate dispositis) inter tessellas et ambulacra interpositâ; tessella madreporiformis super marginem in angulo interradiali locata.

This remarkable form of Starfishes is perhaps more nearly related to the genus *Luidia* than to any other. It differs from that genus, however, in the lateral spinulose narrow plates not reaching quite to the ambulacra, but leaving a narrow intermediate space occupied by groups of spines in line with the plates, which gradually becomes wider towards the base (in *Luidia* the plates extend to the ambulacra); and also it has no elongated spines, and the body is larger in proportion to the rays than it is in that genus.

Leptoptychaster kerguelenensis, *Smith*.

(Plate XVII., fig. 2.)

Leptychaster kerguelenensis, Smith, Ann. & Mag. Nat. Hist. 1876, xvii., p. 110.

Archaster excavatus, Wyville-Thomson, Journ. Linn. Soc. 1876, xiii., p. 72, fig. 10.

Discus quinque-radiatus, modice amplus, depressus, supra et infra planus, in latitudine ad radii longitudinem circiter ⅗-æqualis; radii apud bases haud admodum lati, apices versus sensim attenuati, spinis ambulacralibus gracilibus, transverse (ratione ad sulcos habitâ) quaternis vel quinis positis, duabus interioribus longioribus, cæterisque gradatim brevioribus; iidem ad latera subtusque, tessellis minute spinulosis (spinulis ipsis scabris et apud radiorum bases longissimis, indeque apices versus paulatim abbreviatis) angustis, transversis, lamelliformibus, obtecti; aliqui ex his apices radiorum versus usque ad sulcos ambulacrales pæne attingent, bases vero versus ita gradatim ab his recedunt ut apud angulos interambulacrales aream trigonalem utrinque relinquant; inter tessellas et spinas ambulacrales fere per totam radii longitudinem, in serie unicâ, fasciculi parvi spinarum brevium interponuntur, et præter hanc juxta basin, in areis trigonalibus supradictis, series aliæ quatuor singulatim per vices intercalantur, omnes fasciculi cum tessellis ordinibus collocantes; anguli oris interradiales acutis spinis utrinque quaternis vel quinis muniti. Superficies dorsalis fasciculis pedunculatis spinarum brevium confertim serta; tessella madreporiformis modice magna, subovalis, fasciculis spinarum obsita, ad angulum inter-radialem super marginem posita.

Disci diam. 23, crass. 8; radiorum longit. 38, ad bases lat. 13 mm.

Hab.—On roots of *Macrocystis*, Observatory Bay; also obtained in the island by the Antarctic Expedition; and "Off Cape Maclear, south-east coast of Kerguelen Land, from a muddy bottom at a depth of 30 fathoms" (*Wy. Thomson*).

The small fascicles of little spines on the dorsal surface are borne on short peduncles of skin, and are so closely packed together that their apices constitute an even surface; and the madreporic plate being covered with spines similar to those of the rest of the surface is concealed from view.

OPHIURIDÆ.

Ophiacantha vivipara.

(Plate XVII., fig. 3 a-c.)

Ljungman, Öfvers. K. Vetensk. Akad. Forhandl. 1870, p. 471; Smith, Ann. & Mag. Nat. Hist. 1876, xvii., p. 110.

Ophiocoma didelphys, Wyville Thomson, Journ. Linn. Soc. 1876, xiii., p. 78, fig. 13.

Examples from Kerguelen Island differ slightly from specimens of *O. vivipara* found at Patagonia, in the oral shields, the adorals, the upper arm-plates, and the uppermost arm-spine.

(Kerguelen Island form).—Disk covered on both sides with conical scabrous granules (as described by the author); oral shields only slightly longer than broad; adorals *quadrangular* (not trigonal), about ⅔ as large as the orals, almost touching one another on the inner side, their lower margins the longest; 3 or 4 oral papillæ on each side of an angle, of which the outermost one (or sometimes two) is broad and flattened; arms, side and lower arm shields normal (*i.e.* according well with Ljungman's description); arm-spines slender, 9 or 10 in number, the uppermost one is shorter than the next spine, and after this the remaining spines gradually diminish in length; the upper arm-plates are not so wide as in the typical form.

Diam. disk 12; length of the longest arm-spine 3½ mm.

Hab.—Common on the roots of kelp in Observatory and Swain's Bays (*Eaton*); entrance to Royal Sound (*Wyville Thomson*); also Patagonia.

The differences pointed out above are hardly of sufficient importance to constitute specific distinction between the Kerguelen Island and the Patagonian forms. Patagonia, as Lütken suggested (Zoological Record 1872, p. 448), is most probably the true habitat of the typical specimens, and not Altata on the W. coast of Mexico (the locality cited by Ljungman), which is doubtless erroneous.

The geographical range of *Ophiocantha* is very extensive: 5 of its species are from the northern seas, 1 from the Mediterranean, 1 is Portuguese, 1 West Indian, 1 Javanese, and lastly there is this 1 from Patagonia and Kerguelen Island.

Ophioglypha hexactis.

(Plate XVII., fig. 4 a–c.)

Smith, Ann. & Mag. Nat. Hist. 1876, Feb., xvii., p. 3; Verrill, Bulletin U. S. Nat. Mus. 1876, May, iii., p. 72.

Discus hexagonalis, angulis propter radios interruptis, lateribusque leviter concavis, depressus; papillæ orales apud angulos septenæ vel octonæ, acute conicæ, una ad apicem longissima, cæteræ utrinque gradatim breviores; scuta oralia pali- vel ligoni-formia, parva, manubrio latissimo brevissimo ab ore remoto, et laminâ cordatâ ad latera utrinque leviter concaviusculâ; scuta adoralia angustissime linearia, oralibus adjacentia; intra angulos orales prope bases sunt scuta subovalia-oblonga bina, et pone hos alterum minus transversum apud angulorum apices situs. Radii sex, graciles, circiter triplo-longiores latitudine disci; *scutorum inferiorum* sextum post basin transversum, latissimum, margine externâ in medio levissime angulatâ, lateribus brevissimis rotundato—truncatis, et margine orali externæ simillimâ, nisi angulo magis prominenti; *scuta lateralia* subtus conjuncta, commissuris apicem radii versus gradatim in longitudinem excrescentibus; *scutorum superiora* aliqua circiter radii medium subquadrata, paulo ab apicibus bases versus angustata, postice arcuata, ad latera recta, antice concava; aliqua ad discum propiora subovalia, et his valde disparia, gradatim minora breviora atque latiora, scutellis parvis multis formâ et amplitudine diversis, ad scuta lateralia conjuncta; *scuta radialia* subparva, anguste subovalia, sese longe remota; papillæ ad latera inscisuræ disci per-minutæ atque numerosæ (circiter 40), supra que juxta basin brachii aliæ sex adversus sex illorum in ordine obstant ita ut fimbria faciant. Spinæ brachiales tres, breves, crassiusculæ, haud per-acutæ, pallidæ, æquilongæ (nisi supremâ sæpe paulo longiori); papillæ ambulacrales super foramina ultima (infra-brachialia) et fissuris oris conjunctæ, quatuor, super foramen penultimum tres; super pauca sequentium binæ, et super cætera unicæ, omnes diverse formatæ, aliæ breves, compressæ et squamuli- vel scuti-formes, aliæque spinis brachialibus subsimiles nisi paulo breviores. Discus diverse minuteque squamulosus. Color supra ubique purpureo-niger, subtus sordide albidus.

Disci diam. 21 mm.

Hab.—Very common on roots of *Macrocystis*, Observatory Bay; also Patagonia. They require to be killed as soon as they are taken out of the sea, as otherwise they rapidly break themselves up.

O. hexactis cannot be confounded with any other species hitherto described. The number of arms, the peculiar shape of the oral shields, and the peculiarities of the ray-shields and spines distinguish it at once. The ventral portion of the disk visible between the arms is rather large; and the oral shields are only half as long as the space between them and the sides of the disk. In colour and general appearance it approaches *O. Sarsii*, Lütken, of the Greenland coast, which seems to be its nearest ally; but these species are so different from one another in detail, that it is needless to specify their distinctions.

The genus *Ophioglypha* is very widely distributed; 2 of its species are from arctic or the northern seas, 4 from the Mediterranean, 2 from between Cuba and Florida, 1 from China, 1 from Puget Sound, 1 from Patagonia, 2 from Kerguelen Island, and 2 from Sydney.

Ophioglypha brevispina.

(Plate XVII,., fig. 5 a–c.)

Smith, Ann. & Mag. Nat. Hist. 1876, xvii., p. 112.

Discus in latitudine ad radii longitudinem circiter $\frac{1}{3}$ æqualis, depressus, ad latera rotundatus; papillæ orales ad angulos oris septenæ vel octonæ, quarum unæ ad apices cæteris longiores, et extremæ cæteris latiores; dentes compressi, hastati, ad latera curvati, quatuor; scuta oralia paulo longiora quam lata, trigono-obcordata, angulis exterioribus rotundatis, et apice subacuto; scuta adoralia per-angusta, intus scutis oralibus subtus conjuncta, ad juncturas latiora quam extus; infra scuta adoralia alia duo eis paulo ampliora (duplo quidem latiora) fiunt anguli oris. Radii quinque, modice longi, paulo latiores quam crassi; scutorum inferiorum sextum post basin transversum, margine externâ leviter curvatâ atque in medio parum acutâ, marginibus lateralibus rectis per-brevibus, marginibus interioribus leviter excavatis atque apicem acutum versus convergentibus; scutorum primum (*i.e.*, basale vel intimum) aliis major et dissimile, margine posticâ excavatâ in loco anguli levis, et margine anticâ haud acuminatâ,—secundum tertio amplius, cæteraque usque ad apicem gradatim minora tandemque ibi per-minuta; scutorum lateralium subtus quinque vel sex priora disjuncta, cætera tamen per transversum conjuncta, commissuris apicem radii versus gradatim in longitudinem excrescentibus,—supra septendecim priora disjuncta, cætera super dorsum conjuncta; scutorum superiorum sextum post basin parum transversum, margine externâ rotundatâ, lateribus antice recte convergentibus, et margine anticâ concavi-truncato,—cætera radii apicem versus gradatim angustata, præsertim a fronte, itaque tandem quodque in angulo acuto ante producitur; squamæ disci diversæ et impares, unâ in medio aliisque paucis parum remotis circumpositis quam cæteris majoribus; scuta radialia diversa, contigua, præcedentibus subæqualia; papillarum ad latera inscisuræ disci (in exempl. max. 22, in exempl. minoribus circiter 16–17) sex vel septem superiora cæteris ampliora; spinæ brachiales tres, incrassatæ, per-breves, paulo tantum squamis ambulacralibus longiores, quarum binæ sunt tertio, quarto, atque interdum quinto scutorum, atque cæteris unicæ; papillæ super margines primi foraminum ambulacralium (infra-brachiale) utrinque quatuor vel quinque. Color albidus.

Disci diam. 9 mm.

Hab.—Observatory Bay.

Several species inhabiting the seas of the North bear a superficial resemblance to this form:—such are *O. albida*, Forbes, *O. robusta*, Ayres, and *O. nodosa*, Lütken. And besides these *O. Lymani*, Ljungman, from Patagonia, is very like it. *O. brevispina* resembles this last species in having very short arm-spines scarcely longer than the ambulacral papillæ; but differences in their size, and in the relative length of their arms, and in the form of their radial shields, &c., afford good specific characters for their distinction from one another.

NOTE ON THE ACTINOZOA.—*By the Rev. A. E. Eaton.*

At Kerguelen Island, as in all cold seas, *Actiniidæ* are not usually found between tide-marks, but occupy sites constantly under water. In high northern latitudes they seldom occur in situations less than a fathom in depth, and are usually of a red colour. At Kerguelen Island the genus *Actinia* was represented by one if not by two red species frequenting rocks in shallow water just below the lowest limits of the tide, and on the roots of *Macrocystis* in four or five fathoms. They did not seem to be identical with any of the species figured by Gosse in his History of British Sea Anemonies. An *Ilyanthus* was obtained sparingly by the dredge on mud in Observatory Bay; but the specimens unfortunately perished during my temporary absence from the station at the time of the Transit.

HYDROIDA.—*By Professor Allman, M.D., LL.D., F.R.S., P.L.S.*

(Plate XVIII.)

The species of Hydroida collected at Kerguelen Island and placed in my hands for determination amount in number to seven. Of these, one cannot be separated from the widely distributed *Sertularella polyzonias*;* the others were before unknown.

Among them the gymnoblastic hydroids are represented by a single species only, a *Coryne* or *Syncoryne*, whose nearer determination in the absence of the gonosome is impossible.

The new calyptoblastic forms are represented by five species. Of these one constitutes the type of a new genus (Hypanthea). The others belong to the genera *Sertularella*, *Halecium*, and *Campanularia* (provisional), all of which are well represented by other species in our own latitudes.

The species which has been referred to *Campanularia* (though in the absence of a fuller knowledge of its gonosome, only provisionally) cannot be specifically distinguished from a hydroid obtained last autumn by H.M.S. "Valorous" in Baffin's Bay. It belongs to a common group of campanularian forms; but yet the fact of identical forms occurring in such widely separated localities, though under conditions probably very similar, is one of great interest and significance, more especially as the distribution can hardly be explained, as in certain other cases, by the transporting agency of ships' bottoms.

On the whole the hydroid fauna of Kerguelen Island, so far as it is represented by this collection, exhibits little which can be referred to as impressing on it anything of a special or characteristic facies. The only unusual type is offered by the form for which I have constituted the new genus *Hypanthea*. All the others, notwithstanding the specific peculiarities by which most of them are distinguished from forms occurring elsewhere, belong to well-known and widely distributed types.

The following are the diagnoses of the new Kerguelen Island Hydroida.

HYDROIDA CALYPTOBLASTEA.

SERTULARELLA.

Sertularella unilateralis.

(Plate XVIII., figs. 10, 11.)

Allman, Ann. & Mag. of Nat. Hist., 4th ser., 1876, xvii., p. 114.

Trophosome.—Hydrocaulus attaining a height of about an inch and a half, alternately and pinnately branched, monosiphonic; Hydrothecæ deep, divergent and somewhat tumid below, slightly curving towards the stem above, strongly four-toothed, all of them deflected towards one side of the stem and branches.

---

* I do not now regard the *Sertularella kerguelensis* of the preliminary notice (Ann. & Mag. Nat. Hist. 1876) as sufficiently distinct from S. polyzonias to justify its separation from that species.

Gonosome.—Gonangia arising just below the base of a hydrotheca, ovoid, with a four-toothed terminal orifice; distal portion marked with wide annulations which become obsolete towards the proximal end.

Dredged in Swain's Bay.

The mode in which the hydrothecæ, though springing from opposite sides of the stem and branches, are all deflected towards one side, causes them to appear to be monostichous, and thus gives to the species a peculiar and well-marked character.

Sertularella lagena.

Allman, Ann. & Mag. of Nat. Hist., 4th ser., 1876, xvii., p. 114.

Trophosome.—Hydrocaulus springing from a creeping stolon, attaining a height of about an inch, slightly branched; internodes much attenuated towards their proximal ends, and there furnished with two or three well marked oblique annulations. Hydrothecæ rather distant, borne by the internode close to its distal end, tumid below, becoming narrow towards the orifice which is distinctly four-toothed.

Gonosome.—Not known.

Dredged in Observatory Bay, Kerguelen Island.

This is a small species distinguished chiefly by its flask-shaped hydrothecæ, and proximately attenuated internodes with oblique annulations.

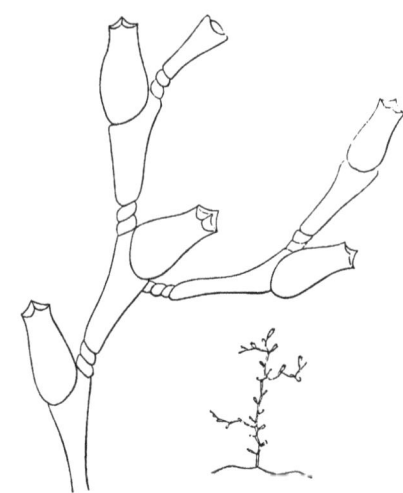

Sertularella lagena.

HALECIUM.

Halecium mutilum.

(Plate XVIII., figs. 8, 9.)

Allman, Ann. & Mag. of Nat. Hist., 4th ser. 1876, xvii., p. 114.

Trophosome.—Hydrocaulus attaining a height of about an inch, irregularly branched; branches with two or three oblique annulations at their origin; internodes short, each carrying for the support of the hydranth close to its distal end a bracket-shaped process which is not produced into a tube, and which is surrounded by a narrow membranous punctate margin.

Gonosome.—Not known.

Dredged in Observatory Bay, Royal Sound.

This species like *H. macrocephalum*, Allman, from the Western part of the Gulf Stream, and *H. sessile*, Norman, from the Hebrides, is remarkable for the utter absence of the tubular prolongation of the lateral orifice of the internode which gives support to the hydranth in most of the species of *Halecium*.

CAMPANULARIA.

Campanularia ? cylindrica.

(Plate XVIII., figs. 4, 5.)

Allman, Ann. & Mag. of Nat. Hist., 4th ser., 1876, xvii., p. 114.

Trophosome.—Hydroid attaining a height of about a quarter of an inch; peduncles springing from a creeping filiform stolon, marked with several annulations at their proximal ends, and having at their distal ends each a single globular annulation which immediately supports the hydrotheca; in their intermediate portion they are slightly corrugated. Hydrothecæ deep, cylindrical, with the margin deeply cut into about twelve strong teeth.

Gonosome.—Gonangia on very short peduncles which spring from the creeping stolon, cylindrical above with a flat summit, and then tapering below towards the peduncle.

Dredged in Swain's Bay.

Beyond our knowledge of the situation and the external form of the gonangia, we know nothing of the gonosome, and therefore the reference of this species to the genus *Campanularia* is merely provisional. A form which cannot be distinguished specifically from this, has more recently been dredged by H.M.S. "Valorous" from 60 fathoms in Baffin's Bay.

HYPANTHEA.

Trophosome.—Hydrothecæ pedunculate, inoperculate, with walls enormously thickened and so far encroaching upon the cavity as to render impossible the complete retraction of the hydranth.

Gonosome.—Gonangia enclosing fixed sporosacs.

The genus *Hypanthea* is very remarkable from having the place of the ordinary hydrothecæ occupied by bodies which may be said to support rather than contain the hydranths, which are thus almost as incapable of being withdrawn into a protective receptacle as are the hydranths of the various species of *Halecium*,—a most exceptional condition among the calyptoblastic hydroids.

Hypanthea repens.

(Plate XVIII., figs. 6, 7.)

Allman, Ann. & Mag. of Nat. Hist., 4th ser., 1876, xvii., p. 115.

Trophosome.—Peduncles attaining a height of about a quarter of an inch, springing at intervals from a creeping stolon, having each a globular annulus just below the hydrotheca, destitute of annulations in the remainder of their extent. Hydrothecæ obconical with very oblique margin, their cavity forming distally a shallow cup, whence is prolonged a narrow cylindrical tube backwards through the axis of the hydrotheca.

Gonosome.—Gonangia elongated, narrow, passing gradually into a short peduncle which springs from the creeping stolon; colonies monœcious; the male gonangia exceeding in height the peduncles of the hydrothecæ, fusiform, opening on the summit by a narrow circular orifice; the females shorter than the males, scarcely narrowing towards their distal extremity where they open by a wide orifice.

Dredged in Swain's Bay.

The singular form of the hydrothecæ, the large naked hydranths, and the greatly elongated gonangia, give to *Hypanthea repens* a striking physiognomy. The great development of the chitinous perisarc is shown not only in the hydrothecæ but also in the peduncles on which these are borne; for the perisarc of the peduncles like that of the hydrothecæ attains a great thickness and narrows their cavity in a manner similar to that by which the walls of the hydrothecæ contract the space contained by them. The lips of the orifice of the female gonangia are inverted for a short distance. In each gonangium of either sex one sporosac only is developed. This forms a greatly elongated sac which occupies almost the whole of the cavity of the gonangium. The presence in this hydroid of male and female gonangia in the same colony is another very exceptional feature.

Several of the hydranths in the specimen were sufficiently well preserved to admit of their form and their relations to the other parts being determined.

HYDROIDA GYMNOBLASTEA.

CORYNE.

Coryne (?) conferta.

(Pl. XVIII., figs. 1–3.)

Allman, Ann. & Mag. of Nat. Hist., 4th ser., 1876, xvii., p. 115.

Trophosome.—Hydrocaulus attaining a height of about an inch and a half; much and irregularly branched, forming dense tufts; stems and branches distinctly and regularly annulated. Hydranths with about 20 tentacles.

Gonosome.—Not known.

Gathered on *Mytilus* at or near low water mark in Observatory Bay (the tide falls only about 2 feet). Scarce.

The absence of the gonosome renders the reference of this species to *Coryne* provisional. It might with equal reason be regarded as a *Syncoryne*. Its densely tufted stems with their strongly annulated perisarc confer upon it a well marked character.

This is the only Gymnoblastic species in the collection.

Spongiidæ.—*By H. J. Carter, F.R.S.*

The collection of Sponges from Kerguelen Island is very limited in extent. So far as it goes, it may be said to present a European, and more especially a British facies. Half of the species at the fewest, may be picked up at any time on the beach of South Devon:—viz. *Isodictya rosea, Halichondria plumosa, H. carnosa,* and *H. sanguinea*. To these we might add a fifth species, *H. panicea,* for the Kerguelen variety differs from the normal British form only in the possession of spicules twice the size of those of the latter. Of the three species remaining *Thalysias* is common to the Mediterranean and the seas between the Americas; the *Ute* occurs on the N.W. coast of Spain and in the Mediterranean; and one only, the *Tethya,* is decidedly antarctic. This last was the only specimen obtained from a considerable depth; all of the others were either collected with the grapple within the Laminarian zone, or were the produce of shore-collecting between tide-marks or amidst the refuse of the beach. Probably more extended research would have brought to light divers of the many peculiar forms which abound in the Cape seas and in those of the southern part of Australia.

In the course of my examination I have met with very few *Foraminifera,* no *Globigerina,* and no *Coccoliths*.

Halichondria panicea.

Johnst. Brit. Spong. 1842, p. 114, pl. x. & xi. 5.

Var.—Dry specimen: Irregularly globose, 4 inches in diameter; colour on the surface white, interiorly light sponge yellow. Structure felt-like, fibreless: texture light, delicate, sub-compressible. Spicules of one form only, viz.:—long, smooth, acerate, curved and often slightly bent in the centre; average maximum dimensions $\frac{100 \times 4}{6000}$ to $\frac{120 \times 4}{6000}$ inch.

This specimen was picked up dry on the shore, apparently after it had been for a considerable period rolled about in the sea. It afforded accommodation to an extensive menagerie of animals, belonging to *Crustacea, Annelida,* and *Spongida*. An amorphous fragment of the same species gathered when fresh was preserved in spirits; it is of a light yellowish sponge-colour.

Hab.—(Var.) Royal Sound, Kerguelen Island; (normal form) British coast.

Isodyctya rosea.

Bowerbank, Monogr. Brit. Spong., vol. ii. 282.

Amorphous, encrusting the malleoloid expansions of a *Melobesia*. Colour pinkish.

Structure isodictyal (vide illustr. in Bowerbk., op. cit. pl. xx., 309). Texture paniceous, friable. Spicules of one form only, viz. :—rather short, smooth, acerate, curved, and often slightly bent in the middle. Average maximum dimensions $\frac{40 \times 2}{6000}$ inch.

The largest specimen appears to have been picked up on the shore after many years exposure. Among other examples are some amorphous fragments in spirits, gathered when fresh, of a light sponge-colour; also some dry pieces still retaining the characteristic roseate tint presented by this species.

Hab.—Royal Sound, Kerguelen Island; also British coast.

Thalysias.

Thalysias, Duch. de Fonbressin et G. Michelotti, Spongiares de la Mer Caraïbe, Harlem, 1864, p. 85, pl. xvii., 1, &c. *Schmidtia*, Balsamo-Crivelli, Atti d. Soc. Ital. de Scienza, 1863 (*Schmidtia ficiformis*), vol. v., taf. iv., 2, &c. Gulf of Naples.

Thalysias subtriangularis, *D. & M.*

Mammiferous sessile form. Dry specimen. Mammiform processes rising out of a procumbent mass. Surface even, uniformly and closely pierced with small holes. Colour yellowish brown. Vents large, single, each at the end of a mammiform process. Pores originally in sarcode tympanising the little holes on the surface. Structure compact. Texture close, uniformly firm. Spicules of one form only, viz. :—smooth, acerate, curved, and often slightly bent in the middle : average maximum measurements $\frac{45 \times 3}{6000}$ inch.

Hab.—Royal Sound, Kerguelen Island; also the Mediterranean, and seas of Central America.

Halichondria plumosa, *Johnston*.

Hab.—Royal Sound, Kerguelen Island, on *Halichondria panicea*: also British coast.

Halichondria carnosa, *Johnston*.

With spicules of one form only, viz. :—pin-like, smooth, with globular terminal head, slightly fusiform shaft, and fine point : average maximum dimensions $\frac{100 \times 3}{6000}$ inch.

Hab.—The same as that of the preceding species.

Halichondria sanguinea, *Johnston*.

With spicules of one form only, viz. :—smooth arcuate and curved : average maximum dimensions $\frac{50 \times 2}{6000}$ to $\frac{60 \times 2}{6000}$ of an inch.

Tethya antarctica.

Carter, Ann. et Mag. of Nat. Hist., 4th ser., 1872, ix. 412, pl. xx.

Part of a large specimen; in appearance very like grey hair matted together with sand and mud, [in form similar to a small cocoa-nut prolonged at the base into a short cylindrical peduncle as thick as one's wrist]. Spicules, long large stout acerates, mixed with apparently an equal quantity of delicate anchor-and-fork-headed long shafts. The acerates are much larger and longer than those of the specimen described by me in the Annals & Magazine of Natural History (loc. cit.) averaging in their maximum dimensions $\frac{7}{24}$ by $\frac{5}{1800}$ inch.

Hab.—Royal Sound, 40 fathoms, brought up by anchor, Eaton; also dredged in lat. 74° 30′ S., long. (obliterated), from a depth of 300 fathoms, and in lat. 77° 30′ S., long. 175° 2′ E., from 206 fathoms, Antarctic Expedition.

Ute capillosa.

Schmidt, Spong. Adriat. Meeres, p. 17, taf. i., 6–6 a–b.

White; long and compressed, sac-like with narrowed aperture; 2 inches long, $\frac{3}{4}$ inch broad, $\frac{2}{3}$ inch thick. Spicules tri-radiate of two sizes, mixed with thick, stout, rudely formed, acerates, which project hair-like above the surface. (In spirits; gathered in the fresh state.)

Hab.—Royal Sound, Kerguelen Island: also Vigo Bay, dredged by Mr. Saville Kent; and the Mediterranean.

THE COLLECTIONS

FROM

RODRIGUEZ.

INTRODUCTORY NOTES.

I. THE PHYSICAL FEATURES OF RODRIGUEZ.—*By Is. Bayley Balfour, Sc.D.*

SITUATED in the Indian Ocean, 300 miles eastwards from Mauritius, the island of Rodriguez, like the sister Mascarene Islands, is a mass of volcanic rock. A fringing reef of coral, studded with islets, skirts it on every side, extending on the west about three miles from the land, but at the eastern end the edge of the reef is within about a hundred yards of the beach. The island consists of a series of hills. Its extreme length, from a little north of east to a little south of west, is about 11 miles and its breadth from north to south about 5 miles. Within this base the land rises towards the centre of the island, where are several peaks, none attaining any great elevation, the highest point, Mount Limon, being only 1,300 feet above the sea level. A main ridge runs along the island in a direction parallel with its greatest diameter, and rather nearer the southern shore. Its slopes rise with some abruptness from the sea on the eastern side, but on the west extend more gradually seawards, and terminate in a wide coralline limestone plain studded with elevations between Baie Topaze and Anse du Peril. The sides of the ridge as they stretch to the sea are deeply cut into ravines. The slopes on the southern side are shorter, and the ravines deeper and more numerous, than on the north. In their upper parts these ravines are bordered by lofty and inaccessible cliffs, upon which the volcanic structure of the island is well marked, and coulée is seen to succeed coulée, separated only by thin beds of cinder, agglomerate, or variously coloured clays. In some instances these cliffs are 300 feet high, and as many as 12 successive coulées may be counted on one cliff. Through these ravines the streams as they descend form in their upper parts a series of cascades, and sometimes high falls. One of the finest is the Cascade Victoire at the head of the Rivière Poursuite, where it falls over a cliff more than a hundred feet high. As the sea is approached, the ravines expand into wide valleys flanked by gently sloping and terraced ridges. These ridges too are in some places marked by lofty cliffs, on which the columnar basaltic character of the rocks is well seen. A splendid example of this is Tonnerre Cliff in the valley Rivière aux huitres, a bold perpendicular face of prismatic columns 200 feet high. At Pointe la Fouche also this structure is conspicuous.

On the south-west, the central volcanic ridge gradually descends, the ravines become less deep, and the ground spreads out into a large coralline limestone plain. The demarcation betwixt the limestone and the volcanic rock is very sharp, but isolated patches of limestone are met with on the surface of the volcanic region, in the vicinity of the main mass. The caves from which the bones of the Solitaire and other extinct birds have been obtained occur in this limestone plain. Some of them extend for a great distance through the rock, and are rich in stalactites and stalagmites; others again are mere small holes. The whole plain is riddled with these caves, and on walking over it one constantly passes small apertures and fissures, evidently "blow-holes" of some subterranean cavern. Wide and deep hollows are also met with, on the floor of which large fragments of limestone lie in confused heaps. These are apparently old caves of which the roofs have fallen in, and the continuation of the cavern may be found at either extremity. The floor of these hollows is composed of volcanic soil, often with large masses of volcanic rock on the surface, and commonly clothed richly with vegetation. It is in such places that many of the largest trees on the island are now to be seen. The limestone is not found along the northern or southern shores, until we near the eastern extremity, where patches occur at the mouths of the valleys, and even at some distance from the shore. One mass I discovered in valley Rivière de l'Est, more than a mile from the sea. It is not so abundant at this end of the island.

On the southern shore between Rivière Palmiste and Rivière Poursuite, indications of raised beaches are seen, reaching about 20 feet above the sea level.

The existence of these masses of coralline limestone indicates clearly a former lower level of the island, and the evidence of raised beaches confirms this. But a consideration of the coral reefs points as clearly to a time when the island stood at a higher level. The present coral reef fringes the coast, extending, as I have mentioned, about three miles on the south-west side, but coming close inshore on the east. An older reef, however, exists now quite submerged in some places to a depth of over 90 fathoms. Upon it the present reef rests, and it extends westwards nearly 15 miles from the present coast, while on the east it stretches about six miles.

We have thus proofs of great and intermittent oscillations of the level of the island.

Of the islets scattered over the reef, some are volcanic, and the others are composed of coralline limestone and sand. They are all within the compass of the present reef, and only occur on its wider parts; consequently there are none east of Port Mathurin on the north, and of Port Sud Est on the south. Eight islets are of volcanic origin, Diamond, Booby, Katrine, Marianne, Desinée, Frigate, Crab, and Hermitage. Only the last mentioned is on the south reef, the rest range round from the south-west to the north. The coralline limestone and sand islets are more numerous, and are confined to the southern and western reefs, none occur on the north. Ten of them receive names, as follows:—Gombrani, Pierrots, Platte, Pantad, Pianqui, Misel, Chat, Zozo, Coco, and Sable. The last two are mere accumulations of sand and coral débris close to the western edge of the reef, the others are all on the south. It will be seen, then, that whilst the volcanic islets are chiefly on the north and west, the limestone ones are on the south.

The prevailing type of rock composing the island is a dolerite rich in olivine, in many places greatly decomposed. Interspersed with the coulées of rock are found extensive beds of clay. These, which possibly result from the decomposition in situ of the dolerite, are highly coloured, usually bright red or ochre, and form a prominent feature in the landscape. The lava coulées seem to have flowed with great regularity, but there have been marked periods of intermission of volcanic energy. Here and there dykes are seen, one specially well-marked occurs at the seaward end of the Charpentier ridge of the valley Rivière Bouteille, forming a conical projection through the layers of rock; and it belongs to the same period of formation as do the higher parts of the island,—apparently the last outburst of volcanic activity.

The exact position of old craters is difficult to determine. Probably there were many foci, but the main ones seem to have been situated about the Grande Montagne and Mount Malartic. Many of the small conical isolated hills, such as Montagne du Nord, scattered over the island, no doubt also mark the site of old foci.

The time that has elapsed since the last exhibition of activity has been sufficient to allow of a considerable amount of denudation, as is evidenced by the fragments of rock and débris with which the more level ground is strewn, and which cover the gentle hill slopes, rendering progression a matter of difficulty; and the smooth and rounded outline of the hills, only occasionally interrupted by a projecting torr or pinnacle, as well as the deep ravines, testify to the same.

A curious feature deserves notice. In the Baie aux Huitres are found masses, about a foot in diameter, of shells of a mollusc embedded in or rather cemented by a material resembling tuff. Unfortunately the specimens were lost in transmission, and thus a further determination of their nature is prevented.

The island is comparatively dry, the soil is parched and arid, and during the warm season many of the streams are dried up. But the size of the water-courses and the enormous boulders filling their beds, indicate large torrents in the rainy season. In some places, issuing from the clay, springs occur, of which the water is brackish, has a very disagreeable taste, and is slightly tepid, but has no smell. As a rule the water of the streams is good and safe to drink, but that of some rivers, notably the Rivère Saumâtre, is most unpalatable and apt to cause slight catharsis.

The climate is much like that of Mauritius, where the average annual temperature is 78° Fahr. During the north-west monsoon from November to April, the weather is wet and warm, and frequently in the first months of the year the island is visited by severe hurricanes. From May to October the south-east monsoon prevails, and then the weather is cool and dry. The rainfall is exceedingly irregular, the hills being hardly high enough and not sufficiently wooded to arrest cloud; hence also fogs are rare.

II. REPORTS OF PROCEEDINGS OF THE NATURALISTS.

1. REPORT OF DR. IS. B. BALFOUR.

I LEFT England upon the 1st July 1874, and embarked at Marseilles on board the Messageries Maritimes steamer. I reached Mauritius with my confrères on the 4th of August. On our arrival there we met Lieut. Neates, the head of the expedition to Rodriguez, who had come by the troopship "Elizabeth Martin" some days previously. H.M.S. "Shearwater" with the other astronomers and the astronomical instruments arrived at Mauritius on the 6th, and as there was no cause for delay there, Captain Wharton decided to sail for Rodriguez on the 11th. This was carried out, and after a week's voyage the ship anchored in Mathurin Bay, Rodriguez, on the afternoon of Tuesday the 18th. Owing to the difficulty of carrying heavy baggage over the reefs, rendered doubly so by the neap tides which then occurred, some days elapsed ere we could take up our abode on the island. These days I spent in making an examination of the general features of the island. On the 28th of August I quitted the ship, and went to live on shore, and at once commenced my botanical and geological work.

As I have already explained in my letters to the Royal Society,* the flora is very scant owing to the frequency of the fires which have swept across the island. Indeed fully one half of the island is entirely bare of vegetation, save the few social weeds which now occupy the place of the destroyed indigenous plants. The only parts of the island where any amount of vegetation is to be found are the upper regions of some of the valleys, round which the fires have evidently passed, leaving them comparatively unscathed; but even here the plants are few and the dryness of the soil precludes any luxuriance. I explored thoroughly all those valleys which promised any botanical treasures, and I regret to say the result is very poor.

In accordance with the wishes of Captain Wharton, I had all my specimens packed and on board the "Shearwater" by the day of the Transit. On the 15th of December we quitted Rodriguez and sighted Mauritius on the 18th.

I remained in Mauritius until the 5th of February 1875. During my stay I spent much time at the botanical gardens examining the native flora, and I also made excursions over the island for the purpose of investigating the species of indigenous Pandani.

Specimens, the result of my work, I have transmitted to England.

Leaving Mauritius by the mail steamer on the 5th February, I reached Bourbon next morning. Here I remained investigating its rich and little known flora. Unfortunately bad weather during my stay rendered botanical and geological explorations both unpleasant and difficult. I amassed a considerable number of specimens illustrating the vegetation of the island, and I specially devoted some time to the examination of the Pandani.

On the 6th March the mail came in, and I had reluctantly to leave Bourbon. I reached Marseilles on the 31st March and London on April 2nd.

Is. BAYLEY BALFOUR.

* See Proc. Roy. Soc. 1874, p. 135.

2. Report of Henry H. Slater, Esq., B.A.

I LANDED in Mauritius August 4th, 1874. The Hon. Edward Newton kindly asked me to stay with him. My instructions directed me to take his advice as to my mode of procedure in Rodriguez, and accordingly I requested George Jenner, Esq., Director of Emigrants for the Grand Port District, to select for me nine good men and a cook.

This took some time and I regretted much that I was not able to leave for Rodriguez until September 9th, by the second trip of the "Shearwater." I was wishing at first, nevertheless, to sail without my men, had not Mr. Newton and Mr. Jenner (10 years magistrate of Rodriguez) assured me that I could do nothing without my men, and that the natives of Rodriguez were quite unreliable.

We landed in Rodriguez on September 14th, with our 10 men. Settled in Port Mathurin and stayed there for a day to allow the men to recover from the voyage, which had been stormy; Government House had been allotted to the naturalists as a residence and head-quarters, but two tents had been supplied from the "Shearwater," one of which was to be my habitation at the caverns. After this the weather was stormy for some days, so much so that it would not have been prudent to attempt the sea voyage to the caves, and no native would pilot one there; but afterwards the weather changed and I secured two boats, in which, upon the 18th of September, I carried my men, tent, stores, and tools down to the caverns. Owing to the roughness of the country, which is very hilly, it would have been impossible to have carried the stores, &c. overland, without many porters and great expense.

The 19th was occupied in establishing camp, in building a hut for the men, and in prospecting for caves, of which I found several. Next morning we started work: the caves had all been previously dug over, but I determined to dig them again, and to a depth of three feet, and we found a good many bones (mostly tortoise, but with some of Pezophaps) though they were in a bad state of preservation.

I used to go out every evening to take my daily exercise and to shoot game (no fowls or goats could be got at the caverns, which part of the island is nearly uninhabited, so that it was necessary to shoot a supply of game), and on these walks I found all the caves.

This sort of work, *i.e.*, digging over caves second hand, continued till the 6th of October. I used to go home always on Saturday afternoon with the week's bones, stay over Sunday, and leave Government House for the caves on Monday morning in my canoe, carrying a week's rations for the men and my own supplies.

On October 7th I found some new caves. These formed a branch of a cavern already dug, but as entrance to them was extremely difficult, they had hitherto escaped notice. In these we found a large quantity of Solitaire bones and the almost perfect skeletons of a male and female mingled, including the rings of the trachea; very few tortoise bones were intermixed, and I found it an almost invariable rule, that where Solitaire bones were found in large numbers and apparently occupying the spots where they died, there were few or no tortoise bones amongst them.

After this I had varying success, sometimes getting a good haul and making valuable notes, sometimes doing nothing in either particular till October 23rd. On that day I was waiting for my head man near a curious ravine which we called "the Gorge," intending to make a survey of it with a view to new caves, when I noticed a small dark hole behind a huge block of coralline limestone. I peered in and found it to be a large cave which had been obliterated by a fall of rock; so when my Sirdar came, a thoroughly intelligent and enthusiastic man, we got a large mallet, and after reducing some huge blocks to movable fragments, gained entrance and found many bones, including one skeleton of Pezophaps which had evidently fallen into a cleft and had been unable to extricate itself; the bones being not quite covered with earth some had become quite decayed; indeed, had it not been that bones lying on the surface were always almost or quite decayed, I should have many skeletons of Pezophaps perfect.

On November 6th we found another virgin cave; it yielded about 24 crania, several quite perfect and with the beaks still attached, maxillæ, mandibles, and quadrates in their right places; and other bones in proportion, including many furcula, the auditory ossicles, plates of sclerotic, and the first phalange of wing, one of the few bones not already found.

After this, although vigorous search was continued, no caves were found of any value, all were disturbed by the action of running water, for some of the caves occasionally served as beds for subterranean rivers.

We dug for three days in a neighbouring marsh, about 450 feet above the sea level. Hitherto all the bones of Didus ineptus recently found have been discovered in the marshes of Mauritius, and I hoped that some result might reward a similar search in Rodriguez, but our work there was entirely ineffectual.

I never could find any traces of the stones mentioned by Lequat as invariably occurring in the stomach of Pezophaps, although I sought for them diligently; we used to pass all the soil found in the bottom of the more fertile caves through sieves, and nothing of the sort was found. Surely a stone which in 1693 was sufficiently hard and close grained to serve as a good whetstone for knives, should have survived as long as the fragile ribs and crania of the bird!

The last week I was in Rodriguez (December 9–15) was employed in packing up my numerous specimens. It may be imagined that so many fragile bones would need a large quantity of some soft and elastic medium in which to pack them, but there was very little straw, no grass was to be had in the island, and I was reduced to great shifts. I finally put them temporarily in barrels, but when I unpacked them in Mauritius I was horrified to find that some of the most valuable specimens of all, which had been placed in sawdust, had, owing to the displacement of the sawdust, got mixed one layer with another; but I do not see that this will cause any real inconvenience.

I left Rodriguez on December 15th, and arrived in Mauritius December 19th; left that island for Réunion January 9th; Réunion for Europe February 6th; and landed in England March 10th.

---

3. Report of George Gulliver, Esq., B.A.

After my arrival in Rodriguez, in company with Mr. H. H. Slater, as stated in his report, I immediately proceeded in accordance with my instructions to make as complete a collection as possible of the land and freshwater animals of the island. As directed by my instructions I paid no special attention to marine zoology. I left Rodriguez immediately after the Transit of Venus, the period of my stay there being accordingly about three months.

Here my colleagues and I must express our special thanks to Mr. Newton, C.M.G., the Colonial Secretary at Mauritius, to Commander Wharton, R.N., and the officers of H.M.S. "Shearwater," to Mr. Bell the present, and Mr. Turner the late magistrate at Rodriguez, and to the residents in Mauritius for the assistance they constantly rendered and the kindness they always displayed to us. Personally I must again make acknowledgments, more especially to Commander Wharton, and the officers of the "Shearwater," and to Mr. Bewsher, of the Oriental Bank Corporation. Finally it would be difficult for me sufficiently to thank Professor Rolleston for the kindness which he has constantly shown to me, both before my departure from, and since my return to England.

PETROLOGY.

By N. S. Maskelyne, F.R.S.

The numerous specimens illustrating the rock formations of the Isle of Rodriguez, collected by Mr. I. Bayley Balfour from different localities, need only a cursory inspection to attest the volcanic character of the whole mass of the island.

Rodriguez, in fact, consists of Doleritic lavas that appear to have been poured out at a considerable number of volcanic orifices at successive periods. It would be difficult, without more minute description of the physical geography of the island than is accessible, to assign any precise date or duration to these volcanic eruptions, or to trace with any certainty the degree to which and the mode in which subsequent denudation has helped in giving the island its present remarkable aspect.

But the fact of that denudation and the degree to which alteration has proceeded in affecting the minerals composing rocks that by their position must have been among the later of the out-poured lavas, would point to a remote date, possibly to one contemporary with the tertiary period, as that of the volcanic activity of Rodriguez.

A central ridge, lofty for an island of only about 13 miles in length by 5 in width, divides the north from the south, except that a break occurs at about two-thirds of the distance from east to west. A succession of peaks on this ridge range from the Mount Quatre-vents, which forms the western bastion, 1,120 feet high, of the central mass of the island, to the Grande Montagne, which is 1,140 feet in height, and looks into the steep valleys of the rivers Grévier, Palmiste, de l'Est, and Bouteille, that lead down from it in a semicircle to the sea around the eastern end of the island.

Between these points, or in positions a little in advance of the general line of mountains, Mount Limon, 1,300 feet, Le Pitou, 1,160 feet, and another height between it and Mount Limon, 1,240 feet, are among the peaks that crown this volcanic mass of mountain. The steep flanks of these high lava ridges assume in some cases columnar forms, the basaltic columns of the cliffs, at Mount Tonnerre for instance, rising to 200 perpendicular feet.

It would seem, however, from the contour of the island and from the nature of the rock composing it, that the steep valleys which groove the sides of the chief mountain mass have been mainly engineered by the brooks that find their way to the sea, and meet the innumerable little bays and inlets that indent its shores on every side.

It is especially on the east and round the eastern end that the mountain range is steepest, for there the coral reef comes nearly close to the shore, and the island would seem to rise rather abruptly from the depths of the ocean.

The rock specimens which Mr. Balfour collected appear to have been brought together with a naturalist's discrimination as regards their variety, and with considerable care as regards their representation of the different localities.

The petrological study of them brings out the fact that, while externally they differ considerably in their mechanical texture and appearance, they all, as has been above stated, belong to the doleritic and basaltic group of rocks, and especially to a variety of these rich in Olivine, and free from any appreciable amount of glassy ground-mass.

The coarser grained and more strictly doleritic varieties of this rock (strictly, perhaps, they should be called Anamesite) are illustrated by various specimens taken from Mount Grenade, from the summit of Mount Pitou, from the summit and also from a valley south of the Grande Montagne; likewise in some of the carefully collected specimens, of which further mention will be made, from the Cascade valley, a valley carved out by a stream which, rising between Mount Pitou and Mount Limon, almost exactly in the centre of the island, enters the sea at the back of the little town of Mathurin, on the northern shore. The cliff at the back of this town, the valley running up from the east of English Bay Point, a hill overhanging the mouth of the point and bay of Banane, and a steep ridge above the Anse (or bay of) Bouteille, are other localities taken in topographical order from W. to E., from which the coarser-grained lavas were collected.

These coarser or more doleritic kinds vary in tint, but are generally of light grey colour, the larger grained kinds tending to a somewhat violet grey, the finer to a bluer and more slaty hue. They are by no means hard in texture; some of them are compact, others more or less vesicular, and occasionally carry zeolitic minerals, or aragonite.

Viewed microscopically, the white felspar is seen to be mottled with dark grey minerals, among which conspicuous crystals of orange-yellow Olivine stand out, and are generally decomposed to a considerable degree, and here and there small grains of magnetite can be recognised.

In thin sections these dolerites of coarser grain present in the microscope a beautiful assemblage of interlaced felspar crystals, which in polarised light exhibit the twinned and ribboned stripes that characterise the anorthic felspars. This felspar is in fact Labradorite. That it is so is confirmed by the analysis of the crystals picked with great care from the crushed rock; a specimen quite characteristic in its felspar being selected from those from the river Palmiste.

In some of these dolerites, as in that from Mount Grenade and from Pitou, the crystals of felspar are long and blade-like in form, while in those from some of the other localities the crystals are comparatively short in proportion to their length;

they do not, however, appear to belong to different varieties, for they pass through intermediate stages. The felspar in most of the rocks becomes prominent in individual crystals, and imparts a micro-porphyritic character to the rock, independently of the Olivine. Of ground-mass, strictly speaking, these dolerites are devoid. The mass of the rock is made up of the interlaced felspar crystals, the interstices being filled with a brown, and in finer grained varieties a nearly opaque, mineral, which sometimes crystallises in distinct forms, and in its less distinct variety is only to be recognised as crystalline by examination with a high power between crossed tourmalines. It is undoubtedly Augite. In some instances the Augite is to be recognised in quite distinct and transparent short pale greenish brown crystals, quite unattacked by decomposition.

The Olivine which often porphyrises the rock by its abundance and distinctness of crystallisation rises into an important ingredient of these dolerites. Some of them might indeed be termed Olivine dolerites, from the abundance of this mineral.

The Olivine belongs to a variety rich in iron, and has undergone the usual characteristic decomposition; the crystals being encrusted by, and sometimes entirely converted into, a dark reddish brown mineral, which has also penetrated the crystals wherever fissures have afforded lines of entrance by which acid or thermal waters could attack them. In one or two of the rocks of finer grain a mineral closely resembling Chlorophæite can be recognised.

Minute grains of magnetite, the rectangular or nearly rectangular sections of which can often be recognised, are abundant in some specimens of the rock, and are seen very evenly distributed in the microscopic sections.

A very characteristic accompaniment of the felspar is Apatite, a mineral which presents itself in long transparent needles and microlites. The presence in some little abundance of phosphoric acid is attested by treating a solution of the mineral, freed from its Silica, in nitric acid by Ammonium Molybdate, confirmatory of the anticipations formed with the microscope.

The Mount Grenade Dolerite is highly vesicular, and like that from the east side of English Bay contains more Augite than that from Pitou, or from what would seem to be the older lavas from the bottom of the Cascade Valley.

Passing to the other extreme of the rock series, that, namely, in which the grain is finest and the texture most dense, we have a few specimens which, as regards these characters, may be classed as somewhat coarse basalt, or as fine-grained anamesite. Their colour is generally dark in contrast with the coarser grained lavas. The eye detects no distinguishable minerals when aided by an ordinary lens, except that a few crystals of felspar and numerous crystals of olivine are to be seen enclosed in the formless mass, and that in one dense rock from the Tonnerre Cliff a certain silky sheen can be detected in the fractured surface, which is due to a stream-like distribution of the felspar crystals.

In the microscope the appearance of those finer rocks is quite similar to that of the dolerites. The Tonnerre Cliff rock, dark grey in hue, is under the microscope a beautiful congeries of minute blades of Labradorite streaming in directions that run along and round the large crystals of Olivine, and a great number of very small black points of magnetite mingled with some pale yellowish green Augite.

These basalts, which are at once the finest in grain and hardest in texture of all the Rodriguez rocks, seem to form the material of the columnar masses of the island, though not confined to these. The localities noted by Mr. Balfour on some of the more characteristic of them are, besides Tonnerre Cliff, the locality of the finest columns in the island; Beline high fall, the rock in the bed of the cascade brook, a columnar mass in the Cascade Valley, and the valley on the E. side of English Bay Point.

The banks of the river Poursuite yield also a fine basaltic rock; sometimes vesicular and charged with Aragonite, a Magnesian Lime-carbonate and Chlorophæite in fissures, and Zeolites, among which striated little crystals of Chabazite are distinctly recognisable.

The felspar in this basalt offers indications of commencing decomposition.

And similar minerals occur in other vesicular varieties, chiefly of the finer-grained rocks.

A curious feature of some of the Rodriguez rocks (and among those collected it is confined to the Anamesites and fine-grained and more strictly basaltic varieties) is seen in the tendency of the rock to break into unevenly rounded fragments; this variety of rock possessing but little tenacity, and being without difficulty separated into knob-like pieces. This nodular texture is traceable in the more pronounced cases to a great number of curved fissures traversing the mass, which are perhaps the result of the conditions under which it cooled. Not improbably they may be due to an imperfect development of a kind of pyromerid structure, in the production of which the felspar may have been the formative ingredient.

Another peculiarity of certain of these Rodriguez Anamesites is probably due to, and in part only a differently developed effect of, a similar cause. These show a spotted or variolitic surface, in which very light spots are seen on the darker slaty gray ground. Usually they are about an eighth to a quarter of an inch in diameter, and are rarely confluent. Microscopically examined the rock is seen at the spots to be nearly devoid of Olivine, whereas this mineral is very plentiful in the other parts: while microscopical inspection reveals in some of the spots a peculiar structure in the felspar, which is here and there grouped with some of its crystals rudely tangential to a circle, at some distance from a larger crystal or crystals of the same mineral, which forms as it were the centre or nucleus of the spot. It would seem that here also the spot is due to the tendency of the felspar to originate an orbicular rock structure, analogous to, but by no means resembling, in the grouping of its crystals, that of the orbicular corsite. It is to be noted also, as bearing on this point, that the spots in the true variolite seem to be due to a peculiar segregation of

Labradorite. The variolitic structure is especially noticeable in a rock from the Grenade Bay.

A variety of the basalt from the bed of the torrent in the Cascade Valley is remarkable for a peculiarity converse to that just described. Here the base of the rock presents the pale grey hue of that of the spots in the rock from the Grenade Bay, but is flecked with tortuous stripes, and with spots of darker hue. At these parts it is seen in a microscopic section that the felspar is less, and the augite more abundant in the dark spots and stripes than in the paler ground. The felspar is here in somewhat larger crystals than in that of the Tonnerre basalt, but presents a similar stream structure.

There remain for consideration certain substances varying greatly in colour from a dull brick red to a liver-brown, and presenting the features of earthy clays, of which Mr. Balfour collected a very complete assortment.

It would seem that they form intermediate beds, intervening between two layers of regular basalt; as in the case of the Cascade Valley, where the most complete series of them was made. The overlying basalt has evidently exercised on the upper layers of this stratum the colouring action to be expected from a molten lava acting on a clay bed; it has given the clay a brick-red colour, and so far hardened it. The lowest bed is of a brown hue, and appears unchanged by heat. The intermediate beds vary in hue from reddish yellow and grey to a deep chocolate or burnt sienna brown.

That these clay strata are the result of the action of water on the rocks of the island seems borne out by their composition, which is shown by a qualitative analysis to be that of a ferric-aluminous silicate, almost entirely free from calcium or magnesium compounds; a residue, in short, left after these earthy and the alkaline bases had been removed by the action of carbonated waters. In this respect they no doubt correspond to the red earth of Mauritius, the history of which would appear from the scanty notices (Mauritius, by Rev. F. Flemyng, pp. 18 and 19) of it, to be that of a disintegration of the volcanic rocks of that island; though, from the description, it would seem that it has a more pisolitic structure, and is richer in iron oxide than the corresponding earth in Rodriguez. Remotely, too, it may be compared with some kinds of laterite.

Mr. Bayley Balfour made some careful observations on this earth with a view to establish its petrological relations and history; and the following extract from his journal will show the sort of evidence which he derived on the spot in favour of its being a product of the alteration of the lava-rock of the island :—

"This red bed, as I call it, is evidently the same as the one which I noticed further down, about a mile from the town, and on the opposite side of the valley. I could trace that one a good way up the valley; but eventually it became hidden by the boulders and scrub, but its presence was indicated by the fragments at the side of the burn below. This bed, at the point where these specimens were taken, is about a foot thick, and is well seen on both sides of the valley, the intervening mass which has been washed away being seen on the face of the Fall Cliff, the distance being about 30 yards. It is to this soft bed that the fall is due, as it lies below a harder rock, of which 64 is a specimen. My specimens show more or less of a gradation from normal basalt to this clay. This clay may be the same as the Terre rouge of Mauritius, but as I have not seen the latter, I cannot say whether it be so or no. It is evidently decomposed lava, the contained crystals of minerals being numerous; but I have not yet analyzed them. I cannot think of any cause for this decomposition. I observed at the place about a mile from Mathurin, where I first picked this red clay, a spring of water rising just at the top of this red bed, and coming along a tunnel of about 4 inches diameter out of the side of the hill. The opening being about 3 feet above the burn level, and about 4 feet from the water as it then was; but in wet seasons the burn must rise fully a long way above its opening. However, the water of the burn was good to drink, and beyond the usual earthy taste, possessed more or less by all the water here, it had no taste; but this spring water was abominable, warm and brackish, specially noticeable on contrasting it with the burn water. It may be that this spring, containing as it clearly does some chemical substances in solution, in addition to the usual constituents of water, may have some influence and should; as it seems probable from facts I have learnt from inhabitants, as also from observation, that many such exist. We may have here a cause of the rapid disintegration. At the Cascade itself, on the E. side, where I picked my specimens, I noticed a hole in the rock just above the red bed similar in form to that through which, below, the spring water emerged, but there was no sign of any moisture—contrarily, the rocks here were very dry."

That it is the result of water-worked disintegration of the basalts and dolerites, the analysis renders more than probable. But whether it does not in fact mark an epoch of pause in the volcanic activity of the island, or at least in the N.E. end of it, during which a long process of denudation and water action supervened over a then submerged area, would need a more minute and extensive topographical study of the district than has yet been given to it. There are veins of a white waxy Aluminium silicate traversing the brown or yellowish brown clays, which probably fill fissures produced by the drying of the original mass, which in some cases, as in that of certain specimens of the red baked upper layer, seems to have assumed a quasi-columnar structure.

BOTANY.

By Is. Bayley Balfour.

Introductory Remarks.

WHEN in 1691 Leguat sighted the island of Rodriguez his eyes fell on mountains " richly spread with great and tall trees," over which flowed streams with banks " adorned with forests," and altogether the scene was such as to call forth from him the designations of " a lovely isle," " an earthly paradise," " a little Eden." He speaks of it thus (New Voyage to the East Indies, p. 248) :—" 'Tis, as I have hinted, composed of lovely hills, covered all with fine trees, whose perpetual verdure is " entirely charming ;" and " between these great and tall trees one may walk at " ease and find such refreshing coolness in their shade at noon, so sweet, so healthy, " that 'twould give life to those that are dying. Their spreading and tufty tops, " which are almost all of an equal height, joyn together like so many canopys " and umbrellos, and jointly make a ceiling of an eternal verdure, supported by " natural pillars which raise and nourish them." Such is his picture of the aspect of the vegetation of Rodriguez. Is such its character now ? No. The great and tall trees have now almost entirely disappeared, the eternally verdant canopy formed by their boughs no longer exists, and the "little Eden" is now a dry and comparatively barren spot, clothed with a vegetation mainly of social weeds, and destitute of any forest growth save in unfrequented and more inaccessible parts in the recesses of the valleys; and, we may ask, what has wrought this change ?

We find in the history of the vegetation of Rodriguez a case similar to that of St. Helena. The same causes which destroyed the peculiar and most interesting Flora of that fertile island have operated, and I regret to say still operate, in Rodriguez, and have effectually changed the face of the island.

Goats, introduced long ago, are now found in enormous numbers, eating the young shoots and leaves of any herb, shrub, or tree within their reach ; and now too several thousand head of cattle graze on the island, and effectually keep down the vegetation of the spots they frequent. Fires have occurred with great frequency, and every now and then at present sweep across parts of the island, destroying everything within reach. And then there are introduced foreign plants. These are now in great abundance, and in many cases completely occupy the ground to the exclusion of the native vegetation, which is driven to the secluded parts of the island. Perhaps one of the best instances of their power in this respect is found in the case of " *L'Acacie* " (*Lucœna glauca*). This plant, introduced about 30 years ago, is now found covering the ground for acres, forming so dense a scrub that it is impossible to penetrate it, and beneath which nothing will grow. Originally planted in the valley near Port Mathurin, it is now found in almost every valley in the island, spreading from the banks of the streams up the sides of the valley. It owes its spreading in a great measure to the cattle and goats, which are exceedingly fond of the leaves and pods, and thus the seeds are carried about. Finally, a certain amount of destructive influence is attributable to the settlers indiscriminately cutting down the trees over large tracts. This, however, has now received a check, as the cutting of timber is forbidden by law. These agencies, then, have directly effected the destruction of a great part of the vegetation of the island, so that over large areas hardly a tree or shrub is seen, and the ground is covered by only a scanty clothing of grass and tropical weeds.

But fires and the hand of man, through the alteration in the climate consequent on their destruction of the forests, have effected indirectly a more permanent injury on the Flora; for now we have a bare, parched volcanic pile, with deep streamcourses for the most part dry, in place of the verdant well-watered island of 200 years ago.

Can we wonder, then, that we find but a remnant of what we may consider the old vegetation still extant in Rodriguez? That the island had originally a rich Flora there can be little doubt, judging from its position and from analogy with the sister islands. Unfortunately Leguat, from whose account we derive all our early information regarding the island, does not enter in any great detail into the native plants of the island at the time of his visit, occupying himself more with those which he and his companions found useful, and with those they introduced into the island, and we thus have really no record of the exact nature of the primitive luxuriant vegetation. He only mentions 10 plants as found on the island, and these, though often rather curtly described, I have been able to identify in every instance save one, at least generically. The following is the list of plants he mentions, and alongside of each is placed the name of the plant with which I believe they can be identified:—

Purslain	*Portulaca oleracea.*
Tree with fruit like olive	*Elæodendron orientale.*
Nasty Tree	*Clerodendron laciniatum.*
Pepper	*Capsicum frutescens.*
Ebony	*Diospyros diversifolia.*
Plantane	*Latania Verschaffeltii.*
Palm Tree	{ *Dictyosperma alba* var. *aurea.* *Hyophorbe Verschaffeltii.*
Pavilion	*Pandanus.*
Rodrigo Kesta	*Ficus.*)

Flower white as a lily and like a jessamine.

I know of no plant on the island which answers the description of the one last mentioned, which runs as follows:—" There's a certain admirable flower in this " island which I should prefer to Spanish jessamine, 'tis as white as a lily and " shaped some think like common jessamine. It grows particularly out of the trunks " of rotten trees, when they are almost reduced to the substance of mould. The " odour of this flower strikes one agreeably at a hundred paces distance."

I can only suppose it to be some species of Orchid. The comparative absence of Orchids, as I shall show, is a notable feature on the island, and it is probable that Leguat refers to a species now passed away.

The vegetation of Rodriguez at the present time is thus very different from what it was at a comparatively recent period, and a very potent influence in altering its character has been cultivation. The island, though an outlier of the Mascarene group, at a great distance from the other members, and, as it were, out of the way, has seen many changes in this respect. Leguat gives a curious account of the cultivation of the soil by himself and his companions, and of the plants they grew. He says they sowed Water-melons, Ordinary Melons, Succory, Wheat, Artichokes, Purslain, Turnips, Mustard, Gillyflower, Clover-grass. Some of these seem to have thriven, but others did not, and it is curious to read of the wheat, " Of three grains of wheat that came up, we could preserve but one plant; it had above 200 ears, " and we were full of hopes that it would come to something, but it produced " only a sort of tares, which very much troubled us as you may imagine. How- " ever, we should not from hence conclude that wheat corn will always turn to " tares here, since in Europe such like degenerations are often to be met with." His observation rather militates against the plant being true Wheat,—more probably Millet. Some of these plants are now cultivated or occur spontaneously on the island. When the island was in the possession of the French many settlers lived and cultivated large estates, but with the liberation of the slaves cultivation decreased in amount. It thus happens that at the present day only a relatively small acreage of land is cultivated. The staple of the cultivation now is the Patate, or Sweet Potato, which is grown very widely, and in almost equal quantity is the Manioc, whilst of other roots Yams are chiefly grown. Of the cereals fair crops of Maize and Millet are obtained, and Rice also grows very well, but is not cultivated in quantity sufficient for the use of the inhabitants. Wheat formerly largely sown is now seldom seen, and this mainly because of the parroquets and Java sparrows which abound. Beans (*Phaseolus lunatus*), Lentils (*Ervum lens*), Gram (*Cicer arietinum*), Dholl (*Cajanus Indicus*), Pistache (*Arachis hypogea*) are all grown to a certain extent, though the rats are great enemies. Of other vegetables Garlic (*Allium sativum*), Giroumon (*Cucurbita Pepo*), Margose (*Momordica balsamina*), Melon d'eau (*Citrullus vulgaris*), Oignon (*Allium Cepa*), Papingaye (*Loffa acutangula*), and Patole (*Trichosanthes anguina*) are the most common. Ginger, Safran (Turmeric), and Arrowroot are also cultivated. Of economic plants

Coffee has formerly been largely grown, but now is never cultivated, as the hurricanes so frequently destroy the crops. Fair-sized trees, the remnant of old plantations, are now found fruiting freely. Vanilla grows well, and is tried in some spots. The want of water is now a great difficulty in the way of manipulating the Sugar-cane, which is therefore not much grown. Formerly Indigo plantations covered some of the central portions of the island, but now its cultivation has ceased.

That the soil of the island is good, and that water was formerly abundant, the fact of so many plants being cultivated clearly proves; but at the same time their cultivation must have acted very prejudicially on the indigenous vegetation.

From all these causes then indigenous plants have suffered, and the aspect of the vegetation now is a very peculiar one.

The elevation in the island is not sufficient to render possible any marked difference betwixt the vegetation of the higher and of the lower parts, and the relative amount of moisture is about the same in the two regions, as the hill-tops are seldom enveloped in mists. But we do find some plants which only occur in the upper parts, while others are found only on the lower districts or on the shore. Although altitude does not much affect the vegetation, difference of soil does so to a great extent. In most places the soil is volcanic, but there are many wide expanses of upraised coral reef forming a limestone soil, and, as we might expect, there are very marked differences between the vegetation on them.

Commencing at the shore we find first of all that we have representatives of a Phænogamic Flora below high-water mark in two species of *Halophila*, abundant on the reefs, and which are also found in many other tropical islands; but we miss the Mangrove which is found in Mauritius. At the mouth of some of the rivers *Ruppia maritima* and *Zannichellia palustris* occur. The shore at high-water mark is freely strewn with *Sesuvium portulacastrum*, while *Ipomœa pes-capræ*, *Canavalia obtusifolia*, and *Zoysia pungens*, are found carpeting sandy flats. Of other shore plants *Psiadia Coronopus* is sparingly found, and so is *Phyllanthus dumetosus*, both only on the south side of the island, whilst *Clitoria Ternatea*, *Teramnus labialis*, *Boerhaavia diffusa*, and *Achyranthes aspera* are also found abundantly on the shore. Of trees *Hibiscus tiliaceus* forms dense thickets close to the sea, and with it *Thespesia populnea* and *Pisonia viscosa* are also found. Where coralline limestone exists on the shores and on the coral islets on the reef such plants, as *Suriana maritima*, *Pemphis acidula*, *Oldenlandia Sieberi*, *Tournefortia argentea*, *Ipomœa fragrans*, *I. leucantha*, *I. nil*, *Lycium tenue*, and *Myoporum mauritianum*, specially occur.

As we pass inland we meet in the valleys at the embouchures of the rivers with the following plants,—*Cardiospermum microcarpum*, *Cæsalpinia Bonducella*,

Physalis peruviana, Datura alba, Ricinus communis, Erythrina indica, Carica Papaya, and *Coix Lachryma,* in abundance; and, on continuing up the rivers, *Nasturtium officinale, Herpestis Monnieria, Alocasia macrorhiza, Colocasia antiquorum, Chara Commersoni,* and such *Algæ* as species of *Batrachospermum, Cladophora,* and *Conferva* are found in their waters. The banks of the streams in most valleys are covered, especially in the lower parts, for about twenty yards on each side of the stream by a dense thicket of *Leucæna glauca,* giving place in the upper parts in many valleys to *Eugenia Jambos,* and close to the stream may be found *Oxalis corymbosa, Hydrocotyle bonariensis, Salvia coccinea, Plantago major,* and *Rumex crispus.* On moist rocks at the tops of valleys *Lobelia vagans* and *Pilea Balfouri* are abundant, and many Mosses and *Algæ* are to be found coating the rocks along with *Hepaticæ* and *Trichomanes cuspidatum.*

The undergrowth is very rank in many places, and the plants which most commonly contribute to its formation are such as *Malvastrum tricuspidatum, Sida carpinifolia, Abutilon indicum, Urena lobata, Gossypium barbadense, Melochia pyramidata, Corchorus trilocularis, Triumfetta glandulosa, Oxalis corniculata, Crotalaria retusa, Atylosia scarabœoides, Rhynchosia minima, Rubus rosæfolius, Ageratum conyzoides, Vinca rosea, Trichodesma zeylanicum, Stachytarpheta indica, Achyranthes aspera, Cassytha filiformis, Commelyna communis, Nephrolepis acuta,* and species of *Cyperaceæ* and *Gramineæ.* Rocks and stones are everywhere covered with Lichens, chiefly species of *Lecanora, Lecidea,* and *Pertusaria.*

On the slopes of the valley are found occasionally such plants as *Toddalia aculeata, Gouania retinaria, Scutia Commersoni, Indigofera argentea, Tephrosia purpurea, Canavalia ensiformis, Daucus Carota, Danais corymbosa, Eupatorium cannabinum, Plumbago zeylanica, Tanulepis sphenophylla, Heliotropium indicum, Solanum sanctum, Barleria Prionitis, Agave americana, Fourcroya gigantea,* and *Aloe lomatophylloides.*

The commonest tree intermixed with these is *Pandanus heterocarpus,* and on the higher parts of the island *P. tenuifolius;* but the following trees and shrubs are also very common: *Pittosporum Senacia, Quivisia laciniata, Elæodendron orientale, Albizzia Lebbek, Terminalia Benzoin, T. Catappa, Fœtidia mauritiana, Mathurina penduliflora, Fernelia buxifolia, Pyrostria trilocularis, Scyphochlamys revoluta, Carissa Xylopicron, Ardisia sp., Olea lancea, Securinega durissima, Ficus rubra, F. consimilis, Dracæna reflexa, Dodonæa viscosa, Eugenia uniflora, E. cotinifolia, Punica Granatum,* and *Phyllanthus Casticum.*

Confined to limited areas in unfrequented spots a few plants such as the following are found: *Aphloia mauritiana,* var. *theæformis, Dombeya ferruginea D. acutangula, Zanthoxylum paniculatum, Allophylus Cobbe, Sclerocarya castanea, Eugenia Balfouri, Randia heterophylla, Psychotria lanceolata, Psiadia rodriguesiana, Sideroxylon sp., Buddleia madagascariensis, Hypoestes rodriguesiana, Obetia ficifolia, Peperomia Rodriguezi, P. hirta, Viscum tænioides, Oberonia brevifolia,* and *Bulbophyllum incurvum.*

In the vicinity of habitations or old plantations we usually meet with such plants as:

Anona muricata.
Argemone mexicana.
Brassica juncea.
Gynandropsis pentaphylla.
Moringa pterygosperma.
Eriodendron anfractuosum.
Triphasia trifoliata.
Citrus decumana.
Indigofera tinctoria.
Abrus precatorius.
Hæmatoxylon Campechianum.
Poinciana regia.
Phaseolus lunatus.
Cajanus indicus.
Acacia Farnesiana.
Prunus communis.
Eugenia Jambolana.
Lagenaria vulgaris.
Momordica balsamina.
Citrullus vulgaris.
Opuntia Tuna.
Coffea arabica.
Parthenium Hysterophorus.
Siegesbeckia orientalis.
Bidens pilosa.
Lobelia Cliffortiana.
Ipomæa Batatas.
Lycopersicum Galeni.
Solanum Melongena.
Nicotiana Tabacum.
Mirabilis Jalapa.
Leonurus sibiricus.
Amaranthus tristis.
Chenopodium ambrosioides.
Basella rubra.
Persea gratissima.
Tetranthera laurifolia.
Euphorbia peploides.
Phyllanthus Niruri.
Manihot utilissima.
Musa paradisiaca.
Ravenala madagascariensis.
Dioscorea alata?

On looking at the vegetation as it now clothes the island, one perceives at once that a line may be drawn across the island which will separate portions very different in aspect. Thus starting from the mouth of the Rivière Saumatre, and passing up to the head of the valley, thence striking somewhat south-west to a point on the opposite side of the island about the mouth of Rivière Coco, we have on the east a lofty district intersected by many deep ravines, the slopes coming somewhat abruptly down to the sea. The sides of the hills are in this region covered with a thick undergrowth and scrub, often in great part of Ferns, and dotted over them are a fair number of small shrubs and trees, notably abundance of Screw-pines. The valleys themselves in their upper parts are here filled with a tolerably dense growth of trees and shrubs. To the west of the line stretches a hilly country of lower altitude, sloping gradually to the south-west, cut by ravines, which are not so deep and whose sides are not so steep as those on the east. The higher land on this side is covered with a great number of small stones and débris of volcanic rocks, and is quite barren of any trees or shrubs, save perhaps a stray stunted Vacoa, Palmiste, Latanier, or Citron, and the ground is over wide areas coated by but a scant covering of *Cyperaceæ* and Grasses mixed with social weeds; whilst the valleys are as a rule quite destitute throughout of a covering of tree or shrub, except it may be a few stragglers at the margins of the streams. As we pass south-west, however, we come, on the banks of the Rivière Quitorze, to an abrupt line of demarcation betwixt this bare district and one on which vegetation is relatively more abundant and varied. Crossing the line we enter on an extensive coralline limestone plain occupying the whole south-west end from Anse Peril to Anse Topaze. This is covered with a poor vegetation, but one that is very characteristic, such herbs abounding as *Tridax procumbens, Premna serratifolia, Senecio linearis, Sarcostemma viminale, Cassytha filiformis, Ipomœa leucantha,* and *Boerhaavia diffusa;* whilst of trees and shrubs *Ludia sessiliflora, Terminalia Benzoin,* and *Antirrhœa frangulacea,* are specially abundant, and more rarely *Gastonia cutispongia;* and sparingly such herbs as *Nesogenes decumbens, Dichondra repens, Hypoestes inconspicua,* and *Selaginella Balfouri,* occur. On the north-west where none of this coralline limestone is found and the volcanic rock passes directly into the sea the barrenness continues.

The difference in landscape offered by the two regions, eastern and western, of the island are very striking, and the abruptness of the line of demarcation is very remarkable. On sailing round from Port Mathurin to the Anse Topaze, as we did in going to the caverns where the bones of Solitaire are found, we had occasion frequently to observe this. In the valley of Rivière Saumatre the features are very well marked, for on one side the valley is thickly covered with vegetation, whilst the opposite side is quite barren. The difference is no doubt due to the fires, lighted to clear portions of land for cultivation without due care being taken to prevent their spreading and fanned by the prevailing south-east wind, sweeping across the south and north-west of the island more especially; they have thus converted the paradise into a wilderness.

When we come to consider more closely the Flora of Rodriguez we find that it is composed of 470 species and varieties, belonging to 293 genera included in 85 natural orders. This then is a proportion of about $5\frac{1}{2}$ species to each family, and hardly 2 species to each genus.

This ratio is, however, brought about by the comparatively large number of Cryptogams, for the 470 species comprise no less than 173 Cryptogams. There are therefore 297 Phænogams. Speaking of the Flora we shall deal separately with the two divisions.

Of Phænogamic plants then we have 297 species belonging to 214 genera included in 75 natural orders. An analysis of these shows us the extraordinary fact that 108 species or over one-third are introduced plants, in many cases relics of former or escapes from present cultivation. These I have thought it well to retain in the list and record as having been found on the island however sparsely, as at some future period any one of them may occupy a more prominent position in the Flora. In taking cognizance, however, of the extent of the Flora at present they must be expunged. From the total number we must also subtract 14 species which are not finally determinable, owing to imperfect specimens having been obtained. And here I may state that from the same cause, in some cases, the determination of species has been very difficult and can only be considered provisional, for our visit to the island happened unfortunately at a time of year when few plants were in flower, and it was just when we were leaving the island that plants were coming into bloom. In all cases of difficulty, however, I have indicated it under the specific name. Of the 14 species I here allude to some most probably are endemic plants whilst others are perhaps widely spread. The following is a list of them so far as can be determined.

Desmodium, sp.   *Asclepiads,* 2 *sp.*
Mucuna, sp.   *Stachys, sp.*
Eugenia, sp.   *Asparagus,* 2 *sp.*
Olea, sp.   *Angræcum, sp.*
Ardisia, sp.   *Cyperus, sp.*
Sideroxylon, sp.   *Bambusa, sp.*

This leaves 175 Phænogams belonging to 119 genera and 57 orders which we consider the indigenous Flora, though even some of the plants in this number may have to be excluded. We have thus a ratio of about 3 species to each order and about 2 to each genus,—a very small proportion. Of the total number, Monocotyledons constitute 49 or about two-sevenths, a comparatively large proportion for a tropical island. Endemic species number 35 or one-fifth of the whole, 6 of them being Monocotyledons; 31 species or about two-elevenths are peculiarly Mascarene, of which number one-fifth are Monocotyledons; and of the rest, 8 species or more than one-twentieth are African plants which do not occur in Asia, whilst 14 species or nearly one-twelfth are found in Asia, but do not reach Africa. The remaining 88 species comprise a certain number which are widely spread in the tropics of the old world, 22 species in all or one-eighth of the Flora being of this nature, whilst 66 species or three-eighths are universal tropical weeds. Thus half the indigenous Phænogamic vegetation consists of common weeds of the tropics.

Amongst indigenous Dicotyledons the prevailing orders are, *Gramineæ*, represented by 21 species; *Leguminosæ*, by 14 species; *Convolvulaceæ*, 11; *Malvaceæ*, 9; *Rubiaceæ, Cyperaceæ,* and *Euphorbiaceæ*, each 8; *Liliaceæ*, 6; *Compositæ* and *Amaranthaceæ*, each 5. These, it will be observed, are all orders which as a rule compose a great part of the vegetation of any tropical island reached by civilization; but the number of *Rubiaceæ* is specially worthy of attention, more especially when compared with the number of *Compositæ*.

Gramineæ constitutes then three-sevenths of the whole indigenous Monocotyledons, and about three twenty-fifths of the whole Phænogams, a fairly large number compared with continental areas,—a usual characteristic of tropical islands. The species call for no special mention; none are peculiar, though a form of *Panicum* is Mascarene. The majority are widely spread, cosmopolitan, or old world species, a few being Asiatic.

Leguminosæ, represented by fourteen species, forms two-twenty-fifths of the whole Flora. It contains as usual a great number of species found along or near the shores, such as *Canavalia obtusifolia*, and *Clitoria Ternatea*, and includes many of the commonest plants. Very few plants of Rodriguez besides the endemic ones are absent from Mauritius, but curiously this family contains three species not found in Mauritius; these are *Canavalia ensiformis*, *Mucuna gigantea*, and *Rhynchosia minima*. Species of the genera of the two former are found, but not of the latter. Most of the Rodriguez plants of this family are universal in the tropics.

Convolvulaceæ is next most numerously represented, chiefly by species of *Ipomœa*, of which eleven are known. Many of the species are littoral, such as *Ipomœa pescapræ, Ip. leucantha, Ip. nil. Ip. fragrans*. The three last mentioned do not occur in Mauritius.

Malvaceæ, with nine species, includes, with one exception (*Hibiscus liliiflorus*), widely spread tropical plants.

Cyperaceæ and *Amaranthaceæ* call for no special notice, and the other families are noticed in the account of the endemic Flora.

The endemic Flora consists of 35 species, or one-fifth of the whole Phænogamic plants. Of this number 29 are Dicotyledons, and 6 are Monocotyledons.

The following is a list of the species:—

Zanthoxylum paniculatum.	*Diospyros diversifolia.*
Quivisia laciniata.	*Tanulepis sphenophylla.*
Sclerocarya castanea.	*Sarcostemma Odontolepis.*
Eugenia Balfouri.	*Hypoestes rodriguesiana.*
Mathurina penduliflora.	*Hypoestes inconspicua.*
Danais corymbosa.	*Nesogenes decumbens.*
Randia heterophylla.	*Clerodendron laciniatum.*
Pyrostria trilocularis.	*Pisonia viscosa.*
Scyphochlamys revoluta.	*Ærua congesta.*
Psychotria lanceolata.	*Pilea Balfouri.*
Psiadia Coronopus.	*Peperomia hirta.*
Psiadia rodriguesiana.	*Peperomia reticulata.*
Abrotanella rhynchocarpa.	*Peperomia Rodriguezi.*
Lobelia vagans.	*Euphorbia daphnoides.*

Phyllanthus dumetosus.
Listrostachys Aphrodite.
Aloe lomatophylloides.
Latania Verschaffelti.

Hyophorbe Verschaffelti.
Pandanus heterocarpus.
Pandanus tenuifolius.

Of the Dicotyledonous orders in which endemic species occur *Rubiaceæ* is the most remarkable, presenting the greatest amount of peculiarity. Of the eight species it contains, five are endemic, and they all belong to different genera. One of these, *Scyphochlamys*, is peculiar to Rodriguez, and the genera of two others— *Pyrostria* and *Danais*—are essentially Mascarene, the latter extending into Madagascar. The two remaining species belong to the widely-spread genera *Randia* and *Psychotria*, of which the former is unknown in the other Mascarene Islands, but may have a representative there in the nearly allied *Gardenia*, which is absent from Rodriguez.

Three species of *Rubiaceæ* are peculiarly Mascarene, two belonging to genera of wide range, *Antirrhœa* and *Oldenlandia*, but the Mauritian type of the species of *Oldenlandia* is modified into a distinct variety in Rodriguez; the third is a species of *Fernelia*, a Mascarene genus.

This comparative abundance and peculiarity of the *Rubiaceæ* is very interesting when compared with similar features seen in other oceanic islands of like nature; and it is further interesting to note that in *Euphorbiacæ* we have an order co-extensive, though not presenting so great peculiarity. The eight species composing it belong mainly to *Phyllanthus* and *Euphorbia*, of each of which there is a peculiar species. There is also a Mascarene species of *Phyllanthus* which derives interest as one of the plants collected by Commerson, and originally described from his specimens; in addition there is a species of *Claoxylon*, and one of *Securinega*, both Mascarene.

Compositæ and *Piperaceæ* are equally numerously represented by peculiar species, each containing three. In the first-named order, of which five species in all occur, the endemic plants belong to two genera. *Psiadia*, a Mascarene type of *Astereæ* extending into Africa, but not Asiatic, includes two of the peculiar species. Both are suffruticose, one, *Ps. rodriguesiana*, growing only on the higher levels, and there sparingly, has the velutinous character so well developed in many of the Bourbon species, the other, *Ps. Coronopus* is glutinous, occurring only on the shores in small quantity, and is interesting as being one of the few plants brought from Rodriguez by Commerson, and originally described as *Sarcanthemum Coronopus* by Cassini. The third peculiar Composite is a species of *Abrotanella*, a small genus of *Cotuleæ*, ranging through Australia, New Zealand, and some Antarctic islands, and unknown in any other of the Mascarene group. A fourth Composite is a *Senecio* confined to the Mascarene Islands, very variable in its character, and the remaining one is the world wide *Sonchus oleraceus.*

In the *Piperaceæ* we have three peculiar species of *Peperomia* with a strong East Indian affinity.

Asclepiadaceæ, Acanthaceæ, and *Verbenaceæ* each contain two endemic plants. The Asclepiadaceous *Sarcostemma* comprises two species, of which one is a novelty, and there is a climber, common on the island, the type of a new genus *Tanulepis* closely allied to the East Indian *Brachylepis*. The two peculiar Acanthads are species of *Hypoestes*, and in *Verbenaceæ* one of the novelties is a species of the hitherto monotypic *Nesogenes*, a genus confined to a few of the Polynesian Islands, whilst the other is a species of the common tropical *Clerodendron*.

Of other Dicotyledonous orders none contain more than one peculiar species. *Anacardiaceæ* has a single species of the African genus *Sclerocarya*, which differs, however, very considerably from the generic type. Of three *Myrtaceæ* one species of *Eugenia* is endemic, whilst another is mainly Mascarene, though occurring elsewhere in the tropics, and the third is the Mascarene *Fœtidia mauritiana*.

I shall only specially mention another order, *Turneraceæ*, of which one generically peculiar plant is found, *Mathurina penduliflora*, whose nearest congener is a central American genus *Erblichia*, and it is worthy of note that this family has hitherto been unknown in the Mascarene Islands. The remaining families with endemic species are *Rutaceæ, Meliaceæ, Campanulaceæ, Ebenaceæ, Nyctaginaceæ, Amaranthaceæ*, and *Urticaceæ*.

Looking next at the 6 peculiar Monocotyledons we find that they are included in 5 genera belonging to 4 orders. The Monocotyledons from this island are specially interesting, as they include those plants which give a character to the vegetation. Every visitor to Rodriguez will be struck at once by the peculiar features impressed on the landscape by the prevalence of the Screw-pines. They are indeed the physiognomic plants, and far outstrip in numbers any other species; but it is remarkable that though individually so numerous, specifically the family is not rich, there being only two species of *Pandanus* on the island, *P. heterocarpus* and *P. tenuifolius*, both peculiar and very distinct from any other of the indigenous Mascarene forms. Three other species have been registered by various authorities, they are *P. odoratissimus*, *P. utilis*, and *P. muricatus*. None of these are Mauritian or Bourbon species, the first being a native of Asia and the two latter both Madagascar species, and the evidence of their occurrence in Rodriguez is faulty.

Next in interest amongst the Monocotyledons come the Palms, and they are very peculiar. We find three species indigenous, and these belong to different genera which are all Mascarene. They are *Latania Verschaffelti, Hyophorbe Verschaffelti*, and *Dictyosperma album*, var. *aureum*. *Latania Vershaffelti* has been for some time known to European nurserymen as *Latania aurea*, and *Hyophorbe Verschaf-*

felti has also been known to horticulturists under the name of *Areca Verschaffelti*. Each of the genera, *Latania* and *Hyophorbe*, includes three species, which present a remarkable correspondence in distribution.

The *Latania Verschaffelti* of Rodriguez is represented in Mauritius by *L. Commersoni*, which grows also in Bourbon, and on Round Island a third species, *L. Loddigesi*, is found. Of *Hyophorbe*, the Rodriguez plant *H. Verschaffelti*, is represented in Mauritius and Bourbon by *Hyophorbe indica*, and on Round Island *H. amaricaulis* occurs.

The genus *Dictyosperma* is Mascarene, and has been created by Wendland to include the type of Palm, originally described by Bory St. Vincent (Voy. I. 306) as *Areca alba*. The Mascarene palms, formerly described as species of Areca, are all removed from that genus, the non-spiny forms now constituting *Dictyosperma*, and the spiny forms, *Areca crinita* and *A. rubra*, combining to make the genus *Acanthophœnix*. *Dictyosperma* is montoypic, but the species is very variable, and in Rodriguez assumes a very graceful and delicate habit; the characters, however, are hardly specific, and it is therefore merely a variety. This palm is the well-known *Areca aurea* of nurserymen.

Of the other endemic Monocotyledons one is a species of *Aloe* which is very distinct. *Liliaceæ* altogether constitutes one-twenty-ninth of the Flora, and in addition to this peculiar Aloe comprises a Mascarene species of *Asparagus*. An *Asphodelus* which has a tolerably wide range is to be noted as occurring only on two of the small coral islets on the southern reef, Gombrani and Pierrots, and is absent entirely from the main island.

A marked feature in the Flora is the paucity of *Orchidaceæ*. Only four species have been determined; a fifth, a species of *Angræcum*, was found, but in too imperfect a state for identification. These belong to genera two of which range into Africa and not into Asia, and one is Asian but not African. One species of *Listrostachys* is peculiar, an *Oberonia* and a *Bulbophyllum* are Mascarene, and a distinct variety of a Mauritian *Aeranthus* occurs. The dryness of the soil and climate no doubt have to do with the scarcity of the family, and also to a large extent the destruction of the old forests, as no epiphytic forms now exist. This paucity is the more remarkable when contrasted with the profusion of this family in the sister islands.

Analysing the relationship of the endemic plants, we find that three genera are endemic, of these one, *Mathurina*, has a near American affinity, another, *Tanulepis*, has a close Asiatic connexion, whilst the third, *Scyphochlamys*, has its nearest congener peculiarly Mascarene. Five are Mascarene genera,—*Quivisia*, *Danais*, *Pyrostria*, *Latania*, and *Hyophorbe*. In the case of four, *Sclerocarya*, *Psiadia*, *Listrostachys*, and *Aloe*,—we have genera of peculiarly African range; and of the rest, *Abrotanella* is an Antarctic genus and *Nesogenes* is Polynesian, the remainder being four old world and twelve generally distributed tropical genera. We may tabulate these as follows:—

ENDEMIC.	MASCARENE.	AFRICAN.	OLD WORLD.	COSMOPOLITAN IN TROPICS.
Mathurina.	*Quivisia.*	*Sclerocarya.*	*Sarcostemma.*	*Zanthoxylum.*
Scyhochlamys.	*Danais.*	*Psiadia.*	*Hypoestes.*	*Eugenia.*
Tanulepis.	*Pyrostria.*	*Listrostachys.*	*Ærua.*	*Randia.*
	Latania.	*Aloe.*	*Pandanus.*	*Psychotria.*
	Hyophorbe.			*Lobelia.*
ANTARCTIC.				*Diospyros.*
				Clerodendron.
Abrotanella.	POLYNESIAN.			*Pisonia.*
				Pilea.
	Nesogenes.			*Peperomia.*
				Euphorbia.
				Phyllanthus.

The African affinity of the endemic Flora thus becomes manifest.

We have stated that 31 species of Phænogamic plants in the Flora are peculiarly Mascarene, of which 25 are Dicotyledons, and the rest Monocotyledons. They are,—

Pittosporum Senacia.
Aphloia mauritiana.
Hibiscus liliiflorus.
Dombeya acutangula.
Dombeya ferruginea.
Toddalia paniculata.
Elæodendron orientale.
Gouania retinaria.
Fœtidia mauritiana.
Terminalia Benzoin.
Gastonia cutispongia.
Oldenlandia Sieberi.
Fernelia buxifolia.
Antirrhœa frangulacea.
Senecio linearis.
Olea lancea.

Carissa Xylopicron.
Buddleia madagascariensis.
Solanum macrocarpum.
Myoporum mauritianum.
Obetia ficifolia.
Phyllanthus Casticum.
Securinega durissima.
Claoxylon parviflorum.
Viscum tænioides.
Oberonia brevifolia.
Bulbophyllum incurvum.
Aeranthus arachnites.
Asparagus umbellulatus.
Dictyosperma album.
Andropogon foliatus.

Several of these I have already referred to, and I now mention specially the occurrence of the curious *Myoporum mauritianiam*, very sparingly found on the island, which differs very markedly from the type of *Myoporineæ*, and is perhaps endemic in Rodriguez, as the evidence of its occurrence in other Mascarene Islands is faulty.

An analysis of the relationships of these plants shows us that,—

Six belong to genera which are essentially Mascarene; these are *Aphloia, Fœtidia, Gastonia, Fernelia, Obetia,* and *Dictyosperma,* and two, *Dombeya* and *Aeranthus,* are essentially African. Ten of the remainder are old world genera, and the remaining twelve are spread tropical genera, but curiously one of these, *Antirrhœa,* is not African. Here we have evidence of the individuality of the Mascarene Flora and also of its primary relationship with the African rather than the Indian type.

Tabulating these we have:—

MASCARENE.	AFRICAN.	OLD WORLD.	COSMOPOLITAN IN TROPICS.
Aphloia.	*Dombeya.*	*Pittosporum.*	
Fœtidia.	*Aeranthus.*	*Toddalia.*	*Hibiscus.*
Gastonia.		*Olea.*	*Elæodendron.*
Fernelia.		*Carissa.*	*Gouania.*
Obetia.		*Myoporum.*	*Terminalia.*
Dictyosperma.		*Claoxylon.*	*Oldenlandia.*
		Viscum.	*Antirrhœa.*
		Oberonia.	*Senecio.*
		Bulbophyllum.	*Buddleia.*
		Asparagus.	*Solanum.*
			Phyllanthus.
			Securinega.
			Andropogon.

The connections exhibited by the plants of other countries indigenous in Rodriguez are not very strong. There are a few, eight, species which have specially an African distribution, and there are fourteen of specially Asiatic range which do not reach Africa: that is, more than one-twentieth of the whole Flora have African and nearly one-twelfth Asiatic distribution.

The African plants are,—

Ludia sessiliflora.
Desmodium mauritianum.
Desmodium incanum.
Sarcostemma viminale.
Lycium tenue.
Ipomœa fragans.
Dracæna reflexa.
Andropogon finitimus.

The Asiatic and non-African are,—

Calophyllum Inophyllum.
Toddalia aculeata.
Allophylus Cobbe.
Erythrina indica.
Mucuna gigantea.
Atylosia scarabœoides.
Eugenia cotinifolia.
Ipomœa peltata.
Cassytha filiformis.
Dracæna angustifolia.
Carex gracilis.
Panicum Balfouri.
Stenotaphrum subulatum.
Zoysia pungens.

It will be observed that the African species are all of a more restricted range than the Asiatic.

Of the 88 generally distributed tropical weeds I need say nothing. They will be seen to be such as one would expect to meet in an island which has passed through such vicissitudes as Rodriguez.

Of the introduced Phænogamic plants, 108 in number, of 90 genera included in 56 natural orders, the prevailing families are—

Leguminosæ	15 species
Solanaceæ	9 ,,
Gramineæ	6 ,,
Compositæ	6 ,,
Myrtaceæ	5 ,,
Convolvulaceæ	
Cucurbitaceæ	4 ,,
Rutaceæ	

orders commonly and widely distributed, and including those social weeds which follow man's footsteps over the world. I need not further refer to them.

Besides those Phænogams I have recorded as occurring on the island, there are a few others reported as existing, but of none of them have I seen authentic specimens from the island, and as I did not find them there, I do not include them in the list. They are the following:—

Achillea, sp.

Lantana, sp.

Ficus, sp.

Pandanus muricatus.

Pandanus odoratissimus.

Pandanus utilis.

In estimating the relationship of the Cryptogamic Flora of Rodriguez, we find more difficulties to encounter from the fact that excepting the Vascular Cryptogams, little comparatively has been done amongst the Cryptogams of the other Mascarene Islands. The lower forms from Rodriguez have, however, been carefully worked out, and I find that there are now known from the island, 173 species of Cryptogams (this number excluding marine *Algæ*). Thus the Phænogams and the Cryptogams are almost equal in number.

Of this number, 26 species or about one-seventh are *Filices, Ophioglossaceæ*, and *Lycopodiaceæ*. *Musci* constitute about two-elevenths, numbering 33 species, whilst of *Hepaticæ* I have 18 species, and there is one species of *Chara*. *Lichens* are most abundant of all, there being 75 species, or over three-sevenths of the whole Cryptogams; whilst *Fungi* number 8 species, freshwater *Algæ* 13, and I have only 39 species of marine *Algæ*.

In Vascular Cryptogams, the Rodriguez Flora contrasts very unfavourably with that of the other Mascarene Islands. The 25 species known, of which 20 are Ferns, is a number relatively and absolutely very small. The scarcity of this group of plants is accounted for by the dryness of the island, and in confirmation of this fact, we observe the Tree Ferns of the other Mascarene Islands have here no representatives, and of the large moisture loving genera *Trichomanes* and *Hymenophyllum*, so abundant elsewhere in the group, the former is here typified in one species only, *T. cuspidatum*, whilst the latter is absent. And when we consider the nature of the species which do occur, they are those characteristic of dry and arid regions, such for example as *Adiantum caudatum, Asplenium furcatum, Asplenium falcatum, Nephrodium unitum, N. molle, Polypodium phymatodes, Nephrolepis acuta.*

None of the species are novelties; all occur in the Mascarene Islands, and three of them, *Trichomanes cuspidatum, Davallia mauritiana*, and *Nephrodium crinitum*, are peculiarly Mascarene, but most are widely spread tropical species. Whilst some, as *Pellœa hastata, Pteris flabellata*, and *Nephrodium elatum*, are found in Africa, but not in Asia; on the other hand we have the Asian but not African *Asplenium hirtum*, and *Polypodium adnascens*. Three Ferns, *Lindsaya acutifolia*, another *Nephrolepis* and another *Lastrea* are reported from Rodriguez, but I have seen no authentic specimens, and as I did not find them myself, I have not included them in my list.

The *Lycopodiaceæ* further exemplify the dry character of the island. Only four species are known. Two are widely spread *Lycopodium Phlegmaria* and *Psilotum triquetrum*, whilst two species of *Sclaginella, S. Balfouri* and *S. rodriguesiana*, are novelties.

Mosses may be considered as fairly represented in Rodriguez by 33 species, though this is a small number compared with those found in Mauritius; the number of species in that island, according to the latest enumeration I have seen, being 104. Of the 33 species found, 17 are peculiar; and of the remainder, 13 occur in the other Mascarene Islands, or in Africa, a few corresponding with those from the western coast of tropical Africa; and it appears from what little is known of the Mosses of the eastern coast, that some species have an enormous range on the African continent. Two other species are found in Asia, and one, *Weisia controversa*, is cosmopolitan.

Of the Rodriguez species all except one belong to genera which are represented in the Mascarene Islands. The genus *Ectropothecium* is most abundantly represented by four species, and they are perhaps the commonest on the island. It is curious that *Hypnum*, which occurs so extensively in the sister islands, should include but one species in Rodriguez. Of all species, the most interesting is *Orthotrichum plicatum*, which is not uncommon on the island, previously known only from the specimens gathered in Bourbon by Du Petit Thouars, and described and well figured by Schwaegrichen; it is undoubtedly a member of the family of the *Orthotricha*, but its characters as a genus are, according to Mitten, intermediate, and its position in that family unique. One of the most elegant as it is one of the rarest species is the endemic *Macromitrium astroideum*, found trailing over boulders at the top of the valley of the Rivière aux Huitres, and along with it is also found *M. aciculare*. *Octoblepheum albidum* is by no means common, growing in tufts on decayed branches of trees and rich vegetable humus in the same valley, and this is also the only station for *Neckera lepineana* and *Meteorium involutifolium*, and sparingly on trees also occurs *Pterogonium curvifolium*. On the moist clay rocks at the sides of the beds of streams in shady places, the various species of *Ectropothocium* are found most abundantly, *E. doleare* especially forming a thick covering, and this is probably the commonest Moss on the island; with it also species of *Sematophyllum*, notably *S. incurvifolium*, are commonly found as also *Callicostella læviuscula*, *Rhacophilum africanum*, and occasionally species of *Fissidens*. On moist rocks near the stream sources we find species of *Bartramia*; *Bryum* occurs commonly over the island, and *Weisia* and *Calyperes* are also frequently met with.

The island possesses 18 species of *Hepaticæ* included in 6 genera, and of these 13 are novelties. Of the remainder, three, *Chiloscyphus oblongifolius*, *Frullania squarrosa*, *Anthoceros fuciformis*, are Mascarene or African, one, *Lejeunia minutissima*, is found in the Eastern Archipelago, and *Anthoceros lævis* is the only cosmopolitan species. *Lejeunia* is the most extensive genus, embracing seven of the total number of species, and six of these are endemic. Of the species, *Lejeunia Balfouri* is one of the commonest, occurring on moist clay rocks at the sides of streams, and in similar situations with it *Chilocyphus oblongifolius* and *Radula appressa* are found, and in great abundance *Anthoceros lævis* and *A. fuciformis*. *Lejeunia furva* is rare, but in similar situations. The stems of the Screw-pines are frequently clothed with *Frullania squarrosa*, *Fr. Apicalis*, and *Fr. obscurifolia*, though they also occur more sparingly on other trees, as does *Phragmicoma carinata*.

One species of *Chara* is found in many of the rivers, and it also occurs in Mauritius and Bourbon.

Looking now at the Lichens, we find that of all Cryptogams they occur most abundantly. Altogether, 76 species and varieties were collected in a determinable state. Of these, the large number of 35 or nearly one half are novelties, whilst of the remainder, 11 are known from Mauritius, and 7 or 8 from Bourbon. This large number of Lichens is very remarkable when compared with the number in Mauritius and Bourbon. From the former island, 89 species and varieties are enumerated by Weddell (Trans. Roy. Soc. Arts and Scien. Maur. vii. 163) and Nylander (Ann. Sc. Nat. 4th ser. xi. 248) determines 112 species in Bourbon. But I think this relatively large proportion is due rather to our imperfect knowledge of the Lichen Flora of these islands than to their poverty in species as compared with Rodriguez. In addition to the species determined, there are fragments of other species, mainly crustaceous, in my collection which are quite undeterminable, being either sterile or with imperfectly developed apothecia.

Of the Rodriguez species, all save two belong to genera represented in the other Mascarene Islands. The two exceptions are *Heppia*, of which we have a single endemic species found very sparingly on the island, and *Pyrenastrum*, a single species of this genus, *P. Americanum*, occurring very abundantly on the bark of trees.

The most abundantly represented genus is *Lecanora*, which includes twenty-one species and varieties, and fourteen of these are peculiar; and next to it comes *Lecidea*, with nine species, of which six are novelties. Of the other genera *Ramalina* is most numerously represented, having five species and varieties, of which two are peculiar. *Arthonia* has two peculiar species, and *Cladonia*, *Pyxine*, and *Opegrapha* have each a single species, which is endemic. The scarcity of species of *Cladonia* is curious, as the genus is well developed in the other Mascarene Islands.

Of the species, one, which will at once attract the attention of any who visit the island, is the beautiful *Usnea dasypogioides*, only occurring in the higher parts of the island; it there hangs beard-like in great abundance from the tree branches, and along with it *Ramalina subfraxinea* is usually found. Close to these may be seen on the rocks the small tufted *Ramalina gracilenta* and *R. gracilenta* f. *nodulosa* intermingling with white patches of *Lecanora atra* f. *succedanea*, with its large black apothecia. *Parmelia latissima* is a not uncommon species in dry spots along with *Sticta aurata*, and on the boulders *Physcia speciosa* is found in abundance. The stems of the Screw-pines are invariably dotted over with stellate patches of *Physcia picta*, and many species of *Verrucaria* and *Graphis* also find thereon a suitable nidus, as well as *Lecanora achroa*; whilst their withered and dried leaves give a home to *Arthonia phylloica* and *A. dendritella*.

The features imposed on the rocks in many places by certain Lichens is very striking. Where such species as *Lecanora obliquans*, *L. conizopta*, *Pertusaria impallescens*, *Lecidea continens*, and *L. configurans* are abundant, one might suppose that the rocks had been whitewashed, this character being visible at a long distance. Again, on the more decomposed or cindery rocks, *Lecanora subfusca* f. *pumicicola*, *L. apostatica*, *L. cinnabarina*, *Lecidea spuria*, *L. achroopholis*, and *L. immutans* unite to give a dull mottled and variegated aspect to the rocks.

The stems of other trees besides the Screw-pines are favourite sites for certain species, notably we may mention Bois Gandine and Bois Puant; of the former, more especially, the bark is usually quite concealed by lichens, giving it a very white or grey appearance. On it alone *Lecanora leucoxantha* occurs with its orange apothecia, and also the pure white *Pertusaria velata*; *Lecanora punicea* and *Coccocarpa molybdæa* are found on trees with rough bark, and so is the scarlet *Trypethelium cruentulum*; whilst on decayed and decorticated wood forms such as *Opegrapha difficilior*, *Lecanora conizœa*, and *Pannaria rubiginosa* grow in profusion.

Fungi are poorly represented in my collection by 8 species. I found a few more species on the island, but in course of transmission to Britain I regret they have disappeared. The family is not however abundant on the island. Of the eight species which have been determined, three turn out to be novelties. One of them, *Polyporus aspidolopus,* is perhaps the commonest form on the island, growing abundantly on the stems of trees, but apparently with a preference for Screw-pines. The widely spread *Schizophyllum commune* and the common British *Hirneola Auricula Judæ* occur sparingly on trees in the island; and on the barren ground, towards the south-west, *Bovista lilacina* grows in considerable quantity.

The short time devoted to the collection of marine forms accounts for the smallness of the collection of *Algæ* from Rodriguez. Of the 52 species and varieties enumerated, 39 are marine, a number so small that it evidently cannot be considered as a fair representation of the marine Algal Flora. None of them call for special mention, as they are all widely spread in the Indian Ocean and South Seas. The remaining 13 are fresh-water forms. Six of these have a general range throughout Europe, and some of the six are quite cosmopolitan. The rest are more restricted in distribution. Two species only, *Thorea violacea* and *Cladophora pannosa,* are peculiarly Mascarene; *Chantransia cærulescens* is curiously enough a Cayenne species, whilst *Conferva Ansoni* and *C. Moluccaæ* are each confined to a single island in the Eastern Archipelago, and there is also *Cladophora Rottleri,* a restricted Indian species. In addition to those enumerated in the list, fragments of *Vaucheria* and *Odontidium* occur amongst the specimens, but in too imperfect a state for determination.

The variability of species in the Rodriguez Flora is a very remarkable characteristic. The number of genera, to species of which it is difficult to assign a limit, is indeed not very large; but the number of genera, species of which exhibit variations in size and form during their stages of growth, and the amount of this variation, is remarkable, and probably is not exceeded in any Flora of similar extent.

Of the genera whose species vary greatly we find, leaving out *Scutia, Eugenia, Senecio, Lobelia, Achyranthes, Cyperus,* and such like widely distributed genera, which are variable wherever found, and which present their ordinary variations in Rodriguez, several genera endemic or of limited extent, such as *Aphloia, Danais, Dombeya, Quivisia, Sideroxylon,* and *Psiadia,* which in the Mascarene group exhibit great variation in their species. In Rodriguez, where we have so small a Flora, and they are represented by usually a single species, the limit of variation is not wide, and specific characters are fairly precise and easily discerned. In only three genera, *Oldenlandia, Dictyosperma,* and *Aeranthus,* do we find a Mascarene species so altered in Rodriguez as to be recognisable as a distinct variety. But in the genus *Pandanus* we have examples in Rodriguez of exceedingly variable species, and it is necessary to have a very large series of forms before determining specific limits. We can recognise that there are two distinct species, but these vary so much, and approach one another so nearly, that of some forms it is almost impossible to say to which species they ought to be referred. The variation occurs in every part of the plant, but more specially in the fruits. Whether hybridisation has to do with this or not is a matter of dispute, but I am inclined to allow that this is a cause in some instances.

But the more marked feature of variability in the vegetation, and one deserving great attention, is the diversity in form, size, and habit exhibited in the leaves of many plants at different periods of their growth. The variation is confined almost absolutely to small trees or shrubby plants, the only exception being the small Composite *Abrotanella*. In species exhibiting this heteromorphism the young plant produces leaves of, as it were, a lower stage of development than the adult, and as the individual increases in age the leaves successively produced approach more nearly the mature, or as we must consider it, the type form, until at a certain stage of its growth only the typical leaves of the adult are found; and once this stage is reached all the leaves produced on the branches of the tree are of the typical adult form. But should any adventitious shoots develope from the base of the trunk, or appear on the stem anywhere below its first branching, these always have the juvenile and not the adult form of leaf. And, as may be supposed, if a tree be blown or cut down and from the stumps young shoots develope, those always bear juvenile leaves. An interesting point to determine would be whether shoots arising from a branch so treated would produce juvenile or adult leaves, and to what extent variations in foliage might be so produced. It seems to me there is room here for some interesting observations and experiments which I would press on the attention of those who have opportunity of making such investigation. In the following 17 species this heterophylly is extremely marked.

Ludia sessiliflora.
Aphloia mauritiana.
Hibiscus liliiflorus.
Dombeya ferruginea.
Quivisia laciniata.
Elæodendron orientale.
Terminalia Benzoin.
Fœtidia mauritiana.
Mathurina penduliflora.

Randia heterophylla.
Fernelia buxifolia.
Pyrostria trilocularis.
Scyphochlamys revoluta.
Abrotanella rhyncocarpa.
Diospyros diversifolia.
Carissa Xylopicron.
Clerodendron laciniatum.

In all the heterophylly is not of the same kind or to the same extent, but whilst the heteromorphism of the vegetation of the island as a whole varies greatly both in degree and kind, each species presents variations always of the same kind, and this holds true of a species if it grows also on the sister islands. Whether all the species of one genus in the island exhibit the same kind of variation is a point I had no opportunity of determining, as in no case did I find two species of one genus heterophyllous; but it is certain that the representatives in adjacent islands of heterophyllous species do not when they produce diverse formed leaves always have the same kind of variation. In the Mascarene Islands I only know of two genera possessing representative species which exhibit heteromorphism. In one of these, *Quivisia*, the type of variation is the same in both species; in the other, *Clerodendron*, it is different. The phenomenon is confined to no special order, though I may note that in four species of *Rubiaceæ* it is very well marked. For the sake of clearness I shall consider the kinds of variation observed as of three types.

1st. Variation dependent on imperfect or arrested development of the whole leaf in the young plant, the lamina developing equally.

In plants which exhibit this type the leaves of the juveniles are to a certain extent miniatures of the adult. They are very small, but possess the same or almost the same relative proportion of length of lamina and breadth thereof as is seen in the adult. The lamina is developed equally and is not lobed, and the margin may be entire, but sometimes the parenchyma towards the edge of the lamina is somewhat deficient, and thus, the veins being left prominent, the leaf has a spinose margin. With advance in age of the plant the leaves both increase in absolute and relative size, and also the edges of the lamina fill up in those cases where the margin was spinose, so that the spinoseness disappears. Sometimes, however, there is a tendency to the perpetuation of a certain amount of the spinoseness in the adult, especially at the apex of the leaves, which may be hard and sharp pointed. Three of the species have this type of heterophylly,—

Ludia sessiliflora.
Fernelia buxifolia.
Carissa Xylopicron.

These are all species which in habit somewhat resemble the Box tree (*Buxus sempervirens*) and their leaves are small and coriaceous. The resemblance between the two latter species, when adult, is very close, but the first and last resemble each other most nearly when young. *Fernelia buxifolia* has the leaves in the young plant entire and not spiny. In the other two they are spiny. The three are Mascarene species, but I have record of *Ludia sessiliflora* only as presenting heterophylly in the sister islands.

2nd. Variation arising from non-development of the young leaf in one direction, the transverse, usually but not always accompanied by an increase in the other direction, the longitudinal, the lamina developing equally.

In plants which have variation after this type, and it is the commonest, the young leaves are usually greatly elongated, frequently being two or three times as long as the adult. Thus in *Randia heterophylla* the young are usually over a foot long, the adult vary from 2½ to 6 inches. More rarely the juvenile leaves are not elongated and may even be shorter than the adult, but it is only in species in which the adult leaves are not large and have a firm and coriaceous consistence that this is observed, a curious point when considered along with what I have noted in the first type of variation. But whether the leaves be elongated or short, the relative proportion betwixt the length and breadth of the lamina in the juvenile is vastly different from what is observed in the adult. The juvenile are usually linear, always very greatly narrowed, often only one-twentieth the breadth of the adult, and the contrast in such a case is, as may be imagined, very striking. For example, take *Scyphochlamys revoluta*, the juvenile form is only $\frac{1}{8}$th of an inch in breadth, while the adult averages $1\frac{1}{2}$ to $2\frac{1}{2}$ inches. The lamina developes equally at the margins, which are never spiny, and though any crenatures or dentations which characterise the mature form, may be represented, the lamina is never lobed or deeply cleft. As I have said this is the commonest type of variation, occurring in no less than nine, that is, in more than half of the heterophyllous species, these are:—

Dombeya ferruginea.
Elæodendron orientale.
Terminalia Benzoin.
Fœtidia mauritiana.
Mathurina penduliflora.
Randia heterophylla.
Pyrostria trilocularis.
Scyphochlamys revoluta.
Diospyros diversifolia.

The first four of these are peculiar Mascarene species, the remainder are endemic. *Elæodendron orientale* presents the same variations in the sister islands, and judging from a remark of Cavanilles (Diss. III., 121), *Dombeya ferruginea* is also heteromorphic; I have no information regarding variation in the other two species. Of the endemics, the *Pyrostria* and *Diospyros* are the only two species in which the linear juvenile leaves are shorter than or at least do not exceed the adult. Another species *Eugenia cotinifolia* is probably heterophyllous after this type, but the specimens I found represented differences only to so slight an extent as hardly to warrant my including it in this list.

3rd. Variation due to unequal development of the lamina in the young leaves.

In the plants included in this group the young leaves are not much less, may even be greater in absolute size than the adult, and the relative proportion of length to breadth of the lamina is the same, but the lamina developes unequally, so that a lobed or cleft, sometimes very deeply cleft leaf is produced. The lobation gradually disappears in the older leaves, though frequently slight traces of it remain, especially in an emargination of the apex. If the mature form of leaf has a pinnately arranged venation, then the young leaves are of the pinnatifid type; if the veins are radiate the palmatifid type is seen. The amount of lobation in the juvenile greatly varies, and the primary lobes are sometimes again once or twice cleft. Four species exhibit this variation,—

Aphloia mauritiana.
Hibiscus liliiflorus.
Quivisia laciniata.
Clerodendron laciniatum.

Of these *Hibiscus liliiflorus* has palmatifid young leaves; in all the others they are pinnatifid. The first two are Mascarene species, the others are endemic.

Aphloia Mauritiana is as variable in Mauritius, and so also probably *Hibiscus liliiflorus*. *Quivisia laciniata* and *Clerodendron laciniatum* are each represented in Mauritius by a heterophyllous species. But whilst *Quivisia heterophylla* of Mauritius varies in the same manner as the Rodriguez plant, *Clerodendron heterophyllum* has the second type of variation developed.

Abrotanella rhyncocarpa I may mention here; it is heterophyllous, but after a different manner. It is a small tufted herb, with very persistent leaves. The young leaves, those at the top of the shoots, are entire and oblanceolate, but the older leaves below are markedly pinnatifid. We have thus the converse, as it were, of this last type of variation.

Some species, in addition to the differences in form and size, exhibit variations in habit of the leaves at different periods of their growth. This is most marked in *Dombeya ferruginea*, of which the young leaves are quite green and glabrous on both sides, but in the adult are clothed on the under surface with a dense brown tomentum. The converse is seen in some to a slight extent, thus *Clerodendron laciniatum*, *Terminalia Benzoin*, and *Randia heterophylla* have pubescent young leaves, while in the adult the leaves are glabrous.

As might be expected, when the leaves of young plants are not so perfectly developed as in the adult, they are often of a more firm and rigid consistence. This is specially apparent in species varying according to the first type, in which there is deficiency in parenchymatous tissue, for instance, in *Toddia sessiliflora* and *Carissa Xylopicron*; but it is also apparent in plants varying after different types.

In 17 species then, belonging to as many genera, of 13 natural Orders, the heterophylly exists; and it is a fact of great significance that every * one of them is either endemic or Mascarene only, that is, one fourth of the whole endemic and Mascarene species on the island. Further, on considering the genera, we find that two of them are endemic, four are Mascarene alone, three extend to Africa, and of the remainder seven are cosmopolitan, and one Polynesian.

Our information regarding the variability of the plants of Mauritius and Bourbon is too slight to enable us to say whether an equal amount of variation occurs there. But certainly some of the Mascarene species, which vary in Rodriguez, present, as I have already noticed, the same amount and kind of variation in these islands though in other cases the heterophylly may not be to such an extent. So that I think there can be little doubt that the heteromorphism in foliage is a feature of the whole Mascarene Flora.

In concluding this subject it is of interest to note that *Hibiscus* (*Paritium*) *tiliaceus* is not heterophyllous in Rodriguez, though it is recorded as such from Africa (Oliv. Flor. Afr. Trop. i., p. 208). A plant of *H. tricuspis*, a South Sea Islands species growing in the Saharampore gardens, is described † as having sent up from a

---

* I must qualify this. One species, *Ludia sessiliflora*, passes into Africa.
† Bell in Trans. Bot. Soc. Edinb. VII. 565. King in Linn. Soc. Journ. XV. 83.

rooting decumbent branch a plant with entire leaves very different from the parent, and indeed like those of *H. tiliaceus*. Prof. Thiselton Dyer suggests that *H. tricuspis* is an insular form of *H. tiliaceus*, which I think is not at all improbable. But it is strange that in Rodriguez, where heteromorphism is so prominent a feature, *H. tiliaceus* does not exhibit the peculiarity; although I may remark there are two distinct varieties of the tree on the island, which I have noticed under the species.

Such is the aspect and nature of the vegetation of Rodriguez, and a consideration of the Flora leads us, I think, to the following conclusions :—

1. It is a small Flora, and fragmentary.
2. It is that of a dry rather than of a moist region, as is exemplified in the paucity and nature of some groups, such as Ferns, Orchids, and the abundance of such others as Lichens.
3. It is an insular Flora, as indicated by—
    a. The relative proportion of species, genera, and orders.
    b. The almost total absence of indigenous annuals.
4. Its facies is tropical.
5. It is essentially Mascarene, though possessing a fair amount of individuality.
6. It presents affinities with the Floras of many other parts of the globe. Its strongest relationships are with the African, but it has also very strong Eastern connections, and some close American and Polynesian affinities.
7. Many species exhibit a great amount of variation, but within certain sharply defined limits.

Our knowledge of the Flora of all the other Mascarene Islands is as yet too scant to allow us to adjudge the exact extent of the affinities subsisting between the Floras of the individual islands of the group. But enough may be learned from what I have indicated of the vegetation of this single island to point strongly in the direction of their being fragments of a once more extensive Flora, which has been gradually broken up by geological and climatic changes. How far the geological evidence bears this out is indicated elsewhere.

The following is a list, in detail, of the plants which compose the Flora:—

ANONACEÆ.

Anona muricata, *Linn.; DC. Prod.* i. 84. Nom. vulg. Corosol.
Cultivated.

Anona squamosa, *Linn.; DC. Prod.* i. 85; *Bot. Mag.* t. 3095. Nom. vulg. Atte.
Established, and grows freely in many parts of the island, where well sheltered by trees.

PAPAVERACEÆ.

Argemone mexicana, *Linn.; DC. Prod.* i. 120; *Bot. Mag.* t. 243. Nom. vulg. Chardon.
A frequent weed near habitations. This plant is much used medicinally by the inhabitants.

CRUCIFERÆ.

Nasturtium officinale, *R. Br.; DC. Prod.* i. 137; *Eng. Bot.* t. 125. Nom. vulg. Brède cresson.
Grows abundantly in Rivière Pistache. There is some dispute as to whom belongs the credit of introducing this plant into the Mascarene Islands. By some M. de Reine, a captain of infantry, is allowed the honour. He, during his sojourn in Mauritius, having tried ineffectually to obtain seeds from France, on his return to the mother country sent some seed to père André at Pamplemousses in Mauritius, who successfully grew it. Others consider that Fusée Aublet, author of the Flora of French Guiana, introduced it about 1760. I have no record of its introduction into Rodriguez, but it now grows very abundantly in some of the rivers.

Brassica juncea, *Hook. fil. and Thoms. Fl. Ind.* i. 157.
Is cultivated, and is an escape in a few places.

CAPPARIDACEÆ.

Gynandropsis pentaphylla, *DC. Prod.* i. 238; Cleome pentaphylla, *Bot. Mag.* t. 1681. Nom. vulg. Brède caya.
A few plants found in waste ground near Mathurin. Is eaten in Mauritius as brède, but is too scarce in Rodriguez.

MORINGACEÆ.

Moringa pterygosperma, *Gärtn. Fruct.* ii. 314, t. 147, f. 2; *Wt. Ill.* t. 77. Nom. vulg. Brède mouroungue.
Cultivated at Port-Mathurin and also on Frigate Island for its leaves, which are

boiled and eaten like spinach. The pods also, when young, are eaten as brède and in curries. Many and various medicinal virtues are ascribed to this plant in Mauritius—laxative, antispasmodic, anthelmintic, antiseptic, &c.—the leaves, bark, and the juice of the root being used; but in Rodriguez the plants are few in number, and it is not much employed.

PITTOSPORACEÆ.

Pittosporum Senacia, *Putterl.; Walp. Rep.* i. 250; Senacia undulata, *Lam.* Nom. vulg. Bois malabar.

Is a common tree in the valleys. The wood is white and fine-grained, and is used for making handles for implements.

BIXACEÆ.

Ludia sessiliflora, *Lam.; DC. Prod.* i. 261. Nom. vulg. Goyave marron.

This Mascarene plant in Rodriguez is found growing only on the limestone plains of upraised coral reef, and as these only occur at the east and west ends of the island, there only is the plant found. The wood is hard and fine, but brittle, very durable, and is used in making "pirogues." Like many others on the island, this tree is heterophyllous. The leaves in young plants and on adventitious shoots of old trees are small, ovate acute, $\frac{1}{4}$–$\frac{1}{3}$ in. long, about $\frac{1}{12}$ in. across, with very short petiole, and the margin spinose dentate. The leaves in the adult, on the other hand, are obovate or elliptical, with a cuneate base, and quite obtuse, sometimes even retuse, about 2 in. long and an inch broad, with a distinct petiole. Lamarck, Dict. iii. 613, cites this from Mauritius, and specimens are in Kew Herbarium sent by Bouton, which exhibit the heterophyllous character. Lamarck records also a species *L. myrtifolia* from Bourbon, but this seems to be merely a form of *L. sessiliflora*, Lam.—the curved style, its only distinction, being of no specific value. *L. heterophylla*, Lam. another species recorded from Mauritius, is also identical with our plant. Clos in Ann. Sc. Nat. 4th ser. viii. 244, describes a species, *L. bivalvis*, from Mauritius on specimens in "Herb. Delessert." It is evidently the plant here mentioned. The genus Ludia then is monotypic, but the species is not confined to the Mascarene Islands, but occurs also at Zanquebar.

Aphloia mauritiana, var. theæformis, *Baker Flor. Maur. Seych.* 12; Aphloia theæformis, *Benn.;* Prockia theæformis, *Willd.; DC. Prod.* i. 261; P. serrata, *Poir. (non Willd.)*. Aphloia madagascariensis, var. seychellensis, *Clos in Ann. Sc. Nat.* 4th ser. viii. 274, seems to be the same. Nom. vulg. Bois d'anémone.

I follow Baker in reducing to one species the several forms of this endemic Mascarene genus, and of these forms *theæformis* alone occurs in Rodriguez, though not very abundantly. It is heterophyllous. The specimens of this tree, however, which I gathered do not exhibit the heterophylly in such a marked degree as do some specimens (in herb. Kew) of the tree from Mauritius, where the young leaves are distinctly pinnatifid, thence they pass through stages of greater or less dentations up to the adult form. Bory de St. Vincent, Voy. t. 24, gives a figure of a plant under the name of *Ludia heterophylla,* Lam. There is a mistake here as to the name, the plant represented being this *Aphloia mauritiana,* var. *theæformis,* Baker, and not a *Ludia.*

PORTULACEÆ.

Portulaca oleracea, *Linn.; DC. Prod.* iii. 353. Nom. vulg. Pourpier.

Common in waste ground and specially abundant on the barren ground towards the west of the island about Mount Pourpier, of which the name is probably derived from the plant. Is not infrequently eaten as a brède or salade.

It is interesting to read in Leguat's account of his sojourn in Rodriguez (p. 64) " that he did not find in this island any plant, tree, shrub, or herb, which grows " naturally in any part of Europe, that was known to us, except Purslain, which " is small and green. There's plenty of it in some places of the valleys, and that " which we sow'd, having brought some of the seed from the Cape, came up exactly " like the Purslain of the Island;" and he records how, when a green caterpillar appeared after a hurricane and destroyed the greater part of their crops, the Purslain was untouched.

GUTTIFERÆ.

Calophyllum Inophyllum, *Linn.; Planch. et Trian. Mon. Guttif.* 254; *Wt. Ic.* t. 77. Nom. vulg. Bois tatamaka malgache.

Only a few trees of this found in the higher parts of the island. Yields a soft gum-resinous wood.

MALVACEÆ.

Malvastrum tricuspidatum, *A. Gray Pl. Wright* 16; Malva borbonica, *Willd.; DC. Prod.* i. 430.

A very common plant.

Sida angustifolia, *Lam.; DC. Prod.* i. 459.

This may be merely a variety of *S. spinosa,* Linn.; *DC. Prod.* i. 460. It was found only at one spot on the island, at the mouth of a valley east of English Bay point.

Sida carpinifolia, *Linn. fil.; DC. Prod.* i. 461. Nom. vulg. Herbe à paniers.

Common everywhere, and assuming very various forms according to its position; from a dwarfed and stunted plant, with few and small leaves, to a small, freely branching under-shrub, about 5 feet high, clothed with large leaves.

Sida cordifolia, *Linn.; DC. Prod.* i. 464. Nom. vulg. Mauve.

Common everywhere. This, like the other species of *Sida,* is often used as a demulcent.

Abutilon indicum, *G. Don Gen. Syst.* i. 504; *Wt. Ic.* t. 12; Sida indica, *Linn.*; *DC. Prod.* i. 471. Nom. vulg. La mauve.

A very common weed.

Abutilon graveolens, *W. and A. Prod.* i. 56; *Hook. Comp. Bot. Mag.* t. ii.; Sida graveolens, *Roxb.*; *DC. Prod.* i. 473.

Only a few plants on the shore in Oyster Bay close to habitations.

Urena lobata, *Linn.*; *DC. Prod.* i. 441; *Bot. Mag.* t. 3043. Nom. vulg. Herbe à paniers.

A common weed.

Hibiscus liliiflorus, *Cav.*; *DC. Prod.* i. 446; H. fragilis, *DC. Prod.* i. 446; H. Genevii, *Bojer in Bot. Mag.* t. 3144.

This endemic Mascarene plant is not abundant on the island. It is remarkable from the variation in form of its leaves. On the youngest trees I met with, the leaves were rounded at the base, deeply trifid, the lobes being linear acuminate. In older specimens the lobation of the leaf gradually disappears, and we have ovate acute leaves. Finally, by a gradual transition of forms, we reach the leaves of mature plants, which are obovate, obtuse or deeply emarginate, and cuneate at the base. The venation also becomes more distinct in the adult leaves, and they are quintuplinerved. A hybrid from this plant, the male being *H. Rosa-sinensis*, Linn., is figured, Bot. Mag. t. 2891, from Mauritius, and it seems to possess the variable foliage of one parent; its flowers, though similar to, are much larger than those of *Hibiscus liliiflorus*. *H. Genevii*, Boj. Hort. Maur. 28; Bot. Mag. t. 3144, is probably only a form of this plant with more dentate leaves.

Hibiscus tiliaceus, *Linn.*; *DC. Prod.* i. 454; Paritium tiliaceum, *Wt. Ic.* t. 7. Nom. vulg. Vaur or Var.

Everywhere along the shores.

Of this plant there are two distinct varieties on the island which are recognised by the inhabitants. Of these (A.) *Var blanc* is the more scarce, at least I met with it less frequently. It forms a tree about 25 to 30 feet high, but I am informed also in some places forms a thicket, though I did not meet with it in such condition. The wood is very hard, is heavy and close grained, and makes a good timber, though difficult to work from its hardness. The bark of the trunk and large branches is thick, quite smooth, and does not split or crack. The bast layers are light-coloured, and make a capital cordage, which is greatly used, as cattle will not gnaw or eat it. The leaves are pilose at the junction of the lamina and petiole, and the veins are puberulous. The calyx lobes are eglandulose. (B.) *Var rouge*.—This seemed to me the commonest variety, and forms dense tickets, and I do not recollect seeing it as a large tree. The wood is soft, porous, and very light, comparatively useless for carpentry, and only fit for burning. The bark is not so thick as in *Var blanc*, is rough, cracking, and splitting after the manner of that of an Oak. The bast layers have a reddish tinge, and do not make so good a cordage as those of *Var blanc*. The leaves are less coriaceous than those of *Var blanc*, and rather velutinous than pilose. The epicalyx seems more deeply cleft, and each lobe of the calyx has a linear median dorsal gland which is very conspicuous.

The forms are very easily distinguished at first sight by the bark. In no instance did I find any trace of heteromorphism in the leaves. This is interesting in connexion with an observation made in the Saharunpore gardens by Bell in 1863 (Trans. Bot. Soc. Edin. vii. 565), and again by King in 1868 (Journ. Linn. Soc. xv. 101) on *Hibiscus* (*Paritium*) *tricuspis*, Cav. A lateral branch of this tree curving downwards " entered the soil, and re-appearing about 2 feet from its point of entrance, gave rise to a large leafy bush." The leaves of this sport, unlike the parent, were not trifid, but like those of *H.* (*Paritium*) *tiliaceus*; and in addition to some other peculiarities, the calyx segments had each " a large oblong gland full of a viscid secretion on the back." It must therefore approach the *Var rouge* of Rodriguez. Thiselton Dyer, in a note appended to King's remarks, suggests that *H. tricuspis*, which is a South Sea Island plant, may be a local form derived from *H. tiliaceus*, and this I think is not at all improbable. It is, however, curious that in an island where heterophylly is so marked a character in many of the trees, no trace of it is observable in this species, although a heterophyllous variety is noted by Masters in Oliv..Flor. Afric. Trop. i. 208, as occurring in Africa.

Thespesia populnea, *Corr.; DC. Prod.* i. 456; *Wt. Ic.* t. 8; Hibiscus populneus, *Linn. Sp. Pl.* 976. Nom. vulg. Mahoe.

Frequently met with on the shores, specially towards the western end of the island.

Gossypium herbaceum, *Linn.; DC. Prod.* i. 456; *Wt. Ic.* tt. 9, 10, 11. Nom. vulg. Coton.

Gossypium barbadense, *Linn. DC. Prod.* i. 456; *Wt. Ill.* tt. 28^a, 28^b, 28^c. Nom. vulg. Coton.

Both species of *Gossypium* are widely distributed over the island; *G. barbadense* is found most abundantly on Ile Pierrot, a small coral islet near the edge of the reef on the south side, covering it almost entirely to the exclusion of other plants. I believe it was introduced from an American ship, which was wrecked some years ago on the reef. The cotton produced is of very good quality, but the inhabitants are too lazy to pick it clean and make use of it.

Eriodendron anfractuosum, *DC. Prod.* i. 479; *Wt. Ic.* t. 400; *Bot. Mag.* t. 3360. Nom. vulg. Ouat.

A few trees of this are found planted about the habitations (now ruins) of the earliest settlers at the top of Soupir valley.

STERCULIACEÆ.

Dombeya acutangula, *Cav. Diss.* t. 38, f. 2; *DC. Prod.* i. 498.

An endemic Mascarene species of which I found only one specimen in leaf at the mouth of the Rivière de l'Est.

Dombeya ferruginea, *Cav. Diss.* t. 42, f. 2; *DC. Prod.* i. 499. Nom. vulg. Bois pipe.

Is met with in the upper parts of the valleys. The wood is heavy, and as the popular name implies is used for making pipes. The leaves are hetermorphic. The adult leaves frequently are more cordate and less crenate than in the type, but all are densely ferrugineo-tomentose beneath. In young trees the leaves are lanceolate and taper to the base, and the under surface is pale green and glabrous, not in the least tomentose. Cavanilles l.c. says of the species " Folia sunt certe diversa," but this he applies only to the amount of lobation; and then, again, he says, " tomento rufescente, in junioribus albicante." His specimens were derived from Mauritius, so that in that island the species apparently varies in the same manner as in Rodriguez. This species is endemic in the Mascarene islands.

Melochia pyramidata, *Linn.*; *DC. Prod.* i. 490.

Is naturalised and grows everywhere.

TILIACEÆ.

Corchorus trilocularis, *Linn.*; *DC. Prod.* i. 504.

Established, and grows in abundance on the barren slopes of the western end of the island. The Rodriguez plant differs from the type slightly, the pods dehiscing usually by four or five valves.

Triumfetta glandulosa, *Lam.*; *DC. Prod.* i. 506. Nom. vulg. Herbe à paniers.

A common weed. An infusion of the leaves is a favourite tisane.

GERANIACEÆ.

Oxalis corymbosa, *DC. Prod.* i. 696. Nom. vulg. Alleluia or Oseille.

A plant I found in leaf in Rivière des Acacies is referred to this species. The leaves closely resemble specimens of a plant so named sent from Mauritius by Bouton; and there is also a close resemblance to the figure of *O. Martiana,* Zucc., in Bot. Mag. t. 3938, a typical American species. As I got no flowers it is impossible to identify the plant with certainty.

Oxalis corniculata, *Linn.*; *DC. Prod.* i. 692; *Wt. Ic.* t. 18. Nom. vulg. Petite oseille.

Very common everywhere.

RUTACEÆ.

Zanthoxylum paniculatum, *Balf. fil.*

Arborea, ramulis validis teretibus, spinis paucis nigrescentibus uncinatis armatis; foliis imparipinnatis, 15–19 foliolatis, ad apicem ramulorum confertis, breviter petiolatis; foliolis oppositis, sessilibus, subcordato-oblongis, obtusis, basi inequaliter cordatis, glabris, subcoriaceis, supra nitidis, subtus pallidioribus, costa venisque prominentibus; paniculis subsessilibus, patentibus, folio brevioribus, ramis puberulis, pedicellis brevibus; capsula globosa, bivalvi, glabra minute tuberculata, breviter stipitata.

Arbor glabra usque ad 20 pedes alta, cortice albido. *Folia* alterna, 5–8 poll. longa; foliola superiora sæpe majora, 2–3½ poll. longa, ⅚ poll. lata; gemmarum tegmenta extus glutinosa; petiolus communis brevis, teres, subpuberulus. *Flores* ignoti. *Capsula* fusco-nigrescens, ¼ poll. diam., usque ad basim in valvas duas fissa, stipite $\frac{1}{24}$–$\frac{1}{12}$ poll. longo. *Semina* ignota.

Nom. vulg. Bois Pasner.

I only found one or two trees of this near the shore of Anse Quitorze. The wood is white, fine grained, and very hard. It is not far removed from *Z. Budrunga*, DC. Prod. i. 728, but differs in the want of large glands at the leaf crenatures. *Z. tomentellum*, Hook. fil. Fl. Brit. Ind., i. 493, has some resemblance, but is more spiny.

Toddalia aculeata, *Pers.; DC. Prod.* ii. 83; *Wt. Ill.* t. 66.

Only in the higher parts of the island.

Toddalia paniculata, *Lam.; DC. Prod.* ii. 83.

A species confined to Mauritius and Rodriguez. Is not common. Only in secluded parts of the island. In Mauritius, infusions of the leaves of this and of the foregoing species are much used as expectorants, whilst the bark is said to be astringent.

Citrus medica, *Linn.*; var. **medica proper,** *Hook. fil. Fl. Brit. Ind.* i. 514. Nom. vulg. Citron.

Grows spontaneously everywhere, forming in many places impenetrable thickets. It fruits very freely, and the fruit is in great demand. The leaves and rind of the fruit are used in preparing tisanes for various maladies. The Citron is distinguished by the inhabitants from the real sour Citron, which, however, I was never fortunate enough to find, but I am informed it grows abundantly in Mauritius. I suppose this latter is the *var. acida,* Hook. fil. l.c. The natives say that if the seed of the sour Citron of Mauritius be sown in Rodriguez, it produces a less sour fruit and one like the Citron, and in fact becomes it.

Citrus Aurantium, *Linn.*; var. *a.* **Aurantium proper,** *Hook. fil. Fl. Brit. Ind.* i. 515. Nom. vulg. Oranger.

Only a few of these trees occur which fruit very freely.

Citrus Aurantium, Var. *b.* **Bigaradia,** *Hook. fil.* l. c.; *Wt. Ic.* 957. Nom. vulg. Bigarade.

This is the commonest *Citrus* on the island, and is very abundant, forming along with *C. medica* close and dense thickets. The fruit is not eaten raw, but is preserved in various ways; and the leaves and the rind of the fruit are used for tisanes in shiverings and colic.

Citrus decumana, *Linn.; Hook. fil. Fl. Brit. Ind.* i. 516. Nom. vulg. Pamplemousses.

One or two trees of this grow near some of the oldest habitations.

Besides those species and varieties mentioned the inhabitants speak of the " Vangasaille " which was described to me as about the size of a mandarin Orange, and the "Limon," said to be a little smaller than the Citron. I did not meet with these forms.

Triphasia trifoliata, *DC. Prod.* i. 536. Nom. vulg. Orangine.

Not common, and only near the dwellings of the early settlers, where many introduced plants occur.

SIMARUBACEÆ.

Suriana maritima, *Linn.; DC. Prod.* ii. 91.

Common along the shores where is coralline limestone and on all the coral islets of the reef.

MELIACEÆ.

Quivisia laciniata, *Balf. fil.* (Plate XIX.).

Frutex glaber, ramosissimus; foliis oppositis v. suboppositis, subsessilibus, nitidis, rigide coriaceis, reticulato-venulosis; adultis obovato-cuneatis, obtusis v. retusis, marginibus integris, junioribus profunde pinnatifidis, lobis 3-5 obtusis erectopatentibus, in formam adultam gradatim transeuntibus; floribus in cymas axillares bifloras brevissime pedunculatas collectis, pedicellis brevibus validisque, erectis v. suberectis; calyce cotyliformi, minute 4-dentato, strigoso, dentibus deltoideis; petalis 4 oblongo-ellipticis obtusis, patentibus; staminibus 8, tubo brevi; ovario dense strigoso; stylo versus apicem subito incrassato ibique strigoso nec non constricto; stigmate capitato, leviter umbilicato, obscure 4-lobato, lævi.

Frutex habitu *Buxi*. *Folia* diversiformia, $\frac{3}{4}$-1 poll. longa, $\frac{1}{4}$-$\frac{1}{3}$ poll. lata. *Pedicellus* $\frac{1}{8}$-$\frac{1}{6}$ poll. longus, leviter strigosus. *Alabastrus* globosus, subtrigonus. *Calyx* $\frac{1}{12}$ poll. diam. *Petala* flava, $\frac{1}{6}$-$\frac{1}{8}$ poll. longa, extus in medio strigosa, ad margines glabra. *Staminum* tubus $\frac{1}{16}$ poll. longus; anthera ovoidea, filamentis dimidio breviora. *Fruct.* ign.

Nom. vulg. Bois balais.

This handsome member of an endemic Mascarene genus is very abundant in Rodriguez. It produces a hard fine-grained wood which is greatly used for making spoons. It is nearly allied to *Q. filipes*, Baker Fl. Maur. Seych. 46, a Mauritian species, but is sufficiently distinguished by the short, thick pedicels, the larger size of the flowers, and the heteromorphic leaves.

Plate XIX. Fig. 1. Twig from adult plant. 2. From a younger plant showing the pinnatifid juvenile leaves. 3. From a plant younger than that of 2. 4. Flower bud. 5. Expanded flower. 6. Corolla and androecium spread out. 7. Detached petal. 8. Four stamens detached. 9. Gynæcium enclosed in calyx. 10. Transverse section of nearly mature ovary. 11. Vertical section of nearly mature ovary. Figs. 1, 2, 3, about nat. size. Rest magnified.

An introduced species of Melia, called by the inhabitants Lilas, which furnishes a wood, termed *Bois de Singapore* is found planted in a few places on the island.

CELASTRACEÆ.

Elæodendron orientale, *Jacq. Ic.* t. 48; *DC. Prod.* ii. 10. Nom. Vulg. Bois d'olive.

This is the most frequently met with of all the trees in the island, and many specimens, where far removed from dwellings, have attained a large size. The leaves are heteromorphic. The young ones are linear and acute, about 8 in. long by $\frac{1}{6}$ in. broad, with no petiole, or an exceedingly short one, the midrib and veins being of a bright red colour, and the margins faintly undulate. From this we trace a succession of forms to the most mature, which are obovate or elliptico-oblong, obtuse or retuse, 3 to $3\frac{1}{2}$ in. long by an inch or more broad, and with a distinct petiole $\frac{1}{3}$ to $\frac{1}{4}$ in. long, the midrib and veins usually green and the margins distinctly crenate.

The inhabitants distinguish two varieties of this tree. One they call "*rouge*" the other "*blanc*" according as the wood of the tree is red or not red, and the "*blanc*" is the harder wood of the two. But it is impossible to recognise these varieties. The wood of the young plant is usually reddish, and as the plant grows the new wood tends to become paler, until in mature trees the new wood may be hardly red, or it may have a distinct red tinge. And so it happens that in every tree of this species the wood in some part of the diameter of the trunk has a reddish or pink colour. The wood is tough, and is used, more than any other tree, in carpentry and for making "pirogues." I think this must be the tree to which Leguat refers when he says, "The Tree bore a fruit something like an olive, and the parrots "lov'd the nuts of it mightily." From this tree exudes an enormous quantity of gum in the form of tears, which soon harden and form large masses in crevices of the stem or on the ground around.

RHAMNACEÆ.

Zizyphus Jujuba, *Lam.*; *DC. Prod.* ii. 21; *Wt. Ic.* t. 99. Nom. vulg. Masson.

Is occasionally met with on the island. The infusion of the leaves is frequently used for cough and cold.

Scutia Commersonii, *Brong. in Ann. Sc. Nat.* x. 363, t. 15, *f.* 1; Sc. indica, *Brong.* l.c. Nom. vulg. Bois senti.

Frequently met with in the less frequented valleys, where it is very annoying from its recurved spines. It also occurs on some of the most elevated hill slopes. The bark has great repute in Mauritius as an astringent.

Gouania Retinaria, *DC. Prod.* ii. 40 ; Retinaria scandens (volubilis *in icon.*), *Gärtn. Fruct.* ii. 187. t. 120, f. 4.

This endemic Mascarene twiner is not common. I only found it in two localities, both in the higher parts of the island. Baker (Fl. Maur. Seych. 52), following Bojer (Hort. Maur. 77), makes this species a synonym of *G. tiliæfolia*, Lam., a Bourbon plant according to De Candolle, Prod. l.c.; but I cannot identify them. The description of *G. Retinaria* "fructibus alato-triquetris" seems to me irreconcilable with that of *G. tiliæfolia* "fructibus subovatis apteris."

SAPINDACEÆ.

Cardiospermum Halicacabum, *Linn.*, var. **microcarpum**; C. microcarpum, *H.B.K.*; *DC. Prod.* i. 601. Type figured *Bot. Mag.* t. 1049; *Wt. Ic.* t. 508. Nom. vulg. Bonnet des prêtres.

Is a common twiner.

Allophyllus Cobbe, *Blume;* Hiern in Hook. fil. Fl. Brit. Ind. i. 673; Schmidelia Cobbe, *Wt. Ill.* t. 141.

This is a glabrous form of the species and the leaves are almost entire. It is not common, only found in one or two of the valleys.

Sapindus trifoliatus, *Linn.;* Hiern in Hook. fil. Fl. Brit. Ind. i. 682; S. emarginatus, *Vahl;* Wt. Ill. t. 51; DC. Prod. i. 608. Nom. Vulg. Bois savon.

A few trees are found on the shore a little east of Venus Point.

Dodonœa viscosa, *Linn.; DC. Prod.* i. 616; D. Burmanniana, *DC. Prod.* i. 616; *Wt. Ill.* t. 52. Nom. vulg. Bois gournable.

Very commmon on the hill slopes.

ANACARDIACEÆ.

Sclerocarya castanea, *Baker Fl. Maur. Seych.* 63. (Plate XX.)

Arbor ramulis validissimis, teretibus; foliis imparipinnatis 7–11-foliolatis, ad apicem ramulorum confertis, petiolatis; foliolis oppositis, sessilibus, rarius brevissime petiolulatis, oblongis v. ovatis, acutis v. breviter acuminatis, basi inæqualiter rotundatis, obscure crenulatis, submembranaceis, supra nitidis glabrisque, subtus medio nervo subhirsutis; floribus dioicis?, breviter pedicellatis, in racemis brevibus sessilibus v. breviter pedunculatis solitariis in axillis foliorum terminalium dense confertis; ♀ sepalis 5 minutis, rotundatis, ciliatis; petalis 5 oblongis, obtusis, reflexis, imbricatis; disco crenato-lobato; staminibus 10, partim anantheris; ovario oblongo, glabro, biloculari (?); stylis 5 validis, distantibus, brevibus, erectis, divaricatis, sub apicem ovarii sitis; stigmatibus capitatis, spongiosis; ovulis solitariis, pendulis.

Arbor parva, glabra, usque ad 30 pedes alta. *Folia* 6–9 poll. longa; foliola 2½–3½ poll. longa, ¾–1½ poll. lata, inferiora minora, latioraque; petiolus communis hispidulus, 1½–2 poll. longus. *Racemus* petiolum subœquans. *Petala* albida, $\frac{1}{12}$ poll. longa, sepalis triplo longiora. *Stamina* basi disci inserta, alterne breviora. *Styli* $\frac{1}{24}$ poll. longi. *Fructus* ignotus.

In the valley of the Rivière Palmiste and at the top of the valley of the Rivière Mouruc.

This is a rare tree 20–30 ft. high, and I found it only in the two localities mentioned. I have followed Baker in describing it as a species of *Sclerocarya*; but its five-symmetrical flowers and crenated disk are marked points of difference from the description of that genus. Of the many points of agreement I specially would note the mode of attachment of the ovule to the placenta, which is very peculiar. It answers much more closely the description of *Harpephyllum*, an imperfectly known monotypic Cape genus; the male flowers of which are alone described in Bentham and Hooker's Genera Plantarum (see also Harv. and Sond. Fl. Cap. I. 525), but of the female flowers I have seen a MS. description at Kew. Unfortunately the male flowers of my plant are unknown; but the female flowers correspond so closely with the description that I should have been inclined rather to place it in this genus had Baker not previously described it as *Sclerocarya*; and as our information regarding the genera is imperfect I prefer to leave it so. May not this species bring *Sclerocarya* and *Harpephyllum* into one genus? It seems to me to point in that direction, but with such fragmentary material as we possess it is impossible to decide.

Plate XX. Fig. 1. Terminal portion of twig with leaves and inflorescence. 2. Flower bud. 3. Expanded flower. 4. Vertical section of flower. 5. Detached stamen. 6. Style detached with stigma. 7. Branchlet bearing fruit. 8. Transverse section of unripe fruit. Fig. 1 nat. size. Rest magnified.

Mangifera indica, *Linn.; DC. Prod.* ii. 63; *Bot. Mag.* t. 4510. Nom. vulg. Le mangue.

Many forms of this tree occur on the island. The inhabitants told me that the seed of the Mango in Rodriguez never contains a small grub which is always, or nearly always, present in the Mango in Mauritius. The trees not being in fruit during my stay I had no opportunity of verifying the statement. But supposing it true, it seems to point to the probable absence from Rodriguez of some insect which is present in Mauritius and visits the Mango flower. May this have anything to do with the fertilisation of the Mango? I may mention in this connexion that a great number of Mango trees which flowered most profusely showed no signs of producing fruit.

LEGUMINOSÆ.

Crotalaria retusa, *Linn.; DC. Prod.* ii. 125; *Bot. Mag.* t. 2561. Nom. vulg. Casse-cavelle.

This is found under two forms on the island. One, an erect branching plant with bright green leaves nearly two inches long, grows in the valleys, and places where the scrub has not been shortened. The second form is a dwarfed and stunted irregularly spreading plant with smaller leaves, which are more silky; this grows on the barren plains and open ground, where the vegetation is kept short by the cattle.

Arachis hypogæa, *Linn.; DC. Prod.* ii. 474; *Mart. Fl. Bras.* xv. pt. 1, t. 23, f. 1. Nom. vulg. Pistache.

Cultivated as an article of diet.

Indigofera argentea, *Linn.; DC. Prod.* ii. 224. Nom. vulg. Indigo batat.

Is found in many places.

Indigofera tinctoria, *Linn.; DC. Prod.* ii. 224; *Wt. Ic.* t. 365.

Formerly was cultivated largely on the island, especially on the higher parts, and it has escaped in several places and is now naturalised.

Tephrosia purpurea, *Pers.; DC. Prod.* ii. 251. Nom. vulg. Indigo sauvage.

A common weed on the island.

Desmodium incanum, *DC. Prod.* ii. 332. Nom. vulg. Gros treff.

Grows everywhere.

Desmodium mauritianum, *DC. Prod.* ii. 334. Nom. vulg. Petit treff.

Everywhere on the grass slopes.

Desmodium triflorum, *DC. Prod.* ii. 334; *Wt. Ic.* t. 292. Nom. vulg. Petite oseille marron.

Common.

Desmodium, sp.

I got a single specimen of a *Desmodium*, but not in flower or fruit, which resembles a specimen in like condition gathered by Horne in Seychelles, and which Baker (Fl. Maur. Seych. 75), considers as most likely *D. adscendens*, DC. Prod. ii. 332; Bot. Reg. t. 815.

Abrus precatorius, *Linn.; DC. Prod.* ii. 381. Nom. vulg. Reglise.

A few plants only near Mathurin.

Clitoria Ternatea, *Linn.; DC. Prod.* ii. 233; *Bot. Mag.* t. 1542. Nom. vulg. Ambrevade marron.

Not common. Only near the shore at English Bay Point.

Teramnus labialis, *Spreng. Syst. Veg.* iii. 235.

Common on the hill slopes.

Erythrina indica, *Lam.; DC. Prod.* ii. 412; *Wt. Ic.* t. 58. Nom. vulg. Mouruc.

This tree is occasionally found in most parts of the island, but is specially abundant at the mouth of the valley Rivière Mouruc, on the south side of the island, to which it gives the name.

Atylosia scarabæoides, *Benth. Pl. Jungh,* 242.

Common near the shore.

Rhynchosia minima, *DC. Prod.* ii. 385.

Common everywhere. This plant is not reported from Mauritius nor from Seychelles.

Cæsalpinia Bonducella, *Flem. in Asiat. Res.* xi. 159; Guilandina Bonduc, *Boj. Hort. Maur.* 117. Nom. vulg. Cadoc.

Common everywhere.

Cæsalpinia sepiaria, *Roxb. Fl. Ind.* ii. 360; *Wt. Ic.,* t. 37. Nom. vulg. Cassie.

Is not common, but is met with in a few valleys.

Hæmatoxylon campechianum, *Linn.; DC. Prod.* ii. 485; *Bentl. and Trim. Med. Pl.* t. 8. Nom. vulg. Bois campêche.

This tree is planted as a hedge around gardens in the vicinity of Mathurin.

Poinciana regia, *Boj. Hort. Maur.* 119; *Bot. Mag.* t. 2884. Nom. vulg. Flamboyant.

A few trees of this are found planted on the links at Port Mathurin, in front of Government House, where they flower and fruit very freely.

Cassia occidentalis, *Linn.; DC. Prod.* ii. 497; *Bot. Reg.* t. 83. Nom. vulg. Casse puante.

A few plants of this are found upon the island. It is often used medicinally.

Tamarindus indica, *Linn.; DC. Prod.* ii. 488; *Bentl. and Trim. Med. Pl.* t. 92; T. officinalis, *Hook. Bot. Mag.* t. 4563. Nom. vulg. Tamarin.

Is found scattered over the island. It is said that the early Dutch settlers introduced this tree to Mauritius. It is used very extensively by the Creoles in treating disease. The bark is said to be astringent and tonic. The pulp of the fruit is well known as a mild laxative.

Mucuna gigantea, *DC. Prod.* ii. 405; *Wt. Illustr. in Hook. Bot. Misc.* ii. 351. *Suppl.* t. 14. Nom. vulg. Mort aux Rats.

Common in many places. The popular name indicates the power ascribed to it by the natives.

Mucuna, sp.

I have the leaves of another *Mucuna* which I have not been able to identify with any species. The leaves are thick and coriaceous, and are more oblong-oval than those of *M. gigantea*. The flower and fruit I have not seen. It is a common climber in the valleys, forming very thick festoons from tree to tree.

Canavalia obtusifolia, *DC. Prod.* ii. 404.

Grows on the shore at English Bay, interlacing with *Ipomæa pes-capræ, Roth.*

Canavalia ensiformis, *DC. Prod.* ii. 404; *Bot. Mag.* t. 4027. Nom. vulg. Cocorico.

Only found in the upper part of Rivière Cascade valley.

Phaseolus lunatus, *Linn.; DC. Prod.* ii. 393; *Wt. Ic.* t. 755. Nom. vulg. Haricot vert.

Is cultivated on the island, and is occasionally found as an escape. Many varieties of Haricot were formerly cultivated, but lately a caterpillar has appeared and so damaged the crops that they are now less commonly cultivated.

Cajanus indicus, *Spreng Syst.* iii. 248; Cajanus bicolor, *DC. Prod.* ii. 406; *Bot. Reg.* t. 31. Nom. vulg. Ambrevade.

This is cultivated occasionally, and the seeds are used as Dholl. In some places it has escaped and grows spontaneously. This plant is reputed most efficacious medicinally as diuretic.

Desmanthus virgatus, *Willd.; DC. Prod.* ii. 445; *Bot. Mag.* t. 2454.

Grows abundantly in the neighbourhood of Port Mathurin.

Leucæna glauca, *Benth. in Hook. Lond. Journ. Bot.* (1842) iv. 416. Nom. vulg. Acacie.

This plant was introduced into the island about 30 years ago, and now has spread everywhere, filling up completely many of the valleys, and destroying the indigenous vegetation. The young twigs are a favourite food for the goats, and the straight stems of the young trees are used as poles for propelling "pirogues."

Acacia farnesiana, *Willd.; DC. Prod.* ii. 461; *Rchb. Fl. Germ. Ic.* t. 2052.

A few plants of this occur planted as hedges along with *Hæmatoxylon campechianum* L., near Mathurin.

Albizzia Lebbek, *Benth. in Hook. Lond. Journ. Bot.* (1844) iii. 87. Nom. vulg. Bois noir.

Is found abundantly on the island. This tree is said to have been introduced into Mauritius about 1767, from Bengal, by Cossigny, but I have no record of its reaching Rodriguez.

ROSACEÆ.

Prunus communis, *Benth. et Hook. f. Gen. Plant.* i. 610.; Amygdalus communis, *Linn.; DC. Prod.* ii. 530. Nom. vulg. La pêche.

Is naturalised.

Rubus rosæfolius, *Smith; DC. Prod.* ii. 556; *Hook. Ic. Pl.* iii. t. 349. Nom. vulg. Framboise.

Everywhere on the island.

CRASSULACEÆ.

Bryophyllum calycinum, *Salisb.; DC. Prod.* iii. 396; *Bot. Mag.* t. 1409. Nom. vulg. Soutu fafan.

Not common on the island. Is used as an application to bruises.

COMBRETACEÆ.

Terminalia Benzoin, *Linn. fil Suppl.* 434 (*excl. syn. et loc.*); T. mauritiana, *Lam.; DC. Prod.* iii. 11. Nom. vulg. Bois charron.

This endemic Mascarene species is one of the heterophyllous trees of the island. It occurs abundantly. The contrast between the leaves of young plants and the adult form is so great that it was some time ere I could convince myself they belonged to the same species. The young leaves are linear and about 2 inches long, and $\frac{6}{8}$ in. broad, very shortly petiolate, clustered at the ends of the branchlets, densely pubescent with undulated and recurved margins. The adult leaves are quite glabrous, with long petioles almost equalling the lamina, which is over 2 inches long and nearly an inch or more broad; oval-oblong with a crenate margin and coriaceous. *T. angustifolia,* Jacq. Hort. Vind. iii. t. 100, is a form of this with leaves narrower than usual. The wood of the tree is very hard, and is the best for the purpose of wheelwrights, hence the common name. The bark is supposed to be a good astringent.

Terminalia Catappa, *Linn.; DC. Prod.* iii. 11; *Bot. Mag.* t. 3004. Nom. vulg. Badamier.

Common on the island.

MYRTACEÆ.

Psidium pomiferum, *Linn.; DC. Prod.* iii. 234; P. pyriferum, *DC.* iii. 233. Nom. vulg. Goyave.

Frequent. The fruits are frequently preserved by the inhabitants.

Psidium Cattleianum, *Sabine; Bot. Reg.* t. 622. Nom. vulg. Goyave de Chine.

I doubt very much if this is really my plant. The fruit in the figure is purple, and is quite globular. Mine has a pyriform fruit which is bright yellow. It may be, however, a variety such as we have in the case of *Ps. pomiferum L.*

Eugenia uniflora, *Linn. Sp. Pl.* 673; E. Michelii, *Lam.; DC. Prod.* iii. 263; Plinia pedunculata, *Bot. Mag.* t. 473. Nom. vulg. Roussaille.

Often met near habitations.

Eugenia Jambos, *Linn. Sp. Pl.* 672; *Bot. Mag.* t. 1696; Jambosa vulgaris, *DC. Prod.* iii. 286. Nom. vulg. Jamrosa. Jamrose.

Very common in the valleys. I am told that in the seed of this fruit in Mauritius a grub is always found, just as in the Mango, but it is absent in the fruit as grown in Rodriguez.

Eugenia Jambolana, *Lam. Encyc.* iii. 198; *Wt. Ic.* t. 535; Syzygium Jambolanum, *DC. Prod.* iii. 259. Nom. vulg. Jamlongue.

A few trees near habitations at Oyster Bay.

Eugenia Balfourii, *Baker Fl. Maur. Seych.* 116.

Arborea, ramosissima, ramulis glabris tetragonis; foliis breviter petiolatis, oblongis v. ovali-oblongis v. oblanceolatis acutis, basi cuneatis, glabris, rigide subcoriaceis, pellucido-punctatis, penninerviis, nervis tenuibus plurimis arcte positis; paniculis longe pedunculatis, paucifloris, axillaribus, folia æquantibus, glabris, ramis late patentibus; floribus sessilibus v. brevissime pedicellatis, paucis; calyce obscure dentato, dentibus deltoideis.

Arbor parva, glabra, usque ad 15 pedes alta, cortice albido corrugato. *Folia* opposita, $2\frac{1}{2}$–3 poll. longa; petiolus $1\frac{1}{2}$ poll. longus. *Panicula* $1\frac{1}{2}$–2 poll. lata. *Calyx* $\frac{1}{4}$ poll. longus. *Fructus* $\frac{1}{2}$ poll. diam., globosus, ruber, lobis calycis persistentibus coronatus.

Nom. vulg. Bois clou.

" Near E. Jambolana, from which it differs by its smaller leaves, with much closer veining and fewer larger flowers, Baker, l. c." This is a small tree which I found growing in the higher part of the island, which has a remarkably white bark, and Baker has considered it a new species. It agrees very well, however, with the descriptions of *Syzygium paniculatum*, DC. Prod. iii. 259, collected by Commerson, in Bourbon, where it is known as *Bois à écorce blanche*, which is *Eugenia paniculata*, Lam. Dict. iii. 199. Unfortunately my specimens are very imperfect, wanting flower and perfect fruit, and I have found no specimens of Commerson's plant in the Kew herbarium with which to compare it, and it is therefore very difficult to determine the point, but I do not think it improbable that my plant is this species. In the meantime I have followed Baker.

Eugenia sp.

Growing on the slopes of the Grande Montagne I found another species of *Eugenia*, only in leaf, which is not far removed from the foregoing, but it is impossible to determine it.

Eugenia cotinifolia, *Jacq. Obs.* iii. t. 53. Nom. vulg. Bois de fer.

Specimens of a small shrub, about 12 feet high, with a habit very like a Holly, growing very abundantly on the island, but of which I neither got flowers or fruit, has been referred by Baker l. c. to the above species. The wood of the tree is very hard and heavy, hence its popular name. I think this species is probably heterophyllous, for I have found bushes with leaves much less rounded than the adult, but have no positive evidence.

Fœtidia mauritiana, *Lam. Ill.* t. 419; *DC. Prod.* iii. 295. Nom. vulg. Bois puant.

This peculiar Mascarene plant is very common. The tree is heterophyllous, but the specimens of heteromorphism I got show by no means so extensive a variation as do many other trees. The heterophylly, so fas as I observed it, is most marked when the the plant is growing on the seashore, the young leaves then being more elongated and approaching a linear form. The plant receives its name Bois puant on account of the sickening and disgusting odour exhaled from the leaves when the sun shines on it. The wood is very good, and is often used for making "pirogues."

LYTHRACEÆ.

Pemphis acidula, *Forst.; DC. Prod.* iii. 89. Nom. vulg. Bois matelot.

Grows abundantly on the shore where there is coralline limestone.

Punica Granatum, *Linn.; DC. Prod.* iii. 3; *Bot. Mag.* 1832, A and B. Nom. vulg. Grenade.

Is sometimes found naturalized. Of it the inhabitants distinguish two varieties, *Grenade rouge* and *Grenade blanc*, the difference lying in the colour of the endocarp of the mature fruit, which in one case is tinged with red. These would correspond respectively to the varieties *a, rubrum,* and *b, albescens,* distinguished by De Candolle, l. c. I cannot say that I satisfied myself of the validity of the distinction. Used as a powerful astringent.

TURNERACEÆ.

Mathurina, *Balf. fil. in Linn. Soc. Journ.* xv. 159. (Plate XXI.)

Sepala 5, ovato-lanceolata v. elliptico-oblonga acuminata, costa prominente, glandula magna bilobata intus basi adnata. *Petala* 5, subhypogyna, obcuneata v. obovata acuta, nuda, subunguiculata, basi sepalis leviter adnata, reclinata et corrugata. *Stamina* 5, subhypogyna, exserta; filamentis subulatis calycis glandulis in fundo imo vix adhærentibus; antheris lineari-oblongis introrsis. *Ovarium* sessile, uniloculare, liberum, oblongum, glabrum; ovula adscendentia; styli 3, filiformes, terminales stigmatibus dilatatis subfimbriatis. *Capsula* oblonga, triquetra, glabra, 3-valvis, poly sperma. *Semina* obovoideo-cylindracea, lente curvata, funiculo brevi, arillo longe piloso-sericeo basim seminis circumdante, testa crustacea extus foveolata, albumine carnosa; embryo axilis, rectus, cotyledonibus ovatis plano-convexis, radicula tereti.

Arbor parva. *Folia* alterna, petiolata, lanceolata, sæpe obovata, acuta, crenatoserrata; petiolus 2-glandulosus; stipulae glanduliformes, deciduæ. *Flores* magni, pedunculati, vulgo solitarii, axillares, nonnunquam in cymas trifloras dispositi, albi; pedunculi infra medium articulati, 2-bracteolati; bracteolæ subfoliaceæ, serratæ v. crenatæ, lineares.

M. penduliflora (species unica).

Arbor parva, usque ad 20 pedes alta, ramis erectis, foliorum cicatricibus magnis. *Folia* lanceolata v. obovata v. obcuneata, acuta, 3-4 poll. longa, 1 poll. lata, penninervia, nervis tenuibus nervulis intramarginalibus conjunctis, sed per adolescentiam linearia vel ligulata, $\frac{1}{8}$ poll. lata, in formam adultam gradatim transeuntia; petiolo brevi, margine utroque versus medium glandula instructo, lamina decurrente. *Flores* albidi; pedunculi 1–2 poll. longi; bracteolæ lineares. *Glandulæ* sepalorum intus sulcatæ, pubescentes, apice emarginatæ. *Sepala* petalaque 1 poll. longa. *Stamina* perianthium dimidio excedentia. *Ovarium* glabrum, oblongum, 1 poll. longum; stylis incurvatis.

Nom. vulg. Bois gandine.

Usually found on the higher parts of the island. The stem is usually thickly clad with lichens, and the wood is light coloured and fine grained. The tree is heterophyllous. The young leaves are quite linear, about $\frac{1}{8}$ inch broad, with slight widely separated serrations, but the adult leaves are usually obovate or obcuneate, almost an inch broad, and with very marked crenatures. The nearest affinity is with the monotypic genus *Erblichia* of Seemann, a native of Panama, from which, however, it is distinguished by the sepaline gland, absence of petaline fringes, stigmas, and the arillate seeds.

Plate XXI. Fig. 1. From a photograph. 2. Twig from adult with typical leaves. 3. Leaf from a young tree, more linear and elongated. 4. Twig with leaves from a very young plant or adventitious shoot. 5. Flower spread out. 6. Detached sepal. 7. Detached petal. 8. Gynæcium. 9. Apex of style. 10. Fruit dehiscing. 11. Transverse section of fruit. 12. Seed (nat. size). 13. Seed magnified. 14. Embryo.

PASSIFLORACEÆ.

Carica Papaya, *Linn. Sp. Pl.* 1466; *Bot. Mag.* tt. 2898, 2899; Papaya vulgaris, *DC. Prod.* xv. 1,414. Nom. vulg. Papaye.

Grows now spontaneously in several places. The juice is used most extensively, specially as an anthelmintic, and also for several other diseases. The inhabitants hold the common idea that fresh killed meat if hung up under this tree for an hour or two becomes quite tender.

CUCURBITACEÆ.

Lagenaria vulgaris, *Ser.; DC. Prod.* iii. 299; *Wt. Ill.* t. 1057.

Is cultivated.

Momordica balsamina, *Linn.; DC. Prod.* iii. 311. Nom. vulg. Margose.

Cultivated and occasionally an escape.

Citrullus vulgaris, *Schrad. in Eckl. et Zeyh. Enum.* 279. Nom. vulg. Melon d'eau.

Cultivated and sometimes found as an escape.

The Water melon was one of the plants introduced by Leguat. He says they brought five seeds from the Cape of Good Hope, and the plants springing from those he describes thus :—" Among our five plants of water melons there were two sorts, " red and white; the first were the best. The rind was green and the inside red; " they are very refreshing, and never do any hurt, no more than the others (*i.e.* " ordinary melons): they are so full of water that one may easily go without drink " when they are eaten; sometimes they were so big that all eight of us could " hardly eat up one of them. These several kinds of melons grew without taking " pains about them, as I have said already, and produced fruit in great abundance. " When we mingled a little ashes with the earth in the place where they were sown " it made 'em grow and fructify extraordinarily, and the fruit was more than " ordinarily delicate."

Citrullus Colocynthis, *Schrad.; Naud. in Ann. Sc. Nat.* 4th ser. xii. 99; *Wt. Ic.* t. 498; Cucumis Colocynthis, *Linn.*; *DC. Prod.* iii. 302.

A plant which is provisionally referred to this species is found in several places on the island, usually on coral or on sandy soil. The specimens are too imperfect for absolute determination. Seeds are, however, sown at Kew.

CACTACEÆ.

Opuntia Tuna, *Mill.; DC. Prod.* iii. 472. Nom. vulg. Raquette.
Occurs near habitations, often planted as a hedge.

FICOIDEÆ.

Sesuvium Portulacastrum, *Linn.; DC. Prod.* iii. 453; *Bot. Mag.* t. 1701.
Everywhere on the shore about high water mark.

UMBELLIFERÆ.

Hydrocotyle bonariensis, *Lam.; DC. Prod.* iv. 60.
Grows in many valleys.

Daucus Carota, *Linn.; DC. Prod.* iv. 211; *Eng. Bot.* t. 515. Nom. vulg. Carotte sauvage.
Common on the hills.

ARALIACEÆ.

Gastonia cutispongia, *Lam.; DC. Prod.* iv. 256; Polyscias cutispongia, *Baker Fl. Maur. Seych.* 127; Polyscias repanda, *Baker,* pars quo ad habitat Rodriguez. Nom. vulg. Bois blanc.

A scarce tree only growing on coralline limestone. I have referred my plant to this, the solitary species of an endemic Bourbon genus, but the Rodriguez plant is not typical. The leaves are more rounded at the base, are less coriaceous and have distinct petioles; the calyx is smaller; the fruit also is more globular, the style disk longer and style branches shorter and more recurved, and the whole more deeply umbilicated than in the type form. These variations might almost be considered specific, but as my specimens are not perfect, I think it is better to include it under this species, which seems to be very variable, until more complete specimens are obtained.

Baker, Fl. Maur. Seych. 126, unites *Gastonia* with the genus *Polyscias* and refers my plant to *Polyscias repanda*, Baker, to which he also refers *Gilibertia repanda*, D.C. I cannot agree with him. *Polyscias* and *Gastonia* are very closely allied, but the articulated pedicels and the calyculus of the former are very characteristic, as also the few-celled ovary, and keep them sufficiently distinct. *P. repanda*, Baker, so far as the description applies to the Mauritian plant is a true *Polyscias*; but the Rodriguez plant, which he also includes, is a *Gastonia*, and if not a variety of, is very nearly allied to, *Gastonia cutispongia*, Lam., as above mentioned.

RUBIACEÆ.

Danais corymbosa, *Balf. fil.*

Herba scandens volubilisve, ramulis tenuibus, glabris, tetragonis; foliis oppositis, breviter petiolatis, oblongis v. lanceolatis acutis v. acuminatis, basi cuneatis, integris, glabris, subcoriaceis, reticulato-venulosis, subtus pallidis; stipulis minutis; cymis corymbosis axillaribus in axillis foliorum terminalium, densifloris, breviter pedunculatis, pedicellis erectis, tenuibus, brevibus, bracteolis minutissimis; calyce 5-dentato, tubo campanulato, dentibus lanceolatis; corolla hypocrateriformi, segmentis oblongo-spathulatis acutis, patentibus, tubo dimidio brevioribus, fauce dense villosa; florum brevistylium staminibus longe exsertis; stylo furcato tubum corollæ æquante, ramulis teretibus clavatis; capsula globosa, glabra.

Herba lignosa, late scandens. *Folia* pallide-virescentia, ad extremitatem utramque attenuata, 2–4 poll. longa, $\frac{1}{2}$–$\frac{1}{6}$ poll. lata, marginibus siccitate revolutis sub-repandis; petiolus $\frac{1}{6}$–$\frac{1}{4}$ poll. longus; stipulæ deltoideæ, $\frac{1}{12}$ poll. longæ. *Corymbus* 1$\frac{1}{2}$–2 poll. diam; pedunculus $\frac{1}{2}$ poll. longus; pedicelli sub-puberuli, $\frac{1}{12}$–$\frac{1}{3}$ poll. longi. *Corolla* $\frac{1}{12}$ poll. longa, calyce triplo-longior. *Anthera* $\frac{1}{12}$ poll. longa. *Capsula* profunde loculicida $\frac{1}{6}$ poll. diam. *Semina* plurima, minuta.

This species is not common on the island and only occurs in the higher districts. DeCandolle, Prod. iv. 361, records four species of this Mascarene genus, three from Mauritius and one common to Mauritius and Bourbon. Of these, three, namely, *D. fragrans*, Comm., *D. rotundifolia*, Poir., and *D. laxiflora*, DC., are merely forms of one species, and they have been all reduced to one *D. fragrans*, Comm., by Cordemoy in Adansonia x. 356, whom Baker follows, Fl. Maur. Seych. 137. *D. sulcata*, Pers., the fourth species mentioned by De Candolle, is probably also a form of *D. fragrans*, Comm. The Rodriguez plant is not unlike some of the forms of *D. fragrans*, Comm., but differs conspicuously in the form and long petiolation of its leaves and its longer paniculate inflorescence, the rachis exceeding considerably the petiole. The flowers in this genus formerly considered dioecious have been shown by Cordemoy l.c. to be really dimorphic. I only collected the short-styled form in Rodriguez.

Oldenlandia Sieberi, *Baker,* var **congesta.**

Herba perennis dense cæspitosa, caule brevi, ramis confertis stellatim patentibus, tetragonis, subalatis; foliis oppositis $\frac{1}{6}$ poll. longis, ovatis v. oblongo-ovatis v. obovatis obtusis, inferne in petiolum brevem attenuatis, glabris incrassatis, coriaceis, nitidis.

This plant grows only in tufts on the coralline limestones along with a small species of *Ærua*. Baker, Fl. Maur. Seych. 138, considers it a distinct species, but I do not think the characters are sufficient to separate it from the Mauritian *O. Sieberi*, Baker. Its congested habit, which is the only marked point of distinction, is quite accounted for by its habitat on dry limestone soil; the type *O. Sieberi*, Baker, being a plant of roadsides and damp ground. *O. callipes*, Griseb. of Coll. Wright, Pl. Cub. n. 2678 in Kew herbarium seems also a very close ally.

Randia heterophylla, *Balf. fil.* Plate XXII.

Suffrutex glaber, ramulis tetragonis; foliis oppositis, breviter petiolatis, rigide coriaceis, glabris, supra nitidis, adultis oblongis v. ellipticis, obtusis mucronatis v. emarginatis, ad extremitatem utramque rotundatis, vel sæpe lanceolatis et versus extremitates attenuatis, juvenilibus lineari-lanceolatis, elongatis, acutis, hispidulis, gradatim in formam adultam transeuntibus; stipulis brevibus, connatis, subtruncatis; cymis solitariis, extra-axillaribus, patentibus, 1–5-floris, pedunculis glabris petiolum longe excedentibus, bracteolis fere obsoletis; floribus sessilibus v. brevissime pedicellatis, erectis; calyce anguste-infundibuliformi, minute 5-dentato; corolla hypocrateriformi, fauce breviter villosa, segmentis lanceolatis; antheris partim exsertis, ligulatis, acutis; ovario 5-gono; fructu ovoideo-oblongo 5-angulato.

Suffrutex inermis. *Folia* heteromorpha, opposita, adulta $2\frac{1}{2}$–6 poll. longa, $1\frac{1}{2}$–$2\frac{1}{2}$ poll. lata, juniora pedem excedentia vixque poll. lata; petiolus $\frac{1}{6}$ poll. longus; stipulæ $\frac{1}{6}$ poll. longæ, extus glabræ, intus piloso-sericeæ. *Pedunculus* glaber $\frac{3}{4}$–1 poll. longus. *Calyx* angulatus, dentibus obscure deltoideis, ciliatis. *Corollæ* tubus $\frac{1}{2}$ poll. longus, calycem sextuplo excedens, segmentis tubo longioribus. *Discus* pulvinaris. *Anthera* $\frac{1}{3}$ poll. longa. *Fructus* costatus, coriaceus, $1\frac{1}{2}$–2 poll. longus.

Nom. vulg. Café marron.

This is one of the prettiest and most interesting plants from Rodriguez, belonging, as it does, to a genus hitherto unknown in the Mascarene Islands, though abundantly represented in Africa. It is one of the few relics of the old Flora of the island, and is only found in most unfrequented spots at the heads of the valleys. It is heterophyllous as is represented in the plate. I am inclined to think there are two species on the island, one of which in the adult has elliptical leaves or leaves rounded at both ends, and the other with leaves narrowing to both ends; but my specimens are not sufficient to determine the point, and where we find, as we do in this flora, such variations in individual characters amongst so many species, we must allow a very wide range of specific variation. The parts of this plant are not put to any use by the natives, and indeed it is so scarce many of them are unaware of its existence. The wood is hard and white.

Plate XXII. Fig. 1. Twig from a young plant, with narrow but not very long leaves. 2. Leaf from a younger plant. 3. Leaf from an adult, typical form. 4. Flower bud. 5. Flower expanded. 6. Stamens detached. 7. Vertical section of ovary, with style attached. 8. Transverse section of ovary. 9. Fruit not mature. 1, 2, 3, 4, 9, nat. size. Rest magnified.

Fernelia buxifolia, *Lam.; DC. Prod.* iv. 398. Plate XXIII. Nom. vulg. Bois bouteille.

This small tree or shrub is very common on the island, and exhibits a very marked heterophylly; the leaves or young plants and on adventitious shoots being very minute, oval, and rigid, but not spiny. In the adult they become almost orbicular and lose much of their rigidity. This is one form of what I have referred to in my introductory remarks as the first type of heterophylly. To illustrate this a figure of the plant is given, but the heterophylly is not so clearly marked in the plate as I could wish for. This variation accounts for the multiplication of species in this genus. In DC. Prod. iv. 398, there are two described in addition to *F. buxifolia*, Lam., viz., *F. obovata*, Lam., and *F. pedunculata*, Gärtn , but these have rightly, I think, been reduced by Baker, Fl. Maur. Seych. 142, to the type species, *F. buxifolia*, Lam. So that we have in *Fernelia* a Mascarene endemic monotypic genus.

Some confusion as to the popular name of this tree has arisen, and it is often referred to as Bois de ronde. This is the name of *Psiadia rodriguesiana*, Balf. fil., a Composite plant. But the name Bois de ronde is often erroneously given to *Carissa Xylopicron*, Pet. Th., an Apocynaceous plant, of which the leaves, and specially the young leaves, are very like those of *Fernelia buxifolia*, Lam., and hence the name Bois de ronde has been sometimes associated with *F. buxifolia*, Lam. The wood is hard, but not used for any special purpose.

Plate XXIII. Fig. 1. Twig from an adult plant. 2. Twig from a young plant. 3 and 4. Leaves of a different form from an adult plant. 5. Flower expanded. 6. Flower in vertical section. 7. Fruit. 8. Fruit in vertical, 9. in transverse section. 10. Seed in vertical section. 11. Embryo. Figs. 3–6 from Mauritius specimens.

Antirrhœa frangulacea, *DC. Prod.* iv. 460. Nom. vulg. Bois goudron.

A shrub 9-10 feet high I found growing on the coralline limestone at the southwest end of the island, unfortunately not in flower or fruit, resembles in foliage this endemic Mascarene species. The wood is bright yellow.

Vangueria edulis, *Vahl;* *DC. Prod.* iv. 454. Nom. vulg. Vavangue, Voavang.
A common plant on the island.

Pyrostria trilocularis, *Balf. fil.;* Plectronia ? trilocularis, *Baker Fl. Maur. Seych.* 147. Plate XXIV.

Frutex ramosus glaber, ramulis tetragonis; foliis rigide coriaceis, marginibus leviter revolutis, glabris, nitidis, penninerviis, subtus costa prominente, adultis breviter petiolatis, oblongo-ovalibus v. oblongis v. fere obovatis acutis, juvenilibus sub-sessilibus, linearibus, mucronatis; stipulis connatis glabris, deciduis, longe cuspidatis; floribus axillaribus; fructu 3-rarius 2-loculari, globoso, subtrigono, breviter stipitato.

Folia heteromorpha, 2-5 poll. longa, 1-1½ poll. lata, petiolus ¼ poll. longus; juvenilia 1-1½ poll. longa, ⅙ poll. lata; stipulæ cuspide laminam æquante, ¼ poll. longæ. *Drupa* rubra, $\frac{5}{12}$ poll. diam., pyrenis ⅙ poll. longis; seminibus infra medium affixis; embryone axili, curvo; cotyledonibus cordatis, compressis; radicula tereti, inferiore.

Nom. vulg. Bois chauve souris.

This is a small heterophyllous tree or shrub about 15-25 ft. high, of very erect habit, the young leaves being narrow and linear and acute, becoming in the adult broader and more obtuse and usually shorter. It is very abundant and bats feed greatly on its fruit, hence its common name.

I have had considerable difficulty in determining to what genus to refer this plant, and its position in the genus *Pyrostria* must only be regarded at present as provisional, as my specimens are imperfect, I have no flowers. With *Pyrostria* it agrees in most points; the fruit is, however, fewer-celled than is common in that genus. The embryo of the Rodriguez plant is slightly curved, and has an inferior radicle. Of *Pyrostria* the embryo is at present unknown, but the genus is located amongst the *Vangueriæ*, Benth. et Hook. Gen. Plant. ii. 22, a group characterised by a superior radicle. An important result then of the confirmation of this plant as a *Pyrostria* will be the removal of that genus from its present position to one in the tribe *Ixoreæ*, where the radicle is inferior when it would come near *Myonima* and *Coffea*. With *Myonima* the plant has many points of resemblance, more especially with a species, *M. heterophylla*, Boj. Hort. Maur. 169, but it differs therefrom in its axillary clustered inflorescence, not terminal panicles. With *Coffea* also it has certain resemblances, but there are sufficient characters to keep it out of that genus. Baker refers it doubtfully to *Plectronia*, but it is impossible to put it there, as in that genus the radicle is superior.

Plate XXIV. Fig. 1. Fruit bearing twig from an adult tree with type leaves. 2. Twig from a younger tree with narrower leaves. 3. From a still younger plant, or from an adventitious shoot. 4. Fruit. 5. Fruit in vertical section. 6. Fruit in transverse section. 7. Pyrene in vertical, 8. in transverse section. 9. Seed in profile. 10. Seed from the side of attachment. 11. Embryo. Figures of leaves natural size. Rest magnified.

Scyphochlamys, *Balf. fil.*

Calycis tubus campanulatus ; limbus truncatus v. obscure 4-lobatus, coriaceus, persistens. *Corolla* infundibularis, tubo brevi, fauce villosa ; limbi lobi 4–5(6 ?) lanceolato-triquetri, acuti, incrassati, valvati. *Stamina* 4–5(6?) ore corollæ inserta, filamentis brevibus, antheris dorso infra medium affixis, vix exsertis, lanceolato-acutis. *Discus* annularis v. pulvinaris. *Ovarium* 4–5(6 ?) loculare ; stylus validus, profunde 4-fidus ; ovula in loculis solitaria, infra apicem loculi pendula. *Fructus* pyriformis, costatus . . . . (maturum non vidi). *Semina* pendula, non compressa.

Arbores parvæ, glaberrimæ, ramulis subtetragonis. *Folia* opposita, breviter petiolata, rigide-coriacea, penninervia; stipulæ interpetiolares, latæ, connatæ, coriaceæ, persistentes. *Flores* in capitulum 6–12-florum densum pedunculatum congesti. *Capitula* intra bracteas duas oppositas conniventes crasse coriaceas persistentes in involucrum cyathiforme connatas inclusa.

S. revoluta, *Balf. fil.* Plate XXV.

Arbor parva, ramosa ; foliis heteromorphis, adultis breviter petiolatis, ellipticis v. late ovato-oblongis, nitidis, venulosis, marginibus revolutis, $3\frac{1}{2}$–4 poll. longis, $1\frac{1}{2}$–$2\frac{1}{4}$ poll. latis (petiolo excluso $\frac{1}{6}$–$\frac{1}{4}$ poll. longo) ; juvenilibus linearibus 4–6 poll. longis $\frac{1}{8}$ poll. latis ; stipulis $\frac{1}{3}$ poll. longis ; capitulis solitariis, axillaribus, erecto patentibus ; pedunculis petiolum excedentibus, $\frac{1}{3}$–$\frac{5}{12}$ poll. longis ; involucro $\frac{1}{2}$ poll. diam., basi intus tuberculis appressis brevibus teretibus linearibus dense vestito ; corolla $\frac{1}{4}$ poll. longa, calyce triplo longiore ; fructu in singulo involucro solitario.

Nom. vulg. Bois mangue.

This small tree grows commonly along with *Randia heterophylla*, Balf. fil. in secluded parts of the island, and is not common. I am unable to refer it to any known genus, and have taken it as the type of a new one. Its nearest affinity is with *Pyrostria*, but it is distinguished by the sessile anthers, which are included or sub-included ; the style, which is thick and is deeply 4-cleft ; but the most prominent feature is the large bracts, which form a complete involucre to the flowers. These bracts are clothed at the base within by many short conical adpressed processes, which may represent abortive flowers. The embryo in the single immature fruit obtained is unfortunately too decayed to determine its relations. This tree is heterophyllous. The leaves of young plants being linear pass through a gradation of forms to the adult type, which in many cases is almost orbicular.

Plate XXV. Fig. 1. Twig from an adult tree, with type form of leaf. 2. From a younger plant. 3. From a very young plant or adventitious shoot. 4. One half of involucre. 5. Flower bud removed from involucre. 6. Flower expanded. 7. Corolla and andrœcium opened out. 8. Stamen detached. 9. Gynæcium with portion of calyx attached. 10. Transverse section of ovary. 11. Vertical section of ovary, with portion of calyx attached. 12. Single not ripe fruit enclosed in involucre. 13. Vertical section of unripe fruit. 1, 2, 3, 4, 12, 13, nat. size. Rest magnified.

Coffea arabica, *Linn.; DC. Prod.* iv. 499; *Bot. Mag.* t. 1303. Nom. vulg. Café.

Has been largely cultivated on the island, but now seldom; but many good-sized trees are found in the vicinity of old plantations.

Psychotria ? lanceolata, *Balf. fil.*

Frutex glaber, ramulis tetragonis; foliis oppositis, lanceolatis, acutis, inferne in petiolum brevem gradatim attenuatis, coriaceis; stipulis deciduis; floribus minutis, breviter pedicellatis in paniculam corymbosam pedunculatam ebracteatam v. minutissime bracteatam dispositis, paniculis in axillis foliorum superiorum congestis; calyce late cupuliformi, truncato, obscure 5-dentato; corolla subcampanulata, lobis lanceolatis incrassatis tubo brevioribus, fauce dense pilosa; antheris dorso affixis, filamentis brevibus; disco magno; ovario 2-loculari; ovulo oblongo-ovato, compresso, in singulo loculo solitario, erecto, basilari; stylo furcato, ramis brevibus. Fruct. ign.

Frutex erectus. *Folia* opposita, 3–4 poll. longa, $\frac{3}{4}$–1 poll. lata, internodiis brevibus; petiolus $\frac{1}{3}$ poll. longus. *Pedunculus* petiolo triplo longior, pedicelli petiolo breviores. *Corolla* $\frac{1}{6}$ poll. longa, calycem triplo excedens, tubo $\frac{1}{8}$ poll. longo. *Antheræ* $\frac{1}{12}$ poll. longæ.

Nom. vulg. Bois lubine.

This small shrub is exceedingly rare. I only know of one plant in the island on the slopes of the Grande Montagne. My specimens are very imperfect, as I have only flower-buds, and no fruit. It is difficult therefore to be certain of the genus, but the 2-celled ovary, with a solitary basilar erect ovule in each loculus, places it distinctly in the tribe *Psychotriæ*, and it seems to fall most naturally into the genus *Psychotria*.

COMPOSITÆ.

Ageratum conyzoides, *Linn.; DC. Prod.* v. 108; *Hook. Exot. Flor.* i. 15. Nom. vulg. Abgrat.

A common weed.

Eupatorium triplinerve, *Vahl. Symb.* iii. 97; E. Ayapana, *Vent.; DC. Prod.* v. 169. Nom. vulg. Ayapana.

Frequently met with. A reputed panacea. Was introduced into Mauritius from Rio Janeiro in 1797 by Captain Augustin Baudin. The story is, that he having heard at that port of its medicinal virtues, and having endeavoured in vain to obtain living plants, the night before leaving stole a plant from the window of one of the houses, which he carried to Mauritius.

Psiadia rodriguesiana, *Balf. fil.*

Suffrutex, ramulis teretibus pubescentibus; foliis lanceolatis, acutis, superne profunde-serratis, inferne cuneatis integrisque in petiolum pilosum brevem attenuatis, dense pilosis, nervo centrali subtus prominente nervos 7-9 adscendentes parallelos subtus prominulos emittente; capitulis breviter pedunculatis, in corymbos laxos axillares patentes dispositis; bracteis involucri glabri interioribus lanceolatis v. oblongis acutis margine submembranaceis, exterioribus brevioribus post anthesin patentibus; achæniis radii compressis angulatis, rugis calvis, inter rugas strigosis, pappo rigido setoso-scabriusculo.

Folia 3-4 poll. longa, $\frac{3}{4}$-1 poll. lata; petiolus $\frac{1}{4}$-$\frac{1}{3}$ poll. longus. *Cymæ* oligocephalæ; pedunculi breves, puberuli; bracteæ interiores $\frac{1}{12}$ poll. longæ.

Nom. vulg. Bois de ronde.

This is now a very rare plant in Rodriguez. It must have existed formerly in great abundance, as the limestone plains are in many places thickly strewn with fragments of branches and stems. The wood is very hard, and has a dark ochry colour. It differs from most species of *Psiadia* in Mauritius in being velutino-pubescent, not glutinous, and resembles many of the Bourbon species, originally described as species of *Conyza*, but now referred to *Psiadia*.

Psiadia Coronopus, *Benth. et Hook. Gen. Plant.* ii. 285; P. trinervia, *Willd.* var. macrodon, *Baker Fl. Maur. Seych.* 172; Sarcanthemum Coronopus, *Cass.;* *DC. Prod.* v. 367. Plate XXVI.

Suffrutex ramulis teretibus glutinosis; foliis anguste oblongo-lanceolatis, basi attenuatis, apice serratis, triplinerviis, coriaceis; capitulis breviter pedunculatis in corymbos terminales dispositis; involucris hemisphæricis floribus paulo brevibus; receptaculo paleaceo fimbrillifero; florum tubo crasso ampliato; achæniis radii glabris compresso-obovatis pappo setoso paleaceo.

This is a very interesting plant. It is very rare on the island. I only found it at one point, on the shore, near the mouth of the Rivière Poursuite. It differs from all other species of *Psiadia* in the presence of scales on the receptacle, and in the dilatation at the base of the corolla tube.

The plant was originally described by Lamarck, Encyc. ii. 89, as *Conyza Coronopus*, from specimens brought by Commerson from Rodriguez. I have compared my plants with the original specimens of Commerson in the Museum at Paris, and have confirmed their identity. Cassini in Bull. Philom. 1818, p. 74 (*see also* Dict. Sc. Nat. xlvii. 349, and DC. Prod. v. 367), created a new genus, *Sarcanthemum*, for the plant, distinguishing it by the receptacle and dilated corolla tube. But Bentham and Hooker, Gen. Plant. l.c., place *Sarcanthemum* under *Psiadia*, remarking " planta Mascarensis a nobis non visa ex charactere dato a *Psiadia* non differt nisi " corollis basi crassis ampliatis." The specimens brought now from Rodriguez admit of a full examination of the plant, with the result that it is confirmed as a species of *Psiadia*, although it differs slightly from that genus in the corolla, and also in the fimbrilliferous centre of the receptacle.

Baker, Fl. Maur. Seych. 173, refers it as a variety to *P. trinervia*, Willd., but this it is certainly not. Cordemoy in Adansonia x. 21, supposes Commerson's plant may be *P. retusa*, Lam., but this also is erroneous. As to the distribution of the plant, I believe it to be endemic in Rodriguez. De Candolle l.c. puts Mauritius as a locality on the authority of Sprengel, but querries it. Sprengel himself, Syst. Veg. iii. 510, only says, " Ins. Mascaren," which does not necessarily refer to Mauritius. And neither Bojer, Hort. Maur. 179, nor Bouton, knew the plant there. Bourbon is also given by De Candolle l.c., on the authority of Bory, from whom he received specimens, but I am inclined to think there has been some confusion as to the locality. Cordemoy knows nothing of the plant in Bourbon.

Plate XXVI. Fig. 1. Capitulum isolated. 2. Capitulum in vertical section. 3. Flower of ray with an involucral scale. 4. Flower of disk with basal scale. 5. Stamens and style from flower of disk.

Parthenium Hysterophorus, *Linn.; DC. Prod.* v. 532; *Bot. Mag.* t. 2275. Nom. vulg. Herbe blanche.

A common weed in waste ground near habitations. An infusion of this plant is a favourite tisane.

Siegesbeckia orientalis, *Linn.; DC. Prod.* v. 495; *Wt. Ic. t.* 1103. Nom. vulg Herbe de flacq.

A common weed near habitations.

Bidens pilosa, *Linn.; DC. Prod.* v. 597. Nom. vulg. Ville bague.
Common near habitations.

Tridax procumbens, *Linn.; DC. Prod.* v. 679.
Abundant on the coralline limestone of Rodriguez and on the coral islets around.

Abrotanella rhynchocarpa, *Balf. fil.* Plate XXVIIA.

Herba pusilla, pulvinata, glabra, caulibus dense confertis; foliis imbricatis, stellatim patentibus recurvatis, late amplexicaulibus, lyrato-pinnatifidis v. pinnatipartitis, rarius et solum versus apicem ramulorum simplicibus spathulatisque, acutis, coriaceis, uninerviis, nervo prominente; capitulis solitariis sessilibus, singulo 6–10-floro; involucro 6-phyllo, bracteis ciliatis, exterioribus plurinerviis latioribus, interioribus 2–3-nerviis angustioribus; receptaculo foveolato; floribus 3–4 exterioribus femineis, interioribus hermaphroditis fertilibus; feminea corolla basi globosa, 3-dentata, stylo exserto breviter bifido; hermaphrodita corolla 4-dentata; antheris inclusis, ovoideis acutis, inappendiculatis; stylo valido tubo corollæ subæquilongo, alte bifido, ramis ciliatis; achæniis obovatis, compressis, calvis, lente curvis, obscure 4–5 angulatis, apice in rostrum primùm tortum demum rectum elongato.

Caules cæspitosi, 1–1½ poll. longi, inferne foliis vetustioribus dense obsiti. *Folia* ¼–⅜ poll. longa, $\frac{1}{16}$ poll. lata, lobis erectis, rotundatis, nervo prominente. *Capitula* ¼ poll. lata. *Bracteæ* ⅑ poll. longæ. *Flores* flavi. *Corollæ* dentibus tubus quadruplo longior. *Achænia* ⅛ poll. longa.

This interesting little plant grows only on the coralline limestone, and there in no great abundance. The young leaves are undivided and gradually become pinnatifid as they get older. This is hardly to be considered a form of heterophylly comparable to what is so common in many of the plants from the island.

The genus *Abrotanella*, to which it is referred, originally founded by Cassini for a species *A. emarginata*, brought by Gaudichand from the Falkland Islands, now contains about nine species, and these are all Australian, New Zealand, or Antarctic forms. The Rodriguez plant, while it closely resembles the other species of the genus, its nearest ally apparently being *A. (Ceratella) rosulata*, Hook. fil. Fl. Antarct. i. 25, a species from Campbell's islands, possesses a very remarkable peculiarity in the ovary and fruit. The apex of the ovary and of the young fruit is prolonged into a beak, which coils on itself upon the top of the ovary, making usually one complete turn, and to its extremity is jointed the corolla tube. In the mature fruit this beak becomes straightened out into a long process. The genus is unknown in the other Mascarene islands.

Plate XXVIIA. Fig. 1. Capitulum enclosed in leaves. 2. Flower of ray with outer involucral scale. 3. Flower of disk with inner involucral scale. 4. Flower of disk in vertical section. 5. Achene with beak still coiled on the summit. 6. Achene with beak straightened.

Senecio linearis, *DC. Prod.* vi. 375; S. Lingua, *Poir.; DC.* l. c.; S. salicifolius, *Pers.; DC.* l.c.; S. Boutoni, *Baker Fl. Maur. Seych.* 181.

This plant is common on the island, but under two forms. When growing near the sea the leaves and stems usually become more succulent and fleshy than they are when the plant grows inland.

The species was founded by De Candolle on specimens from Mauritius in the herbarium of the Museum at Paris. The type specimens I could not find for comparison, nor did I find named specimens of *S. Lingua*, Poir., and *S. salicifolius*, Pers. There are, however, specimens unnamed, both in the general herbarium and in Jussieu's herbarium from Mauritius and Bourbon, collected by Commerson, which are identifiable with the descriptions of the above-mentioned species, *S. linearis*, DC., *S. Lingua*, Poir., and *S. salicifolius*, Pers., and which are manifestly mere insular variations of one species, to which the Rodriguez plant may also be referred. There is also a specimen at Paris from Bourbon, collected by Abbé Pourret and named in MS. *S. borbonicus*, which is clearly this plant; and a specimen of Commerson's from Madagascar is probably the same. I have no hesitation in uniting the four species in one. Baker, l.c., from want of sufficient information as to *S. linearis*, DC., describes the Rodriguez plant as *S. Boutoni*. The species is exclusively Mascarene.

Sonchus oleraceus, *Linn. sp. Plant.* 1116; *Eng. Bot.* t. 810. Nom. vulg. Lastron.

Common. Two forms of this occur. When grown on coralline limestone it becomes much stunted, and the leaves are exceedingly narrow and congested. The leaves of this plant are frequently eaten as salad, and as a brède.

In a letter to me at Rodriguez Mr. Horne mentioned that a collector in 1864 brought to Mauritius a yellow flowered *Achillea* from Rodriguez, but I do not know it.

CAMPANULACEÆ.

Lobelia vagans, *Balf. fil.*

Annua; caulibus repentibus, tenuibus, glabris; foliis membranaceis, inferioribus ovato-lanceolatis v. elliptico-ovato-acutis v. obovatis, basi cuneatis integrisque, apice crenatis, breviter petiolatis, superioribus linearibus acutis v. obtusis retusis v. submucronatis subsessilibus crenato-serratis; floribus solitariis in axillis foliorum superiorum, pedicellatis, pedicellis bracteis brevioribus; calycis tubo obconico laciniis subulatis sub-æquilongo v. breviter excedente; corollæ tubo calycis lobis quadruplo longiore, lobo inferiore obovato-oblongo acuto, superiore lineari; antheris 2 inferioribus apice barbatis, superioribus pilosiusculis; stylo longo, stigmate bilobato; capsula obovoidea.

Caulis late patens, gracilis. *Folia* 1–2 poll. longa (petiolo incluso $\frac{1}{4}$–$\frac{1}{12}$ poll.), $\frac{1}{4}$–$\frac{2}{3}$ poll. lata, superiora $\frac{1}{6}$ poll. lata. *Pedicelli* adscendentes, $\frac{1}{4}$–$\frac{1}{2}$ poll. longi. *Corolla* alba, $\frac{1}{2}$ poll. longa, labio inferiore profunde inciso. *Capsula* tubo calycis triplo-longior, $\frac{1}{5}$ poll. longa.

Grows abundantly in the upper and shady parts of the valleys. Is very closely allied to *L. serpens*, Lam.; DC. Prod. vii. 368, and which I take to be the same as *L. filiformis*, Lam.; DC. Prod. vii. 368. It is distinguished, however, from these species by the larger size of the flowers and its distinctly two-lobed stigma and the much broader leaves.

Lobelia Cliffortiana, *Willd.; DC. Prod.* vii. 372.

Only a few plants on the shore at the mouth of the Rivière Grande.

PLUMBAGINEÆ.

Plumbago zeylanica, *Linn.; DC. Prod.* xii. 692; *Bot. Reg.* t. 23.

Very common everywhere.

MYRSINACEÆ.

Ardisia, sp. Nom. vulg. Bois de bœuf.

There are several forms of *Ardisia* growing in Rodriguez. Unfortunately they were only in fruit during my visit, and my material is not sufficient for the determination of species. But there is certainly more than one.

SAPOTACEÆ.

Sideroxylon, sp. Nom. vulg. Bois des pommes.

This tree I obtained in fruit only. It is undoubtedly a *Sideroxylon*, and is not improbably a new species. It has some resemblance to the Mauritian *S. Bojeranum*, DC. Prod. viii. 179, but differs from all the Mauritian species by its very large fruit and seed, the former being as large as a pigeon's egg.

EBENACEÆ.

Diospyros diversifolia, *Hiern in Trim. Journ. Bot.* iv. 353. t. 172. Nom. vulg. Bois d'ebène.

A not uncommon endemic heterophyllous tree, usually about 16–20 feet high. The young leaves being narrow and linear, the older much broader. This heterophylly in *Diospyros* is quite a novelty. Hiern l. c. remarks, "I am acquainted "with no parallel to this diversity of foliage in the case of any other species "throughout the order." One seldom meets with a large tree. In all the large ones I met with the dark heart wood was quite decayed, leaving a shell of newer wood outside. I was told by inhabitants that this rotting of the centre always takes place in this species after it attains a certain height.

OLEACEÆ.

Olea lancea, *Lam.; DC. Prod.* viii. 286. Nom. vulg. Bois tambalacoq.

This Mascarene species is a common tree in Rodriguez.

In addition to the fore-mentioned, I have leaves of another very abundant small tree, which is probably an *Olea*, and Baker, Fl. Maur. Seych., makes it a variety *latifolia* of *Olea lancea*, Lam. I doubt its being so, as the tree differs in habit and general facies from *Olea lancea*, Lam., but it is impossible to decide what it is from my materials. It is known on the island as Bois d'olive petites feuilles.

APOCYNACEÆ.

Carissa Xylopicron, *Pet. Th.; DC. Prod.* viii. 333. Plate XXVIIB. Nom. vulg. Bois sandal.

This Mascarene tree is remarkably heterophyllous. The young leaves are small, about $\frac{1}{4}$–$\frac{1}{2}$ in. long, spinose dentate on the margins, and with the apex obtuse or slightly retuse; also the branchlets bear a number of interpetiolar spines, one between each pair of leaves. In the adult the leaves are 1–2 in. long by $\frac{2}{3}$–1 in. broad, with entire recurved margins and the apex deeply emarginate or obtuse, and there are no spines or rarely on the branchlets. Between these two forms all stages are found. This belongs to the first type of heterophylly referred to in my introductory remarks, and is well represented in the plate.

The tree is very common on the island. The wood is very hard and dense, quite like boxwood, and is much used for making netting needles and such like articles.

The wood is not unlike that of the Bois de ronde (*Psiadia rodriguesiana*, Balf. fil.), but the heart wood is of a much brighter yellow. The plant is abundant still on the island, but formerly existed in enormous quantity, for the limestone plains are in many places strewn with dead branches and stems of it mixed with those of Bois de ronde.

Plate XXVIIB. Fig. 1. Twig from an adult plant with type form of leaf. 2. Twig from a young plant with small and spinose leaves. 3. Twig from a young shoot with smaller leaves. 4. Flower bud. 5. Flower expanded. 6. Corolla spread out with stamens included. 7. Style. 8. Ovary in transverse section. 1, 2, 3, nat. size. Rest magnified. Fig. 3 and the analyses are of Mauritian specimens.

Vinca rosea, *Linn.; DC. Prod.* viii. 382; *Bot. Mag.* t. 248. Nom. vulg. Chaponière.

Of this there are the two varieties *rosea* and *alba*, known to the natives as *Chaponière rouge* and *C. blanc* respectively—both abundant. The leaves are said to make an excellent infusion for bronchitis and other chest affections.

ASCLEPIADACEÆ.

Tanulepis, *Balf. fil.*

Calyx minutus, 5-partitus, basi intus 5-glandulosus. *Corolla* subrotata, profunde 5-fida, subvalvata. *Coronæ* squamæ 5, lineari-lanceolatæ, elongatæ, petalis subæquilongæ, antheras breviter excedentes, subincrassatæ, basi connatæ, gynostegioque adhærentes. *Stamina* filamentis latis brevibusque basi annulo coronæ affixa; antheræ apice cuspidatæ, conniventes; pollen granulosum, in quoque loculo in massas duas subcohærens, appendicibus latis orbicularibus corpusculorum longe stipitatorum applicitum. *Stigma* 5-gonum. *Folliculi* divaricati, tenues, subteretes, subcostati. *Semina* comosa.—*Suffrutex* volubilis, glaber. *Folia* opposita, nitidula. *Cymæ* laxe ramosæ, ad apices ramorum breviter pedunculatæ. *Flores* parvi, pedicellati. *Bracteæ* minutissimæ.

T. Sphenophylla, *Balf. fil.* Plate XXVIII.

Caulibus tenuibus; foliis membranaceis, breviter petiolatis, lanceolatis acuminatis v. oblanceolatis, basi cuneatis, $1\frac{1}{2}$–$2\frac{1}{2}$ poll. longis (petiolo excluso $\frac{3}{8}$ poll.), $\frac{1}{3}$–$\frac{1}{2}$ poll. latis, integris, nitidis, subtus pallidioribus, penninerviis; cymis paucifloris, pedunculis $\frac{1}{16}$ poll. longis, bracteis deciduis, pedicellis $\frac{1}{3}$–$\frac{1}{2}$ poll. longis; corollæ lobis $\frac{1}{5}$ poll. longis; stipite corpusculorum appendicem duplo excedente; folliculis fusco-nigris, glabris, 2–3 poll. longis, $\frac{1}{12}$–$\frac{1}{8}$ poll. latis; seminibus compressis, bisulcatis, $\frac{1}{8}$ poll. longis.

This twiner is fairly abundant, and is also found on Frigate Island. Its nearest congener is *Brachylepis*, W. and A., a monotypic East Indian genus, from which the elongated corona scales, its inflorescence, and the arrangements of the stamens exclude it. The Madagascar monotypic *Harpanema*, Dcne, is also not far removed, but the same characters separate them. I only obtained the plant in fruit, and the flowers are described from specimens sent home by Bouton.

Plate XXVIII. Fig. 1. Flower bud opened. 2. Flower expanded with a portion of one of the corona scales removed. 3. Stamen seen from the inner side. 4. Flower with corolla, corona, and stamens removed. 5. Corpuscles. 6. Follicles.

Sarcostemma viminale, *R. Br.; DC. Prod.* viii. 538. Nom. vulg. Liane calé. Plate XXIX. Figs. 1-3.

Very abundant. The Rodriguez plant differs slightly from the type in having puberulous not glabrous receptacles.

Plate XXIX. Fig. 1. Portion of a stem with inflorescence. 2. Flower expanded. 3. Flower still more magnified and with corolla and portion of one of the scales of the inner corona removed.

Sarcostemma Odontolepis, *Balf. fil.* Plate XXIX. Figs. 4-10.

Planta aphylla scandens v. decumbens, caulibus teretibus, glabris, carnosis, ad nodos articulatis ibique squamis minutis oppositis in loco foliorum instructis; umbellis solitariis ad nodos sessilibus breviterve pedunculatis v. terminalibus; receptaculo paleis vestito; floribus breviter pedicellatis; calycis segmentis ovatis v. suborbicularibus, concavis, margine ciliatis, extus puberulis; corolla profunde lobata, laciniis oblongo-ellipticis glabris; corona exteriore corollæ adnata profunde 5-partita, lobis tridentatis, singulo squamæ interiori adhærente, dente centrali maximo; corona interiore exteriorem triplo excedente, segmentis superne saccatis dolabriformibus gynostegio subæquilongis; stigmate apiculato, obscure bilobato, umbilicato.

Pedicelli subpuberuli, $\frac{1}{6}$-$\frac{3}{8}$ poll. longi: *Corollæ* laciniæ $\frac{1}{6}$ poll. longæ, calyce quadruplo longiores.

Nom. vulg. Liane calé.

This plant resembles the foregoing species in almost every point save in the flower, and therein it differs in the corona. In place of the sinuose-dentate cyathiform outer corona, characteristic of *Sarcostemma*, and well seen in *S. viminale*, R. Br., the outer corona consists of 5 scales, distinct except at the very base, each tridentate, the central tooth being largest. These are connected with the inner corona scales, which are twice as long as the outer scales, but shorter than the gynostegium, whereas in *S. viminale*, R. Br., the inner scales are thrice the length of the outer, and are longer than the gynostegium. This is a very important difference in this family. But on examining a large series of flowers, I discovered one in which there was an arrangement of the corona exactly half-way between the two forms I have described. The cyathiform outer corona of *S. viminale*, R. Br., was divided about half-way down into five tridentate lobes, and the relations of length between outer and inner scales was intermediate. This then connects the two forms. It is then a question, have we to deal with two or one species? I rather incline to the former view, that there are two distinct species, for we find certain other minute differences between them, and at Rodriguez I was led to suspect there might be two species, from a certain difference in facies of the plant in different situations. However we consider them, the character of the genus must be slightly emended. Should they turn out to be one species, we have here a species varying beyond the generic limits. Acting on the opinion expressed, I have described a new species of *Sarcostemma*.

Plate XXIX. Fig. 4. Portion of a stem bearing inflorescence. 5. Unopened flower bud. 6. Flower expanded and magnified. 7. Flower still more magnified and with corolla and portion of the scales of inner corona removed. 8. Pollen-masses. 9. Gynostegium. 10. Ovary in transverse section.

Two species of Asclepiadaceous twiners are occasionally met with, but as I only obtained them in leaf, and they are unlike any known Mascarene or Seychelles species, it is impossible to determine them.

LOGANIACEÆ.

Buddleia madagascariensis, *Lam.; DC. Prod.* x. 447; *Bot. Mag.* t. 2824.

The leaves of a plant, which seems to be this Mascarene species, occurs near the dwellings of the early settlers.

BORRAGINACEÆ.

Tournefortia argentea, *Linn. f.; DC. Prod.* ix. 514. Nom. vulg. Voultie.

Common on the coralline limestone, on the shore of Rodriguez, and on most of the coral islets.

Heliotropium (Heliophytum) indicum, *Linn.; DC. Prod.* ix. 556; *Bot. Mag.* t. 1837. Nom. vulg. Herbe à papillons.

An occasional weed. The juice of this is said to be an excellent vulnerary, and also the infusion as a wash for ulcers.

Trichodesma zeylanicum, *R. Br.; DC. Prod.* x. 172; *Bot. Mag.* t. 4820. Nom. vulg. Madame Tombé.

A common weed.

CONVOLVULACEÆ.

Argyreia tiliæfolia, *Benth. et Hook. Gen. Pl.* ii. 869.

Is very common.

Ipomœa paniculata, *R. Br. Prod.* 486; *Bot. Reg.* t. 62; Batatas paniculata, *Choisy in DC. Prod.* ix. 339.

A plant I obtained in leaf, which seems to be this, is common.

Ipomœa purpurea, *Lam.;* Pharbitis hispida, *Choisy in DC. Prod.* ix. 341.

Is not common.

Ipomœa fragrans, *Boj.* MSS.; Pharbitis fragrans, *Boj.*; *Choisy in DC. Prod.* ix. 341.

A plant in leaf resembling this occurs on Gombrani and some of the other coral islets. I never found this on the main island.

Ipomœa Nil, *Roth.*; Pharbitis Nil, *Choisy in DC. Prod.* ix. 343.

On the coast in many places, and on the coral islets.

Ipomœa (Calonyction) muricata, *Roxb. Fl. Ind.* 499 (non *Cav.*); *Choisy in DC. Prod.* ix. 353.

Very common.

Ipomœa Batatas, *Lam.* Nom. vulg. Patate or Batat.

Many varieties are cultivated, and it is the chief staple of food on the island, being easily grown, and not affected by hurricanes.

Ipomœa Pes-Capræ, *Sw.*; *Choisy in DC. Prod.* ix. 349. Nom. vulg. Batatran.

Very common on the shores. In English Bay it grows mixed with *Canavalia obtusifolia*, DC., and it is difficult at sight to discover the one from the other. A curious example of mimicry.

Ipomœa peltata, *Choisy in DC. Prod.* ix. 359.

Only found at the top of valley of Rivière Coco, where it grows in great profusion, covering the trees.

Ipomœa Turpethum, *R. Br.*; *Choisy in DC. Prod.* ix. 360; *Bot. Reg.* t. 279.

Occurs pretty frequently in the valleys.

Ipomœa tuberosa, *Linn.*; *Choisy in DC. Prod.* ix. 362; *Bot. Reg.* t. 768.

I have leaves of a twining plant very probably this species. It grows frequently in the valleys.

Ipomœa obscura, *Ker*; *Choisy in DC. Prod.* ix. 370; *Bot. Reg.* t. 239.

Very common on the shore.

Ipomœa leucantha, *Jacq.*; *Choisy in DC. Prod.* ix. 382.

Very common on the coralline limestone.

Ipomœa palmata, *Forsk.*; *Choisy in DC. Prod.* ix. 386.

Common in the valleys.

Dichondra repens, *Forst.*; *Choisy in DC. Prod.* ix. 451.

Only on coralline limestone towards the south-west of the island. Usually along with *Hypoestes inconspicua*, Balf. fil. and *Selaginella Balfouri*, Baker.

SOLANACEÆ.

Lycopersicum cerasiforme, *Dunal in DC. Prod.* xiii. 1. 26. Nom. vulg. Pomme d'amour.

Is cultivated and is seen near habitations occasionally.

Solanum nigrum, *Linn.; Dunal in DC. Prod.* xiii. 1. 50. Nom. vulg. Brède martin.

A very common plant, much used by the inhabitants as a brède.

Solanum indicum, *Linn.; Dunal in DC. Prod.* xiii. 1. 309; *Wt. Ic.* t. 346. Nom. vulg. Petite anghive.

Is not common.

Solanum macrocarpon, *Linn.; Dunal in DC. Prod.* xiii. 1. 353. Nom. vulg. Bringelle.

A specially Mascarene form not very common on the island.

Solanum Melongena, *Linn.; Wt. Ill.* t. 166. Nom. vulg. Gros bringelle.

In the vicinity of dwellings.

Solanum sanctum, *Linn.; Dunal in DC. Prod.* xiii. 1. 369. Nom. vulg. Bringelle marron.

Very common in the vicinity of dwellings.

Physalis Peruviana, *Linn.; Dunal in DC. Prod.* xiii. 1. 440. Nom. vulg. Pocke-pocke.

Common.

Capsicum frutescens, *Linn.; Dunal in DC. Prod.* xiii. 1. 413. Nom. vulg. Petit piment.

Abundant everywhere.

Capsicum cordiforme, *Mill.; Dunal in DC. Prod.* xiii. 1. 427. Nom. vulg. Gros piment.

Not common.

Leguat speaks of certain trees " that bear a sort of pepper, and are not a little " like plum trees of a moderate size; their leaves are much like that of the jessa- " mine; they bear their fruit in little bunches, and it did very well in our sauces." I suppose it is to one of these species of *Capsicum* he refers.

Lycium tenue, *Willd.*, var. *Sieberi, Dunal in DC. Prod.* xiii. 1. 515.

Usually on the coralline limestone near the sea. Specially abundant on the coral islets. In one spot at the top of the valley Rivière de l'Est, I found this plant inland, and it there had lost its stunted, short-branched, rigid character, and formed a widely-spreading, long and slenderly-branched plant. The leaves also were larger and less fleshy. The description of this species answers my specimens, but they differ rather from the specimens named *L. tenue*, Willd., in the herbarium at Kew. It is a Cape species, but the variety *Sieberi* is Mauritian. I have seen no type specimen of the variety. The inland form of the Rodriguez plant most resembles the Kew specimens of *L. tenue*, Willd.

Datura alba, *Nees; Dunal in DC. Prod.* xiii. 1. 541; *Wt. Ic.* t. 852. Nom. vulg. Herbe du diable.

A frequent weed.

Nicotiana Tabacum, *Linn.; Dunal in DC. Prod.* xiii. 1. 557; *Hayne Gewächse* 12. t. 41. Nom. vulg. Tabac.

Cultivated and in some places now grows spontaneously.

SCROPHULARIACEÆ.

Herpestis Monnieria, *H.B.K.; Benth. in DC. Prod.* x. 400; *Bot. Mag.* t. 2557.

Common in the streams towards the west end of the island where the woods have been destroyed and the streams pass over barren plain.

ACANTHACEÆ.

Barleria cristata, *Linn.; Nees in DC. Prod.* xi. 229; *Wt. Ic.* t. 453; *Bot. Mag.* t. 1615.

Not at all uncommon near habitations and also in some of the more frequented valleys.

Barleria Prionitis, *Linn.; Nees in DC. Prod.* xi. 237; *Wt. Ic.* t. 452.

Only found in the Champ de Roi near Port Mathurin, but there abundant.

Hypoestes rodriguesiana, *Balf. fil.* Plate XXX.

Herba parva, ramosissima, glabra, ramis crassis lignosis; foliis longe petiolatis ovato-lanceolatis, inferne deltoideis, integris, subcoriaceis, substrigulosis; racemis breviter pedunculatis in axillis foliorum confertis, bracteis obovatis v. oblanceolatis capitulis unifloris oppositis arcte dispositis; involucro tubuloso, subventricoso, tetraphyllo, ad medium discreto, lobis inæqualibus, exterioribus longioribus ovato-acutis, extus plaga lineari lateraliter notatis, interioribus lanceolatis brevioribus; calyce brevi, extus hispidulo, laciniis tubo brevioribus, ciliatis; corolla profunde bifida, pilosa, labiis tubo subæquilongis, superiore oblongo, inferiore tridentato v. trifido; staminibus 2, filamentis exsertis; ovario oblongo, compresso, glabro; stylo filiformi, exserto.

Herba humilis, cortice albido. *Folia* 1–1½ poll. longa; petiolus ¼–½ poll. longus. *Capitula* uniflora, opposita decussataque, brevissime pedunculata; involucrum ¾ poll. longum, lobis integris hispidulis. *Calyx* ⅙ poll. longus. *Corolla* uncia longior, eburnea, labio superiore integro obtuso v. acuto. *Filamenta* glandulosa; anthera oblonga.

A rare plant and not far removed from the Madagascar *H. Bojeriana*, DC. Prod. xi. 507, but distinguished by its densely clustered racemes, unequally lobed involucre, and large flowers. I only found one plant on the top of Malartic, one of the highest points on the island, growing in a very exposed situation, which may account for the dwarfed growth, for it did not rise more than 6 inches from the ground.

Plate XXX. Fig. 1. Flower bud enclosed in involucre. 2. Involucre opened out with calyx included. 3. Calyx spread out. 4. Flower expanded. 5. Corolla opened out with androecium and gynæcium. 6. Stamen detached. 7. Transverse section of ovary. 8. Vertical section of ovary.

Hypoestes inconspicua, *Balf. fil.*

Herba tenuis, caule repente valde articulato, ad nodos radicante; foliis oppositis distantibus, ovalibus v. obovatis, obtusis v. subacutis, strigosis, integris, inferne in petiolum subæquilongum longioremve attenuatis; capitulis solitariis axillaribus subsessilibus, bracteis nullis; involucro unifloro, strigoso tetraphyllo, laciniis lineari-lanceolatis inæqualibus usque ad medium connatis, exterioribus dimidio longioribus; calyce brevissimo, lobis lanceolatis acuminatis, involucro multo breviore; corolla involucro longiore; cæt. ign.

Caulis ramosissimus, glaber sed extremitate strigosus. *Folia* $\frac{1}{6}-\frac{1}{4}$ poll. longa; *Petiolus* $\frac{1}{4}-\frac{5}{12}$ poll. longus. *Involucrum* $\frac{1}{4}$ poll. longum. *Calyx* $\frac{1}{24}$ poll. longus.

A very small plant discovered on one small patch of coralline limestone at the western end of the island growing along with *Dichondra repens*, Forst., and *Selaginella, Balfouri*, Baker. It has some affinity with *H. serpens*, R. Br.; Nees in DC. Prod. xi. 501, and *H. Alsine*, Nees in DC. Prod. xi. 502, but is distinguished by its non-pubescent involucre and very minute calyx.

MYOPORINEÆ.

Myoporum mauritianum, *A. DC. Prod.* xi. 711.

A very curious Mascareno plant. Rare in Rodriguez, only a few plants having been seen on the coralline limestone on the shore in Anse Coton at the east end of the island. A. De Candolle described the species from Mauritian specimens in Sieber's herbarium, including it doubtfully in *Myoporum*. For whilst most characters place it there it is peculiar in having a distinctly 4-lobed stigma. The Rodriguez plant differs from De Candolle's description in its 5, not 4-lobed corolla.

VERBENACEÆ.

Nesogenes decumbens, *Balf. fil.* Plate XXXI.

Herba diffusa perennis, ramis laxis, oppositis v. suboppositis, teretibus, tenuibus, divaricatis, humifusis, glabris sed versus extremitatem biseriatim pilosis; foliis oppositis, lanceolatis v. rhomboideis acutis, integris, margine ciliatis, basi in petiolum brevem attenuatis, siccitate non nigrescentibus; floribus in cymas 1–2-floras axillares dispositis, pedicellis brevissimis pubescentibus; calyce amplo, $\frac{1}{4}$ poll. longo, alte 5-dentato, dentibus acutissimis crenatis, extus intusque glandulosis; corollæ lobis subæqualibus, rotundatis, tubo quintuplo v. sextuplo brevioribus; antheris oblongis; fructu lævi compresso, apice hispido, basi glabro, stylo mucronato calyce cincto.

Caulis ab basi ramosus. *Folia* breviora $\frac{1}{4}-\frac{3}{4}$ poll. longa, $\frac{1}{4}-\frac{1}{3}$ poll. lata; petiolus $\frac{1}{3}-\frac{1}{6}$ poll. longus. *Calyx* $\frac{1}{4}$ poll. longus, laciniis tubo æquilongis. *Anthera* oblonga. *Ovarium* ovoideum, stylo filiformi incluso.

Only found in one place, on a patch of coralline limestone about ¾ of a mile from the sea, at the west end of the island, close to where *Hypoestes inconspicua*, Balf. fil., and *Dichondra repens*, Forst., were found.

A most interesting plant belonging to a hitherto Polynesian monotypic genus. The other species, *N. Euphrasioides*, Hook. and Arn., was first described from Whitsunday Island in Botany of Beechey's Voyage 67, and there doubtfully referred to *Myoporum*. De Candolle, Prod. xi. 703, constituted the genus *Nesogenes* for that species including it in *Myoporaceæ*. But it is now referred by Bentham and Hooker, Gen. Plant iii. 1141, to *Verbenaceæ* and placed near to *Spartothamnus*. In Kew herbarium are specimens from Sow Island, Chain Island, and one or two islands of the Dangerous Archipelago, so that the species is probably common there. But it is curious that another species should occur so sparingly in Rodriguez, an island almost antipodal.

Plate XXXII. Fig. 1. Expanded flower. 2. Corolla and andrœcium spread out. 3. Detached stamens. 4. Gynæcium. 5. Apex of style. 6. Fruit enclosed in calyx. 7. Fruit removed from calyx. 8. Transverse section of fruit. 9. Vertical section of fruit. All magnified.

A species of *Lantana* was shown to me in Mauritius by Mr. Horne, who informed me it came from Rodriguez. I found no such plant there, which is strange, as species of *Lantana* usually grow so freely and spread so widely it is not likely it should be overlooked.

Stachytarpheta indica, *Vahl; Schauer in DC. Prod.* xi. 564. Nom. vulg. Queue des rats.

A very common weed. The juice is said to have a styptic action, and is often used for wounds.

Premna serratifolia, *Linn.; Schauer in DC. Prod.* xi. 632. Nom. vulg. Bois sureau.

Not at all uncommon.

Clerodendron laciniatum, *Balf. fil.* Plate XXXII.

Frutex v. arbor parva; foliis oppositis, petiolatis, coriaceo-membranaceis, poroso-punctatis, adultis ovatis v. ovato-oblongis, acutis, basi subdeltoideis, integris, glabris, subtus pallidis, juvenilibus filiforme-tripinnatipartitis segmentis distantibus puberulis ligulatis obtusis, per formas varias in adultam transeuntibus; cymis axillaribus, paucifloris, bis trifidis, patentibus, breviter pedunculatis, bracteolis minutissimis; calyce cupuliformi truncato, ore integro v. obscure lobato, glabro, demum patulo; corolla infundibuliformi calycem sextuplo excedente, tubo intus resino-papillato, laciniis obtusis tubo brevioribus; genitalibus breviter exsertis.

Rami teretes, cinerei, apice puberuli. *Folia* heteromorpha graveolentia, 2–3 poll. longa; petiolus $\frac{1}{2}$–1 poll. longus. *Cymi* corymbosi, 2–3 poll. diam.; pedunculi $\frac{1}{6}$–$\frac{5}{12}$ poll. longi; pedicelli breves $\frac{1}{12}$–$\frac{1}{8}$ poll. longi. *Calyx* $\frac{1}{12}$ poll. longus.

Nom. vulg. Bois cabri.

A small endemic fairly abundant tree exhibiting an extreme form of heterophylly well represented in the plate. It is very closely allied to the Mauritian and Bourbon species, *C. heterophyllum*, R. Br., but its larger pinnatipartite leaves on the young plant and relative size of the parts of the flower sufficiently distinguish them.

This small tree is easily recognised by its disagreeable odour, which has occasioned its popular name. The wood is very white and close-grained, but is not put to great use, save for burning, probably on account of the odour. I have no doubt this is the tree to which Leguat refers when he says, "There is a tree we call the " *Nasty* tree because it stunk. 'Tis the best wood of all for carpenter's use, but " 'twas of no service to for us it stinks so, that it makes all the places about " it smell of it, and the smell is very offensive." Some have thought he refers to the Bois puant (*Foetidia mauritiana, Lam.*), but the odour of that tree though exceedingly objectionable is only apparent when the sun shines upon it, and is evanescent, and not at all like the persistent odour which the Bois cabri emits.

Plate XXXII. Fig. 1. Twig with flowers. 2. Leaves from adult, typical form. 3. Leaf from a young tree. 4. Leaf from a still younger tree. 5. Twig with leaves from a very young tree, or from an adventitious shoot. 6. Flower bud. 7. Expanded flower. 8. Corolla and androecium opened out. 9. Anthers detached. 10. Gynaecium enclosed in calyx. 11. Transverse section of immature fruit. 12. Vertical section of immature fruit. 1, 2, 3, 4, 5, natural size. Rest magnified.

LABIATÆ.

Ocimum canum, *Sims; Benth. in DC. Prod.* xii. 32 ; *Bot. Mag.* t. 2452.

A few plants in the vicinity of Port Mathurin.

Salvia coccinea, *Linn.; Benth. in DC. Prod.* xii. 343.

Is very common in the valleys.

Stachys, sp.

A plant which is apparently a species of *Stachys* was found in leaf only in the valley of Rivière Grande des Bamboux (Cascade).

Leonurus Sibiricus, *Linn.; Benth. in DC. Prod.* xii. 501.

Frequently found near dwellings.

PLANTAGINEÆ.

Plantago major, *Linn.; Dcne in DC. Prod.* xiii. 1. 694; *Eng. Bot.* 2nd edit. t. 1162.

This occurs abundantly near the top of Valley Rivière Baleine on the south side of the island along with a species of *Rumex*.

NYCTAGINACEÆ.

Mirabilis Jalapa, *Linn.; Choisy in DC. Prod.* xiii. 2. 427; *Bot. Mag.* t. 371.
Nom. vulg. Belle de nuit.
Occasionally found near habitations.

Pisonia viscosa, *Balf. fil.* Plate XXXIII.

Arbor inermis; foliis petiolatis, ovatis v. obovatis, obtusis v. acutis v. breviter cuspidatis, basi deltoideis, membranaceis, subrepandis, primum subpuberulis demum scaberulis, siccitate nigrescentibus; petiolo tenui; ☿ panicula corymbosa in axillis foliorum terminalium oriente, brunneo-pubescente, densa, pedunculo compresso, bracteolis minutis deciduis; perianthio infundibuliformi, extus piloso, breviter 5 dentato, dentibus reflexis; staminibus 10 exsertis; ♀ panicula laxe lateque patenti, pedicellis erecto-patentibus; perigonio elongato, 5-angulato, angulo singulo tuberculis brevibus uncinatis glandulosis seriatim instructo.

Arbor parva, ramis validis. *Folia* 3–4 poll. longa, 1½–2 poll. lata; petiolus ¼–½ poll. longus. ☿ *Panicula* 1½–2½ poll. diam., folia superans; pedunculus ⅙ poll. longus; bracteolæ subulatæ glanduloso-pubescentes; flores subsessiles. *Perianthium* ¼ poll. longum, tubo extus papillis subrigidis erectis 5-seriatis instructo. *Stamina* basi coalita; anthera rotundata. *Ovarium* breviter stipitatum; stylus crassus; stigma capitatum subfimbriatum. ♀ *Panicula* 6–8 poll. diam.; pedicelli ⅙–½ poll. longi. *Perigonium* ½ poll. longum. *Fructûs* loculus semine impletus.

Nom. vulg. Bois mapou.

This tree is very abundant on a small ledge of coralline limestone on the west side of Frigate Island, where it is the favourite nesting place of the Fou, but it is not so abundant on the main island. The wood is very soft, and of no practical value. The plant has some affinity with *P. ovalifolia*, DC. Prod. xiii. 2. 441, a Mauritian species of which the Mauritian *P. lanceolata*, DC. Prod. xiii. 2. 442, is also probably a form, but is distinguished by the structure of the flowers, and the bristly fruit with the seed filling up the whole cavity.

Plate XXXIII. Fig. 1. Male flowers. 2. The same cut open. 3. Portion of female inflorescence. 4. Single female flower. 5. The same more enlarged in transverse section. 6. Embryo. Fig. 3 nat. size. Rest enlarged.

Bœrhaavia diffusa, *Linn.; Choisy in DC. Prod.* xiii. 2. 452.

Widely spread in the island; when growing on the coralline limestone it has a more stunted and woody habit.

POLYGONACEÆ.

Rumex crispus, *Linn.; Meissn. in DC. Prod.* xv. 44.

A plant exceedingly closely allied to this, if it be not identical, grows abundantly with *Plantago major*, Linn., in the valley of Rivière Baleine.

AMARANTHACEÆ.

Amaranthus tristis, *Linn.; Moq. in DC. Prod.* xiii. 2. 260; *Wt. Ic.* tt. 514, 713. Nom. vulg. Brède malabar.

Very common near habitations, and is very much used as a brède by the inhabitants.

Ærua congesta, *Balf. fil.*

Herba perennis, dense cæspitosa, ramosissima, pulvinata, ramis firmis lignosis humifusis; foliis alternis petiolatis, obovato-spathulatis, obtusis v. acutis, subcoriaceis, glabris, junioribus subtus pilosis; spicis oblongis v. subrotundatis, in axillis villosis foliorum superiorum sessilibus solitariis erectis, bracteolis deltoideis perianthio brevioribus, glabris; perianthii segmentis oblongo-lanceolatis uninerviis, exterioribus duobus latioribus pilosiusculis, interioribus tribus angustioribus denseque pilosis; antheris rotundis; staminodiis minutissimis; utriculo compresso suborbiculari glabro; semine inæqualiter reniformi, minute tuberculato, margine obtuso.

Herba pusilla, ramis patentibus usque ad 2-3 poll. *Folia* $\frac{1}{8}-\frac{1}{4}$ poll. longa. *Spicæ* $\frac{1}{8}$ poll. longæ. *Perianthii* segmenta $\frac{1}{24}$ poll. longa bracteolas albidas dimidio excedentia. *Semen* nitidum nigrum.

A small tufted plant growing only on coralline limestone, along with such plants as *Abrotanella rhynchocarpa*, Balf. fil., and *Oldenlandia Sieberi*, Baker var. *congesta*, and frequent on the coral islets of the reef.

Achyranthes aspera, *Linn.; Moq. in DC. Prod.* xiii. 2. 314; *Wt. Ic.* 1777. Nom. vulg. Herbe l'argent.

Very common weed.

Achyranthes argentea, *Lam.; Moq. in DC. Prod.* xiii. 2. 315; *Sibth. Fl. Græc.* t. 244. Nom. vulg. Herbe l'argent.

Occurs on the coralline limestone near the shore and on the coral islets. Perhaps is a mere form of the foregoing. An infusion of the root of Herbe l'argent is said to be of great service as a cure for cough. The juice of the leaf is a vulnerary of great repute.

Alternanthera sessilis, *R. Br.; Moq. in DC. Prod.* xiii. 2. 357; *Wt. Ic.* t. 727.

A common weed.

CHENOPODIACEÆ.

Chenopodium ambrosioides, *Linn.; Moq. in DC. Prod.* xiii. 2. 72. *Wt. Ic.* t. 1786.

Is an occasional weed in waste ground in the vicinity of Port Mathurin.

BASELLACEÆ.

Basella rubra, *Linn.; Moq. in DC. Prod.* xiii. 2. 222. Nom. vulg. Brède d'angole.
Is cultivated and used much as a brède.

LAURACEÆ.

Persea gratissima, *Gärtn.; Meissn. in DC. Prod.* xv. 1. 52; *Wt. Ic.* t. 1823; *Bot. Mag.* t. 4580. Nom. vulg. Avoca.

A few trees at the top of valley Rivière Palmiste, near the site of some old dwellings.

Tetranthera laurifolia, *Jacq. Hort. Schœn.* t. 113; *Meissn. in DC. Prod.* xv. 1. 178; T. apetala, *Roxb. Cor.* t. 147. Nom. vulg. Bois Zozo.

Planted near Port Mathurin.

Cassytha filiformis, *Linn.; Meissn. in DC. Prod.* xv. 1. 255; *Wt. Ic.* t. 1847. Nom. vulg. Liane sans fin.

Very abundant, covering the ground and trees in many places.

URTICACEÆ.

Obetia ficifolia, *Gaud. Atl. Bon.* t. 82. Nom. vulg. Figue marron.

Occasionally found in the upper parts of the valleys.

Pilea Balfouri, *Baker Fl. Maur. Seych.* 276. Plate XXXIV.

Herba parvula monoica perennis glaberrima, caule brevi 4-gono basi cæspitose ramoso, ramis patentibus oppositis, exsiccatis cystolithiferis; stipulis deltoideis minutissimis; foliis oblongis rhomboideis, acutis v. acuminatis, basi cuneatis integrisque, apice profunde inciso-crenatis, triplinerviis, nervis usque ad apicem productis, laminis exsiccatis membranaceis, paginis utrisque cystolithis linearibus suffultis; cymis patentibus, multifloris, breviter pedunculatis, axillaribus, petiolo brevioribus; floribus breviter pedicellatis; ♂ perianthio bilobato, lobis concavis subcucullatis acutis; ♀ achæniis lævibus, acute marginatis, compressis, ovoideis, vix segmentum intermedium perigonii superantibus.

Rami late patentes. *Folia* 1–4 poll. longa, $\frac{3}{4}-\frac{1}{4}$ poll. lata, opposita, internodiis longis usque ad 2 poll.; petiolus $\frac{3}{4}-1\frac{1}{4}$ poll. longus. ♂ *Alabastrus* $\frac{1}{16}$ poll. longus, exsiccatus cystolithiferus. *Calycis* lobi $\frac{1}{24}$ poll. longi.

Common in the shady and moist spots in the upper part of valleys. This is a near ally of *P. cuneiformis*, Wedd. in DC. Prod. xvi. 1. 133, a Mauritian species, but is distinguished by its habit, longly petiolate leaves, and spreading inflorescence.

Plate XXXIV. Fig. 1. Portion of male inflorescence. 2. Male flower opened. 3. Perianth of female flower. 4. Female flower.

MOREÆ.

Artocarpus integrifolia, *Linn. Suppl.* 412; *Roxb. Cor.* t. 250; *Bot. Mag.* tt. 2833, 2834. Nom. vulg. Jacque.

Frequent near habitations. There are said to be two varieties, *Jacque labou* and *Jacque blanc*, but I never found any plant which could be considered a variety.

Ficus consimilis, *Baker Fl. Maur. Seych.* 286. Nom. vulg. La fouche.

Is very common. The bark is said to be astringent and the juice good for warts. The bast layers make excellent cordage which is much used. The fruit is not edible. The wood is very hard and tough.

Ficus rubra, *Vahl.* var. **amblyphylla,** *Baker Fl. Maur. Seych.* 285. Nom. vulg. La fouche rouge or La fouche petite feuille.

Common. The fruit of this tree is said to be edible. I have followed Baker, l. c., regarding this plant and the preceding. My specimens are not sufficient to allow of a very satisfactory determination.

Leguat speaks of "a wonderful tree whose branches are so round and so thick " 'tis impossible for the sunbeams to penetrate through it. Some of these trees " are so big that two or three hundred people may stand under them and be sheltered " from the sun or the weather. The vast extent of it is occasioned thus. Some " of the great branches naturally tend downwards, and reaching the ground, take " root and become new trunks themselves which make a sort of little forest."

He refers evidently to a species of *Ficus*, and gives a figure of it, but I do not know to what species, certainly neither of those above mentioned, for he describes it more particularly :—

"The *Rodrigo Kastas* (for I sought to keep the *Indian* name at least in the *Indies*) bear leaves as broad as one's hand, pretty thick and somewhat like that of a lilach or heart in shape, they are softer than satin to touch. Their flower is white and smells well. Their fruit is red and round, and as big as a black damask plum. Their skin is hard and within it is a thin seed, a little like that of a fig. The fruit is not prejudicial to health, but 'tis insipid. The batts commonly feed on it, and multitudes of them nest in the tufted branches of the tree."

I found no species which would answer this description, which if accurate indicates a species formerly extant, now extinct, or at least of which all the large individuals are destroyed, only young and inconspicuous ones left.

Mr. Horne showed me in the Botanic Gardens at Pamplemousses, Mauritius, a third species of *Ficus* which he said Mr. Duncan got from Rodriguez, but I do not know what this is.

EUPHORBIACEÆ.

Euphorbia pilulifera, *Linn.; Boiss. in DC. Prod.* xv. 2. 21.

A common weed.

Euphorbia thymifolia, *Burm.; Boiss. in DC. Prod.* xv. 2. 47.

Common specially on coralline limestone.

Euphorbia daphnoides, *Balf. fil.*

Suffrutex caule tenui, ramulis lignosis teretibus; foliis ad apicem ramulorum confertis, breviter petiolatis, oblanceolatis v. anguste oblongis, obtusis, mucronatis, tenuiter coriaceis, glabris, cito deciduis, exstipulatis; cymis terminalibus, pedunculatis, bracteis 2 magnis coriaceis ovalibus v. suborbicularibus mucronatis v. retusis v. emarginatis; involucris campanulatis, breviter pedicellatis, glabris, lobis brevibus, glandulis 5 sæpe rubris rotundatis v. lunatis poroso-punctatis integris; stylo brevi; capsula glabra sub-depressa trisulcata transverso diametro majore, coccis paullum compressis; semine glabro, irregulariter papillato, ovoideo.

Suffrutex glaber, habitu *Daphnes*. *Folia* 2-4 poll. longa, $\frac{1}{2}$-$\frac{3}{4}$ poll. lata. *Bracteæ* ultimæ $\frac{1}{4}$-$\frac{1}{8}$ poll. longæ; *pedicelli* glabri, $\frac{1}{4}$ poll. longi. *Involucrum* $\frac{1}{2}$ poll. longum. *Capsula* $\frac{1}{4}$ poll. diam. *Stylus* $\frac{1}{12}$ poll. longus. *Semen* $\frac{1}{12}$ poll. longum.

Not a common plant. Only found in the valley Rivière de l'Est, and on the flanks of the Mount au Sel. It forms a very handsome undershrub and of it there are two varieties, one with red glands in the flower, whilst in the other they are uncoloured.

Euphorbia peploides, *Gouan; Boiss. in DC. Prod.* xv. 2. 141.

In waste ground near dwellings.

Securinega durissima, *Gmel.; Mull. Arg. in DC. Prod.* xv. 2. 447.

Nom. vulg. Bois dur.

A tree, apparently referable to this Mascarene species is very abundant on the island. I only obtained specimens in leaf, which are therefore not fully determinable.

Phyllanthus Niruri, *Linn.; Mull. Arg. in DC. Prod.* xv. 2. 406.

Nom. vulg. Ananellé.

Common around Port Mathurin.

Phyllanthus dumetosus, *Poir.; Mull. Arg. in DC. Prod.* xv. 2. 398.

Abundant in Anse Baleine. This is one of the plants Commerson brought from Rodriguez, where it is endemic.

Phyllanthus Casticum, *Willem.; Mull. Arg. in DC. Prod.* xv. 2. 348.

Nom. vulg. Castique.

Not uncommon in many places. Many forms of this Mascarene tree occur, varying slightly in the leaves, but all referable to the one type. Is said to be a powerful astringent.

Manihot utilissima, *Pohl.; Mull. Arg. in DC. Prod.* xv. 2. 1064; Jatropha Manihot, *Bot. Mag.* 3071.

Nom. vulg. Manioc. Cassava.

Commonly cultivated. There are several varieties which grow well when planted in a dry soil, otherwise the roots rot. It is always planted in sheltered situations on the hill slopes. Some varieties ripen in 3 months, others not for 14. The roots are boiled whole, or ground down and made into small round cakes known as *Coup de poing manioc*, or large flat ones called *Gullet manioc*.

Ricinus communis, *Linn.; Mull. Arg. in DC. Prod.* xv. 2. 1017; *Bot. Mag.* t. 2209. Nom. vulg. Tang-Tang.

Forms dense thickets in many places, and in some places produces a spiny fruit; in other places the fruit is quite smooth.

Claoxylon parviflorum, *A. Juss.; Mull. Arg. in DC. Prod.* xv. 2. 785.

A tree greatly resembling this Mascarene species occurs on the island, but as I only obtained it in leaf, it is not fully determinable.

PIPERACEÆ.

Peperomia hirta, *Balf. fil.*

Herba repens pilosa, caule simplici v. ramoso, e nodis radicante; foliis oppositis, petiolatis, ellipticis v. oblongo-ellipticis v. obovatis, 5-nerviis, utrinque villosis, nervulo obscuro juxta marginem currente, petiolo villoso. Cæt. ign.

Caules pedali minores $\frac{1}{12}$ poll. crassi. *Folia* $\frac{1}{4}$–$\frac{3}{4}$ poll. longa; petiolus $\frac{1}{3}$–$\frac{1}{4}$ poll. longus.

Discovered within a few yards of the summit of Mount Limon, and nowhere else seen. Its nearest affinity is with *P. elliptica,* Dietr.; C.DC. in DC. Prod. xvi. 1. 440, a Mauritian species from which its pilose character sufficiently separates it. Of this plant I have only leaves, and I am indebted to M. Casimir de Candolle for the specific determination.

Peperomia reticulata, *Balf. fil.* Plate XXXV.

Herba carnosa repens, caule simplici v. uniramoso, in parte inferiore e nodis radicante, versus apicem adscendente; foliis oppositis petiolatis, summis ternis, elliptico-rhomboideis, apice emarginulatis v. acutis, utrinque glabris, 5-nerviis, et reticulato-venulosis, nervulo obscuro juxta marginem currente; amentis solitariis, axillaribus, folia terminalia superantibus, breviter pedunculatis, bracteis subrotundato-peltatis, breviter stipitatis; ovario globoso immerso, stigmate umbilicato, prominente, glabro.

Caules $\frac{1}{2}$–1 ped. longi, $\frac{1}{3}$–$\frac{1}{6}$ poll. crassi. *Folia* 1–2 poll. longa; petiolus $\frac{1}{6}$–$\frac{5}{12}$ poll. longus. *Amenta* 2–4 poll. longa, $\frac{1}{8}$ poll. crassa; pedunculus glabrus petiolum superans. *Fructus* $\frac{1}{24}$ poll. diam.

A small trailing plant not uncommon in the shady parts of the valleys. It is nearly allied to an Indian species, *P. dindygulensis,* Miq.; C.DC. in DC. Prod. xvi. 1. 442, but differing in the glabrous leaves, shorter petioles, and non-puberulous stigma.

Plate XXXV. Fig. 1. Portion of spike. 2. Bract. 3. Flower. 4. Stamen detached. 5. Fruit in vertical section. All magnified.

Peperomia Rodriguezi, *Balf. fil.*

Herba ramosissima, carnosa, repens, caulibus ad nodos radicantibus subpilosis;

foliis oppositis, petiolatis, obovato-ellipticis, basi cuneatis, obtusis utrinque glabris, ciliolatis, subtus albido-pallidis subtiliter 3–5 nerviis enervulosis, petiolo piloso. Cæt. ign.

Caules subpedales, $\frac{1}{12}$ poll. crassi. *Folia* $\frac{1}{2}$–$3\frac{1}{4}$ poll. longa; petiolus $\frac{1}{5}$–$\frac{1}{2}$ poll. longus.

This small species I found only on the southern slopes of Mount Piton creeping over the surface of large boulders. Its nearest ally is *P. Ventenati*, Miq.; C.DC. in DC. Prod. xvi. 1. 446, an East Indian species. I only obtained the plant in leaf, and should not myself have ventured on a specific determination; but M. Casimir de Candolle has kindly examined the specimens, and pronounced it a novelty.

LORANTHACEÆ.

Viscum tænioides, *Comm.; DC. Prod.* iv. 283.

Only in the valley of Rivière Baleine, and abundant there on the branches of *Fernelia buxifolia*, Lam. This plant is peculiar to the Mascarene Islands.

MUSACEÆ.

Musa paradisiaca, *Linn.; Trew Ehret.* tt. 18–20; *Red. Lil.* t. 443. Nom. vulg. Banane. Cultivated.

Musa sapientum, *Linn.; Trew Ehret.* tt. 21–23. Nom. vulg. Banane. Cultivated:

Ravenala madagascariensis, *Sonnerat Voy.* ii. 223. tt. 124–6; *Jacq. Hort. Schœn.* t. 93; Urania speciosa, *Willd.; Boj. Hort. Maur.* 333. Nom. vulg. Ravenal.

A few trees near dwellings of old settlers.

ORCHIDACEÆ.

Mr. S. Le M. Moore has kindly determined the Orchids.

Oberonia brevifolia, *Lindl. Gen. and Sp. Orch.* 16; *Fol. Orch.* No. 36.

Not very common. Only found in valley Rivière Baleine on *Fernelia buxifolia*, Lam., along with *Viscum tænioides*, Comm.

Bulbophyllum incurvum, *Thouars Orch. Afr.* t. 95.

A Mascarene plant not uncommon on the branches of trees.

Aeranthus arachnites, *Lindl. Bot. Reg.* sub. t. 817; *Bot. Mag.* t. 6034. Var. Balfourii. Leaves 10 in. long; lateral sepals $1\frac{1}{4}$ in. long.

A distinct variety of this Mascarene species. It is very common on the branches of trees.

Listrostachys Aphrodite, *Balf. fil. and S. Moore in Baker Fl. Maur. Seych.* 354. Plate XXXVI.

Caulibus erectis, validis, $\frac{1}{2}$–1 ped. altis; foliis subimbricatis, carnosis, lineari-oblongis, oblique-emarginatis, $2\frac{1}{2}$–3 poll. longis, $\frac{1}{3}$ poll. latis v. latioribus; racemis adscendentibus foliis oppositis, fere 5 poll. longis, squamis nonnullis lagis lentis scariosis vaginantibus infra flores vestitis; bracteis rotundatis, $\frac{1}{6}$ poll. longis; sepalis lanceolatis, acuminatis, fere $\frac{1}{2}$ poll. longis; petalis brevioribus, linearibus, labello ovato-rotundato, 3-lobato, lobis lateralibus crenulatis, infra circum columnam convolutis, lobo centrali auguste-lineari, integro vix $\frac{1}{6}$ poll. longo; calcare $\frac{1}{12}$ poll. longo fere recto; polliniis oblongo-ovoideis, caudiculis linearibus leviter in glandulam ovatam attenuatis; lobis rostelli verticaliter resupinatis.

Nom. vulg. Faname.

An endemic plant only occasionally met with on stems and branches of trees.

Plate XXXVI. Fig. 1. Flower bud partly open. 2. Front view of expanded flower. 3. Side view of column.

Angræcum, sp. near *A. caulescens, Thouars.*

Too far advanced for description. A very common plant.

AMARYLLIDACEÆ.

Crinum asiaticum, *Linn.; Kunth. Enum.* v. 547; *Bot. Mag.* t. 1073. Nom. vulg. Fleur de lis.

Very abundant at the mouth of the Rivière aux Huitres, and also near Mont Plaisir in the centre of the island.

Agave americana, *Linn.; Kunth. Enum.* v. 819; *Bot. Mag.* t. 3654 is a variety. Nom. vulg. Aloe.

Very common, especially on the slopes of Mount Piton.

Fourcroya gigantea, *Vent.; Kunth. Enum.* v. 841; *Bot. Mag.* t. 2250. Nom. vulg. Aloe vert.

Equally common with the last. The fibres of this are much used for cordage.

DIOSCOREACEÆ.

Dioscorea sativa, *Linn.; Kunth. Enum.* v. 340. Nom. vulg. Cambare.

Cultivated. Formerly the yam was largely grown, but since whalers ceased to visit the island regularly the cultivation has greatly decreased.

Dioscorea alata, *Linn.; Kunth. Enum.* v. 387.

I have the leaves of a plant which resembles this somewhat. I found it in waste ground near dwellings.

BROMELIACEÆ.

Ananassa sativa, *Lindl. in Bot. Reg.* sub. t. 1068; Bromelia Ananas, *Linn.; Bot. Mag.* t. 1554. Nom. vulg. Ananas.

Very common in many places.

LILIACEÆ.

Aloe lomatophylloides, *Baker in Fl. Maur. Seych.* 372.

Subacaulescens, foliis paucifariis stellatim patentibus, ensiformibus versus apicem gradatim attenuatis, acutis deltoideis dentibus paullum incurvatis subrigidis armatis, carnosis, subtus convexis, supra concavis, paginis lævibus atroviridibus; scapo ramoso striato compresso inferne subalato, alis obscure dentatis, ramulis 2–3 adscendentibus racemos densos gerentibus, bracteis lanceolatis membranaceis; floribus breviter pedicellatis; perianthio obclavato, segmentis lanceolatis, tubum oblongum rectum dilatatumque dimidio superantibus; staminibus perianthio subæquilongis, filamentis membranaceis alatis, basi dilatatis; capsula angulari ovoidea.

Folia $1\frac{1}{2}$ ped. longa, 3 poll. lata. *Scapus* $\frac{1}{2}$ ped. longus, ramuli 3–6 poll. longi; bracteæ $\frac{1}{8}-\frac{1}{6}$ poll. longæ; pedicelli apice articulati, primum $\frac{1}{4}-\frac{1}{2}$ poll. longi, demum $\frac{2}{3}-\frac{3}{4}$ poll. longi. *Perianthium* $\frac{2}{3}-\frac{3}{4}$ poll. longum. *Capsula* $\frac{1}{2}-\frac{3}{4}$ poll. longa.

Nom. vulg. Ananas marron.

A very common and very distinct species, characterised by the smooth non-spotted leaves and the compressed rachis of the flower scapes.

Asphodelus fistulosus, *Linn.*; *Kunth Enum.* iv. 557, var. **tenuifolius**; A. tenuifolius, *Cav.*; *Kunth En.* iv. 558.

This plant I did not find on the main island, but only upon two of the coral islets, Gombrani and Pierrots, on the southern reef.

Dracæna angustifolia, *Roxb.*; *Baker in Linn. Journ.* xiv. 526. Nom. vulg. Bois chandelle.

Very common, attaining a height of about 14 feet or more, and frequently with the aërial roots at the base very numerous.

Dracæna reflexa, *Lam.*, var. **angustifolia,** *Baker in Linn. Journ.* xiv. 531. Nom. vulg. Bois chandelle.

Not common only at the top of valley Rivière aux Huitres.

Asparagus umbellulatus, *Sieber*; *Baker in Linn. Journ.* xiv. 611. Nom. vulg. L'Asperge.

Common.

Asparagus racemosus, *Willd.*; *Baker in Linn. Journ.* xiv. 623. Nom. vulg. L'Asperge.

Common.

Two other species of *Asparagus* occur, but I obtained them only in leaf and the material is too imperfect for identification.

COMMELYNACEÆ.

Commelyna communis, *Linn.*; *Kunth Enum.* iv. 36; *C. B. Clarke Commelyn. Beng.* t. 1. Nom. vulg. Herbe à cochons.

A common plant. The leaves and young shoots are eaten as a salad.

PALMÆ.

Latania Verschaffelti, *Ch. Lem. Ill. Hort.* t. 229. Nom. vulg. Latanier.

Found in every part of the island. Perhaps no plant on the island is put to more uses than the Latanier. The wood is very hard and durable, of a rich mottled black appearance, and used for building huts, though now this is interdicted as the trees are becoming more scarce. The leaves are chiefly used for thatching huts and are also made into baskets. The fibres of the petiole form a very excellent material for cordage and the reticulum is also put to various uses.

Male and female trees exist in about equal numbers. This is the *Latania aurea* of horticulturists, and has been known in Europe for some years. The original description of the plant in Ill. Hort. l.c. errs in the description of the pyrenæ, which are represented as inverted, the apex being described as the base and *vice versâ*.

Leguat speaks of this tree as the "plantane," and thus quaintly describes it,—

"The plantane is a sort of palm tree, and the arborists place it in the same class. Our plantanes have a straight trunk, which seems to be formed of large rings at an equal distance. They have no such prickly scales as I have talked of in the palm tree. At the top of the trunk is a cabbage, very like to that of the palm tree. At the foot of this cabbage instead of palm boughs are broad leaves, with stalks about six or seven foot long; these leaves are strong and thick, and like a fan when it is open, the sticks of which come a little out of the circumference, and are pointed at the end. Some of these leaves are eight foot diameter, insomuch that they serv'd to make rare coverings for our cabbins. We cut 'em out into little pieces, and made hatts and umbrello's of them. The stalk is four inches broad, an inch thick, and a little roundish at the sides; at the bottom where it joyns to the tree, it widens and grows like a flat shell which sticks to the trunk, and in part embraces it. This wide and hollow plate is sometimes above a foot diameter, and of the thickness of a crown piece. We made use of it for dishes, plates, and sawcers. The first rind of the stalk served us instead of ropes, and the fibres of the second made good thread to sow with. One might have wove stuffs with it, had it been prepar'd.

"We cou'd not perceive any difference in the tast, or in any other qualities of the palm tree or plantane. This liquor is whitish like white whey, and so sweet, that no other sweetness, if I may judge of it, can compare to it: the newer it is, the more agreeable. In three or fuor days it begins to turn sowre, and in seven or eight, 'tis as sharp as the strongest vinegar, without changing its colour.

"The dates of the plantane are bigger than those of the palm tree. Having abundance of better things to feed on, fish and flesh, fruits, &c., we left the dates for the turtles and other birds, particularly the *Solitaires*, of which we shall hereafter make mention.

"About the cabbage of the plantane, near the bottom of it, and between the stalks of its broad leaves, is a sort of cotton of a limon colour, which all thro'

India is known by the name of *capoc*. We made very good quilts of it. It may be wove and manufactur'd for all the uses that cotton is put to. Perhaps we might have thought of making a sort of stuff, both of the capoc and the fibres of our plantane leaves; but we had stuff enough of our own to serve a long time, and the air is so mild, so sweet, that we did not make much use of our cloaths."

The genus *Latania* is Mascarene and is represented by three species. Of these one here mentioned is endemic in Rodriguez. Another, *Latania Loddigesi*, Mart. Hist. Palm. iii. 224. t. 161. f. ii. 10–14, the *L. glaucophylla* of horticulturists, is found only on Round Island near Mauritius, while the third species, *L. Commersoni*, Gmel. Syst. viii. 1035, and which includes *L. rubra*, Jacq. Frag. Bot. 13. t. 8, and *L. borbonica*, Lam. Encyc. iii. 427, is distributed in both Mauritius and Bourbon. The three species are easily distinguished specially by the pyrenæ. It is curious to note the similarity of constitution and distribution of this genus with another Mascarene one, *Hyophorbe*.

Phœnix dactylifera, *Linn*, Nom. vulg. Le Datte.

Is not abundant and occurs occasionally on some of the coral islets.

Hyophorbe Verschaffelti, *Wendl. in Ill. Hort.* tt. 462, 463. Nom. vulg. Palmiste marron.

A palm spread over the whole island, but never occurring on the coralline limestone. It is of a very striking appearance by reason of the bulging which takes place in the stem towards the middle, the stem on both sides of the swelling decreasing in size. If the tree be lofty, there may be a second ventricosity. But the tree seems rarely to attain an altitude above 20–25 feet. The external hard part of the stem is very thin, not more than an inch thick, and within it is a soft succulent mass of cellular tissue and fibro-vascular bundles. The juice from this tree is said by the inhabitants not only to be unwholesome, but even poisonous, causing, if taken in small quantities, severe emesis. The leaves have an exceedingly plumose appearance, and with their yellow stripe beneath are very picturesque. The parts of the tree are put to no use.

Hyophorbe, to which the species belongs, is a Mascarene genus represented by only three species. Of these the one here mentioned is endemic in Rodriguez. *H. amaricaulis*, Mart. Hist. Palm. iii. 309, formerly cultivated in Europe under the name *Areca speciosa*, is a second species endemic in Round Island, about 6 miles from Mauritius. This, from its habit, is not unfrequently termed in Mauritius the "bottle palm," and hence it has for long been confounded with the Chilian bottle palm, *Jubæa spectabilis*, with which, however, it has no connexion. The Rodriguez palm I should have said has also been confounded with *Jubæa*. The Round Island palm is very distinct from the Rodriguez plant. The third species is the most delicate. Originally described as *Hyophorbe indica* by Gärtn. de Fruct. ii. 186, the name *H. Commersoniana* was substituted by Martius Hist. Palm. iii. 164.

There seems, however, no sufficient ground for the alteration. Bory St. Vincent, Voy. ii. 296, mentions and describes this palm as *Areca lutescens*, under which name it is frequently and most commonly met with in gardens. The species has a wider distribution than the others, occurring in both Mauritius and Bourbon. It is in these islands confined to the shady parts of the woods and valleys and is now extremely rare. It differs in habit from the other species in having a slender tapering stem not dilated and with no ventricosities.

Dictyosperma album, *Wendl. in Linnæa* xxxix. 181. var. aureum; Areca alba, *Bory Voy.* i. 306.

Palma typo minor tenuiorque usque ad viginta pedes alta; folia 4–8 ped. longa; petiolus communis curtus, 8 poll. longus, intus paullo-convexis squamulis adspersis munitis; vagina 1–2 ped. longa; pinnæ lineari-lanceolatæ, anguste acuminatæ, 1½–2 ped. longæ, 1 poll. latæ, venulis secundariis obscuris; spadix fastigiatus ramulosus, ramulis rigidis erectis 9–11 poll. longis, omnino rectis v. ad basim tortilibus; flores illis typi dimidio minores; fructus violaceus, $\frac{2}{3}$–$\frac{3}{4}$ poll. longus, $\frac{1}{3}$–$\frac{5}{12}$ poll. latus, cylindro-conicus; semina $\frac{1}{2}$ poll. longa. Plantæ juveniles aurantiaceæ, pinnis fere linearibus, spinulis vestitæ.

Nom. vulg. Palmiste bon.

This palm is very abundant in Rodriguez, growing freely both on the volcanic soil and on the coralline limestone. It has for many years been cultivated in the gardens of Europe as *Areca aurea*.

The genus *Areca* has long been a receptacle for many species of doubtful affinity, but Wendland has recently revised the genus and removed therefrom many of its hitherto contained species. As a result of his revision the genus is unrepresented in the Mascarene islands. Some Mascarene species are referred, as already noted, to *Hyophorbe*; two others, *Areca rubra*, Bory, and *A. crinita*, Bory, constitute the endemic Mascarene genus *Acanthophœnix*, and the old *Areca alba*, Bory, is the type of a new genus, *Dictyosperma*. This is a very variable plant, and by reason of this several garden names have been given to its forms under cultivation. Thus we have *Dictyosperma furfuraceum*, *D. rubrum*, and *D. aureum*. These are, however, all varieties of the one palm, *Dictyosperma album*, Wendl., and the last mentioned is that form which occurs in Rodriguez.

Leguat mentions only one species of "palm tree" in Rodriguez, and it is difficult to determine whether he refers to *Hyophorbe* or *Dictyosperma*. He thus describes it.—

"Our palm trees are commonly 30–40 ft. high, their trunk is straight and without leaves, but 'tis cover'd with a sort of prickly scales, whose prickles stand out a little. Some have a smoother bark than others. On the top of the trunks grow those boughs of palm, of which no man ever saw a lively picture. These boughs form a great knot and fall down all about it in plumes. Below these boughs, or rather below the trunk from which they grow, are produc'd long bunches, each fruit or grain as big as a hen's egg, and of the same shape, known by the name of *dates*.

"In the center of this great knot, and at the summet of the trunk, is what we call the cabbidge. One cannot see it, being hidden by the boughs that rise a little all about it. This summet consists of tender leaves which closely embrace each other, joyn together, and form a mass something like that of a cabbage lettice, or common cabbage; 'tis about two foot high if the tree is large, and of the same bigness with the trunk. The large outside leaves of this mass are white, soft, pliable, and as strong as buff, which it resembles. They will serve also for linen, satin, for napkins, table-cloths, and any thing what you please. The membrana, or inside leaves, are tender and brittle, like the heart of a lettice. They are good to eat raw, and tast like a filberd; but we made an admirable ragout of them when we fricasi'd them with the fat and liver of a turtle. We put them also in our soops.

"We come now to the liquor, or rather the nectar of the Isle of Rodrigo. 'Tis call'd palm-wine all over the Indies. There are two ways of drawing out the juice. We make a hole in the trunk of the tree at about a man's height, as big as one's two fists. We presently put a pot or other vessel there to receive the delicious liquor, which runs out fast enough, otherwise we dig the cabbage, and make a little cistern at its head. We need only go twice or thrice a day and draw this rare wine at the fountain head, and we may be abundantly supply'd with it. The wine of the trunk, and that of the cabbage, are in my opinion of an equal goodness.

But those who would be good husbands of their trees (for as for us we were lavish enough of them), the first way is the best, because after the cabbage has yielded its liquor for about a month, it withers, and the tree decays and dies. 'Tis the same thing if you tear off the cabbage, when its head and brain are gone it dies almost suddenly.

"Whereas if you only pierce its side the tree do's not die, provided the wound is not too deep; but the liquor will not run out at that hole above four days. The wounded tree must afterward have time to recover. I do not know what is done elsewhere, but I can tell by experience what I have said here, we having made trial of it daily for two years together. The bark of this tree is very hard, 'tis an inch thick, porous and tender in the inside. If one make the hole on the side of the trunk too wide, there is reason to fear 'twill weaken the tree there, and that then the next hurricane will break it."

"The fruit or grain as big as a hen's egg" makes it difficult to refer his palm to either *Hyophorbe* or *Dictyosperma*, and points rather to the Latanier. And then the trunk "covered with a sort of prickly scales whose prickles stand out a little," does not agree well with either, and least with *Hyophorbe*. But then he says afterwards, "The bark of this tree is very hard, 'tis an inch thick, porous and

3 B

" tender in the inside," which more nearly corresponds with *Hyophorbe* than with *Dictyosperma*. But, on the other hand, the inhabitants say the juice of *Hyophorbe Verschaffelti*, Ch. Lem., is poisonous, whilst Leguat and his companions seem to have used it much. It is then a difficult matter to decide, but if we discard the statement as to the size of the fruit, I think we may most safely consider that Leguat failed to diagnose two species of Palm tree, and in his description includes both species. And as pointing in this direction I think we may take his remark " Some have a smoother bark than others ; " for the bark of *Hyophorbe Verschaffelti*, Ch. Lem. is much smoother than that of *Dictyosperma album*, Wendl.

Cocos nucifera, *Linn*.

A few trees of this occur in the compound of the Government House at Port Mathurin, and one or two trees are found on the south side of the island. The introduction of this plant was on this wise as described by Leguat :—" The sea " having thrown us up some Cocos which began to bud, we planted some of that " fruit some months after our arrival, and when we left the place, the trees were " four foot high."

PANDANACEÆ.

Pandanus heterocarpus, *Balf. fil.*

Arbor ramosa, caudice lævi, radicibus aëreis plurimis, ramis patentibus ; foliis lanceolatis acuminatis, basi amplexicaulibus dilatatis, erecto patentibus, coriaceis, strictis, viridescentibus basi sæpe glaucescentibus, subplanis, marginibus per totum costaque subtus a medio distanter rubrospinosis, spinulis subincurvatis ; ♂ spadicibus laxe racemosis odoris 15-spicatis, spathis subito acuminatis, carinatis, basi latis, spicas trigono-oblongas excedentibus, marginibus costaque subtus per totum spinescentibus ; staminibus laxe dispositis, filamento communi brevi, 5–10 ramoso, ramulis patentibus, antheris lineari-oblongis mucronatis ; ♀ capitulis solitariis globosis subdepressis v. elliptico-oblongis nunc longe nunc breviter pedunculatis, pendulis v. inclinatis, spathis pluribus brevibus deciduis ; drupis 5-locularibus, obpyramidalibus, non compressis, 5–6-gonis, fere per totum coadunatis, apice humiliter pyramidale rarissime convexo sæpius applanato v. umbilicato, distincte 5–6 angulato, areola centrali 2–5 stigmata plana sessilia reniformia suberosa cingente instructo.

Arbor umbraculiformis, usque ad 20 pedes alta. *Caudices* fuscobrunnei, 5–7 poll. diam., læves, erecti, oblique annulati, ramosissimi, ramulis terminalibus non adscendentibus, radices aereas plures seriatim spinescente-tuberculatas emittentes. *Folia* ensiformia, recta v. rarius recurvata v. dependentia, 1½–3 ped. longa, 1½–2½ poll. lata, supra non sulcata, non reduplicata, pallide v. atro-virentia, supra nitida, subtus opaca venulis lateralibus subtiliter striata, marginibus incarnatis a basi spinosis, spinis sæpe in medio deficientibus acutis rubris subincurvatis adscendentibus, costa subtus pallidiori subrosea a medio spinescente, spinis in margine majoribus intervalloque longiore dispositis. *Spadices* masculi 1 ped. longi 15-spicati penduli, spicæ basales

3 poll. longæ, spathæ sursum sensim minores cymbiformes cuspidatæ abrupte angustateque acuminatæ basi submembranaceæ latæ, spicas amplectentes pollibicus pluribus superantes, marginibus carinaque subtus per totum spinulescentibus. *Staminum* filamentum commune $\frac{1}{6}$ poll. longum, ramuli singuli breves patentes $\frac{1}{24}$ poll. longi; antheræ albidæ $\frac{1}{8}$ poll. longæ. *Capitula* 60–70-drupata, nunc 35, rarius 100, $4\frac{1}{2}$–$6\frac{1}{2}$ poll. longa, $4\frac{1}{2}$–$5\frac{1}{2}$ poll. lata; pedunculus trigonus plerumque 8–12 poll. rarius 3 v. 15 longus, $\frac{7}{12}$–$\frac{11}{12}$ poll. crassus. *Drupæ* $\frac{11}{12}$–$1\frac{1}{6}$ poll. longæ, 1–$1\frac{1}{2}$ poll. latæ horizontaliter, $\frac{3}{4}$–1 poll. in diam. verticaliter, usque ad $\frac{5}{6}$–$\frac{7}{8}$ a basi coadunatæ, parte conjuncta maturitate rubra v. flava, apicis margine nonnunquam ruguloso; areola centrali distincte marginata; stigmatibus $\frac{1}{8}$–$\frac{1}{4}$ poll. diam. reniformibus, pelvis centrum versus spectante; putamine rotundato lateribus lævibus apice applanato processibus 2–5 instructo; mesocarpio spongioso, amplo; semine $\frac{1}{3}$ poll. longo.

Nom. vulg. Vacoa calé rouge, V. calé blanc, V. sac, V. poteau, V. parasol, V. mâle.

It is a very variable tree, and the popular names indicate this. It occurs very abundantly everywhere on the island from the seashore to the highest points. And according to its situation, its habit and appearance vary. Thus when on sites exposed to the wind it has a stunted habit. The branches are few, thick, and short, and the leaves are also short and are erect. In such situations it is known as the *Vacoa calé*. The inhabitants make a distinction between two varieties of this according as the head, peduncle, and united parts of the drupes are red, or are greenish yellow or yellow. The former they style *rouge* the latter *blanc*. The first of the characters on which they base their varieties by the colour of the head and peduncle is worthless, and the difference in colour of the bases of the drupes is found in all the forms of the species, but is not sufficient to characterise varieties. If the tree grows in suitable soil, and in a sheltered position where it has room to develop its branches properly, then it forms a dense and compact dome and the branches may droop downwards, so far as almost to conceal the stem, and is then known as *Vacoa parasol*.

When in any situation the tree develops a trunk of good size, and is allowed to grow until the wood is hard and firm to the centre, and is capable of being used as a post for a house, then the tree is called *V. poteau*. The name *V. sac* is given to the young plants when the leaves are long and broad, and may be made into bags or sacs. *V. mâle* is of course the male tree, known by the inhabitants as not bearing fruit.

It is not only in habit that the tree varies, but few plants exhibit such an amount of variation in the fruit and it well deserves its specific designation. It is impossible here to enter into a consideration of these which will form the subject of a separate memoir. This species belongs to the *Utilis* group of the species of *Pandanus*. It has a close affinity with *P. utilis*, but is sufficiently distinguished in habit and in fruit.

Leguat gives a curious description of the Screw pines of Rodriguez as follows:—
" Among the great number and variety of trees in this island planted by nature,
" there is one which is wonderful and worthy our particular observation for its
" beauty, bigness, roundness, and the rare symmetry of its exact branches, the ends
" of which are very much tufted, and its leaves so great and thick that they fall
" down almost to the ground all about it, so that come which way you will at this
" tree, you can perceive but a small part of its trunk, and that at the bottom of it,
" and sometimes you can see nothing at all of it. It being as one may imagine, all
" shady in the middle; the branches are within-side like dry poles, which seem to
" be the work of a carpenter, and set there to bear up the plumes or branches which
" are quite about it, and thus make a sort of cage or tent of the tree. 'Tis true, the
" greatest beauty of this tent, is in its charming outside, though the coolness and
" shelter of the inside have also their charms: 'twas unhappy that its fruit was not
" good to eat. Those of us who had the curiosity to eat it, found it sour, and knew
" by experience that was all the hurt that was in it. It had the smell of a very
" fine quince. 'Tis a sort of a grape, the seeds of which are close and altogether;
" it looked at a distance like the fruit of the ananas, for which reason we us'd to
" call these trees Ananas, tho' there's a great difference between the two plants:
" as for me I was calling it the *pavilion* or tent. The leaves are of an admirable
" green, and the stalks of them are so short, that one wou'd think they grew imme-
" diately to the wood. The greatest are four or five inches broad, sharp at the end,
" and about five inches long. They form a great bunch, and here and there one
" may see the grapes, which are of divers colours, according as they are more or less
" ripe. I have often taken pleasure to survey these natural palaces, and was
" equally ravish'd with its largeness and singular beauty."

Pandanus tenuifolius, *Balf. fil.*

Arbor parva, caudice laevi, radicibus aëreis paucis, ramis erectis crassis; foliis angustate lineari-lanceolatis acuminatissimis, apice subflagellatis, basi dilatatis lateque amplexicaulibus, coriaceis recurvatis pendulisque, subplanis atro-viridibus saepe basi dealbatis, marginibus subincrassatis per totum costaque a medio rubro-spinosis, spinis brevibus acutis incurvatis adscendentibus; capitulis solitariis pauci-drupatis subglobosis, pedunculo curvato pendulosis; spathis pluribus deciduis; drupis subpyriformibus 2-5-locularibus, non compressis, usque ad $\frac{2}{3}-\frac{3}{4}$ a basi coadunatis, parte conjuncta 5-6-gona obpyramidali, apice libero rubiginoso rotundato umbraculiformi obscure 5-6-angulato, vertice applanato v. depresso, areola marginata 2-5 stigmata plana sessilia reniformia cingente instructo.

Arbor usque ad 15 pedes alta plerumque magnum spatium instar arbustum occupans, rare solitaris. *Caudices* fusci, laeves, 3 poll. diam., ramos erectos validos abrupte terminantes angulo acuto emittentes. *Folia* 8 poll.—$2\frac{1}{4}$ ped. longa, $\frac{1}{4}-\frac{7}{12}$ poll. lata, supra non sulcata, subplana, non reduplicata, supra nitida, infra opaca, lateralibus nervulis subtiliter striata, marginibus subincrassatis a basi spinosis, spinis sæpe in medio deficientibus, rubris, arcte dispositis, incurvatis v. subappressis, adscendentibus, costa prominente rubra, a medio spinis distantibus acutis longioribus instructa. *Capitula* 20–40-drupata, $3-5\frac{1}{4}$ poll. diam.; pedunculus trigonus, curvatus, 5–11 poll. longus, $\frac{1}{3}-\frac{5}{12}$ crassus. *Drupa* breviter pyriformes, $1\frac{1}{2}$ poll. longæ, parte superiore $\frac{1}{3}-\frac{1}{4}$ libera, horizontaliter $1\frac{1}{4}-1\frac{1}{2}$ poll. latæ, verticaliter $1\frac{2}{3}-1\frac{5}{6}$ poll. latæ, parte conjuncta maturitate rubra; stigmatibus $\frac{1}{8}-\frac{1}{3}$ poll. diam., reniformibus, pelvi centrum versus spectante; putamine rotundato lateribus lævibus apice depresso 2–6 processus gerente; mesocarpio crasso spongioso; seminibus $\frac{1}{3}$ poll. longis.

Nom. vulg. Vacoa chevron.

A small and very graceful tree, usually growing in clusters and forming frequently a very regular dome. It is confined to the higher parts of the island and the upper parts of valleys. Though nearly allied to, it is easily distinguished from the foregoing by its arching narrow delicate leaves, and its few large druped fruits. It is also a very much smaller plant. But certain trees on the island seem to possess characters intermediate between these two species and may be hybrids. This is, however, a difficult matter to settle. Its local name is derived from its common use as a rafter for huts, for which its close-grained and hard wood when mature well fits it.

These are the only two species I found on the island, but Sir Henry Barkly has sent home figures of two (and of one a specimen) other species from the island. These are *P. utilis*, Bory, and *P. odoratissimus*, Linn. f. His figures and specimen of the former are most distinctly those of *P. utilis*, Bory, but whilst I do not dispute the fact that this plant may have existed or may exist on the island, I may notice that Sir Henry Barkly's drawing and specimen are of a tree grown in the Botanic Garden in Mauritius from seeds which Mr. Duncan is said to have brought from Rodriguez. I think this takes away considerably from the value of the observation, especially as the trees Mr. Horne pointed out to me as from Rodriguez are growing closely surrounded by Mauritius ones. At the same time I may remark that in the young state, before many branches have been given off, *P. utilis*, Bory, and *P. heterocarpus*, Balf. fil., are so much alike as to be almost undistinguishable, and I may have in that way missed *P. utilis*, Bory, in Rodriguez. As to *P. odoratissimus*, Linn. f., I can only say I never saw it, and it must be very scarce, if there at all; and it seems to me very strange that such a very marked form should be unknown by the inhabitants, especially as several of them were employed by Mr. Jenner, the magistrate of the island, at the request of Sir Henry Barkly, to collect Pandani for him. Mr. Horne showed me trees of this in the Botanic Gardens at Mauritius, said to have grown from seeds brought from Rodriguez by Mr. Duncan, and as this may have been the source of Sir H. Barkly's figures the same objection may be urged as in the case of *P. utilis*, Bory. It seems to me very strange that no mention is made by Sir H. Barkly of *P. tenuifolius*, Balf. fil., from Rodriguez, which is very abundant, whilst *P. utilis*, Bory, and *P. odora-*

tissimus, Linn. f. must be very scarce, if there. Bojer in his Hortus Mauritianus says of *P. muricatus*, Pet. Thouars, that it grows naturally at Rodriguez. I believe Bojer visited Rodriguez, and he ought to have known the Pandani, but he says nothing about *P. heterocarpus*, Balf. fil. or *P. tenuifolius*, Balf. fil., which are so common on the island. He appends to his notice of *P. muricatus*, Pet. Thouars, that some trunks are to be found at Grand Port. To recognise a species of *Pandanus* by leaves, fruit, and stem is in many cases difficult enough, but to discriminate species by stems only is very fallacious, if not impossible. I am convinced the species does not now exist on the island; and, indeed, Petit Thouars only recorded it from Madagascar. In the Botanic Gardens in Mauritius a species was shown me by Mr. Horne which he supposed might be *P. muricatus*, Pet. Thouars, but this is merely a form of *P. utilis*, Bory, with drupes much divided at the apex.

AROIDEÆ.

Colocasia antiquorum, *Schott Prod.* 138; Arum colocasia, *Linn.*; *Wt. Ic.* t. 786; Caladium esculentum and nympheæfolium, *Vent.* Nom. vulg. Songe rouge.

Grows very abundantly in the streams. Is eaten in times of scarcity by the inhabitants, but is said to produce a severe form of skin eruption if much eaten. It is chiefly used to feed pigs when they can get nothing else.

Alocasia macrorhiza, *Schott Prod.* 146; Arum macrorhizum, *Linn.*; *Boj. Hort. Maur.* 356. Nom. vulg. Songe blanc.

Grows abundantly with the former, and is the better to eat of the two.

NAIADACEÆ.

Ruppia maritima, *Linn.*; *Kunth Enum.* iii. 123; R. rostellata, *Koch*; *Reich. Ic.* ii. 66. t. 174.

Abundant at the mouth of English Bay River.

Zannichellia palustris, *Linn.*; *Kunth Enum.* iii. 124.

Abundant, especially in Rivières Bouteille and De l'Est. This plant is not recorded from Mauritius.

Halophila ovalis, *Hook. fil. Fl. Tasman.* ii. 45; H. ovata, *Gaud. in Freyc. Voy.* 429. t. 40. f. 1.

Grows abundantly on the sandy flats on the reefs, and is left uncovered at the fall of the tide.

Halophila stipulacea, *Aschers. in Linnæa* xxxv. 172; Thalassia stipulacea, *König*; *Kunth Enum.* iv. 120.

Abundant on the sandy flats on the reefs, specially where there is a current, and does not seem to be left uncovered by the tide.

CYPERACEÆ.

Cyperus lævigatus, *Linn.;* var. albidus, *Beklr. in Linnæa,* xxxv. 487.

Very common species, occurring under several forms, according to situation.

Cyperus polystachyus, *Rottb.; Kunth Enum.* ii. 13; *Beklr. in Linnæa,* xxxv. 477.

Not uncommon on the island.

Cyperus dubius, *Rottb.; Beklr. in Linnæa,* xxxvi. 336. Nom. vulg. Herbe à oignon.

Common everywhere, and is a great pest.

Cyperus distans, *Linn. fil.; Kunth Enum.* ii. 93; *Beklr. in Linnæa,* xxxv. 612.

Frequently found.

Cyperus, sp.

A species without flower I found close to Gabrielle village.

Kyllingia monocephala, *Rottb.; Beklr. in Linnæa,* xxxv. 427.

Very common.

Fimbristylis glomerata, *Nees; Beklr. in Linnæa,* xxxvii. 47.

Common, especially on coralline limestone.

Fimbristylis communis, *Kunth Enum.* ii. 235.

Common.

Carex gracilis, *R. Br.; Kunth Enum.* ii. 513; *Boott Car.* i. 59. t. 154–156. Nom. vulg. Herbe Madame Pitelle.

Very common.

GRAMINEÆ.

Coix Lachryma, *Linn.; Kunth Enum.* i. 20; *Bot. Mag.* t. 2479. Nom. vulg. Larme de Job.

Very abundant, especially near habitations on the banks of streams.

Zea Mays, *Linn.; Kunth Enum.* i. 19; *Mart. Flor. Bras.* ii. 2. t. 11.

Cultivated, but not so much as it might be. Is exported to Mauritius. A good head is supposed to yield 640 grains, and each plant bears usually two heads. In planting three seeds are put in each hole, and five heads are expected from the three plants.

Paspalum scrobiculatum, *Linn.; Kunth Enum.* i. 53; *Trin. Sp. Gram.* ii. t. 143. Nom. vulg. Herbe filé.

Very common.

Panicum brizoides, *Linn.; Kunth Enum.* i. 78; *Trin. Sp. Gram.* ii. tt. 158–9.

Common.

Panicum fluitans, *Retz.; Kunth Enum.* i. 78.

Panicum maximum, *Jacq. Ic.* t. 13.

Panicum plicatum, *Lam.;* var. **P. costatum,** *Roxb. Fl. Ind.* i. 314; *Kunth Enum.* i. 94.

Commonest grass on the island.

Panicum serpens, *Kunth Enum.* i. 117.

Panicum Balfourii, *Baker Fl. Maur. Seych.* 438.

Frequently met with in the valleys. Closely allied to *P. undulatifolium,* Ard.

Stenotaphrum complanatum, *Schrank.; Kunth Enum.* i. 137.
Common.

Stenotaphrum subulatum, *Trin. Sp. Gram.* iii. t. 360.
Only growing on Gombrani and Pierrot islets.

Cenchrus echinatus, *Linn.; Kunth Enum.* i. 166. Nom. vulg. Herbe cateaux.
Is widely dispersed.

Zoysia pungens, *Willd.; Kunth Enum.* i. 471.
Very common on the sandy and coral islets on the reef.

Saccharum officinarum, *Linn.; Kunth Enum.* i. 474; *Hook. Bot. Misc.* t. xxvi.
Formerly much cultivated. It grows well, but is not much cultivated now, on account of the difficulties in manipulation from want of water.

Andropogon contortus, *Linn., Kunth Enum.* i. 480.

Andropogon muricatus, *Retz.; Kunth Enum.* i. 505. Nom. vulg. Vetivert or Cuscus.

Grows in abundance on the slopes of Mount Malartic, and the roots used much by the natives for scenting their cabinets and apparel.

Andropogon Schœnanthus, *Linn.; Kunth Enum.* i. 493. Nom. vulg. Citronelle.
An infusion of this is a very favourite tisane. It grows abundantly on the slopes of Mount Limon.

Andropogon foliatus, *Steud. Syn. Gram.* 389. Probably a variety of *A. Schœnanthus.*

Andropogon finitimus, *Hochst.; Steud. Syn. Gram.* 385.

Andropogon halepensis, *Sibth.; Kunth Enum.* 502. Nom. vulg. Petit mille.
Cultivated.

Sporobolus virginicus, *Kunth Enum.* i. 210.

Aristida adscensionis, *Linn.; Kunth Enum.* i. 190.

Cynodon Dactylon, *Pers.; Kunth Enum.* i. 259. Nom. vulg. Chiendent.
On the shores where sandy, and on most of the coral islets.

Eleusine indica, *Gärtn.; Kunth Enum.* i. 272; *Trin. Sp. Gram.* i. t. 71.
Everywhere on the island.

Chloris barbata, *Sw.; Kunth Enum.* i. 264; *Trin. Sp. Gram.* iii. t. 306. Nom. vulg. Esquine.

Very common.

Phragmites communis, *Trin.; Kunth Enum.* i. 251.

On the banks of the stream in valley Rivière Grande des Bamboux.

Bambusa, sp.

Abundant on the banks of the streams.

LYCOPODIACEÆ.

Lycopodium Phlegmaria, *Linn.; Spring. Monog. Lycop.* i. 63.

Not common. Only at the top of the valley of the Rivière aux Huitres.

Psilotum triquetrum, *Sw.; Boj. Hort. Maur.* 423.

Very common.

Selaginella Balfouri, *Baker Fl. Maur. Seych.* 522.

Caulibus decumbentibus dense cæspitosis ad 3–4 poll. procumbentibus, ramulis compositis plurimis curtis adscendentibus instructis; foliis majoribus obliquis, oblongis, obtusis, patentibus, $\frac{1}{16}$ poll. longis, albo-viridibus, membranaceis, ciliatis, summis marginibus conniventibus; foliis minoribus obovatis, mucrone longa armatis; spicis $\frac{1}{4}$–$\frac{1}{2}$ poll. longis, tetragonis; bracteis uniformibus, deltoideis, cuspidatis.

The specimens from which this species is described are from two localities. On the coralline limestone growing along with *Dichondra repens*, Forst. and *Hypoestes inconspicua*, Balf. fil., a plant of which the leaves have a reddish and dried appearance was found. Other specimens of a pale green and fresher look were obtained in the valley Rivière Grande des Bamboux (Cascade).

Baker l. c. considers the plants from both localities one species as above described. Professor MacNab informs me that he believes the two forms belong to different species, and he thus describes them:—

Form 1. Growing on coralline limestone. Large leaves *ovate, rather obtuse,* $\frac{35}{1000}$ *long,* $\frac{20}{1000}$ *broad, margin serrated, midrib central, base wide, lower side rounded not produced, upper side produced not so short, denticulate auricle;* small leaves $\frac{25}{1000}$ long, $\frac{12}{1000}$ broad, contracted below, widest above middle, then narrowed into an acute point, margin of leaf denticulate, base with lower side produced into a long denticulate auricle, inner margin with a very minute auricle.

Form 2. Growing in moist localities in valley Rivière Grande des Bamboux (Cascade). Large leaves elongate, widest in middle, narrowed towards each end, apex rather blunt yet pointed, $\frac{54}{1000}$ long, $\frac{25}{1000}$ broad, *upper side of leaf with a rounded outline, lower side nearly straight and less serrated, vein nearer lower than upper margin, base auricled, upper auricle long curved and almost ciliate, lower ciliate and denticulate;* small leaves $\frac{45}{1000}$ long and $\frac{15}{1000}$ broad, contracted below, widest above middle, then narrowed into an acute point, margin denticulate, base with outer margin produced into a large denticulate auricle, inner margin with a very minute auricle.

Selaginella rodriguesiana, *Baker Fl. Maur. Seych.* 523.

Caulibus decumbentibus, non-articulatis, ad pedem procumbentibus, ramulis compositis plurimis distantibus adscendentibus instructis; foliis majoribus obliquis, oblongis, obtusis, falcatis, $\frac{1}{12}$ poll. longis, viridibus, membranaceis, integris obscure ciliatis, summis marginibus conniventibus; foliis minoribus obovatis, longe mucronatis; spicis $\frac{1}{4}$–$\frac{1}{2}$ poll. longis, tetragonis; bracteis uniformibus, deltoideis, valide cuspidatis.

Very abundant. A species not far removed from *S. concinna*, Spring. Monog. Lycop. ii. 199. Professor MacNab says of this plant, "the form of the large leaves "is like *S. concinna*, Spring., but the expression 'foliis rigidis longe biauriculatis' "can hardly apply, the leaves being soft, thin, and with very small auricles. Then "the small leaves are much more oblique, and by no means 'aristato-mucronatis.' "The species comes near one I have from Kew Gardens cultivated as *S. concinna*, "but locality unknown to me."

OPHIOGLOSSACEÆ.

Ophioglossum reticulatum, *Linn.; Hook. & Baker Syn. Fil.* 446.
Amongst the grass on the slopes in a few of the valleys.

FILICES.

Trichomanes cuspidatum, *Willd.; Hook. & Baker Syn. Fil.* 73.
On the rocks at the sides of streams, only near their source. A Mascarene plant. Not very abundant.

Davallia mauritiana, *Hook. & Baker Syn. Fil.* 96.
A Mascarene species common on the dry slopes of the valleys.

Adiantum caudatum, *Linn.; Hook. & Baker Syn. Fil.* 115. Nom. vulg. La capillaire.
Very common everywhere. An infusion of this is used as a tisane in fevers.

Adiantum Capillus-Veneris, *Linn.; Hook. & Baker Syn. Fil.* 123.
Not common only in the valleys of Rivière Cascade and Rivière Poursite, a few plants growing on moist rocks.

Adiantum hispidulum, *Swartz; Hook. & Baker Syn. Fil.* 126.
Common in the upper parts of valleys.

Pellæa (Platyloma) hastata, *Link.; Hook. & Baker Syn. Fil.* 152.
Very abundant everywhere.

Pteris flabellata, *Thunb.; Hook. & Baker Syn. Fil.* 161.
Not very common. Grows in shady places in the upper parts of valleys.

Asplenium (Thamnopteris) Nidus, *Linn.; Hook. & Baker Syn. Fil.* 190. Nom. vulg. Langue de bœuf.

Common on trees in the valleys.

Asplenium hirtum, *Kaulf.; Hook. & Baker Syn. Fil.* 205.

Common on the banks of streams.

Asplenium falcatum, *Lam.; Hook. & Baker Syn. Fil.* 208.

Not common. Growing on dry stony slopes of valleys.

Asplenium furcatum, *Thunb.; Hook. & Baker Syn. Fil.* 214.

Common on banks of streams.

Asplenium (Anisogonium) decussatum, *Sw.; Hook. & Baker Syn. Fil.* 243.

Only found at the foot of Cascade Victoire.

Nephrodium (Lastrea) crinitum, *Desv.; Hook. & Baker Syn. Fil.* 265.

A Mascarene species abundant in all the valleys near the bed of the stream.

Nephrodium unitum, *R. Br. (non Sieb.); Hook. & Baker Syn. Fil.* 289.

Grows in large patches on the dry slopes of some valleys.

Nephrodium molle, *Desv.; Hook. & Baker Syn. Fil.* 293.

Very abundant in all valleys.

Nephrodium elatum, *Baker; Hook. & Baker Syn. Fil.* 502.

Only at the foot of the Cascade Victoire with *Asplenium decussatum*.

Nephrolepis exaltata, *Schott.; Hook. & Baker Syn. Fil.* 301.

Everywhere.

Nephrolepis acuta, *Presl.; Hook. & Baker Syn. Fil.* 301.

Everywhere. This and the foregoing species in many places cover the ground for several acres, notably on the eastern side.

Polypodium (Niphobolus) adnascens, *Sw.; Hook. & Baker Syn. Fil.* 349.

Abundant on dry rocks near the mouths of the valleys.

Polypodium (Phymatodes) phymatodes, *Linn.; Hook. & Baker Syn. Fil.* 364. Nom. vulg. Polypod.

Everywhere abundant. An infusion of the rhizome is an excellent tisane for cough. Mr. Horne showed me in the Herbarium of the Botanic Gardens, Mauritius, specimens of a crested form of this from Rodriguez.

There are specimens of a *Nephrodium (Lastrea)* grown by Lady Barkly at Reduit, which is said to have come originally from Rodriguez, in the Herbarium at the Botanic Gardens, Mauritius. I do not know what species, and I did not gather it.

In a letter to me at Rodriguez, Mr. Horne mentions that in 1864 a collector brought from Rodriguez *Nephrolepis splendens*, but there must be some mistake as to the name, as there is no such species.

Lindsaya acutifolia, Desv., is also reported from Rodriguez, but I have not seen specimens.

MUSCI.

By William Mitten, A.L.S.

The number of the mosses as yet known to occur in the island of Rodriguez is thirty-three, of these ten were gathered in a fertile state, the remaining twenty-three being barren. With one exception all the genera are the same as those known from Africa or the Mascarene Islands, and thirteen of the species are identical with those found in those countries. Three species are found also in India and the Indian Archipelago, and seventeen are considered to be new, of these eleven are without fruit, and six fertile.

The most interesting moss is the *Orthotrichum plicatum*, Beauv., before known only from the specimens gathered in Bourbon by Du Petit Thouars, described and well figured by Schwægrichen; it is undoubtedly a member of the family of the *Orthotricha*, but its characters as a genus are intermediate, and its position in that family unique. A few species correspond with those from the western coast of tropical Africa, and it appears from what little is yet known of the mosses of the eastern coast that some species have an enormous range in the African continent.

Cosmopolite mosses are represented solely by *Weisia controversa*.

Octoblepharum albidum, *Lin.* (Bryum).

Not a common species. Found on decayed branches of trees and vegetable humus.

Calymperes (Hyophilina) lævifolium, *Mitt.*

Caulis ascendens ramosus. *Folia* dense inserta e basi erecta dilatata late obovata margine minute crenulata intus cellulis hyalinis areolata exinde patentia lineali ligulata apice lato acuta subapiculatave margine incurvo canaliculata integerrima cellulis minutis sublævibus obscuris et ad margines folii basis erectioris limbum formantibus descendentibus areolata, vitta intramarginalis nulla. Folia anomala longiora erecta attenuata apice propagulis glomerulatis coronata.

Stems about half an inch high, with the leaves a line and a half wide. The uppermost leaves green, the lower all brown, about a line long, when dry, curved and loosely contorted.

Not uncommon.

In its leaves this moss comes nearest the South American *C. Richardi*, C. Mull. and *C. parasiticum*, Sw., and differs from the African *C. Afzelii*, Sw. in the absence of the marginal vitta.

Calymperes pallidum, *Mitt.*

Caulis brevis subsimplex. *Folia* a basi parum latiore erecta cellulis hyalinis areolata margine superiore crenulata exinde patentia oblongo-lanceolata, cellulis rotundatis vix obscuris papillis brevibus aspera, nervo crassiusculo concolori intus extusque scabro percursa, vitta angusta sub-intra-marginali lutescente circumdata. Folia anomala angustiora ex apice nervi glomerula corpuscularum viridium fusiformium ferentia.

Stems about two lines high. Leaves a line long, pale glaucous green when dry, loosely incurved and contorted.

Frequent.

In the structure of the leaf nearest to *C. varium* from Ceylon and Moulmein, in the glaucous appearance of the upper cells like *C. Dozyanum*, Mitt. (*C. Moluccense*, Bryol. Jav. t. 37; but not of Schwægrichen), so widely spread in the Indian Archipelago and Pacific Islands.

Weisia (Euweisia) controversa, *Hedw. Fund. Musc. Frond.* iii. 12 t. 5.

Everywhere on the banks of streams.

Weisia (Hyophila) Roscheri, *Lorentz.*

Found with other species of Weisia.

Small barren stems, which agree pretty well with others from Zanzibar. The species is very similar to *W. (H.) involuta*, so common in India.

Weisia (Tortularia) incerta, *Mitt.*

Caulis brevis ramosus. *Folia* patentia patenti-recurvaque lineali ligulata apice obtusa, nervo crasso rufescente in mucronem brevem excurrente, marginibus superioribus inflexis integerrimis, cellulis superioribus rotundatis obscuris basalibus paucis oblongis parvis rectangulatis hyalinis.

Stem a quarter of an inch high. *Leaves* one line long, dark green, when old of a rusty brown colour.

Common on banks of streams.

This appears to be a moss very nearly resembling *W. navicularis*, Mitt. Musc. Austr. Amer. 139, but it is a species of which too little is known to render its position here well ascertained.

Tortula (Planbelia) mutica, *Mitt.*

Humilis, subsimplex. *Folia* erecto-patentia ovato-oblonga obtusissima, nervo valido concolori sub apice abrupto dorso superne aspero inferne lævi carinata, integerrima, cellulis superioribus parvis fere obscuris inferioribus basin versus rotundo-quadratis parvis pellucidis.

Stems two lines high, red. *Leaves* one third of a line long, glaucous green, when dry contorted.

Along with species of *Weisia*, but more scarce.

Very near to *T. orientalis*, Willd., and to *Pottia (Hyophila) papillinervis*, Lorentz, from Zanzibar, which is doubtless a species of this group, being very similar to *T. orientalis* in size, habit, and in the structure of its leaves.

Leiomitrium, genus nov.

Caulis repens ramis fertilibus erectis. *Theca* exserta. *Peristomium* duplex externum octodentatum (dentibus bigeminatis) internum octociliatum. *Calyptra* campanulata plicis nullis.

L. plicatum, *Beauv. Prodr.* 81; *Schwægr.* t. 52 (Orthotrichum).

Not uncommon.

This curious and interesting species has been well figured and described by Schwægrichen, its habit is that of a *Macromitrium*, its leaves like those of the species of *Zygodon* which are allied to *Z. viridissimus*, and its capsule and peristome as in *Ulota*; there is, however, no species of *Orthotrichum* or *Ulota* to which it has any resemblance, except in its capsule. In the non-plicate calyptra this genus agrees with *Schlotheimia*, without, however, having any other near affinity or resemblance.

Macromitrium astroideum, *Mitt.*

Caulis repens prostratus, foliis e basi patente divergentibus apice incurvis hastato-lanceolatis acuminatis apice angustis cellulis inferne oblongis nonnullis curvulis superioribus rotundatis obscuris areolata, rami dense dispositi breves erecti densifolii, foliis linealibus e basi suboblonga erecta cellulis pellucidis angustis oblongis paucis curvatis papillis elevatis intus asperis, exinde patentibus apice obtuse acutis incurvis cellulis rotundis densis obscuris vix papillosis areolatis costa superne concolori inferne flavescente percursa carinatis.

Stems creeping in a centrifugal manner with closely set branches about a line high. *Leaves* one line long, when dry curled and contorted, forming globular tufts.

Rare. Only at the top of Oyster River valley, trailing over boulders.

This appears to be near *M. fimbriatum*, Schw., but has narrower leaves.

Macromitrium aciculare, *Brid.; Schwægr.* t. 111.

Occurs more commonly than the foregoing.

Stems all barren, but agree with Mauritian specimens. It appears to be one of the most abundant amongst a group of very closely allied species found in Mauritius.

Bartramia (Philonotis) pungens, *Mitt.*

Dioica. *Caulis* erectus ascendensve late cæspitosus gracilis rufus infra perichetium ramis plerumque ad unam lateram decurvis ramosus. *Folia* laxe imbricata erecto-patentia lanceolata sensim angustata acutissima costa angusta dorso denticulata percursa margine parum recurva serrulata e cellulis oblongis laxis pellucide areolata, perichætialia e basi dilatata ovata subulata nervis lævioribus. *Theca* in pedunculo rubro sæpe sinuoso parva globoso-pyriformis inclinata sicca plicata operculo convexo breviter mamillata peristomio normali. *Flos masculus* gemmiformis rufus foliis perigonialibus e basi dilatata excavata vaginante sensim angustatis elongatis remotius serrulatis.

Stems half an inch or more high, with the leaves half a line wide. *Leaves* when dry appressed, rather soft in substance, the upper of a pale sub-glaucous green, the lower pellucid, with the nerves and margin pale brown. *Seta* three-quarters of an inch high, shining. *Capsule* when dry and deoperculate curved and plicate, of a red brown colour.

On moist rocks near the sources of all streams.

Very nearly resembling the Ceylon *B. imbricatula*, Mitt. Musc. Ind. Or., but with narrower and more lax areolation. The male flower is like that of *B. rufiflora*, Hornsch. in Mart. Flor. Brasil.

Plate 37 A. 1. Plants of the natural size. 2. Leaves. 3. Portion of perichætium. 4. Capsule mature with operculum. 5. Older collapsed capsule. 6. Male flower: all magnified.

Bartramia (Philonotis) flaccidifolia, *Mitt.*

Gracilis, subsimplex. *Folia* caulina laxe inserta apicalia subsecunda planiuscula late lanceolata acuta nervo dorso lævi in acumen pungentem dissoluto margine parum recurva serrulata, cellulis oblongis laxis pellucidis areolata. Folia in ramulorum inferiorum nonnulla parva obtusa.

Found with the foregoing.

Larger than *B. pungens*, with leaves twice as wide, but in their substance less different, although more flaccid. It is possible that this may be only an altered state of *B. pungens* by some peculiarity of locality, but its appearance is different.

Brachymenium pulchrum, *Hook. Bot. Miscell.* i. t. 38.

On the stems of many trees.

A small quantity of plants which appear to be a young state of this or some nearly allied species. When dry the foliage is much twisted, but the form of the leaf is as in the S. African moss.

Bryum erythrocaulon, *Schwægr.* t. lxxx.

Widely spread on the island.

The specimens agree well with barren examples from Mauritius and Natal. The species appears to have softer and narrower leaves than is usual in *B. leptothecium*, Tayl., which in other respects is very closely allied.

Bryum megalacrion, *Schwægr.* t. 71.

Common on trees.

Specimens barren.

Bryum pachypoma, *Mont. in Ann. Sc. Nat.* 3d ser. iv. 104.

Rare. Only in a few places on rotten wood.

Stems mostly barren, all without fruit. They agree with specimens from Ceylon and Java.

Rhacopilum Africanum, *Mitt. in Journ. Linn. Soc.* vii. 162.

Found intermingled with species of Ectropothecium on moist rocks in the bed of streams.

Specimens barren, but they appear to be in no way different from those from the western coast of tropical Africa. The species differs from *R. mucronatum*, Beauv. Prodr. 66, in the cells of its leaves being larger and easily defined with a lens that leaves those of *R. mucronatum* obscure.

Callicostella læviuscula, *Mitt.*

Synoica. *Caulis* procumbens. *Folia* compressa, media late ovali oblonga apice obtusa late sub-apiculata, intermedia longiora apice truncato obtusa, lateralia apice late acuta, omnia costis divergentibus sub apice abruptis dorso parum erosis marginibus versus apicem crenulatis cellulis parvis rotundis lævibus areolata; perichætialia erecta vaginulam vix tegentia. *Theca* in pedunculo rubro lævi horizontalis decurvave ovalis operculo longe subulato calyptra ad medium thecæ descendente apice scabra, peristomii dentibus crassiusculis solidis cæterum normalibus.

Abundant.

One of a large genus, most of the species of which to the naked eye present scarcely any obvious differences; on a close examination, however, of the areolation of the leaves considerable differences are apparent in the form of the cells, in the presence or absence of the papillæ, also in the position of the papillæ on the cells; due consideration being given to this, *C. læviuscula* is allied in structure to *C. papillata*, Mont., so widely distributed in the Pacific islands, to *C. Africana*, Mitt. in Linn. Trans. xxiii. 53. t. 5, f. ix., from tropical Africa, to *C. depressa*, Swartz, from the West Indies, and to *C. Martiana* and *C. Merkeli* and *C. microcarpa*, Hornsch., all from S. America, but appears to differ from all in its shorter leaves which are not evidently papillose.

Plate 37 B.—1, stem of the natural size; 2, a leaf from the middle series of the upper side of the stem; 3, a leaf of the intermediate series on the same side; 4, a lateral leaf; 5, perichætium; 6, capsule with calyptra: all magnified.

Meteorium involutifolium, *Mitt. in Journ. Linn. Soc.* vii. 156.

Not common. At the top of Oyster River Valley.

Specimens showing the moss in a young or incomplete state, with simple stems about an inch long, arising from a creeping rhizoma like stem denuded of leaves. On the ascending branches the leaves are imbricated, and at their points are congested into a cuspidate bud, they are inserted in five twisted rows. This species is found also in Madagascar as well as on the Western Coast. Fertile stems are frond-like, and show the species to be nearly related to *M. Mauineuse*, Sull., and *Cryptotheca cochlearifolia*, Hornsch.

Pterogonium curvifolium, Mitt.

Monoicum. *Caulis* primarius repens radicellis purpureis apice ascendens arcuatus. Rami laxe dispositi ascendentes simplici vel interdum ramosi. *Folia* ramea patentia incurva parum compressa sicca tereti imbricata, late ovato acuminata concava margine recto ob prominentia cellularum parietum minute crenulata nervo furcato brevi subobsoleto, cellulis superioribus anguste elongate oblongis apicibus in papillas elevatis basin versus ad lateras densis minutis transverse oblongo-rotundis obscurioribus, perichætialia duplo majora patentia ovato-lanceolata læviora. *Flos masculus* exilissimus.

Stems about an inch long. *Branches* three to four lines long, when wet straight, when dry curved, with the leaves one fourth of a line wide.

Found sparingly on trees in Oyster Bay Valley.

This pretty moss forms extensive rather loosely interwoven patches. In miniature it has the habit of *P. gracile*; it differs from *Clasmatodon*, *Habrodon*, and *Anisodon* in the almost obsolete nerve and papillose leaves.

Plate 37 C.—1, stem of the natural size; 2, a portion with leaves; 3 and 4, leaves detached: all magnified.

Anomodon exilis, *Mitt. in Journ. Linn. Soc.* xiii. 309.

Rare.

A single stem amongst *Pterogonium curvifolium* agrees well with the South African specimen. The species is allied to *A. triste*, Cesati, but has the foliage of its branches compressed.

Neckera Lepineana, *Mont. in Ann. Sc. Nat.* 3ᵈ ser. x. 107.

Only in Valley Rivière des Huitres, trailing over the boulders.

Agrees with specimens from Madagascar and from Johanna, where it was gathered by Speke. The species is found in Ceylon, the Indian Archipelago, and the Pacific islands. Its fruit is rare.

Sematophyllum fulvifolium, *Mitt.*

Monoicum. *Caulis* decumbens ramosus. *Folia* compressa interdum secunda laxe imbricata ovata ovaliaque parum acuminata acuta concava, margine late subplana integerrima breviter obsoleteque binervia cellulis oblongis distinctis obscuriusculis alaribus utrinque tribus majoribus pellucidis areolata, perichætialia erecta longiora acuminata. *Theca* in pedunculo gracili rubro ovalis suberecta inclinatave basi callosa, operculo convexo tenui curvirostro, peristomii dentibus luteis inferne latis crassis.

Stems an inch long, with the leaves scarcely a line wide. *Leaves* green, almost glossy. Seta four lines long. *Capsule* greenish brown, after the fall of the red operculum, much contracted below the mouth and urceolate, at its base where contracted into the seta is a callous ring.

Not uncommon.

The specimens appear to be specifically identical with a moss gathered in Mauritius by Telfair, Bewsher, and Pike, in which the leaves are usually fulvous. *S. brachycarpum*, Hampe Ic. xi., has narrower leaves with a different areolation.

In the obscure cells of the leaf *S. fulvifolium* agrees with the South American *S. Kegelianum*, C. Muller.

Plate 37 D.—1, stem of the natural size; 2, leaves from the same; 3, perichætium and male flower; 4, capsule with operculum: all magnified.

Sematophyllum incurvum, *Mitt*.

Monoicum. *Caulis* repens pinnatus ramis assurgentibus incurvis. *Folia* caulina ovato-lanceolata acuminata concava nervis binis brevibus uno validiore vel obsoletis, margine obsolete serrulata ramea undique patentia sicca sursum secunda elliptico-lanceolata concava margine recurva integerrima enervia e cellulis elongatis angustis basi ad angulos rotundatis incrassatis alaribus oblongis hyalinis: perichætialia erecta caulinis similia duplo longiora. *Theca* in pedunculo breviusculo rubro ovalis horizontalis pendulave, basi apophyse calloso, operculo convexo longe oblique rostrato.

The prostrate *stems* are about an inch long, the branches from two to three lines long, and with the leaves about two-thirds of a line wide. The *leaves* glossy green and straw coloured. *Seta* three lines long.

Common.

This differs greatly in the substance of its leaves from *S. fulvifolium*, and agrees more nearly with *S. brachycarpum*, from which it may be distinguished by the form of its leaves, which are all directed upwards on the incurved branches.

Plate 38 A.—1, stem of the natural size; 2, leaves detached; 3, perichætium and male flower; 4 and 5, capsules, one with calyptra.

Ectropothecium scaturagineum, *Brid*. ii. 418 (Hypnum).

Perhaps the commonest species on the island.

Specimens all without fruit, varying greatly in appearance, but not much in the form of the leaves.

Ectropothecium subenerve, *Mitt*.

Caulis procumbens pinnatus ramis assurgentibus in cæspitem intertextus. *Folia* caulina compressa subsecunda ovata acuminata integerrima, nervis obsoletis, caviuscula ad margines vix implana, e cellulis elongatis angustis firmis areolata.

Branches from three to six lines long, with the leaves three-fourths of a line wide.

Along with the foregoing.

Ectropothecium doliare, *Mitt*.

Monoicum. *Caulis* procumbens intricatus foliis a symmetricis ovato-lanceolatis. *Rami* ascendentes in cæspitem laxum congesti foliis distichaceis compressis apicibus decurvis subfalcatis fulcatisve anguste lanceolatis omnibus margine apicem versus minute serrulatis, nervis binis brevibus subobsoletis, cellulis elongatis angustis lævibus areolatis; perichætialibus longioribus subulato-angustatis. *Theca* in pedunculo gracili breviter ovalis pendula operculo convexo brevirostro, peristomio e dentibus latis crassis interno processibus et ciliis in uno coalitis in membranam alte exsertam insidentibus.

Branches with the leaves about two-thirds of a line wide. *Leaves* pale green, glossy. *Seta* half an inch long. *Capsule* small and short.

Very common on clay rocks on the banks of streams. Forming extensive soft loose patches, in which the ramification, so frequently elegantly pinnate in allied species, is in the present scarcely perceptible.

Plate 38 B.—1, a stem of the natural size; 2, a leaf from the middle series on the upper side of the stem; 3, a leaf from the intermediate series of the same side; 4, a lateral leaf: 5, perichætium and male flower; 6, capsule.

Ectropothecium subulosum, *Mitt.*

Caulis procumbens radicans ramis brevibus pinnatim ramosus. *Folia* subcompressa lateralia patentia apicalia subfalcata, omnia lanceolato-subulata, caulina integerrima, ramea apicem versus sub-serrulata, costis binis brevibus inconspicuis, subobsoletis, cellulis angustis firmis areolata.

Branches about two lines long ascending, with the leaves two-thirds of a line wide. *Leaves* green, shining, scarcely altering when dry.

Frequent along with the preceding.

This moss nearly resembles a Mauritian species in its narrow leaves, but seems a little larger and of a firmer substance.

Hypnum (Rhynchostegium) pectinatum, *Mitt.*

Caulis procumbens ramis assurgentibus. *Folia* laxe inserta patula compressa, sicca pectinatim patula, ovata acuta vel breviter acuminata argute serrulata, costa tenui ultra medium evanida, cellulis ubique elongatis teneris mollibus areolata.

Stems with branches an inch long, slender, with leaves one line long, subpellucid, thin and glossy.

Not common: on dry places.

Nearly resembling a species from Natal, and also the S. African *H. rhaphidorhynchum*, C. Muller, but not exactly agreeing with any.

Fissidens procumbens, *Mitt.*

Caulis elongatus inferne procumbens crassiusculus simplex. *Folia* alterna patentia lineari-lanceolata acuta, costa vix pellucida in apice evanida, lamina vera ultra medium producta acuta, dorsali angusta haud decurrente, ubique integerrima e cellulis minutis obscuris densis areolata.

Stems an inch or more long with the leaves a line wide.

Occurs in tufts occasionally.

A species to all appearance different from any of the large group to which it belongs by its immarginate obscure leaves, which are not crenulate from the prominent marginal cells, as is so frequently observable among the species destitute of the hyaline limb.

Fissidens brevifrons, *Mitt.*

Caulis brevis oblique ascendens. *Folia* circiter sex juga longitudine increscentia frondem sub-flabelliformem sistentia, superiora anguste linealia apice latiuscule acuta nervo valido pallidiore percursa, lamina vera vix mediam folii attingente apice acuta parum inæquale, dorsali ad nervi basin anguste continuata, e cellulis minutis fere opacis areolata integerrima.

Stem a line long. The upper *leaves* one and a half line long, of a glaucous olive green, becoming brown.

On clay rocks. Rare.

Allied to the S. African *F. plumosus*, Hornsch., but with leaves narrower at the points and more opaque.

Fissidens flaccidus, *Mitt. in Trans. Linn. Soc.* xxiii. 56. t. 6, f. xviii.

Sparingly on clay rocks.

The specimens, all of a dark brown colour, agree with those from Western Africa. In this species the nerve vanishes some distance below the point of the leaf, an uncommon circumstance in the genus.

Fissidens Vogelianus, *Mitt. in Trans. Linn. Soc.* xxiii. 54. t. 5, f. x.

With *Anthoceros fuciformis*.

The specimens have the leaves a little wider than in those from the Western Coast of Africa. In other respects they agree closely.

HEPATICÆ.

By WILLIAM MITTEN, A.L.S.

The entire number of the Hepaticæ is eighteen, of which thirteen were obtained with more or less complete fructification and five only were barren. As in the Mosses the genera are all African, but only three of the species are identical with African or Mascarene forms; one only is also found in the Eastern Archipelago, and one *Anthoceros lævis* may be said to be cosmopolitan.

Thirteen of the species appear to be new, of these nine are fertile and four barren.

Lophocolea subrotunda, *Mitt.*

Caulis repens pallidus. *Folia* imbricata alterna explanata marginibus recurvis convexa vel sursum secunda apicibus recurvis ovato-rotunda, apice rotundata subretusave, integerrima; amphigastria parva caulem vix latitudine superantia quadrifida, basi uno latere cum folio subjacente anguste connexa, cellulis parvis rotundatis pellucidis areolata, perianthium triquetrum superne latiore, labiis longis spinosis;

angulis ala inconspicua. Folia involucralia apice bidentata cæterum integerrima, basi cum amphigastrio magno dentibus pluribus spinosis elongatis connexa.

Not common. On decayed wood and peaty soil.

Allied to *L. pallida*, Mitt. from Lord Auckland's Islands, in its entire leaves, but although less in size more nearly resembling *L. heterophylla*. No bidentate leaves are observable except the two next the perianth.

Plate 38 C.—1. Plant of the natural size. 2. A portion with leaves and stipule as seen on the dorsal side. 3. Perianth with involucral leaves and stipule: all magnified.

Lophocolea opposita, *Mitt.*

Caulis procumbens. *Folia* explanata imbricata marginibus recurvis convexula, oblongo-ovata, apice truncata sinu obtuso breviter bidentata integerrima, basi antice sæpius cum folio opposito decurrente connexa postice cum amphigastrio quadridentato caule latiore utrinque coalita, e cellulis parvis rotundis areolata, involucralia conformia margine dorsali ventralique dentibus 2-3 brevibus armata amphigastrio ovali apice breviter bi-quadridentato, perianthium triquetrum ala indistincta labiis spinoso-dentatis.

Along with and in same situations as the last. Not common.

In size similar to *L. semiteres* (*Chiloscyphus*), Synops. Hepat. 190, but more pellucid and of a pale whitish green. Among the small species of this genus, to which the present must be placed, it is remarkable for its leaves being united in front, or, as usually written, on the dorsal side of the stems.

Plate 38 D.—1. Plant of natural size. 2. Portion of stem with leaves and stipule as seen on the dorsal side. 3. Involucral leaves and combined stipule removed from the stem. 4. Perianth as seen on the side: all magnified.

Chiloscyphus oblongifolius, *Mitt. in Trans. Linn. Soc.* xxiii. 58. t. 6, f. xxii.

Common on the moist banks of streams.

Stems all barren, a little more robust than the specimens from the Niger; otherwise without difference. This species is found also in Mauritius.

Radula appressa, *Mitt.*

Caulis procumbens parum ramosus. *Folia* dorso imbricata divergentia subobovata ob marginem inferiorem medio sinuatum apice rotundato decurva integerrima cellulis parvis densis obscuris dorso minute papulosis areolata, lobulo subquadrato appresso apice obtuso ad medium usque folii latitudinis protracta basi ad insertionem caulem ad medium tegente.

Very abundant.

A small species resembling *R. buccinifera*, Taylor, and *R. Novæ Hollandiæ*, Hampe.

Phragmicoma emergens, *Mitt.*

Caulis repens ramosus. *Folia* humida fere verticalia divergentia sicca imbricata ambitu explanata ovato-orbiculata, margine ventrali incurva lobulo semiorbiculari plerumque tridenticulato appresso, cellulis rotundo-ovalibus exterioribus rotundioribus carnosulis areolata, amphigastria transverse oblonga caulem duplo latiora apice late truncata obtusa. *Folia involucralia* magna lobuloque apice rotundato una cum amphigastrio obovato rotundato integerrima. *Perianthium* obovato-oblongum emergens sub-compressum, dorso et ventre tricarinatum, igitur octoplicatum.

Stems and *branches* with the leaves about a line wide. Leaves brown, with cells twice the size of *P. carinata*.

Only in a few places on branches of trees.

Plate 39 A. 1. Portion of plant of the natural size. 2. Part of the stem with leaves and stipule as seen on the ventral side. 3. The same as seen on the dorsal side. 4. A leaf removed, as seen on the under side. 5. Perianth with capsule, with involucral leaves and stipule, as seen on the ventral side. 6. Transverse outline of the perianth: all magnified.

Phragmicoma carinata, *Mitt.*

Caulis repens ramis pluribus irregularibus ramosus. *Folia* subverticaliter patentia sicca imbricata, ambitu explanata ovato-rotundato margine ventrali incurva, lobulo lato appresso semiorbiculari plerumque unidentato, e cellulis ovalibus carnosulis chlorophylloso obscuratis marginem versus rotundioribus areolata, amphigastria caulem duplo latiora subrotunda apice lata subtruncata margine recurva subretusata. *Folia involucralia* majora obovata obtusa angulatave lobulo oblongo, amphigastrium magnum ovale apice obtusum integerrimum margine in medio apicis recurvum. *Perianthium* emergens oblongum dorso planum ventre alte carinatum, igitur triplicatum.

Stems and *branches* with the leaves two-thirds of a line wide. *Leaves* obscure, brown. When dry the older parts are ochraceous brown.

Common on trees. Often on the stem of *Pandanus*.

Plate 39 B. 1. Stem of the natural size. 2. A portion seen on the ventral side with leaves and stipules. 3. A leaf as seen on the under side. 4. Perianth with involucral leaves and stipule: all magnified.

Lejeunia Balfouri, *Mitt.*

Caulis procumbens parum ramosus subsimplex. *Folia* a dorso visa imbricata convexula divergentia oblongo-ovalia rotundata integerrima, lobulo ventrali obsoleto, cellulis rotundis parvis crassiuscule limitatis laevibus areolata. *Amphigastria* cauli triplo latiora rotunda integerrima. *Perianthium* in ramo brevissimo subsessile late obcordatum applanatum laeve marginibus superioribus undulato sinuatis rugosis. *Folia involucralia* conformia lobulo parvo transverse oblongo apice obtuso, amphigastrium ovali-rotundum integerrimum.

The *stems* with the leaves a line wide.

Common on the moist clay rocks at the sides of streams. In dull olive brown or blackish patches.

Like *L. Hobsoniana,* Lindenb, from the West Indian Islands, but perianth destitute of wing or teeth, owing to its position near the base of the stems it is very inconspicuous.

Plate 39 C. 1. Branch of the natural size. 2. A portion of the same seen on the ventral side, with leaves and stipule. 3. Perianth with involucral leaves and stipule seen on the ventral side: all magnified.

Lejeunia abortiva, *Mitt.*

Caulis vage pinnatim ramosus procumbens. *Folia* divergentia dorso imbricata ovali orbiculata apicibus incurvis planisve integerrima, e cellulis rotundo-hexagonis parietibus angustis nigro-fuscis areolata lobulo obsoleto. Amphigastria parva rotunda integerrima caulem vix duplo latiora. *Folia involucralia* amphigastriumque majora conformia. *Perianthium* obovatum obtusum, dorso planum, ventre bicarinatum. Carinis vetralibus lateralibusque angulato erosis.

Stems with the leaves about half a line wide. Entire plant blackish brown.

Grows in tufts on the clay rocks.

Allied to *L. subfusca,* Nees ab E., but the carinæ of the perianth are not alate nor with elongated teeth. It may be the species mentioned in the Synopsis, p. 315, under *L. subfusca,* as from Mauritius, in which the lobule of the involucral leaves is stated to be wanting.

Plate 39 D. 1. Plant of the natural size. 2. A portion of the stem with leaves and stipules seen on the ventral side. 3. Perianth with subjacent leaves and stipules as seen on ventral side. 4. Transverse outline of perianth: all magnified.

Lejeunia furva, *Mitt.*

Caulis procumbens ramosus. *Folia* subimbricata directione patula rotundata integerrima cellulis rotundatis parietibus crassis fuscoviridibus areolata, lobulo minuto quadrato vel omnino obsoleto. Amphigastria rotunda parva caulem vix duplo latiora bifida laciniis acutis. *Fructus* ad ramulorum latera. *Folia involucralia* parva lobulo oblongo acuta. *Perianthium* obovatum obtusum æqualiter quinqueplicatum.

Stem with the leaves half a line wide. Entire plant dull brown, not altered in drying.

Along with Mosses on the moist clays at the sides of streams.

A species with the habit of the European *L. serpyllifolia,* but with round leaves.

Plate 40 A. 1. Plant of the natural size. 2. Portion of the stem with leaves and stipules, with perianth as usual arising from the side of lateral branch, also with the capsule past maturity and empty, seen from the ventral side: all magnified.

Lejeunia surrepens, *Mitt.*

Caulis repens. *Folia* imbricata divergentia ovali orbiculata lobulo oblongo apice rectangulato ad medium marginis ventralis producta e cellulis rotundo-hexagonis pellucidis æqualibus areolata. *Amphigastria* rotundata caulem triplo superantia, apice fissa, rarius indivisa.

Branches half a line wide. *Foliage* whitish green.

Not common on tree stems.

This appears to be not different from a specimen, also barren, from Mauritius. The length the lobule continued along the ventral margin renders this immediately distinguishable from the numerous species resembling *L. serpyllifolia*.

Lejeunia angulifolia, *Mitt.*

Caulis repens ramosus. *Folia* alterna patenti divergentia subovata apice angulata subacuta integerrima, lobulo subrotundo saccato, cellulis prominulis papuloso apice acuto, cellulæ foliorum rotundatæ pellucidæ læviæ parietibus crassis obscuris viridibus. *Amphigastria* minuta bifida caulem vix latiora. *Folia involucralia* acutiora lobulo ovato acuto. *Amphigastrium* ovale profunde fissum laciniis acutis.

Stems with the leaves about one third of a line wide. *Leaves* all angled at the apices, and occasionally with one or more angles on the dorsal margin.

On tree stems. Not uncommon.

A small species, less than *L serpyllifolia*, of a pale green colour. Allied to *L. Nilgiriana*, Mont., but more tender and with a more lax areolation.

Lejeunia minutissima, *Sm.*

Gathered sparingly on stems of trees.

Lejeunia subciliata, *Mitt.*

Caulis repens appressus ramosus. *Folia* imbricata, dorso planiuscula, late ovata apice rotundata, margine dorsali cellulis elongatis hyalinis tenerrimis scarioso marginata, cellulis reliquis rotundo-hexagonis parietibus crassiusculis areolata, lobulo tumido ovale apice apiculo spiniformi terminato. *Amphigastria* nulla.

Nearly related to *L. planissima*, Mitt., from Ceylon, which has the same kind of scarious margin, but is much more closely adherent to bark.

Stem with the leaves scarcely more than half a line wide. *Leaves* whitish green, closely appressed to the bark on which the specimens were growing. The scarious margin is composed of a single row of elongate cells placed side by side, and an irregularity in their length causes the edge to be erose; this hyaline limb of diverse cells is here and there broken off, leaving the margin of the leaf itself entire.

Rare on stems.

Plate 40 B.—1. Stem of the natural size. 2. A portion of the same as seen on the ventral side. 3. Portion of the edge of the leaf on the dorsal edge: all magnified.

Frullania obscurifolia, *Mitt.*

Caulis procumbens repensve ramis vagis subpinnatim ramosus. *Folia* imbricata orbiculata e cellulis angulato rotundis parietibus angustis teneris sed chlorophylloso obscuris, lobulo ovali dimidium folii longitudinis tegente compresso apice paululum a cauli divergente interdum explanato lacinia minuta interjecta. *Amphigastria* caulem duplo latiora rotundata apice breviter bifida laciniis extus unidentatis. *Folia involucralia* ovalia lobulo lato ovato-lanceolato acuto inferne lacinia dentiformi instructa amphigastrioque profunde bifido laciniis acutis integerrimis. *Perianthium* obovatum obtusum, apiculo brevi, compressum, dorso uni ventre bicarinatum.

Common on trees.

Similar to *F. trinervis*, Lehm. et Lindenb., from S. Africa, but differing in its obscure areolation and the longer lobule of its leaves.

Plate 40 C.—1. Plant of natural size. 2. A portion of the stem with leaves and stipules on the ventral side. 3. Perianth with involucral leaves and stipule. 4. Transverse section of perianth: all magnified.

Frullania squarrosa, *Nees ab E.*
Very common on trees.

Frullania apicalis, *Mitt.*
Caulis gracilis pinnatim ramosus. *Folia* dorso convexa divergentia orbiculata integerrima cellulis parvis rotundatis distinctis sed utriculo collapso obscuratis areolata, lobulo ovali-cylindraceo erecto cauli parallello lacinia parva interjecta, in ramulorum apicibus lobulo apice ad caulem incumbente, amphigastria obovata bifida laciniis extus unidentatis, folia involucralia acuta lobulo ovato acutiore amphigastrioque marginibus dentatis. *Perianthium* oblongo-obovatum, dorso plano ventre unicarinato.

Common on trees, specially on Pandanus.

In size and colour, as well as in the form and imbrication of its leaves, nearly resembling *F. Capensis*, Gottsche Syn. Hep. 449, but with more the habit of *F. gracilis*, and the leaves of the involucre dentate.

Plate 40 D.—1. Stems of the natural size. 2. Portion with leaves and stipules as seen from the ventral side. 3. Perianth with involucral leaves and stipule: all magnified.

Anthoceros lævis, *Linn.*
Common on moist clay rocks on the banks of streams.

Anthoceros fuciformis, *Mont. in Ann. Sc. Nat.* 2d ser. xx. 296.
Growing along with the foregoing.

This appears to be the same as the specimen from Bourbon. The spores are opaque and echinulate.

CHARACEÆ.

Chara Commersoni, *A. Braun.*
In many streams.

LICHENES.
By Rev. J. M. Crombie, F.L.S.

Family COLLEMACEI.
Tribe Collemei.

Collema, sp. ——?

Several specimens of a *Collema* occur which externally bears some resemblance to young states of *C. flaccidum*, but sterile and with the thallus discoloured so as to be quite indeterminable.

On trunks of trees sparingly.

Leptogium tremelloides, *Ach*.

Sufficiently typical, and also occasionally somewhat atypical.

Common, on the trunks of trees; fertile.

Family LICHENACEI.
Tribe Cladoniei.

Cladonia Balfouri, Cromb.

Thallus minor, basi squamulosus, squamulis parvis, albo-glaucescentibus, subfirmulis, crenatis, podetiis simplicibus, brevissimis, apice raro divisis, albo-pulverulentis (vel pulvere subvirescenti-albo obductis), K—, contra squamis K+ (flaventibus); apothecia fusca (ut videtur), rite evoluta non visa.

The general aspect of this plant is sufficiently peculiar to warrant its being regarded as a distinct species, notwithstanding that the specimens seen are probably not fully developed as to thallus. Only a few faint traces of young apothecia are visible, which would refer it to the Phæocarpæ, where it seems to have its affinity in the section of *C. pyxidata*.

On dead stumps of trees; not common.

Tribe Roccellei.

Roccella Montagnei, *Bel.*, f. teretior, *Cromb*.

This form is distinguished by the narrow rounded thallus, which gives it more the aspect of *R. phycopsis* than of *R. fuciformis*, which typical specimens more closely resemble.

On rocks; infertile.

Various and evidently young states, which seem referable to this form, are also present.

Tribe Usneei.

Usnea florida (*L.*), *Ach.*
On the branches of trees; sparingly and infertile.

Usnea dasypogoides, *Nyl.*
"Subsimilis *Usneæ dasypogæ*, sed thallo subflavido, lævi, glabro, nec nisi in ramulis vel fibrillis sæpius leviter sparsim setuloso (quibus setulis etiam distinguitur ab *U. trichodea*, Ach.). Apothecia ignota."—*Nyl. in litt.*

These specimens vary considerably in length, the largest being upwards of a foot long. Unfortunately none of them are fertile, though scattered "cephalodia" not unfrequently occur.

Abundant on the branches of trees in the higher parts of the island.

Tribe Ramalinei.

Ramalina gracilenta, *Ach.*
On rocks and trees in the higher parts of the island; fertile.

Ramalina gracilenta, *Ach.*, f. **nodulosa**, *Cromb.*
Thallus minute, densely branched and fastigiate, forming more or less rounded nodules.

This is evidently a young or rather stunted condition of the plant, presenting a somewhat singular appearance. In one specimen a single minute apothecium was visible.

Rare; on rocks with the type.

Ramalina farinacea, *Ach.*, *Nyl.*
On the branchlets of trees; not common. In higher parts of island. Also found in Mauritius.

Ramalina farinacea, *Ach.*, *Nyl.*, f. **pendula** (*Schrad.*).
On the branchlets of trees, with the type.

Ramalina subfraxinea, *Nyl.*
Common on the branches of trees in the higher parts, along with *Usnea dasypogoides*; fertile.

Ramalina canaliculata, *Tayl.*
On the branches of trees; fertile.

Ramalina canaliculata, *Tayl.*, f. **brevior**, *Cromb.*
Thallus shorter, $\frac{1}{2}$-1 inch; laciniæ rather broader.
On the branches of trees; fertile, but like the type.

Ramalina intermedia, *Del.*, *Nyl.*
On rocks: not uncommon, but sterile.

Tribe PARMELIEI.

Parmelia latissima, *Del.*

Very common on trees and rocks; sterile.

The thallus is often discoloured, probably from the action of salt water.

Parmelia subconspersa, *Nyl.*

Common on rocks; fertile.

Parmelia atrichoides, *Nyl.*

On rocks; sparingly and infertile.

Tribe STICTEI.

Stictina dissimilis, *Nyl.*

On trees; sparingly and infertile.

Stictina agyracea, f. insidiata, *Nyl.*

Thallus somewhat densely covered with isidia.

On trees; sparingly and infertile.

Sticta aurata, *Ach.*

On trunks of trees and dry places, with *Parmelia latissima;* infertile.

Tribe PHYSCIEI.

Physcia flavicans, f. crocea (*Ach.*).

Rare; on the branches of trees. Only on Mount Piton, near south end; sterile.

Physcia speciosa (*Wulf.*).

Abundant, on trees and on boulders; fertile.

Physcia ægialita (*Ach. Syn.* p. 179, sub *Lecanora*).

On rocks; fertile, but very sparingly gathered, only a single small specimen being present.

Physcia picta (*Sw.*).

On bark of Pandanus, not uncommon.

Tribe PYXINEI.

Pyxine petricola, *Nyl.*

"Thallus glaucescens, laciniis subintricato-contiguis, adnatis (latit. 1 millim. vel angustioribus), late diffusus; apothecia nigra, planiuscula, lecideina, marginata (latit. 1 millim. vel minora), vel margine extus epithallino obducta, intus obscura; sporæ 8næ, nigrescentes, oblongæ, biloculares, longit. 0·011-0·015 millim., crassit. 0·0045 millim.; epithecium cæruleo-nigricans; paraphyses graciles; hypothecium sub-incolor. Iodo gelatina hymenialis bene cærulescens.

"Species certe bene distincta notis allatis. Thallus K=."---*Nyl. in litt.*

On rocks; sparingly.

Tribe LECANOREI.

Pannaria rubiginosa, f. cinerascens, *Nyl.*

Thallus greyish; otherwise similar to the type.

On rocks. This species occurs only in the higher parts of island. Only two apothecia visible on the specimen.

Two other sterile specimens may belong also to this form.

Pannaria rubiginosa, var. dispartita, *Nyl.*

" Forsan propria species, thallo tenuius diviso, effuso. Apothecia non visa."— *Nyl. in litt.*

On rotten stumps on the ground; sparingly.

Pannaria luridula, *Nyl.*

" Thallus luridus, squamulosus, squamulis crenatis vel crenato-lobatulis, subintricatis; apothecia testaceo-rufa vel testaceo-rufescentia (latit. circiter 1 millim. vel minora), bene lecanorina, margine thallino crassiusculo crenato; sporæ ellipsoideæ, longit. 0·016–0·025 millim., crassit. 0·009–0·011 millim. Iodo gelatina hymenialis cærulescens.

" Affinis *P. immixtæ*, Nyl., sed color thalli luridus et hypothallus niger ullus vix visibilis etc."—*Nyl. in litt.*

On the ground; rare.

Coccocarpia molybdæa, *Pers.*

Not uncommon on bark of trees; infertile.

Heppia Rodriguezi, *Cromb.*

Thallus cervino-luridus, adnato-squamulosus, squamulis planiusculis vel alibi convexiusculis, subanguloso-rotundatis (latit. 1–2 millim. vel sæpius minoribus, crassit. circiter 0·25 millim.), contiguis aut subdispersis; apothecia rubello-carnea, concaviuscula, innata (latit. 0·2–0·3 millim.); thecæ polysporæ; sporæ breviter ellipsoideæ vel oblongo-ellipsoideæ, longit. 0·004–0·006 millim., crassit. 0·0025–0·0035 millim., paraphyses gracilescentes. Iodo gelatina hymenialis cærulescens, dein thecæ præsertim violacee rubescentes.

This is a fine species, which is at once distinguished by the colour of the apothecia. The syngonimia are usually sufficiently small, and then contain but few gonimia. With the exception of the colour of the apothecia, it presents almost the appearance of *Lecanora cervina*, Pers. (*squamulosa*, Schrad.).

On rocks; very rare in the Cascade Valley; only a few apothecia visible.

Lecanora murorum, var. lobulata, (*Smrf.*).

As the thallus is not sufficiently developed in the specimen seen, the plant is somewhat uncertain, although the apothecia and spores agree with this variety. Spores 0·010–0·013 millim. long, 0·006–0·007 millim. thick.

On rocks; not common.

Lecanora cinnabarina, (*Sw.*).

Frequent on rocks; in the specimens seen associated with other lichens, and not very characteristic.

Lecanora aurantiaca, (*Lightft.*).

On the bark of trees; sufficiently typical, but very sparingly gathered. Spores 0·011–0·016 millim. long. 0·006–0·010 millim. thick.

Lecanora aurantiaca, var. **isidiosella.**

A somewhat peculiar variety, which is well distinguished by the thallus being more or less covered with subconcolorous isidia. The apothecia are also smaller and only sparingly present in the few fragments seen.

On the bark of trees, associated with other lichens.

Lecanora aurantiella, *Nyl.*

" Subsimilis *L. aurantiacæ* et *erythrellæ*, Ach., sed apotheciis minoribus et innatis, sporis minoribus (longit. 0·010–0·012 millim., crassit. 0·006–0·008 millim.)."—*Nyl. in litt.*

On rocks in the Cascade Valley, associated with other lichens.

Lecanora leucoxantha, (*Spr.*).

On the bark of *Mathurina penduliflora*; found in fine fruit.

Lecanora glaucofuscula, *Nyl.*

" Thallus glauco-cinerascens (vel humidus cinereo-virescens), tenuis, continuus, passim subrimosus; apothecia fusca, subinnata, plana (latit. circiter 0·05 millim.), margine thallino tenui vel parum distincto cincta; sporæ 8næ, incolores, placodinæ, longit. 0·009–0·014 millim., crassit. 0·006–0·007 millim.; epithecium (in lamina tenui) lutescenti-inspersum (K purpurascens); hypothecium incolor. Iodo gelatina hymenialis intensive cærulescens. Nulli cognitæ affinis, vergens autem versus *L. encephalarti* (Kphb)."—*Nyl. in litt.*

On rocks; sparingly.

Lecanora glaucofuscula, f. **biatoroidea.**

This differs from the type only in the apothecia having a more distinctly biatorine appearance.

On rocks; like the type, very sparingly.

Lecanora apostatica, *Nyl.*

" Thallus albus vel albidus, tenuis, areolato-rimosus, hypothallo cærulescenti-nigricante passim visibili; apothecia nigra, lecideina, plana, marginata (latit. 0·05–0·07 millim.), intus albida; sporæ 8næ, incolores, placodinæ, longit. 0·011–0·015 millim., crassit. 0·005–0·008 millim.; paraphyses mediocres; epithecium et perithecium cærulescenti-nigrescentia; hypothecium incolor. Iodo gelatina hymenialis intensive cærulescens.

" Est species e stirpe *Lecanoræ diphyodis*, proxima *L. Kurzii* (Kphb.). Affinitas videtur inter *Lecanoras*, etiamsi apothecia lecideina. Thallus K flavens. Spermogonia arthrosterigmatibus munita; spermatia minutula, oblongo-cylindrica, longit. 0·002 millim., crassit. 0·0006 millim."—*Nyl. in litt.*

Not uncommon on cindery rocks.

Lecanora obliquans, *Nyl.*

" Differt thallo cinerascente tenuissimo continuo. Occurrit quoque epithecio albo-suffuso. Forsan propria species."—*Nyl. in litt.*

Very common on rocks, giving them a white appearance.

Lecanora carneofusca, *Nyl.*

" Thallus glauco-cinerascens vel glauco-albidus, tenuis, inæqualis, areolato-rimulosus; apothecia carneo-fusca vel rufescenti-fusca, lecanorina (latit. 1 millim. vel minora), margine thallino integro cincta; sporæ 8næ, incolores, placodinæ, longit. 0·011–0·018 millim., crassit. 0·007–0·009 millim.; paraphyses fere mediocres; epithecium sordide sublutescens. Iodo gelatina hymenialis intensive cærulescens.

" Species accedens ad *Lecanoram camptidiam,* Tuck., a qua thallo areolato rimuloso et sporis nonnihil majoribus præsertim differt."—*Nyl. in litt.*

On rocks. Common.

Lecanora subfusca, f. pumicicola, *Nyl.*

" Differt a f. *campestri,* Schær. præsertim thallo subdisperso, e granulis sat tenuibus constans vel variantibus convexiusculis, ambitu subcrenatis (latit. 1 millim. vel minoribus), passim contiguis. Apothecia fusca (latit. 1 millim. vel minora); margine thallino integro aut obsolete subcrenulato; sporæ longit. 0·011–0·012 millim., crassit. 0·006–0·007 millim.; paraphyses tenues, molles; epithecium rufescenti-fuscum. Iodo gelatina hymenialis cærulescens, dein violaceo tincta.—*Nyl. in litt.*

This very distinct form may easily be recognised by the above characters of the thallus, &c.

Common on cindery rocks.

Lecanora subflavicans, *Nyl.*

" Thallus albidus vel albido-flavescens, tenuis, granuloso-inæqualis (K+, flavens); apothecia testaceo-lutescentia vel testaceo-pallida (latit. (0·5–0·8 millim.), margine crassiusculo-ruguloso-inæquali vel demum integro cincta; sporæ 8næ, incolores, ellipsoidæ, longit. 0·010–0·012 millim., crassit. 0·006–0·008 millim.; paraphyses gracilescentes; epithecium inspersum. Iodo gelatina hymenialis cærulescens, dein thecæ violaceo tinctæ.

" Est species quasi intermedia inter *Lecanoram subgranulatam* et *L. albellam.*" *Nyl. in litt.*

On the bark of trees; common.

Lecanora achroa, *Nyl.*

" Thallus glaucescens, tenuis, rugulosus, demum rimosus, nigricanti-limitatus (K flavens); apothecia lutescenti-pailida, plana (latit. 0·5–0·6 millim.), superficialia, margine thallino subcrenato vel subintegro cincta; sporæ 8næ, ellipsoidæ, longit. 0·010–0·016 millim., crassit. 0·006–0·008 millim.; paraphyses graciles; epithecium granuloso-inspersum. Iodo gelatina hymenialis cærulescens, dein thecæ obscure nonnihil violacee tinctæ.

"Ex affinitate videtur *Lecanoræ chlaronæ*. Spermatia vulgo leviter arcuata, longit. 0·018-0·023 millim., crassit. 0·0005 millim. (minus arcuata quam in comparanda *L. achroella*, quæ simul aliis notis differt."—*Nyl. in litt.*

On bark of trees, specially on *Pandanus*.

Lecanora achroella, *Nyl.*

"Thallus glaucescens, tenuis, sublævigatus, rimulosus, nigricanti-limitatus (K flavescens); apothecia lutescenti-pallida, plana, subinnata, minuta (latit. circiter 0·3 millim.), margine thallino tenui integro cincta; sporæ 8næ, ellipsoideæ, longit. 0·009-0·014 millim., crassit. 0·006-0·007 millim.; paraphyses graciles; epithecium vix vel parum inspersum. Iodo gelatina hymenialis cærulescens, dein thecæ subviolaceo tinctæ vel violaceo-fulvescentes.

"Ex affinitate videtur *Lecanoræ chlaronæ*. Spermatia arcuata, longit. 0·015-0·018 millim., crassit. 0·0005 millim."—*Nyl. in litt.*

On the trunks of trees, associated with the preceding.

Lecanora perlutescens, *Nyl.*

"Thallus flavidus vel albido-flavescens, tenuis, subleprosus, determinatus; apothecia cerino-lutea vel luteo-pallescentia (latit. 0·4-0·6 millim.), plana, margine thallino tenui integro vel evanescente cincta; sporæ 8næ, oblongæ vel ellipsoideæ, longit. 0·009-0·013 millim.; crassit. 0·0035 millim.; paraphyses gracilescentes; epithecium inspersum. Iodo gelatina hymenialis cærulescens, dein lutescens.

"Affinis *Lecanoræ lutescenti* (DC) sed minor, magis determinata, sporæ minores, etc. Thallus et nonnihil apothecia CaCl erythrinose vel aurantiace tincta. Spermatia arcuata, longit. circiter 0·016 millim., crassit. 0·0005 millim."—*Nyl. in litt.*

On the bark of trees.

Lecanora conizæa (*Ach.*), *Nyl.*

Entirely similar to our European plant. Thallus CaCl—. Spores 0·010-0·014 millim. long, 0·003-0·004 millim. thick.

On decayed wood, not common.

Lecanora conizopta, *Nyl.*

"Thallus albidus, tenuis, opacus, passim rimulosus (K flavens); apothecia carneo-lutea vel luteo-rubella, subinnata, plana (latit. 0·2-0·4 millim.), margine thallino depresso inæquali cincta; sporæ 8næ, ellipsoideæ, simplices, longit. 0·010-0·012 millim., crassit. 0·005-0·006 millim.; paraphyses gracilescentes. Iodo gelatina hymenialis cærulescens.

"Videtur e stirpe *Lecanoræ conizææ*, et notis datis facile agnoscitur."—*Nyl. in litt.*

Common on rocks, and very abundant in many places.

Lecanora parella * phlœoleuca, *Nyl.*

"Thallo tenuissimo albo præsertim differens a typo, et facie apotheciorum subpertusariodea. Sporæ longit. 0·038–0·050 millim., crassit. 0·023–0·025 millim. Fortasse propria species."—*Nyl. in litt.*

On the bark of trees.

Lecanora atra, f. succedanea, *Nyl.*

"Differt a *L. atra* europæa saltem spermatiis nonnihil brevioribus (longit. 0·007–0·011 millim., crassit. fere 0·001 millim.)."

On many rocks in the higher parts of the island; more rarely on dead wood and bark of trees.

Lecanora punicea, *Ach.*

On the bark of trees; common, as it is in Mauritius.

Pertusaria velata (*Turn.*)

On the bark of trees, specially *Mathurina penduliflora*; quite similar to our British plant.

Pertusaria impallescens, *Nyl.*

"Thallus albidus, opacus, continuus, tenuis (crassit. circiter 0·2 millim.), sublævis, passim rimosus vel rimulosus (K—); apothecia incoloria in protuberantiis thallinis depresso-convexulis innata, 1–4 in quavis tali protuberantia, ostiolis nigris, punctiformibus, vix prominulis; sporæ 8næ, incolores, solitæ generis, longit. 0·070–0·082 millim., crassit. 0·032–0·040 millim. Iodo gelatina hymenialis intensive cærulescens.

"Notis datis satis distincta species, nec ulli cognitæ affinis. Spermatia peculiaria, aciculari-fusiformia (vel fusiformia utroque apice longe aciculari-producto), longit. 0·007–0·009 millim., crassitie medio vix 0·001 millim."—*Nyl. in litt.*

On rocks. One of the commonest lichens on the island, giving a white appearance to the rocks.

Tribe LECIDEEI.

Lecidea achroopholis, *Nyl.*

"Thallus albidus vel pallescenti-albidus, squamulosus, squamulis (latit. circiter 1–2 millim.) adnatis, planis, contiguis vel subdispersis, ambitu sublobatulis vel interdum subcrenatis, subtus nigris; apothecia nigra, planiuscula, marginata (latit. 0·7–1·5 millim.), intus obscura; sporæ 8næ, incolores, oblongæ, simplices (vel obsolete vel spurie 1-septatæ), longit. 0·009–0·014 millim., crassit. 0·0035–0·0045 millim.; paraphyses mediocres apice subpurpurascenti-fusco; epithecium et hypothecium subpurpurascenti-infuscata (vel hoc parte supera rufescente). Iodo gelatina hymenialis cærulescens, dein subfulvescens.

"Species insignis in stirpe *Lecideæ decipientis*, ut videtur. Thallus firmus K non reagens. Thalamium totum dilute subpurpurascens et K simul cum epithecio et perithecio et hypothecio inferiore purpurascens. Spermogonia supra nigella, intus incoloria; spermatia tenuiter bacillaria nonnihil vel obsolete curvula (quare subrecta apparent, licet vix unquam sint vere recta), longit. 0·005–0·007 millim., crassit. 0·0006 millim."—*Nyl. in litt.*

On rocks, usually of a vesicular character.

Lecidea coccocarpioides, *Nyl.*

"Thallus albido-glaucescens, squamaceus, subadnatus, squamis (latit. 2–5 millim.) inciso-lobatis, incisuris clausis (nec hiantibus) et nigricantibus, ambitu subcrenulato, pagina infera nigra; apothecia nigra, innata, planiuscula, immarginata (latit. 1–2 millim.), interdum 2–3 confluentia, intus obscura; sporæ 8næ, ellipsoideæ vel oblongæ, 1-septatæ (septo saltem spurio), longit. 0·006–0·012 millim., crassit. circiter 0·0035 millim.; paraphyses mediocres; epithecium cærulescens; hypothecium obscure rufescens. Iodo gelatina hymenialis bene cærulescens.

"Species omnino peculiaris et propriæ stirpis. Facie fere *Coccocarpiæ molybdææ*, sed thallus subnitidiusculus, nihil superficie concentrici habet, et ceteroquin est gonidicus."—*Nyl. in litt.*

On rocks.

Lecidea melopta, *Nyl.*

"Thallus albus, tenuis, areolatus, in hypothallo nigro subprædominante instratus; apothecia nigra, in areolis vulgo demum singulatim innata, plana (latit. 0·5–0·7 millim.), immarginata; sporæ 8næ, incolores, ellipsoideæ, longit. 0·008–0·010 millim., crassit. 0·004 millim.; paraphyses non distinctæ; epithecium cærulescens; hypothecium incolor. Iodo gelatina hymenialis bene cærulescens.

"Species bene distincta, et fere urceolariiformis, inter species stirpis *Lecideæ planæ*. Thallus K flavens (I demum nonnihil subroseo tinctus, hæc reactio vero passim obsoleta). Variat thallus crassior (crassit. 0·5 millim.) et pluribus apotheciis in quavis areola (areolis circumcirca lateribus nigris), apotheciis impressiusculis."— *Nyl. in litt.*

On rocks; common.

Lecidea disciformis (*Fr.*), *Nyl.*

On the bark of trees specially *Fœtidia mauritiana,* Lam.

Lecidea triphragmia, *Nyl.*

On the bark of trees, specially of *Ficus.*

Lecidea spuria (*Schær.*)

On decomposed rocks.

Lecidea immutans, *Nyl.*

Thallus ochraceo-rosellus (verisimiliter autem typice albidus), areolato-diffractus (crassit. circiter 0·2 millim.), determinatus (K +, flavens); apothecia nigra, superficialia, plana, obtuse marginata (latit. 0·5–0·9 millim.), intus concoloria; sporæ 8næ, fuscæ, ellipsoideæ, 1-septatæ, longit. 0·016–0·023 millim., crassit. 0·008–0·010 millim., leviter superficie granuloso-inæquales; paraphyses gracilescentes (non bene discretæ); epithecium et hypotheclum luteo-fuscescentia (aut hoc magis infuscatum). Iodo gelatina hymenialis intensive cærulescens.

"Species bene distincta in stirpe *Lecideæ spuriæ*. Thallus I-. Spermatia breviter bacillaria, longit. 0·004 millim., crassit. 0·0006 millim."—*Nyl. in litt.*

On the rocks with the foregoing.

Lecidea continens, *Nyl.*

"Thallus lacteus vel sublutescenti-albidus, sat tenuis, continuus, superficie areolato-rimulosus (K flavens, medulla I cæruleo-obscurata); apothecia nigra, plana marginata (latit. 0·5–0·6 millim.), margine fuscescente, intus obscura; sporæ 8næ, fuscæ, ellipsoideæ, 1-septatæ, longit. 0·011–0·014 millim., crassit. 0·006–0·007 millim.; paraphyses fere mediocres; epithecium et hypothecium fuscescentia. Iodo gelatina hymenialis intensive cærulescens.

"Ad *Lecideam spuriam* quodammodo accedens et reactionibus conveniens, at thallus alius. Variat perithecium fusco-pallescens. Spermogonia non visa."

Not uncommon on rocks.

Lecidea configurans, *Nyl.*

"Thallus lutescenti-glauco-albidus, lævis (crassit. 0·5 millim. vel tenuior), subrimulosus, determinatus; apothecia fusco-nigra vel fusca, innata, plana (latit. 0·3–0·5 millim.), immarginata, circumscissa, intus albido-incoloria, demum 2–6 submoniliformi-confluentia (tum quasi lomentaceo-linearia et varie flexuosa); sporæ 8næ, incolores, oblongæ, simplices, longit. 0·011–0·014 millim., crassit. 0·004–0·005 millim.; paraphyses graciles; epithecium sublutescens aut dilute glauco-fuscescens; hypothecium incolor. Iodo gelatina hymenialis cærulescens, dein violascens.

"E stirpe *Lecideæ lævigatæ* species, simplex, facie fere *Lecideæ phæopis*. Thallus K flavens."—*Nyl. in litt.*

Frequent on rocks with the foregoing.

Tribe GRAPHIDEI.

Opegrapha difficilior, *Nyl.*

"Thallus macula albescente indeterminata vel obsoleta indicatus; apothecia nigra linearia (longit. 1–2 millim. vel breviora, crassit. fere 0·25 millim.), subsimplicia, recta aut subflexuosa, superficialia; epithecium rimiforme demum arcte angustatum; sporæ 8næ, fusiformes (vetustate obscuratæ), 5–7-septatæ, longit. 0·027–0·033 millim., crassit. 0·004–0·006 millim., hypothecium nigricans. Iodo gelatina hymenialis vinose rubens.

"Sat affinis *Opegraphæ hapaleoidi*, sed thallo, apotheciis et sporis differens. Apothecia sæpius conferta. Spermatia tenuia, rectiuscula, longit. 0·004–0·005 millim., crassit. 0·0005 millim. Comparanda, quoad faciem, cum *O. difficili*, Duf. (quæ non est nisi *O. cinerea*, Chev.! lignicola), spermatiis aliis mox divergens."—*Nyl. in litt.*

On decorticated stumps of trees; not common.

Arthonia phylloica, *Nyl.*

"Thallus macula alba indicatus; apothecia nigra tenuia erumpentia linearia (longit. 1 millim. vel nonnihil longiora aut breviora, latit. parum 1 millim. excedentia), simpliciuscula aut ramulum unum alterumve emittentia, variantia subastroidea; sporæ 8næ, incolores, oblongæ, 3-septatæ, minutæ, longit. circiter 0·011 millim., crassit. 0·003–0·004 millim.; epithecium subcærulescens. Iodo gelatina hymenialis vinose rubens (præcedente cærulescentia)."—*Nyl. in litt.*

Common on decayed leaves of *Pandanus*.

Arthonia dendritella, *Nyl.*

"Thallus macula albo-cinerascente subdeterminata indicatus; apothecia nigricantia, tenuia, innata (latit. 0·1 millim. vel tenuiora), subdendritico-divisa aut simpliciora, sat conferta; sporæ 6–8næ, incolores, oviformi-oblongæ, 3-septatæ, longit. 0·010–0·014 millim., crassit. 0·004–0·005 millim.; epithecium cærulescenti-nigrescens; hypothecium incolor. Iodo gelatina hymenialis subincolor et thecæ vinose violascentes.

"Species e stirpe *Arthroniæ palmicolæ*, præcipue facie nebuloso-petricola et reactione iodo effecta distincta."—*Nyl. in litt.*

On decayed leaves of *Pandanus*. Frequent.

Graphis scalpturata, *Ach.*

On bark of trees.

Graphis assimilis, *Nyl. Prodr. Gall.* p. 150 (*N. Gran.* p. 74).

On the bark of trees.

Graphis analoga, *Nyl.*

On the bark of trees, associated with other lichens.

Glyphis cicatricosa, var. **favulosa** (*Ach.*) *Nyl. N. Gran.* p. 108.

On bark of trees, not uncommon.

Glyphis tricosula, *Nyl.*

"Thallus albus vel albidus, opacus, tenuissimus, tenuiter nigricanti-limitatus; apothecia nigra, subcæsio-pruinosa, dendroideo-divisa, sparsa, parum ramosa (divisionibus latit. 0·2 millim. vel tenuioribus); sporæ 8næ, oblongæ, 4-loculares, longit. 0·011–0·012 millim., crassit. 0·005–0·006 millim. (demum infuscatæ); hypothecium fuscum.

"Est *Glyphis* facie *Medusulæ tricosæ* simplicioris, notis allatis satis tute distinguenda."—*Nyl. in litt.*

On stems of *Pandanus*.

Tribe PYRENOCARPEI.

Verrucaria 5-septatula, *Nyl.*

"Thallus macula albida indistincta vel nigricanti-limitata indicatus; apothecia pyrenio integre nigro, supra denudata convexula (latit. 0·2–0·5 millim.); sporæ 8næ, incolores, fusiformes 5-septatæ, parvæ, longit. 0·013–0·018 millim., crassit. 0·004–0·005 millim.; paraphyses sat graciles, non confertæ. Iodo gelatina hymenialis non tincta.

" Species videtur e stirpe *Verrucariæ chloroticæ*, sed paraphyses minus confertæ, quamobrem sit potius stirpis *Verrucariæ 5-septatæ*, Nyl. Pyrenoc. p. 58."

On *Pandanus* stems, not uncommon.

Verrucaria chlorotica, *Ach.*

On rocks; probably not uncommon, and quite similar to the European plant. Spores 0·018 millim. long, 0·0045 millim. thick.

On rocks; common.

Verrucaria denudata, *Nyl. Pyrenoc.* p. 49.

On the bark of trees.

Trypethelium cruentum, *Mnt.*

On the bark of trees. Spores 0·026–0·036 millim. long., 0·012–0·015 millim. thick.

Trypethelium cruentulum, *Nyl.*

" Forsan subspecies *Trypethelii cruenti*, Mnt., Nyl. Pyrenoc. p. 73, a quo differt præcipue sporis minoribus (longit. 0·018–0·025 millim., crassit, 0·008–0·011 millim." —*Nyl. in litt.*

On the rough bark of trees; notably on *Fœtidia mauritiana*.

Pyrenastrum americanum, *Spr.*

On the bark of trees; frequent on *Pandanus*.

FUNGI.

By the Rev. M. J. Berkeley, M.A., F.L.S.

Agaricus (Pleurotus) Balfouri, *Berk.*

Imbricatus, conchiformis; pileo primitus subtiliter pulverulento pallide umbrino, demum glabrescente ruguloso spadiceo, margine involuto, lamellis in basin incrassatam decurrentibus, primum venosis obtusis, antice reticulatis, demum latis acie crenulatis dentatisve concoloribus.

Pilei 1½ inch wide when fully developed, but often much less, especially when densely imbricated; occasionally the pileus is merely excentric, the gills running down to a distinct attenuated stem, with the gills reticulated above; in highly developed specimens they are two lines broad.

Is not at all common. Endemic.

Schizophyllum commune, *Fr. Ep.* p. 403.

On decayed branches of *Leucæna glauca* in one or two localities.

Hirneola polytricha, *Mont. Bél. Voy.* p. 154, sub *Exidia.*

Rare; only in the Cascade Valley.

Hirneola Auricula Judæ, *Fr. Hym. Eur.* p. 695, var. lævior, *Berk. Handb.* tab. 18, fig. 7.

Along with the foregoing.

Polyporus (Pleuropus) aspidopus, *Berk.*

Pileo dimidiato infundi buliformi e basi orbiculari oriundo, cito glabrescente rugosiusculo umbrino, stipite brevissimo vel omnino obsoleto cum basi confuso hymenio concolore, margine sterili; poris rotundis minimis, acie obtusis.

About 1 inch in diameter, generally dimidiate, or, from confluence of the margins behind, infundibuliform, in all cases arising from a distinct orbicular more or less hispid base; pores perfectly round, $\frac{1}{108}$ inch in diameter, with an obtuse margin; occasionally, especially below, the dessepiments are elongated into hydniform aculei, and the disc has raised processes on its surface.

On many trees; very common on *Pandanus.* Endemic.

Polyporus (Placodermei) microphlebius, *Berk.*

Ochroleucus; pileo dimidiato postice decurrente laccato zonato, zonis hic illic elevatis vel elongatis, scabroso-venulosis; hymenio concavo; poris minutissimis acie hispidulis

Pileus 2 inches wide, nearly 1 inch long, reniform; the zones sometimes slightly elevated, sometimes projecting, their surface very minutely venulose as if there had been a contraction of the laccate substance when first developed. Pores $\frac{1}{150}$ inch in diameter. In external appearance the species, of which there is a single specimen only, has some resemblance to one of the forms of *Polyporus zonalis,* B.

Not common. Endemic.

Trametes occidentalis, *Fr. Ep.* 491; *Klotzch Linn.* viii. p. 486 *sub Polyporo.*

Not common.

Bovista lilacina, *B. & M. in Hook. Lond. Journ.* iv. p. 64.

A single specimen only, in very bad condition. The spores are smooth as described in the place quoted above, but it is possible that they may become hispid as stated by Montagne, Sylloge, p. 287.

On the plains towards the south-west of the island.

ALGÆ.

By George Dickie, M.D., F.L.S.

FUCACEÆ.

Sargassum vulgare, *Ag.*
Geog. Dist.—West Indies; Senegambia; Spain and Portugal; Australia.

Sargassum ilicifolium, *Turner.*
Geog. Dist.—Cape Comorin; Philippines; Borneo; Sumatra.

Turbinaria ornata, *Turner.*
Geog. Dist.—Ceylon; Philippines; Tahiti; New Zealand.

SPOROCHNACEÆ.

Chroospora implexa, *Hering.*
Geog. Dist.—Red Sea; Ceylon, &c.

DICTYOTACEÆ.

Dictyota dichotoma, *Huds.* A narrow form.
Geog. Dist.—North and South temperate Oceans, less general in warmer seas.

Padina Pavonia, *L.*
Geog. Dist.—Found in most of the warmer seas. South coast of Britain; West, South, and East coasts of Australia.

Zonaria nigrescens, *Sonder.*
A few small examples about an inch in length. Prof. J. Agardh considers it as scarcely distinct from *Z. variegata*.
Geog. Dist.—Coasts of Australia.

Asperococcus sinuosus, *Roth.*
Geog. Dist.—King George's Sound; Tasmania; Mediterranean and Adriatic; Gulf of Mexico; Indian Ocean; Japan; Cape of Good Hope; New Zealand; Falkland Islands.

Hydroclathrus cancellatus, *Bory.*
Geog. Dist.—Tropical and subtropical seas.

CHORDARIACEÆ.

Mesogloia virescens, *Carm.* ?, var. gracilis ?.
Imperfect, and therefore doubtful.
Geog. Dist.—N. Europe; Australia; Tasmania; Friendly Islands; Red Sea; Japan; Loo-Choo.

ECTOCARPACEÆ.

Sphacelaria rigida, *Hering*.
Upon *Digenea*.
Geog. Dist.—Red Sea.

RHODOMELACEÆ.

Digenea simplex, *Wulf*.
Geog. Dist.—West Indies; East Coast of North America; Mediterranean and Adriatic; Red Sea; Ceylon; Japan.

LAURENCIACEÆ.

Laurencia papillosa, *Fork*.
Geog. Dist.—Warmer parts of Atlantic; on African shores; Gulf of Mexico; Mediterranean; Adriatic; Red Sea; Mauritius; Sandwich Islands; New Zealand.

Laurencia obtusa, *Hudson*.
Geog. Dist.—Temperate shores of Atlantic; Mediterranean; Adriatic; Ceylon; Mauritius; Philippines; Tahiti, &c.

Laurencia flagellifera, *J. Ag*.
Geog. Dist.—Shores of Hindostan.

CORALLINACEÆ.

Amphiroa fragilissima, *L*.
Geog. Dist.—Mediterranean; shores of Hindostan.

Lithothamnion mamillare, *Harv*.
Geog. Dist.—Shores of Brazil; Cape Verde; Algoa Bay.

SPHÆROCOCCOIDEÆ.

Gracilaria corticata, *J. Ag.*, var. linearis.
Geog. Dist.—Shores of Hindostan; Ceylon; Mauritius.

GELIDIACEÆ.

Hypnea hamulosa, *Turner*.
Geog. Dist.—Red Sea; Cape of Good Hope.

Hypnea Valentiæ, *Wulf*.
Geog. Dist.—Red Sea; shores of Hindostan; Philippines.

Gelidium rigidium, *Vahl*.
Geog. Dist.—Mauritius; Ceylon; Philippines; Tahiti; Gulf of Mexico.

HELMINTHOCLADIEÆ.

Liagora viscida, *Fork*.
Geog. Dist.—Mediterranean; warmer parts of Atlantic; Red Sea; Indian Ocean; Australia.

CRYPTONEMIACEÆ.

Chylocladia vigens, *Ag.*
Geog. Dist.—West Indies; Marian Islands; Japan; Philippines.

SIPHONACEÆ.

Caulerpa ericifolia, *Ag.*
Geog. Dist.—West Indies; Caroline Islands (Polynesia).

Caulerpa Boryana, *J. Ag.*
Geog. Dist.—Warmer parts of Pacific.

Caulerpa plumaris, *Vahl.*
Geog. Dist.—Ceylon; Red Sea; Philippines; Australia.

Caulerpa mamillosa, *Mont.*
Geog. Dist.—Indian Ocean; Gambier Islands.

Halimeda Opuntia, *Lamx.*
Geog. Dist.—Warmer parts of Atlantic; Mediterranean; Red Sea; Philippines, &c.

Halimeda Tuna, *Lamx.*
Geog. Dist.—Mediterranean; Philippines, &c.

Codium tomentosum, *Stackh.*
Geog. Dist.—Tropical and temperate oceans of both hemispheres.

Codium adhærens, *Ag.*
Geog. Dist.—Coasts of Europe; Ceylon; Cape of Good Hope; Friendly Islands; Loo Choo.

VALONIACEÆ.

Valonia Ægagrophila, *Ag.*
Geog. Dist.—Adriatic; Rawak Island.

Microdictyon Agardhianum, *Dne.*
Geog. Dist.—Mediterranean; Red Sea; Australia; Sandwich Islands.

Dictyosphæria favulosa, *Dne.*
Geog. Dist.—Red Sea; Indian Ocean; Sandwich Islands.

ULVACEÆ.

Ulva latissima, *L.*
Geog. Dist.—General in both hemispheres.

Enteromorpha intestinalis, *Link.*
Geog. Dist.—Widely distributed in both hemispheres.

Enteromorpha compressa, *Grev.* ?
The specimens rather imperfect, and therefore doubtful.
Geog. Dist.—Like the former, extensive in its range.

Enteromorpha ramulosa, *E. B.*
Geog. Dist.—European shores; coast of Amboina.

BATRACHOSPERMEÆ.

Batrachospermum moniliforme, *Ag.*, var. equinoctiale, *Bory.*
Geog. Dist.—Its various forms have an extensive range in the fresh waters of Europe, America, and Africa.

Thorea violacea, *Bory.*
Geog. Dist.—Mauritius.

CONFERVACEÆ.

Conferva moluccana, *Ag.* ?
The few examples are referred with some doubt to this species.
Geog. Dist.—Island of Pisang, one of the Moluccas.

Conferva bombycina, *Ag.*, var. inequalis, *Ktz.*
Geog. Dist.—Various parts of Europe.

Conferva Ansoni, *Ag.*
Geog. Dist.—" In fonte Ansonii insulæ Tinian " (Gaudichaud).

Cladophora Roettleri, *Ktz.*
Geog. Dist.—Tranquebar.

Cladophora pannosa, *Dickie.*
The specimens seem identical with the plant from Mauritius, which in a notice of the Algæ of that island was considered a new species.

Chantransia cærulescens, *Mont.*
Geog. Dist.—Cayenne.

Œdogonium fonticola, *A. Braun.* ?
Agrees with the species in the oogonia and proportions of the individual joints, but the characters of the basal and terminal cells could not be seen.
Geog. Dist.—Various parts of Europe, and found also in the island of Madeira.

ZYGNEMACEÆ.

Rhynconema angulare, *Hass.*
The few specimens observed agree in general characters with this species.
Geog. Dist.—In various parts of Europe.

Staurospermum capucinum, *Bory.* ?
The specimens agree in the proportions of the individual cells as to length and breadth; the latter = ·00055, but in the absence of zygospores the plant is doubtful.
Geog. Dist.—Various parts of Europe.

OSCILLARIACEÆ.

Lyngbya majuscula, *Dillw.*
Geog. Dist.—European shores.

Leptothrix lutea, *Ktz.*
Geog. Dist.—" In thermis Enganearum" (Meneghini).

NOSTOCHINEÆ.

Nostoc commune, *Vauch.*
Geog. Dist.—Europe; America; Java.

ZOOLOGY.

EXTINCT FAUNA.

OBSERVATIONS ON THE BONE CAVES OF RODRIGUEZ.—*By Henry H. Slater, B.A., Naturalist to the "Transit of Venus" Expedition.*

I.—The Cave Region of Rodriguez.

The cave tract in Rodriguez is situated about the S.W. side of the island, and is of a very curious nature. We find there 10 or 12 patches of limestone scattered upon the basalt which forms the island; these patches are of irregular form, and usually terminated by an escarpment of various height, from 3 to 10 or 12 feet, which marks their juncture with the basalt. On examination, these patches are found to consist of marine coral upheaved with the basalt.

Sometimes the limestone preserves its original structure or grain which it had in life; sometimes this grain is brought out by weathering, the outer walls or thecæ being harder than the rest, and standing out in relief when the rest is denuded away; sometimes, but more rarely, the septa are visible. But, as a rule, the limestone is much altered: whether by pressure during or previous to upheaval, or whether by an infiltration of acidulous water with carbonate of lime in solution, I must leave to more able geologists than myself to settle upon an investigation of my colleague, Mr. Balfour's, collection of rocks; but the stone is usually rendered very hard and solid, though always very porous. Sometimes weathering produces a quasi-oolitic grain on the surface.

It is in these coralline limestone patches solely that the caves are situated; in Bourbon, on the contrary, the only caves were in the igneous rock itself, and appeared to be the effect of rapid cooling.

The caves have been, if not originated, at least much enlarged by water, of which many bear abundant traces, and in the rainy season some are evidently the courses of subterranean streams.

Another proof of the marine origin of the caves is found in the presence therein of several marine shells, as *Pleurotoma* sp., 150 yards above the sea, and 2,500 yards from the coast line; and *Terebellum* sp. In places where the coral is at its minimum of alteration, perforations were seen in it, which from their slightly oval shape, and gradually increasing diameter, seemed identical with those of the *Lithodomi*, at present common on the reefs in Rodriguez. I quote in allusion from Sir C. Lyell's "Principles:" "Perforations of Lithodomi in limestone cliffs, and in the

" Temple of Serapis at Puteoli, afford conclusive evidence of change in the sea level
" of coasts in modern times."

There exists near my camp at the caves a sort of ravine, terminated at each end by a cavern, and having others opening into it. The terminal caves and precipitous sides at once determined me that this has been a vast cavern, the roof of the greater part of which has fallen in; this fall has left a sort of ravine with a level bottom surrounded on all sides by precipices, nearly, if not quite, perpendicular, and having a height of from 30 to 90 feet; the bottom is now covered with earth and full of large trees, the tops of which rise to the level of the cliffs. Descent can only be effected with ease in two places, where two heaps of limestone blocks rear themselves against the precipices. There is no reason to believe that water ever accumulates in the caverns opening into this "Gorge," and in these caves most of my specimens of any value were found, whilst the trees in the Gorge, being the abode of guinea-fowl, supplied my table. There were, judging from the hollow sound of the coral in various places under my boots, other caves with no mouths at all, but I did not judge it needful to open them. Capt. Wharton, of H.M.S. "Shearwater," opened one about a mile and a half W.S.W. of the Gorge, and descended by means of a rope, but nothing was to be seen but stalactites, nothing like earth on the floor.*

II.—*Bone Earths and Stalagmites.*

The cave-earth was always similar in colour to the reddish brown earth found on the basaltic parts of the island, but differed from it in being generally mixed with a greater or smaller per-centage of coralline sand, formed from the crumbling of the interior of the cave, or washed in. The earth itself was probably washed into such caves as were liable to the influx of water; into the dry caves it seemed to have been blown in in the form of dust, which form it often kept.

Mixed with the cave-earth was a large admixture of land shells, including—

1. *Cyclostoma articulatum*, now living in the island abundantly.
2. *Cyclostoma hæmastoma*, much smaller than above, and also living in the island.
3. *Helix bewsheriana*, $1\frac{3}{4}$ inches in diameter across the whorls.
4. *Pupa sp.*, 1 inch in length, and $\frac{1}{3}$ inch in diameter.

The two latter are now extinct, and only found in the caves; there are also a small *Pupa*, and two small *Bulimi* now living in the island, and an *Achatina* (extinct), which occur in the cave-earth, with remains of a huge *Julus*, much larger than those now in the island, equalling the species in the Seychelles.

I never met with any human remains, either bones or implements, in the cave-earth.

* I was surprised when I made my preliminary survey of the cave tract in Rodriguez to discover the amount of work already done. Out of the first 13 caves, I found 12 to bear unmistakeable signs of previous research. Nevertheless, I deemed it best to dig them over again, a proceeding which elicited a certain number of bones, though it gave no information whatever as to their deposition.

The depth of the bone-earth is very variable; in some caves we find it with a depth of from six inches to three feet; in others, however, it varies from four to nine feet in depth. Below about two feet I never found many bones, which makes me believe that the agency which deposited the bones in the caverns, never operated until the later days of the existence of the Solitaire. The bones might certainly have decayed, but yet I usually found that the bones which were well covered with earth were in much better preservation than those near or upon the surface, which were usually much decayed. This makes me think that the Solitaire resorted to the caverns in case of fire in the island, which has been known to have denuded it several times of its trees; more so, as in several cases I found nearly perfect skeletons, which lay evidently as they died; this precludes the idea that they were carried there by wild cats. Again, in the bottom of a cleft near the mouth of a cave, I found the greater part of the skeletons of a male and female Solitaire; they had clearly fallen in and were unable to extricate themselves, but the bones being but imperfectly covered, many bones were so much decayed as to prevent their removal. But I could not doubt that they arrived there alive, and if not driven into the neighbouring hole as a shelter under some alarm, what could have allured them there? Nor were any bones found in the caves at any distance from the mouths, at least, if there were, they were single, and looked as if they had been washed there. Land crabs inhabit some of the caves, and may have conveyed the dead Solitaires into the inmost recesses, when decay enabled them to remove portions of the carcases at a time.

Stalagmite of any thickness is uncommon in the caves; in three caves only did it occur in anything like extensive sheets, but it often occurred in small patches of a yard or so in diameter under some long stalactite. But in one long cavern, near a fishing station called Patates, there was evidence of there having been rather an extensive system of stalagmite layers. This cave was nearly a mile in length, and ran downhill, having an elevation of about 200 feet above the sea level at one mouth, and not more than 25 at the other; moreover, it was clearly occasionally the bed of a stream. In one place in this cave there was an interesting relic in the shape of a sort of column composed of alternate floors of stalagmite and cave-earth. It occurred in an angle of the cave where the force of a stream coming down from the higher mouth would just have missed it. It was about 8 feet in height, and there were three or four layers of stalagmite in it. At one period, before the streams began to flow through the cave, this system of alternate layers had existed all over the cavern, but had been broken up by the stream, which would wash out the earth, when the stalagmite would fall in. Evidences of the stream were to be found in the trunks of trees and beds of leaves, with smooth ripple-marked sand in the floor of the caves.

THE EXTINCT BIRDS OF RODRIGUEZ.*—*By Dr. A. Günther, F.R.S., and Edward Newton, C.M.G., M.A., F.L.S.*

(Plates XLI.–XLIII.)

The following notes are the results of our examination of two sets of materials, one collected by Mr. H. H. Slater, one of the naturalists of the Transit of Venus Expedition, the other obtained previously by Mr. George Jenner when magistrate of Rodriguez. The majority of the remains brought home by Mr. Slater have been deposited in the British Museum, whilst the bulk of the latter set has been deposited in the Museum of Zoology of the University of Cambridge. As far as practicable, and without detracting from the value of each set, the two sets have been mutually supplemented by an exchange of duplicate specimens, but originally they consisted of the following remains:—

Carine murivora.

	T. of V.	Jenner.		T. of V.	Jenner.
Cranium	—	1	Metacarpus	1	2
Mandible	1	1	Coracoid	—	1
Pelvis	2	3	Femur	3	9
Humerus	4	5	Tibia	2	10
Ulna	—	2	Metatarsus	3	6

Necropsar rodericanus.

	T. of V.	Jenner.		T. of V.	Jenner.
Cranium	1	1	Ulna	8	—
Mandible	1	2	Metacarpus	1	—
Sternum	—	1	Femur	2	2
Coracoid	1	—	Tibia	15	23
Humerus	5	7	Metatarsus	11	5

Turtur picturatus.

	T. of V.	Jenner.		T. of V.	Jenner.
Sterna	3	—	Tibia	4	—
Humerus	1	—	Metatarsus	1	—

Necropsittacus rodericanus.

	T. of V.	Jenner.		T. of V.	Jenner.
Upper part of cranium	—	1	Ulna	2	1
Mandible	1	1	Femur	2	1
Coracoid	2	—	Tibia	2	2
Humerus	6	4	Tarsus	1	—

Palæornis exsul.

	T. of V.	Jenner.		T. of V.	Jenner.
Humerus	—	6	Femur	—	1
Ulna	—	1	Tibia	—	5

* With the exception of the Solitaire, which is the subject of a subsequent separate paper.

Erythromachus leguati.

	T. of V.	Jenner.			T. of V.	Jenner.
Cranium	1	2	Ulna		8	4
Bill	5	2	Radius		2	—
Mandible	—	5	Metacarpus		2	—
Pelvis	5	6	Femur		13	15
Sternum	2	—	Tibia		16	18
Scapula	1	—	Tarsus		10	17
Humerus	3	12				

Ardea megacephala.

	T. of V.	Jenner.		T. of V.	Jenner.
Cranium with bill	—	1	Ulna	—	2
Tympanic	—	1	Radius	—	1
Mandible	2	1	Metacarpus	1	2
Pelvis	—	1	Phalanx	—	1
Vertebra	—	3	Femur	2	1
Sternum	—	1	Tibia	2	2½
Coracoid	—	2	Fibula	—	1
Scapula	—	2	Metatarsus	4	4
Humerus	2	4	Phalanx	—	3

Carine murivora (A. M.-Edw.)

(Plate XLI., Figs. A.–F.)

M. A. Milne-Edwards (Ann. Sc. Nat. Zool. t. xix., art. 3, p. 12) recognized two tibiæ and one metatarsus as belonging to two species of Owls, one of which he named *Strix (Athene) murivora*, the other (represented by one tibia only) he left unnamed. This last was 6 mm. longer than that of *C. murivora*, and said to be distinguished by a shorter and less projecting peroneal crest. Having this individual bone before us, together with a considerable number of tibiæ of the shorter and longer kind, we have been able to ascertain that a great portion of the peroneal crest is broken away, that in perfect specimens the crest of the longer bones is as much developed as in the shorter ones, and that, in fact, the former are evidently those of females and the latter of males, according to the difference in size which obtains between the sexes of nearly all the nocturnal birds of prey.

After having satisfied ourselves that all the Strigine remains from Rodriguez are referable to one species only, *Carine murivora*, we hoped to be able, with the aid of so much more material than was at the disposal of M. Milne-Edwards, to throw more light on its generic affinities. In this we were not successful, chiefly for the reason that the genera of Owls adopted by some ornithologists do not by any means coincide with osteological modifications, and that the Rodriguez Owl shows a combination of osteological characters which we have not found in any of the numerous skeletons of *Striges* with which we have compared it (*Bubo, Scops, Surnia, Carine, Ninox, Glaucidium, Asio*, &c.).

Of the *cranium* (Fig. A.), the frontal region, with a portion of the parietal region, has been preserved. The frontal bone is not swollen at the root of the bill, as, for instance, in *Carine boobook* or *C. hirsuta*, but rather concave, as in *C. infuscata* and *C. noctua*. The orbit appears to have been of an unusually short diameter, and the eye is not protected by the long supraorbital process of *C. noctua* and *C. infuscata*, this process being reduced to a very short prominence, as in *C. novæ-zelandiæ*, *C. boobook*, and *C. hirsuta*. The margin of the orbit in front of the supra-orbital protuberance is longitudinally grooved. Width of anterior part of frontal region, 13 mm.; least width of interorbital space, 15½ mm.

The *mandible* (Fig. B.) does not show any peculiarity; one of the rami is 40 mm. long.

The form of the *pelvis* (Fig. E.), particularly of the postacetabular portion, differs from that of all other Owls with which we are acquainted. The anterior iliac blades are of moderate width, presenting a deeply excavated outer surface, the lower portion of which is nearly horizontal, a small pointed tubercle projecting on its outer margin (as in *Scops rutilus*). The convergent dorsal edges of this part of the ilium do not coalesce into a crest, but are connected by a flattened lamina, which forms the roof over the large longitudinal vacuities on each side of the sacral crest. In its middle the pelvis is but moderately constricted. The postacetabular portion is broad, chiefly owing to the great width of the sacrum, which is of rhombic shape, dorsally slightly depressed in the median line. The post-iliac crest is unusually prominent and more strongly curved, in an S-shaped line, than in any other Owl with which we are acquainted.

There exist considerable modifications of the configuration and arrangement of the foramina for the sacral nerves in the various species of Owls, even in such as are referred to the same genus. The Rodriguez Owl resembles in this respect *Carine noctua*, and still more *Scops rutilus*, the three anterior foramina being equidistant from each other, of nearly the same width, and separated by the strong pleurapophyses, all of which are perpendicular to the axis of the vertebral column. The fourth foramen is much constricted by pleurapophyses, which are much stronger than the preceding, and situated at a different level, opposite to the point of the greatest constriction of the pelvis. The deep anterior and posterior renal excavations are separated from each other by a pair of slender pleurapophyses. On the whole, the pelvis of this bird is of an unusually strong build, and well adapted to form the base of the powerful muscles of the trunk and hind limbs by which this small-bodied bird was enabled to keep down the rats with which the Island was infested, as Leguat informs us, and later to overpower " les petits oiseau et les petits lézards " as the author of an old manuscript says it did.*

---

* Professor Newton has published the abstract of this MS. " Relation de l'Ile Rodrigue," in Proc. Zool. Soc. 1875, p. 39; and the entire MS. has been published by Professor A. Milne-Edwards in Ann. Sci. Nat. (6) ii. art. 4, pp. 20.

The passage runs as follows :—" On voit une oiseau qui est à peu près comme la chouette, et qui mange " les petits oiseau et les petits lézards."

The pelvis does not appear to vary much in size.

	Carine murivora.	Carine boobok.	Asio accipitrinus.
Length of pelvis (to end of sacrum)	36 mm.	34 mm.	40 mm.
Greatest width of anterior part	21 ,,	22 ,,	21 ,,
Least width of anterior part	14 ,,	14 ,,	13½ ,,
Greatest width of posterior part	29 ,,	23 ,,	25 ,,

The single *limb-bones* do not show any characteristic features, and very little need be said about them beyond the general conclusion at which we have arrived from comparative measurements of those of other Owls. The *coracoid* (Fig. C.) is a little weaker than in *C. boobook*, and 30 mm. long, its sternal extremity being 10 mm. broad. The *foramen nutritium* of the shaft of the *humerus* (Fig. D.) is generally (but not constantly) nearer to the distal than to the proximal extremity. The *ulna* shows elevations for the secondaries. In the *femur* (Fig. F.) all the lines and grooves for muscular insertions are strongly marked; the outer inferior condyle is very prominent, and provided with a deep groove for the tendon behind. The *tibia* and *metatarsus* have been described by M. Milne-Edwards.

The discovery of a part of the cranium and the pelvis enabled us to obtain a pretty accurate idea of the size of the Rodriguez Owl. Its body was rather larger than that of *C. boobook*, but considerably smaller than that of *Asio accipitrinus*. The wing was clearly somewhat shorter and a little less developed than in either of those Owls, though the power of flight of this bird cannot have been much impaired. On the other hand, the length and strength of its legs, as compared to the size of the body, and especially to the length of the wing, was much more than in any of the species of Owls, the skeletons of which we have been able to compare, as will be seen from some of the measurements which we have taken, and which we tabulate thus:—

	Humerus.	Ulna.	Metacarpus.	Femur.	Tibia.	Metatarsus.
	longest. shortest.			longest. shortest.	long. short.	long. short.
Carine murivora	69 × 64	74	36	54 × 50	76 × 69	46 × 41
,, boobook	69	76	36	48	74	41
,, infumata	49	28	58	39	58	36
,, noctua	51	28	56	41	59	37
,, hirsuta	56	30	64	39	60	32
,, castaneoptera	48	55	24	33	49	23
,, connivens	115	137	60	65	109	—
Spiloglaux novæ-zelandiæ	58	30	65	41	64	35
Asio accipitrinus	81	90	42	54	77	40
Scops rutilus	47	54	23	32	50	28
Bubo virginianus	163	178	85	101	146	75
Ketupa zeylonensis	76	121	68	120	129	54

Necropsar rodericanus, Slater.

(Pl. XLII., Figs. A.–G.)

The discovery of an extinct Starling in Rodriguez, allied to the *Fregilupus* of Réunion, which appears to have held out a little longer in struggling for its existence within so narrow limits, is undoubtedly one of the most interesting results of Mr. Slater's labours. He recognised and determined the affinities of the remains found by him, and says, in preliminary manuscript notes appended to his general report: "Professor Newton prophesied, as it were, the existence of this bird (Proc. "Zool. Soc. 1875, p. 41) when, alluding to a bird mentioned in a manuscript "'Relation de l'Ile Rodrigue,' he said, 'I am at a loss to conjecture what these birds "were, unless possibly of some form allied to *Fregilupus*.' *Necropsar* is altogether a "smaller bird than *Fregilupus*, to which it is most closely allied. If it were a "mammal, I should not make a separate genus of it; but as it is a bird, I think I "cannot do less."

After a close scrutiny of all circumstances, we quite agree with the discoverer of this bird in questioning the propriety of generically separating two species, on what appear to us very slight modifications of the osteological frame; and we retain the name *Necropsar* merely from the wish of conforming with the present ornithological practice.

In the following notes more especially those points will be mentioned in which the Rodriguez Starling differs from that of Réunion, the latter having been fully described in Dr. Murie's excellent paper in Proc. Zool. Soc. 1874, p. 474, to which we have to refer for other details. The skeleton of *Fregilupus* itself, which is unique, has been lent to us by Professor Newton.

Of the *skull* of *Necropsar* (Fig. A.) the cranium with the maxilla and mandible has been preserved; but all the bones of the palatal region and the tympanic have been lost. It is extremely similar to the skull of *Fregilupus*, and more especially possesses a postorbital muscular impression above the temporal fossa, which is very conspicuous in *Fregilupus* and *Acridotheres* (*cristatellus*), and faintly visible also in the common Starling (*Sturnus*). But it differs from *Fregilupus* in having not only the supraoccipital ridge conspicuously more developed, but also in showing, besides, an additional line of muscular insertion (biventer *) running across the parietal region, subconcentric with the supraoccipital ridge. The development of these ridges indicates a corresponding increase of strength in the muscles of the neck and articular end of mandible. The maxilla is a little shorter, and decidedly less curved, and less slender at the extremity than in *Fregilupus*, the nasal aperture being longer and generally larger. The mandible † is on

---

* See Owen, Anat. Vert. II., p. 85, fig. 35. 18.

† The articular facet of the mandible is figured on Pl. XLII., Figs. *a* and *a*¹, half as large again as the natural size. Fig. *a* represents its configuration as it appears in two of the specimens before us; but in

the whole of a stouter build, and about the middle of its length (at the angle of the mouth) more bent, in the fashion of the corresponding part of the Starling. Peculiarities all of which indicate that *Necropsar* had to use greater force than *Fregilupus* in searching and perhaps digging for its food, and that it possessed in an eminent degree the capability found in several species of Starlings and Crows, of removing objects or forcing an entrance in their search for food, by first inserting their wedge-shaped bill and then opening the mandibles.

The greater strength of the bill would also be fully accounted for, if *Necropsar* be the bird described in the old manuscript previously mentioned as feeding on eggs and dead tortoises " qu'ils savent assez bien déchirer."*

The fenestra in the interorbital septum is wider than in *Fregilupus*.

	Fregilupus.	Necropsar.			
		A.	B.	C.	D.
Length of cranium (measured from occipital condyle)	26 mm.	—	—	28 mm. or	29 mm.
Greatest width of cranium	23 ,,	—	—	21 ,,	21 ,,
Width of interorbital space	7½ ,,	—	—	8 ,,	8 ,,
Length of bill	41 ,,	—	—	39 ,,	36 ,,
,, nasal aperture	11 ,,	—	—	13 ,,	12 ,,
,, mandible	58 ,,	59	53	53 ,,	—
Greatest depth of mandible	5 ,,	5	5	4½ ,,	—

Of a *sternum* only the fore part has been preserved, which does not show any difference from the corresponding bone of *Fregilupus*, except in the foremost part of the keel being 1 mm. lower.

The *coracoid* (Fig. B.), 27½ mm. long, is otherwise identical with that of *Fregilupus* (29 mm.).

Such of the wing and leg bones as have been preserved differ nowise from those of *Fregilupus*, *Acridotheres*, or *Sturnus*. Figures, half as large again as the natural size, are given of the *humerus* (C.), *metacarpus* (D.), *femur* (E.), *tibia* (F.), and *metatarsus* (G.). The ulna is distinctly marked with elevations for the secondary quills. All these bones vary in size; but as far as we can judge (with any amount of

the third specimen (Fig. *a*) the angular developes a transversely extended process directed inwards, which is entirely absent in the two other specimens. One of the authors (E. N.) is inclined to consider this process as an individual malformation.

* The passage runs as follows:—

" On trouve un petit oiseau qui n'est pas fort commun, car il ne se trouve pas sur la grande terre ; on en voit su l'ile au Mât, qui est au sud de la grande terre, et je crois qu'ils se tient sur cette île à cause des oiseaux de proie qui sont à la grande terre, comme aussi pour vivre avec plus de facilité des œufs de ces oiseaux de pêchent qui y pondent, car ils ne mangent autre chose que les œufs ou quelques tortues mortes de faim qu'ils savent assez bien déchirer. Ces oiseaux sont un peu plus gros qu'un merle et ont le plumage blanc, une partie des aîles et de la queue noire, le bec jaune aussi bein que les pattes, et ont un ramage merveilleux ; je dis un ramage quoiqu'ils en aient plusieurs, et tous différents, et chacun des plus jolis. Nous en avons nourri quelques uns avec de la viande cuite hachée bien menu, qu'ils mangeaient préférablement aux graines de bois."—Proc. Zool. Soc. 1875, p. 41.

safety) by comparing their average length with that of *Fregilupus*, the length of the forearm of the latter was relatively to that of the humerus somewhat greater than in *Necropsar*, while the proportionate lengths of *tibia* and *metatarsus* were exactly the same, as may be ascertained from the following measurements :—

	Necropsar.		Fregilupus.	Sturnus vulgaris.
	longest	shortest		
Length of humerus	35	32	38	27
,, ulna	40	37	47	34
,, metacarpus	22½	—	24	—
,, femur	34	32	37	26
,, tibia	59	52	65	46
,, metatarsus	41	36	45	—

Necropsittacus rodericanus, M.-E.

(Pl. XLII., Figs. H.—N.)

As more parts of the skeleton of this extinct bird are discovered, its affinity to *Palæornis* (which has been maintained by M. Milne-Edwards, in his second treatise on this bird, p. 134,) becomes more and more apparent. The most peculiar portion of the skeleton (as far as we know it at present) is the cranium (Fig. H.), which is singularly depressed, the occipitals being low, the occipital foramen wider than deep, the parietals but slightly convex, and gently sloping behind towards the occipital region. This depressed form of skull we have not observed in any other genus of parrots except in *Ara*, but the birds of that genus have a complete suborbital arch, while in *Necropsittacus*, as in *Palæornis*, the uncinate bone is not anchylosed to the squamosal process. To judge from the figure of *Lophopsittacus mauritianus* (Encyl. Brit. Ed. 9, iii., p. 732), that bird may have had a similarly flat and depressed cranium.

	Necropsittacus.	Palæornis (alexandr).	Ara militaris.
Length of base of cranium (without bill).	50 mm.	35 mm.	59 mm.
Width of cranium	38 ,,	28 ,,	48 ,,
Depth of ,,	24 ,,	21 ,,	34 ,,

The several limb-bones vary considerably in size and length, in a similar manner as those of the other extinct birds of Rodriguez; but this variation does not seem to indicate a sexual difference, as there are intermediate specimens between the longest and shortest. Such of the bones as have been preserved do not show any noteworthy feature; we have given figures of the natural size of the *coracoid* (J.), *humerus* (K.), *femur* (L.), *tibia* (M.), and *metatarsus* (N.) A comparison of their relative lengths with those of *Ara* and *Palæornis* appears to be of direct interest.

	Necropsittacus.		Palæornis (alexandri).	Ara militaris.
	longest	shortest		
Length of coracoid	—	35	31	53
,, humerus	53	49	43	79
,, ulna	—	57	52	99
,, femur	49	46	37	60
,, tibia	63	59	50	83
,, metatarsus	—	22	18	31

If we considered the skull as a safe guide to obtain an estimate as to the size of the bird, we should infer that the length of the wing as well as of the leg had been diminished in this bird, as compared with *Palæornis* and *Ara*. But we believe that the pelvis affords a much surer basis for such calculation, and are, therefore, contented to leave this question undecided for the present, merely maintaining that in the extinct Rodriguez Parrot all the limbs were smaller relatively to the size of the head than in *Ara* and *Palæornis*, and that the fore-limbs were relatively to the hind-limbs not less developed in the Rodriguez bird than in those two genera.

Palæornis exsul, A. Newton.

The fragments of the skeleton collected by Messrs. Jenner and Slater have been mentioned at the beginning of this paper. As, however, a nearly complete skeleton of a freshly killed bird has been obtained, a description of the isolated bones in their collection would be obviously superfluous.

Columba rodericana, M.-E.

M. Milne-Edwards has described this Pigeon from a single sternum so different from that of other members of this family (as far as they are known), that we cannot refer it, at present, to any particular genus. A metatarsus seemed to him to afford evidence of the existence of a second species, which he believed to be *Turtur picturatus* of Madagascar, Réunion, and Mauritius. We have before us two metatarsi, not differing from that examined by M. Milne-Edwards, and do not see the reason why they should be separated from *C. rodericana*, to which, evidently, all the other bones of our series belong.

All the limb-bones (*humerus, femur, tibia*, and *metatarsus*) vary not inconsiderably in size, but there is a gradual passage from the shortest specimen to the longest, so that this difference is not likely to be sexual. The only evidence we have found of the possible existence of a second larger species of Pigeon is a femur in Mr. Jenner's collection which is 42 mm. long, and comparatively more slender than that of *C. rodericana*. The line of muscle-insertion on the front of the shaft starts from the very margin of the bone below the trochanter, as in a

Pigeon, and not inwards of the margin as in a Parrot. But no definite conclusion can be safely arrived at from this single bone.

The limb-bones of the Rodriguez Pigeon do not show any perceptible noteworthy peculiarities. The *humerus* has a deep semi-globular pneumatic cavity, and a small tubercle on the ulnar margin just above the condyle. The *femur* is slightly bent forwards, with the line of muscle-insertion starting from the margin of the bone, and running inwards along the middle line of the front side of the shaft. The following are the measurements:—

	Shortest.	Longest.
Length of humerus	34 mm.	37 mm.
,, femur	36 ,,	37 ,,
,, tibia	44 ,,	48 ,,
,, metatarsus	25 ,,	26 ,,

Aphanapteryx and Erythromachus.

(Plate XLIII.)

The classical paper "Sur les affinités zoologiques de *l'Aphanapteryx*," in which M. A. Milne-Edwards, with the most convincing precision, demonstrates the Ralline characters of a few ornithic remains sent from Mauritius, recognizing their connexion with the picture so fortunately and opportunely discovered by Von Frauenfeld, offers such a sure base for all succeeding workers that scarcely more remains to be done than to supplement his memoir by description of those parts of the skeleton with which he was not acquainted.

At a somewhat later period M. Milne-Edwards received a metatarsus and sternum from Rodriguez, in which he recognized the representative of the Mauritian *Aphanapteryx*, but misled (as we think) by Leguat's description, he believed that the Rodriguez bird differed by having a straight bill and shorter legs, referring it to a distinct genus which he named *Erythromachus* (Ann. des Sc. Nat. Zool. t. xix. art. 3, p. 6). These differences do not in reality exist, the bill of the Rodriguez bird is as much curved as in the Mauritian, and their legs have the same proportion. The wings were clearly more developed in the species inhabiting the smaller island, and certainly lack the power of flight, but this can scarcely be relied upon as a generic distinction, and, therefore, we prefer looking upon the Rodriguez bird as a smaller species of *Aphanapteryx* with somewhat longer wings, and to treat of it as *Aphanapteryx leguati*.

The search for further evidence of *Aphanapteryx* in Mauritius, subsequent to M. Milne-Edwards's memoir, has resulted only in the discovery of some more or less broken metatarsals, besides fragments of other portions of the leg which do not admit of determination. From the small size of the former bones, it is clear that *Aphanapteryx broecki* varied considerably in size, or that two races existed, which is not very probable.

The *cranium* of *A. leguati* (Fig. A.) is, in general appearance, extremely similar to that of *Rallus* (*aquaticus*) and *Ocydromus*, being slightly elongate, equally convex in every direction, and with the processes and ridges but slightly projecting. The orbital edge of *Ocydromus* is truncate nearly in its whole circumference (perhaps for the attachment of a glandular band), but in *A. leguati* it is rounded off, as in *Rallus*. The frontal region is narrow, long, with the lateral margins slightly diverging in front. All the bones belonging to the palate and the lachrymals are lost, but so much of the jaws has been preserved as to give a perfect idea of the form of the bill, which is curved in exactly the same manner and to the same degree as in *A. broecki*. It is very feeble, and varies extraordinarily in length, being in some specimens one third shorter than in others. The extremity of the longer beaks (Fig. A.) is more conspicuously curved than that of the shorter ones (Fig. B.) Besides, we may state at once that throughout the series of bones at our disposal a marked difference in size may be traced; but whether the short bills belong to the smaller individuals, or whether the difference in length of beak and in size generally is attributable to sex, we are unable to say. The culmen of the bill above the nostril is almost straight; the nasal aperture of extraordinary length, slightly longer than the cranium. The extremity of the bill is provided with a deep and narrow groove along its under side.

The *mandible* (Figs. A., C., C¹) is identical with that of the Mauritian species in every respect, except in size, more especially it has the three postarticulary apophyses (C.) arranged in exactly the same manner, but less projecting, as may be expected in the smaller species:—

Length of cranium measured from occipital condyle	38 mm.
Greatest width of cranium	20 ,,
Least width of frontals	6 5 ,,
Length of bill	77 ,,
,, nasal aperture	43 ,,
,, lower jaw	98 ,,
Greatest depth of lower jaw	8 ,,

The *pelvis* (Fig. D.) is large, very strongly built relatively to the size of the bird, and has thoroughly Ralline characters; its præacetabular portion is narrow, with the iliac blades scarcely divergent, and dorsally confluent into a high crest, the profile of which is much more strongly curved than in *Ocydromus*; a strong crest runs along the median line of the back of the sacrum. The postacetabular portion is dilated, much broader than in *Ocydromus* or *Rallus*, with deep and capacious renal excavations, and with the outer descending surface of the os ischium formed into a deep broad short hollow, bordered by an upper (postiliac) projecting crest, and by the lower reverted margin of the bone. The postiliac crest is not continuous, but interrupted above the ischiatic foramen as in *Rallus* and *Ocydromus*.

The ischiatic foramen is circular, about equal in extent to the acetabulum. The slender pubic rods are not dilated posteriorly or attached to the ischium, so far as we can judge from a single specimen, in which the pubic bone is preserved, with only a very small portion of its extremity broken off.

Length of pelvis to end of sacrum - - - 57 mm.
Greatest width of anterior part - - - - 20 ,,
Least width of anterior part - - - - 13 ,,
Greatest width of posterior part - - - - 36 ,,

A single *scapula* (Fig. E.), which we are inclined to refer to this bird, is imperceptibly dilated behind the middle of its length, and tapering in its posterior third. The articular surface for the shoulder-joint is very small. The whole bone is 45 mm. long, with a greatest width of 4 mm.

The *humerus* (Fig. F.) resembles much that of *Ocydromus*, its shaft being straight, somewhat compressed, and moderately slender. The projecting process behind the head is almost hamate; and there are no pneumatic foramina. Notable is the great difference in the size of this bone, the largest specimens being 50 mm., the shortest 45 mm. long. The larger and smaller specimens are represented in about equal numbers in the series before us; and several are intermediate between the extremes mentioned.

The *ulna* is considerably bent, in some specimens rather more so than others; relatively to its length it is stout, with a conspicuous ridge running from the inner side of the elbow joint along the greater part of its length; some faint elevations for the attachments of the secondary quills are visible. As regards length, it varies to a similar extent as the humerus, the longest specimen being 42 mm., the shortest 37 mm., long.

The *radius* is much more slender and weaker than in *Ocydromus*.

In the *metacarpus* (Fig. G.), the imperfect development of the wing is still more apparent than in the proximal bones; its constituent parts being very slender, and the middle and annular bones anchylosed at the extremities only, leaving a large vacuity between them. A minute tubercle is situated on the side of the middle bone which is opposed to the annular. The two specimens before us measure 16 and 14 mm. in length, the larger being as long as the metacarpal of *Ocydromus*, and not much longer than that of *Rallus aquaticus*.

The *femur* (Fig. H.) is, relatively to its length, stouter than in *Ocydromus*; the trocharterian ridge well marked. The shaft is slightly bent, with a faint ridge for the insertion of muscles along its anterior and posterior side. As regards length, the specimens vary from 56 to 63 mill.

The *tibia* (Fig. I.) and *metatarsus* (Fig. K.) are, in every detail of their conformation, identical with those of *A. broecki*, so that any description would be merely a repetition of that given by M. Milne-Edwards. The shaft of the tibia seems to be a little more elliptical in a cross section in the Rodriguez bird; and its middle trochlea projects somewhat less beyond the outer ones. The great extent of variation in size of the Rodriguez Rail is clearly shown by our series of these two bones; specimen A (from Mr. Slater's collection) must have belonged to an individual of unusually large size; but the passage from specimen B to C is very gradual, as is also the case between the extreme sizes of the metatarsus.

	Spec. A.	B.	C.
Length of tibia	101 mm.	96 mm.	84 mm.
Width of middle of tibia	6 ,,	5½ ,,	4½ ,,

	Longest.	Smallest.
Length of metatarsus	60	52
Width of middle of metatarsus	5½	5
Width of trochlear extremity	11½	11

It is not by any means certain that the tibia and metatarsus of *A. broecki* described by M. Milne-Edwards belonged to the same individual, although the state of their preservation, the same deep black colour with which they are stained, the circumstance that they were found in the same locality, and the corresponding size of their condyle surfaces, would seem to lead to that conclusion. Neither have the femur and metatarsus of the same individual of *A. leguati* been found, or at least preserved. It is therefore impossible to state with precision the relative lengths of these bones in either species. But taking such examples of bones of *A. leguati* as appear to be of an average size, and comparing them with those of *A. broecki*, we come to the conclusion that in all probability the Rodriguez species did not differ from the Mauritian as regards the relative lengths of the long leg bones.*

Nycticorax megacephalus, (*Ardea megacephala*, M.-E.).†

(Pl. XLI., Fig. G.)

Researches made subsequently to M. Milne-Edwards's description of this bird, have brought to light several bones with which that author was not acquainted, viz., the penultimate and antepenultimate cervical vertebræ, the fifth dorsal vertebra, the pelvis, the scapula, ulna, and radius, the second phalanx of the inner toe, and the first of the hind toe. These bones need scarcely any description, as they are

---

* NOTE ON *Fulica newtoni*, M.-E.

Like the other birds indigenous to the Mascarene Islands, the Coot of Mauritius varied considerably in size. The tibia described and figured by M. Milne-Edwards is 144 mm. long. Two others which are now before us, and which like the former were found in the Mare aux Songes, and are now in possession of Mr. Lucas, of Upper Tooting, measure 140 and 124 mm., the shortest specimen being that of a full-grown bird.

† Ann. des sc. nat. Zool. t. xix. art. 3, p. 10.

of the same form as those of the birds of this family, and especially of *Nycticorax*, but they are important additions, assisting us essentially in arriving at safe conclusions as to the relative development of the various parts of the skeleton, and the real affinities of this bird.

M. Milne-Edwards had considered it to be a true *Ardea;* finding the fore part of its frontal region flat as in a Heron, whilst this part is concave in *Nycticorax*, he dismissed the question of the affinity to the latter which to us appears to be undeniable. The distinctive features of the skull common to the European and Rodriguez Night-Herons are:—first the great width of the occipital region; then the mastoid processes, which are as distant from each other as the temporal: the relative distances between the mastoid, temporal, and postorbital processes are the same in both birds. The temporal fossa is nearly of the same width; the foramen occipitale is broader than deep, more as in the Rodriguez species than in *N. griseus;* the arch of the supraorbital margin is in both much more open than in the Heron; the præantenasal groove is equally deep, and extending equally far forwards; the bill of the Rodriguez species is not less curved downwards than in *N. griseus*, though the bill is equally slight in either. The bill of *N. megacephalus* is much stronger than in the European bird, but not more so than in other species of this genus. The principal difference between these skulls is that that of the Rodriguez species is much more depressed, with scarcely any transverse and longitudinal depression near the base of the bill; it is also a little longer.

The evidence gathered from a comparison of the *pelves* leads to the same conclusion. The præacetabular portion is nearly equally narrow and constricted; the anterior iliac blades coalesce for a short distance only, leaving a great part of the sacral crest uncovered. The width of the postacetabular half, and the arrangement of the foramina, pleurapophyses, &c. are nearly the same.

	N. megacephalus.	N. griseus.
Length of pelvis	63 mm.	61 mm.
Width in its narrowest part	15 ,,	14 ,,
Greatest width above acetabulum	35 ,,	33 ,,

The pelvis of *Ardea cinerea* has a length of 82 mm., and a greatest width of 39 mm., and is therefore in general shape (as well as in other details of configuration) widely different from that of *Nycticorax*.

Taking the *pelvis* as guide, the body of the Rodriguez Night-Heron was of nearly the same size as that of *N. griseus;* the cranium also was nearly of the same size, whilst the bill and mandible were much stronger, and in accordance with this powerful development of the maxillary apparatus, the cervical portion of the vertebral column was proportionally stouter than in *N. griseus*.

The reduction of the power of flight has been already demonstrated by M. Milne-

Edwards, and we are able to corroborate his opinion by completing the measurements of the sternum and wing in comparison with the European Night-Heron:—

	N. megaceph.	N. griseus.
Length of sternum	64 mm.	69 mm.
Breadth of sternum	34 ,,	37 ,,
Greatest depth of keel	14 ,,	20 ,,
Length of scapula	72 ,,	72 ,,
,, coracoid	55 ,,	60 ,,
,, humerus	114 ,,	126 ,,
,, ulna	121 ,,	139 ,,
,, radius	117 ,,	132 ,,
,, metacarpus	62 ,,	70 ,,

It will be observed from this table that whilst all the bones of the wing have been reduced in length (and strength), the scapula has not been affected by the diminished amount of exercise taken by this bird. As regards form, it is exactly the same as in other Herons, viz., tapering behind, without dilatation.

With regard to the leg, M. Milne-Edwards, guided in his estimate of the general size of the bird by the length of the skull and femur, inferred that this part of the osseous frame was much reduced in length. Having shown from the pelvis, with which M. Milne-Edwards was not acquainted, that the body of this bird was considerably less in size than he supposed, in fact equal to that of the European Night-Heron, we arrive at the opposite conclusion, viz., that the leg is proportionally much more developed in length and strength. And this will be readily perceived from the following table:—

	N. megaceph.	N. griseus.	Ardea cinerea.
Length of femur	86 mm.	82 mm.	89 mm.
,, tibia	136 ,,	136 ,,	185 ,,
,, metatarsus	93 ,,	89 ,,	139 ,,
Width of metatarsus	$6\frac{1}{2}$,,	4 ,,	—
Length of 2nd phalanx of inner toe	20 ,,	19 ,,	26 ,,
,, 1st ,, hind toe	30 ,,	28 ,,	35 ,,

In this table we have added also the corresponding measurements of the Heron, in order to show that the Rodriguez bird agrees with *Nycticorax*, and differs from *Ardea* in the length of the femur as compared with that of tibia and metatarsus. The metatarsus and the phalanges are the parts in which the greatest development has taken place, the thickness of these bones being nearly twice as great as in *N. griseus*, the bird having been clearly of much more cursorial habits than its congeners, chasing rather terrestrial animals (lizards) than aquatic.

Thus the effect of the prolonged isolation on the two vertebrate-hunting birds of Rodriguez, the Owl and the Night-Heron, was precisely the same. Without losing the power of flight, they became brevipennate; but the increased development of the legs compensated for the reduction of this power, and enabled the one to destroy animals of larger size when the smaller kinds became scarcer, and the other to chase its swift-running prey. In the Night Heron the increase of development was confined to the legs in conformity with its acquired habit; and it was principally the metatarsus which became enlarged to receive and form a base for the tendons of the foot. But the Owl required additional strength for the purpose of mastering and tearing its prey, not only in the muscles of the lower leg, but also in those attached to the trunk, and hence we find in this bird the greatest development in the femur and pelvis.

Explanation of the Plates.

Plate XLI. :—

    A.—F. *Carine murivora* (nat. size).

A. Cranium.	B. Mandible.
C. Coracoid.	D. Humerus.
E. Pelvis.	F. Femur.

    G. Pelvis of *Nycticorax megacephalus* (nat. size).

    H. Pelvis of *Nycticorax griseus* (nat. size).

Plate XLII. :—

    A.—G. *Necropsar rodericanus*.

    All these figures are half as large again as the natural size, with exception of A. (skull), which is of the natural size.

A. Cranium.	a, a^1. Articular surfaces of mandible.
B. Coracoid.	C. Humerus.
D. Metacarpus.	E. Femur.
F. Tibia.	G. Metatarsus.

    H.—N. *Necropsittacus rodericanus* (nat. size).

H. Cranium.	I. Coracoid.
K. Humerus.	L. Femur.
M. Tibia.	N. Metatarsus.

Plate XLIII. :—

    Aphanapteryx leguati.

    All the figures are of the natural size, with exception of C. and C^1, which are three times enlarged.

        A. and B. Cranium of long and short billed forms.

        C. Articular surface of mandible.

        C^3. Posterior aspect of end of mandible.

D. Pelvis.	E. Scapula.
F. Humerus.	G. Metacarpus.
H. Femur.	I. Tibia.
K. Metatarsus.	

On the Osteology of the Solitaire (*Pezophaps solitaria*, Gmel.).—By Edward Newton, C.M.G., M.A., F.L.S., and John Willis Clark, M.A., Superintendent of the Museum of Zoology and Comparative Anatomy in the University of Cambridge.

(Plates XLIV.–L.)

In the Memoir on the Osteology of the Solitaire in the Philosophical Transactions for 1869, by Messrs. Alfred and Edward Newton, the authors entered fully into the history of the Bird, and recounted the circumstances under which a large collection of its remains had been discovered in the caverns of the island of Rodriguez.

That island having been selected as one of the stations from which the Transit of Venus in 1874 should be observed, it was suggested that a thorough examination of the caves should be instituted in the hope of obtaining those portions of the skeleton which the previous researches had failed to discover. The naturalist appointed to that station, Mr. H. H. Slater, sent home a large series of bones, out of which several male and female skeletons, almost complete, have been sorted. The major part of this collection is now in the British Museum.

Previous to this expedition Mr. Edward Newton had requested Mr. George Jenner, resident magistrate of Rodriguez, who had before been good enough to interest himself in the search, to make a fresh examination of the caves. The result was a large collection of bones, supplementing many deficiencies; but unfortunately, neither in Mr. Slater's collection (when it arrived in England) nor in that of Mr. Jenner were there any remains which could be proved to be associated. Mr. Jenner's collection, found in January and February 1871, has been deposited in the Museum of the University of Cambridge. He accompanied his collection by a most interesting report, in which he described the localities where the bones had been found. This it was our wish to have been allowed to print in connexion with this paper. As, however, Mr. Slater had gone over the same ground, and composed a report of a similar character to that by Mr. Jenner, we were reluctantly compelled to acquiesce in the suppression of the latter.

In the following Paper we have naturally drawn our descriptions mainly from specimens in the collection that was most accessible to us; but the Council of the Royal Society having placed Mr. Slater's collection at our disposal for description, we have availed ourselves of that series of specimens when it appeared to us necessary to do so.

Vertebræ.

In the former Memoir the difficulty of coming to any accurate conclusion respecting the number of vertebræ was felt to be so great that the authors stated that "it is beyond our power to determine precisely the number which the skeleton " contained." The assumption that nineteen vertebræ intervened between the skull and the last dorsal (that which is anchylosed to the pelvis) was made doubtfully, and with the full expectation that more extended study might reverse it. It was based on the probability that *Pezophaps* would possess as many vertebræ as *Didus*, and on the fact that Prof. Owen's artist had assigned that number to the latter.*

The former collection contained one hundred and sixty-one vertebræ. Those received in the recent collection of Mr. Jenner alone augment this number to three hundred and thirty-two assignable to the cervical region, which we conceive should be distributed as follows:—

	I.	II.	III.	IV.	V.	VI.	VII.	VIII.	IX.	X.	XI.	XII.	XIII.	Total.
♂	2	10	11	12	16	14	13	13	13	14	15	11	10	154
♀	1	12	13	13	17	16	18	22	14	18	12	10	12	178

The penultimate free dorsal, the eighteenth of the series according to our present view, is represented by ten female and eleven male examples. We are, therefore, now enabled to give a complete figure of this vertebra, Plate xliv. figs. 1, 2, of which the former collection possessed only more or less broken examples. The neural spine is developed so as to attain a height equal to that of the three anchylosed vertebræ that precede it, and as in them its summit is marked by a ridge. The other parts of the vertebra were sufficiently described in the former paper (p. 332, Plate xv. figs. 56–59).

The fourteenth vertebra (figured in the former paper, Plate xv. figs. 48–50) has its neural spine prolonged into a thick broad ridge, which does not reach quite so high as the coalesced spines of the three succeeding vertebræ. It sends forwards a long blunted process extending rather beyond the anterior edge of the centrum.

The thirteenth vertebra, Plate xliv. figs. 3, 4, has its neural spine developed in a similar manner to the fourteenth, but not to so great an extent. The walls of the

---

* Phil. Trans. 1869, p. 332. The number nineteen is not given in the text, but nineteen vertebræ are figured (Plate xv.), and in the "Description of the Plates," p. 359, the words occur, "Twelfth (and last?) " cervical vertebra," "Penultimate (seventh?) dorsal vertebra."

neural canal slope inwards, and the roof curves downwards in the centre, so that the outline becomes sub-cordate. The anterior zygapophyses are elliptical, and nearly flat; their surfaces slope inwards and upwards. The posterior are slightly concave and slope upwards and outwards. The transverse processes are broad and flat and inclined downwards and backwards. In some specimens of this vertebra in the collection the canal for the vertebral artery is developed, but in only one instance on both sides. The hypapophysis resembles that of the fourteenth vertebra, but is smaller.

In the twelfth vertebra, Plate xliv. figs. 5, 6, the neural processes have not united into a spine, but are present as two thin plates of bone, including a canal. The transverse processes are extremely wide and strong, and develop broad processes which extend downwards and backwards to meet the process sent up from the anterior edge of the centrum, and so form a large arterial canal.

Ribs.

Six appears to be the normal number of dorsal ribs, though in some examples, as in the large male skeleton, of which the pelvis is here figured, Plate xlvi. fig. 1, a seventh appears.

The first of these articulates with the fourteenth vertebra: the sixth with the nineteenth vertebra, which coalesces with the pelvis. The first and second are without any corresponding sternal ribs, but the third, fourth, fifth, and sixth, had corresponding sternal appendages. The seventh, when present, had a similar appendage, which failed to reach the sternum, but was connected by ligament with the sterno-costal rib immediately preceding. Only three or four examples of these ribs, attached in this manner to each other, have been found, even in the combined collections, and then only on one side or the other. The figure, Plate xlv. fig. 1, from a specimen in the British Museum, represents the structure in a very old individual, where the two sternal ribs have become anchylosed.

In the former paper it was said (p. 334), "There appears to have been eight "pairs of dorsal and four of sternal ribs. The first probably articulated with the "thirteenth vertebra, the last or eighth with the twentieth." This statement we now hold to be partially incorrect.

It was formerly estimated that nineteen vertebræ were interposed between the cranium and the last dorsal vertebra, that which is anchylosed to the pelvis. We now incline to consider that there are eighteen only, of which we assign thirteen to the cervical region, and five to the dorsal; so that there are six dorsal vertebræ in all, for we cannot count as normally dorsal that which we have found in a few examples only, as above stated, with a rib attached to it.

In *Didus* the number of ribs appears to be eight. The first two, as in *Pezophaps*, have no sternal appendages; the next five have sternal appendages articulating with the sternum, and the last an appendage which fails to reach the sternum, but articulates with the sterno-costal appendage preceding it. This description is derived from a study of Professor OWEN's figure.*

In *Didunculus* there are only six ribs; the first is free; the second has a much attenuated sternal appendage corresponding to it, which articulates with the sternum, but fails to meet the rib by a slight interval; the next three have sternal appendages articulating both to them and to the sternum, and the last has a sternal appendage attached by ligament to the similar appendage of the preceding rib.

In the following Pigeons of which we have been able to examine specimens, the dorsal rib, articulating with and anchylosed to the first pelvic vertebra, has no proper sterno-costal rib, but only an appendage which joins its extremity, and is attached to the sterno-costal rib immediately preceding by ligament. The species are *Phaps chalcoptera*, *Phaps elegans*, *Phlogœnas crinigera*, *Phlogœnas cruentata*, *Carphophaga microcerca*, *Calœnas nicobarica*, *Geotrygon violacea*, *Starnœnas cyanocephala*, *Treron calva*, *Columba œnas*, *Didunculus strigirostris*. In *Goura* and *Pezophaps*, however, this rib has its own sterno-costal rib. Out of all these species *Pezophaps*,† *Phaps chalcoptera*, and *Goura coronata*, alone present any evidence of a rib articulating with the second vertebra of the pelvis.

We figure (Plate xlv. figs. 2–6), the vertebral ribs of the right side of a female skeleton. The uncinate processes are somewhat more slender and recurved than those of *Didus*, but in other respects the ribs bear a close general resemblance to those of that bird. We also figure three of the sternal ribs on the same plate. Of these, the first and second, figs. 7, 8, are of the left side; the third, fig. 9, of the right. The surface articulating with the sternum is marked *a*; that with the rib is marked *b*.

Pelvis.

The present collection formed by Mr. Jenner contains eight examples, three of which were probably males, and five females, in which the sacro-caudal vertebræ are complete. We find the number of these, as stated before, to be eighteen.

The anterior portion of the pelvis was so complete in the former specimens, that any further description is unnecessary. We can now, however, describe and figure the posterior extremities of the ilium, ischium, and pubes, which before were unknown.

The last six sacro-caudal vertebræ are of uniform width. The first three of these are separated from the ilia by a depression on the superior surface in older specimens,

---

* Transactions of the Zoological Society of London, vol. vi. pl. xv.

† It will be observed that the first pelvic vertebra is the eighteenth of the series in *Goura* and *Phaps elegans*, the nineteenth in *Calœnas nicobarica*, *Phlogœnas cruentata*, *Carpophaga violacea*, *Phaps chalcoptera* and *Pezophaps*, and the twentieth in *Phlogœnas crinigera*.

and by a slight interval in younger ones. The last three are quite free, and the inner edge of the posterior portion of the ilium bends away from them, first outwards, then inwards, then outwards again, and forms a blunted process at the termination of the ridge which separates the ilium from the ischium, Plate xlv. fig. 2.

The extremity of the ischium was figured before, (Plate xviii. fig. 70); and the perfect specimens now before us show that the extremity, as there drawn, was hardly, if at all, broken. The assertion there made (p. 336), as to the "lower margin "sloping downward and outward, as if to pass and avoid the pubic style" is fully borne out by extended observation. In one specimen only, a very large male,* (Plate xlvi. fig. 1) they might possibly have met, as there is a trace of an articular surface on each bone (Plate xlv. fig. 13). The extremity of the pubic bone, perfect in two specimens, a male and a female, would appear to vary considerably (Plate xlv. fig. 13, Plate xlvii. fig. 1).

In the former (the specimen mentioned above) the extremity turns round the posterior process of the ischium, and terminates bluntly at about a quarter of an inch beyond it. In the latter it turns inwards and slightly upwards on reaching the same parts, making a hook-like process about half an inch in length.

The collection contains six caudal vertebræ, including the coccyx. It is of course quite impossible to say to how many individuals these belong. Their most noticeable peculiarity is the feeble development of the neural ridge, as might have been expected from the similar conditions of the sacro-caudal vertebræ, and the extraordinary shape of the coccyx (Plate xlvi. fig. 2.) This bone is nearly rectangular, and tapers very slightly from before backwards. Its length is nearly equal to its height. From its inferior surface it sends forward a small process to pass under the centrum of the preceding vertebra.

The drawing of a restored pelvis (Plate xlvi.) will show how completely the tail must have fallen within a line joining the pelvic bones. This must have been the case in *Didus* also; but in that bird the ilium was suturally connected with the entire series of sacro-caudal vertebræ, so that the caudal alone were free.†

Sternum.

We are now able to figure nearly complete specimens of this bone. In the former paper the anterior portions of the "pleurosteon" were described and figured,‡ but the posterior portion was broken in all the specimens. Immediately behind the articulations for the ribs is a short broad "metosteon," subject to considerable indi-

---

* This pelvis has been placed on the mounted skeleton in the Museum of the University of Cambridge.
† Owen, *ut supra*, Plate xix.
‡ Plate xviii. Figs. 71-74.

vidual variation, but generally projecting outwards and backwards. Behind this, the edge of the bone curves gradually inwards, until near the posterior extremity, when it curves outwards again. There is a short blunt process on either side of the extremity of the posterior edge, and a more or less deep notch in the centre, between the two ridges that mark the origin of the keel. Between these processes and the central notch the edge curves outwards. The form of these parts will be best understood from the figures of a male and a female sternum, here given, Plates xlviii., xlix., figs. 1, 2. Individual variations, probably depending to some extent on age and sex, seem to be very great, and the median notch is in some specimens wholly absent. The costal border is, moreover, perfect in a sufficient number of specimens to place beyond all doubt or question the existence of *four* articular surfaces only.*

Scapular Arch.

The present collection, made by Mr. Jenner, contains three complete furculæ (Plate xlvii. fig. 3, Plate, l. fig. 7) and the fragments of three others. We are thus enabled to decide that this bone is most thoroughly Columbine in form. We have nothing to add to the description before given of the scapula, coracoid, and the bones of the wing, for unfortunately no phalanges have yet been found.

Bones of the Leg.

The femur, tibia, and metatarsal have been already so fully described by Dr. Melville, Mr. Strickland, and Messrs. Newton, that we need not enter upon that part of the subject in this place. As, however, it is on the difference of size in these bones that especial stress has been laid by Strickland, and more lately by Professor Owen,† in their attempt to prove that there were two species of Solitaire, which they designate respectively *Pezophaps solitaria*, and *P. minor*, we have carefully measured a very considerable number of specimens, as will be seen by the following table:—

	Femur.		Tibia.		Metatarse.	
	R.	L.	R.	L.	R.	L.
Presumed Males	41	28	42	41	36	34
Presumed Females	31	38	40	42	29	33

* Compare Newton, p. 338, with Owen, Trans. Zool. Soc. vii., p. 514.

† P.S.—Aug. 1878.—Still more recently (Ann. and Mag. Nat. Hist., January 1878), Professor Owen has tacitly admitted the error he had espoused.

In this number of bones we have before us the remains of at least 42 specimens of each sex.

The dimensions of these are as follows, in inches:—

	Femur.		Tibia.		Metatarse.	
	Largest.	Smallest.	Largest.	Smallest.	Largest.	Smallest.
Presumed Males	7·28	6·60	10·90	10·20	7·45	6·80
Presumed Females	6·10	5·55	9·30	8·35	6·14	5·40

	Femur.	Tibia.	Metatarse.
Difference in length between largest and smallest presumed male.	·68	·70	·65
Difference in length between largest and smallest presumed female.	·55	·95	·74
Difference in length between smallest male and largest female	·50	·90	·66

From this it will be seen that though the differences in length between the largest and smallest of each supposed sex are generally greater than the difference between the smallest male and the largest female, and though there are bones of length intermediate between the largest and smallest of one sex, there are no bones of a length intermediate between the smallest male and the largest female, as would doubtless have been the case had there been more than one species on the island.

The present collection contains twenty-eight ungual phalanges which agree in character with the three figures referred to above, but we consider that it is almost if not quite impossible to determine their exact position on the foot. We believe, however, that the present collection contains those phalanges which were noted in the former paper as wanting, and we have accordingly reconstructed the foot of *Pezophaps* (Plate i., fig. 7) after Strickland's figure of that of *Didus*.*

Skull.

Of the cranial portion of the skull, the collection includes seventeen specimens, together with fragments of eight others. Moreover, we had nine submitted to us by the authorities of the British Museum. Of the twenty-six specimens that are more or less complete, twelve may be assigned to females, and nine to males.

In all these, the depression of the central tract noticed in the former Memoir

* We have come to the conclusion that some of the ungual phalanges formerly figured (Plate xx., Figs. 113, 114, 125, 127) were incorrectly referred to *Pezophaps*. They belong to some species of *Chelonia*. On that plate, Figs. 108, 109, and 115, alone refer to *Pezophaps*.

(p. 246), is most marked, subject, of course, to individual variations, being in some quite flat, in others slightly concave, or divided into two portions by a low transverse tumescence. The occipital and frontal ridges are well marked in all; but the degree of their elevation, especially in the case of the former, varies with age and sex; being occasionally raised into warty protuberances separated by deep depressions: and the severance of the occipital from the ex-occipital portion is always observed. The frontal ridge, or boundary of the central tract, is less elevated than the occipital. It is always divided into a more or less distinct central portion, forming the forehead of the bird (Plate l., fig. 1, a.), and two lateral protuberances marked by more or less prominent exostoses (Ibid., b.), directed forwards and inwards, and in some cases even meeting in the median line. In old specimens, the maseteric ridge always exists as an elevated line, forming the lateral boundary of the central tract behind the orbits. This ridge is well shown in the specimen figured in the former Memoir (Plate xxi., fig. 142, Plate xxii., fig. 149).

On making a vertical and longitudinal section of the skull of the Solitaire, the extent of cancellous structure disposed between the external and internal walls of the cranium is seen to be of considerable extent, but developed very differently from the same tissue in *Didus;* as is seen by looking at the figures of Prof. Owen (Plate xxiii., figs. 1, 5). The enormous lofty and rounded elevation of the skull between the orbits is there seen to be due to a mass of pneumatic diploë exceeding "the longitudinal " diameter of the cavity containing the cerebral hemispheres." This is wholly wanting in *Pezophaps*. We find the cancellous structure to be thinnest precisely where it is thickest in *Didus*, namely, immediately above the centre of the cerebrum (Plate l., fig. 2 a); while in front, the forehead consists of a mass of the same extending between the orbits, and round the nasal passages (Ibid., b.). Professor Owen, who had already noticed the flatness of the frontals between the orbits in *Pezophaps* (Trans. Zool. Soc., Vol. vii. p. 517), thus wrote of this part of the skull in 1871:—
" I suspect that when the part of the skull of the Solitaire may be found, supplying
" what is wanting in the specimens figured in figs. 149, 150, pl. 22 (N.), there will
" be a depression or concavity in the profile contour between the fore part of the
" frontals, and the naso-premaxillaries, which will suggest the presence of a 'frontal
" 'protuberance,' differing only in degree from that so called in *Didus*."

The present series of specimens, while revealing (as the study of a long series alone can do with certainty) the normal form of the forehead, of which the section shows the internal extent, do not in reality add anything to our knowledge of the essential information of that part, which Professor Owen might have found delineated in the very figure (Plate xxii., fig. 149), to which he has made reference above. Not only is there no depression or concavity as anticipated by him, but the anterior aspect of the conjoined frontals slopes gradually forwards, without either elevation or depression. It was, no doubt, to this that the frontlet was attached in the female birds, of which Leguat says:—"Elles ont une espèce de "bandeau comme un bandeau de veuves au haut du bec qui est de couleur tanée."

If the sections of the two crania be placed side by side, so that the basi-cranial axis may be horizontal, it will be seen that in *Pezophaps* the base of the brain lies in the main parallel to it, while in *Didus* it is thrown so far backwards and upwards that it makes a considerable angle with it, as in the Common Pigeon.

The cancellous structure varies considerably in different parts of the cranium. On the upper surface, and above the frontals, it is of close, firm texture, consisting of small sub-equal and sub-spherical cells; but above and below the nasal passages, and in the body of the basi-sphenoid, the cells become larger, and the bone more light and spongy. The united naso-maxillaries form a solid mass of bone, contrasting with the porous structure of this part in *Didus*: while a section of the beak shows one long elliptical cell (Plate l., fig. 2, *c.*), surrounded by smaller ones of similar shape. The maxillary branch of the premaxillary, on the contrary, the peculiar structure of which was dwelt upon in the former memoir (p. 347), commences as a solid bone, which, as it widens, is lightened by the expansion of large cells within its mass, the walls of which are pierced by smaller openings, and the whole included within extremely thin outer walls of bone.

In the former paper, it was stated of the maxilla:—"There is a remarkable "variation in the size of the upper mandible in different individuals to the extent "of very nearly one-half the linear dimensions between the largest and smallest "specimens, of which the collection contains thirteen in all, some of which, how-"ever, are merely fragmentary, and the best exceedingly imperfect."

This passage was cited by Prof. Owen (Trans. Soc. Zool. VII., p. 517), with the question appended, "Is there an intermediate gradational series ? May this difference "of length of beak concur with that pointed out by Strickland in length of leg ?"

The present collection, which includes fifteen perfect specimens, enables us to answer this question in the affirmative. The following tables of admeasurements, taken along the lines *a—b, c—d,* (Plate 1. fig. 5) show that while the difference in size between the largest and the smallest beaks is not so remarkable as that observable in the legs, yet it is entirely of the same character and as fully dependent on sex:—

—	Presumed Adult Males.					Presumed Adult Females.									
	1.	2.	3.	4.	5.	6.	7.	8.	9.	10.	11.	12.	13.	14.	15.
Length (in inches)	3·40	3·40	3·36	3·35	3·33	3·10	3·07	3·07	3·06	3·05	3·05	3·00	2·95	2·88	2·87
Height (in inches)	·95	·92	·91	·90	·90	·79	·81	·76	·78	·80	·80	·75	·70	·74	·70

From the above table we learn that the greatest difference between the skull of the largest and smallest presumed males is in length ·13 of an inch, and in height ·05; that between the largest and smallest presumed females is in length ·23, in height ·11; and that between the smallest presumed male and the largest presumed female is in length ·23, and in height ·11.

The variation in the size of the skull is not so marked, and the 19 specimens which could be fairly measured shewed the following result:—

—	Presumed Adult Males.							Presumed Adult Females.											
	1.	2.	3.	4.	5.	6.	7.	8.	9.	10.	11.	12.	13.	14.	15.	16.	17.	18.	19.
Length from forehead to base of occiput.	3·36	3·36	3·36	3·35	3·33	3·33	3·30	3·13	3·08	3·07	3·05	3·04	3·02	3·00	3·00	2·96	2·95	2·95	2·92
Extreme breadth of skull.	2·80	2·72	broken	2·76	2·80	2·71	2·69	broken	2·37	broken	broken	2·38	2·35	2·40	2·30	broken	broken	2·38	2·39

It will be seen from this table that the greatest difference between the largest and smallest presumed adult male in length of skull is ·06, in breadth ·11; the greatest difference between the largest and smallest females is in length ·21, in breadth ·1; and the greatest difference between the smallest male and the largest female is in length ·17, and in breadth ·11.

There is not only considerable variation in the degree in which the tip of the beak is hooked, as has been already shewn in the former memoir (Plate xxii., figs. 155-157) but the beak is occasionally almost straight. The posterior portion, that which abuts upon the frontals, was in the former collection either absent or imperfect. The present collection includes six specimens which show this part (three males and three females).

The nasals (Plate l. fig. 4, *na.*), and the premaxillaries (Ibid. *pmx*) are of nearly equal width at the point of junction with the cranium; an equality which is preserved for some distance. That portion of the nasals which appears on the superior surface of the beak is never anchylosed to the premaxillaries. There is even occasionally an interval left between them. The premaxillaries also are not only not confluent throughout, even in adult specimens, but for a distance equal to about one-fourth of their entire length from the cranium even include a narrow space between them. In *Didus* a certain resemblance to *Pezophaps* is observable; the premaxillaries being broad and flat, and the nasals free for a short distance (STRICKLAND, Plate ix.); but the proportions are quite different. In *Goura*, on the other hand, each nasal is three times the width of the united premaxillaries.

The palatals, which we are now enabled to describe and figure for the first time (Plate l., figs. 4, 5, 6 *pl*), are complete in four specimens. Their posterior attachments, which abut against the rostrum of the basi-sphenoid (Ibid., fig. 4, *c.*), are broad and slightly concave, but there is no reason for supposing that they meet in the median line. The hinder border is thin and abruptly truncate. The horizontal plate of the bone bends abruptly outwards, then forwards and inwards to join the maxillary; the inner edge, which is more vertical, is produced at its anterior extremity into a very delicate and sharp point (Ibid. *a*.).

The maxillo-palatals (Ibid., figs. 2, 5 *mxp*) are two elongated processes of a light spongy texture. Their shape is constant in the seven specimens we have examined; but the degree of cancellous structure varies in individuals, and even on the opposite sides of the skull of the same bird. There is usually a deep pit hollowed out at the end nearest the nasal passages and extending right through the substance of the bone to its inner including wall. This is sometimes simple in outline, but more frequently subdivided by thin spicules of bone stretching across it in different directions, so as to subdivide it into numerous small cells.

The jugals, of which there are four specimens, are thin and flat, with a feebly marked process on the superior surface at their proximal end. The squamous suture, by which they are joined to the maxilla, extends for nearly two-thirds of the length of the bone. One of these is figured (Plate xlv., fig. 12).

The lachrymal (Plate l., fig. 3 *la*), forms the anterior, and, to a certain extent, the inferior, edge of the orbit. It is attached above to the prefrontal by a long articular surface distinguishable only in young specimens. It sends out forwards a long pointed process (Ibid. *a*.) varying somewhat in shape in different individuals, which approaches the boundary wall of the nasal passage, but neither unites with it, nor advances sufficiently far forward to join the nasals. Below this the lachrymal sends out backwards a sub-conical recurved process (Ibid. *b*.) spongy in texture. This is met and supported by a thin process developed from the upper part of the rostrum of the basi-sphenoid and forming part of the outer wall of the nasal passage.

The pterygoid, of which the collection contains nine specimens, is a short, stout, pillar of bone, nearly straight. The process by which it articulates with the quadrate, makes nearly a right angle with the plane of the bone and is bounded by a thin expanded surface. At the distal end it abuts against the sphenoid by a level triangular surface of considerable extent, bounded distally by a sharp edge, which meets, without distinct articular facets, the thin hinder edge of the palatals (Plate xlv., figs. 10, 11).

Brain.

We have had a cast made of the brain-cavity, by the help of which we can form a tolerably accurate idea of the size and shape of the brain. It seems to have been rather small in proportion to the size of the bird; but comparison with that of a tame Pigeon (*Columbia livia* var. *domestica*) shews that it closely resembles it in general shape, and especially in the great breadth of the central lobes. It is figured of the natural size (Plate xlv. fig. 14).

Stone swallowed by the Solitaire.

In his description of the hen Solitaire, LEGUAT says :—

"On leur trouve toujours dans le gésier (aussi bien qu'aux mâles) une pierre brune de la grosseur d'un œuf de poule; elle est un peu raboteuse, platte d'un côté et arrondie de l'autre, fort pesante, et fort dure. Nous avons jugé que cette pierre naît avec eux; parce que quelque jeunes qu'ils soient, ils en ont toujours, et n'en ont jamais qu'une; et qu'outre cela, le canal qui va du jabot au gésier, est trop étroit de moitié pour donner passage à une pareille masse. Nous nous en servions préférablement à aucune autre pierre, pour aiguiser nos couteaux."*

The attention of Mr. H. H. Slater was specially drawn to this statement before he went to Rodriguez; but notwithstanding his careful examination of the caves of that island, he was unsuccessful in finding anything bearing out Leguat's statement. Shortly after his return, Mr. Caldwell, of Mauritius, Corresponding Member of the Zoological Society of London, visited Rodriguez.† He was more fortunate, and obtained three of what he believes to be the stones mentioned by Leguat, which he describes as follows :—

"Of the three stones found *in situ* A and B were taken by Sergeant Morris [of the Police], in my presence, and C, the stone I have given you, by myself; but all in my presence.

"The first found was B. Morris had called me to see the leg bones protruding from the dry powdery soil in the upper or entrance chamber of the cavern, and we began carefully to remove the earth. We then came to the sternum (which was in every case keel above), when Morris put his fingers under it to lift it entire, and said 'there is a round stone under it.' The earth was then more carefully removed, the sternum lifted, and the stone close up to it. No other stone was found until we reached the floor about 20 inches below the surface, but even the fragments occasionally found in no way resembled this, or the other two subsequently found, being coralline. We did not at first pay much attention to the stone; I had merely put it in my pocket, being puzzled to account for a rounded basalt pebble being found in the dry earthy dust of a cavern in the face of a low cliff above the large main cavern we had explored in vain. It should be noted that in this chamber there was plenty of light.

"Stone A was found in the lower chamber of the cavern by Morris. When he called to me that there appeared to be an entire bird embedded, I came up. There was a very bright lamp with us and each bone was carefully uncovered and put aside. Here again the bird was on its back buried in the loose dusty soil, and the sternum consequently uppermost. It was very carefully raised by Morris, when stone A was found in exactly a similar position to the last. It was then Morris remarked 'it would be curious if this were the stone the bird was said to have in its belly.'

---

* Voyages et Avantures de François Leguat, &c. Londres: MDCCVIII., vol. i. p. 100.
† Proc. Zool. Soc. Lond., 1875, pp. 644–647.

"The third stone was found in the upper chamber by me, and also under the fragment of the sternum which accompanies the stone, but the bird was very incomplete, and apparently young.

"J. CALDWELL, 6 Dec. 1877.

"P.S.—There are no stones of similar composition to these in the neighbourhood of any of the caverns where Solitaire bones have been found. I should think the nearest place where fragments of basalt could be found would be at least two if not three miles from the cavern where I found them."

One of these stones he was so good as to give to Mr. Edward Newton. It is here figured in three aspects (Plate xlvii. figs. 4, 5, 6). It weighed, before it was cut for the purpose of determining its mineralogical nature, a little over $1\frac{3}{4}$ oz. It is brown, somewhat rough, heavy, and hard. It is hardly, however, to be called "flat on one side," as Leguat describes those he had the opportunity of observing. In connexion with this fact, however, it may be remarked that the bird with whose remains it was associated appears to have been young.*

We have to thank Professor Bonney, M.A., F.R.S., for the following report on the microscopic structure and mineralogical composition of the stone:—

"This rock externally presents considerable resemblance to a dolerite. Such it is proved to be on microscopic examination; the slide shewing a crystalline mixture of plagioclase felspar, augite, olivine, and a peroxide of iron (? hematite). The plagioclase is well preserved; the sections are commonly about six times as long as wide, and exhibit the characteristic twinning; probably it is labradorite. Enclosures of opacite, augite (?), and other microliths, with minute gas-cavities, are frequent in some crystals, rare in others; colours with polarizing apparatus fairly bright. The augite in the sections is of a pale puce-brown colour, rather rough in texture, and with the nicols fairly rich coloured. The olivine also shews brilliant colours; with ordinary light it is nearly colourless, except where stained a warm brown through incipient decomposition; some crystals are thus rendered almost opaque. For this reason, and the absence of serpentine, I conclude this to be a ferriferous variety approaching hyalosiderite. The grains of iron peroxide are not very numerous and are rather irregular in form. It seems most probable that they are hematite. These minerals are enumerated in order of frequency. The felspar, as is not unfrequent in doleritic rocks, is pierced in places by long acicular microliths, nearly colourless; some of which may possibly be apatite."

* P.S., August 1878.—The stone was exhibited to the Zoological Society, 5th March 1878, by Professor Newton. The preceding description is borrowed from the remarks made by him on that occasion. (Proc. Zool. Soc. Lond., 1878, p. 291.)

Concluding Observations.

The minute examination of the enormous series of specimens at our disposal enables us confidently to affirm the "bold" statement made in the former Memoir (p. 330):

"There does not seem to be a single bone in the skeleton of *Pezophaps solitaria* which is not liable to greater or less individual variation of some kind or other; . . . . . the individual variation is not at all confined to absolute size; it extends to the relative proportion of divers parts of the bone, to processes or depressions upon them such as are commonly held to be specifically characteristic, so that it is often utterly impossible to predicate any definite limits of individual modification."

We are not aware that the osteology of any vertebrate, other than man, has been studied with the same wealth of materials as that of the Solitaire. About the wonderful variability of this particular species no doubt can exist. The bearings of this fact on the theory of Evolution will not here be entered into; but that they will of necessity prove to be most important in future discussions of that theory, and of many of the questions arising from it, can hardly fail to be admitted. We have deviated so far from the example set in the former Memoir as in certain cases to give precise measurements of certain bones; but it must especially be borne in mind that variations in length, breadth, and thickness form but a comparatively immaterial portion of the variations which exist. These are often, nay almost always, of a kind which cannot possibly be expressed in words, and could only be represented by a series of figures almost equal in number to the specimens.

The age of these bones has not hitherto been taken into consideration. Some of those found on a former occasion were considered by Professor Steenstrup (Proc. Zool. Soc. 1855, p. 718) to bear traces of having been broken by man, or by some predatory animal, in order to extract the marrow. In the present collection, however, we cannot find any examples of this kind. There is no evidence by which we can determine the age of the bones; and the collection may include the remains of birds which lived at a very remote as well as at a very recent period. It is to this cause that we ought perhaps to ascribe some portion of the wonderful variability observable in the remains of *Pezophaps solitaria*.

P.S., August 1878.—In his recently published paper (Ann. and Mag. Nat. Hist., January 1878, p. 87), Prof. Owen remarks of the præsacral vertebræ of *Pezophaps*:—
"So much of the vertebral formula thus accords with that of *Didunculus*." This is a tribute to the fidelity with which the late Mr. JAMES FLOWER obeyed the directions of Professor FLOWER, which were that the skeleton of *Pezophaps*, at first deposited and mounted at the Royal College of Surgeons, and subsequently transferred to the British Museum, where it was examined by Professor OWEN, should be articulated in accordance with the skeleton of *Didunculus*, in which of course the number of vertebræ is known.

The Extinct Reptiles of Rodriguez.
By Dr. A. Günther, F.R.S.

The earliest notice of the Tortoises and Lizards of Rodriguez we find in LEGUAT (Voyages et Avantures). He says "that there are such plenty of Land-Turtles in "Rodriguez, that sometimes you see two or three thousand of them in a flock, so "that you may go above a hundred paces on their backs." According to Admiral KEMPINFELT, who visited the island in 1761 (see Grant's Maurit. p. 100), small vessels were constantly employed in transporting these animals by thousands to Mauritius for the service of the hospital. But early in the present century the work of extermination appears to have been accomplished, and there is, at present, of the Rodriguez Tortoise not a single living example in the island, or in any other locality.

Remains of this Tortoise had been discovered and had reached Europe many years ago, but no particular attention was paid to them. M. J. DESJARDINS, one of the first explorers of the fauna of Mauritius, sent a bone of a Tortoise found in 1786 in a cave in Rodriguez, with some remains of the Solitaire, to Paris,* where they were examined by CUVIER and BLAINVILLE, who erroneously stated them to have been recently found under a bed of lava in Mauritius.† Another Mauritian naturalist, C. TELFAIR, in searching, in 1832, for bones of the Solitaire in Rodriguez, succeeded in obtaining "numerous bones of the extremities of one or more large species of Tortoise," which were presented to the Zoological Society of London, and exhibited at one of the meetings.‡ These bones were still in the possession of the Society three or four years before the publication of Strickland and Melville's memoir on the Dodo (1848); but no further attention being paid to them they were lost. Another portion of TELFAIR's collection was presented by him to the Andersonian Museum at Glasgow, where they are still preserved.

Some well-preserved bones, kindly sent to the writer by M. BOUTON, of Port Louis, in 1872, satisfactorily proved that the Tortoise of Rodriguez is distinguished from all its congeners by well-marked characters (Ann. & Mag. Nat. Hist., 1873, xi. p. 397); but it was only when these remains were supplemented by those preserved in the Andersonian Museum at Glasgow, and entrusted to me by the curators of that institution for examination, and when, finally, the extensive series collected during the Transit-of-Venus Expedition arrived, that our knowledge of its specific characters became tolerably complete. No further important additions can be expected from Rodriguez, with the exception of the small bones of the foot and caudal vertebræ;

---

* Proc. Comm. Zool. Soc. ii. p. 111. Strickland and Melville, "The Dodo," pp. 51–53.
† Edinb. Journ. Nat. Sci. iii. p. 30. ‡ Proc. Zool. Soc. 1833, p. 31.

and these will be but of small value, unless they be found in their natural connexion.

With the aid of the carapaces brought home by Mr. SLATER we are now enabled to recognize the Rodriguez Tortoise in some carapaces which reached Europe in the last century, probably during the lifetime of the species, and which we find noticed by the following herpetologists:—

1. SCHOEPFF (Histor. Testud. 1792, p. 103, pl. 22, fig. B.) has reproduced a sketch of a Tortoise $2\frac{3}{4}$ feet long, which was communicated to him by Vosmaer, who examined the specimen which then was in " Museo Principis Arausionensis " in the Hague. This seems to have been a male, with a carapace very similar in form to that of the male described below; its front and hind margins, being still provided with the epidermoid scutes, have an undulated outline. Schoepff was informed by Vosmaer that the carapace had been brought from the Cape of Good Hope; and expressing himself uncertain whether it should be considered a distinct species, or a sexual, local, or individual variety of the Tortoise described by Perrault, he named it " *Testudo Indica, Vosmaeri.*"

2. DUMÉRIL and BIBRON recognized Schoepff's Tortoise in a skeleton with complete carapace in the Paris Museum. The description of the specimen, whose shell measured 75 centims. over the curvature, again perfectly agrees with our male specimen, and supplies a detailed account of the outer epidermoid covering. The authors adopt the binominal term, " *Testudo vosmæri,*" which, of course, supersedes that proposed by myself (*Testudo rodericensis*, Ann. & Mag. Nat. Hist., 1873, xi. p. 397). By the singular resemblance of the general form of the male of this species to that of some of the Galapagos Tortoises, they were led into the error of supposing that *T. vosmæri* came from the Galapagos Islands (Erpétol. Génér. ii. p. 140).

3. A second specimen, probably a young female, likewise in the Paris Museum, and without known history, was considered by the French herpetologists a distinct species, *Testudo peltastes* (ibid. p. 138). This description agrees in every respect with our young carapaces from Rodriguez.

The Rodriguez Tortoise* differs from the Mauritius and Galapagos Tortoises by the more slender build of all the various parts of its skeleton; its neck must have been capable of still greater flexion, as is evidenced by the deep postapophysial impressions or actual perforations of the cervical vertebræ. Although careful comparative measurement show beyond doubt that this Tortoise had longer limbs and a longer neck than even some of the Galapagos Tortoises, yet, taking also into consideration the extreme thinness and fragility of its carapace, we must infer that this general slenderness of the bones must have been partially due to the same cause, probably a diminished supply of the calcareous salts, or a diminished power of assimilation of them.

* For the detailed illustrated description of these remains, I refer to " Gigantic Land-Tortoises (Living and Extinct). Lond. 1877. 4to."

The bones collected by the Naturalist of the Transit-of-Venus Expedition belonged to several hundred individuals; and there are in some cases as many as 40 specimens of one and the same bone in the collection; yet no variation in structure equivalent to that observed among the Galapagos, Aldabra, and Mauritius Tortoises could be detected, so that evidently in this small island there was room for one species only. The only variation which is worth recording is one which can be explained as a sexual difference, the female having been of a smaller size and somewhat stouter form than the male, as is the case in the other Gigantic Tortoises.

Bones far exceeding in size the majority of their kind are not rare, and prove that the Rodriguez Tortoise was quite equal in bulk to *Testudo elephantina*, many (probably male) individuals having had a carapace $4\frac{1}{2}$ feet in length. From Duméril and Bibron's descriptions we learn that the scutes were perfectly smooth or nearly so, and that the shell of the adult was black, whilst the young were of a lighter brown colour, the sternum being dotted with yellow.

Two kinds of *Lizards* appear to have been known to LEGUAT. He says (p. 108), "Les Palmiers et les Lataniers sont tous chargez de Lézards de la longueur d'un "pied: on ne sçauroit se lasser d'en considerer la beauté. Il y en a de noirs, de "bleus, de verds, de rouges, de gris, et de tout cela du plus vif, et du plus éclatant. "Leur nourriture la plus ordinaire est le fruit du Palmier. Ils ne sont nullement "malicieuse, et sont si familiers qu'ils venoient souvent manger nos melons sur la "table en nôtre presence, et mêmes entre nos mains. Ils servent souvent de proye "aux oiseaux, sur tout aux butors. Quand nous les faisions tomber des arbres, "avec une perche, ces oiseaux accouroient et venoient les engloutir devant nous, "quoi que nous pussions faire pour les en empêcher; et lors que nous en faisions "seulement le semblant, ils venoient de la même manière, et nous suivoient "toûjours."

"Il y a une autre espece de Lézards nocturnes, de couleur grisâtre, dont la figure "est fort vilaine: ils sont gros et longs comme le bras, et la chair n'est pas mauvaise. "Ils aiment beaucoup les Lataniers."

The former of these Lizards may have possibly a species of *Phelsuma*, a genus which is well represented in the Mascarene region, and the species of which are subject to great variation in colour. The latter is probably the species of which Mr. Slater collected, with remains of the Solitaire and Tortoise, several bones. He recognized them as the remains of a Lizard, possibly belonging to the family of Skinks. In my opinion it is a Geckoid Lizard, which, as far as the evidence before us goes, cannot be separated from the genus *Gecko*; but the species from Rodriguez appears to have attained a much larger size than *Gecko verus* (to which it is very similar), or than any other Geckoid known. Referring, then, this Lizard to the genus mentioned, I concur in Mr. Slater's proposal of naming it after Mr. E. Newton,— *Gecko newtonii*.

The bones collected consist of two parietals, posterior half of right ramus of lower jaw, right humerus, right half of pelvis, five left and two right femurs, and therefore must have belonged to at least five individuals, of which the one indicated by the pelvis was the largest. In the following description these bones have been compared with the skeleton of a *Gecko verus*, the vertebral column of which is 100 millims. long (exclusive of the caudal vertebræ), and the skull 45 millims.

The *parietal* agrees in size and shape entirely with that of *G. verus*, in which the two long processes into which this bone bifurcates behind are separated by a large vacuity from the paroccipital. In *Phelsuma* (which genus is so well represented in these islands and on the coasts of this geographical region, and which might have been expected to occur in Rodriguez) the parietal has quite a different shape (*P. seychellense*), and its posterior processes are addressed to the paroccipital.

Parietal bone of *Gecko newtonii*. Upper aspect and of natural size.

The *articulary* piece of the mandible differs nowise from that of *G. verus*; like the latter it is produced behind the condyle into a hamate grooved process, which, however, is much more concave on its upper surface in the Rodriguez species than in *G. verus*.

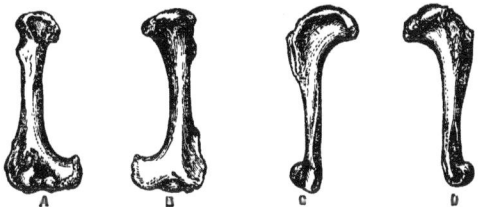

Humerus of *Gecko newtonii*, nat. size. A, anterior, B, posterior, C, ulnar, and D, radial aspects of bone.

The *humerus* offers a more striking difference from *G. verus* than the preceding bones; it is much stronger, and especially its extremities are comparatively much more dilated. Its head is transversly elongated, passing into a curved and projecting prominence, which answers to the ulnar tuberosity. The radial crest is strongly developed, and does not extend beyond the proximal third of the length of the bone. The transverse diameter of the distal extremity is nearly rectangular to that of the proximal. The whole of this part of the bone is much dilated, particularly by a broad trenchant crest running along the radial border of the bone. Of the two condyles the radial one is much the more prominent one and projects towards the anterior side of the bone. In all these particulars *G. newtonii* resembles *G. verus*, all the ridges and prominences being, however, much more developed.

	G. verus. millims.	*G. newtonii.* millims.
Length of the humerus	22½	25
Least width of the shaft	2	3
Transverse diameter of proximal end	5½	9
Transverse diameter of distal end	6½	10

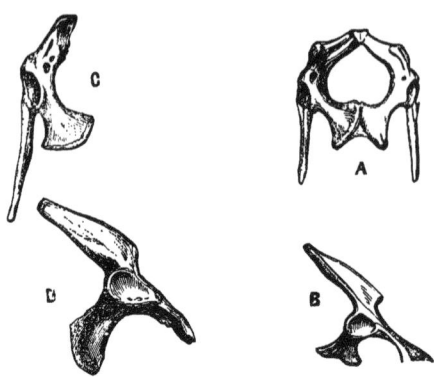

A. Inferior surface of the pelvis of *Gecko verus*. B. The outer aspect of the right os innominatum of the same animal. C. Lower surface of the right half of the pelvis of *G. newtonii*, and D, its exterior surface. All the figures are of natural size.

The *pelvis* and *femur* are so similar to those of *G. verus* that the accompanying figures and statements of measurements will suffice to give a perfect idea of those bones.

	G. verus. millims.	G. newtonii. millims.
Length of the os ilium (from acetabulum)	11	13
Greatest width of os ilium (from acetabulum)	4	5½
Length of os pubis (from acetabulum)	11	(injured)
Width of os pubis at its base	3	5½
Length of os ischii (from acetabulum)	6½	11
Least width of the os ischii	2½	11
Length of femur	26	30
Width in the middle	1½	3
Width of its lower extremity	5½	7½

The femur of *Gecko newtonii* in four different aspects and of natural dimension.

THE EXTINCT MOLLUSCA OF RODRIGUEZ.

The only extinct form as yet discovered is a land-shell (*Helix bewsheriana*), which was obtained in a sub-fossil condition in the bone caves by Mr. Slater. This species will be again referred to in the account of the recent mollusca of the Island.

RECENT FAUNA.

MAMMALIA.—*By G. E. Dobson, M.A., M.B., etc.*

The only indigenous mammal consists of a species of frugivorous Bat, described below, which appears to have hitherto escaped the notice of zoologists.* Bones of the rabbit, rat (*Mus decumanus*), mouse, and pig (introduced by the settlers), were brought home by the collectors, but they possess no special interest.

Pteropus rodericensis.

Dobson, Catal. Chiropt. Brit. Mus. p. 36.

Ears slightly longer than the muzzle, but projecting by their upper third only from the long fur surrounding them; ear-conch moderately broad in lower three-fourths, very abruptly narrowed above by flattening of the upper fourth of the inner margin and a deep concavity of the corresponding part of the outer margin, terminating in a narrow acute tip; more than two-thirds of the concave surface of the conch is well clothed with moderately long hairs, the upper fourth and the back of the ear naked or with a few short hairs only.

Fur long and dense, in quality intermediate between that of *Pt. edwardsii* and *Pt. vulgaris*; that covering the back of the head, neck, and shoulders very long, on the back shorter but not appressed, the hairs directed backwards, extending thickly upon the wings almost as far outwards as a line drawn from the proximal third of the humerus to the knee, upon the femur, knee, and adjoining wing-membrane along the proximal third of the tibiæ, about two-thirds of which are thickly covered with straight hairs directed backwards, their distal thirds being nearly naked as well as the feet; and, although the fur of the back extends upon the interfemoral membrane, it does not conceal its posterior margin, and its lower third, supported by the calcanea, is naked. Beneath, the antebrachial membrane is covered with long thinly-spread hairs, and similar hairs clothe the wing-membrane between the humerus and femur, and extend outwards in a broad band behind the forearm. Face reddish brown, with a few greyish or shining hairs; chin and throat darker brown; top of the head and nape reddish or yellowish brown, passing into a band of bright yellow, which extends across the back of the neck from shoulder to shoulder, and downwards on the sides of the neck and thorax, limited by a longitudinal band of

---

* Mr. Gulliver, who obtained some of the specimens of this Bat, remarks:—" Insectivorous Bats are " entirely absent in Rodriguez, though they exist in Mauritius. At the latter place these Bats may be seen " in the caves, but in the caves at Rodriguez I never observed a trace of them, nor did Mr. Slater, who, " from the constancy of his occupation there, could hardly have failed to see them had they existed."

dark fur passing backwards from the chin along the thorax to the abdomen, which is clothed with dark brown fur, of which some hairs have greyish or shining extremities; fur of the back behind the shoulders dark brown, the extremities of the hairs more or less yellowish. In the male the extremities of the hairs on the head and back are much brighter coloured than in the female.

First upper premolar deciduous; the second upper premolar and the second and third lower, also the first and second upper molars and the first lower, have each a small but distinct posterior basal cusp; even the small first lower premolar has an indication of a posterior basal cusp; last upper molar very small and circular, not as large as the first lower premolar, last lower molar slightly larger than the first lower premolar.

Length (of an adult ♂ preserved in alcohol): head and body 7″; head 2″·2; ear 0″·9; ear from tip of nostril 2″; eye from tip of nostril, 0″·8; forearm 4″·9; thumb-metacarp, 0″·35; ph. and claw, 1″·6; third finger-metacarp, 3″·3; 1st ph. 2″·4, 2nd ph. 3″·55; fourth finger-metacarp, 3″·25; 1st ph. 2″, 2nd ph. 1″·85; fifth finger-metacarp, 3″·35, 1st ph. 1″·5, 2nd ph. 1″·45; tibia, 2″·25; calcaneum, 0″·6; foot, 1″·15.

This small species resembles *Pteropus rubricollis*, Geoffr., from the island of Mauritius, in size only. The ears project by their extremities beyond the fur of the head, and their margins are quite naked. In *Pt. rubricollis* they are covered by long hairs and quite concealed by the fur; the fur of the body is also quite different in texture, being much coarser and not in the least degree woolly, and the lower third of the tibia is naked.

BIRDS.—*By R. Bowdler Sharpe, F.L.S., F.Z.S., etc., Assistant in the Zoological Department, British Museum.*

Our knowledge of the recent Avifauna of the island of Rodriguez has hitherto been confined to a few memoirs published by Mr. Edward Newton, and Professor Alfred Newton of Cambridge. The former has given us an account of a hurried trip of a few days duration made by him in 1864 (Ibis, 1865, pp. 146-154), and it says much for his zealous collecting that the collections now received add very little to his account of the ornithology of the island, and confirm in nearly every instance the correctness of his identifications. Two indigenous species were then discovered by him, both of them proving new to science, and these were described by Professor Newton as *Foudia flavicans* and *Drymœca? rodericana*. More recently he has described a *Palœornis* from this island as *Palœornis exsul* (Ibis, 1872, p. 33), and in 1875 a figure of this interesting bird was published (Ibis, 1875, pl. vii.). The naturalists attached to the present expedition did not procure a specimen, but Mr. Henry Slater saw one on one occasion only :—" This was on the 30th of September, " towards the south-western end of the island, where there is a good deal of wood, " and he could have shot the bird if he had had a gun with him; but neither of his " companions was favoured with a sight of this expiring species, and no further " information about it could be obtained from the Creoles." (Newton, Ibis, 1875, p. 343.) Of the collections brought home by the naturalists to the Expedition, that of Mr. Slater is the largest, as he procured not only several specimens of the *Foudia* and *Drymœca*, but also a number of waders and sea-birds which are found on the island. Mr. Gulliver's specimens of the two indigenous species were very well preserved, and he also succeeded in obtaining their eggs.

Among Mr. Slater's series were examples of the following introduced species, *Psittacula cana, Acridotheres tristis, Francolinus ponticerianus,* and *Numida coronata.*

1. Bradypterus rodericanus.

Drymæca (?) rodericana, Newton, P. Z. S. 1865, p. 47, pl. 1, fig. 3.

Drymoica rodericana, Gray, Handl. B. i. p. 199, No. 2771.

Adult male. General colour above olivaceous brown, very little paler on the upper tail-coverts; lores pale yellow, extending above the eye but not forming a distinct eyebrow; round the eye a ring of pale yellow feathers; cheeks and sides of face pale yellow, the ear-coverts olivaceous; under surface of body pale yellow from chin to vent, the sides of the neck and sides of breast olivaceous brown; under wing-coverts light yellow; wing-coverts above brown washed with olivaceous like the back; quills dark brown, the secondaries edged with olivaceous, inclining to yellowish white on the outer margin of the primaries; tail-feathers brown, narrowly margined with olivaceous, the outer ones slightly tipped with white; bill horn-brown in skin, the lower mandible yellow; feet fleshy brown in skin.

The following are the measurements of the specimens sent, seven in number:—

	Total Length.	Culmen.	Wing.	Tail.	Tarsus.
a. ♂ ad. Rodriguez, Nov. 1874 (Gulliver)	6·3	0·65	2·45	3·0	0·95
b. ♀ ad. ,, ,, ,,	6·3	0·6	2·35	2·95	0·85
c. ♂ ad. ,, ,, (Slater)	6·2	0·65	2·45	2·85	0·95
d. ♂ ad. ,, ,, ,,	5·8	0·65	2·45	2·9	0·95
e. ♂ ad. ,, ,, ,,	6·2	0·65	2·45	2·95	0·9
f. ♀ ad. ,, ,, ,,	6·0	—	2·35	2·9	0·85
g. ad. ,, ,, ,,	6·0	0·65	2·25	2·8	0·85

As is usual among the Warblers the female is smaller than the male. The difference in coloration is not very great, ranging from a dark olive-brown to a clear olive or greyish brown; the shade of yellow also varies in individuals.

The present bird is a difficult one to place in the natural system. In general plumage it is unlike a *Drymœca*, but, as Professor Newton observes (*l.c.*), it cannot be associated with the *Cisticolæ*, which have ten tail-feathers. Its affinities must, therefore, be with *Drymœca*, *Orthotomus*, *Prinia*, &c., or with the Reed-Warblers, *Acrocephalus*, *Bradypterus*, &c. It cannot be referred to *Orthotomus*, as all the members of the latter genus have the tail shorter than the wing, or about equal to it in length. In true *Drymœca* the tail is always longer than the wing, and so it is in *Prinia*, but in *Apalis* the wings and tail are about equal, so that I can hardly understand Mr. G. R. Gray's suggestion that *D. rodericana* resembled *Apalis*. The shorter and broader bill, the minuteness of the rictal bristles, and the fact of the first primary being equal to the inner secondaries in length, all show that *Apalis* is not the genus to which the present bird could be referred, though in the "Handlist" a clean sweep is made of all the Grass-Warblers and their allies, which are sunk into one enormous genus *Drymoica*. It remains therefore clear that the Rodriguez Warbler must be referred either to *Prinia*, to *Drymœca*, or to the immediate vicinity of the Reed-Warblers. After comparing them I have come to the conclusion that it should be referred to the last-mentioned group of Warblers, as its abnormally long bill and strong rictal bristles separate it off from *Drymœca* and *Prinia*. In the two latter genera the bill, though long, never exceeds in length the hind toe and claw, and is only equal to about half the tarsus. The well developed first primary prevents the admission of *D. rodericana* into *Acrocephalus*, and it therefore must be referred to the genus *Bradypterus*, of which it exhibits the generic characters, but of which genus it is rather an aberrant member, on account of its long bill and yellow coloration.

The nests of this species, obtained by both naturalists, are very similar in structure, but one is rather larger than the others. They are both cup-shaped and neatly constructed, but firmly built, and the walls of the nest decidedly thick. The principal material used in their construction appears to be fibre with a few grasses interwoven; some leaves and shreds of cotton are also added.

Mr. Gulliver brought an egg of this Warbler. It is something of a Whitethroat type, being white, thickly spotted, more particularly at the obtuse end. All over the surface of the egg are small underlying spots of pale purple, and towards the thicker end of the egg are larger clay-brown spots, and also good-sized spots of reddish brown. Length 0·7; diam. 0·5.

2. Foudia flavicans.

Foudia flavicans, Newton, P.Z.S., 1865, p. 47, pl. 1, figs. 1, 2.
Ploceus flavicans, Gray, Handl. B. ii., p. 45, No. 6624.

Adult Male. Above brown, slightly washed with olivaceous, the back broadly streaked with dark brown, the lower back uniform, the rump yellow, forming a band across it; upper tail-coverts light olivaceous brown; wing-coverts blackish brown, with greyish olive edgings to the lesser series, the greater and median coverts broadly-edged with whitish: quills dark brown, the primaries narrowly edged with yellowish, the secondaries more broadly margined with whity brown, lighter on the innermost; tail-feathers ashy brown, all edged with olive yellow; head and neck all round bright yellow, as also the breast; the forehead, lores, sides of face, and cheeks orange scarlet, the throat also washed with the latter colour; feathers in front of and round the eye black; rest of under surface of body light yellow, the flanks brownish; under wing-coverts light ashy-brown. Total length 5·2 inches; culmen 0·6; wing 2·7; tail 2·1; tarsus 0·8.

The series of male specimens is very complete, showing gradual transitional changes from the time when the back is like that of the female without any trace of the yellow band on the rump, the orange colouring on the head and neck being apparently gained by a moult. The yellow band across the rump appears to be assumed after the yellow head and neck are fully coloured. Of two males killed by Mr. Gulliver in November 1874, one has the band tolerably well defined, the other only shows its approach by the presence of a single yellow feather.

Adult Female. General colour above brown, the head washed with olive, all the feathers mesially centred with dark brown, more distinct on the back and scapulars; rump uniform brown; wing-coverts blackish brown, with greyish edgings to the least series, the median and greater coverts rather broadly edged with white; quills dark brown, narrowly margined with olive yellow, the secondaries with lighter brown; tail-feathers dark brown with narrow olive-yellow margins; lores yellowish; ear-coverts olivaceous brown with paler shaft-streaks; cheeks and under surface of body light yellow, the sides of the neck ashy, the sides of the body light brown; under wing-coverts light ashy brown; bill brown. Total length, 5 inches; culmen, 0·55; wing, 2·65; tail, 2·1; tarsus, 0·85.

The size of the bill in this species varies very much in length and also in stoutness. In some examples of the brown plumaged birds the beak is black, and these I take to be young males, as the black bill is accompanied by a certain bright yellow shade about the face and throat, and the under surface is rather brighter yellow.

Three nests collected are very interesting. That obtained by Mr. Slater is a large domed structure, composed entirely of fibre, with a few leaves and a little moss. The two obtained by Mr. Gulliver are firmly fixed in the small twigs of a tree, and in addition to the domed structure which is exemplified in Mr. Slater's specimen, these two have a doorway leading into the nest, and the probability is, therefore, that the nest which Mr. Slater found is not finished. With the exception of a few shreds of cotton and scraps of lichens the nest is entirely constructed of fibres.

Two eggs taken by Mr. Gulliver in December 1874, were of a clear pale blue. Length, 0·75 inch; diam. 0·55.

3. Ægialitis geoffroyi.

Ægialitis geoffroyi (Wagl.); Harting, Ibis, 1870, p. 378, pl. xi.

Mr. Slater obtained a single female of this Plover, in winter plumage. It measures as follows:—Total length, 8·3 inches; culmen, 0·9; wing, 5·55; tail, 2·2; tarsus, 1·35.

4. Strepsilas interpres.

Strepsilas interpres (L.); Finsch. & Hartl. Vög. Ostafr. p. 662.

Six specimens collected by Mr. Slater. They are all in winter plumage, with here and there an appearance of the summer dress being donned.

5. Numenius phæopus.

Numenius phæopus (L.); Finsch. & Hartl. Vög. Ostafr. p. 739.

The vernacular name of this bird in Rodriguez is "Corbiseau," according to Mr. Slater, who obtained one specimen in full winter plumage.

6. Butorides atricapilla.

Ardea atricapilla (Afzel.); Finsch. & Hartl. Vög. Ostafr. p. 701.

Two specimens were collected by Mr. Slater, an old female and a young bird, the latter in the usual mottled plumage peculiar to these Herons. The old bird agrees with the African *B. atricapilla*, and I must say that it surprises me to see that some authors would unite this African species to the Indian *B. javanica*. The latter is much darker, and is ashy-brownish where the African species is light grey, while the green shade of the upper plumage is always more bronzy.

Mr. Slater also procured an egg of this Heron. It is light greenish blue, and is not quite so brightly coloured as one procured by Mr. Monteiro from Angola. Length, 1·7 in.; diam. 1·2.

7. Sterna bernsteini.

Sterna bernsteini, Schl. Mus. P. B. Sterna, p. 9; Gray, Handl. B. iii. p. 120, No. 11,060; Saunders, P. Z. S. 1876, p. 657.

Thalasseus bernsteini, Blasius, J. f. O. 1866, p. 81.

Adult in winter plumage. General colour above bright silvery grey, the upper tail-coverts white; forehead, sides of face, sides of neck, and hinder neck pure white, forming an indication of a collar on the latter; crown of head mixed black and white, the plumes of the nape long and pointed, forming a crest; the ground colour of the feathers of the crown white, with a greater or less amount of black on them, sometimes forming only a dusky black streak, sometimes occupying nearly the entire feather, evidently indicating a change of plumage; wing-coverts light silvery grey like the back; quills slightly darker, silvery grey, white along the inner web, this colour more extended on the secondaries, occupying nearly the whole of the inner web of the innermost, and extending round the outer web of the shorter secondaries; primaries with the shafts pure white, the first quill blackish grey on the *outer* web near the base, shading gradually off into lighter grey towards the end of the feather, the other long primaries with a broad line of dark grey extending alongside of the shaft on the inner web, and widening out towards the apex of the same web; edge of the wing pure white, as are also the outer webs and tips of the thumb-coverts (bastard wing); tail light silvery grey, rather whiter on the inner webs of the feathers, the centre rectrices whitish; entire under surface of body, including the thighs and under wing and tail-coverts and axillaries pure white; bill (in skin) yellow, lead coloured towards the base of both mandibles; feet blackish brown in skin. Total length, 16·8 inches; culmen, 2·3; wing, 13·5; tail, 3·5, to outermost feathers, 5·5; tarsus, 1·0.

The description is taken from a bird obtained by Mr. Slater.

Another specimen obtained by the same gentleman is not quite in such good condition. It is marked a male, and has the following measurements:—Total length, 17·5 inches; culmen, 2·4; wing, 13·0; tail, 4·0, to tip of outermost feathers, 6·0; tarsus, 1·05. Probably the dimensions of the tail-feathers may be a shade wrong, as both the birds are moulting. A misprint has evidently occurred in the dimensions of the bill (1·8 inch) given by Mr. Howard Saunders (*l.c.*).

The species seems to be a very distinct one and apparently finds its nearest ally in *Sterna bergii*. From this, however, it differs in many important particulars:—

Sterna bergii.	Sterna bernsteini.
Upper surface dark slaty grey.	*Upper surface* light silvery grey.
Upper tail-coverts and *tail* uniform with back.	*Upper tail-coverts* and *tail* much lighter than the back, and nearly white.
Plumes of the bastard wing uniform with the other coverts.	Plumes of the bastard wing tipped with white.
Primaries externally blackish, but with a broad grey line intervening between the shaft and the black outer web.	*Primaries* externally blackish or grey, perfectly uniform, without any lighter shade intervening between the outer web and the shaft.

Mr. Slater says that the vernacular name for this species in Rodriguez is "Goilon."

8. Sterna dougalli.

Sterna dougalli (Mont.); Saunders, P. Z. S., 1876, p. 653.

Two specimens obtained by Mr. Slater have the bill perfectly black, and measure as follows:—

	Total Length	Culmen.	Wing.	Tail.	Outer Tail-feathers.	Tarsus.
a. Rodriguez (*Slater*)	15·0	1·5	8·05	3·0	6·0	0·7
b. ,, ,,	15·0	1·55	8·9	2·9	6·5	0·7

9. Sterna fuliginosa.

Sterna fuliginosa (Gm.); Saunders, P. Z. S., 1876, p. 666.

A pair of the large Sooty Tern were collected by Mr. Slater, and agree with the other skins from West Africa and other parts of the Southern Ocean. The measurements are as follows:—

	Total Length.	Culmen.	Wing.	Tail.	Long Tail-feathers.	Tarsus.
a. Rodriguez (*Slater*)	16·8	1·65	11·6	3·4	6·1	0·85
b. ,, ,,	16·7	1·7	11·6	4·1	6·8	0·8

10. Gygis candida.

Gygis candida (Gm.); Saunders, P. Z. S., 1876, p. 667.

A pair of birds were collected by Mr. Slater, who gives the vernacular name as "Goilette blanche." The following are the measurements of these birds:—

	Total Length.	Culmen.	Wing.	Tail.	Tarsus.
a. ♂ Rodriguez (*Slater*)	12·2	1·75	9·2	4·3	0·45
b. ♀ „ „	12·3	1·6	9·25	4·3	0·45

It will be seen from the above that the sexes in this Tern are very similar in size, but the female has a slightly smaller bill. They do not, however, show any approach to the slender-billed species *G. microrhyncha* of Mr. Saunders, of which the Museum has a specimen presented by Admiral Sir E. Belcher, and collected somewhere in the eastern seas. The white shafts to the quills and tail feathers in addition to the small size of the bird seem to be adequate distinctions for *G. microrhyncha*.

Four eggs of Gygis were obtained by Mr. Slater, but the dates of capture are not recorded, nor the circumstances of nesting. Mr. Saunders writes (*l. c.* p. 669):—
"The nesting of Gygis is peculiar, the single egg of clay-white mottled with brown
"being placed on the cavity of the branch of a tree, or in a fork of two branches,
"and on the points of the coral reefs—anywhere, in fact, where it will lie." The fact of there being but one egg laid by these birds will account partly for the difference in type exhibited by the four specimens obtained by Mr. Slater.

No. 1. Ground colour clay-white, sparsely spotted all over, the underlying spots and blotches being of a pale purple, the overlying marks being larger and of a dark brown colour, deepening on some to blackish brown; the larger blotches seem to be congregated more about the larger end and centre of the egg. Length, 1·55; diam. 1·2.

No. 2. Very different from the foregoing, being of a dull clay colour, the underlying purple blotches being very large, and distributed nearly equally over the whole egg, the brown mottlings taking the form of scribblings and wavy lines, confluent and forming blotches near the top and centre of the egg. Length, 1·5 inch; diam. 1·05.

No. 3. The ground colour of this egg is dull clay-colour like the preceding, but the whole surface is covered with small underlying spots of purple, with surface scribblings of dark brown all over the egg; in no place are there any large confluent blotches. Length, 1·65 inch; diam. 1·2.

No. 4. The ground colour in this specimen is clay-white as in the first described, but the markings are most like those of the last mentioned, being a collection of innumerable small spots and blotches of pale purple as the underlying markings, with an equally confused mass of dark brown spots and scribbled lines on the upper surface. There are a few broad blackish scribblings near the upper end of the egg. Length, 1·55 inch; diam. 1·2.

11. Anous tenuirostris.

Anous tenuirostris (Temm.); Saunders, P.Z.S., 1876, p. 670.

Adult Female. General colour above sooty-brown, rather darker on the rump and upper tail-coverts; wing-coverts sooty brown like the back, the least series blacker, as also those near the edge of the wings; quills black with an obscure gloss of sooty-brown on the outer webs, the inner secondaries sooty-brown like the back; tail uniform dark sooty-brown; crown of head light grey, gradually shading off into the ashy chocolate colour of the entire hinder part and sides of neck; feathers between the bill and the eye light grey; in front of the eye a black spot; above the hinder part of the eyes and below the same a narrow line of whitish feathers on the eyelid; sides of face chocolate brown, shaded with grey as they approach the sides of the neck; entire under surface of body sooty-brown; lower surface of quills ashy chocolate brown. Total length, 13·4 inches; culmen, 1·4; wing, 8·7; tail, 4·4; tarsus, 0·75.

The name of this species at Rodriguez is "Malanne" according to Mr. Slater, who collected a single specimen of this small Noddy. It bears out the distinctions of Mr. Saunders' paper (*l.c.*), but as in the latter essay the characters do not seem very easily made out, I have described the present species in detail from Mr. Slater's Rodriguez example.

12. Anous stolidus.

Anous stolidus (L.); Saunders, P. Z. S. 1876, p. 669.

One specimen collected by Mr. Slater (*vide infrà*). Like the *A. tenuirostris* it is called "Malanne." Mr. Slater has also sent six eggs of this species.

No. 1 is white, with faint purple spots and blotches distributed over its surface, but only distinct at the larger end, which is further slightly clouded by large blotches and spots of chesnut, with a few spots of black intermixed.

No. 2 is very similar, but the chesnut spots are not distributed so thickly round the thicker end of the egg, but there is a large blotch of dark chesnut at the apex.

No. 3. Very like No. 1, but more yellowish white, and having the upper end thickly clouded with chesnut and black spots, the underlying purplish blotches being rather more distinct.

No. 4. Creamy white, the blotches large but sparingly distributed round the larger end of the egg in a ring, leaving the apex unspotted.

No. 5. White with the blotches extremely large and confluent round the larger end of the egg, the chesnut colour being rather pale and blending with the underlying purple colour.

No. 6. White with very few markings, the underlying purple spots as well as the larger chesnut marks being distributed all over the egg.

13. Puffinus chlororhynchus.

Puffinus chlororhynchus (Less.); Gray, Handl. B. iii., p. 103, No. 10,844.

♀. Total length, 17; culmen, 1·4; wing, 10·1; tail, 5·4; tarsus, 1·7.
♂.   ,,   16; ,,   1·45; ,, 11·2; ,, 5·2; ,, 1·8.

14. Phaeton flavirostris, Brandt; Gray, *t. c.* p. 124, No. 11,096.

An adult female procured by Mr. Slater.

15. Sula piscator, (L.); Gray, *t. c.* p. 126, No. 11,109.

An adult female procured by Mr. Slater. "Booby."

Note on Anous.—The following would seem to be the differences between the members of the genus *Anous*, of which the British Museum has a good series. I give herewith a list of the specimens at present in the national collection, with the dimensions of the wings in each case:—

KEY TO THE SPECIES.

a. Crown of the head grey or whitish, not uniform with the back.
  a^1. Lores and feathers in front of the eye sooty-brown or blackish like the sides of the face.
    a^{11}. Forehead and crown grey, gradually shading off into ashy-brown or chocolate on the neck.
      a^{111}. Very much larger; wing, 10·2–11·5 inches.
        a^4. Crown of head rather clear grey, whiter on the forehead and over the eye; lores and sides of face greyish chocolate . . *stolidus*.
        b^4. Similar to *A. stolidus*, the crown greyish brown, with a narrow superciliary line of white in strong contrast; lores and sides of face dull chocolate brown . . . . . . *superciliosus*.
      b^{111}. Very much smaller; feathers in front of the eye black; wing not exceeding 9·5 inches.
        c^4. Cheeks and throat all blackish chocolate, darker than the head and sides of the face . . . . . . *melanogenys*.
        d^4. Cheeks and throat sooty-grey, like the neck . . *plumbeigularis*.
    b^{11}. Crown and nape pure white, forming a cap . . . . *leucocapillus*.
  b^1. Lores and feathers in front of the eye grey like the crown.
    c^{11}. Underneath sooty-brown, as also the sides of the face . . . *tenuirostris*.
    d^{11}. Underneath silvery grey or greyish white . . . . . $\begin{cases} \textit{cæruleus.} \\ \textit{cinereus.} \end{cases}$
b. Crown of the head uniform brown like the rest of the upper surface . . *galapagensis*.

1. Anous stolidus, L.; Saunders, P.Z.S., 1876, p. 669.

a. ad. Africa. Lieut. Friend, R.N. Wing, 10·2.
b. ad. Gold Coast. R. B. Sharpe, Esq. Wing, 10·7.
c. ad. Cameroons. Capt. R. Burton. Wing, 10·9.

d. e. ad. Cape Seas. Capt. Thomas Row Harry.
f. ad Reunion. Herr. J. C. Van Dam. Wing, 10·9.
g. ad. „ „ „ Wing, 11·2.
h. ♂ ad. Rodriguez. H. H. Slater, Esq. Wing, 11·15.
i. ad. S. Paul's Rock. Antarctic Expedition. Wing, 11·5
k. ad. Morty Island. A. R. Wallace, Esq. Wing, 11·2.
l. ad. Raines' Islet. J. B. Jukes, Esq. Wing, 11·2.
m. ad. At sea between New Zealand and Australia. E. Newman, Esq. Wing, 11·25.
n. ad. Huaheine. J. H. Gurney, Esq. Wing, 11·0.
o. ad. Navigator's Islands. Rev. J. B. Stair. Wing, 10·85.
p. q. ad. Ellice Islands. Rev. S. J. Whitmee.
r. ad. North America. H.R.H. The Duke of Sussex. Wing, 10·6.
s. ad. North America. Purchased. Wing, 10·2.
t. ad. Western Coast of America. Sir W. Burnett and Admiral Fitzroy. Wing, 11·3.

2. Anous superciliosus, sp. n.

A. similis *A. stolido* sed pileo summo cinerascenti-brunneo, lineâ superciliari albâ distinctâ distinguendus. Long. tot., 12·2 ; culm., 1·6 ; alæ, 10·7 ; caudæ, 5·3 ; tarsi, 0·95.

Hab.—Coast of Central America and the Antilles.

a. ad. Coast of Mexico. G. Barclay, Esq. Wing, 10·75.
b. ad. S. Nevis. W. Cottle, Esq. Wing, 9·7.
c. ad. Jamaica. Purchased. Wing, 10·7.
d. ad. "Caught at sea in long. 91 E., lat. 8 N." R. B. Sharpe, Esq. Wing, 9·7.

3. Anous melanogenys, Gray; Saunders, *t. c.,* p. 670.

a. Locality unknown. Captain Lord Byron. Wing, 8·9.
b. ♀. Moreton Bay. J. Macgillivray, Esq. Wing, 8·95.

4. Anous plumbeigularis, sp. n.

A. ubique cinerascenti-chocolatinus ; capite summo canescente ; maculâ supra-et infraoculari albidâ ; plumis anteocularibus nigris ; loris, facie laterali totâ clarè cinerascenti-chocolatinis, corpore reliquo subtus chocolatino-brunneo ; subalaribus cinerascenti-chocolatinis ; tectricibus alarum superioribus chocolatinis, remigibus rectricibusque nigricantibus. Long. tot., 14·4 ; culm., 1·55 ; alæ, 9·8 ; caudæ, 5·8 ; tarsi, 0·9.

Hab.—Red Sea.

Mr. Saunders hints that the Noddy of the Red Sea is probably incorrectly determined, and I have therefore examined a specimen which we have in the Museum from that locality. I find that the bird is apparently distinct from *A. melanogenys*, differing in its clearer grey face and throat, which are not so black; the wing is also an inch longer, the back rather lighter brown, and the tail chocolate brown without any grey shade. The size of the bird is not nearly large enough to allow of its being confounded with *A. stolidus*.

a. ad. Red Sea. J. Middleton, Esq.

5. Anous leucocapillus, *Gould ; Saunders, t. c.* p. 670, pl. LXI,, fig. 3.

a. Off Bristow Island, S. coast of New Guinea. J. Macgillivray. Wing, 8·5.
b. Off Barnard Islands. J. Macgillivray. Wing, 8·6.

c. Locality unknown. Wing, 9·3.
d.   „   R. B. Sharpe, Esq.*
e. Off Western coast of South America. R. B. Sharpe, Esq.
f. Ellice Islands. Rev. S. J. Whitmee.

6. **Anous tenuirostris** (*Temm*); *Saunders, t. c.* p. 670, pl. LXI., fig. 1.

a. ad. Mauritius. R. B. Sharpe, Esq. Wing, 8·15.
b. ad. Rodriguez. H. H. Slater, Esq. Wing, 8·7.
c. ♂ ad. South Island, Houtmann's Abrolhos. J. Gould, Esq. Wing, 8·7.

7. **Anous cinereus,** *Gould*; *Sharpe*, P.Z.S. 1878, p. 271.

a. b. ad. Australia. Sir T. Mitchell. Wing, 8·45.

8. **Anous cæruleus** (*Bennett*); *Sharpe*, P.Z.S. 1878, p. 272.

a. b. ad. Ellice Islands. Rev. S. J. Whitmee.

9. **Anous galapagensis,** sp. n.

A. similis A. stolido, sed rostro minore et capite summo fuliginoso unicolori distinguendus Long. tot. 13·4; culmen, 1·4; alæ, 10·65; caudæ, 5·5; tarsi, 0·85.

Hab.—Galapagos Archipelago.

a. ad. Dalrymple Rock, Chatham Island.

Mr. Saunders thinks that this may be a young *A. stolidus*. I believe the bird to be quite adult, and to constitute a distinct species.

* Said to have been procured off the Gold Coast, but the locality is probably erroneous.

REPTILES.—*By Dr. Albert Günther, F.R.S.*

Numerous specimens of a Gecko, *Peropus mutilatus*, were collected by Mr. Gulliver. This small species is not only generally distributed in the Mascarene Islands, but occurs also in numerous islands of the East Indian Archipelago.

Mr. E. Newton has received quite recently a specimen of *Phelsuma cepedianum*, which was caught by Mr. H. E. Desmarais on the 1st of July 1876, in the jungle of Rodriguez. This species is common in Madagascar and almost all the Mascarene Islands, but seems to be very scarce in Rodriguez.

FISHES.—*By Dr. Albert Günther, F.R.S.*

A.—MARINE SPECIES.

The marine fish fauna of Rodriguez does not differ in any respect from the well-known fauna of Mauritius, the Seychelles, or Zanzibar; and the species collected by the naturalists of the Expedition being not uncommon species, it will be sufficient to enumerate them in the following list:—

Holocentrum diadema, Lacép.
    „   *spiniferum*, Forsk.
Serranus hexagonatus, Forst.
    „   *flavocœrulens*, Lacép.
Mesoprion fulviflamma, Forsk.
Dules marginatus, C. & V.
Pentapus aurolineatus, Lacép.
Mulloides flavolineatus, Lacép.
Lethrinus mahsena, Forsk.
    „   *nebulosus*, Forsk.
Chrysophrys bifasciata, Forsk.
Chætodon setifer, Bl.
    „   *lunula*, Lacép.
    „   *vagabundus*, L.
    „   *lineolatus*, C. & V.
Holacanthus alternans, C. & V.
Heniochus monoceros, C. & V.
Psettus argenteus, L.
Caranx crumenophthalmus, Bl.
    „   *melampygus*, C. & V.
Gobius albopunctatus, C. & V.
Salarias quadricornis, C. & V.
Teuthis nebulosa, Q. & G.
Naseus unicornis, Forsk.
Acanthurus desjardinii, Benn.
    „   *matoides* (*blochii*), C. & V.
    „   *strigosus*, Benn.
    „   *triostegus*, L.
Atherina pinguis, Lacép.
Pomacentrus trilineatus, C. & V.
Glyphidodon sparoides, C. & V.
    „   *cœlestinus*, C. & V.
Dascyllus aruanus, L.
Chilinus punctatus, Benn.
Anampses diadematus, Rüpp.
Pseudoscarus harid, Forsk.
Plotosus anguillaris, Bl.
Tetrodon nigropunctatus, Bl.
Balistes aculeatus, L.
    „   *rectangulus*, Bl.
Ostracion cubicus, L.
Diodon hystrix, L.

B. Freshwater Species.

Also the freshwater fishes (which were collected by Mr. Gulliver) are species well known to inhabit or to freely enter fresh waters of the Mascarenes and other parts of the western shores of the Indian Ocean. Two Grey Mullets, however, differ sufficiently from those of the neighbouring islands to be entitled to specific distinction.

1. List of the Species.

Dules marginatus, C. & V.
„ *rupestris*, Lacép.
Chrysophrys sarba, Forsk.
Psettus argenteus, L.
Gobius albopunctatus, C. & V.
„ *ocellaris*, Brouss.
Eleotris monteiri, O'Shaughn.

Mugil rodericensis, sp. n.
Myxus cæcutiens, sp. n.
Megalops cyprinoides, Brouss.
Chanos salmonoides, Bl.
Anguilla mauritiana, Benn.
„ *virescens*, Ptrs.

2. Description of the Grey Mullets.

Mugil rodericensis.

Allied to *M. troschelii*, *smithii* and *compressus*, but with a narrower and longer caudal peduncle.

D. $4\frac{1}{8}$, A. $\frac{3}{9}$, L. lat. 30. L. transv. 11.

The height of the body is a little more than the length of the head, and one fourth of the total (without caudal); the head is two-thirds as high as long; the diameter of the eye is one-fourth of the length of the head, and contained once and three-fourths in the width of the interorbital space. Eyelids not developed. Upper profile of the head and nape nearly straight. Præorbital scaly, emarginate, and finely denticulated. Snout a little shorter than the eye; the maxillary extends beyond the præorbital, and its extremity is uncovered; upper lip thin. There are twenty-one scales between the snout and the dorsal fin. The least depth of the tail is less than one-half of the length of the head, and the caudal peduncle is considerably longer than deep. The spinous dorsal is higher than the soft; the spines are strong, the length of the first-exceeding that of the postorbital portion of the head; the base of the first spine a little nearer to the root of the caudal than to the end of the snout. The soft dorsal and anal nearly entirely scaleless; the anterior third of the anal in advance of the origin of the soft dorsal.

Caudal fin forked. Pectoral shorter than the head. Coloration uniform.

One specimen 10 inches long, and several young ones were collected by Mr. Gulliver.

Myxus cœcutiens.

D. $4\frac{1}{8}$, A. $\frac{3}{8}$, L. lat. 43. L. transv. 14.

Teeth very small, moveable, bent, those of the upper jaw in a single series; a notch in the middle of the upper jaw to receive the mandibular symphysis. Lower jaw with a similar series of horizontal teeth; other smaller teeth behind appear to be destined to replace those in function. Lower surface of the mandible without transverse folds. Two round, hard, apparently toothless, naked patches on each side of the palate. The maxillary does not quite extend to the front margin of the eye. Snout obtuse, shorter than the eye, which is two-sevenths of the length of the head, and two-thirds of the width of the interorbital space. Eye with a broad anterior and posterior adipose eyelid. The depth of the body is nearly equal to the length of the head, which is two-sevenths of the total length (without caudal). Pectoral extending to the commencement of the spinous dorsal, which corresponds to the thirteenth scale of the lateral line. Dorsal spine moderately strong, the length of the first being more than half the length of the head. Caudal fin deeply emarginate. Coloration uniform.

Two specimens, 5 inches long, were collected by Mr. Gulliver.

MOLLUSCA.—*By Edgar A. Smith, F.Z.S., Assistant Zoological Department, British Museum.*

(Plate LI.)

The following list comprises all the species of mollusca collected at the Island of Rodriguez by Messrs. G. Gulliver and H. H. Slater. The marine shells, with one exception (*Rhizocheilus monodonta*), were obtained by the latter.

They are for the most part very common forms, which are found abundantly in the Indian and Pacific Oceans, on the shores of the Mauritius, Ceylon, Andaman, and Seychelles Islands, Philippines, North Australia, Samoa, Fiji, &c.

The terrestrial and fluviatile molluscan fauna, as might be expected, and as shown by M. Crosse, assimilates most closely to that of the most adjacent land, the Mauritius, Bourbon, Madagascar, and the Seychelles, a few of the species being common to Rodriguez and these localities.

The living terrestrial and fluviatile mollusca hitherto recorded from this island are 25 in number. To these are now added five others, and a species of *Bulimus* which in all probability never lived on Rodriguez, but has been conveyed there by chance, or may have been cast overboard from a ship and washed on shore, where it was picked up I believe, since it was packed up with marine shells collected on the beach.

It will be noticed how few of the species are quoted as Madagascarian; this no doubt simply arises from the fact that this country has been so little investigated in regard to its molluscan fauna.

The whole collection consists of 78 marine species, represented by 260 specimens; 16 terrestrial forms, consisting of 1,149 examples; and six fluviatile species, which include 574 specimens.

The following species not obtained by the Transit of Venus Expedition are mentioned in M. Crosse's paper, entitled "Faune malacologique terrestre et fluviatile de l'île Rodriguez," published in the Journal de Conchyliologie, 1874, Vol. XXII., pp. 221-242:—1. *Pupa (Gonospira) metableta*, Crosse; 2. *Omphalotropis tæniata*, Crosse; 3. *O. littorinula*, Crosse; 4. *Truncatella teres*, Pfeiffer; 5. *Melania commersoni*, Morelet; 6. *Melania tuberculata*, Müller; 7. *Neritina caffra*, Gray; 8. *N. mauritiana*, Morelet; 9. *Streptaxis distortus*, Jonas; 10. *S. piriformis*, Pfeiffer.

List of the Species.

I. Marine.

	Island of Rodriguez only.	Previously found elsewhere.						
		Mauritius.	Seychelles.	Bourbon.	Madagascar.	Andaman Islands.	Ceylon.	Other Localities.
Conus hebræus, Linn.						×	×	Cape of Good Hope, Zanzibar, China Sea, N. Australia, Fiji, Sandwich Islands, Samoa, Society Islands, Aneiteum, Tonga Islands.
„ *tessellatus*, Born.		×			×		×	Mozambique, Red Sea, Philippines, N. Australia, Fiji.
„ *catus*, Hwass		×						Tahiti, Hardy's Island.
„ *miles*, Linn.			×				×	Philippines, Solomon Islands, Fiji.
„ *pennaceus*, Born.		×			×	×		Persian Gulf, Philippines, Moluccas, Sandwich Islands, Tahiti, Society Islands.
„ *lividus*, Hwass		×						Cape of Good Hope, Red Sea, Philippines, Sandwich Islands, Solomon Islands, Fiji, New Caledonia.
„ *rattus*, Hwass			×					N. Australia, Annaa Island.
„ *balteatus*, Sowerby								Philippines.
„ *ceylanensis*, Hwass							×	Red Sea, Sandwich Islands, Lord Hood Island, Niue, Red Sea, Natal.
Sistrum arachnoides, Lamarck			×			×		Lord Hood's Island, Samoa, N. E. Australia, Sandwich Islands.
„ *morus*, Lamarck								Lord Hood's Island, Cape of Good Hope.
„ *muticum*, Lamarck		×						Mozambique.
„ *horridum*, Lamarck						×		Philippines, Port Essington, Society Islands.
Rhizochilus (*Coralliophila*) *monodonta*, Quoy & Gaimard								Fiji Islands, "Tonga Tabu" (*Quoy*).
Rhizochilus (*Coralliophila*) *squamosissimus*, sp. nov.	×							
Harpa imperialis, Linn.		×						
„ *minor*, Lamarck					×			Red Sea, Mozambique, Philippines, Society, and Navigator's Islands.
Fasciolaria trapezium, Linn.			×			×		Red Sea, Philippines, Solomon Islands.
Latirus polygonus, Gmelin			×					Mozambique, Philippines, N. Australia, Tonga Islands.
„ *nassatulus*, Lamarck						×		Cape of Good Hope, Philippines, Port Essington, Fiji Islands.
Strombus gibberulus, Linn.			×			×		Red Sea, Cape of Good Hope, Zanzibar, Philippines, New Guinea, Swan River, Fiji, New Zealand, Society Islands.
„ *floridus*, Lamarck			×			×		Philippines, Society Islands, Sandwich Islands, Red Sea, Cape of Good Hope, Zanzibar.
„ *elegans*, Sowerby								Red Sea, Philippines.
„ *samarensis*, Chemnitz.						×		Philippines.

	Island of Rodriguez only.	Previously found elsewhere.					
		Mauritius.	Seychelles.	Bourbon.	Madagascar.	Andaman Islands.	Ceylon.	Other Localities.
Pterocera chiragra, Linn.						×		Timor, Australia, Solomon Islands.
Tritonium cynocephalum, Lamarck.								Philippines.
„ *tuberosum*, Lamarck						×		Sandwich Islands, Annaa, Fiji.
Cassis vibex, Linn.								Philippines, Samoa, Pacific Islands, Annaa, Society Islands, Cape of Good Hope.
Cerithium echinatum, Lamarck								
„ (*Vertagus*) *asperum*, Linn.		×						Red Sea, Mozambique, N. Australia, Fiji, Louisiade Archipelago.
Natica maroccana, var. Chemnitz.								Natal, Mozambique, Red Sea, N. Australia, Philippines, Fiji, Society Islands, W. Africa, W. Indies, California.
N. (*Ruma*) *melanostoma*, Lamarck.			×					Red Sea, Mozambique, Philippines, Darnley Island.
„ *simiæ*, Chemnitz								Philippines.
Janthina communis ? Lamarck								
Nerita plicata, Linn.							×	Natal, Mozambique, Zanzibar, Round Island, N. Australia, New Guinea, Fiji.
„ *polita*, Linn.		×	×			×	×	Red Sea, Cape of Good Hope, Zanzibar, Philippine Islands, Sandwich and Solomon Islands.
„ *albicilla*, Linn.		×	×					Hardy's Island, Zanzibar, Cape of Good Hope, Philippines.
Fenella reticulata, A. Adams								Gulf of Suez, Japan.
Cypræa tigris, Linn.		×						Mozambique, Samoa, Tongatabu, New Caledonia.
„ *arabica*, Linn.			×			×		Natal, Mozambique, Red Sea, N. Australia, Upolu.
„ *mappa*, Linn.								Annaa.
„ *mauritiana*, Linn.		×				×	×	Navigator's Island, Tongatabu.
„ *vitellus*, Linn.			×			×		Fiji Islands, Cape of Good Hope.
„ *carneola*, Linn.			×			×		Natal, N. Australia, Fiji, Sandwich Islands.
„ *isabella*, Linn.		×	×			×		Sandwich Islands.
„ *caput-serpentis*, Linn.			×			×		Cape of Good Hope, Mozambique, New Zealand, Tahiti, Savage Island.
„ *nucleus*, Linn.								Philippines, Borneo, Savage Island.
„ *helvola*, Linn.			×				×	Mozambique, Sandwich Islands, Cape of Good Hope.
„ *lamarckii*, Gray								Mozambique.
„ *moneta*, Linn.			×					Fiji, Sandwich Islands.
„ *annulus*, Linn.			×					Sandwich Islands, Cape of Good Hope.
Dolium perdix, Linn.						×		Samoa Island, "Florida" (*Green*).
Amalthea acuta, Quoy & Gaimard.								"New Ireland" (*Quoy*), "Natal," *Krauss*.
Turbo margaritaceus, Linn.			×					Philippines, Tongatabu.
Trochus (*Cardinalia*) *virgatus*, Gmelin.								Red Sea, Philippines.
Trochus (*Infundibulum*) *delicatulus*, Philippi.							×	India.

	Island of Rodriguez only.	Previously found elsewhere.						
		Mauritius.	Seychelles.	Bourbon.	Madagascar.	Andaman Islands.	Ceylon.	Other Localities.
Haliotis venusta, Adams & Reeve.								Eastern Seas.
Hexabranchus mauritiensis? Abraham.		×						
Bulla ampulla, Linn.		×						Red Sea, Philippine Is., Tahiti, Fiji, New Zealand.
Dione læta, Linn.								Philippines.
Circe pectinata, Linn.						×		Upolu, Australia, Philippine Islands, Red Sea.
„ (*Leioconcha*) *picta*, Lamarck.							×	Indian Ocean, China.
Tridacna gigas, Linn.								Indian and Pacific Oceans, Philippines.
Lucina fibula, Reeve								Red Sea, Philippines, China Sea, Port Essington [West Columbia, *Rve.*].
Diplodonta lateralis, sp. nov.	×							
Cardium rugosum, Lamarck					×			Zanzibar, Nicobar Islands, Australia.
„ (*Fulvia*) *pulchrum* Reeve.		×						China "Cape of Good Hope" (*Jay*).
Lasea australis, Sowerbie								"New Caledonia" (*Souverbie*).
Avicula (*margaritifera*) *margaritifera*, Linn.							×	Cape of Good Hope (*Krauss*). N. Australia, Solomon Islands, Persian Gulf, Society and Sandwich Islands.
Avicula fusco-purpurea, sp. nov.	×							
Ostrea, sp.								
Chama ruppellii, Reeve								Red Sea.
Modiola auriculata, Krauss								Red Sea, East Africa, Zanzibar.
Mytilus variabilis, Krauss								Natal, Zanzibar, Red Sea.
Pinna muricata (Linn.), Reeve.								[W. Indies, *Rve.*]?
Pecten imbricatus, var. Gmelin.								Red Sea.

II. Fluviatile.

	Island of Rodriguez only.	Mauritius.	Seychelles.	Bourbon.	Madagascar.	Andaman Islands.	Ceylon.	Other Localities.
Melania scabra (Müller), Reeve.	×	×	×					
Melania rodericensis, sp. nov.	×							
Planorbis rodericensis, Crosse								
Niritina gagates, Lamarck		×		×				Sumatra.
„ *longispina*, Recluz		×						
Navicella porcellana, var., Linn.		×		×	×			Philippine Islands.

III. TERRESTRIAL.

	Island of Rodriguez only.	Previously found elsewhere.						
		Mauritius.	Seychelles.	Bourbon.	Madagascar.	Andaman Islands.	Ceylon.	Other Localities.
Helix bewsheriana, Morelet	×							
„ *rodericensis*, Crosse	×							
„ sp.								
„ *similaris*, Férussac		×	×					Brazil, Cuba, Natal, Bengal, China, Java, Japan, Sandwich Islands.
Bulimus fibratus var., Martyn								New Caledonia.
„ (*Stenogyra*) *gracilis*, Hutton.								India, Siam, Archipelago.
Pupa lienardiana, Crosse	×							
„ *desmazuresi*, Crosse	×							
„ (*Gibbus*) *rodericensis*, Crosse.	×							
Pupa (*Gibbus*), *chloris*, Crosse	×							
„ („), *palanga*, Lesson.		×						
Succinea nevilli, Crosse		×		×				
Vaginula rodericensis, sp. nov.	×							
Cyclostoma articulatum, Gray					×?			
„ *hæmastoma* var. Anton.		×						Round Island.
Omphalotropis littorinula, Crosse.	×							

REMARKS ON VARIOUS SPECIES.

Conus lividus. Both forms of this species, the coronated (*lividus*) and non-coronated (*C. flavidus* Lamarck), were collected at Rodriguez. Of the former the specimens are strongly tubercled and broad-shouldered, clothed with a brownish olive epidermis and encircled around the lower half of the whorl by about eight nodulous, but only faintly elevated lirations. The latter variety also has similar liræ, and agrees precisely with certain specimens in the British Museum, from the Fiji and Solomon Islands and Port Essington.

Conus ceylanensis. The single specimen of this species is in very bad condition, but is interesting on account of its unusually broad form, very similar to that of *C. sponsalis*. It has a purple aperture, and is painted exteriorly with spiral series of reddish dots.

Conus balteatus. Plate LI., fig. 12. The shells which I have referred to this species display considerable difference in form from the normal type. They are much shorter and consequently more briefly conical and even somewhat pyriform. The tubercles of the spire are also more remote from one another. The mouth is deep purple within, and white at the upper and middle parts of the lip. The exterior surface is obscurely two-banded with brown, variegated with a few opake white dots here and there, and is clothed with a very thin yellowish-olive epidermis.

Conus rattus. Plate LI., fig. 11. Of this species, besides two fairly typical specimens, there is a third one in the collection which is a very curious short shell apparently belonging to, although differing considerably from, this species. It is 30 millimeters long, and 18 in breadth and partakes much of the outline of *C. sponsalis*. It also shows considerable affinity to *C. balteatus* and, indeed, it is possible that it may be a variety of it. From this species it differs in having the spire blotched with brown and not wholly white, and it lacks the rosy apex. It is, however, coronated, which is very unusual in *C. rattus*. The body-whorl is broadly two-banded with brown and dotted with opake white and the aperture is deep purple brown.

Conus catus. Plate LI., fig. 13. A single specimen of this species is remarkable on account of its uniformly pink colour.

Latirus nassatulus. The pretty example of this species agrees in all respects with the normal form, with the exception of colour being in every part perfectly white.

Sistrum arachnoides. The only specimen of this species is a very fine large representative of it. It has fine large bold spines with the interrupted band on the labrum and columella of a rich orange.

Sistrum horridum. This shell also, like the specimens of *S. morus*, is thickly coated with lime, and is remarkable for the thickness of the labrum and the development of the teeth, which form an almost uninterrupted series of eight, which are about equal in size, and the four on the columella are also very stout, and the aperture is consequently very narrow.

Fenella reticulata. Plate LI., fig. 6. Besides the three lirations which encircle the upper whorls and are nodulous on crossing the longitudinal costæ, which are about 14 in number in each whorl, there is a fourth fine thread-like lira around the upper part which borders the suture and is only slightly nodose. The body-whorl below the four above-mentioned lirations is ornamented with four or five others ("liris numerosis!" Adams' Annals Nat. Hist. 1860, VI., p. 422) which generally are not nodulous like the upper ones, as the longitudinal ribs usually terminate at the fourth lira from the suture. The largest example in the British Museum is from Japan, and has a length of 4 millim, and its last whorl is $\frac{1}{3}$ broad; aperture 1 long and $\frac{2}{3}$ wide. Specimens from Suez, the Persian Gulf, and Rodriguez, have invariably rather smaller dimensions than the preceding.

Janthina communis? The specimens of this genus from Rodriguez have a rather elevated form, being well represented by the figure 1a in Reeve's Monograph in the Conchologia Iconica, to which he has given the name *J. roseola*, which, it is impossible to say definitely, may or may not be distinct from the common species.

Amalthea acuta. The single shell from Rodriguez which appears to be but a variety of this species, differs slightly from the description and figure given by Quoy in the Voyage of the "Astrolabe." It is slightly more elevated and the margin is more or less stained with brown. It is a very stout, solid shell, and coarsely radiately ridged exteriorly, white and stained with reddish-brown, especially towards the apex.

Natica maroccana. One specimen which is in a somewhat broken condition agrees almost precisely with others in the Museum from the Red Sea and Sandwich Islands. The sutural wrinkles are well marked. It is of a pale livid grey colour with a white zone beneath the suture and another above the middle of the whorl, the base of which is also white. The aperture is brown banded.

Strombus elegans. A curious example of this species was obtained by Mr. Slater. It is a small shell about an inch in length, whitish, with a few brown spottings; the columella is deep red, and the upper and lower parts of the labrum are stained with the same colour. On the last whorl the longitudinal plicæ become very strongly nodulous on the dorsal region, and are not prolonged downwards, and the transverse striation is particularly well defined.

Strombus floridus. There are two dwarf specimens of this shell, only ten lines long, having a very thick labrum, and the thick callous of the columella is scarcely striated at all. They are white with only a few brownish markings.

Pinna muricata, *Reeve*. The specimens of this species from Rodriguez do not appear to differ in any very material respects from that said to come from the West Indies, and described by Reeve as *muricata* of Linneus. However, there are one or two slight distinguishing characters in them, but not sufficient to separate them specifically. The ribs on the ventral half of the valves which are scale-bearing, do not extend so close to the margin; the dorsal edge is more arcuated and the base is also slightly sinuated and not convex, as in the shell figured by Reeve in the Conchologia Iconica, fig. 23.

Helix rodericensis. The specimen of this species originally described was not in a perfect condition, as no mention is made of the beautiful epidermis which should adorn it. Two of the best preserved shells collected by Mr. Gulliver display a very pretty brown epidermis, which is thick and almost lamellar, being elevated into close-set oblique shaggy laminæ, which have here and there one or two somewhat pointed elevations, which produce the appearance of the whorl being spirally ridged. On the body-whorl there are about seven of these pseudo-ridges, three above and four beneath the periphery, and the oblique riblets or laminæ do not extend very far over the lower surface of the whorl. The height of the spire in this species is subject to variation, as some specimens are much more depressed than that depicted in the Journal de Conchyliologie, which also is rather a small one. The largest living shell at hand measures $13\frac{1}{2}$ mill. at its greatest diameter and some of the semi-fossil specimens which were obtained from the bone-caves are still larger, having a width of 17 mill.

Helix bewsheriana. This sub-fossil species was collected by Mr. Slater, in the bone-caves.

Helix sp. ? This species is represented in the collection by a single specimen, obtained by Mr. Slater. It is very small, measuring only $1\frac{1}{2}$ millim. at its greatest diameter, depressed, discoid, and consists of $3\frac{1}{2}$ whorls which are obliquely, finely lirate. The umbilicus is rather large; aperture lunate circular.

As there is but a single specimen, and that possibly immature, I have not thought it advisable to designate it with a specific name. Several closely allied species inhabit New Zealand and the islands of the Pacific.

Helix similaris. This almost cosmopolitan shell was "very common" on the island according to the observations of Mr. Gulliver.

Bulimus fibratus, *var.* The presence of this New Caledonian species in the island of Rodriguez is very surprising. Only one specimen was in the collection made by Mr. Slater. It seems impossible that it can be truly indigenous to the island, for, as is well-known, land-shells are very local in their geographical distribution except in a few instances, such as in the case of the preceding ubiquitous species, *H. cellaria* and one or two others. It appears to me therefore, that in all probability this specimen must have been conveyed to Rodriguez Island by human agency, and may have been picked up on the shore, having been thrown overboard from a vessel and washed on to the beach. The specimen varies somewhat from the typical *fibratus* and seems to be a connecting link between it and B. Souvillei, Morelet.

Bulimus (Stenogyra) gracilis, Hutton. On comparing the numerous examples from Rodriguez, with Indian specimens of this species, some of them appear to be quite identical. It is a very variable species with regard to the thickness of the spire, and if the two extreme forms be separated from the others they appear to be distinct species; but in a long series of specimens such as that before me, all the intermediate subtle gradations clearly show that they are but variations of one and the same species. *B. (Stenogyra) subula* Pfr. and *oparanus* Pfr. are scarcely separable from this species.

Cyclostoma articulatum, Gray. "This species is especially abundant on the Coral-"line limestone, the shells, chiefly dead, occurring in vast numbers. The living "animals are to be found for the most part on trees "—(*Gulliver*).

The description of *Cy. desmazuresi* given by Crosse (Journal de Conch. 1874. XXII., p. 237 pl. VII., f. 12) agrees so exactly with the young stage of *articulatum*, that there can be no doubt that his shells are not a distinct species, but merely immature examples of the present one.

Cyclostoma hæmastoma, *var.* The variety of this species which dwells on Rodriguez has been styled *var. rodriguezensis* by Crosse, who points out the following distinctions. It is "much smaller than the typical form, possessing only four and a half (instead

" of five) whorls and presenting a smaller number of spiral sulci (12 instead of 14 " on the penultimate whorl). It occurs everywhere in abundance."—(*Gulliver*.)

Omphalotropis hamiliana. The coloration of this species varies somewhat, the reddish brown colour differing in intensity, and the white band around the periphery sometimes being almost obsolete. The spiral striation will be found to be minutely punctured under the microscope.

Omphalotropis littorinula. Crosse is very closely allied to, if not the same as, certain varieties of this species.

Descriptions of New Species.

Melania rodericensis.

Plate LI., Figs. 9 and 10.

Annals and Mag. Nat. Hist. 1876, May, p. 404.

Testa subulata, pallide olivacea, strigis longitudinalibus purpureo-fuscis irregulariter picta, et circa basim anfr. ultimi fascia lata coloris saturatioris cincta, frequenter corio limoso rubro induta; anfractus 11, convexiusculi, sutura leviter obliqua discreti, liris transversis tenuibus plicisque longitudinalibus supra liras paululum granosis ornati; apertura ovata, superne acuminata; columella arcuata, alba.

Long. 16 mill. diam. 5; apertura long. 5, diam. 2½. var. major, long. 23 mill. diam. 7.

Shell small, subulate, thinnish, of a pale olive or yellowish olive colour, ornamented with longitudinal little undulating purplish-brown or red stripes, which frequently stop short of, and do not flow into, the broad dark purplish-brown band encircling the base of the body-whorl; frequently the whole surface of the shell is covered with a reddish earthy deposit; whorls 11 moderately convex, encircled with fine subæqual spiral lirations (9 or 10 on the upper whorls and about 20), which are crossed by slightly oblique longitudinal plications, which are slightly granulous at the points of contact with the transverse ridges and scarcely extend to the suture beneath; they are slightly flexuous, about 17 in number and on the last whorl they become obsolete about the middle. The aperture is ovate and acuminated at the upper extremity, and has much of the tint of the exterior of the shell, the longitudinal flames and the dark basal band being quite distinctly visible; the columella is oblique, not much arched superiorly but more so below at its junction with the outer lip where it is white.

Variety. The shell is rather larger, rather more ventricose and has the sculpture less pronounced.

The dark band, encircling the base of the body-whorl, is very constant in all specimens. This species was collected by Mr. Gulliver, both in fresh and brackish water, but appeared most abundant in the latter.

There are several species of *Melaniæ*, which partake very much of the forms of the present species.

M. fasciolata of Olivier appears to exhibit the greatest affinity to it. The former is however larger and the sculpture coarser, both as regards the longitudinal ribs and the transverse sulcations.

Vaginula rodericensis.
Plate LI., Figs. 4 and 4a.

Annals et Mag. Mat. Hist., 1876, May, p. 405.

Corpus elongatum, utrinque rotundatum, postice leviter angustatum et acuminatum, superne rotundatum, lateraliter carinatum, pallium supra et infra minutissime granulatum, testaceum, irregulariter confertim nigro tessellatum, vel punctatum infra paulo pallidius, marginibus lateralibus haud nigro-punctatis; pes augustus latitudinis corporis $\frac{1}{4}$ adequans, testaceus, usque ad extremitatem corporis fere productus; tentacula oculifera nigrescentia; caput tentaculaque buccalia flavo-testacea; orificium fermineum paulo pone medium corporis situm.

Long, 30 mill., diam. 10 (specim. in alcohol. serv.).

In one or two specimens there is the faintest trace of a mesial dorsal narrow yellowish stripe. *V. maculata* of Templeton from Ceylon appears to be the nearest ally of this species, which is not at all common at Rodriguez according to the observations of Mr. Gulliver.

Diplodonta lateralis.
Plate LI., Figs. 7 and 7a.

Annals and Mag. Nat. Hist. 1876, May, p. 405.

Testa alba, transverse globoso-ovata, antice augustata, postice latissime rotundata, valde inæquilateralis; margo dorsalis utrinque paululum declivis et rectiusculus; valvæ tenuiter concentrice striatæ, hic illic fortius; umbones haud prominentes, conspicue excentrici; impressio muscularis antica elongata augustiuscula, possica aliquanto latior.

Long. 18 mill., latit. $20\frac{1}{2}$, crass. 11.

Shell pure white, transversely globosely ovate, a little narrowed anteriorly and posteriorly broadly rounded, very inequilateral, equivalved. The dorsal margin sloping on both sides of the umbones, the posterior slope is straight and rather short, the anterior, just a little excavated near the umbo; the valves are concentrically striated, the striæ being very fine with here and there the intervals of growth distinctly marked; the umbone are small and not prominent and very eccentric, the anterior muscular scar is elongate and narrowish, the posterior broader. The linear fossa which holds the exterior hinge ligament is about 5 millim. in length.

D. coreensis *Ad. and Rve.* resembles the present species somewhat as regards outline. In that species, however, the umbones are decidedly more central and the anterior dorsal slope is consequently longer. The anterior end of the ventral margin is very faintly angled at its juncture with the anterior lateral margin; but in *D. lateralis* no trace of such an angulation is observable, and it is also a more compressed species than *coreensis*.

Within the valves of the single specimen at hand parallel with the palial impression on the side towards the umbones, there is a second impression, which is punctured with small pits; but probably this is merely an individual peculiarity.

Avicula fuscopurpurea.

Plate LI., Figs. 5 and 5*a*.

Annals and Mag. Nat. Hist. 1876, May, p. 405.

Testa inæquivalvis, perobliqua, postice peroblique alata, striis concentricis confertis regulariter sculpta; color externus fusco-purpureus, antice interdum pallescens; rostrum valvæ dextræ medriocriter breve, paululum declive, confertim rugose lamellosum, valvæ sinistræ angustum, magnum; cauda inconspicua; ala perobliqua, imo recurva, in margine postico valde sinuata; margo valvarum cardinalis rectus; margarita pulcherrime iridescens, a limbo fusco-purpureo, antice pallidiore, posticeque latissimo cincta.

Long. max. 55 mill., diam. 35.

This species is more nearly allied to *A. spadicea* of Dunker (Küster's Conchylien-Cabinet, *Avicula* p. 31, pl. 10 f. 1 and 8) than to any other, however it is more deeply sinuated on the posterior margin.

The obliquity of the shell varies with age, for young specimens do not exhibit nearly so much of a wing as older ones. The colour of the exterior surface is a dark brownish or blackish purple which is of the darkest tint posteriorly, and beneath the rostrums, or on the anterior portion of the valves, it is generally paler.

This is also the case with the inner surface, the anterior part of the limbus surrounding the pearl being of a dirty yellow colour gradually blending into the dark blackish purple tint of the posterior margin.

A. malleoides, *Reeve*, agrees somewhat with the present species in form. It is however more oblique and produced, and lacks colouration.

Rhizochilus (Coralliophila) squamosissimus.

Plate LI., Figs. 8 and 8*a*.

Annals and Mag. Nat. Hist. 1876, May, p. 404.

Testa ovato-fusiformis, aliquanto umbilicata, alba; spira elevata, turrita; anfractus 7 convexi, costis spiralibus inæqualibus pulcherrime squamatis, irregulariter alternatim majoribus cincti, et plicis longitudinalibus obliquis clathrati; anfr.

ultimus costis spiralibus circiter 26 ornatus, ventricosus, inferne angustatus; apertura ovato-pyriformis, alba, longitudinis totius ⅗ vix æquans; labrum margine crenulato; columella alba, callo lævi induta; canalis angustus, obliquus, leviter recurvus.

Long. 31 mill., diam 16; apertura long. 19, diam. 9.

Shell ovately fusiform, narrowly umbilicated, entirely white; spire elevated, turrited, with the converging outlines almost rectilinear; whorls 7 convex, gradually increasing, encircled by numerous beautifully and very closely imbricated lirations which are more or less regularly alternately stouter and finer and are about 12 in number on the penultimate whorl and about twice that number on the last. The whorls are also ornamented with obliquely longitudinal costations or plications, which are more conspicuous and closer together on the upper whorls than on the ultimate; they number about 12 on the penultimate and the same on the last whorl; the latter is ventricose above and narrowed inferiorly; aperture white, ovately pear-shaped, scarcely equalling ⅗ of the whole length of the shell; the outer lip is crenulated on the edge; columella covered with a thin white callosity, only slightly arcuate in the middle; canal oblique, rather narrow and a little recurved.

Fusus inflatus, Dunker, in Philippi's Abbildungen neuer Conchylien, pl. iv., f. 2, must be allied to this species, but the sculpture and proportion of the aperture to the shell is very different. Perhaps *R. exaratus*, of Pease, which is found in the Sandwich Islands, has more the form of this species than any other in the genus. It is, however, at once distinguishable on account of its very distinct sculpture.

Explanation of Plate LI.

(Figs. 4 to 13—Mollusca.)

Fig. 4. Vaginula rodericensis, dorsal view.
„ 4a. „ „ ventral surface.
„ 4b. „ „ orificium femineum.
„ 5. Avicula fusco-purpurea, upper side.
„ 5a. „ „ interior of the same valve.
„ 6. Fenella reticulata (much enlarged).
„ 7–7a. Diplodonta lateralis.
„ 8. Rhizochilus (Coralliophila) squamosissimus.
„ 8a. „ „ portion of the surface magnified.
„ 9. Melania rodericensis.
„ 10. „ „ var.
„ 11. Conus rattus, var.
„ 12. „ balteatus, var.
„ 13. „ catus, var.

CRUSTACEA.—*By Edward J. Miers, F.L.S., F.Z.S., Assistant in the Zoological Department, British Museum.*

The Crustacea collected by Messrs. G. Gulliver and H. H. Slater amount in all to 189 specimens, representing 35 species. All of these are forms that are widely distributed throughout the Indo-Pacific or Oriental Region (which includes the eastern coast of Africa, the south and east of Asia and islands adjacent, Australia, and the islands of Polynesia), with the following exceptions:—*Atergatopsis signatus* (hitherto only known from the Mauritius), *Caridina typus* (original locality not known), *Palæmon dispar* (hitherto recorded only from the Malayan Archipelago), *Palæmon hirtimanus* (from Mauritius, Réunion, and the Indian Ocean), *P. debilis* (from Amboina and the Sandwich Islands), and the new species of *Talitrus* (*T. gulliveri*), which is described below. With two exceptions all the species in the collection belong to the *Podophthalmia*. The following are the sub-tribes represented, with the number of species belonging to each:—

Sub-Tribe.	No. of Species.	Sub-Tribe.	No. of Species.
Oxyrhyncha	2	*Caridea*	6
Cyclometopa	8	*Squillidea*	2
Catometopa	9	*Oniscoidea*	1
Oxystomata	1	*Gammaridea*	1
Paguridea	5		

The Crustacea inhabiting the Red Sea have been made the subject of special study by Rüppell and Heller, those of Madagascar and the islands adjacent by Hoffmann, of Mauritius and Réunion by Alphonse Milne-Edwards, and of the South African coast by M'Leay and Krauss. Valuable additions to our knowledge of the Crustacea of the East African coast have been published by Hilgendorf, in Van der Decken's "Reisen in Ost-Afrika," where will also be found a conspectus of all the known species of East African Crustacea by Von Martens. So far as I am aware, however, no species have hitherto been recorded as inhabiting the Island of Rodriguez.

PODOPHTHALMIA.

Decapoda.

Menæthius monoceros, Latr.

A single specimen, an adult male, was collected by Mr. Slater. There is probably no species of the *Oxyrhyncha* subject to greater variations, both in the form and length of the rostrum and anterior legs and in the tuberculation of the carapace. Nearly a dozen supposed species of the genus *Menæthius* have been described, all of which are united by M. A. Milne-Edwards under the single name of *M. monoceros*; and although one or two of these forms (as, for example, the *M. tuberculatus*, Ad. and White, from the Mauritius) appear at first sight to present marked distinctive peculiarities, it is probable that these are no more than individual variations. The specimen from Rodriguez is of the normal form, in which the carapace is indistinctly tuberculated, the legs nearly smooth, the palm slender, and about twice as long as the fingers, which are not strongly arcuate. This variety is widely distributed throughout the Indo-Pacific region. Specimens are in the British Museum from the Gulf of Suez, Mauritius, Fiji Islands, Philippines, and Eastern Seas.

Parthenope horrida, Linn.

Two specimens, males, of this common species were collected by Mr. Slater. It is probably distributed throughout the Indo-Pacific region, and M. Milne-Edwards records its occurrence in the Atlantic Ocean. Specimens are in the British Museum from the Red Sea and Mauritius. This species is not to be confounded with *P. spinosissima* of M. Alph. Milne-Edwards (Faune Carcinologique, in Maillard, Notes sur l'ile de la Réunion. Annexe F., p. 8, pl. xviii., fig. 1).

Atergatopsis signatus, Ad. & White.

A single specimen, a male, was obtained by Mr. Slater. This specimen is a little smaller than the typical example of this species in the British Museum Collection, from the Mauritius, hitherto its only recorded habitat.

Zozymus æneus, Linn.

One example, a male, of this beautiful but very common species is in the collection obtained by Mr. Slater. It is found in the Red Sea, at the Seychelles, the Mauritius, Réunion, on the shores of the Indian Ocean, and in the Loo-Choo Islands, Australian Seas, and South Pacific (Samoa and Fiji Islands and Paumotu Archipelago).

Actæodes tomentosus, M.-Edw.

Two specimens, male and female, of this species were collected. It is found at the Mauritius, and is distributed throughout the Indo-Pacific region; its occurrence having been recorded from the Red Sea, Mozambique, the Indian Ocean, Nicobars, Sumatra, Sulu Archipelago, Samoa, and Fiji Islands; and specimens are in the Collection of the British Museum from the Seychelles, Philippines, Sandwich Islands, and Minerva Reef. *A. affinis*, Dana (U. S. Expl. Exp. xiii., Crust. i. p. 198, pl. xi. fig. 3), from the Paumotu or Society Islands, differs only in its more distinctly areolated and minutely granulous carapace, and must be united with this species, as specimens in the British Museum Collection presented by the Smithsonian Institution prove.

Daira perlata, Herbst.

Three examples of this species, a male and two females, were collected. It is probably distributed throughout the whole Indo-Pacific Region. Specimens are in the British Museum from the Mauritius, Philippine Islands, Australian Seas, and Minerva reef. Milne-Edwards mentions its occurrence in the Atlantic Ocean and on the Coast of Brittany, and it has been recorded from the Samoa Islands, New Caledonia, Loo-Choo Islands, and Auckland.

Cymo andreyossii, Audouin.

A single specimen of this curious species was obtained by Mr. Slater. Its geographical range is probably as extended as that of the preceding species. Specimens are in the British Museum Collection from the Red Sea and Fiji Islands, and it has been recorded from the Loo-Choo and Samoa Islands, Tahiti, and New Caledonia. The *C. melanodactylus*, of De Haan (Faun. Japon. Crust. p. 22), from Java, as described and figured by Dana (U.S. Expl. Exp. Crust. p. 225, pl. xiii. fig. 1) from specimens collected at the Fiji Islands, is probably as M. A. Milne-Edwards has stated, a variety of this species. The colour of the fingers is not a constant character in *C. andreyossii*, but there appear to be two distinct forms; in one of which the carapace and anterior legs are nearly smooth, and the front nearly entire; in the other the sides of the carapace and anterior legs are conspicuously granulated, and the front denticulated on its anterior margin. The form of the carapace is very variable; it is usually nearly orbiculate, but sometimes longer and broadest in front. These characters, however, do not appear to be of specific value, as intermediate varieties occur, and forms in all respects identical are found at widely separated localities.

Trapezia rufo-punctata, Herbst., var. *maculata*, M'Leay.

One example is in the collection, a male, obtained by Mr. Slater. Specimens are in the Collection of the British Museum from the Mauritius, Red Sea, and Ceylon, and its occurrence is recorded at Zanzibar, the Cape of Good Hope, Tahiti, the Sandwich Islands, New Caledonia.

The specimen from Rodriguez is of the form figured by Dana (U.S. Expl. Exp. Crust. i. p. 256, pl. xv. fig. 4), as *T. maculata*, of M'Leay (*Grapsillus maculatus*, M'Leay, in Smith Zool. S. Africa, Crust. p. 67). This variety is distinguished from the typical *T. rufo-punctata* by the form of the frontal lobes, which are obscurely defined and obtuse or rounded, not strongly prominent and acute. The two or three anterior teeth upon the anterior margin of the arm are truncate or slightly emarginate at their apex, the red spots, with which the whole animal is covered, are smaller and fewer in number than in the typical variety.

Tetralia cavimana, Heller.

Three specimens, two males and a female, were collected by Mr. Slater. Examples of this species are in the Collection of the British Museum from the Fiji Islands and Australian Seas; and it occurs also at Tahiti and the Red Sea and New Caledonia.

Achelous granulatus, M.-Edw.

One example, a male, was collected by Mr. Gulliver. This species is very commonly and widely distributed throughout the Indo-Pacific Region. Specimens are in the British Museum Collection from the Red Sea, Mauritius, Ceylon, the Philippine Islands, Fiji and Sandwich Islands; and it has been found also at Réunion, Zanzibar, Java, Samarang, and Japan, and New Caledonia.

Gelasimus vocans, Linn.

Four specimens, males, were collected by Mr. Gulliver, who observed that both this and the following species were very common in the island, burrowing in the sands by the sides of the estuaries rather than on the seashore. The prominent triangular lobes or teeth on the inner margin of the immobile finger vary considerably in size, and one or even both may be obsolete, as in the species or varieties described as *G. cultrimanus* by Adams and White (Voy. Samarang, Zool. Crust. p. 49), from the Philippines and Hong-Kong. Specimens of this species are in the British Museum Collection from Mauritius, Singapore, and the Fiji Islands; and it has been recorded from the Islands of Nossy-Faly and Nossy Bé adjacent to Madagascar, Zanzibar, Java, Malabar, the Nicobars, and Loo-Choo.

Gelasimus annulipes, M.-Edw.

Six specimens, including males, females, and young, were obtained by Mr. Gulliver, which I refer to this very variable species. The larger hand and fingers are smooth externally, without indications of longitudinal grooves, or with one obscurely-indicated at the base of the palm, which has a short transverse groove near its anterior margin. On the inner surface of the palm there is a high prominent oblique granulated crest, and two transverse lines of granules near the anterior margin. The fingers are granulated on their inner margins, the upper one arcuate, the lower nearly straight. There is always a tooth near the distal extremity, and usually one at about the middle of the inner margin of the lower finger, sometimes also two or three other tubercles or teeth placed at unequal distances along the inner margins of both the upper and lower fingers; often, however, these are absent. Specimens of *G. annulipes* are in the collection of the British Museum from Zanzibar, Ceylon, the Malayan Archipelago, the Philippines, and the Fiji Islands; and it has been recorded from the Islands of Nossy-Faly and Nossy-Bé, Madras, and the Nicobars.

Ocypode cordimana, Desm.

Two specimens, a male and a female, were collected.

Mr. Gulliver states that the female taken by him was found in a wood, and considers it probable that this species was hibernating in holes at the time of his sojourn on the island. It has been recorded from the Red Sea, Zanzibar, Mauritius, Réunion, Mozambique, Natal, and the Cape of Good Hope, and its range extends over the whole Indo-Pacific Region; as specimens are in the British Museum from Ceylon, China, and the Australian Coast, the Samoa Islands, Fiji Islands, New Hebrides, and it has also been recorded from the Nicobars, Manilla, Japan, Hong Kong, Loo-Choo, and New Caledonia.

The *O. arenaria*, Say, which inhabits the Atlantic Coasts of the American Continent, is very nearly allied to this species, young individuals are indeed hardly to be distinguished, but adult specimens may be recognized by the laterally-compressed ambulatory legs, which are smooth (not granulated as in *O. cordimana*), and thickly clothed on the upper and lower margins with dense fulvous hair.

Grapsus pictus, Latr.

A female example of this very variable, common, and widely distributed species was taken by Mr. Slater. Specimens are in the British Museum from the Cape Verde Islands, St. Helena, S. Africa, the Mauritius, Ceylon, Fiji Islands, Galapagos, Peru, California, and New Zealand, and it also inhabits the West Indies and Coast of Florida.

Geograpsus grayi, M. Edw.

Three specimens, an adult male and two females, were collected. The examples obtained by Mr. Gulliver were found by him at some distance from the seashore. In the younger animal the hands are more granulous, and the series of teeth or spines upon the anterior margin of the arm more acute. In all the adult examples of this species that I have examined the frontal margin is straight or very nearly so, never bi-emarginate as in the specimen figured by Hilgendorf as the adult of *G. rubidus*, Stimpson (Crust. in van der Decken's Reisen in Ost-Afrika, pl. v. fig. 1c.). This species has been recorded from the Red Sea, Mauritius, Réunion, Zanzibar, Madagascar, the Indian Ocean, Australia, Bonin, New Caledonia, and the Sandwich Islands; and specimens from the Fiji Islands are in the Collection of the British Museum.

Metopograpsus messor, Forskal.

Four specimens, including males and females, were collected by Mr. Gulliver, who states that he found this species plentifully near freshwater streams in the woods, but never actually on the seashore. It varies considerably in the marbled coloration of the carapace. It has been recorded from the Red Sea, Zanzibar, Natal, Persian Gulf, Coast of Malabar, Madras, Ceylon, Bonin, Sandwich Islands, Tahiti, and New Caledonia, and specimens are in the British Museum from Mauritius, the Eastern and Western shores of Australia, and the Fiji Islands. The *G. thukujar*, of Owen (Beechey's Voyage, Zool. Crust. p. 80, pl. xxiv. fig. 3), is, I believe, a variety of this species.

Sesarma tetragonum, Fabr.

Three examples, male and female and young, were collected by Mr. Gulliver who found them in burrows near the estuaries. In the adult male the fingers are arcuate, and meet only at the corneous excavated tips; in the young male the fingers are straight. This species has been recorded from the Red Sea, Mauritius, Zanzibar, Natal, Madagascar, Indian Ocean, Chinese Seas, and New Caledonia. Specimens from the Fiji Islands are in the Collection of the British Museum.

Sesarma quadratum, Fabr.

Seven examples, males, of a species of *Sesarma* were collected, which I refer with some hesitation to *S. quadratum*. The animal is of a blackish-brown colour; the carapace has the lateral margins entire, and is slightly granulated on the post-frontal region, with the four truncate frontal lobes well-defined; the arm has a prominent angulated lobe or tooth at the distal extremity of its anterior margin; the wrist is uniformly rugose without spines; the exterior surface of the palm smooth below, slightly granulated toward the superior margin, the upper surface has two small oblique very closely and minutely pectinated ridges, and behind these two or three oblique granulated lines. *Sesarma quadratum* has been recorded from Zanzibar, Natal, Pondicherry, and New California; and there are specimens in the British Museum Collection from Borneo and the Fiji Islands. The species figured by De Haan, (Faun. Japon. Crust. pl. xviii. fig. 5), under the name of *Pachysoma affinis*, is probably identical with this species; also the *Sesarma aspera*, of Heller (Voy. Novara, Crust. p. 63, pl. vi. fig. 1), from Ceylon, Madras, and the Nicobars. Dr. Hilgendorf, basing his observations upon a large series of specimens from the Indian Archipelago, Malacca, Sumatra, Luzon, Flores, and Timor, states that *S. quadratum* varies very considerably in the form of the carapace, frontal lobes, and anterior legs, but it is possible that he has confounded two or more species under the above-mentioned designation, as he neither mentions nor figures the minutely-pectinated ridges on the upper margin of the hand, which are characteristic of *S. quadratum*. (*See* Van der Decken's Reisen in Ost-Afrika, Crust. p. 90, pl. iii. fig. 3c. and pl. iv. fig. 3.)

Cardisoma carnifex, Herbst.

Two examples, a fine adult and a smaller male, were collected by Mr. Gulliver, who found this species burrowing in sandy ground by the estuaries of the streams, but did not observe that it penetrated far inland. There is usually a small tooth behind the external orbital tooth or spine, but this is sometimes rudimentary or entirely absent: it is present in the smaller, wanting in the larger specimen from Rodriguez. As the animal increases in age, the body becomes very much more convex over the branchial regions, and is expanded laterally beyond the marginal line, which is hardly distinguishable in the oldest specimens; one or other of the anterior legs is very greatly developed, and the hair upon the penultimate and antepenultimate joint of the ambulatory legs becomes very long and dense. Specimens of this species are in the British Museum Collection from the Mauritius, Pondicherry, Philippines, the Fiji, Sandwich, and Samoa Islands. It has been recorded from Zanzibar, the Nicobars, and New Caledonia, and a young example obtained at Madeira is referred by Dana to this species.

Calappa hepatica, Linn.

Three specimens, two males and a female, were collected. It is one of the commonest and most widely distributed of the Indo-Pacific *Brachyura*. This species has been recorded from the Red Sea, Zanzibar, Natal, Madagascar, the Mauritius, Indian Ocean, Nicobars, Auckland, the Fiji and Sandwich Islands, and New Caledonia; and there are specimens in the Collection of the British Museum from the Seychelles, Ceylon, the Coast of China, the North-east Coast of Australia, the Loyalty Islands, New Hebrides and Samoa Islands.

Pagurus punctulatus, M. Edw.

Three examples of this species, one of the largest and most beautifully-coloured of the genus *Pagurus*, were collected. One is an adult female, bearing a large quantity of ova. It has been recorded from the Mauritius, Red Sea, Madagascar, Indian Ocean, Nicobars, the China Seas, Tahiti, and the Sandwich Islands; and specimens are in the British Museum from the Philippines, Australia, and the Fiji Islands.

Aniculus typicus, Dana.

Four specimens were collected. The recorded localities of this species are the Mauritius, Réunion, Batavia, Manilla, Japan, Auckland, the Paumotu Archipelago, Wakes Island. Specimens are in the Collection of the British Museum from the Seychelles, Australia, Samoa and Fiji Islands.

Calcinus tibicen, Herbst.

A single specimen inhabiting a shell of a species of *Turbo* was collected by Mr. Slater. This species has been recorded from Zanzibar and the Coast of Natal, the Nicobars, Straits of Balabac, Loo Choo, the Paumotu Archipelago, the Society, Sandwich, Samoan Islands, and Wakes Island. Specimens are in the Collection of the British Museum from the Pelew Islands and Mozambique.

Calcinus elegans, M. Edw.

A female example, with ova, is in the Collection, obtained by Mr. Slater. Both the anterior legs are unfortunately wanting, but the colouration agrees very well with the descriptions of this species; the legs of the second and third pair being beautifully marked with broad alternating bands of crimson and white, the tarsi with crimson spots upon a white ground. This species has been recorded from the Coast of Natal, Loo Choo, the Paumotu Archipelago, Hawaiian Islands, Wakes Island, Tahiti, New Ireland, and Western Patagonia.

Cœnobita rugosa, M. Edw.

Fourteen specimens of this species were collected. The greater number of these inhabit shells of *Nerita polita*, one specimen was found in a shell of the terrestrial *Cyclostoma articulatum*. Mr. Gulliver observes of this species that it was often found in perfectly arid places a mile or so from the sea-shore. It is very commonly and widely distributed throughout the Indo-Pacific Region, and its range even extends to the Western Coast of America. It has been recorded from the Red Sea, Zanzibar, Mozambique, Réunion, Natal, Persian Gulf, Indian Ocean (Ceylon, Madras, Nicobars), Java, China Seas (Loo Choo), Australia (Sidney), the Sooloo Sea, Paumotu Archipelago, Fiji and Samoan Islands, Tahiti, and Panama. To these localities I may add, on the authority of specimens in the Collection of the British Museum, the Mauritius, Coquimbo, and Vancouver Island. M. Milne-Edwards (Hist. Nat. Crust. ii. p. 241) states that a specimen from Messina in the Collection of the Paris Museum appears to belong to this species, but may have been wrongly labelled.

Caridina typus, M. Edw.

To this species I refer a single specimen obtained by Mr. Gulliver, which only differs from Milne-Edwards' description and figure in having two instead of three minute teeth on the inferior margin of the rostrum, which is proportionately somewhat shorter and deeper than that of the example figured by Milne-Edwards. In many of the *Caridea* the form and number of teeth of the rostrum are so variable that these differences can scarcely be considered as of specific importance; and it may be that the *C. exilirostris* of Stimpson (Proc. Ac. Nat. Sci. Phil. p. 29, 1860) from Loo Choo, in which the inferior margin of the rostrum is obsoletely 2–3 dentate, represents an intermediate variety of this species. The habitat of the type-example is unknown.

Alpheus laevis, Randall.

A single example, a female with ova, is in the Collection obtained by Mr. Slater. The antennæ are imperfect, the smaller hand is wanting; the larger hand agrees well with the descriptions and Dana's figure of the species. It has been recorded from the Red Sea, Nicobars, Sydney, Tahiti, and the Sandwich and Fiji or Friendly Islands.

Palaemon ornatus, Olivier.

A large number of specimens of this, one of the commonest, and most universally distributed of the Prawns inhabiting the freshwaters of the Indo-Pacific region, were collected, forty-one in all; of these, however, but few are in a perfect condition, the greater number having lost one or both of the large legs of the second pair; and but two or three examples are of full size. Many of the smaller examples are females bearing ova. The teeth on the rostrum vary between 6–11 on the upper and 2–3 on the lower margin; there are usually 8–9 upper and 3 lower marginal teeth ($\frac{8-9}{3}$). The males of this species are at once recognized by the strong greatly elongated second pair of legs, which are closely beset with numerous minute spines, the palm in the adult animal is much longer than the wrist, the fingers are inflexed at the tips, the upper often considerably arched, the immobile finger has two, the mobile finger a single strong conical tooth on the inner margin. In the females and young these characteristics are not nearly so strongly marked, and perhaps some of the smaller specimens ought to be referred to the following species, (*P. dispar*). *P. reunionensis* and *P. longimanus* of Hoffmann, (Recherches Faune Madagascar, Crust. pp. 33 & 34, pl. ix., fig. 66 & 67, & 68, 69), from Réunion, appear to me to be founded respectively on younger and adult examples of this species. *Palaemon ornatus* has been recorded from Amboina and the Islands of Waigiu and Adenare, the Celebes, Philippines, Tahiti, and Auckland. There are specimens in the Collection of the British Museum from the Mauritius, Australia (Moreton Bay), the Samoa and Fiji Islands, New Hebrides.

Palaemon dispar, von Martens.

To this species is certainly to be referred a male example in the Collection obtained by Mr. Gulliver. When adult, *P. dispar* is distinguished without difficulty from *P. ornatus*, by its smaller size; the straighter rostrum, with more numerous marginal teeth, the very unequal legs of the second pair, and the form of the fingers, which, in the male, each have a series of conical acute spines on their inner margins. Hitherto *P. dispar* has been recorded only from Adenares Island, near Flores, in the Malayan Archipelago, but it may have easily been confounded with other species of this very difficult genus. The *P. alphonsianus* of Hoffmann (Rech. Faune Madagascar, Crust. p. 35, pl. ix., fig. 63–65) from Réunion is in all probability synonymous with it.

Palaemon hirtimanus, Olivier.

One specimen, a male, is in the Collection, obtained by Mr. Gulliver. In this example the large leg of the second pair on the right side has been broken off above the second joint, and its place is supplied by another leg, evidently newly-formed, and very small and weak, but well-proportioned in all its joints. This species has been recorded from Mauritius and Réunion, and the Indian Ocean.

Palaemon (Leander) debilis, Dana.

No fewer than fifty-six specimens were collected, but a considerable number of these are more or less imperfect. Like all the species of this sub-genus, *P. debilis* is of small size, never exceeding 2 in. in length. This species is distinguished by the form of the slender rostrum which is directed upward in its distal half, and bears on the upper margin close to the extremity a small tooth; the apex thus, in some specimens, appearing bifid. The number of teeth upon the upper and lower margins is very variable. Of thirty-nine specimens from Rodriguez in which the rostrum is perfect, fourteen have five teeth upon the upper and six on the lower margins, exclusive of the sub-apical tooth ($\frac{5}{6}2$); but beside this, which appears to be the normal form, there are no fewer than nine varieties, as shown in the following table:—

	Dental formula.	No. of Specimens.		Dental formula.	No. of Specimens.
Pal. debilis	$\frac{5}{6}2$	14	var. *e*	$\frac{4}{5}2$	3
var. *a*	$\frac{4}{6}2$	5	,, *f*	$\frac{6}{6}2$	2
,, *b*	$\frac{5}{7}2$	5	,, *g*	$\frac{6}{5}0$	1
,, *c*	$\frac{5}{7}2$	4	,, *h*	$\frac{5}{4}2$	2
,, *d*	$\frac{4}{7}2$	2	,, *j*	$\frac{4}{3}2$	1

The second pair of legs are very small and slender, resembling those of the first pair. The specimens described by Dana were collected at the Sandwich Islands; Von Martens records the occurrence of this species at Amboina, at the mouth of the Weynitu River, in *salt-water*; Mr. Gulliver, on the other hand, found it in the *freshwater* streams of the Island of Rodriguez. The species of the sub-genus *Leander* are, I believe, generally marine, those of *Palaemon* fluviatile.

Dana has figured a variety of *P. debilis* from the Hawaiian Islands, to which he has applied the name of *P. attenuatus*; having six teeth on the upper and nine on the lower margin of the rostrum.

Stomatopoda.

Lysiosquilla maculata, Lam.

A single large male example (length nearly 13 in.) was obtained by Mr. Slater. This species has been recorded from the seas of Asia and Japan. Specimens are in the Collection of the British Museum from the Indian Ocean, Philippines, Samoa and Fiji Islands. On account of its large size, and the colouration of the segments of the body, which are marked with broad alternating bands of black and yellow, this species is perhaps the most conspicuous of the *Stomatopoda*.

Gonodactylus chiragra, Fabr.

Two specimens of this species were collected. The largest of these, an adult male, has a length of nearly 3½ in. It is a very common and widely distributed species, and its range probably extends throughout the warmer latitudes of the globe. There are specimens in the Collection of the British Museum from the Mediterranean, Red Sea, Zanzibar, Indian Ocean, Ceylon, Western, Southern, and North-eastern Australia, Sir C. Hardy Island, Philippine Islands, New Hebrides, Fiji Islands, and Panama. It has been recorded from Natal and Réunion.

EDRIOPHTHALMIA.

Isopoda.

Oniscus ? (*sp.*)

Five specimens of a species of Terrestrial Isopod, probably belonging to the genus *Oniscus* or *Porcellio*, are in the Collection obtained by Mr. Gulliver, who states that this species is common under stones. In all these specimens both antennæ and uropoda are wanting, and it is therefore impossible to ascertain even the genus to which they ought to be referred with certainty. The posterior margins of the first three segments of the body are straight, the terminal segment almost T-shaped, the posterior half being very narrow. From *P. truncatus*, M. Edwards (Hist. Nat. Crust. iii., p. 171) inhabiting the Mauritius, this species differs in the form of the terminal segment; from a species described under the same name by Brandt (Bull. Mosc. Soc. Nat. vi., p. 181), but placed in a different section of the genus by Milne Edwards (l. c. p. 173), from the Cape of Good Hope, in the form of the head, in which the front is scarcely at all prominent, the antero-lateral lobes large and rounded.

Amphipoda.

Talitrus gulliveri.

T. gulliveri, Miers, Ann. Mag. Nat. Hist. ser. 4, xvii., p. 406 (1876).

Slender, smooth, with the buccal organs considerably projecting, as in *T. locusta*. Head small, with the anterior margin straight, eyes round, black. Second and third segments of the postabdomen with the posterior margins straight, the inferoposterior angle somewhat prominent and acute. Superior antennæ scarcely as long as the head and about reaching to the extremity of the second joint of the inferior antennæ, with seven joints exposed, the basal very short, the two next longer, sub-equal, together about as long as the four last joints, of which three are equal and the last minute. Inferior antennæ slender, not as long as the body, with the basal joint of the peduncle very short, the second joint more than half the length of the third joint, the flagellum about eighteen-jointed and rather longer than the peduncle. First and second pairs of legs small and weak; the first pair with the three last joints short, subequal, and not dilated; the second pair (in some specimens) longer, with the palm slightly dilated and the finger quite rudimentary as in *T. platycheles*. Next three pairs of legs short, sixth and seventh pairs considerably longer, with the second joint moderately developed. Rami of the fourth and fifth pairs of tail-appendages subequal, those of the fourth shorter than those of the fifth pair, sixth pair quite rudimentary. The distal extremities of the joints of the antennæ, the legs, and the rami of the postabdominal appendages are fringed with short hairs. Length about 5 lines.

This species was found by Mr. Gulliver under stones in damp places, but never observed actually in the water.

I feel no hesitation in referring this species to the genus *Talitrus*, as in none of the specimens are either the first or second legs largely developed and subcheliform as in other genera of *Orchestiidæ*. It differs from the common European *T. locusta* in its much shorter and slenderer inferior antennæ, and in the form of the segments of the post-abdomen; the posterior margins not being sinuated as in that species. It seems to be more nearly allied to *T. platycheles*, Guerin. (Exped. Scient. Morée, Zool. Crust. p. 44, pl. xxvii., fig. 4), but differs in the proportionate length of the joints of the peduncles of the antennæ, in the larger second pair of legs, and sixth and seventh pairs of legs, and in the more obtuse infero-posterior angles of the post-abdominal segments, &c.

MYRIOPODA AND ARACHNIDA.—*By Arthur Gardiner Butler, F.L.S., F.Z.S., &c.*
(Plate LII.)

Of the twelve species of *Myriopoda* obtained by Messrs. George Gulliver and H. H. Slater, seven are referable to the *Chilognatha* and five to the *Chilopoda*: all, with the exception of the two species of *Scolopendra* (and possibly *Spirostreptus sorornus* and the *Mecistocephalus*), appear to be new to science.

So few naturalists have interested themselves in the study of these animals, that it is impossible at present even to form an opinion as to the geographical distribution of the species of Rodriguez; all that can be said is that, in all probability, the two forms of *Scolopendra* will be found abundantly in Madagascar.

Of the *Arachnida* twenty-eight belong to the *Araneidea*, one to the *Acaridea*, and one to the *Scorpionidea*, eleven of them being new species. Of the remainder, three have been previously recorded from Mauritius, eight from Bourbon, two from Madagascar, three from Mauritius and Bourbon, and one from Mauritius, Bourbon, and Madagascar. The last-mentioned species (*Olios leucosius*) is very common, and probably has a wide range. Specimens in the Collection of the British Museum are ticketed "Magdeleine."

The species of the greatest interest is a new form of the curious genus *Miagrammopes* of Cambridge, a group of delicate and extremely strange-looking four-eyed spiders, originally discovered by Mr. Thwaites in Ceylon. There is also an abnormal form of *Sphasus*, which in colouring, and general aspect much resembles *Miagrammopes*; unhappily both of these species seem to be very rare.

MYRIOPODA.

Name of Myriopoda found in Rodriguez.	Hitherto found in Rodriguez only.	Previously obtained elsewhere.		
		Mauritius and Bourbon.	Madagascar.	Other Localities.
Strongylosoma errucaria, *Butler*	×			
Cambala nodulosa, *Butler*	×			
Spirostreptus avernus, *Butler*	×			
Spirostreptus sorornus, *Butler*	×			
Spirostreptus gulliveri, *Butler*	×			
Spirostreptus simulans, *Butler*	×			
Spirobolus hecate, *Butler*	×			
Eurylithobius slateri, *Butler*	×			
Scolopendra mossambica, *Peters*				Mozambique.
Scolopendra ferruginea, *Koch*				(Locality previously unknown).
Rhombocephalus smaragdinus, *Butler*	×			
Mecistocephalus gulliveri, *Butler*	×			Réunion?

Polydesmidæ.

Strongylosoma, Brandt.

Strongylosoma errucaria.

Ann. & Mag. Nat. Hist. s. 4, vol. xvii., p. 444 (1876).

Deep chocolate-brown; under surface, legs, lateral wings, the segments pale testaceous; segments glabrous, first dorsal segment with well-defined lateral carina; preanal segment terminally rostriform. Length 8 lines, width 1 line.

(Coll. by Gulliver.)

This species seems to be allied to *S. Guerinii* of Gervais, but it differs in its well-marked lateral carina, smaller size, and deeper coloration; young examples are, however, paler than the type.

Ten specimens were brought home by Mr. Gulliver.

Cambala, Gray.

Cambala nodulosa.

Ann. & Mag. Nat. Hist. s. 4, vol. xvii., p. 444 (1876).

Reddish testaceous; dorsal segments with a lateral grey stripe; first segment longitudinally multisulcate; remaining segments crossed transversely by two rows of tubercles, between which, on each segment, is a depressed line; posterior margins narrowly rebordered. Length 8 lines, width 1 millim.

(Coll. by Gulliver.)

This species has about fifty dorsal segments, but they are difficult to count, owing to its small size and elaborate sculpturing.

Nine examples were obtained.

Julidæ.

Spirostreptus, Brandt.

Spirostreptus avernus.

Ann. & Mag. Nat. Hist. s. 4, vol. xvii., p. 445 (1876).

Black, legs and antennæ castaneous; forty-three dorsal segments; first segment smooth, with narrow marginal ridge; remaining segments rugulose and depressed in front, smooth behind, striated at the sides; preanal segment produced into a point behind. Length 13 lines, width 1 line.

Cascade Valley   (Coll. by Gulliver.)

Eight examples.

Spirostreptus sorornus.

Ann. & Mag. Nat. Hist. s. 4, vol. xvii., p. 445 (1876).

Colour and general structure of the preceding; forty-nine dorsal segments, which are not rugulose in front. Length 1 inch 9 lines, width 3 millims.

(Coll. by Slater.)

This may be the *S. nigerrimus* of Newport, but the latter is only represented by the head, nuchal plate, and two or three of the anterior dorsal segments, in a dried state. Only one specimen was obtained.

Spirostreptus gulliveri.

Ann. & Mag. Nat. Hist. s. 4, vol. xvii., p. 445 (1876).

Olivaceous; dorsal region dark; eyes black; face, margins of first dorsal segment, and hind margins of other segments tawny; legs reddish tawny; fifty-seven dorsal segments, first segment with square anterior angles, above which are four depressed lines; marginal ridge feebly developed; each of the remaining segments crossed by a depressed line, behind which they are swollen; sides striated as usual; preanal segment slightly produced, with bisinuate hind margin. Length 2 inches 4 lines, width 5 millims.

(Coll. by Gulliver.)

Nine examples were collected. It is said to be "common."

Spirostreptus simulans.

Ann. & Mag. Nat. Hist. s. 4, vol. xvii, p. 445 (1876).

Olivaceous; a dorsal series of dark spots in front and a continuous lateral series of blackish spots; first segment clouded with brown behind; forty-three dorsal segments, first segment with marginal ridge feebly developed; remaining segments rugulose and depressed in front; preanal segment regularly arched. Length $10\frac{1}{2}$ lines, width 2 millims.

(Coll. by Gulliver.)

This species looks like a small form of the preceding, but its structure is quite distinct. Eight specimens were obtained.

Spirobolus, *Brandt.*

Spirobolus hecate.

Ann. & Mag. Nat. Hist. s. 4, vol. xvii., p. 445 (1876).

Shining black; front margins of the segments below castaneous; fifty-nine dorsal segments; lateral wings of first segment quadrisulcate in front; remaining segments swollen behind, laterally striated; preanal segment deeply bisinuated behind, the centre of the hind margin being produced. Length, 4 inches 3 lines, width 7 millims.

(Coll. by Gulliver.)

Only one individual of this fine species was obtained; it is "not common."

Lithobiidæ.

Eurylithobius, Butler.

Lithobio affinis, multo latior, segmentis viginti; antennarum articulis septendecim, quorum decem primoribus et postremo distinctis, aliis inconspicuis; oculis utrobique quatuor; pedum paribus uno et viginti, latis, breviusculis.

Gen. typ. *E. slateri.*

Eurylithobius slateri.

Ann. & Mag. Nat. Hist. s. 4, vol. xvii., p. 446 (1876).

Fawn-coloured, legs and ventral surface testaceous; dorsal segments gradually narrowing towards each extremity, with two longitudinal depressed lines and a lateral marginal ridge; terminal segment subconical behind. Length of body 1 inch, including hind legs 1 inch 8 lines; width in the centre $4\frac{1}{2}$ lines.

(Coll. by Slater.)

Only one specimen was obtained.

Scolopendridæ.
Scolopendra, *Linnæus.*
Scolopendra mossambica, Peters.

Twenty-one specimens.

Scolopendra ferruginea, Koch.

Thirty-three specimens.

The two preceding species look much like different ages of the same animal; but Dr. Peters seems to have figured his species from an adult example, which has alone prevented my uniting them.

Rhombocephalus, *Newport.*
Rhombocephalus smaragdinus.

Ann. & Mag. Nat. Hist. s. 4, vol. xvii., p. 446 (1876).

Anterior half of body emerald-green, posterior half olive green; segmental incisions deeper-coloured; legs pale greenish, in twenty pairs; posterior pair without denticles; preanal plate oblong. Length of the body $10\frac{1}{2}$ lines.

(Coll. by Slater.)

One specimen.

Mecistocephalus, *Newport.*
Mecistocephalus gulliveri.

Ann. & Mag. Nat. Hist. s. 4, vol. xvii., p. 446 (1876).

Amber-coloured, legs paler; head reddish castaneous; forceps of mandibles black, the latter with about four teeth; head below deeply excavated; dorsal segments with a deep longitudinal sulcus; forty-eight pairs of legs. Length 1 inch 1 line.

(Coll. by Gulliver.)

This species, of which only one example was found, is evidently nearly allied to the Geophilus insularis of Lucas; indeed it is just possible that the two may be varieties of one animal.

ARACHNIDA.

Names of Species obtained in Rodriguez.	Hitherto found in Rodriguez only.	Mauritius.	Bourbon.	Madagascar.	Other Localities.
Clubiona insularis, *Vinson*	-		×		
Scytodes thoracica, *Latreille*	-		×		
Pholcus elongatus, *Vinson*	-		×		
Pholcus vexillifer, *Butler*	×				
Theridium borbonicum, *Vinson*	-		×		
Theridium diurnum, *Vinson*	-		×		
Latrodectus menavodi, *Vinson*	-			×	
Linyphia argyrodes, *Walckenaer*	-	×			
Epeira opuntiæ, *Dufour*	-	×	×		Spain and Algeria.
Epeira isabella, var., *Vinson*	-		×		
Argiope mauritia, *Walckenaer*	-	×		×	South Africa and Madeira.
Meta undulata, *Vinson*			×		
Meta vacillans, *Butler*	×				
Nephila inaurata, *Walckenaer*	-	×			
Nephila ardentipes, *Butler*	×				
Nephila instigans, *Butler*	×				
Tetragnatha protensa, *Walckenaer*	-	×			
Tetragnatha nero, *Butler*	×				
Gasteracantha mauritia, *Vinson*	-	×	×		
Uloborus vanillarum, *Vinson*			×		
Miagrammopes Gulliveri, *Butler*	×				
Olios leucosius, *Walckenaer*		×	×	×	"Magdeleine."
Olios captiosus, *Walckenaer*		×	×		
Sphasus lucasii, *Vinson*				×	
Sphasus extensipes, *Butler*	×				
Salticus baptizatus, *Butler*	×				
Salticus scabellatus, *Butler*	×				
Salticus rodericensis, *Butler*	×				
Holothyrus testudineus, *Butler*	×				
Tityus marmoreus, *Koch* *	×				

* Locality not previously recorded.

ARANEIDEA.

Clubiona, *Latreille.*

Clubiona insularis, Vinson.

One specimen of this species was obtained by Mr. Gulliver.

Scytodes, *Walckenaer.*

Scytodes thoracica, Latreille.

Seven examples. It is "common under leaves, bark, &c."—*G. Gulliver.*

Pholcus, *Walckenaer.*

Pholcus elongatus, Vinson.

Five examples obtained.

Pholcus vexillifer. (Plate LII., figs. 6, 6a.)

Ann. & Mag. Nat. Hist. s. 4, vol. xvii., p. 441 (1876).

♀. Cephalothorax inverted heart-shaped, waved at the ends, bright reddish castaneous; caput ascending, with the front margin and a central marking yellowish; eyes upon elevations of the surface, those of the lateral groups forming triangles, anterior pair of eyes small; falces blackish, yellowish at base; maxillæ and labrum tawny; pectoral shield pyriform, yellowish; legs pale amber-yellow, clothed with whitish hairs and black spines; palpi testaceous, with piceous tips: abdomen subcylindrical with a wavy outline, above testaceous, with a central stellate streak, several oblique lines, and the borders brown, posterior extremity testaceous; underside the same colour, with the base and a central streak greyish brown. Length 10 millims.; relative length of legs 1, 2, 4, 3.

(Coll. by Gulliver.)

Of the lateral groups of eyes, those nearest the mesial line are smallest, the claws of the legs are black; the spinners are testaceous.

Two examples were obtained.

Theridium, *Walckenaer.*

Theridium borbonicum, Vinson.

One specimen found "on leaves."—*Gulliver.*

Theridium diurnum, Vinson.

One specimen.

Latrodectus, *Walckenaer.*

Latrodectus menavodi, Vinson.

Two specimens.

Linyphia, *Latreille.*

Linyphia argyrodes, Walckenaer.

Four specimens found "on leaves."—*G. Gulliver.*

Epeira, *Walckenaer.*

Epeira opuntiæ, Dufour.

Fifty-eight specimens, exhibiting numerous variations, among which is the form named *var. flava* by M. Vinson. This species according to Mr. Gulliver is "common" and "lives in a web of irregular construction."

Epeira isabella var. Vinson.

Two specimens.

Argiope, *Koch*.

Argiope mauritia, Walckenaer.

One example was obtained.

Meta, *Koch*.

Meta undulata, Vinson.

Four examples were obtained.

Meta vacillans. (Plate LII., figs. 4, 4*a–c*.)

Ann. & Mag. Nat. Hist. s. 4, vol. xvii., p. 441 (1876).

♂ ♀. Cephalothorax smoky testaceous, obovate, truncated at the ends, depressed behind; caput ascending, convex, with a marked depression on each side; a central longitudinal ridge; eyes black, in two slightly arched series: abdomen cylindrical, testaceous, with two central lines and a broad undulated longitudinal dusky band, bounded by three black spots on each side; the sides silvery; underside black, with a silvery line on each side: legs long, slender, testaceous, paler below, blackish at the joints, sparsely dotted with long spines; palpi slender, testaceous; male palpus with a clavus resembling the head of a vulture, covered with long bristles; falces subcylindrical, testaceous, about nine teeth on their inner margins; pectoral plate pentagonal scutiform, testaceous. Length of body 6 millims.; relative length of legs 1, 2, 4, 3.

(Coll. by Gulliver.)

The cephalothorax is slightly convex in front, and concave behind, with the centre of the thoracic region depressed. Eyes black, arranged in two slightly convex series across the anterior part of the caput, those of the posterior series rather larger than those of the anterior. The falces have a long curved terminal movable fang, the maxillæ and labrum are sordid testaceous.

Four examples were collected.

Nephila, *Leach*.

Nephila inaurata, Walckenaer.

Seventeen specimens, including both sexes, were obtained.

Nephila ardentipes. (Plate LII., figs. 1, 1*a–c*.)

Ann. & Mag. Nat. Hist. s. 4, vol. xvii., p. 443 (1876).

♀. Cephalothorax black, hairy, oblong, with convex margins, contracted behind the caput; two central shining black tubercles; a deep excavation at the back of the thoracic region; caput ascending, laterally excavated; oculiferous tubercles shining; palpi cylindrical, mahogany-red, hairy at the base, the last three joints black and hairy: abdomen grayish testaceous, silvery pubescent; the dorsal region subochraceous, enclosing a quadrangle of four black spots in front and two parallel rows of black spots behind, where it is also bounded by six black spots, three on each side; sides covered with parallel, irregular, interrupted black lines; underside brownish, dark towards the base and crossed by a yellow band, beyond which is an area enclosed by black dots and shaped like an heraldic shield: legs robust, hairy, mahogany-red; the apices of the femora, knees, apices of tibiæ, the tarsi, and metatarsi black, the black parts densely hairy; maxillæ black, with reddish margins; pectoral plate heptagonal, with eight tubercles, three of which are prominent and form a triangle; falces black, with three marginal teeth. Length 36 millims.; relative length of legs 1, 2, 4, 3.

(Coll. by Gulliver.)

Most nearly allied to *N. nigra* of Vinson; but with the abdomen smaller and not clouded with black, and the legs altogether redder. I have examined thirty-seven examples, but have seen no variation.

"Web strong and sticky; used by the natives where we use waxed ends."— *G. Gulliver.*

Nephila instigans. (Plate LII., fig. 10.)

Ann. & Mag. Nat. Hist. s. 4, vol. xvii., p. 442 (1876).

♀. Cephalothorax oblong, wider behind than before, laterally convex, contracted behind the caput, dull mahogany red, but the back of thoracic region shining black, with a deep central depression; caput ascending, covered with black bristles, which also form a central longitudinal line; legs bright ochre-yellow, sprinkled with black spines; the tarsi and metatarsi castaneous, clothed with brown hair; proximal extremities of the femora blackish; basal half of palpi yellow, nearly smooth, apical half castaneous, clothed with black bristles; falces dark reddish castaneous, with three teeth on each inner margin; pectoral shield heptagonal subpyriform, black, with central orange longitudinal streak; abdomen sandy brown, with a quadrangle of four punctures; region of epigyne blackish. Length 19 millims.; relative length of legs 1, 2, 4, 3.

(Coll. by Gulliver.)

This species does not possess the usual tufted legs.

Two examples were collected.

Tetragnatha, *Latreille*

Tetragnatha protensa, Walckenaer.

One specimen.

Tetragnatha nero. (Plate LII., figs. 3, 3*a, b.*)

Ann. & Mag. Nat. Hist. s. 4, vol. xvii., p. 442 (1876).

♂ ♀. Cephalothorax fusiform, truncate at each end, narrowed in front, with a slender marginal ridge; caput ascending, bounded behind by one or two depressed diverging lines, and with a central depressed spot; centre of thoracic region depressed; eyes black; palpi long and slender, of the male with a globose clavus with white cap, and held by finger-like processes; falces with ten central teeth on each inner margin, and two terminal teeth, the longer one of which projects forwards on each side of the movable fang, the latter black; but the remainder of the cephalothorax and its members tawny; the palpi, coxæ, and trochanters pale; pectoral plate deep castaneous; abdomen pale brown, irrorated with æneous, subcylindrical, a darker dorsal line. Length of body, ♂ 8 millims., ♀ 12; relative length of legs 1, 4, 2, 3.

(Coll. by Gulliver.)

Three examples were found by Mr. Gulliver, who says that it "lies with legs " extended on a single thread placed over or near streams."

Gastracantha, *Latreille.*

Gastracantha mauritia, Vinson.

Two examples.

Uloborus, *Latreille.*

Uloborus vanillarum, Vinson.

Six specimens of this species were obtained. It "makes a web in crevices of " rocks."—*G. Gulliver.*

Miagrammopes, *Cambridge.*

Miagrammopes gulliveri. (Plate LII., figs. 5, 5*a, b.*)

Ann. & Mag. Nat. Hist. s. 4, vol. xvii., p. 443 (1876).

Allied to *M. thwaitesii* of Cambridge, but differing as follows:—Cephalothorax longer and broader, more angular when seen laterally, and with convex rather than concave margins; caput longer, rather higher than the thoracic region, more convex in front, testaceous, with silvery reflexions and a whitish hind border; eyes black, forming a convex series across the back of the caput; lateral eyes twice as large as the central pair, the latter far apart; legs amber-yellow, the femora notched at the distal extremity, the intermediate joints shorter than in *M. thwaitesii,* the tarsi of the hind legs broader, claws black; palpi short, slender, hairy, testaceous; falces small, movable fang piceous; sternum somewhat different in form; abdomen nearly twice as long as the cephalothorax, testaceous, with a central blackish line wider than in *M. thwaitesii*, and not throwing off downward branches; basal region less contracted, lateral margins subangulated in the middle, whence they converge towards the hinder extremity; ventral surface slightly concave, testaceous, with two longitudinal brown lines; spinners concealed by dense hairs, which converge towards the central line. Length 6 millims.; relative length of legs 1, 4, 2, 3.

Two examples of this rare and interesting species were "swept from grass" by Mr. Gulliver.

Olios, *Walckenaer*.

Olios lencosius, Walckenaer.

This species, of which 20 specimens were obtained, is said by Mr. Gulliver to be "common in and out of doors."

Olios captiosus? Walckenaer.

One specimen, obtained by Mr. Slater.

Sphasus, *Walckenaer*.

Sphasus lucasii, Vinson.

Four specimens.

Sphasus? extensipes. (Plate LII., figs. 2, 2a.)

Ann. & Mag. Nat. Hist. s. 4, vol. xvii., p. 439 (1876).

♀. Cephalothorax testaceous, irregularly pentagonal, almost flat; a longitudinal ridge from behind the eyes to beyond the caput: eyes black, placed upon a whitish elevation on the front of the caput; the anterior pair very minute, in the centre of the face, the others considerably larger and forming a transversely hexagonal figure: front margin of face projecting in the centre, behind the projection a V-shaped depression: abdomen testaceous, subcylindrical, flattened below, tapering slightly towards the posterior extremity: legs pale ochraceous; the tibiæ and metatarsi of the two anterior pairs fringed with brown bristles; front pair very like those of *Miagrammopes*; palpi short, pale ochraceous, with blackish tips; maxillæ and falces pale brown; pectoral plate elongate hexagonal; ventral surface of abdomen with three central longitudinal grey lines. Length 7 millims.; relative length of legs 1, 4, 2, 3.

Mr. Gulliver obtained one specimen of this little species by sweeping grass. It has the general aspect of the genus *Miagrammopes*, but the structure of a *Sphasus*.

Salticus, *Latreille.*

Salticus scabellatus. (Plate LII., fig. 9.)

Ann. & Mag. Nat. Hist., s. 4, vol. xvii., p. 441 (1876).

♀. Oculiferous region of cephalothorax black, bounded by a crescent-shaped reddish castaneous belt; thoracic region dark piceous; eyes opaline; palpi testaceous; falces, labrum, maxillæ, and anterior tibiæ dark piceous; legs above fawn-coloured, pectoral plate and legs below paler; abdomen grey, subpyriform, its base black; a central pale brown longitudinal stripe, bounded laterally by four black spots, which form a square; venter pale brown, speckled with dark brown, a central line and the margins dark grey. Length 8 millims.; relative length of legs 4, 1, 3, 2.

(Coll. by Gulliver.)

This species seems to be allied to *S. exilis* of Cambridge; only two examples were brought home.

Salticus baptizatus. (Plate LII., fig. 7.)

Ann. & Mag. Nat. Hist. s. 4, vol. xvii., p. 440 (1876).

♀. Cephalothorax inverted bell-shaped, dark shining piceous, with a whitish submarginal line, bordered within laterally by a rather wider tawny line; a minute cruciform amber-coloured spot in the centre of the caput, two lateral crescent-shaped fawn-coloured spots (one on each side) clothed with white scales, and a leaf-like fawn-coloured patch in the centre of the thoracic region, the front of which is depressed; the hinder part and apex of the thoracic patch clothed with whitish scales; falces, maxillæ, labrum, femora, tibiæ, and metatarsi of anterior legs castaneous; pectoral plate fusiform, testaceous; last three pairs of legs and tarsi of anterior pair amber-yellow, claws black; palpi whitish: eyes opaline, forming an oblong across the caput; relative sizes as follows—anterior central pair, posterior, lateral anterior, lateral intermediate: abdomen elliptical, dark brown, with a lateral whitish line and a broad, white-edged, irregular, central testaceous streak; ventral surface whitish, striated with brown at the sides; a central ferruginous streak; region of spinnerets testaceous. Length 9 millims.; relative length of legs 1, 4, 2, 3.

(Coll. by Gulliver.)

This species has the general coloration of the much smaller species figured in Lucas's "Algeria" under the name of *S. testaceolineatus*. Only one specimen was obtained.

Salticus rodericensis. (Plate LII., fig. 8.)

Ann. & Mag. Nat. Hist. s. 4, vol. xvii., p. 440 (1876).

♀. Cephalothorax dark piceous, clothed with testaceous hairs; a central longitudinal reddish castaneous streak from the back of the caput to the posterior margin; eyes opaline; falces, labrum, maxillæ, legs above, femora and tibiæ below, and epigyne castaneous, clothed with testaceous hairs; tarsi testaceous, with black claws; abdomen ovate, dark brown, pilose, with a dorsal testaceous streak interrupted by angulated lines at its inferior extremity, and intersected by a black line; two arched black lines spotted with testaceous from the sides of the testaceous streak to the spinnerets; pectoral plate, coxæ, trochanters, and venter fawn-coloured, base of venter white; a central tapering grey steak. Length 10 millims.; relative length of legs 1, 2, 4, 3.

(Coll. by Gulliver.)

Apparently allied to *S. intentus* of Cambridge.

Five specimens were obtained.

ACARIDEA.

Holothyrus, *Gervais*.

Holothyrus? testudineus (Plate LII., figs. 11, 11 *a–c*).

Ann. & Mag. Nat. Hist. s. 4, vol. xvii., p. 444 (1876).

Carapace ovate, convex, conical in front, fawn-coloured, clouded with brown, smooth and shining, with well-defined lateral carina; legs rather short and robust; ventral surface flattened and shining behind, with an apparently hinged oblong projecting anterior plate, wider in front than behind; body in front concave, with *only three pairs of legs*, their relative size 3, 2, 1; antennæ with hairy terminal joint. Length ¾ millim.

(Coll. by Gulliver.)

Although fourteen examples of this species have come, not differing in size or structure, I have been unwilling to erect a new genus for its reception. From many points which it appears to have in common with *H. coccinella* of Mauritius, I think it possible that it may turn out to be an immature condition of that species.

SCORPIONIDEA.

Tityus, *Koch*.

Tityus marmoreus, Koch.

Twenty-four examples were collected by Messrs. Gulliver & Slater.

DESCRIPTION OF PLATE LII.

Fig. 1. *Nephila ardentipes,* Butl. Upper surface.
   1a.   ,,   ,,   ,,   Profile view enlarged.
   1b.   ,,   ,,   ,,   Ventral surface of abdomen, showing scutiform area.
   1c.   ,,   ,,   ,,   Pectoral shield, showing arrangement of tubercles.
   2. *Sphasus extensipes*   ,,   Profile view, enlarged three times.
   2a.   ,,   ,,   ,,   Upper surface of cephalothorax and base of abdomen greatly enlarged.
   3. *Tretragnatha nero*   ,,   Upper surface enlarged.
   3a.   ,,   ,,   ,,   Falx opened to show the dentition.
   3b.   ,,   ,,   ,,   Male palpus.
   4. *Meta vacillans*   ,,   Profile view enlarged three times.
   4a.   ,,   ,,   ,,   View of upper surface without the members.
   4b.   ,,   ,,   ,,   Falx opened to show the dentition.
   4c.   ,,   ,,   ,,   Male palpus, as seen from the side.
   5. *Miagrammopes gulliveri,,* Profile view much enlarged.
   5a.   ,,   ,,   ,,   Upper surface, enlarged about 8 times.
   5b.   ,,   ,,   ,,   Profile view of cephalothorax.
   6. *Pholcus vexillifer*   ,,   Upper surface, enlarged.
   6a.   ,,   ,,   ,,   ,,   ,, enlarged about 5 times to show form, pattern, and arrangement of eyes.
   7. *Salticus baptizatus*   ,,   Upper surface enlarged.
   8.   ,,   *rodericensis*   ,,   ,,   ,,   ,,
   9.   ,,   *scabellatus*   ,,   ,,   ,,   ,,
   10. *Nephila instigans*   ,,   ,,   ,, natural size.
   11. *Holothyrus testudineus* ,,   ,,   ,, greatly magnified.
   11a.   ,,   ,,   ,,   Under surface   ,,   ,,
   11b.   ,,   ,,   ,,   Apparently hinged plate projecting from flattened posterior area of the ventral surface.
   11c.   ,,   ,,   ,,   Leg.

COLEOPTERA.—*By C. O. Waterhouse, Assistant in the Zoological Department, British Museum.*

(Plate LIII.)

The collection of *Coleoptera* contained in all 265 specimens. The number of species obtained is only sixty, and is probably only a small proportion of what could have been discovered had the naturalists been able to devote more time specially to them, and had remained longer in the Island. All the large groups are represented, and in the following proportions:—

Geodephaga	1	*Lamellicornia*	4	*Longicornia*	6
Hydradephaga	4	*Serricornia*	2	*Phytophaga*	1
Brachelytra	3	*Malacodermata*	7	*Pseudotrimera*	
Necrophaga	14	*Heteromera*	2	*Erotylidæ*	1
Palpicornia	3	*Rhynchophora*	11	*Endomychidæ*	1

A detailed account of the species is given below, it may therefore be sufficient to state here that none of the species are apterous, that the number described as new to science is thirty-three, for eight of which new genera had to be established. Of the twenty-seven previously described species eight appear to be cosmopolitan or to have a wide distribution.

The following list of species will best show the localities in which each has been met with:—

	Island of Rodriguez only.	Previously obtained elsewhere.		
		Mauritius and Bourbon.	Madagascar.	Other Localities.
Chlænius olivaceus, *C. W.*	×			
Copelatus pulchellus, *Sturm.*		×		Senegal.
Hydaticus signatipennis, *Lap.*				Senegal, E. Indies.
Rhantus socialis, *C. W.*	×			
Dineutes picipes, *C. W.*	×			
Aleochara parvula, *C. W.*	×			
Homalota destituta, *C. W.*	×			
Lithocharis occulta, *C. W.*	×			
Carpophilus mutilatus, *Er.*				Cosmopolitan.
Carpophilus auropilosus, *Woll.*				Madeira.
Carpophilus humeralis, *Fab.*		×	×	Natal, China, Java, &c.
Microporum (n. g.) nitens, *C. W.*	×			
Probænus (n. g.) longicornis, *C. W.*	×			
Epuræa ophthalmica, *C. W.*	×			
Ascomma (n. g.) horrida, *C. W.*	×			

	Island of Rodriguez only.	Previously obtained elsewhere.		
		Mauritius and Bourbon.	Madagascar.	Other Localities.
Endocoxelus (n. g.) variegatus, *C. W.*	×			
Æschyntelus (n. g.) ater, *C. W.*	×			
Murmidius segregatus, *C. W.*	×			
Cryptamorpha desjardinsii, *Guér.*		×		N. Zealand, Madeira, St. Helena, Japan, N. Amer.
Læmophlœus palpalis, *C. W.*	×			
Dermestes vulpinus, *Fab.*				Cosmopolitan.
Trogoderma, sp.	×			
Berosus mixtus, *C. W.*	×			
Philhydrus melanocephalus, *Ol.*				Europe, &c.
Dactylosternum abdominale, *Fabr.*		×		W. Indies, W. Africa, Natal, Ceylon, Andamans.
Rhyssemus tarsalis, *C. W.*	×			
Ancylonycha gradaria, *C. W.*	×			
„ rodriguezi, *C. W.*	×			
Oryctes minor, *C. W.*	×			
Sponsor desjardinsii, *Guér.*		×		
Adelocera modesta, *Boisd.*		×		Aden, China, Java, Senegal, Guatemala, Oceania.
Malthacodes (n. g.) pictus, *C. W.*	×			
Tarsostenus univittatus, *Spin.*				Europe, Venezuela, Natal, India, Australia.
Xylodes (n. g.) albovarius, *C. W.*	×			
Lyctus rugicollis, *Walker*				Ceylon, China.
„ brunneus, *Steph.*				Europe, Australia, &c.
Cis insularis, *C. W.*	×			
„ sexcarinatus, *C. W.*	×			
Opatrum cœsum, *Stcv.*		×		
Cistela brunnea, *C. W.*	×			
Balanodes (n. g.) tomentosus, *C. W.*	×			
Caranistes annulipes, *C. W.*	×			
Cratopus inornatus, *C. W.*	×			
„ virescens, *C. W.*	×			
„ magnifica, *C. W.*	×			
Sitophilus linearis, *Hbst.*				W. Indies, India, &c.
„ oryzæ, *Linn.*		×		Cosmopolitan.
Phlæophagosoma proximum, *Woll.*				Malay Archipelago.
Pentarthrum rodriguezi, *C. W.*	×			
Cossonus marginalis, *Schönb.*		×		
Tomicus ferrugineus, *Fabr.*				Rio Janeiro, Ceylon, Penang, Celebes.
Macrotoma simplex, *C. W.*	×			
Xystrocera globosa, *Ol.*		×	×	Round Is.
Stromatium barbatum, *F.*		×		India, Penang.
Phoracantha semipunctata, *F.*				Australia.
Coptops bidens, *Fabr.*		×		India, Natal, W. Africa.
Micracantha dentata, *Ol.*		×		
Cryptonychus limbatus, *C. W.*	×			
Euxestus parkii, *Woll.*				Madeira.
Trochoideus desjardinsii, *Guér.*		×	×	Siam, Java, Phil. Is.

GEODEPHAGA.

Carabidæ.

Chlænius olivaceus.

Ann. & Mag. Nat. Hist., 1876, xviii., p. 106.

C. capite thoraceque obscure viridi-æneis; capite obsolete subtiliter punctulato; thorace sat crebre fortiter punctato; elytris obscure olivaceis, striatis, interstitiis sat crebre distincte punctatis; antennis articulis tribus basalibus pedibusque flavo-ferrugineis, tarsis obscurioribus. ♂. Long. 5¼ lin., lat. 2 lin.

Somewhat resembles *C. nigricornis*, Fab., in form; but the thorax is relatively a little narrower, the sides are a little more rounded, the posterior angles more rounded; the elytra are slightly more attenuated posteriorly. The antennæ are blackish, with the first three joints reddish yellow. The head is sculptured as in *C. nigricornis*, but rather more distinctly, the punctures are more distinct towards the eyes and on the neck. The thorax is one fourth broader than long, gently convex, strongly punctured but not very thickly; the punctures, however, are closer near the suture and towards the hinder margin; there is a single well-defined fovea near the posterior margin, slightly nearer to the side than to the sutural line. Scutellum impunctate. Elytra somewhat broader than the thorax, less parallel at the sides, and more narrowed towards the apex than in *C. nigricornis*; the pubescence is rather less close, the striæ are well marked (rather more so than in *C. nigricornis*), the interstices are thickly and distinctly punctured, but the punctures are not crowded. The underside is black, shining; a few strong punctures are scattered over the prosternum and the metasternum. The legs are reddish yellow; the apices of the tibiæ and the tarsi are pitchy.

Var. Elytra with a reddish-yellow spot near the apex on the third to sixth interstices. ♀.

Of this species, which was new to science, only two specimens were brought.

HYDRADEPHAGA.

Dytiscidæ.

Copelatus pulchellus, Sturm.

Seven specimens of this species were captured.

Hydaticus signatipennis, Lap.

Two examples only.

Colymbetes (Rhantus) socialis.

Ann. & Mag. Nat. Hist., 1876, xviii., p. 106.

C. elongato-ovalis, supra obscure flavicans, infra niger; capite postice nigro, vertice transversim flavo notato; thorace disco guttis parvis duabus approximatis piceis notato; elytris obscurioribus (flavo limbatis); prosterno pallido. Long. $4\frac{1}{2}$ lin., lat. $2\frac{3}{8}$ lin.

Elongate oval, shining. Head yellow, with an oblique spot on each side on the forehead, and the neck black, the black portions uniting at the eyes. Thorax yellow, with the middle of the anterior and posterior margins and two approximate discoidial spots pitchy; very shining, with a line of punctures along the front margin; there is also a line of obscure punctures along the sides and extending a short distance along the posterior margin; the extreme lateral margins are distinctly incrassate. Scutellum pitchy. Elytra shining, with the sutural line and the sides yellowish, the rest closely spotted with small brownish markings as in *C. notatus*, F.; each elytron with three rows of rather large punctures, each row containing about eight or ten punctures. Underside very shining, black, except the prosternum, which is yellow, and the margins of the abdominal segments, which are obscurely pitchy. Legs pitchy yellow; intermediate femora and tibiæ moderately thickly and finely punctured.

This species most nearly resembles *C. notatus*, Fab., but is a trifle narrower; the coloration is nearly the same. The thorax and elytra are smooth, the sides of the former are not so much impressed above, but the edges are more distinctly incrassate. The three rows of punctures on the elytra are very distinct. The under side is very shining and is not frosted as in *C. notatus*; the prosternum is a little less sharply keeled; the intermediate femora and tibiæ are distinctly less strongly and more finely punctured; the anterior and intermediate tarsi are scarcely dilated in the male.

Eight specimens were taken.

Gyrinidæ.

Dineutes picipes.

Ann. & Mag. Nat. Hist., 1876, xviii., p. 107.

D. obovalis, depressiusculus, sat latus, supra nigro-olivaceus, vix cæruleo-cupreo micans, nitidulus; elytris postice rotundatis, pone apicem in mare leviter emarginatis, in fœmina externe oblique truncatis, ad truncaturæ basin dente parvo deflexo et apice dente acuto armatis; corpore subtus nigro-piceo, ano pedibusque piceis.

♂ long. $6\frac{3}{4}$ lin., lat. $4\frac{1}{6}$ lin.; ♀ long. 6 lin., lat. $3\frac{1}{2}$ lin.

"Always on the surface of Rivière de l'Est." G. G.

BRACHELYTRA.

Aleocharidæ.

Aleochara parvula.

Ann. & Mag. Nat. Hist., 1876, xviii., p. 107.

A. statura fere *A. mœrentis*, at parva, nitida, parcius pubescens; antennis pedibusque obscure testaceis antennarum articulo ultimo tibiarumque basi vix picescentibus; elytris thorace vix longioribus, cum thorace discrete distincte punctatis; abdomine discrete distincte punctato, ano piceo-testaceo. Long. ⅚ lin.

Antennæ stout, scarcely longer than the head and thorax taken together; the first, second, and third joints elongate, subequal, the fourth joint a little broader than long; the fifth to tenth joints becoming gradually broader but not longer; the eleventh joint as long as the two preceding taken together, bluntly acuminate. Head rather broad, sparingly and scarcely visibly punctured. Thorax gently convex, twice as broad as long, narrowed in front, not thickly but distinctly punctured. Elytra a trifle broader than the thorax, and a little longer, distinctly but not thickly asperate-punctate. Abdomen distinctly, evenly but not thickly punctured; the punctuation of the fifth segment is scarcely less distinct than that of the preceding.

The thorax is relatively rather broader than in *A. mœrens*; and the punctuation is less close, as is also that of the elytra.

A single example only.

Homalota destituta.

Ann. & Mag. Nat. Hist., 1876, xviii., p. 108.

H. statura et colore *H. boletobiæ*, Th. (*nigritulæ*, Kz), at antennis paulo brevioribus. Nigra, subnitida, subtiliter punctulata, flavosericea; antennarum basi, elytris pedibusque luteo-testaceis, elytris regione scutellari angulisque posticis obsolete infuscatis, thorace obscure piceo; abdomine discrete subtiliter punctulato, segmentis apice piceis. Long. 1⅖ lin.

Very closely allied to *H. boletobia*, Th., but is a trifle more parallel-sided, the antennæ are a little shorter and stouter, the elytra are relatively a little shorter; the punctuation is throughout slightly finer, especially that of the thorax, which is also less close; that of the abdomen is more even (being scarcely less close on the fifth and sixth segments than on the previous ones). The three basal joints of the antennæ are elongate, subequal, testaceous; the third more slender; the fourth joint is twice as broad as long, shining, pitchy testaceous; the fifth to tenth joints gradually but distinctly broader, transverse; the eleventh joint nearly as long as the two previous joints taken together, obtusely acuminate. Head sparingly and scarcely perceptibly punctured. Thorax pitchy, very nearly as broad as the elytra, not quite twice as broad as long, gently rounded at the sides and base, with a shallow longitudinal impression in front of the scutellum, finely but not thickly punctured. Elytra a trifle longer than the thorax, finely and not thickly punctured. Abdomen with the apex of each segment pitchy, very finely but not closely punctulate; the punctuation is less close and distinctly finer than on the elytra.

This is another of the new species of which a single example was found.

Pæderidæ.

Lithocharis occulta.

Ann. & Mag. Nat. Hist., 1876, xviii., p. 108.

L. ferruginea; antennis, elytris pedibusque flavo-testaceis; capite thoraceque subopacis, sat crebre fortiter punctatis; elytris thorace $\frac{1}{5}$ longioribus, nitidis, sat crebre punctatis; abdomine nitido segmentis 4 basalibus sat parce subtilissime punctulatis, 5° lævi. Long. 1$\frac{1}{6}$ lin.

Sparingly clothed with pubescence. Antennæ one fourth longer than the head; first joint elongate, second a little longer than broad, third not longer than the second but narrower, the fourth as broad as long; fifth to tenth joints transverse and gradually becoming broader; the eleventh joint ovate, sub-acuminate. Head quadrate, convex, subopaque, finely rugulose and somewhat thickly and strongly punctured, straight behind, sides parallel, the extreme angles blunt; the portion of the head behind the eyes is considerable longer than the diameter of the eye. Thorax at its broadest part scarcely as wide as the head, gently narrowed behind, scarcely as long as broad, subopaque, finely rugulose and moderately thickly and strongly punctured. Elytra a little broader than the head, one fifth longer than broad, shining, punctured as the thorax (but not rugulose). The punctuation of the abdomen is very fine and obscure, gradually disappearing towards the apex. Tarsi long and slender.

The portion of the head in front of the eyes resembles in form that generally seen in *Sunius*, that is to say, it is somewhat broad, slightly impressed, covering the base of the antennæ, with the front margin arcuate; the general structure, however, of the tarsi and maxillary palpi, so far as I am able to examine them without dissection, appears to be that of *Lithocharis*.

A unique example of this interesting species was brought by Mr. Slater.

NECROPHAGA.

Nitidulidæ.

Carpophilus mutilatus, Er.

Eleven specimens of this species were brought; nine being flavous and two nearly black varieties.

Carpophilus auropilosus, Woll.

A single specimen which I could not separate from *C. auropilosus*, taken by Mr. Wollaston in Maderia.

Carpophilus humeralis, Fabr.

Three specimens. The punctuation (especially of the thorax) is very strong, much more so than examples from Natal which are in the National Collection, but the form is rather that of *C. humeralis* than *C. foveicollis*; the punctuation is also even more strong than in this latter, I therefore prefer considering the specimens from Rodriguez mere varieties of *C. humeralis*.

Microporum, C. Waterh.

Ann. & Mag. N. Hist., 1876, xviii., p. 109.

Mentum tranverse, much narrower in front; ligula short, the anterior angles produced. Labial palpi short, the apical joint large. Lobe of the maxillæ elongate, slender, ciliate; maxillary palpi with penultimate joint quadrate, the apical joint twice as long as the previous joint, slightly attenuated. Mandibles strongly arched, terminating in a sharp point. Head transverse, narrowed in front of the eyes, terminating in a short muzzle. Eyes prominent. Antennæ inserted immediately in front of the eyes, moderately short; the first joint stout, cylindrical, the rest much more slender; the second to fifth subequal, elongate; the sixth and seventh a little shorter, the eighth as broad as long, narrowed at the base; the ninth to eleventh forming an oblong-ovate club. Thorax twice as broad as long, gently arched in front, arcuata at the sides, broadest at the posterior angles, which are acute and slightly directed backwards. Elytra oblong, leaving two segments of the abdomen uncovered, as wide as the thorax, margined at the sides, the apex of each elytron slightly rounded, especially at the outer angle. Legs moderately stout; tarsi with the basal joints somewhat dilated, ciliated.

The insect which appears most nearly to approach the one here described is *Idocolastus*, and I have accordingly placed it next to that genus.

Microporum nitens (Plate LIII., fig. 2).

C. Waterh., l.c., p. 109.

M. oblongum, leviter convexum, nitidum, piceo-testaceum, crebre evidenter punctatum, elytris pedibusque testaceis; capite transverso; oculis prominentibus, nigris; antennis capite dimidio longioribus; thorace longitudine duplo latiore, convexo, angulis anticis deflexis rotundatis, lateribus arcuatis, angulis posticis acutis retrorsum directis; elytris thorace haud angustioribus at duplo longioribus, humeris obtusis, lateribus arcuatis, angulis apicalibus exterioribus bene rotundatis apicibus arcuatis.

Long. 1⅙ lin.

The punctuation although somewhat fine is distinct, especially on the elytra, and the punctures are rather close together. The eyes (which are coarsely granular) are oblique and very prominent, especially behind. The thorax is very little narrowed in front, the sides are marginal and gently arcuate, the greatest width is at the posterior angles, which are acute and slightly lap over the shoulders of the elytra. The elytra are very nearly twice as long as the thorax, and their width slightly narrowed towards the apex; on each side of the suture there is a lightly impressed stria, joining the suture at the apex, but not reaching the scutellum.

The head has on the forehead between the antennæ a minute central tubercle, and on the vertex a very shallow (scarcely perceptible) fovea; both these characters are only to be seen in certain lights.

One specimen.

Probænus, C. Waterh.

Ann. & Mag. N. Hist., 1876, xviii., p. 110.

Mentum somewhat transverse, suddenly narrowed in front, the extreme apex truncate; on either side of the narrow portion (but on a lower plane) there is a somewhat triangular projection; the ligula is not prominent, but its anterior angles are produced into two delicate, subcylindrical, slightly curved projections, nearly as long as the palpi. Labial palpi short and thick; the basal joint round, scarcely visible; the second and third equal, slightly elongate, subcylindrical. Lobe of the maxillæ slender, ciliated, with a distinct sharp tooth nearly on a level with the basal joint of the palpus. Maxillary palpi with the second joint twice as long as broad, narrowed at its base; the third joint a little shorter, a little longer than broad, the apical joint one third longer than the previous one, slender, a little narrowed at the apex. Mandibles very prominent, completely visible from above, impressed, broad at the base, narrow and pointed at the apex. Head transverse, slightly lobed over the base of the antennæ; eyes prominent. Antennæ very long, basal joint very large; second joint elongate, cylindrical, inserted near the apex of the basal joint; the third to eighth joints a little longer than the second, and of the same form; the ninth rather shorter and broader than the previous joint, the tenth scarcely transverse, the eleventh oblique truncated. Thorax transverse, broader behind, scarcely narrowed in front. Elytra oblong. Pygidium visible from above. Legs robust; tarsi short, the basal joints slightly dilated, strongly ciliated. Pro- and meso-sterna constructed as in *Mystrops*.

This genus is closely allied to *Mystrops*, from which it differs in the form of the mandibles, the form of the head, the basal joint of the antennæ, and in the elytra nearly covering the abdomen.

I am unacquainted with *Mystrops dispar* from Madagascar, except from description; but I think it probable that it may have to be transferred to the present genus.

Probænus longicornis (Plate LIII. fig. 1).

Ann. & Mag. N. Hist., 1876, xxiii., p. 110.

P. oblongus, leviter convexus, testaceus, nitidus, crebre distincte punctatus; capite sat magno transverso, leviter convexo, antice impresso, utrinque supra antennarum basin parum elevato; oculis prominentibus, nigris; mandibulis porrectis, leviter curvatis, apice acuminatis, basi supra parum concavis; antennarum articulo primo magno, elongato; thorace capite ¼ latiore, longitudine fere duplo latiore, leviter convexo, antice vix angustato, angulis anticis obtusis, lateribus levissime arcuatis, marginatis, angulis posticis rectis, basi fere recta; scutello sat magno; elytris thorace vix angustioribus, at ⅓ longioribus, apicem versus parum angustatis, lateribus levissime arcuatis, marginatis, apicibus arcuatim truncatis.

Long. 1¼ lin., lat. ¾ lin.

The punctuation is rather close and distinct, but not very strong, it is rather finer on the head. The mandibles are nearly as long as the head, not much curved, enclosing a space when the points meet. Antennæ long, about twice the length of the thorax, basal joint very long and thick, slightly narrowed at the apex, the second joint elongate, placed near the apex of the first on the hinder side.

One specimen of this singular insect was captured. From the unusual way in which the second joint of the antennæ is articulated with the first, it is difficult to see how the antennæ could ever be directed otherwise than backwards.

Epuræa ophthalmica.

Ann. & Mag. N. Hist., 1876, xviii., p. 111.

E. oblonga, obscure ferruginea, opaca, dense pubescens; capite lato, transverso, ante oculos angustato, subtilissime coriaceo, sat crebre distincte punctato; labro transverso, lateribus rotundatis, antice medio anguste profunde triangulariter exciso; oculis magnis prominentibus, nigris, antennis capite paululo longioribus, clava picea; thorace capite ⅓ latiore, longitudine ⅖ latiore, leviter covexo, antice vix angustato, coriaceo, sat crebre evidenter punctato, margine antico leviter emarginato, angulis anticis obtuse rotundatis, lateribus tenuiter marginatis, levissime arcuatis, angulis posticis fere rectis (summo ipso obtuso); elytris thorace haud angustioribus, vix duplo longioribus, medio vix ampliatis, coriaceis, minus crebre evidenter punctatis, apice truncatis, angulis externis paulo rotundatis; abdomine supra subtiliter haud crebre punctulato, brevissime flavo-pubescente. Long. 1⅓ lin., lat. ⅔ lin.

The eyes in this species are unusually prominent; and there is not the usual impressed line along the inner margin. The thorax is broadest at the posterior angles, which are rectangles, with the extreme point blunted.

A single example.

Colydiidæ.

Ascomma, C. Waterh.

Ann. & Mag. Nat. Hist., 1876, xviii., p. 111.

General build of *Endolphlœus*, but with the head not widened in front of the eyes. Eyes partly clothed with erect scale-like setæ. Mentum square; lingula very nearly as broad as the mentum and about half the length, sides parallel, front margin gently arcuate, the angles consequently slightly obtuse; the front margin fringed with dense stiff hair, so dense that it has the appearance of a piece separate from the ligula; the labial palpi are well separated from each other at the base; the apical joint is elongate, slightly acuminate at the apex, a little swollen at the base. The outer lobe of the maxillæ rather widened and truncate at the apex; the palpi are short and thick and do not differ materially from those of *Endolphlœus*. Antennæ eleven-jointed, the third joint rather elongate; the tenth and eleventh joints form a distinct broadly ovate club.

Ascomma horrida (Plate LIII., fig. 3).

Ann. & Mag. Nat. Hist., 1876, xviii., p. 112.

A. oblonga, covexiuscula, opaca, ferruginea, nigro varia, breviter echinata; capite deplanato, antice rotundato, postice angustiore; oculis nigris, sat prominentibus; antennis capite paulo longioribus, nitidis, piceis, clava fere rotundata; thorace capite fere duplo latiore, longitudine $\frac{1}{4}$ latiore, convexo, postice paulo angustato, margine antico utrinque sat fortiter emarginato, medio lobato, angulis anticis acutiusculis, lateribus deplanatis leviter rotundatis, serratis, angulis ante basin fere rectis, basi medio late lobata, disco impressione oblonga; scutello parvulo, rotundato; elytris basi thorace haud latioribus, postice vix ampliatis, fere parallelis, apice rotundatis, singulis elytris tricostatis; tibiis echinatis.

Long. $2\frac{1}{3}$–3 lin., lat. $\frac{3}{4}$–$1\frac{1}{6}$ lin.

Build of *Endophlœus spinulosus*, but longer; the head is scarcely wider in front of the eyes, so that the eyes (which are furnished with erect black narrow scales) are the most prominent; the antennal groove below the eye is very deep. Antennæ with the first joint short (not visible from above), the second nearly globular, the third elongate, the fourth fifth and sixth becoming shorter, the seventh to ninth nearly globular, the tenth and eleventh forming a somewhat round and distinct club. The thorax has the sides somewhat expanded and reflexed, the margins serrated, but the teeth are difficult to see on account of the short bristles. The whole surface of the insect (with the tibiæ and part of the eyes) is clothed with closely set short erect thick black and ferruginous setæ. The general colour is dull rusty red, the disk of the thorax and a mesial and subapical patch of the elytra are blackish.

Fifteen specimens of this interesting new species were found.
"Under bark," G.G.

Endocoxelus, C. Waterh.

Ann. & Mag. Nat. Hist., 1876, xviii., p. 112.

Mentum slightly transverse, a little narrowed in front; ligula about half the width of the mentum and of the same length, parallel at the sides rounded and ciliated in front; labial palpi somewhat elongate and acuminate; outer lobe of the maxillæ short, traingular, truncate and ciliated at the apex; apical joint of the palpi twice as long as broad, as long as the two preceding joints taken together, truncate at the apex. Head nearly parallel at the sides, gently sinuate at the eyes, and a little narrowed behind; eyes slightly prominent. Antennæ eleven-jointed as long as the head, the first and second joints stouter than those following, scarcely longer than broad; the third to fifth a little elongate, sixth to ninth moniliform, the tenth and eleventh joints forming a distinct club, the eleventh considerably smaller that the tenth. Thorax convex, a little broader than long, slightly narrower behind, distinctly margined all round, but especially at the sides, which are gently arcuate (microscopically serrate) all the angles obtuse. Scutellum very small. Elytra at the base scarcely broader than the thorax, slightly broader posteriorly, twice as long as broad.

Closely allied to *Coxelus*, but with the head not wider in front of the eyes, antennæ with a more distinct club, thorax narrowly but distinctly margined, &c.

Endocoxelus variegatus (Plate LIII., fig. 4).

Ann. & Mag. Nat. Hist., 1876, xviii., p. 112.

E. oblongus, antice paulo angustatus, leviter convexus, opacus, niger, squamulis pallidis adspersus; elytris testaceo-variegatis; tibiis testaceis.

Long. $1\frac{1}{4}$ lin., lat. $\frac{3}{5}$ lin.

Dull pitchy black, the clypeus and the middle of the front and posterior margins of the thorax reddish. Elytra obscure testaceous, each with the scutellar region, a discoidal spot, a subapical patch, and three or four lateral markings blackish brown. Legs (especially the tibiæ) testaceous. The generally decumbent scales which are tolerably thickly scattered over the surface (and on the tibiæ) are very pale yellow. Head flattened, thickly rugulose. Thorax thickly and rugosely punctured, convex, a trifle broader towards the apex, which is not very obtusely rounded, slightly shining, crenate-striate, the interstices scarcely convex, the scales forming rows.

Less convex than *Coxelus pictus*, and with the elytra less narrowed at the base, &c. &c.

Twenty-seven specimens; "Under bark," G.G.

Æschyntelus, C. Waterh.

Ann. & Mag. Nat. Hist., 1876, xviii., p. 113.

Resembles *Bothrideres* except in the form in the head, which is like that of *Deretaphrus erichsoni*. Head bent downwards, much arched, broadest behind the eyes, scarcely narrowed and rounded in front; eyes not prominent, scarcely visible from above. Antennæ as long as the head; the basal joint large, nearly round; the second the same form but smaller; the third, seventh, eighth, and ninth a little longer than broad; the fourth, fifth, and sixth nearly as long as broad; the tenth compressed, cup-shaped; the eleventh much smaller, transverse, truncate and pubescent at the apex. Anterior coxæ very little separated; anterior tibiæ denticulate. Intermediate coxæ moderately separate. Posterior legs very widely separated. Basal segment of the abdomen rather longer than the two following taken together.

Æschyntelus ater (Plate LIII., fig. 5).

C. Waterh., Ann. & Mag. Nat. Hist., 1876, xviii., p. 113.

Æ. elongatus, ater, dorsim depressiusculus; capite crebre fortiter punctato, punctis longitudinaliter confluentibus; antennis piceis, nitidis; thorace antice capite fere duplo latiore, latitudine haud longiore, postice angustato, angulis anticis prominulis obtusiusculis, lateribus antice leviter rotundatis, dein ad basin rectis, medio tuberculo vix perspicuo notatis, basi utrinque levissime sinuata; supra crebre fortiter punctato, medio foveis duabus ovalibus (altera ante, altera post medium) notato; scutello parvo quinquelaterali; elytris thorace paulo latioribus, parallelis, ante apicem solum angustatis, apice truncatis, fortiter punctato-striatis, interstitiis 2° et 3° planis, 4°, 5°, 6°, 7° angustatis costiformibus (plerumque 5° et 7°); pedibus piceis, tibiis anticis quadridenticulatis; corpore subtus sat crebre fortiter punctato.

Long. 2–3⅓ lin., lat. ½–1 lin.

The thorax is more strongly punctured towards the hind margin than in front, the middle of the front margin is rather finely punctured. The punctures in the striæ of the elytra are large and slightly separated, the lateral striæ would be better termed grooves.

Three specimens, "Under bark," G.G.

Murmidius segregatus.

C. Waterh., Ann. & Mag. Nat. Hist., 1876, xviii., p. 114.

M. ovalis, convexus, nitidus, piceo-niger, punctatus; thorace rufopiceo, antennis pedibusque obscure testaceis; capite sat crebre subtilissimo punctulato; thorace crebre subtiliter punctato, circa angulos anticos punctis nonnullis magnis, fortiter transverso, antice leviter emarginato, angulis anticis acutis, sub angulis rotundato-exciso, lateribus arcuatis marginatis, basi marginata utrinque oblique bisinuata; elytris convexis, sat crebre distincte punctatis, postice paulo ampliatis, basi flexuosis, apice rotundatis.

Long. 1¼ mill.

Rather more convex than *M. ovalis,* more rounded at the sides, more narrowed in front, and at once distinguished by the punctures on the elytra not forming lines, but regularly distributed over their surface. The scutellar region is somewhat impressed.

A single example only.

The genus *Murmidius* appears to me to be more conveniently placed at the end of the *Colydiidæ* than with the *Histeridæ.* The genus does not appear in Lacordaire's "Genera," so that we have not that author's opinion as to its affinities.

Cryptamorpha desjardinsi, Guérin.

Psammœcus desjardinsi, Guérin, Ic. Règn. Anim. Ins. 1843, p. 196.

A single example "Under bark," G.G.

This insect was originally described from Mauritius by Guérin under the name *Psammœcus Dejardinsii*. In 1846 Mr. Adam White described the species under the name *Dendrophagus suturalis*, from New Zealand. In 1854 Mr. Wollaston met with it in Maderia, and correctly proposed the new genus *Cryptamorpha* for its reception, naming it *C. musæ*, not observing that it had already received two specific names in genera to which it could not properly be referred. More recently, in 1859, Dr. Leconte seems to have again described the species under the name *Pseudophanus signatus*, from N. America; and I think that in all probability the *Telephanus fasciatus*, of Redtenbacher, from Taïti (1867), will also prove to be the same insect.

Cucujidæ.

Læmophlœus palpalis.

C. Waterh., Ann. & Mag. Nat. Hist., 1876, xviii., p. 114.

L. Elongatus, convexiusculus, niger, nitidus, palpis tarsisque testaceis, tibiis piceis; capite sat convexo, crebre fortiter punctato, post oculos sulco transverso leviter impresso; oculis prominulis; antennis capite $\frac{1}{3}$ longioribus, sat crassis, articulis duobus basalibus crassioribus, 3° minore latitudine paulo longiore, 4° quadrato, 5°–8um moniliformibus, 9° et 10° latioribus, transversis, 11° fere globoso; thorace capite vix latiore, latitudine paulo longiore, coriaceo, crebre fortiter punctato, antice posticeque truncato, basin versus vix angustato, lateribus fere rectis, marginatis, dorso utrinque carina longitudinali; elytris thorace paululo latioribus et fere triplo longioribus, convexiusculis, sat fortiter striatis, humeris obtusis, lateribus vix arcuatis (fere rectis) ad apicem arcuatim angustatis, striis parce obsolete punctatis interstitiis parce punctatis.

Long. $\frac{9}{10}$ lin., lat. $\frac{1}{4}$ in.

This very distinct species has somewhat the appearance of a minute *Passandra*. The punctures on the head and thorax are very strong, those on the head being somewhat confluent, those on the thorax moderately close together but leaving a smooth sutural line. The sides of the thorax are nearly straight, very slightly turned at the angles which are a little obtuse; on each side of the disk near the sides there is a distinct ridge. The interstices of the elytra are slightly coriaceous, each with a row of moderately distinct punctures, those on the sides irregular. The three-jointed club to the antennæ is unusually distinct in this species.

One specimen.

Dermestidæ.

Dermestes vulpinus, Fabr.

Many examples.

Trogoderma spe.—?

Oblong-ovate, shining, brown, clothed with ashy pubescence (which would probably form a pattern, but not bands, on the elytra in a fresh example). Antennæ testaceous, with the basal joint large, a little longer than broad, the second joint globular, third a little elongate and somewhat slender, the fourth to eighth transverse, gradually becoming a little wider, the ninth and tenth joints larger, transverse, the eleventh as long as the two previous joints together, flat, rounded at the apex. Flanks of the thorax beneath excavated to receive the club of the antennæ. Thorax finely and very closely punctured. Elytra a little more distinctly, but rather less closely punctured, a trifle longer than together broad, very slightly narrowed towards the apex.

A single specimen apparently belonging to this genus, but not in sufficiently good condition for me to describe. The above will, however, help to mark the species.

PALPICORNIA.

Hydrophilidæ.

Berosus mixtus.

C. Waterh., Ann. & Mag. Nat. Hist., 1876, xviii., p. 114.

B. statura fere *B. affinis*; oblongus, convexus, sordide flavo-testaceus, vix nitidus, subtus niger; capite thoraceque sat crebre distincte punctatis, clypeo subtilissime punctulato; elytris fortiter striatis, striis crebre punctatis, interstitiis planis sat crebre distincte punctatis, apice truncato, angulo externo breviter unidentato.

Long. $2\frac{1}{2}$ lin., lat. $1\frac{1}{4}$ lin.

Form of *B. affinis* but rather broader behind. Head distinctly and rather closely punctured; clypeus very finely and delicately punctured, a little more distinctly at the sides. Thorax relatively broader than in *B. affinis* and less convex, less deflexed at the sides, the anterior angles much rounded; punctuation very distinct and moderately close. Scutellum with a few fine punctures. Elytra with less prominent shoulders than in *B. affinis*, the striæ are not impressed at the base, but are very deep towards the apex, the interstices are flat (very narrow at the apex), less strongly punctured than in *affinis*, but not more thickly; the extreme apex of each elytron is truncate, and the external angle is marked by a short sharp tooth.

The elytra present six round obscure blackish spots placed as in *B. affinis*.

Half-a-dozen specimens were taken.

Philhydrus melanocephalus, Oliv.

A series of specimens of *Philhydrus* taken by Mr. Gulliver appear to belong to this species. I would note, however, that the average size is rather smaller than that of British examples (4 mill., the largest $4\frac{1}{2}$ mill.), and the punctuation appears a trifle less distinct. Specimens from the Canaries named "*politus*, Küst" by Mr. Wollaston differ from those from Rodriguez in being a little less punctured.

Those from the Cape Verde Group, named "*hesperidum*, Sh." differ in being less bluntly round at the apex of the elytra, and the punctuation is also a little less distinct than in the examples from Rodriguez.

Dactylosternum abdominale, Fab.

Seven examples of this species were obtained.

LAMELLICORNIA.

Aphodiidæ.

Rhyssemus tarsalis.

C. Waterh., Ann. & Mag. Nat. Hist., 1876, xviii., p. 115.

R. fusco-niger, nitidus, fronte granosa, vertice subtilius granuloso; thorace transversim quadricarinato; elytris punctato-striatis, interstitiis biseriatim granulatis.

Long. 1½ lin., lat. ¾ lin.

Extremely close to *R. germanus*, and only differs in being more shining, in having the projection in front of the eye nearly rectangular (scarcely obtuse), the granulation of the elytra a trifle less fine; the basal joint of the posterior tarsi is as long as the spur, whereas in *R. germanus* it appears to be always a little shorter.

Two examples.

Melolonthidæ.

Lachnosterna gradaria.

C. Waterh., Ann. & Mag. Nat. Hist., 1876, xviii., p. 115.

L. oblonga, convexa, brunnea, sat nitida; capite sat magno, collo lævi, fronte planiuscula crebre distincte punctata; clypeo confertim fortius punctato, marginato, medio vix sinuato; thorace longitudine fere duplo latiore, convexo, minus crebre punctato, ante medium paulo angustato, margine antico fere recto, angulis anticis obtusiusculis, posticis obtusis; scutello lævi; elytris basi thoracis latitudine postice paulo ampliatis, convexis, ad apicem rotundatis, haud crebre punctatis, marginibus incrassatis piceis; pectore longe flavo-pubescente; abdomine amplo, parce punctato; pygidio sat crebre fortiter punctato.

Long. 9½ lin., lat. 4½ lin.

Eight examples.

Lachnosterna rodriguezi (Pl. LIII., fig. 6).

C. Waterh., Ann. & Mag. Nat. Hist., 1876, xviii., p. 115.

L. oblonga, leviter convexa, nitida, pallide brunnea, sat lata; capite lato, sat crebre fortiter punctato; clypeo brevi, fortiter transverso, reflexo-marginato, crebre punctato, medio paululo producto, utrinque leviter sinuato; oculis sat magnis; thorace longitudine duplo latiore, leviter convexo, sat crebre distincte punctato, margine antico leviter flexuoso, angulis anticis obtusiusculis, lateribus arcuatis, angulis obtusis, basi utrinque sinuata, medio parum lobata; scutello lævi; elytris thoracis latitudine at $3\frac{1}{4}$ longioribus, post medium paululo ampliatis dorso depressiusculis, ad apicem rotundatis, minus crebre punctulatis, sutura parum elevata.

Long. $10\frac{1}{2}$–$12\frac{1}{2}$ lin., lat. 5–$6\frac{1}{4}$ lin.

Two dead specimens (possibly bleached) found in a spider's web.

Dynastidæ.

Oryctes minor.

C. Waterh., Ann. & Mag. Nat. Hist., 1876, xviii., p. 115.

O. oblongus, niger, nitidus; capite antice angustato, rugoso, medio vix nodoso; thorace longitudine $\frac{1}{3}$ latiore, nitido, parce subtiliter punctulato, antice medio impressione rotundata rugosa et utrinque plaga parva rugosa notato, lateribus leviter rotundatis; elytris thorace haud latioribus, postice ampliatis, fortiter lineato-punctatis; pedibus rufo-piceis.

Long. $11\frac{1}{2}$ lin., lat. $5\frac{1}{2}$ lin.

The elytra are covered with rather large horseshoe punctures, among which may be traced the usual two pairs of punctured lines; the surface between the large punctures has small punctures scattered here and there.

A single specimen found by Mr. Gulliver in a spider's web.

SERRICORNIA.

Buprestidæ.

Sponsor desjardinsii, Guérin.

"Lives in holes in dead wood."—G. G. Two specimens (male and female) taken by Mr. Gulliver. The head and thorax in both specimens are bright green, the elytra in the female are bright purple, those of the male are more golden, reflecting purple in some lights. The male measures 6 mill., the female 8 mill.

These appear to be the only specimens of this genus which have as yet reached this country.

Elateridæ.

Adelocera modesta, Boisd.

Nine specimens. This species has now been found in most of the tropical parts of the world.

MALACODERMATA.

Melyridæ.

Malthacodes, C. Waterh.

Ann. & Mag. Nat. Hist., 1876, xviii., p. 116.

Maxillæ with two lobes, membranous, the internal smaller and narrower than the external; apical joint of the maxillary palpi strongly securiform; mandibles bifid at the apex. Antennæ with the first joint elongate, the second smaller and shorter, the third a little longer than the second but not so long as the first; the fourth to tenth about as broad as long, narrowed at their base; the eleventh oblong. Eyes prominent. Thorax transverse, broadest at the base. Elytra scarcely broader than the thorax and twice and a half as long. Tarsi with the basal joint a little elongate, second to fourth joints subequal, shorter than the first; claws with a membranous lobe beneath each. Body pubescent.

The species upon which this genus is founded resembles *Haplocnemis*, but is of a shorter form, the head is short, and the apical joint of the maxillary palpi is very strongly securiform. *Pelecophorus* is described as having the basal joint of the tarsi very short, shorter than the second; this I cannot apply to the insect here described, or I should have placed it in that genus.

Malthacodes pictus (Pl. LIII., fig. 7).

C. Waterh., Ann. & Mag. Nat. Hist., 1876, xviii., p. 116.

M. oblongus, leviter convexus, nigro-æneus, griseo-pubescens; antennis nigris, articulo basali piceo; thorace longitudine duplo latiore, sat crebre subtiliter punctato, lateribus arcuatis, tenuissime flavo marginatis, basi utrinque leviter sinuato; elytris thorace vix latioribus at $2\frac{1}{2}$ longioribus, crebre sat fortiter punctatis, fasciis duabus flexuosis rufo-testaceis, femoribus piceis, tibiis tarsisque pallide testaceis.

Long. $1\frac{3}{4}$ lin., lat. $\frac{3}{4}$ lin.

A single example only was found of this interesting new genus.

Cleridæ.

Tarsostenus univittatus, Spin.

The single specimen found does not differ from the European examples.

Ptinidæ.

Xylodes, C. Waterh.

Ann. & Mag. Nat. Hist., 1876, xviii., p. 116.

General form nearly that of *Hedobia*. Antennæ rather thick, not approximate at the base; the basal joint oblong, the second shorter and smaller, a little narrowed at its base, the third nearly as long as the first, parallel-sided, the fourth to tenth nearly quadrate, scarcely longer than broad, eleventh nearly twice as long as the preceding, rounded at the apex. Thorax arched in front, as long as broad, slightly narrowed in front and behind. Scutellum pentagonal. Elytra nearly twice as broad as the thorax, and twice and a half as long, parallel, rounded at the apex, punctate-striate. Legs rather short and stout; tarsi rather short and stout, the basal joint scarcely elongate, the second to fourth a little shorter, the fifth elongate and slender. Body velvety.

Xylodes albovaria (Pl. LIII., fig. 8).

C. Waterh., Ann. & Mag. Nat. Hist., 1876, xviii., p. 117.

X. niger, velutinus; thorace supra albo, medio nigro annulato, marginibus nigris; elytris basi fasciaque post medium dentata albis, humeris nigris.

Long. $1\frac{1}{4}$–2 lin., lat. $\frac{2}{3}$–$\frac{4}{5}$ lin.

Six specimens of the pretty new species were found; they vary much in size.

Cioidæ.

Lyctus rugicollis, Walker.

Ditoma rugicollis, Walker, Ann. & Mag. Nat. Hist., 1858, ii., p. 206.
Lyctus rugicollis, C. Waterh., l. c. 1876, xviii., p. 117.

This species, which was originally described from Ceylon by Mr. Walker as a *Ditoma*, is closely allied to *L. obsitus*, Woll. A single specimen was found.

Lyctus brunneus, Steph.

Three examples.

Cis insularis.

C. Waterh., Ann. & Mag. Nat. Hist., 1876, xviii., p. 117.

C. oblongus, convexus, piceo-niger; capiter leviter convexo, crebre sat fortiter punctato; labrum palpisque testaceis; thorace longitudine paululo latiore, convexo, antice paulo angustato, crebre sat fortiter punctato, margine antico leviter arcuato, supra caput vix superante, angulis anticis omnino rotundatis, lateribus leviter arcuatis, reflexo-marginatis, angulis posticis obtuse rotundatis basi marginata; elytris thorace vix latioribus, at $2\frac{1}{4}$ longioribus, convexis, ad apicem arcuatim attenuatis, sat crebre fortiter punctatis; antennis pedibusque piceis, tarcis testaceis.

Long. $1\frac{1}{6}$ lin., lat. $\frac{1}{2}$ lin.

This species has the elytra rather unusually narrowed at the apex; the punctuation is very distinct, thick but not crowded; in the middle of the forehead there is a small, almost imperceptible fovea; the elytra are covered with a slightly metallic bloom.

A single specimen only.

Cis sexcarinatus.

C. Waterh., Ann & Mag. Nat. Hist., 1876, xviii., p. 118.

C. oblongus, ater, vix nitidus, fortiter confertim rugoso-punctatus; capite piceo; antennis testaceis, clava picea; thorace longitudine $\frac{1}{4}$ latiore, lateribus fere parallelis (vix arcuatis), angulis obtusis, basi bisinuata; scutello lævi; elytris basi thorace haud latioribus at duplo longioribus, lateribus fere rectis ad apicem obtuse rotundatis, sutura parum elevata, singulo elytro ad apicem carinis tribus instructo; corpore subtus haud crebre fortiter punctato; pedibus piceis.

Long. 1½ mill., lat. ¾ mill.

Thirteen specimens found in fungus.

HETEROMERA.

Opatrinæ.

Opatrum cæsum, Stev.

Eight examples, which agree with the description of this species.

Cistelidæ.

Cistela brunnea (Pl. LIII., fig. 9).

C. Waterh., Ann. & Mag. Nat. Hist., 1876, xviii., p. 118.

C. oblongo-ovalis, parum convexa, sat nitida, brunnea, breviter aureo-pubescens; capite triangulari, crebre distincte punctato; antennis thorace duplo longioribus; thorace obscuriore, longitudine fere duplo latiore, antice arcuatim parum angustato, confertim fortiter punctato, angulis anticis omnino rotundatis, posticis rectis, basi recto-truncata, medio solum vix lobata; elytris thorace vix latioribus at 3½ longioribus, ad apicem arcuatim attenuatis, irregulariter crebre punctatis.

Long. 4 lin., lat. 1¼ lin.

The elytra are punctate-striate at the apex, the stria next the suture extends for a little distance towards the scutellum.

This species, of which four specimens were brought, may eventually have to be separated from *Cistela* proper.

RHYNCHOPHORA.

Anthribidæ.

Balanodes, C. Waterh.

Ann. & Mag. Nat. Hist., 1876, xviii., p. 118.

Head as long as broad; rostrum a trifle narrower, very short. Antennæ placed close to the eye, a little longer than the head and thorax together, very slender; the first and second joints scarcely stouter than the following; third to seventh a trifle longer than the second, subequal, the eighth rather shorter but a little elongate; the ninth to eleventh subequal, forming an elongate, slender, loose club. Eyes a little prominent, slightly ovate. Thorax nearly semicircular, truncate in front for the width of the head; the basal ridge very slight, scarcely separated from the posterior margin, only visible near the shoulders. Scutellum minute. Elytra at the base scarcely as broad as the thorax, but a little more than twice as long, gradually narrowed towards the apex, which is rounded, punctate-striate. Tibiæ cylindrical, not widened at the apex; the anterior pair rather long and curved beyond the middle, minutely denticulate beneath the base. Tarsi moderately broad, the basal joint a little elongate, the second transverse, the third short, bilobed; claws with a distinct mesial tooth. Body evenly convex, thickly clothed with pubescence.

Closely allied to *Aræocorynus*, but, besides the difference of form it differs in having slightly narrower tarsi and the tibiæ are not widened at the apex.

Balanodes tomentosus (Pl. LIII., fig. 10).

Ann. & Mag. Nat. Hist., 1876, xviii., p. 119.

B. ovalis, convexus, dense breviter flavo-griseo-pubescens, obscure piceus, elytris pedibusque brunneis; thorace creberrime haud fortiter punctato, angulis anticis obliteratis, posticis rectis, basi utrinque vix sinuata; elytris fortiter striato-punctatis, interstitiis alternatis vix latioribus, subtilissime coriaceis; antennis articulis tribus apicalibus piceis.

Long. $2\frac{3}{4}$ lin., lat. $1\frac{1}{2}$ lin.

The pubescence on the elytra is slightly interrupted by the rows of punctures, which gives them a striped appearance.

Another of the interesting new genera, of which only a single example was found.

Caranistes annulipes.

Ann. & Mag. Nat. Hist., 1876, xviii., p. 119.

C. oblongus, nigro-piceus, dense piceo-tomentosus, flavo variegatus; pedibus pallide brunneis nigro annulatis.

Long. $2\frac{1}{4}$–3 lin., lat. $1\frac{1}{6}$–$1\frac{1}{2}$ lin.

Front of the head and the rostrum clothed with yellowish pubescence, that on the vertex brown; eyes rather widely separated above, very slightly emarginate in front. Antennæ obscure testaceous, the club only dark, twice as long as the thorax; the third to eighth joints very slender and gradually diminishing in length; the ninth to eleventh elongate, forming a distinct but slight club. Thorax very finely granular, nearly one-third broader than long, much narrowed in front, the sides nearly straight, rather narrowed behind the basal ridge, which is very distinct, well separated from the base (especially at the sides); the pubescence is brownish. Scutellum yellowish. Elytra as broad as the thorax, one-fourth longer than broad, not very convex, rounded at the apex, rather strongly punctate-striate, the interstices very slightly convex; the pubescence is yellowish, interrupted by small brown square spots; a few of these spots are larger and darker than the others; and the yellow pubescence has a tendency to form a sutural spot behind the middle. The legs are very pale brown; a ring on the femur, two on the tibia, the apical half of the basal tarsal joint, and nearly all the second joint blackish. Some large punctures are scattered over the flanks of the prosternum and the metasternum.

Three examples, found under bark by Mr. Gulliver.

Curculionidæ.

Cratopus inornatus.

Ann. & Mag. Nat. Hist., 1876, xviii., p. 120.

C. elongatus, niger (vel nigro-piceus), nitidus, setis albidis brevissimis parce adspersus; antennis piceis; rostro basi sat crebre, apice parcius lævius punctulato, fronte discrete distincte punctata; oculis oblongis, haud prominentibus; thorace longitudine $\frac{1}{3}$ latiore, antice angustato, vix crebre sat fortiter punctato; scutello lævi; elytris thorace paulo latioribus at $3\frac{1}{2}$ longioribus, ad apicem regulariter acuminatis, fortiter striato-punctatis, interstitiis vix irregularibus, punctis parvis adspersis, apice tuberculis parvis obsitis, marginibus dimidio basili distincte carinatis, ad apicem serrulatis; pectore breviter griseo-pubescente; femoribus anticis subtus dente parvo armatis.

Long. 5–$6\frac{1}{2}$ lin., lat. 2–$2\frac{1}{2}$ lin.

Seven specimens.

Cratopus virescens.

Ann. & Mag. Nat. Hist., 1876, xviii., p. 120.

C. elongatus, piceus, dense virescenti-pubescens; capite rostroque crebre sat fortiter punctato; oculis vix prominulis; thorace longitudine vix latiore, antice angustato, crebre irregulariter granulato, lateribus arcuatis; scutello viridi-albo tecto; elytris thorace paulo latioribus at $3\frac{1}{4}$ longioribus, apice angustatis, sat fortiter striato-punctatis (punctis rotundatis, medio tuberculo minuto instructis), interstitiis haud convexis, parce subtilissime asperato-punctatis, marginibus haud carinatis; femoribus anticis subtus dente magno armatis.

Long. 5–6 lin., lat. 2–$2\frac{2}{5}$ lin.

Nine specimens.

Cratopus magnificus.

Ann. & Mag. Nat. Hist., 1876, xviii., p. 120.

C. elytris piceo-niger, dense squamulis rotundatis parvis tectis; squamulis læte viridi vel aurato vel cupreo nitentibus; sutura lævi.

Long. elytr. $5\frac{1}{2}$ lin.

Only two elytra of this most beautiful insect were found by Mr. Gulliver in a spider's web. The colours are most brilliant, reflecting bright copper, golden, emerald green, or pale bluish-green in stripes, according to the position in which the insect is held.

Mr. Gulliver has this note on the species: "I was told that this beetle occurred very rarely, but is found occasionally concealed in the sheaths of the leaves of the maize plant."

Sitophilus linearis, Hbst.

Five examples.

Sitophilus oryzæ, Linn.

One specimen.

Phlæophagosoma proximum, Woll.

Five specimens, which agreed perfectly with the typical specimen in Mr. Pascoe's collection from Makian.

Pentarthrum rodriguezi.

Ann. & Mag. Nat. Hist., 1876, xviii., p. 120.

P. nigro-piceum (vel. rufo-piceum), convexum, subcylindricum; capite pone oculos lævi, fronte rostroque sat crebre distincte punctatis; thorace longitudine haud longiore, sat crebre distincte punctato, lateribus bene rotundatis; elytris thorace paululo latioribus, at $2\frac{1}{4}$ longioribus, fere parallelis, fortiter striato-punctatis, interstitiis vix convexiusculis uniseriatim parce punctulatis.

Long. $1\frac{3}{4}$ lin. (rostr. incl.).

The second joint of the funiculus is a little longer than the following. The rostrum is about two-thirds the length of the thorax, at the apex very finely punctured. The thorax is a little more narrowed in front than behind.

Five specimens.

Cossonus marginalis, Schonh.

A single specimen, which agrees well with the description of this species.

XYLOPHAGA.

Scolytidæ.

Tomicus ferrugineus, Fabr.

Three specimens, which I think are without doubt to be referred to this species. They also agree well with *Bostrichus testaceus* of Walker from Ceylon.

LONGICORNIA.

Prionidæ.

Macrotoma simplex.

Ann. & Mag. Nat. Hist., 1876, xviii., p. 121.

♀. *M.* fusca, vix nitida; capite piceo-nigro, rugoso-punctato; thorace longitudine duplo latiore, antice paulo angustato, margine antico flexuoso, angulis anticis obtusiusculis, lateribus haud spinosis (medio solum irregulariter denticulato) ante angulos posticos acutos sinuatis, basi utrinque sinuata, disco crebre fortiter punctato (medio plaga parva lævi), lateribus fortiter rugosis; scutello parce subtiliter punctulato; elytris thorace vix latioribus at quintuplo longioribus, basi fortiter punctatis et rugulosis, marginibus plagaque discoidali pallidioribus ruguloso-coriaceis; pectore longe flavo-pubescente.

Long. 18 lin., lat. $6\frac{1}{2}$ lin.

Two specimens, with a portion of a third.

Cerambycidæ.

Xystrocera globosa, Oliv.

The remains of several specimens of this species were found.

Stromatium barbatum, Fabr.

Three specimens of this widely-distributed species were found.

Phoracantha semipunctata, Fabr.

A single specimen of this strictly Australian species was brought; doubtlessly introduced.

Lamiidæ.

Coptops bidens, Fabr.

Six specimens.

Micracantha dentata, Oliv.

Two specimens.

PHYTOPHAGA.

Hispidæ.

Cryptonychus limbatus (Pl. LIII., fig. 11).

Ann. & Mag. Nat. Hist., 1876, xviii., p. 121.

C. elongatus, nitidus, depressus, flavo-testaceus; antennis, thoracis elytrorumque disco, pectoreque piceo-nigris; capite sat parvo fere lævi, medio canaliculato, inter antennas breviter anguste producto; oculis nigris; thorace latitudine haud longiore, antice paululo angustato, post angulos anticos obtusos paulo constricto, margine antico arcuato, angulis posticis vix acutis; elytris basi thorace vix latioribus post medium paulo ampliatis, sat fortiter striato-punctatis, interstitiis subtilissime uniseriatim punctulatis, apice declivi, truncato, utrinque carinato.

Long. $3\frac{1}{2}$ lin., lat. $\frac{5}{6}$ lin.

Five specimens.

PSEUDOTRIMERA.

Erotylidæ.

Euxestus parkii, Woll.

Four examples. I have most carefully compared two of them with the type specimen from Madeira, and I am quite unable to separate them. The two other examples are black with the head pitchy, but I cannot see any other difference except that the punctuation is seen with a little less difficulty.

Endomychidæ.

Trochoidens desjardinsii, Guérin.

Seventeen specimens found in wood.

EXPLANATION OF PLATE LIII.

Fig. 1. Probænus longicornis.
 „ 2. Micraporum nitens.
 „ 3. Ascomma horrida.
 „ 4. Endocoxelus variegatus.
 „ 5. Æschyntelus ater.
 „ 6. Lachnosterna rodriguezi.
 „ 7. Malthacodes pictus.
 „ 8. Xylodes albovaria.
 „ 9. Cistela brunnea.
 „ 10. Balanodes tomentosus.
 „ 11. Cryptonychus limbatus.

HYMENOPTERA, DIPTERA, AND NEUROPTERA.—*By Frederick Smith.*

Fifteen species of *Hymenoptera* were collected by Messrs. G. Gulliver and H. H. Slater, twelve of which belong to the *Aculeata*, the remainder to the *Terebrantia* or Ichneumonidous division; five of the first of these divisions appear to be undescribed. Six species of *Formicidæ* were obtained, four being new to science: the other two have a cosmopolitan range; one is the species described by Heer as the House Ant of Madeira, *Pheidole pusilla*; the other is the *Prenolepis gracillescens* of Nylander, also found in Madeira, but common in conservatories generally; it is found abundantly in the Botanical Gardens both at Kew and Paris. The most interesting species of *Apidæ* is the *Apis unicolor* of Latreille, first discovered in Madagascar, and subsequently in the island of Bourbon. Two species of leaf-cutting Bees were found: both have a wide geographical range, as has also the only species of Wasp, *Polistes hebræus*; this has occurred in India, China, Mauritius, Madagascar, and in several of the islands of the Eastern Archipelago.

Six species of *Diptera* were obtained, two belonging to the *Tipulidæ*; these latter were in a fragmentary condition, and could neither be identified nor described; of the remaining four one is a described species, the others being apparently new.

Of *Neuroptera* three species were found, all of which are described by Rambur from specimens obtained in Mauritius.

HYMENOPTERA.

Name of Species found in Rodriguez.	Hitherto found in Rodriguez only.	Previously obtained elsewhere.		
		Mauritius and Bourbon.	Madagascar.	Other Localities.
Tapinoma pallipes, *Smith*	×			
,, fragile, *Smith*	×			
Monomorium elongatum, *Smith*	×			
,, impressum, *Smith*	×			
Prenolepis gracillescens, *Nyland.*				Cosmopolitan.
Pheidole pusilla, *Heer*				Cosmopolitan.
Scolia (Dielis) rufa, *St. Farg.*		×		
Pelopæus convexus, *Smith*	×			
Megachile rufiventris, *Guér.*		×		India.
,, lanata, *Fabr.*		×		India.
Apis unicolor, *Latr.*		×	×	
Polistes hebræus, *Fabr.*		×	×	India, Java, China.
Evania lævigata, *Latr.*				Australia, Angola, Egypt, Europe, Mexico.
Paniscus perforator, *Smith*	×			
Ophion rufus, *Brullé*		×		Bengal, Java, Africa.

Formicidæ.

Tapinoma, Foerster.

Tapinoma pallipes.

Ann. & Mag. Nat. Hist., 4th ser., vol. xvii., p. 447 (1876).

Worker. Length 1½ line. Head and thorax reddish brown, abdomen dark brown; eyes black; mandibles and antennæ pale testaceous. Thorax rounded anteriorly, slightly narrowed posteriorly, with the metathorax obliquely truncate; the legs, petiole, and squama, pale testaceous, the latter rounded above. Abdomen smooth, shining, and impunctate.

Coll. by Gulliver. 20 specimens taken.

Tapinoma fragile.

Ann. & Mag. Nat. Hist., 4th ser., vol. xvii., p. 447 (1876).

Worker. Length 1 line. Head brown, smooth, and shining; the eyes black; mandibles white and pellucid, antennæ also white. Thorax pale brown, the sides and metathorax darker brown; the prothorax rounded in front; the metathorax oblique and concave; the legs, petiole, and squama, pale testaceous, nearly white. Abdomen brown, smooth, and shining, palest at the base.

Coll. by Gulliver. Three specimens taken.

Found under bark; very like the preceding species, but smaller; the legs and antennæ being pellucid.

Myrmicidæ, *Smith*.

Monomorium, Mayr.

Monomorium impressum.

Ann. & Mag. Nat. Hist., 4th ser., vol. xvii., 447 (1876).

Female. Length 1¾ line. Head and abdomen dark fusco-ferruginous. The clypeus, mandibles, and antennæ flavo-testaceous; ocelli distinct, with an impressed line in front of the anterior one, extending to the insertion of the antennæ; the apex of the scape and the club of the flagellum slightly rufo-fuscous. Thorax narrower than the head and one-third longer; the prothorax rounded in front, the sides nearly parallel, slightly narrowed towards the metathorax; the legs pale testaceous, the femora rufo-testaceous; the tarsi colourless and pellucid; the thorax testaceous, as well as the nodes of the abdomen. Abdomen oblong-ovate, as long as the head and thorax.

Male. Length 1¼ line. Dark fusco-ferruginous; mandibles and antennæ rufo-testaceous; ocelli prominent, with a deeply impressed line in front of the anterior one. Thorax oblong, narrowed posteriorly; the wings colourless hyaline;

legs white, with the femora slightly fuscous in the middle. Abdomen oblong-ovate, smooth, and shining.

Nine specimens found under bark. Coll. by Gulliver.

Monomorium elongatum.

Ann. & Mag. Nat. Hist., 4th ser., vol. vii., p. 448 (1876).

Female. Length 2 lines. Reddish brown; head oblong, slightly narrowed behind the eyes, truncate posteriorly; ocelli glassy, the anterior one placed in a deep fossulet; mandibles and antennæ pale rufo-testaceous. Thorax oblong, narrowed behind; the metathorax truncate; legs pale rufo-testaceous, the femora slightly fuscous. Abdomen oblong-ovate, smooth, and shining, with a thin scattered pubescence.

Worker. Length 1 line. Reddish brown, shining, with the mandibles, antennæ, and legs pale testaceous; the head oblong, the sides nearly parallel, truncate behind, with the angles rounded. Thorax strangulated a little beyond the middle; the metathorax without spines. Abdomen ovate, smooth, and shining.

Fifty-three specimens collected by Gulliver.

SCOLIIDÆ.

Scolia, *Fabr.*

Scolia rufa, St. Fargeau.

Male. Length 7½ lines. Black, the pubescence fulvo-ferruginous; the face densely pubescent; the clypeus and mandibles yellow, the former with a large triangular black macula in the middle, the latter rufo-piceous at their apex and narrowly so on their lower margin; the head pubescent behind. Thorax pubescent, the metathorax densely so; wings fusco-hyaline, the nervures black, slightly iridescent; the anterior legs, with the tips of the femora above, also the tibiæ and tarsi above, yellow; the tips of the joints of the tarsi black; the intermediate and posterior tibiæ yellow above, their tarsi entirely so, with the tips of the joints black; all the calcaria pale testaceous. Abdomen with broad yellow fasciæ, the fasciæ emarginate in the middle and at the sides; the apical segment black, smooth, and shining, with a few strong punctures at the base, the rest of the abdomen with shallow punctures; beneath, the segments have narrow yellow apical marginal fasciæ, and are fringed with fulvous pubescence.

Two specimens. The female only of this species was known, until Mr. Gulliver took both sexes.

Sphegidæ.

Pelopæus, Latreille.

Pelopæus convexus.

Ann. & Mag. Nat. Hist., 4th ser., vol. vii., p. 449 (1876).

Male. Length 5–6¼ lines. Head and thorax blue; abdomen and legs blue black. The face with silvery pubescence; on the cheeks a thin griseous pubescence. The mesothorax evenly punctured and convex, with a central impressed longitudinal line, without any other trace of longitudinal channels; rather strongly but not very closely punctured, on the disk a few transverse irregular carinæ; wings subhyaline, with their apical margins clouded; the thorax has a thin sparing griseous pubescence; the apex of the metathorax with bright silvery pubescence; the tibiæ with the calcaria black.

(Coll. by Slater.)

Three specimens obtained, which differ so much from the males of all the blue species of the genus, in having the thorax very convex, that the male is described as that of a new species: the capture of the other sex will probably prove the correctness of doing so.

Apidæ.

Megachile rufiventris, Guér.

A single specimen taken by Mr. Slater.

Megachile lanata, Fabr.

A single specimen taken by Mr. Gulliver.

Apis unicolor, Latr.

Thirty-two specimens taken by Mr. Gulliver.

Vespidæ.

Polistes hebræus, Linn.

Seven examples taken by Mr. Gulliver.

Ophionidæ.

Ophion rufus, Brullé.

Two specimens taken by Mr. Gulliver.

Paniscus perforator, Smith, Ann. & Mag. Nat. Hist. 4th ser., vol. vii., p. 449.

Female. Length 6 lines. Rufo-ferruginous; the face and scape of the antennæ in front, yellow; the space between the ocelli black. Thorax: the sides, the sutures on the mesothorax, and the scutellum paler than the disk; wings hyaline and iridescent, the stigma and nervures rufous; the legs and abdomen rufous, the latter slightly fuscous at the apex.

A single example taken by Mr. Slater. This species is like *P. melanopus*, Brullé, but it differs in having all the tarsi rufous, as well as the nervures of the wings: in *P. melanopus* they are black.

Evaniidæ.

Evania lævigata, Latr.

A single specimen taken by Mr. Gulliver; the genus Evania being parasitic on Blattidæ, some of the species are conveyed with merchandise to all parts of the world. *E. lævigata* is one of the species that has a wide distribution, and is found occasionally in warehouses in the docks in London.

DIPTERA.

Tabanidæ.

Tabanus sequens, Walker.

A single specimen taken by Mr. Gulliver.

Muscidæ.

Pollenia basalis, Macquar.

Smith, Ann. & Mag. Nat. Hist., 4th ser., p. 449 (1876).

Length 4 lines. Dark blue, with shades of green on the disk of the thorax; the face and cheeks luteous, and clothed with short, dense yellow pubescence, at the end of the facialia a black bristle on each side; antennæ in deep facial grooves, slightly yellow at the base, third joint long, arista longly plumose; palpi luteous, with black bristles; frons black and more than a quarter the width of the head; wings hyaline with their base fuscous; the transverse vein at the end of the wing nearly rectangular below and much curved above, leaving the cell open; the lower transverse vein very oblique. Abdomen with a little short luteous pubescence at the apex and also beneath; alulæ dirty white.

A single example taken by Mr. Slater.

Sarcophaga mutata.

Smith, Ann. and Mag. Nat. Hist., 4th ser., p. 450 (1876).

Female. Length 5 lines. Black, thinly clothed with black hairs and bristles; the face and cheeks with a covering of bright silvery pile; the margin of the facial groove fringed anteriorly with short black bristles; eyes dark reddish brown; the base of the antennæ black, the third joint tawny and with a long plumose black bristle at the base; frons with a broad black stripe running from the antennæ to the vertex. Thorax tawny above and with three black longitudinal stripes; the lateral margins set with black curved bristles; legs black and with a number of black bristles; wings sub-hyaline, the veins black; the tranverse vein at the end of the wing straight above, oblique and curved below; the lower transverse vein evenly waved; the alulæ white. Abdomen tessellated with black and silvery pile, covered with short black hairs, and having a few long black bristles at the apex; there are also four long black bristles on the hinder margin of the scutellum.

Two examples taken by Mr. Gulliver.

Sapromyza squallida.

Smith, Ann. and Mag. Nat. Hist., 4th ser., vol. vii., p. 950 (1876).

Male. Length 2 lines. Pale brown; frons yellow, with a dark brown central line; two long bristles at the hind corner of the eye, and two between them and the antennæ; face pale yellow; peristoma with a row of small bristles. Antennæ yellowish, brown above, the second joint with long black bristles beneath, at the end; arista black, yellow at the base and plumose; the palpi yellow. Thorax with obscure pale (almost white) longitudinal stripes; the scutellum pale brown, its outer margin and a central longitudinal pale yellow line. Legs yellow; the tibiæ have all a preapical bristle; the anterior femora with scattered bristles, the intermediate pair with a row of bristles in front near the tip. Wings faintly yellowish, middle transverse veinlets darkened, and yellowish spots at the tip of the lower transverse veinlet and at the tips of the second and third veins, and also on those veins, some distance from the tips, and on the second vein opposite where the first vein ends in the costa. Abdomen brownish yellow and thinly sprinkled with short black hairs.

A single example taken by Mr. Gulliver.

NEUROPTERA.

Name of Species found in Rodriguez.	Hitherto found in Rodriguez only.	Previously obtained elsewhere.		
		Mauritius and Bourbon.	Madagascar.	Other Localities.
Termes ?	×			
Myrmelion obscurus		×		
Libellula mauritiana		×		
Anax mauritianus		×		
Agrion ferrugineum			×	

Libellulidæ.

Libellula mauritiana, Rambur.

Four specimens obtained by Mr. Gulliver.

Anax mauritianus, Rambur.

Four specimens taken by Mr. Gulliver.

Agrionidæ.

Agrion ferrugineum, Rambur.

A single example only obtained by Mr. Gulliver.

Termes, sp.

One hundred and seven specimens of a small species of Termite were in the collection, they were snow white, but only workers were obtained; in the absence of the females and males it was impossible to draw up distinctive specific characters. Obtained by Mr. Gulliver.

Myrmelion obscurus, Rambur.

A single specimen only obtained by Mr. Gulliver.

LEPIDOPTERA.—*By Arthur Gardiner Butler, F.L.S., F.Z.S., &c.*

Twenty-one species of Lepidoptera were obtained by Mr. George Gulliver; of these, seven are Rhopalocera and fourteen Heterocera; of the latter, four are new and probably endemic forms. Twelve of the species, including the whole of the Butterflies, have been previously known from Mauritius and Bourbon, and eight of them, including six of the Butterflies, from Madagascar. All the genera, with the exception of *Laverna*, and eight of the species, have been recorded from India or Ceylon.

The species here enumerated evidently form but a very small part of the Lepidopterous fauna of Rodriguez: in all probability many of the species which occur at Mauritius and Bourbon will be found to exist there also, of these *Terias floricola* and *Atella columbina* among the Rhopalocera, and *Chærocampa celerio, Daphnis nerii, Protoparce convolvuli, Deiopeia pulchella, Plusia chalcitis*, and *Heliothis armigera*, with several other wide-ranging species, amongst the Heterocera, may be confidently looked for.

It is interesting to find *Acherontia atropos* in Rodriguez; the Mascarene Islands probably form the extreme limit of this species, which (with *Protoparce convolvuli*) seems to be common throughout Europe and Africa, but does not occur in Asia, the Indian representatives of *A. atropos* being all perfectly distinct.

Name of Species found in Rodriguez.	Hitherto found in Rodriguez only.	Previously obtained elsewhere.		
		Mauritius and Bourbon.	Madagascar.	Other Localities.
Danais chrysippus, *Linnæus*	-	×	×	Africa, Asia, S. Europe.
Melanitis ismene var., *Cramer*	-	×	×	Africa, Asia.
Junonia rhadama, *Boisduval*	-	×	×	
Diadema misippus, *Linnæus*	-	×	×	Africa, Asia.
Lampides telicanus, *Herbst*	-	×	×	Africa, S. Europe.
Hesperia forestan, *Cramer*	-	×	×	Natal, E. Africa, Sierra Leone, India.
Pamphila borbonica, *Boisduval*		×		
Acherontia atropos, *Linnæus*	-	×		Europe and Africa.
Argina cribraria, *Clerck*	-	×		India, China, Malaysia.
Laphygma cycloides, *Guénée*	-			S. Africa.
Prodenia littoralis, *Boisduval*		×	×	India, Ceylon.
Caradrina expolita, *Butler*	×			
Cosmophila xanthindyma, *Boisduval*	-	×	×	Asia, ? Australia.
Diomea bryophiloides, *Butler*	×			
Homoptera turbida, *Butler*	×			
Achæa catilla, *Guénée*	-			Senegal, Abyssinia.
Ophiusa properans, *Walker*	-			S. Africa, Natal, W. Africa.
Trigonodes acutata, *Guénée*	-	×		S. Africa, Natal, Senegal, Sierra Leone.
Remigia conveniens, *Walker*	-			Natal, Congo, Sierra Leone.
Asopia vulgalis, *Guénée*	-			Africa, Ceylon, West Indies.
Laverna plumipes, *Butler*	×			

* Identical with *P. testaceoides*, Guénée.

Rhopalocera.

Danais chrysippus, L.

Five specimens, of both sexes, were collected.

Melanitis ismene var., Cramer.

Two specimens.

Junonia rhadama, Boisd.

Fifteen examples of both sexes.

Diadema misippus, L.

Only one male.

Lampides telicanus, Herbst.

One female.

Hesperia forestan, Cramer.

One male.

Pamphila borbonica, Boisd.

One male.

Heterocera.

Acherontia atropos, L.

Six examples of both sexes.

Argina cribraria, Clerck.

Five specimens, one of which is the variety figured by Cramer (pl. 208, fig. C), and another is the *P. astrea* of Drury.

Laphygma cycloides, Guénée.

Three males.

Prodenia littoralis, Boisd.

One male.

Caradrina expolita.

Ann. & Mag. Nat. Hist., s. 4, vol. xvii., p. 407 (1876).

Primaries glossy brown, reniform spot indistinct, a brown streak below it; a waved dark brown transverse line towards the base; a transverse postmedian line, arched, with its lower extremity straight, followed by a series of black points; an irregular dusky discal band, zigzag externally; a marginal series of black dots; the margin indicated by a clay-coloured line; fringe shining brown, darker at base: secondaries shining pale brown, with deeper-coloured outer border; marginal characters and fringe as in primaries: body pale olive-brown, abdomen greyish;

palpi black, with whity-brown tips: wings below shining brownish, a dusky spot on the discocellulars; an oblique black costal litura in primaries, followed by a continuous discal series of black dots upon the nervures; an ochraceous marginal line and a marginal series of black dots; secondaries paler than primaries: body shining whitish. Expanse of wings 1 inch.

(Coll. by Gulliver.)

Allied to *C. conformis.*

Cosmophila xanthindyma, Boisd.

Two specimens.

Diomea bryophiloides.

Ann. & Mag. Nat. Hist., s. 4, vol. xvii., p. 408 (1876).

Primaries rounded at apex, smoky brown; external area darker, crossed by a blackish tapering band bounded externally by whitish dots; several costal spots of the same colour: secondaries pale brown; margin, veins, and a spot on the discocellulars dusky: body smoky brown, abdomen pale; palpi whitish above, blackish below; primaries below paler than above, markings indistinct, a marginal series of black dots; secondaries sordid white; costal area irrorated with brown; basal area ochraceous; a black discocellular spot; two irregular discal brown lines; outer border grey; a marginal series of black dots; legs spotted with ochraceous; coxæ tufted: body below whitish varied with brown, sides of pectus ochraceous. Expanse of wings 11 lines.

(Coll. by Gulliver.)

One example.

Homoptera turbida.

Ann. & Mag. Nat. Hist., s. 4, vol. xvii., p. 408 (1876).

Primaries cinereous, crossed by undulating brown lines; a blackish line at the base, followed by a dark oblique brown band, the outer edge of which is straight, and bounded towards costa by a white streak; reniform spot ill-defined, confounded with a brown costal patch crossed by pale lines; the latter bounded externally by an abbreviated narrow white streak; postmedian line slender, black, sinuated at its upper end, regularly undulated from the first median to the inner margin; outer border broadly grey; a submarginal black line, a marginal series of yellowish spots; margin indicated by a reddish brown line; fringe brown, with basal ochraceous line: secondaries pale buff, internal area (excepting the abdominal region) and the outer border cinereous, crossed by parallel undulated brown bands; two parallel subanal black lines, between which is a brown line; marginal characters as in primaries: head, palpi, and collar pale reddish brown; tegulæ and thorax greyish brown; abdomen pale brown, three dorsal greyish tufts; two subanal dusky bands, the inner one broad; anal tuft tipped with brown: wings and body below pale buff, markings only indicated on the apical half of primaries and costa of secondaries, the latter covered with hair; an apical marginal sinuated black line. Expanse of wings 1 inch 10 lines.

(Coll. by Gulliver.)

Allied to, but quite distinct from, *H. vinsonii* of Guénée.

Achæa catilla, Guénée.

Two specimens, rather more strongly marked than Abyssinian examples (which in all probability would agree with the type from Senegal), but not sufficiently distinct to warrant their separation.

Ophiusa properans, Walker.

One example obtained.

Trigonodes acutata, Guénée.

Five specimens.

Remigia conveniens, Walker.

One specimen.

Asopia vulgalis, Guénée.

One specimen.

Laverna plumipes.

Ann. & Mag. Nat. Hist., s. 4, vol. xvii., p. 409 (1876).

Primaries shining brown, irrorated with black; a subapical black spot; fringe grey: secondaries sericeous pale brown, the apical area darker; fringe pale brown: head and thorax olive-brown, abdomen pale brown: wings below shining pale brown: body and legs below shining whitish; tibiæ of second pair of legs with a long terminal pencil of hair-scales between the spines; tibiæ of hind pair densely clothed with long projecting hairs. Expanse of wings 10 lines.

(Coll. by Gulliver.)

ORTHOPTERA and HEMIPTERA.—*By Arthur Gardiner Butler, F.L.S., F.Z.S., &c.*
(Plate LIV.)

Fifteen species of *Orthoptera* were obtained by Messrs. Gulliver and Slater, one of these was, however, in an immature condition, and therefore only its genus could be ascertained; four of the species appear to be endemic and new to science, the species of most interest being *Phisis spinifera*, a form nearly allied to *P. pectinata* from the Samoa Islands, Molluccas, and Ceylon.

Of the Orthoptera previously described, seven are recorded from Mauritius and Bourbon, and three from Madagascar; two of the latter are, however, cosmopolitan species.

Amongst the *Hemiptera* of which there are 20 species, five are new to science; two are immature and cannot be specifically identified. With the exception of three species, the whole of the recognisable forms are referable to the *Hemiptera Heteroptera*. Their geographical distribution seems to be as follows: five endemic, five occurring in Mauritius and Bourbon, and seven in Madagascar; two of these, *Nezara viridula* and *Conorhinus rubrofasciatus*, appear to be cosmopolitan or nearly so, and one species, *Serinetha lateralis*, seems to be generally distributed over Africa; in addition to these, four species have hitherto only been recorded from South Africa, and two from Algeria.

ORTHOPTERA.

Name of Species found in Rodriguez.	Hitherto found in Rodriguez only.	Previously obtained elsewhere.		
		Mauritius and Bourbon.	Madagascar.	Other Localities.
Forficula auricula, *Linn.*				Cosmopolitan.
Forficula varicornis, *Smith*	×			
Panchlora corticum, *Serville*		×		Congo, Whydah.
Phyllodromia germanica, *Linnæus*		×	×	Cosmopolitan.
Periplaneta americana, *Linnæus*		×	×	Cosmopolitan.
Periplaneta rhombifolia, *Stoll*				China, Brazil.
Polyzosteria latipes, *Walker*				Sierra Leone.
Gryllus capensis, *Fabricius*		×		Almost cosmopolitan.
Mogoplistes sp. (immature)	×			
Nemobius luteolus, *Butler*	×			
Phisis spinifera, *Butler*	×			
Xiphidium iris, *Serville*		×		
Conocephalus differens, *Serville*		×		
Pachytylus cinerascens, *Fabr.*		×		Old World generally.
Epachromia rodericensis, *Butler*	×			
Mantis variegata, *Olivier*			×	Natal and Sierra Leone.
Bacillus incommodus, *Butler*	×			

Forficula auricula, Linn.

A single specimen taken by Mr. Gulliver. This insect is now almost cosmopolitan.

Forficula varicornis.

Smith, Ann. & Mag. Nat. Hist., 4th ser., vol. vii., p. 450 (1876).

Length 4½ lines. Nigro-piceous, shining; the palpi, mouth, and legs pale testaceous, the prothorax rufo-piceous; antennæ 15-jointed, joints 12th and 13th white, five or six of the basal joints rufo-piceous, as are also the palpi; the prothorax oblong-quadrate, narrowly margined, and with a longitudinal impressed line not extending to the posterior margin; apterous; the abdomen smooth and shining; the basal half of the segments with fine shallow punctures; the apical segment with a deeply impressed longitudinal line, its posterior margin emarginate; the forceps very stout, triangular at the base, curved inwardly beyond the middle, acute at the apex, their inner margin crenulated.

Ten specimens obtained by Mr. Gulliver. This species would probably come into the genus Brachylabis of Dohrn.

Panchlora corticum, Serville.

Six examples were obtained by Messrs. Gulliver and Slater. This species has been considered synonymous with *P. Surinamensis*, but the African and Mascarene examples seem to be constantly shorter in the wings than those from Tropical America.

Phyllodromia germanica, L.

One example.

Periplaneta americana, L.

Mr. Slater collected five specimens of this species.

Periplaneta rhombifolia, Stoll.

The four specimens obtained differ in no respect from Chinese or Brazilian examples.

Polyzosteria latipes, Walker.

Eight examples, slightly varying in punctuation.

Gryllus capensis, Fabr.

Ten examples, varying slightly in colour.

Mogoplistes sp. (immature).

Two specimens.

Nemobius luteolus.
(Plate LIV., figs. 2, 2ᵃ.)

Ann. & Mag. Nat. Hist., s. 4, vol. xvii., p. 409 (1876).

Stramineous; vertex of head greyish; pronotum irrorated with brownish, with a lateral slender ridge and a central depressed line; posterior abdominal segments greyish; cerci as long as oviduct, hairy; oviduct castaneous below and at the tip; tegmina nearly extending to the end of abdomen, pale stramineous; tibiæ of hind legs with three pairs of spines, also several terminal shorter spines.

Length of body 5 millims., of oviduct 2.

(Coll. by Gulliver.)

The single example is somewhat shrunken, and may have been killed before attaining its full coloration.

The head is obtusely produced between the antennæ, and is slightly depressed in the centre; the oviduct, as usual, is curved upwards; the wings are whitish and extend when closed to the end of the abdomen.

Phisis spinifera.
(Plate LIV., figs. 1, 1ᵃ.)

Ann. & Mag. Nat. Hist., s. 4, vol. xvii., p. 410 (1876).

Female. Testaceous, pronotum oblong, greyish, with carinated margins; abdomen greyish; oviduct stramineous; the femora of the two anterior pairs of legs with two converging longitudinal ferruginous lines above, of the first pair with six exterior and five interior lateral spines, distal end terminating on each side in a conical denticle; tibiæ with seven spines on each side; femora of second pair with seven exterior short spines, inferior margin denticulated; tibiæ with six exterior and five interior spines, longer than in the femora; femora of hind pair reddish, spinulose below; tibiæ denticulated.

Length of body 19 millims., of oviduct 9.

(Coll. by Gulliver.)

More robust than *P. pectinata*, rather differently coloured; anterior femora with strong terminal conical denticles.

The oviduct is blackish at the tip, compressed, and curved upwards; the tibiæ are more strongly denticulated on their superior than on their inferior surface.

Xiphidium iris, Serville.

Two specimens of this delicate species were collected by Mr. Gulliver.

Conocephalus differens, Serville.

Eight examples. The species obtained under this name from M. Brunner de Wattenwyl, and occurring in the Philippines, is entirely distinct; it is a much more elongated species, with the head much more produced and acuminate. M. Serville says of his type "trouvés à l'Ile-de-France," and his description in all respects agrees with examples from Rodriguez.

Pachytylus cinerascens, Fabr.

Twenty examples in various stages.

Epachromia rodericensis.

(Plate LIV., figs. 3, 3^a.)

Ann. & Mag. Nat. Hist., s. 4, vol. xvii., p. 410 (1876).

Male. Pale carneous; pronotum and head above brown, adorned with an X-shaped marking, intersected by a longitudinal fascia, pale carneous edged with black; pronotum with a lateral oblique black streak, meso- and metanota testaceous, with lateral black spots; knees black at the sides; tibiæ with three black spots below, spines black; tegmina with the basal half coriaceous, brown, inner border whitish, base whitish-speckled, a white subquadrate costal spot; apical half pale brownish hyaline, crossed by ill-defined irregular white bars; veins dark brown; wings hyaline white, veins black.

Length of the body 12 to 18 millims.

(Coll. by Gullivor.)

This species does not vary in coloration.

Eleven specimens in various stages were taken.

Mantis variegata, Olivier.

Fifteen examples in various stages.

Bacillus incommodus.

(Plate LIV., figs. 4, 4[a-c].)

Ann. & Mag. Nat. Hist., s. 4, vol. xvii., p. 410 (1876).

"Green, with red antennæ, when alive."

Ochreous, clouded with purplish brown; eyes black; head truncate-ovate, with central and lateral longitudinal depressed lines, longer than the pronotum; antennæ with twenty-four joints, basal joint broad, depressed; pronotum quadrate, with obtusely carinated margins; two longitudinal series of four obtuse well-developed tubercles; a central depression; mesonotum dorsally carinate, covered with tubercles; a series of lateral tubercles between the pairs of legs; tegmina small; wings minute, coriaceous; anterior abdominal segments roof-like, laterally carinated; anal styles lanceolate; anterior legs considerably the longest, the femora strongly excavated and rugulose at the base, with two internal longitudinal marginal series of denticles; all the legs parallelopipedous, obtusely carinated; posterior femora denticulated below.

Length 3 inches 10 lines.

(Coll. by Gulliver and Slater.)

The pronotum might perhaps be more strictly described as irregularly lobulate, the lobes mammoid, capped with prominent tubercles, and arranged in two longitudinal series of four, between them a depressed mesial line. The general aspect of the species is that of the genus *Lopaphus*, but its structural characters agree in all respects with those of *Bacillus*. Five specimens were found in the island. It is " common on leaves of fan-palm, which it resembles in colour."—*G. Gulliver*.

HEMIPTERA.

Name of Species found in Rodriguez.	Hitherto found in Rodriguez only.	Previously obtained elsewhere.		
		Mauritius and Bourbon.	Madagascar.	Other Localities.
Phricodes hystrix, *Germar*		×	×	Caffraria.
Æthus sp.* (immature)			× ?	
Nezara viridula, var leii, *Hope*		×	×	S. Europe, S. Asia, Malaysia, Columbia.
Serinetha lateralis, *Signoret*		×	×	Caffraria, Senegal, Sierra Leone, Calabar.
Lethæus tartareus, *Stål*				Cape.
Rhyparochromus luctuosus ? *Lucas*				Algeria.
Dysdercus fasciatus, *Signoret*			×	Sierra Leone, Zanzibar.
Mezira caffra, *Stål*				Cape.
Conorhinus rubrofasciatus, *De Geer*		×	×	Sierra Leone, India, Malaysia, China, Brazil.
Oncocephalus cancellatus, *Stål*				Caffraria.
Reduvius laniger, *Butler*	×			
Emesa barbara ? *Lucas*				Algeria.
Velia infernalis, *Butler*	×			
Sigara scutellaris, *Stål*				Country of the River Kuisip, Namaqua Land.
Sigara sulcata, *Signoret*			×	
Sigara felix, *Butler*	×			
Cicada sp. (larva)				
Tylana carinata, *Fabricius*		×		
Coccus ceratiformis, *Butler*	×			
Aptinothrips fasciatus, *Butler*	×			

Hemiptera Heteroptera.

Phricodes hystrix, Germar.

Four specimens were obtained by Mr. Gulliver at Rodriguez.

Æthus sp. (immature).

One specimen.

* May be *Æ. madagascariensis*.

Nezara viridula, Linn., var. *leii*, Hope.

If the various representatives of this species from all parts of the world (differing in form, sculpturing, and colour) be considered conspecific, *Pentatoma plicaticollis* of Lucas may be added to the synonymy.

One example of *N. leii* was taken by Mr. Gulliver "on flowers."

Serinetha lateralis, Signoret.

Twenty-one examples, in various stages of development, were found by Mr. Gulliver "under stones."

Lethæus tartaræus, Stal.

The seven examples obtained by Mr. Gulliver differ in no respect from the Cape form.

Rhyparochromus luctuosus ? Lucas.

Two examples. Judging by the figure I cannot distinguish this species from that described by M. Lucas.

Dysdercus fasciatus, Signoret.

Twenty-five specimens, in various stages of development, found "on Malvaceous trees."—*G. Gulliver*.

Mezira caffra, Stål.

Fifteen examples. "Common under bark of trees."—*G. Gulliver*.

Conorhinus rubrofasciatus, De Geer.

One specimen was obtained by Mr. Slater.

Oncocephalus cancellatus, Stål.

Ten examples were collected.

Reduvius laniger.

(Plate LIV., figs. 6, 6ª.)

Ann. & Mag. Nat. Hist., s. 4, vol. xvii., p. 411 (1876).

Densely clothed with testaceous hair; antennæ slender, the three basal joints amber-yellow, remaining joints blackish; head castaneous; eyes blackish; rostrum pale castaneous, dark at the tip; thorax testaceous, with two nearly parallel longitudinal piceous bands, anterior lobe with a central longitudinal fossa; posterior lobe with a central depression; scutellum piceous, pale at the tip; corium of hemelytra testaceous, with two blackish spots and a cuneiform whitish spot between them; membrane dull purplish black; abdomen above amber-yellow, with marginal brown spots, below piceous, with a yellowish band on each side; pectus shining black; legs amber-yellow; the femora with central and terminal brown bands.

Length 6½ millims.

(Coll. by Gulliver.)

One example brought home; it is said to be "common under cow-dung."— *G. Gulliver*.

R. laniger seems to be of about the same size as *R. nanus*, of which Stål says: "Minima mihi cognita species generis." It cannot, however, be confounded with that insect, as, apart from its different coloration, its hairy character would alone be sufficient to distinguish it.

Emesa barbara? Lucas.

Two specimens, apparently immature. "Common in the hottest time of the year, February and March."—*G. Gulliver*.

Velia infernalis.
(Plate LIV., fig. 5.)

Ann. & Mag. Nat. Hist., s. 4, vol. xvii., p. 411 (1876).

Velvety black above, two grey spots on the front of the thorax, slaty grey below; sides of pectus whitish, anal segments of venter brown; legs black above, brownish below; the base of the antennæ, the coxæ, and the base of the femora orange-yellow.

Length 4 millims.

(Coll. by Gulliver.)

Allied to *V. nigricans* of Burmeister, but distinguished by its dull black colouring, and the two grey thoracic spots above.

Sixteen adult examples and 15 larvæ were obtained.

Sigara scutellaris, Stål.

Two examples.

Sigara sulcata, Signoret.

One specimen only.

Sigara felix.
(Plate LIV., figs. 8, 8ᵃ.)

Ann. & Mag. Nat. Hist., s. 4, vol. xvii., p. 412 (1876).

Ovate, piceous; head rather wider than the thorax; vertex of head and sides of tegmina whitish testaceous, below greyish brown; legs fawn-colour; face rugose, vertex with slender transverse irregular impressions; thorax transversely striated; tegmina punctured, setose.

Length 2–3 millims.

(Coll. by Gulliver.)

Similar in coloration to *S. minutissima*; but smaller and comparatively broader. "Common in streams."—*G. Gulliver.*

Twenty-four specimens were collected.

Hemiptera Homoptera.

Cicada sp. (larva).

Four specimens. "Common."—*G. Gulliver.*

Tylana carinata, Fabricius.

One specimen.

Coccus ceratiformis.

(Plate LIV., fig. 9.)

Ann. & Mag. Nat. Hist., s. 4, vol. xvii., p. 412 (1876).

Female. Irregularly elliptical; above with posterior central depression; thinly covered with a fine powdery, silvery, waxy efflorescence, which conceals the coloration; when this is removed the insect is bright yellow, often with the central area largely piceous; about nine segmentations traceable; under surface somewhat concave; head sinuate in front, with an angular frontal depression, in front of which is an obtuse ridge; rostrum situated in the fore part of a deep obovate depression; antennæ 9-jointed, pale testaceous, sparsely clothed with setæ; legs pale testaceous, rather long (extending beyond the sides of the body in small examples), apparently with three tarsal joints and two terminal claws; pediferous lobes very prominent.

Greatest length 3 millims.

(Coll. by Gulliver.)

I cannot identify this species with any of those described by Messrs. Icery and Signoret.

Fourteen examples.

Physopoda.

Aptinothrips fasciatus.

(Plate LIV., fig. 7.)

Ann. & Mag. Nat. Hist., s. 4, vol. xvii., p. 412 (1876).

Blackish piceous, glabrous; wings hyaline; bases of antennal joints, eyes, ocelli, and five broad abdominal bands crystalline white; frons fulvous; antennæ 7-jointed, basal joint conical, the second to fourth obconical, fifth to seventh fusiform; the terminal joint terminating in an acute point; head rounded, truncate in front and behind, with a central obtuse carina and an oblique stria behind each eye.

Length 3 millims.

(Coll. by Gulliver.)

One example.

ORTHOPTERA AND HEMIPTERA.

Description of Plate LIV

Fig. 1.	Phisis spinifera, *Butl.*		-	Profile view enlarged.
,, 1ª.	,, ,, ,,		-	Upper surface of head and thoracic shield.
,, 2.	Nemobius luteolus, *Butl.*		-	Upper surface enlarged.
,, 2ª.	,, ,, ,,		-	Head and thoracic shield in profile, greatly magnified.
,, 3.	Epachromia rodericensis, *Butl.*		-	Upper surface enlarged.
,, 3ª.	,, ,, ,,		-	Head and thoracic shield in profile, greatly magnified.
,, 4.	Bacillus incommodus	,,	-	Upper surface.
,, 4ª.	,, ,,	,,	-	Head in profile, greatly enlarged.
,, 4ᵇ.	,, ,,	,,	-	Terminal segments of abdomen, ventral surface, enlarged.
,, 4ᶜ.	,, ,,	,,	-	Terminal segments of abdomen, profile view.
,, 5.	Velia infernalis	,,	-	Upper surface, greatly magnified.
,, 6.	Reduvius laniger	,,	-	,, ,,
,, 6ª.	,, ,,	,,	-	Profile of head, ,,
,, 7.	Aptinothrips fasciatus	,,	-	Upper surface, ,,
,, 8.	Sigara felix	,,	-	,, ,,
,, 8ª.	,, ,,	,,	-	Front of head, ,,
,, 9.	Coccus ceratiformis	,,	-	Anterior portion of ventral surface.

ANNELIDA.—*By Professor Ed. Grube.*

The Annelids obtained are few in number and belong to two species:—

Perichaeta, *Schmd.*

Perichaeta rodericensis.

Longius vermiformis, anteriora versus citius magisque attenuata, teres, ex carneo flavescens, læte iricolor, paulo pellucida, *segmentis* fere 110, *ante* 17^{mum} *sitis* majoribus, pæne 2-plo latioribus, *posterioribus* 3-plo fere latioribus quam longis. *Lobus capitalis*, quoad observare licuit, semiovalis, latior quam longus, linea pæne recta a segmento buccali sepositus. *Segmentum buccale* illo ut proximo longius. *Segmentum* 14^{mum} et 15^{mum} (clitellæ) sordide crocea, satis libera ut buccale, nec vero tumida. *Setae* apice quasi curvatæ, segmentorum ante clitellas sitorum paulo majores, 32-nae ad 36-nas, annulis leviter elatis collocatæ, posteriorum minores, sæpe ægre distinguendæ, ad 60-nas, postremorum 40-nae.

Segmentum 17^{mum} subtus pone cingulum setarum utrinque eminentia planiore orbiculari munitum, centro distinctius circumscripto. Aperturæ genitales anteriores haud observatæ.

Longitudo fere 100 mm., partis ante clitellas sitæ 21 mm., clitellarum 2 mm., latitudo maxima (proxime ante clitellas) 4, 5 mm., partis postremæ 3 mm.

Rather common in damp earth and moss.—(*Gulliver*).

Several examples were received, the longest of which measured 110 mm., and the shortest 45 mm.; they were in all cases particularly soft, but, as in but a few cases the integument had pealed off to a very small extent, this did not appear to be due to the mode of their preservation, but to the thinness of the muscular layer. The wall of the body is so transparent that it was possible to see the viscera and large vascular trunks quite distinctly; thus, in the 9th and 10th segments the stomach may be seen shining through the wall of the body; in the 11th and 12th the two pairs of organs to which Léon Vaillant[*] has given the name of testes (l. c. Fig. 4 *h h'*) and in the 16th to the 19th inclusive, as also in the 20th segment, his "Glandes "accessoires du Canal déférent." Dissection revealed the presence in the 18th segment, of a minute C-shaped cæcum; but it was in one example only that I noted the distinct presence of the above-mentioned circular protuberances with their sharply projecting centra (in which I was in no case able to find an orifice). There are no longitudinal costulæ on the most anterior segments; the whole animal is rounded, the hinder end not being either flattened out or

---

[*] Léon Vaillant.—Note sur l'Anatomie de deux Espèces du Genre Perichaeta.—Montpellier, 1869.

dilated, but also well rounded, and gradually tapering off to the extremity of the body. It was impossible to detect the presence of any genital orifices. The cephalic lobe was very indistinct, as the pharynx was partially exserted, but I believe I discerned an almost perfectly straight line of boundary between it and the buccal segment.

The animal under description would seem therefore to belong to the genus *Nitocris*, Kbg.,* inasmuch as the setae on the post-clitellar segments are distinct, and, as a rule, much more numerous than those on the anterior segments; it is true that Kinberg in his diagnosis, says " cingulum nullum "; perhaps, because he regards the presence of a swelling as an essential character of the cingulum (clitellæ); such a swelling is absent also in our *Perichaeta*, but I am, nevertheless, of opinion that the two segments which are so markedly distinguished by their colour, and by the absence of setæ must be regarded as forming a cingulum. As however there is some doubt as to the presence of this organ, I think it best to describe this worm from Rodriguez under the more general name of *Perichaeta* (as defined by Schmarda).

Only one species of *Nitocris* is known, *N. gracilis*, Kbg., from Rio de Janeiro; this has as many as 91 segments, and is from 42–66 mm. long; there is a cephalic lobe which is said to be nearly as long as the buccal segment. In our species it is, I think, shorter; Kinberg does not state the number of setæ on the anterior or posterior segments, or any other distinctive characters; but, in any case, it would be against all experience, if the two forms were identical.

Amphinome, *Brug*.

A. (Eurythoë) pacifica, Kb.

When I was determining the examples of this species I was undecided whether they should be referred to *E. pacifica*, or *E. capensis*, Kbg.; as in both of these species, there are four buccal segments, and the caruncle does not extend to the fifth segment. In *E. capensis*, however, it is said to extend beyond the line which separates the third from the fourth segment, and not to be lobate; in our specimen, the caruncle does not extend so far, and its basal portion is lobed. I did not observe any distinctly finer setæ in the ventral fascicle, where the number of bristles was about 25; the setæ of the superior fascicles are either linear, having their points long and somewhat distinct, or they are much stronger and provided with about 11 denticles at their points, otherwise they are smooth; all white, but those in the ventral bundles are of a more yellowish colour. The dorsal cirrus, which Kinberg states to be shorter than the superior setæ, I find to be of the same, or even of a somewhat greater length. The branchiæ are most intensely coloured, nearly ferrugineous, which I have not observed in any other examples of *A. pacifica*; the filaments are grouped rather in the form of the fan, than in several rows, and, as a

* Kinberg, Annulata nova. Öfvers. Kongl. Vetensk. Akad. Förhandl., 1866, p. 102.

rule, are divided into four branches; the first gill had only four, but the second as many as 15 filaments; the fourth and fifth reach those opposite to them, but this is not the case with the succeeding ones.

The anus is dorsal in position and occupies the three last segments. In one example, in which there are 73 segments, the length is 71 mm., the greatest breadth of the median segments, which are about $4\frac{1}{2}$ or 5 times as broad as long, is on the ventral surface 9 mm., or, adding the pharetrae, 10, or the setæ 13 mm. Kinberg's figure* (l. c. Taf. XII., Fig. 11) represents the animal as much more slender, but in his description no weight is laid on the proportions of the segments, and it is possible that the figure was taken from a living specimen.

The caruncle is 1·4 mm. long; its basal portion has six descending grooves; the upper portion is swollen and undivided; the azygos tentacle is 0·5 mm. long, and somewhat shorter and thinner than the others.

* Eugenies Resa. Annulata.

TURBELLARIA.—By *George Gulliver, B.A.*
(Plate LV.)

Two forms of considerable interest belonging to this group were found to be tolerably abundant. They belong to the genera *Tetrastemma* and *Geoplana*, and I can scarcely entertain a doubt that these animals will finally prove to have a much wider geographical range, or, in other words, that it would be premature to regard them as peculiar to the island of Rodriguez. I may also express it as my belief, that the Land Planarians are much more numerous than has been supposed hitherto; but, unfortunately, with but few exceptions, Zoologists who have visited this region, have paid but little attention to them.

The species of *Tetrastemma* is, I believe, the third Land Nemertean which has been discovered, one having been found by Semper in the Pelew Islands, and another quite recently by the late Dr. Von Willemoes Suhm in the Bermudas, which he has described in "Annals and Magazine of Natural History," 1873, xiii., p. 309. He expresses it there as his opinion that land Nemerteans are much more common on tropical islands than has commonly been supposed, and the discovery of one at Rodriguez goes a long way towards justifying his supposition. Dr. Suhm's, like mine, is a species of *Tetrastemma* (*T. agricola*), but the present species is much more truly a terrestrial species than his, inasmuch as it occurs in woods—in damp rotten wood—and similar situations. This was the case also with that discovered by Semper. The animal was killed almost immediately by immersion in salt-water, for which it showed its distaste by rapidly exserting its proboscis, as it does when placed in alcohol. Fresh water, on the contrary, was by no means so distasteful to the animal, and I have had them remain alive for a whole night in it. It never, however, was observed to attempt to crawl on the upper surface of the water, though it would ascend the sides of the glass, and, if possible, settle itself outside the water. All the specimens seemed to be exceedingly torpid, and showed no disposition to move. If disturbed in their lodgment in rotten wood they appeared to shrink away into some recess, but never attempted to move for any length of time. It would seem to be very probable that this season was one of torpidity with them.

Tetrastemma rodericanum.—(Plate LV., fig. 2.)

Specific character.—Body narrow, elongated: dorsal surface dark green with a narrow white streak along the median line, and one at each side near the ventral surface. Ventral surface white. Head of a lighter green colour than the dorsal surface of the body, and having four white spots on the snout, visible when the animal is looked at from before. Marginal stylet sacs four. Cephalic sacs absent.

Habitat.—On rotten wood, under decayed leaves, &c. in the Island of Rodriguez.

The animal is from one to three inches in length. The dorsal surface is convex, the ventral flat. The colour of the dorsal surface is generally a very dark green, but occasionally the shade is a good deal lighter than usual. The lateral white lines are broader than the median one. The head is bifid anteriorly. Of the four white spots one is situated laterally on each lobe of the snout, and of the remaining two, one corresponds to the aperture of the proboscis, and one to the mouth.

Anatomy.—Tetrastemma Rodericanum corresponds very closely in the arrangement of its internal organs with the other members of the Enoplan division. It presents however certain peculiarities which are worthy of mention.

Cephalic fissures and sacs.—No trace of the cephalic fissures or of the sacs to which they lead, could, after careful examination, be detected. This was also the case in the Bermudian Tetrastemma described by Dr. Suhm.

Proboscis.—As in *Amphiporus hastatus*, there are four marginal stylet sacs situated in the same plane. The structure of the proboscis corresponds very closely with that of other members of the group. There is however one respect in which it appears to be peculiar, namely in the possession of an additional system of circular fibres which traverses the great longitudinal columns which constitute the so-called "beaded layer" (Plate LV., fig. 8). As in the other Enopla, each of the great longitudinal bands is surrounded by fibres which are perhaps derived in part from the external, and in part from the internal, circular coats, but the system of fibres just mentioned is additional to these, though mingled with them. It cuts through the longitudinal fasciculi, dividing each fasiculus into an inner and outer half, constituting a distinct circular coat (Plate LV., fig. 9, F). These fibres are wavy, as are also the others just spoken of as surrounding the longitudinal fasciculi. None of the fibres are distinctly traceable among the bundles of longitudinal muscular fibres. Whilst the internal circular and external circular coats, and the external longitudinal and internal longitudinal muscular coats, are readily stained by carmine, the longitudinal elastic bundles, with the fibres which surround, and those which traverse them, do not take the staining fluid at all.

Nervous system.—The general arrangement of the nervous system presents no peculiarities. The distinction between the fibrous and cellular elements in the lateral nerve trunks is, however, remarkable from its clearness and constancy of the arrangement of the two elements. Well marked groups of nerve-cells are present also in the ganglia, and are continuous with the cellular elements of the lateral cords. Owing to the difficulty of obtaining symmetrical sections of the ganglia, it is very hard to gain anything like a notion of the arrangement of the cells in those bodies. In size and form the ganglionic cells appear to be identical with those of cords, the general arrangement of which can readily be studied in transverse sections. In the present species the cells have a definite arrangement which does not appear to have been the case in any of the species examined by previous writers.

The arrangement is in all the more easy to study, as the cells are deeply stained by carmine or logwood, whilst the fibres are comparatively unstained.

If a transverse section be made through a lateral nerve trunk at any part of the body, appearances similar to those figured in Plate LV. figs. 4, 5, 6, will be seen. The trunk itself is somewhat kidney shaped in section, the convexity being towards the integumental, and the concavity towards the visceral, aspect of the body. It is bound down to the body-wall by well marked fibres. There are seen to be two cell regions, and two fibrillar regions. The uppermost or dorsal cell region (Fig. 6 A.) crosses the cord, obliquely slanting inwards and downwards. Above it, is the upper fibrillar region C., and below it the lower fibrillar region, D. The upper cellular region is, allowing for slight variations, constantly expanded at its external, and to a much less degree at its internal, extremity.

The lower cellular region B. occupies the ventral part of the cord, and is in contact below with the sheath. It has a more or less crescentic form, tending to have its outer horn prolonged to meet the external and lower extremity of the upper cellular region. In some sections the two regions may be seen to meet, but always by a very narrow belt of cells, whilst in most cases the two regions are, as in Fig. 6, separated by the lower fibrillar region. The arrangement here mentioned and figured is so constant that there is never any difficulty in distinguishing the dorsal from the ventral, or the inner from the outer aspect of the cord, in the abscence of any other guide but the arrangement of its cells and fibres. The cells themselves are of small size, granular in structure, and never present an appearance of a distinct nucleus and nucleolus as do those from the ganglia of *Leptoplana* figured by Moseley in the "Phil. Trans.," 1874, Plate XV., nor can any sign of processess arising from them be detected, though in some cases the matrix in which they lie suggests the presence of them from having a fibrillar appearance. Fine bands of fibres, Fig. 6 E., may be constantly seen running into the masses of nerve fibres. They appear to be in a large measure fibrous septa running into these masses, but may in part be fibres derived from the cells, though no distinct evidence of this has been obtained.

With regard to the structure of the fibrillar masses themselves, as might be expected, nothing definite can be made out. In transverse section they appear granular, having the fibres just mentioned running into them at parts.

Eyes.—The eye consists of an anterior clear, and a posterior pigmented, portion. In form it resembles that figured by Leydig as occurring in *Planaria lactea* ("Vom Bau des Thierischen Korpers," pl. I.). In specimens preserved in alcohol, the eye pigment has a tawny brown colour. It was suggested to me by Dr. Rolleston that this might be due to the solvent action of the alcohol, and, accordingly, having examined some specimens which had been preserved in chromic acid, and found the eye pigment of these to be of a deep black colour, it was at once obvious that such was the case. Moseley, *l. c.* p. 121, appears to doubt the solubility of the pigment of Planarians in alcohol, and mentions that it is especially not the case with the eye pigment of *Leptoplana*. Nor is it the case in the planarian, *Geoplana whartoni*, about to be described, nor with regard to the body pigment of the present species of Tetrastemma. It is however most undoubtedly the case that the eye pigment of the latter is soluble in alcohol.

Cephalic Glandular Mass.—Commencing at a very short distance from the snout of the animal, persisting throughout the whole œsophageal region, and, if one may so express it, occupying the space, which after the termination of the œsophagus is filled up by the voluminous digestive tract is a mass, which is here described as glandular, without, it must be confessed, much knowledge as to its true nature. It is figured as it appears under a low power in Pl. LV. fig. 3, F. Under a high power this mass consists of spaces irregular in shape and size, and enclosed by a tissue which is in part homogeneous and in part fibrous. In the spaces thus enclosed there is amorphous granular material, but no distinct appearance of cells.

Connective Tissue.—Dr. Hubrecht has, *loc. cit.*, insisted with great justice on the importance of the "fibro-muscular connective tissue," which serves to support the various internal organs by bands of fibres which pass from the body-wall to these organs. In the present species this system is well developed, and, as in other species, is most marked where spaces intervene between the internal organs and the body-wall. Thus fibres constantly bind down the lateral nerve-trunks to the body wall (Fig. 6, F.), and in like manner pass from the body-wall to the dorsal surfaces of the proboscis sheath and digestive tract. But the connective tissue in this species presents a feature which has not hitherto been described, namely, the presence of cells in addition to fibres. These cells are represented in figures 4 and 5 F., and, highly magnified, in Fig. 6 G. and in Fig. 7. These cells are large and of fairly uniform size, though, scattered amongst them, are, as may be seen in Fig. 7, a few of much less magnitude. At parts these cells present the appearance of being connected with fibres, as is the case with the cell B, in Fig. 7. They present a well-marked nucleus and nucleolus, and might at first in female specimens be mistaken for young ova, as they are especially numerous in the vicinity of the lateral nerve trunks. They may with care be traced all round the body internal to the inner muscular coat, being recognizable in a single layer, and having a flattened appearance, in places where the voluminous digestive tract is in close apposition with the body wall, whereby it makes the recognition of separate fibres out of the question.

Vascular System.—The opacity of this species makes the study of this system very difficult, as no general view can be obtained. In Fig. 3. D. the two lateral vascular trunks are seen in section.

Generative System.—The arrangement of the generative organs presents nothing peculiar. The animal is diœcious. Fig. 4 represents the general arrangement of the testicular masses, and fig. 5 shows an ovum lying in its sac. In all cases the male organs seem to bear a much greater proportion to the size of the body than do the female. In some sections I have observed the appearance of ducts passing from the testis through the body-wall, but not with sufficient constancy or distinctness to justify any lengthened description.

Digestive Tracts.—The digestive tract has the usual arrangement. The mouth and œsophagus present well marked cilia. The contents of the post-œsophageal portion of the digestive tract has a uniform granular appearance, many of the granules being like oil globules. In none of the specimens examined has anything definite been observed that would lead to a knowledge of what the animal feeds upon, and, in fact, at the time when the specimens were collected, it is very probable that the animals were torpid and not feeding at all.

Beside the *Nemertine*, I was fortunate enough to obtain specimens of a small species of Land *Planarian*, which occurs in situations similar to those in which the Nemertine lives, and, indeed, is often found together with it. Wishing to ascertain where this animal should be placed, and referring to the arrangement of Mr. Stimpson (Proc. Acad. Phil., 1857), I found that he states, under the characters of his sub-family *Geoplanidæ*, that the eyes are two or many situated in the anterior part of the body ("Oculi duo vel plurimi in capite dispositi.") This part of the diagnosis Mr. Moseley (*loc. cit.*) has very properly altered in consequence of his discovery of eyes in various parts of the body in *Bipalium*. The genus *Geoplana*, the first under Mr. Stimpson's sub-family, has amongst its other characters this " ocelli numerosi marginales, v. in acervos submarginales, in capite dispositi." This genus is composed of all the species described by Mr. Darwin (Ann. Nat. Hist., Series i. vol. xiv.), except those which have no eyes, and of one species found and described by Mr. Stimpson himself. But there appears nothing in Mr. Darwin's description to justify Mr. Stimpson using the words "in capite dispositi." For instance, in describing *Planaria vaginuloides* (*Geoplana vaginuloides*, Stimpson), Mr. Darwin says, " Ocelli numerous, placed at regular intervals round the anterior extremity of the body and irregularly round the foot." Again, in the description of *Pl. tasmanica*, " Ocelli scattered round the entire margin of the foot, but most frequent at the anterior extremity," Mr. Stimpson's words " in capite dispositi " must mean, " in capite solum dispositi;" at least Mr. Moseley has taken them to mean this, and it would be difficult to put any other construction on them. This explanation is necessary, as the planarian which exists in Rodriguez has, like *Pl. tasmanica*, eyes situated round the entire margin of the foot, but more frequent at the anterior extremity. In other respects this agrees in its characters with the genus Geoplana; and it seems to be necessary then to alter the characters of this genus, as Mr. Moseley has done those of the genus *Bipalium*, in order that the former may include Mr. Darwin's species and the present one. The latter I propose to name *G. whartoni*, after my friend Commander Wharton, R.N., lately in command of H.M.S. "Shearwater."

GEOPLANA (altered from Stimpson).

Corpus depressum, vel depressiusculum, elongatum v. lineare, capite continuo. Ocelli numerosi, marginales, v. submarginales; v. in parte anteriori corporis solum, vel passim circa corpus, singulatim plerumque, nonnunquam in acervos dispositi.

G. whartoni,—Pl. LV., fig. 1.

Body elongate, pointed at both ends. Eyes numerous, placed singly all round the margin of the foot, but most numerous at the anterior extremity. Body cream coloured, and marked with three dark longitudinal stripes, one median and two lateral, which latter converge towards the extremities and meet the median one. Mouth placed behind the centre of the body. Generative orifice nearer to the posterior end of the body than to the mouth. Body one inch to an inch and a half long when extended, much less when at rest.

Habitat.—In the island of Rodriguez on rotten wood. Mr. Darwin mentions (*loc. cit.*) that he found a Land Planarian at Mauritius. In the island of Mahé in the Seychelles group, I observed an animal of similar size and colour, which unfortunately became dry before I had time to examine it.

DESCRIPTION OF PLATE LV.

Fig. 1.—*Geoplana whartoni*, sp. nov., about twice the natural size, from a specimen preserved in alcohol.

Fig. 2.—*Testrastemma rodericanum*, sp. nov., about three times the natural size, from a specimen preserved in alcohol.

Fig. 3.—Transverse section through the anterior part of the body of a specimen of T. Rodericanum. Hartnack's Obj. 2. Drawn with the camera lucida, as were also the succeeding figures.

A. Cuticular and muscular layers of the body-wall. B. Œsophagus. C. Proboscidean sheath. D. Vessels. E. Lateral nerve trunk. F.F. Cephalic glandular mass.

Fig. 4.—Transverse section through the middle part of the body of a male individual of the same species. Hartnack Obj. 4.

A. Cuticle.   B. Circular muscular coat.   C. Longitudinal muscular coat.   E. Testicular masses having well defined walls, enclosing cells and spermatozoa.   F. Connective tissue, consisting of cells and fibres.   G. Portion of intestinal tract with granular contents.

Fig. 5.—Similar section through a portion of the body of a female. A, B, C, D, F, G, as in the preceding figure. E. an ovum contained in its sac, and possessing a well marked nucleus and nucleolus.

Fig. 6.—Transverse section of lateral nerve trunk, highly magnified. Hartnack's Obj. 8.

A. Dorsal cellular region.   B. Ventral cellular region.   C. Dorsal fibrillar region.   D. Ventral fibrillar region.   E.E. Fibres running into the fibrillar masses, probably consisting mainly of connective tissue, but perhaps in part of processes from the cell regions.   F. Fibrous bands binding down the cord.   G. Connective tissue cells.   H. Internal muscular coat.

Fig. 7.—Portion of a transverse section highly magnified, to show the connective tissue.

A. Int. muscular coat shown in outline. The fibres represented passed between it and a portion of the digestive tract.   D. Fibres of connective tissue.   B. A large connective tissue cell, with fibre attached, other cells mostly large, but some small are also seen.

Fig. 8.—Section through a portion of the extended proboscis to show the arrangement of the "beaded layer." Hartnack. Obj. 8.

A. Internal circular muscular coat.   B. External circular muscular coat.   C. Internal longl. muscular coat.   D. External longitudinal muscular coat.   E. One of the great longitudinal bands surrounded by fine wavy fibres, and also divided by similar fibres which pass through it; all these fibres go to constitute the circular elastic coat.   F. The circular elastic coat.

ECHINODERMATA.—*By Edgar A. Smith, F.Z.S.*

(Plate LI., figs. 1-3.)

The collection of Sea-urchins and Starfishes from the island of Rodriguez, although comprising but few species, contains some of considerable interest, among which may be mentioned a very remarkable new *Comatula*, two apparently undescribed species of *Ophiocoma*, and the fine *Ophiomastix venosa*, described by Dr. Peters from Mozambique.

The rest of the forms are well-known inhabitants of the Indian Ocean, certain of them being also common in the Pacific.

Crinoida.

Comatula indica.

(Plate LI., figs. 3-3b.)

Ann. and Mag. Nat. Hist., 1876, xvii., p. 406.

Rays 30; dorsal disk, small, convex, subpentagonal, sculptured with little contiguous and well defined pits which are the sockets of the cirri, these are about 45 in number; radial joints two, the axillary without a syzygium; between this and the next bifurcation are two articulations; after two more joints the two outermost arms bifurcate; none of the brachial axillaries with a syzygium; each third segment above an axillary has a syzygium, and on the arms the next joints with syzygia are at very unequal distances, the most proximate being separated by as many as 20 joints, but usually by about 14; then nearer the extremity of the rays, they are rather closer together, the interlying joints varying in number from 6 to 10.

The second pinnulæ are very long and composed of much elongated joints. The colour of the specimen is pale purplish brown, with the sutures of the ray-articulations blackish brown. Diameter of disk 6 mill., length of rays about 6 inches (153 millim.).

The only example of this interesting species was collected by Mr. Slater. Unfortunately it is imperfect, all the dorsal cirri being wanting.* The remarkable elongate second pinnulæ (17 mill. in length) are situated on the first and second segments above those joints which always have a syzygium, and which are the third joints above the last brachial axillaries, or in other words, they rest upon the fourth and fifth joints above these axillaries. They consist of about 17 joints,

---

* The figured dorsal cirrus was found among *débris* at the bottom of the bottle containing the Comatula, but it is not absolutely certain that it belongs to the present species, although the evidence points to that conclusion.

which are elongate, cylindrical, those at the middle being the longest, and the two or three basal ones the shortest.

In the short diagnosis given in the "Annals and Magazine of Natural History," May 1876, p. 406, by an oversight it is stated that "the two *innermost* arms bifurcate;" it should be *outermost* or *exterior*.

Ophiuroida.

Ophiomastix venosa.

Peters, Wiegm. Arch. Naturgesch., 1852, p. 83; Von Martens, in Von der Decken's Reisen in Ost-Afrika, p. 130; Lyman, Ophiuridæ, p. 92.

Three fine examples of this remarkable species were collected by Mr. Gulliver. It has also been previously found at Mozambique and Zanzibar.

Ophiocoma erinaceus.

Müller and Troschel, System der Asteriden, p. 98; Dujardin and Hupé, Echinodermes, p. 260; Lütken, Additamenta ad Hist. Ophiurid, p. 164; Von Martens, in Von der Decken's Reisen in Ost-Afrika, III., p. 129; Lyman, Ophiuridæ, p. 83.

Seven specimens were obtained by Mr. Gulliver.

This species has a wide geographical range, having been met with in the Red Sea, at Mozambique, Zanzibar, Natal, Seychelles, Mauritius, Sandwich Islands, Kingsmills and Society Islands.

Ophiocoma variegata.
(Plate LI., figs. 2–2d.)

Ann. & Mag. Nat. Hist., 1876, vol. xviii., p. 39.

Disk circular, finely granulated above and beneath; oral shields rather longer than broad, slightly octangular, with the sides rectilinear; adoral shields narrow, extending along the lateral margins of the orals; three mouth-papillæ on each side of each oral angle, the outermost being the largest and somewhat squarish; above the teeth at the apex of the angle is an irregular cluster of about 12 smaller papillæ; teeth four, strong, arcuate at the tips. Arms rather stout, about four times as long as the diameter of the disk; upper plates transversely narrowly oval, with the outer margin faintly angulated in the centre, about twice as broad as long; lower plates squarish, the aboral angles rounded, the side margins excavated; arm-spines four, subequal, the uppermost a trifle the shortest and stout and the lowest but one a little the longest, about as long as the width of the dorsal arm-plates (the tenth); ambulacral scales two in number to the extremity of the arms, short and compressed.

Colour of a specimen in alcohol:—disk above and beneath uniformly purplish brown; upper arm-plates dirty brown, at intervals varied with three or four contiguous pale ones, which are marked with the dark outline of a subquadrate figure, particularly observable towards the *outer* ends of the arms; lower arm-plates pale, blotched with dirty brown, especially towards the outer extremity of the rays; arm-spines of a uniform colour, but lighter than the upper arm-plates; oral shields pale, mottled with dirty brown.

Diameter of disk 38 mill., length of rays 105; width of broadest upper arm-plates 4.

The nearest ally of this species is *O. scolopendrina*, Lamarck, from which it differs in coloration, form of the arm-shields, oral shields and the brachial spines. The dark subquadrate outline on the pale upper arm-plates is very characteristic, as also is the dirty brownish mottling on the oral and lower brachial shields.

Ophiocoma brevispinosa.
(Plate LI., figs. 1–1f.)

Ann. Mag. Nat. Hist., 1876, vol. xviii., p. 40.

Disk sub-circular, flat above, minutely granulated above and beneath; oral shields heart-shaped, broader than long; adoral plates small, crescentiform, bordering the sides of the orals; three or four mouth-papillæ on each side of an oral angle, and a group of about twelve at the apex; teeth four, the two intermediate ones larger than the two exterior. Arms a trifle more than three times as long as the diameter of the disk; upper plates transversely oval, about twice as broad as long; lower plates (twelfth from the base) a little longer than broad; aboral margins arched and a little pointed in the middle, lateral edges rather deeply concave; oral margins somewhat convergently sloping and interrupted by the outer margin of the adjacent preceding plate; tentacle-scales two, short, and compressed; brachial spines short, four (sometimes five on a few plates just beyond the disk), the two upper ones shorter than the others, broad and flattened; the two inferior ones (of which the second or upper is rather the larger) are slightly conical, and not so long as the width of the broadest dorsal arm-shield.

Colour of a specimen in alcohol:—disk dirty white, irregularly mottled with green above and beneath; arms of the same colour as the disk, with a narrow green line above, more or less distinct, down the centre; lower shields, ambulacral scales (and two lower series of arm-spines for the most part) uniformly dirty white, and the two upper series of spines ornamented with one or two narrow greenish rings and dots; oral shields spotted with green. Diameter of disk 17 millim.; length of rays 54. This species is well distinguished by its peculiarity of colouring and the character of the short brachial spines.

Asteroida.

Scytaster variolatus, Retzius.

A single fine large specimen collected by Mr. Slater.

Other localities where this species has been met with are the Red Sea, Réunion, Mauritius, Zanzibar, Bourbon.

Linckia multifora, Lamarck.

All four examples collected by Mr. Gulliver have five rays in different stages of development.

This common starfish also inhabits the Red Sea, the shores of Mozambique, Réunion, Bourbon, Seychelles, Mauritius, New Caledonia, and the Sandwich Islands.

Echinoida.

Phyllacanthus imperialis, Lamarck.

Of this species a fine large example with the ends of the primary spines well fluted, was collected by Mr. Gulliver.

This species is also recorded from the Red Sea, Zanzibar, Aru and Molucca Islands, Sulu Sea, New Holland and Tonga.

Hipponoe variegata, Leske.

The single specimen of this polymorphous and widely-distributed species is of a depressed form, roundly pentagonal, of a brownish colour, and clothed with white spines. It was collected by Mr. Gulliver.

This species appears to be found in all parts of the Indian Ocean, at Zanzibar, Mauritius, Seychelles, Nicobar Islands, in the Red Sea, the Pacific, and at the Philippines, Moluccas, Fiji Islands, &c.

Heterocentrotus mammillatus, Klein.

There are two varieties of this most variable sea-urchin brought home from Rodriguez, one with pale brown spines with dark tips and one or more pale rings, and the other with the primaries of a rich chocolate their entire length, and also having pale annulations near the extremities. Mr. Gulliver reports that this species is "very abundant."

The geographical range of this species is like that of the preceding.

Echinometra lucunter, Leske.

Of this extremely variable species two varieties were met with at Rodriguez; one with almost uniformly violet spines, but palish towards the base, and the other with light greenish-olive spines.

The geographical range of this species extends over the Red Sea, Indian Ocean, Chinese and Japanese Seas, and several parts of the Pacific.

Echinoneus cyclostomus, Leske.

A single example of this species, about an inch in length, was collected by Mr. Slater.

Other localities for the species are Mozambique, Zanzibar, Réunion, Bourbon, Australia, Amboina, Flores, Manila, Hood's Island, and the Kingsmills Islands.

Explanation of Plate LI. (Figs. 1 to 3).

Fig. 1. *Ophicoma brevispinosa*, enlarged view of upper side.
- 1a. " " " " of lower side.
- 1b. " " " " of an oral shield.
- 1c. " " " " of the madreporic shield.
- 1d. " " " " of two upper arm-plates.
- 1e. " " " " of a lower arm-plate.
- 1f. " " " " of an arm-segment (seen from the exterior side and showing the relative length of the spines).

2. *Ophicoma variegata* view of upper side, natural size.
- 2a. " " " lower side "
- 2b. " " enlarged view of an oral angle.
- 2c. " " " of a lower arm-plate and ambulacral papillæ.
- 2d. " " " of an arm-segment (seen from the exterior side and showing the relative size of the spines).

3. *Comatula indica*, enlarged view of the disk and a portion of the rays showing the manner of furcation.
- 3a. " " of a dorsal cirrus (probably belonging to this species).
- 3b. " " lateral view of an outer arm, showing the position and form of the pinnulæ.

CORALS.—*By Dr. F. Brüggemann.**

The specimens of Corals procured at Rodriguez amount to 102 and belong to 49 species. They were taken by Messrs. Slater and Gulliver evidently at moderate depths; at least no deep-sea forms, especially no *Turbinoliidæ* or *Oculinidæ*, are represented amongst them. As might have been anticipated, the Rodriguez coral fauna bears a close affinity to that of Mauritius, Madagascar, and the Seychelles Islands, and there can scarcely be any doubt that the species described as new in the present paper will prove to occur likewise in the localities mentioned as well as in others of the Indo-Pacific Region. Many of the species are common to Rodriguez and the Red Sea; a less considerable number have been recorded from the north-eastern part of the Indian Ocean; some of them extend their range into the Pacific. However, the coral fauna of these regions is hitherto comparatively but little known, and the present collection most likely contains only a small minority of the Rodriguez species; so that it would be premature to offer any further remarks on their geographical distribution.

ALCYONARIA.

Alcyoniidæ.

Alcyonium sphaerophorum.

Lobularia sphaerophora, Ehrenberg, Abhandl. Akad. Wiss. Berl. 1832, I., p. 281 (1834).

Alcyonium sphaerophora, M. Edw. Hist. Nat. Corall, I., p. 119.

Sarcophytum latum.

Alcyonium latum, Dana, U.S. Expl. Exp. Zooph. p. 623, Pl. 58, fig. 6 (1846); M. Edw. Hist. Nat. Corall. I., p. 121.

Milleporidæ.

Millepora verrucosa.

Millepora verrucosa, M. Edw. Hist. Nat. Corall. III., p. 227, Pl. F. 2, fig. 1 a, b, c (1860).

Millepora forskali.

Millepora forskali, M. Edw. Hist. Nat. Corall. III., p. 228.

Several specimens, belonging to different varieties, in one of them the terminal branches being much prolonged, stout and nearly cylindrical. This species I believe to be hardly separable from the West Indian *M. alcicornis*.

---

* This communication on the Corals of Rodriguez was one of the last works of the author, who died before the manuscript passed through the press. Shortly before his death he expressed a wish to make some alterations, but without specifying them. Therefore the editors have considered it best to print the manuscript as it was left by the author.

Millepora gonagra.

Millepora gonagra, M. Edw. Hist. Nat. Corall. III., p. 230, Pl. F. 3, fig. 1 a, b (1860).

ZOANTHARIA.
Actiniidæ.
Palythoa argus.

Palythoa argus, Ehrenb. Abh. Akad. Wiss. Berl. für 1833 I., p. 373 (1834); M. Edw. Hist. Nat. Corall. I., p. 305.

Palythoa violacea.

Corallum thin, explanate, spreading, incrusting coenenchyma rather rigid, coriaceous, its surface covered with numerous grains which are distant by once to twice their diameter. Polyps scattered, small, wholly immersed when contracted, their disk surrounded by six to nine short prominences, a little larger than the small verrucæ on the coenenchyma.

Height of corallum ½ of a mm. Diameter of polyps, about 2 mm. Colour (in spirits), violet; the prominences of the coenenchyma somewhat lighter, whitish.

There is only one small specimen of this apparently new species in the collection, spreading on the surface of a fragment of *Madrepora monticulosa*. It is not well preserved, and, therefore, I am not able to give any details respecting the structure of the polyps.

Stylophoridæ.
Stylophora palmata.

Sideropora palmata, Blainville, Dict. des Scienc. Nat. LX., p. 360 (1830).
Stylophora palmata, M. Edw. & H. Hist. Nat. Corall. II., p. 137.

Astræidæ.
Galaxea laevis.

Surface of corallum slightly convex. Corallites crowded, parallel, subcylindrical, compressed, only slightly angular, united by a rather dense peritheca, the cells of which measure from ½ to 1 mm. Walls comparatively very thick, opake, smooth, with excessively slight indications of costæ for every cycle. Calicular fossæ oblong, sub-angular, deep, Three cycles complete, a fourth one represented only by a few septa in some of the calicles. Septa smooth, unequal, the primary ones very thick, becoming suddenly thin towards the centre, much projecting, regularly truncate at the top; secondary septa a good deal smaller, less exsert; those of the third cycle very thin, narrow and hardly extending beyond the border of the calicle. Greatest diameter of calicular fossæ 8 mm.; distance between the corallites 2 to 5 mm.; height of corallites about 15 mm.

This species comes near to *G. bougainvillei*; it is readily distinguished from all its congeners by its thick smooth walls and septa.

Galaxea fascicularis.

Madrepora fascicularis, Linn. Syst. Nat. Ed. XII., p. 1278 (1767).
Galaxea fascicularis, M. Edw. & H. Hist. Nat. Corall. II., p. 227.

Galaxea laperouseana.

Sarcinula laperousiana, M. Edw. & H. Ann. Scienc. Nat. 3e sér. Zool. X., p. 314, Pl. 6, fig. 5 (1848).
Galaxea laperouseana, M. Edw. & H. Hist. Nat. Corall., II., p. 231.

Mussa umbellata.

Corallites ventricoso-turbinate, arising from a common stem, united till near the summit, then suddenly isolated and much diverging, forming a sort of umbel; few of them remain associated to form short series consisting of two or three corallites. A rudimentary epitheca ascends near to the borders of the calicles. Costae distinct from the base of corallum upwards, projecting but little, smooth for the greater part of their length, then tuberculate, and near the top spinose. Fossae rather shallow, triangular or quadrangular, rounded. Columella well developed, very spongiose. Four cycles complete, a fifth one indicated only in some of the systems. Primary and secondary septa rather thick, inflated, not much projecting; their margins provided with four to six coarse teeth of moderate size, most of which are truncate at the top. Inferior septa much thinner, tapering towards the centre, with delicately dentate margins.

This species is distinguished by its mode of growth as well as by the dentition of the septa. The latter are less projecting than in most of the other members of the genus. Diameter of calicular fossæ 20 to 30 mm., their depth about 7 mm.

Leptoria tenuis.

Maeandrina tenuis, Dana, U.S. Expl. Exp. Zooph. p. 262, Pl. 14, fig. 7 (1846).
Leptoria tenuis, M. Edw. & H. Hist. Nat. Corall. II., p. 407.

Platygyra esperi.

Astroria esperi, M. Edw. & H. Ann. Scienc. Nat, 3e sér. Zool. XI., p. 298 (1849).
Coeloria esperi, M. Edw. & H. Hist. Nat. Corall. II., p. 47.

The two generic names, *Cœloria* and *Astroria* being given by Milne Edwards & Haime to the present genus in 1848 (Compt. Rend. Acad. Sc. XXVII., p. 493) ought to give way to the earlier appellation *Platygyra* of Ehrenberg (Abhandl. Berl. Akad. Wiss. für 1832, I., p. 323, 1834; type: *Madrepora labyrinthica* of Ellis and Solander).

This is a very variable species; there are specimens with short and long, mæandrous and straight valleys; in one nearly all the calicular fossæ are as long as broad, plenty of calicles being thus circumscribed. Walls either thin or thick, even inflated. Septa in most instances very narrow, but in one specimen (the one alluded to above) much projecting towards the centre; their edges faintly or coarsely dentate. Columella rudimentary or well developed, consisting of confluent trabecles or of vertical lamellæ, the latter being placed in the direction of the valleys, and sometimes entirely confluent, so as to render such specimens hardly distinguishable from the genus *Leptoria*. It is true that most of these differences are of what is called " specific " or even " generic " value; but I found it impossible to draw any precise lines of distinction. In the present case the beautiful series procured by the Expedition comprises every stage of transition, and what is most significant, there are frequently different forms represented in different places of the very same corallum. I have made similar observations on the true *Cœloriæ*; in one specimen of this genus taken in the Red Sea by Professor Haeckel, there are really two " species " on the same specimen, one occupying the centre of the surface, the other two opposite sides, and both passing into each other.

Hydnophora microcona.

Monticularia microconos, Lamarck, Hist. Anim. sans Vert. II., p. 251 (1816).
Hydnophora microcona, M. Edw. & H. Hist. Nat. Corall. II., p. 423.

Astræa denticulata.

Madrepora denticulata, Ell. & Sol. Hist. Zooph. p. 166, Pl. XLIX., fig. 2 (1786).
Favia denticulata, M. Edw. & H. Hist. Nat. Corall. II., p. 428.

Astræa affinis.

Parastrea affinis, M. Edw. & H. Ann. Scienc. Nat. 3e Sér. Zool. XII., p. 167 (1850).
Favia affinis, M. Edw. & H. Hist. Nat. Corall. II., p. 429.

Astræa lobata.

Parastrea lobata, M. Edw. & H. Ann. Scienc. Nat. 3e Sér. Zool. XII., p. 171 (1850).
Favia lobata, M. Edw. & H. Hist. Nat. Corall. II., p. 434, pl. D. 8, fig. 3.

Plesiastræa quatrefagesana.

Plesiastræa quatrefagesana, M. Edw. & H. Ann. Scienc. Nat. 3e Sér. Zool. XII., p. 119 (1850); Hist. Nat. Corall. II., p. 491.

The single Rodriguez specimen may prove to belong to a new species; the calicles are remarkably smaller than in the true *Pl. quatrefagesana*, and their borders less well defined.

Acanthastræa angulosa.

Corallum low, nearly explanate, with an even slightly undulate surface. Calicles unequal, irregularly tetragonal or pentagonal, with simple rather thin borders and shallow fossæ. Columella feebly developed, more or less papillose, parietal. Three cycles complete, a fourth one irregular; well developed in several systems of the larger calicles, rudimentary or entirely wanting in others. Septa unequal, thin, slightly granulate, narrow, hardly projecting; their edges provided with long thin spiniform teeth at the summits, and with shorter and feebler ones towards the centre. Greatest diameter of calicular fossæ, 10 to 14 mm.; their depth 4 to 6 mm.

Prionastræa rodericana.

Corallum convex, somewhat lobate. Calicles polygonal, mostly quadrangular, subequal; their fossæ of moderate depth. Walls simple, rather thin. Gemmation takes place at the very borders of the calicles. Columella pretty well developed, trabecular. Three cycles complete, a fourth one very incomplete and obsolete. Septa crowded, unequal, slightly projecting, narrow, thin, smooth, with their edges delicately and equally dentate; those of the first two cycles suddenly enlarging below and there armed with slender pointed, but not paliform, teeth. Greatest diameter of calicles averages from 7 to 8 mm.; depth of fossæ 4 to 5 mm.

Differs from the closely allied *Pr. seychellensis* in its thin walls, the regularly denticulate, smooth septa, the smaller calicles and mode of gemmation. From *Pr. melicerum*, a species I have not yet seen, it may be distinguished by the presence of a fourth cycle, and the form of the columella.

Prionastræa scabra.

Corallum explanate, somewhat convex. Calicles polygonal, slightly irregular; fossæ of medium depth. Walls simple, rather thick, roughly spinose. Columella well developed, distinctly papillose. Three cycles complete, with some additional septa of a fourth one. Septa (about 30 in number) subequal, rather thick, delicately granulate, narrow, hardly projecting; their edges with scarce, spiniform, sharp-pointed teeth. Greatest diameter of calicles, 9 to 12 mm, depth of fossæ, 6 mm.

Very similar to *Pr. tesserifera*, but the columella is crowdedly papillose instead of being spongious, the calicular fossæ are deeper, the septa thicker and their teeth stouter and more pointed, giving the whole corallum a spinose aspect.

Echinopora spinulosa.

Corallum foliaceous, explanate, slightly undulate, unifacial. Costæ indistinct, replaced by irregular series of small, acute, delicately granulate spines. Calicles oblong, rarely circular, moderately crowded, very little prominent, with shallow fossæ. Columella well developed, delicately spongious. Four cycles complete. Septa crowded, altogether thin, spinulous on their surfaces, the primary, secondary, and tertiary ones nearly equal, projecting, unevenly truncate and toothed at the summit, irregularly denticulate and deeply emarginate on the inner edge; those of fourth cycle much thinner, smaller and not prominent. Under surface of corallum faintly striate. Greatest diameter of calicles, 5 to 6 mm.

This species should be placed next to *E. gemmacea* with which it agrees in having the fourth cycle complete and in being partly fissiparous. It differs from this as well as from the other species by the condition of the costæ and by the very slight prominence of the corallites. It bears a general resemblance to *E. ehrenbergi* in which, however, the fourth cycle is entirely wanting, the costæ being well marked and the calicles more projecting. With the coarsely spinose *E. hemprichi* it cannot be confounded; from *E. helli* it is distinguished by the absence of paliform lobes, and by the number and description of the septa.

Fungidæ.

Fungia haimei.

Fungia "discus," M. Edw. Hist. Nat. Corall. III., p. 9.
Fungia haimei, Verill, Bull. Mus. Comp. Zool. I., p. 51 (1864).

Lobactis dentigera.

Fungia dentigera, Leuckart, Zooph. corall. spec. gen. Fung. p. 48, Pl. III. fig. 1, 2 (1841); M. Edw. Hist. Nat. Corall. III., p. 17.

Pavonia cristata.

Madrepora cristata, Ell. & Sol. Hist. Zool. p. 158, Pl. XXXI., fig. 3,4 (1786.)
Lophoseris cristata, M. Edw. Hist. Nat. Corall. III., p. 66.

Eupsammiidæ.

Dendrophyllia ehrenbergiana.

Coenopsammia ehrenbergiana, M. Edw. & H. Ann. Scienc. Nat. 3ᵉ sér. Zool. X., p. 109, Pl. 1, fig. 12 (1848); Hist. Nat. Corall. III., p. 127.

Madreporidæ.

Madrepora pharaonis.

Madrepora pharaonis, M. Edw. Hist. Nat. Corall. III., p. 143 (1860).

Madrepora pustulosa.

Madrepora pustulosa, M. Edw. Hist. Nat. Corall. III., p. 144 (1860).

30. *Madrepora plantaginea.*

Madrepora plantaginia, Lamarck. Hist. Anim. s. Vertebr. II., p. 279 (1816); M. Edw. Hist. Nat. Corall. III., p. 149.

Madrepora acervata, Dana, U. S. Expl. Exp Zooph. p. 460, pl. 34, fig. 4 (1846), M. Edw. Hist. Nat. Corall. III., p. 154.

In two examples the corallum consists of a thin spreading base, covered with numerous low, mammilliform branches. This may be either a younger state or a variety.

Madrepora haimei.

Madrepora haimei, M. Edw. Hist. Nat. Corall. III., p. 151 (1860).

Madrepora gonagra.

Madrepora gonagra, M. Edw. Hist. Nat. Corall. III., p. 151 (1860); Brüggem. Abh. Ver. Brem. V., p. 398.

In the Mauritius specimen mentioned in my paper, the corallum shows a small compact base from which half a dozen of simple, straight, sub-cylindrical branches arise. One of the Rodriguez specimens nearly agrees with it, only the branches are more irregular and partly ramified. The second is apparently very old, the base of the corallum being entirely worn, the branches altogether confluent, and their summits coalescing, forming a nearly continuous plane and bearing the general aspect of the top of a cauliflower. This condition is evidently due to some immediate outer influence; most probably the corallum had approached too near the surface of the water.

Madrepora seriata.

Heteropora seriata, Ehrenb. Abh. Berl. Akad. 1832, I., p. 337 (1834).
Madrepora seriata, M. Edw. Hist. Nat. Coral. III., p. 152.

A low, slender branched variety, in which the seriate disposition of the calicles is not well pronounced.

Madrepora corymbosa.

Madrepora corymbosa, Lamarck, Hist. Anim. sans vert. II., p. 279 (1816); M. Edw. Hist. Nat. Corall. III., p. 154.

In the Rodriguez specimens, the stems and branches show an extreme tendency to coalesce and form bulky masses.

Madrepora flabelliformis.

Madrepora flabelliformis, M. Edw. Hist. Nat. Corall. III., p. 156 (1860).

Madrepora microcladus.

Heteropora microclados, Ehrenb. Abh. Berl. Akad. 1832, I., p. 333 (1834).
Madrepora surculosa, Dana, U.S. Expl. Exp. Zooph. p. 445, Pl. 32, fig. 4 (1846); M. Edw. Hist. Corall. III., p. 158.

Dana considers Ehrenberg's species to be closely allied to, or identical with, *M. spicifera*; Milne Edwards who seems to have examined the type specimen places it near to the *M. millepora*. In my opinion, it should be united to the *M. surculosa*, the descriptions agreeing well with each other.

Madrepora monticulosa.

Corallum solid, explanate, spreading. Surface nearly even, covered for the greater part of its extent with large, stout, rounded, subconical prominences. Coenenchyma faintly developed, spongious. Calicles much crowded, short, sublabellate or nearly tubiform, equal on the cones, intermixed with smaller or immersed ones on the plain parts of the corallum. Walls rather thin, reticulate, slightly echinulate. Septa nearly entirely wanting. Apical calicles always indicated, but not very conspicuous, with thicker walls and more distinct septa, the latter being generally six to eight in number.

Height of corallum, from under surface of base to the summits of largest cones, 00 mm.; greatest height of cones, 30 mm.; greatest diameter at base, 20 mm. Diameter of ordinary calicles 2 mm.; their length the same. Diameter of apical calicles, 3 mm.; height, $1\frac{1}{2}$ to 2 mm.; width of their apertures, 1 mm.

This well-marked species is, in its general aspect, similar to *M. conigera*, but it seems to be always spreading, and therefore, unifacial. Besides, it is distinguished by its lower and everywhere rounded cones, by its much shorter subequal calicles, by the absence of the septa, etc.

Madrepora alces.

Mandrepora alces, Dana, U.S. Expl. Exp. Zooph. p. 437, Pl. 31, fig. 12 (1846); M. Edw. Hist. Nat. Corall. III., p. 160.

One of the specimens is provided with long, erect, cylindrical branches rising at nearly a right angle from the palmato-digitate main stem. There are similar examples in the collection of the British Museum, one of which is stated to be from the Louisiade Archipelago.

Turbinaria mesenterina.

Explanaria mesenterina, Lamarck, Hist. Anim. s. Vert. II., p. 255 (1816).
Turbinaria mesenterina, M. Edw. Hist. Nat. Corall. III., p. 166, pl. E. 1, fig. 1 a, b.

Poritidæ.

Porites lutea.

Porites lutea, M. Edw. & H. Ann. Scienc. Nat. 3e sér. Zool. XVI., p. 28 (1851); M. Edw. Hist. Nat. Corall. III., p. 180.

Milne Edwards quotes as a synonym, the *Porites conglomerata* var. *lutea*, from the "Voyage de l'Astrolabe." Quoy and Gaimard, indeed, mention a variety of *P. conglomerata*, but without naming it.

Porites arenosa.

Madrepora arenosa, Linn. Syst. Nat. Ed. XII., p. 1276, sp. 17 (1767).
Porites arenosa, M. Edw. Hist. Nat. Corall. III., 180.

Montiporidæ.

Montipora divaricata.

Corallum ramose, subarborescent. Branches diverging, more or less angular, enlarged towards the summit; branchlets compressed, obtuse. Calicles immersed, distant by about twice the diameter, subequal, hardly larger on the prominent parts of the corallum. Coenenchyma moderately spongious, with even, delicately echinulate surface. Two cycles; septa subequal, very narrow, rather stout. Diameter of calicular apertures, nearly three quarters of a millimeter.

Allied to *M. rus* and *M. monasteriata*; differing by its mode of growth, and by the absence of the prominent verrucae.

Montipora foliosa.

Madrepora foliosa, Pallas, Elench. Zooph. p. 333 (1766).
Montipora foliosa, M. Edw. Hist. Nat. Corall. III., p. 212.

Montipora incrustans.

Montipora incrustans, Brüggemann, Abhandl. Ver. Brem. V., p. 398 (1877).

This species was established for a specimen from Mauritius. The condition of the coenenchyma and the excessively small calicles render it easily recognizable.

Montipora explanata.

Montipora n. sp., Brüggemann, Abh. Ver. Brem. V., p. 399.

Corallum explanate, spreading, incrusting. Surface uneven, in some places slightly tuberculate. Coenenchyma rather dense, consisting of confluent trabecles covered with small spinulous grains. Calicles immersed, equal, distant by about twice the diameter. Twelve septa, but little developed, thin and narrow, one of the primaries always stouter and larger; the secondary ones sometimes rudimentary.

Height of corallum averages 7 mm.; diameter of calicular apertures, three-quarters of a mm.

Rodriguez (*Slater*); Mauritius (*Brit. Mus., Jena Mus.*).

Distinguished from the preceding by its much larger calicular apertures, by the coarser coenenchymal tissue and less delicately spinulous surface; closely allied to *M. scabricula*, but the latter is a more massive species, with uniformly tuberculate surface. *M. lichen*, with which I was formerly inclined to unite it, has prominent calicles, a character totally absent in *M. explanata*.

Montipora lichen.

Manopora lichen, Dana, U.S. Expl. Exp. Zooph. p. 492 (1846).
Montipora lichen, M. Edw. Hist. Nat. Corall. III., p. 218.

There are several fragments in the collection which seem to belong to the above species. The corallum is explanate, foliaceous, partly bifacial, partly incrusting. Surface more or less even; margin in one fragment reflexed, in another straight and somewhat tumid. Coenenchyma spongious, with most delicately echinulate surface, the minute spines being generally united to small packets. Calicles very small, unequally distant on the upper surface, crowded below, nearly everywhere encircled by a coenenchymal ring, which is either continuous or separated into six rounded papillæ. Primary septa well-developed, thin; secondary ones rudimentary.

Pocilliporidæ.

Pocillipora brevicornis.

Pocillipora brevicornis, Lamarck, Hist. Anim. s. Vert. II., p. 275 (1816); M. Edw. Hist. Nat. Corall. III., p. 304.

The present collection contains the true *P. brevicornis*, as well as a singular variety, which may be termed *P. brevicornis* var. *cerebrum*. The corallum is hemispherical, brain-like. The principal branches are much enlarged and compressed, confluent, mæandrous, subsimple, with nearly smooth summits and obtusely verrucose lateral surfaces. This form would, according to the classification of Milne-Edwards, be placed in quite another section of the genus, viz. close to *P. mæandrina*, etc., and I should never have ventured to regard it as a variety of the *P. brevicornis*, but for the complete series of transitional examples in the collection.

Pocillipora favosa.

Pocillipora favosa, Ehrenb. Abh. Berl. Akad. Wiss. 1832, I., p. 351 (1834); M. Edw. Hist. Nat. Corall. III., p. 305.

Pocillipora grandis.

Pocillopora grandis, Dana, U.S. Expl. Exp. Zooph. p. 533, pl. 51, fig. 2 (1846); M. Edwd. Hist. Nat. Corall. III., p. 307.

The principal branches are, as a rule, more elongate, and their summits more mæandrous than in the specimen figured by Dana. The lateral verrucæ are very variable: either crowded or more remote, or (on the outer surface of lateral branches in some of the coralla) even totally wanting; either appressed, or more horizontal; generally pointed and angular, but in one example obtuse and rounded. One specimen approaches the *P. verrucosa* in having the summits of minor branches covered with verrucæ.

Perhaps the Rodriguez coral is a distinct species; but since I have no specimens of the Pacific *P. grandis* for comparison, I abstain from characterising it under a new specific name. In my opinion, moreover, the species of *Pocillipora* have already been far too much multiplied.

LONDON:
Printed by George Edward Eyre and William Spottiswoode,
Printers to the Queen's most Excellent Majesty.

RANUNCULUS TRULLIFOLIUS, *Hook.f*

Trans of Venus Exped.   KERGUELEN ISLAND.   Botany, Pl. 1

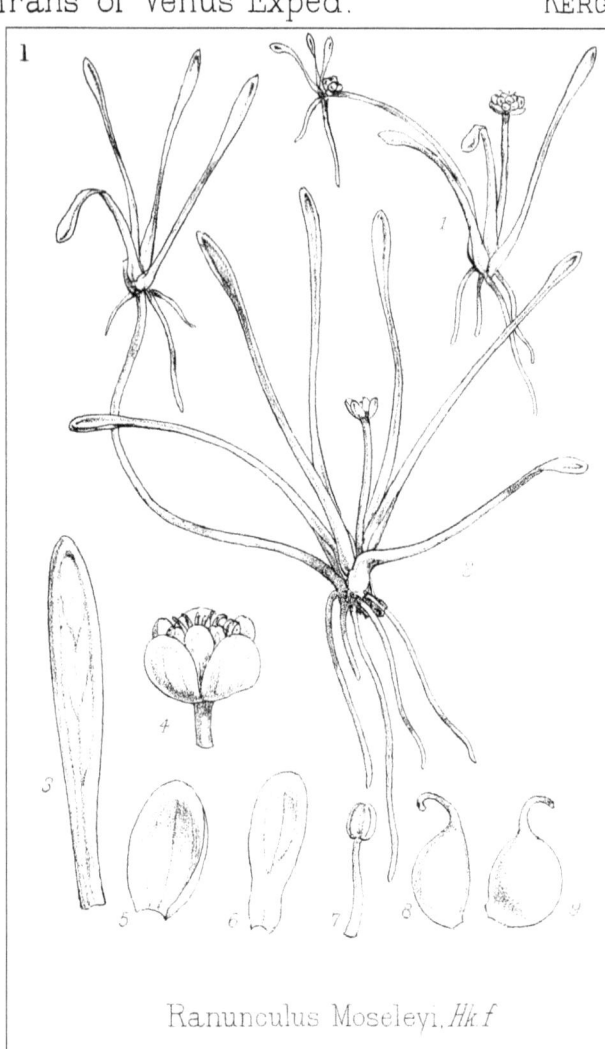

Ranunculus Moseleyi, Hk.f.

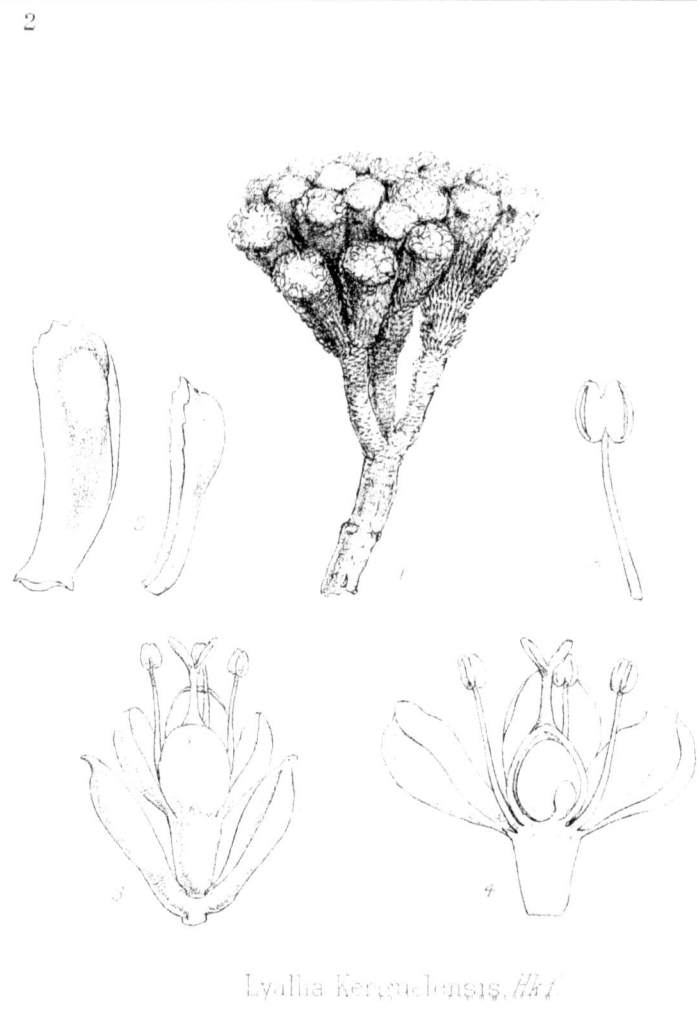

Lyallia Kerguelensis, Hk.f.

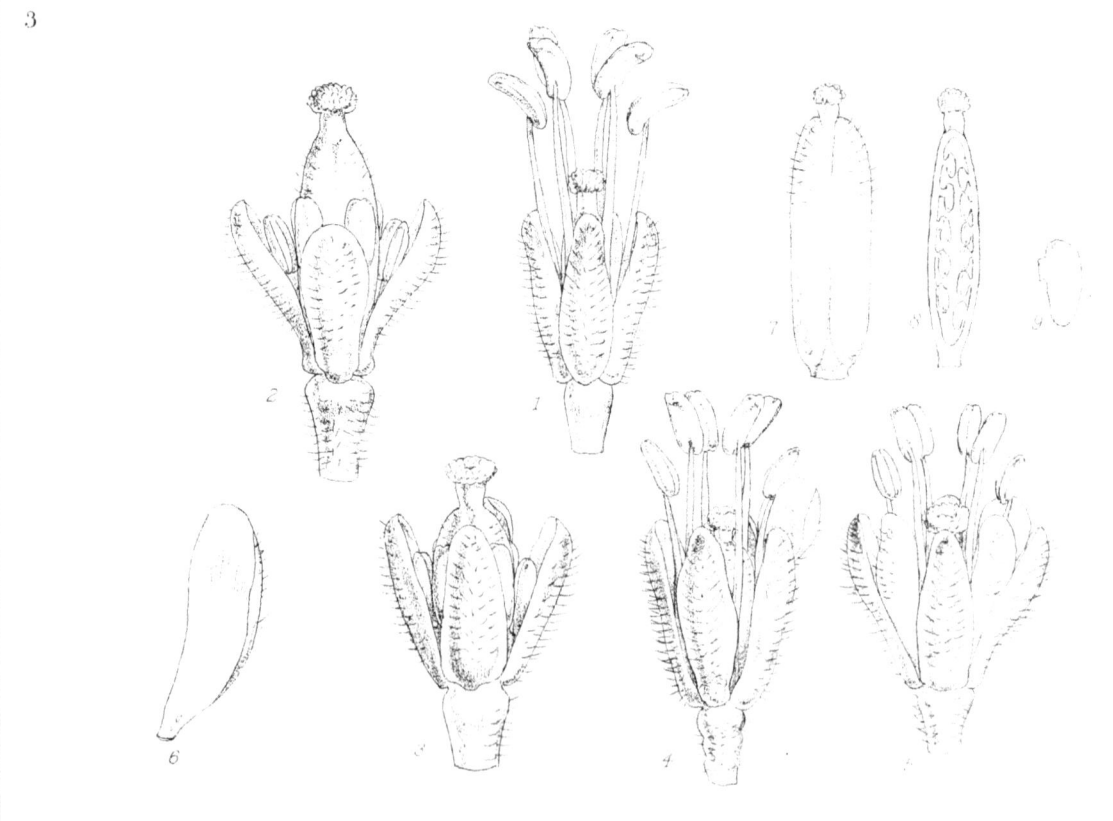

Pringlea antiscorbutica, Hk.f.

W.H.Fitch del. et lith.   M & N Hanhart imp.

Trans of Venus Exped.   Kerguelen Island   Botany. Pl. III.

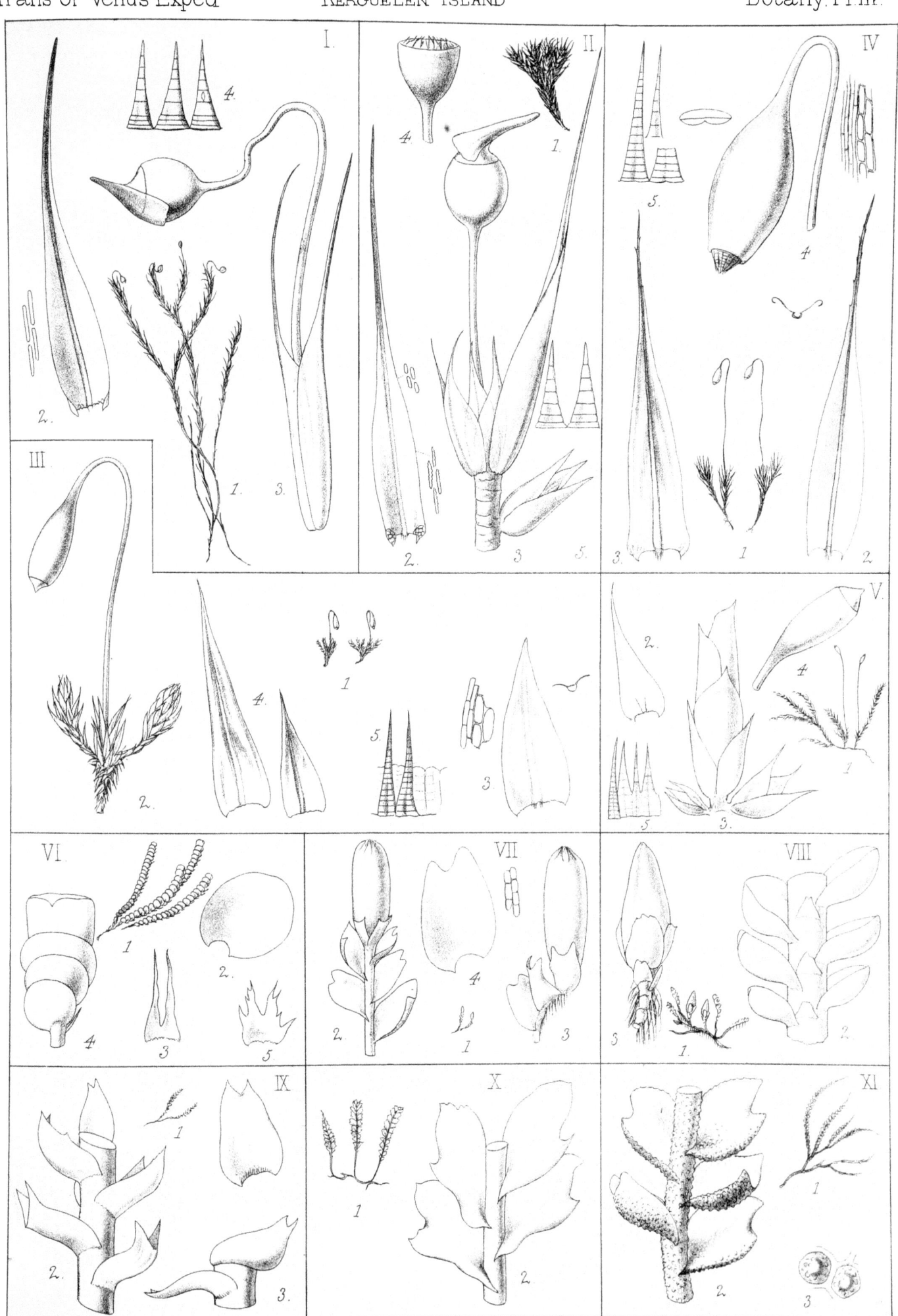

W. Mitten, delt.   J.N. Fitch, lith.

Trans. of Venus Exped.   KERGUELEN ISLAND.   Botany Pl.IV

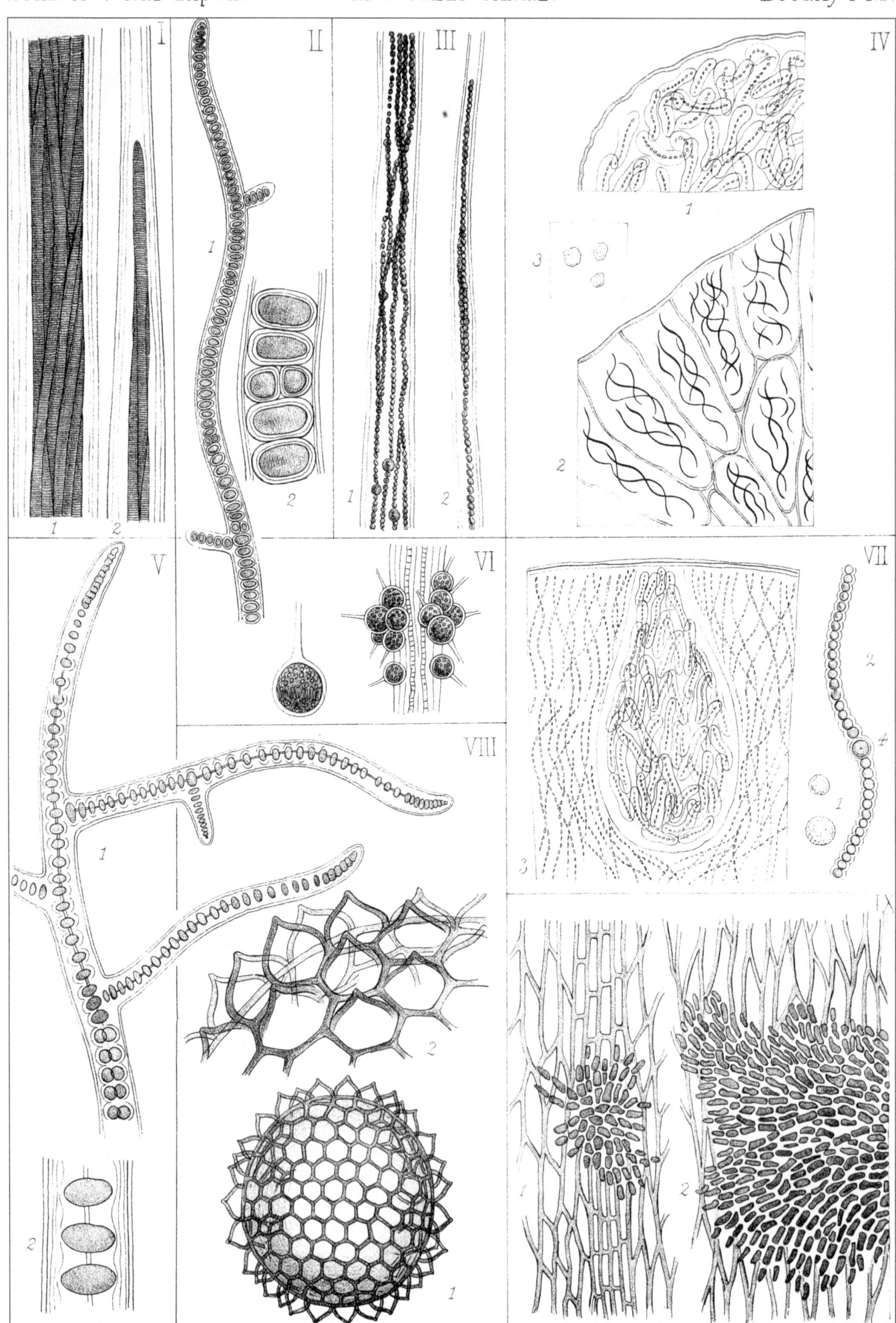

J.N.Fitch lith.   Fitch imp.

Trans. of Venus Exped.   Kerguelen Island.   Botany, Pl. V.

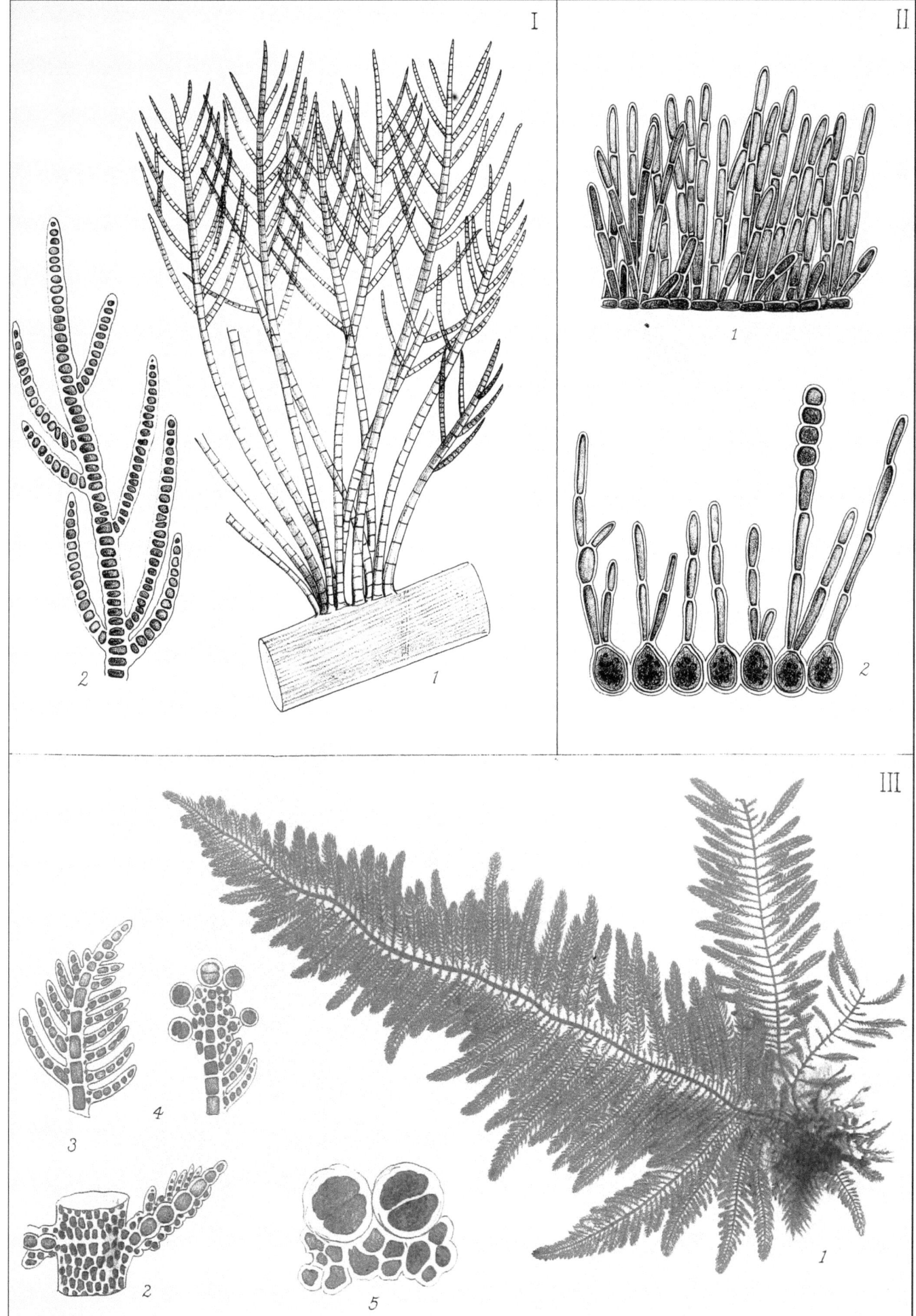

J.N.Fitch lith.   Fitch imp.

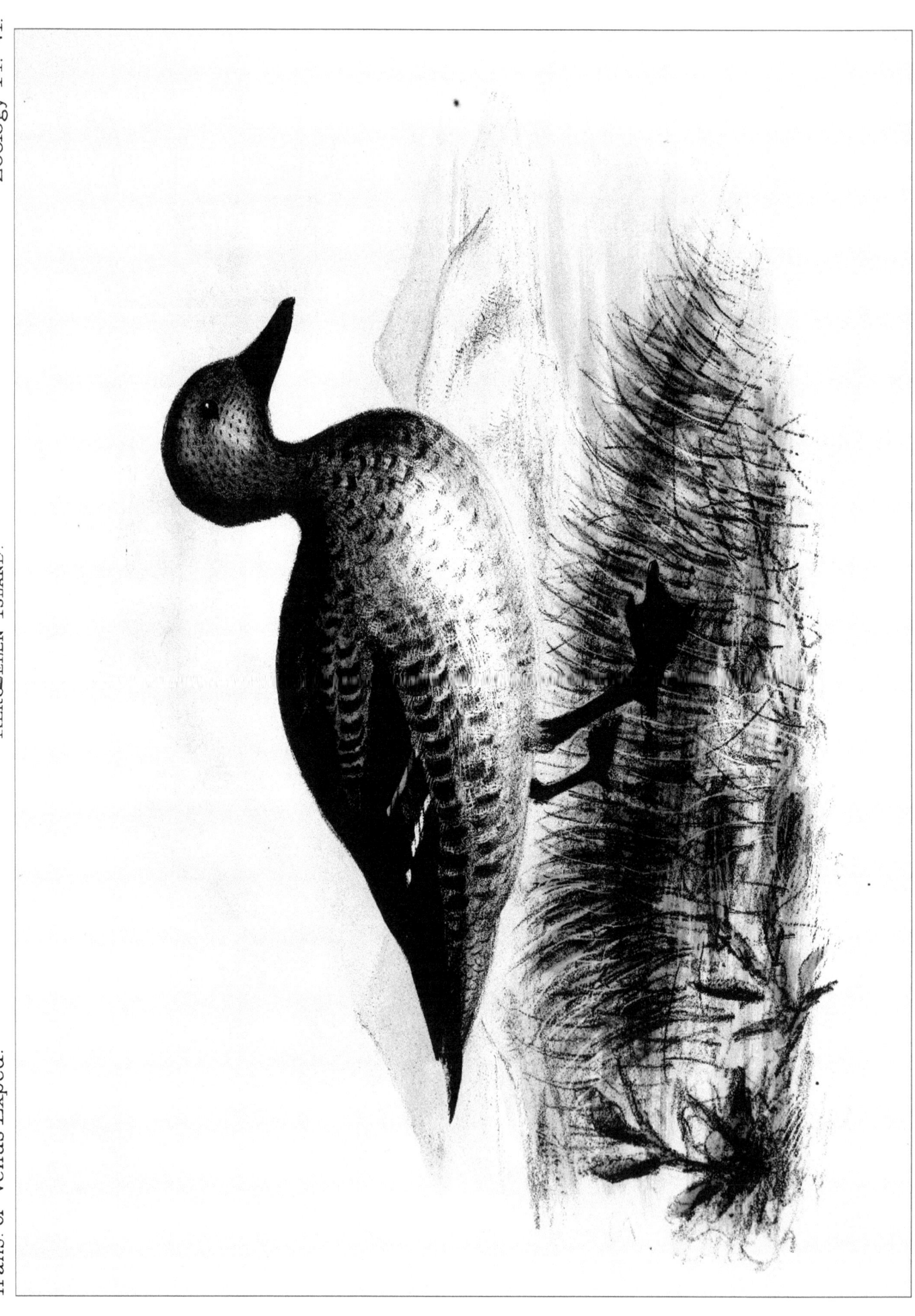

Trans. of Venus Exped.   Kerguelen Island.   Zoology Pl. VI.

Trans. of Venus Exped.   Kerguelen Island   Zoology Pl. VII.

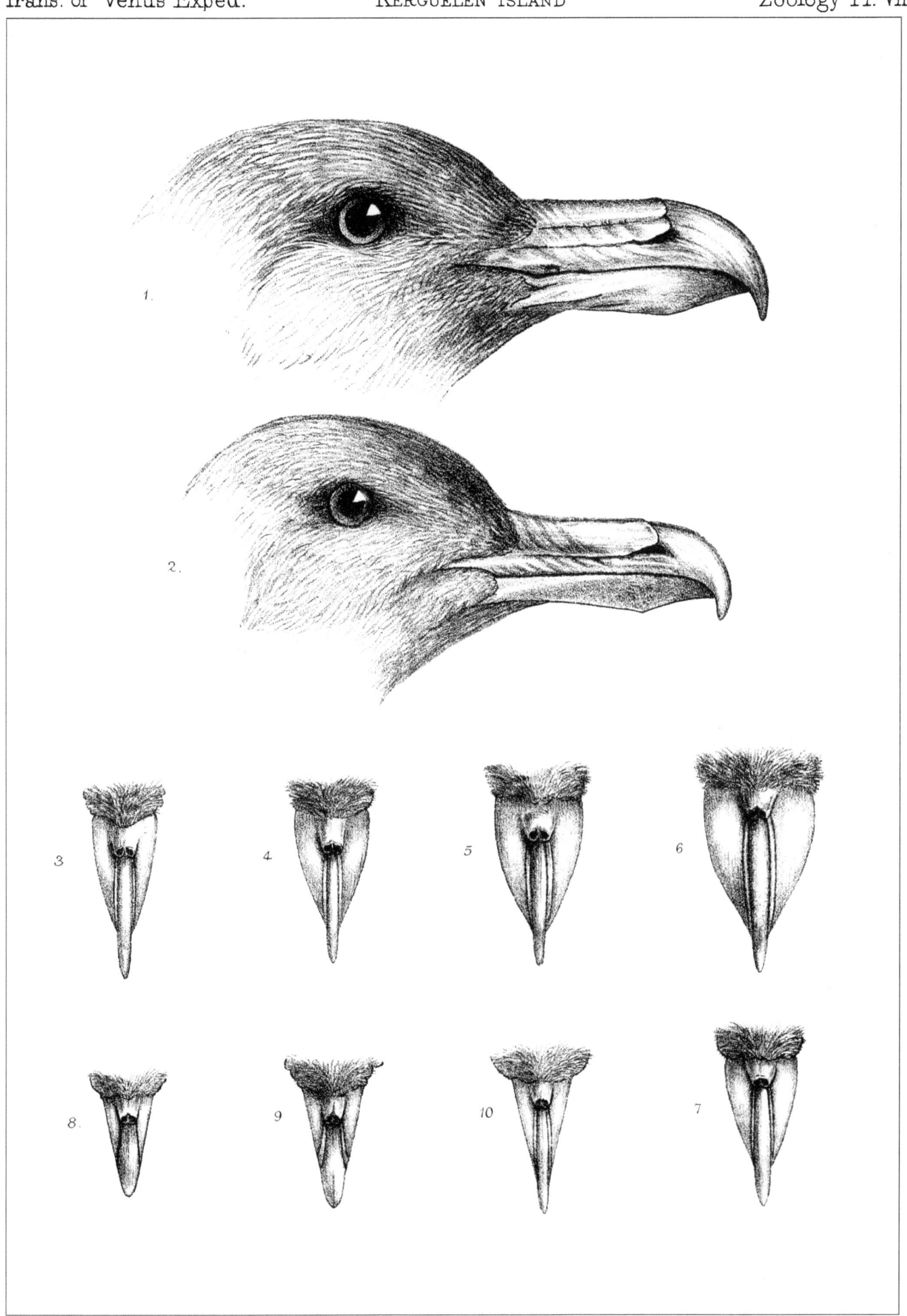

Cyril Davenport del. et lith.   Mintern Bros. imp.

Trans. of Venus Exped.   KERGUELEN ISLAND   Zoology Pl. VIII.

Cyril Davenport del.et lith.   Mintern Bros. imp.

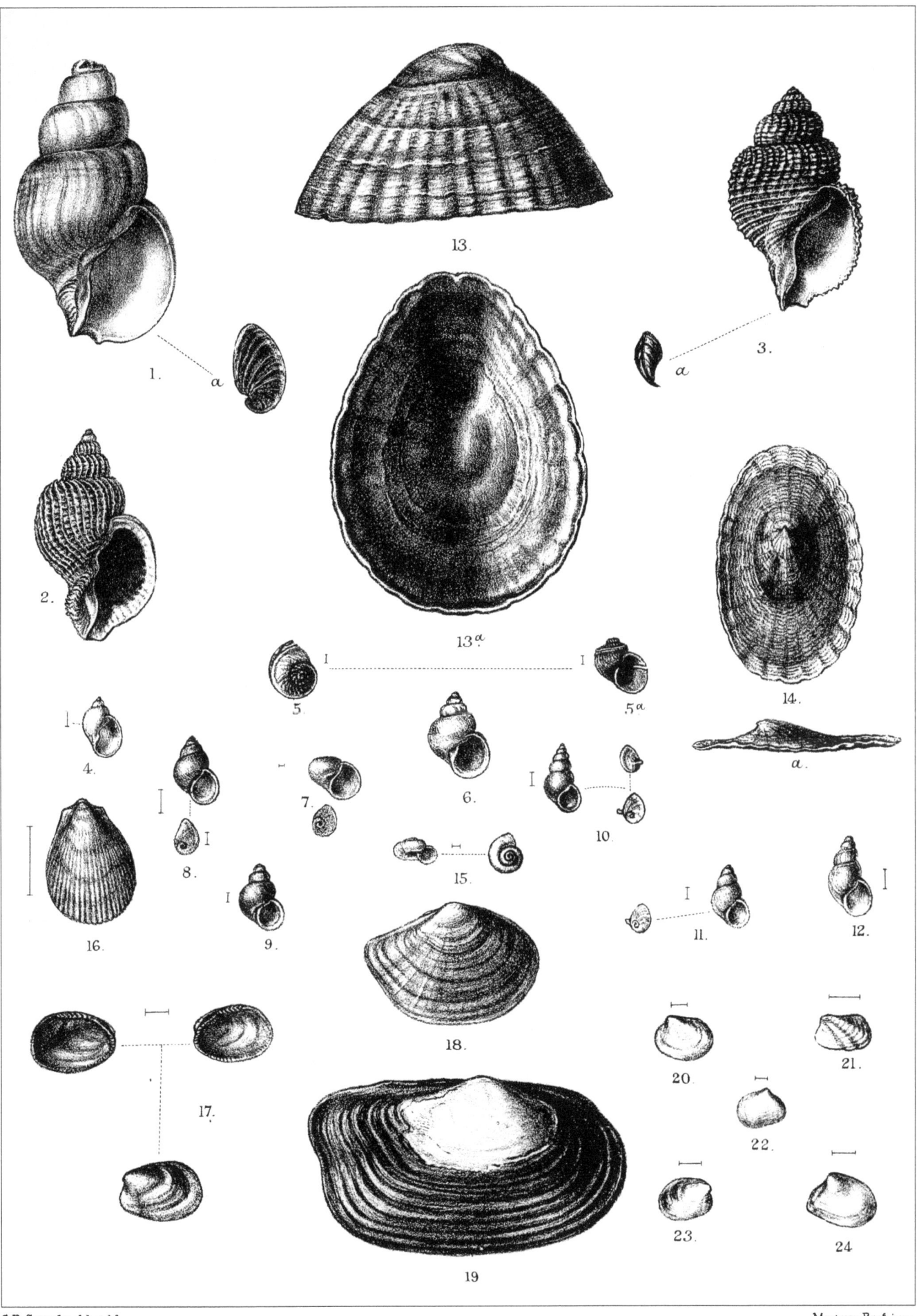

Trans. of Venus Exped.   KERGUELEN ISLAND.   Zoology Pl. XI.

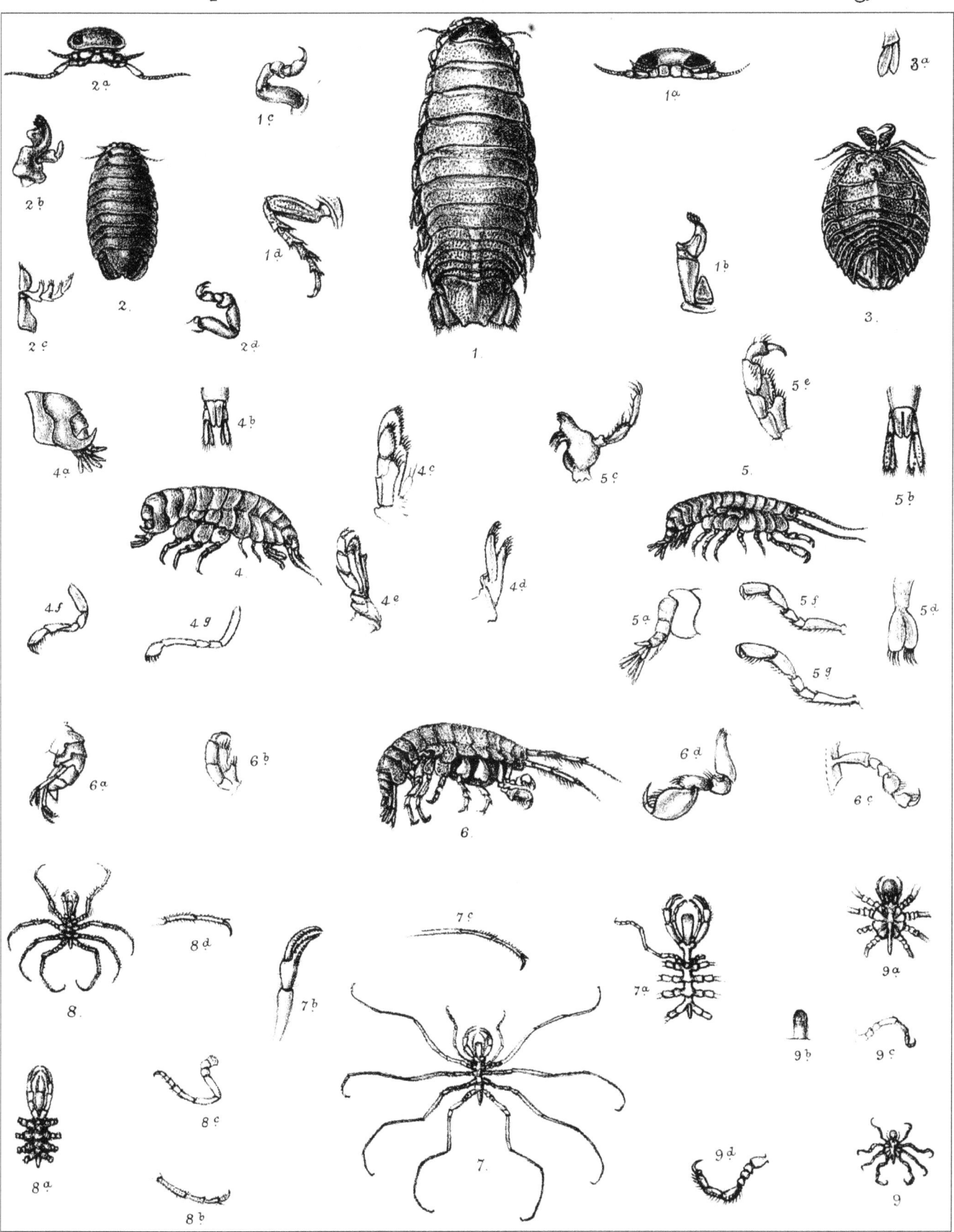

G.B.Sowerby del. et lith.   Mintern Bros imp.

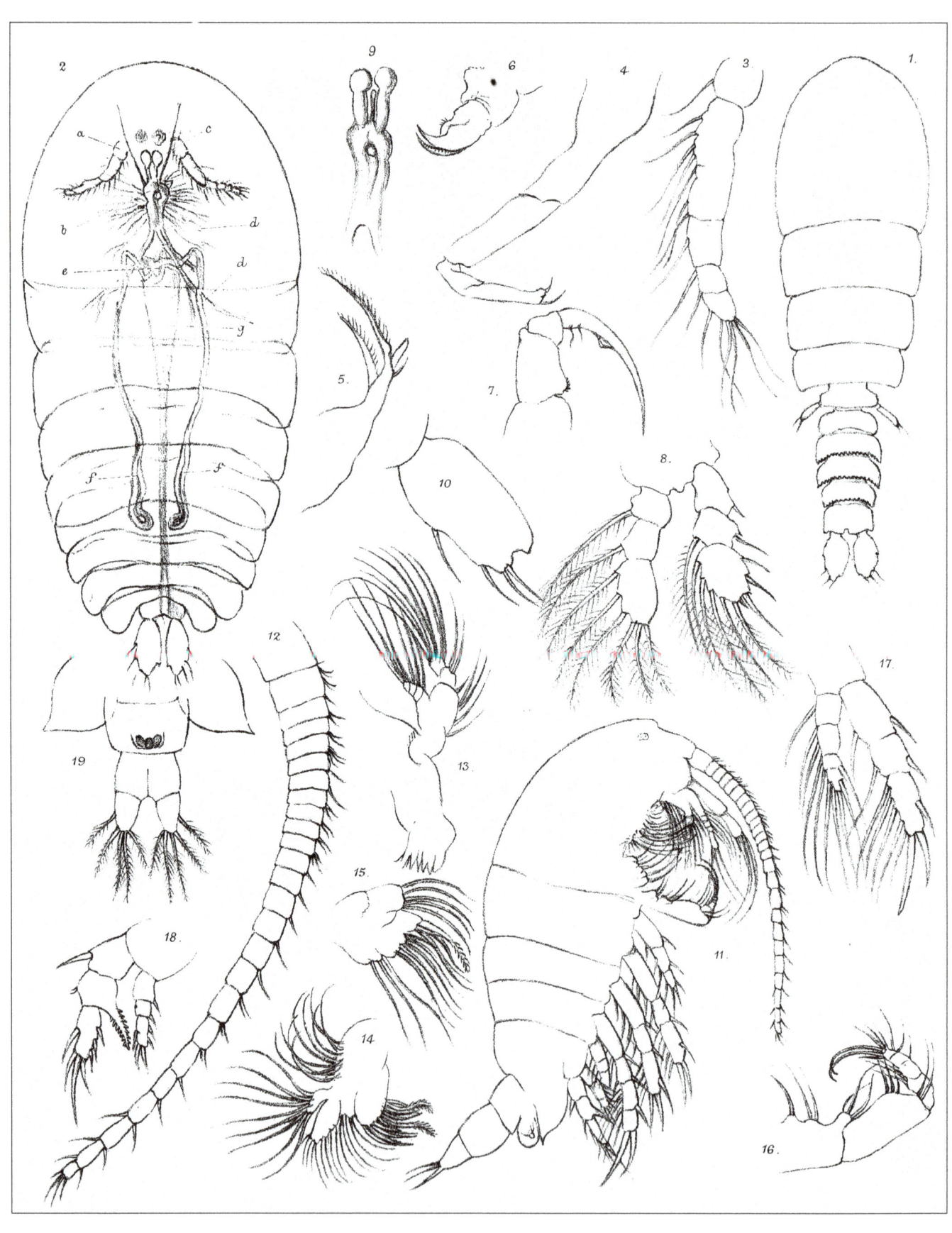

1–10 Sapphirina danæ
11–19 Centropages brevicaudatus

Trans. of Venus Exped.   KERGUELEN ISLAND.   Zoology Pl. XIII

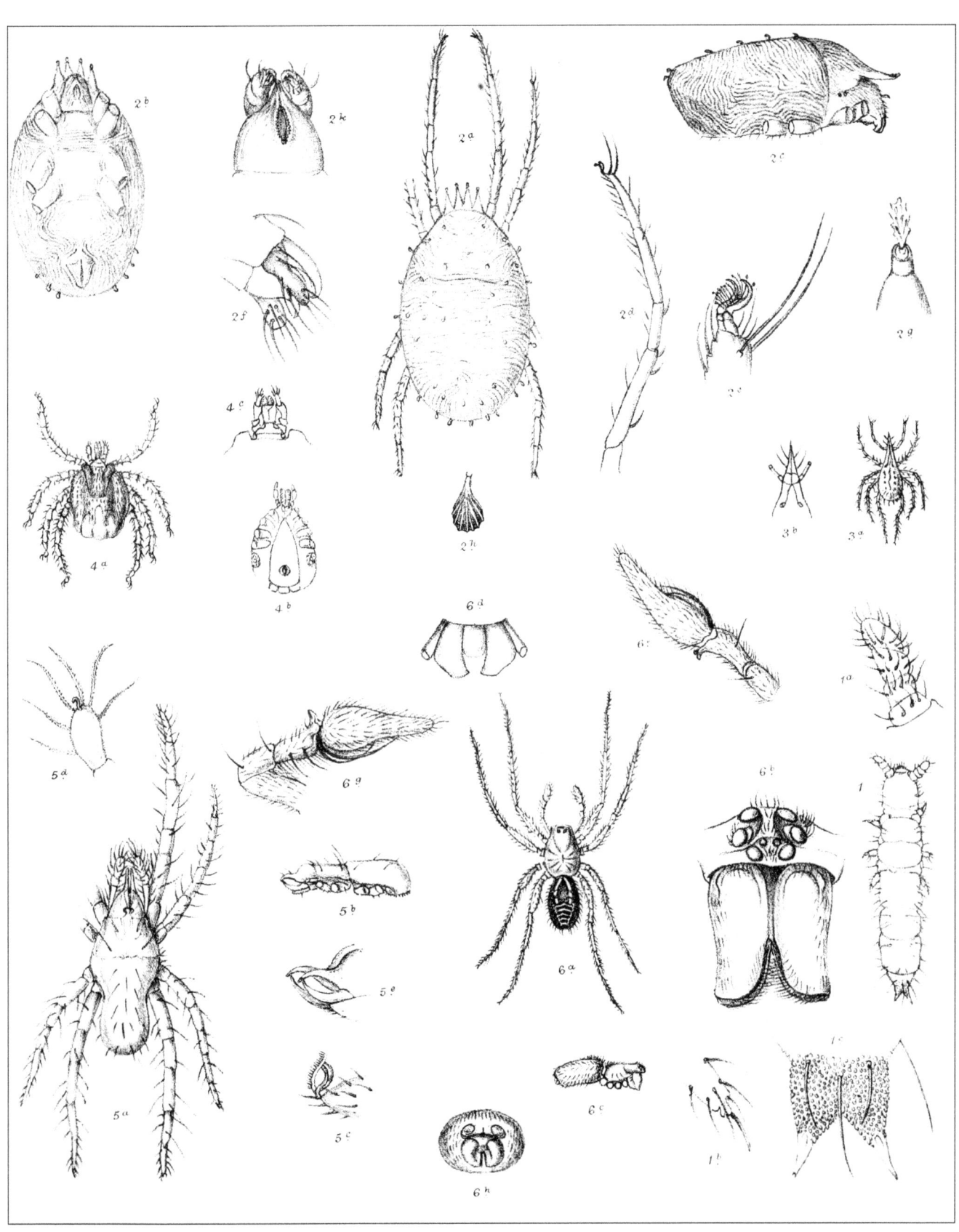

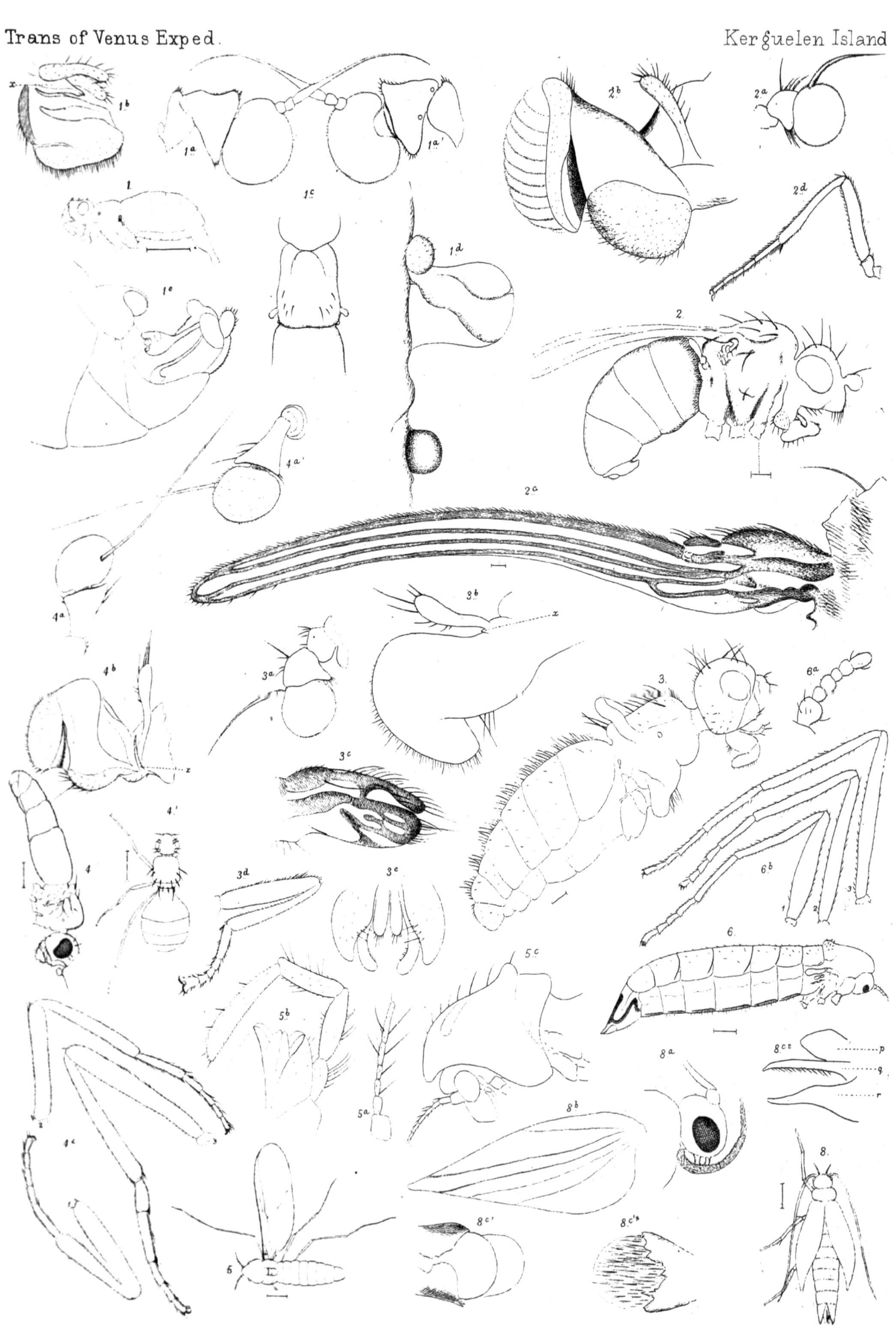

Trans of Venus Exped.   Kerguelen Island

Zoology. Pl. XIV.

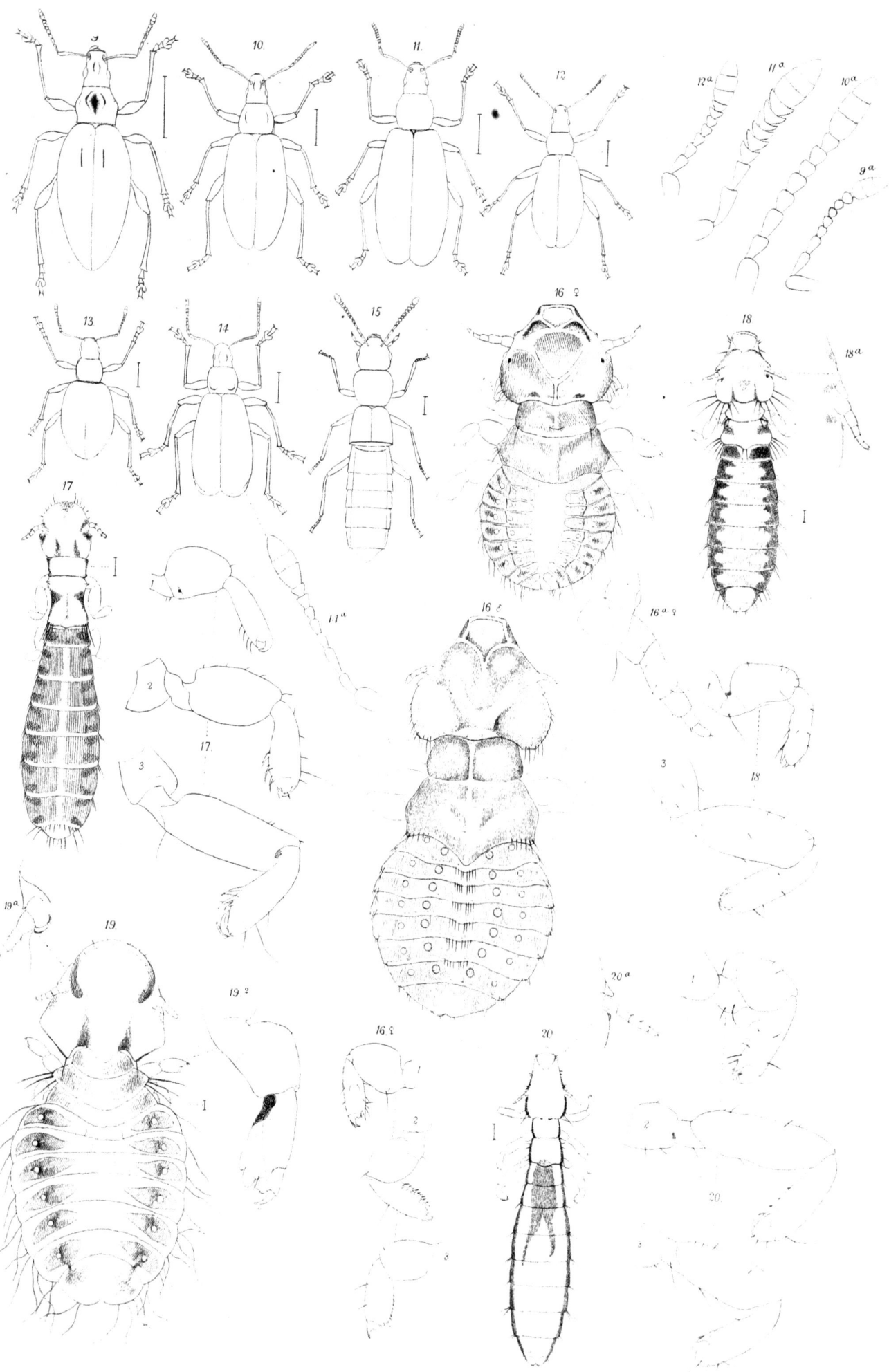

W. West & Co. sc.

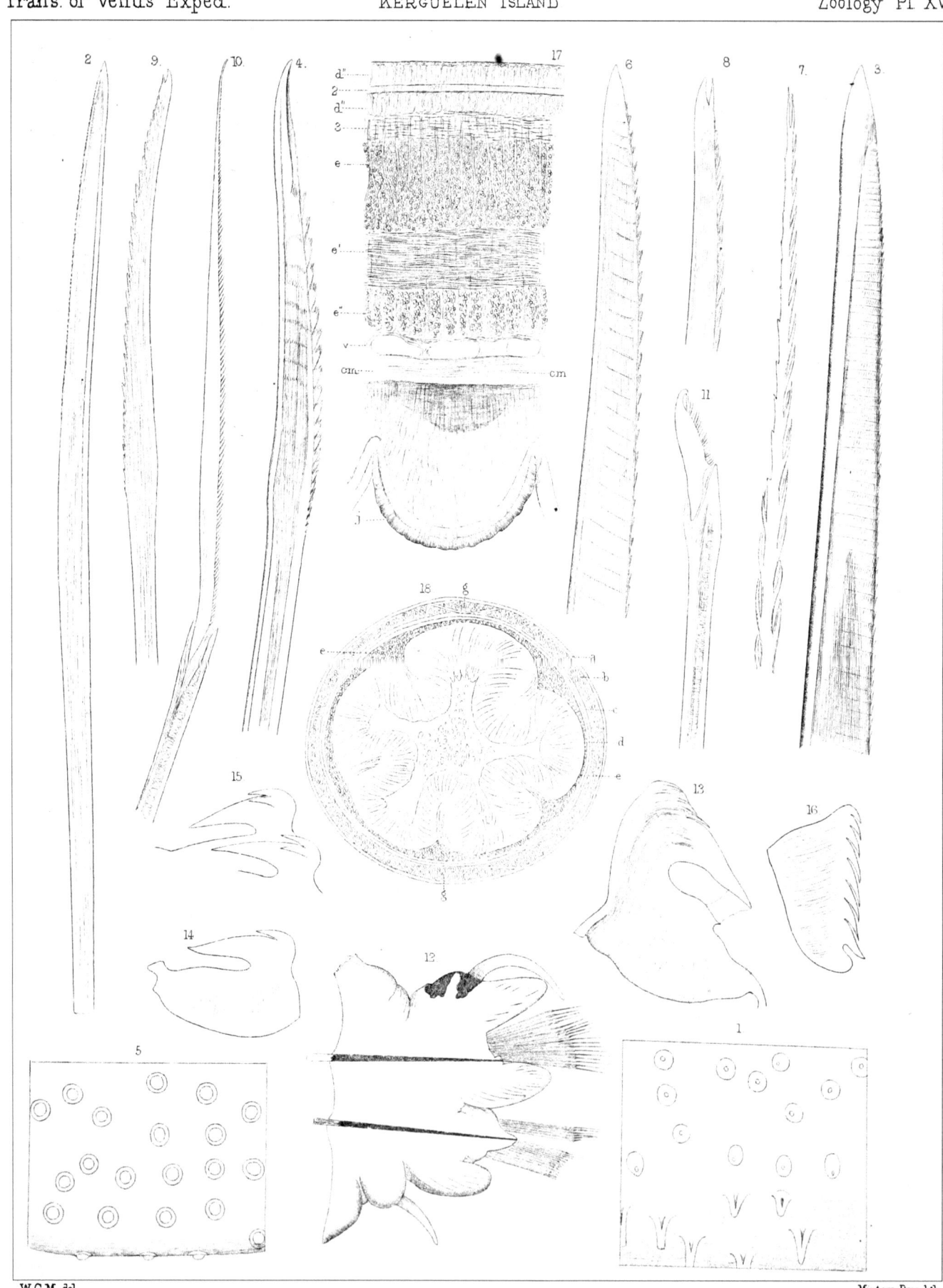

Trans. of Venus Exped.   KERGUELEN ISLAND   Zoology Pl. XVI.

C. Berjeau lith.   Mintern Bros. imp.

Trans. of Venus Exped.   Kerguelen Island   Zoology Pl. XVII.

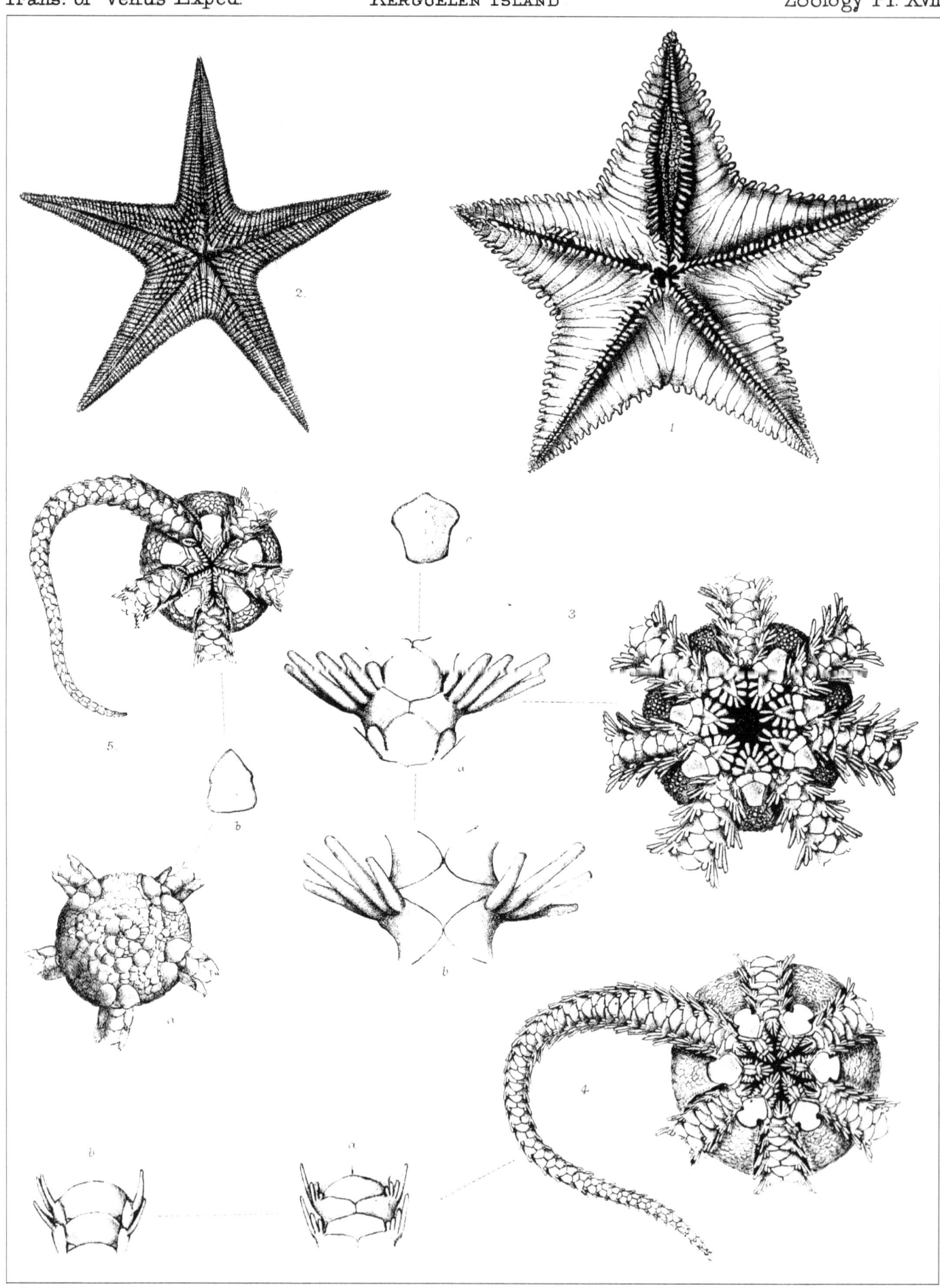

C. Berjeau lith.   Mintern Bros imp.

Trans. of Venus Exped.   KERGUELEN'S ISLAND.   Zoology Pl. XVIII.

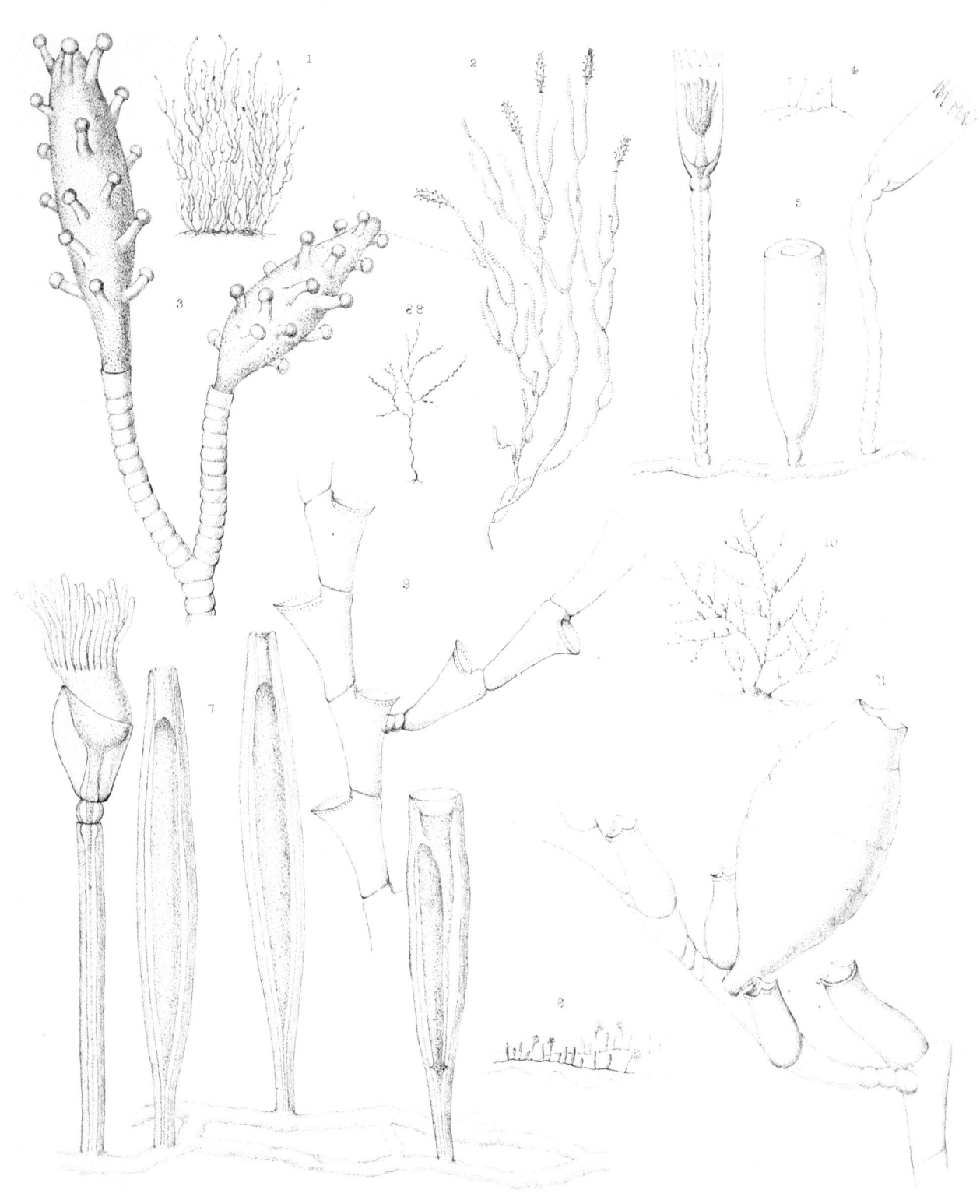

Auctor &T.W.del.TuffenWest sc.   W.West &Co.imp.

QUIVISIA LACINIATA, *Balf. fil.*

SCLEROCARYA CASTANEA, *Baker.*

MATHURINA PENDULIFLORA, Balf. fil.

RANDIA HETEROPHYLLA, *Balf. fil.*

FERNELIA BUXIFOLIA, *Lam.*

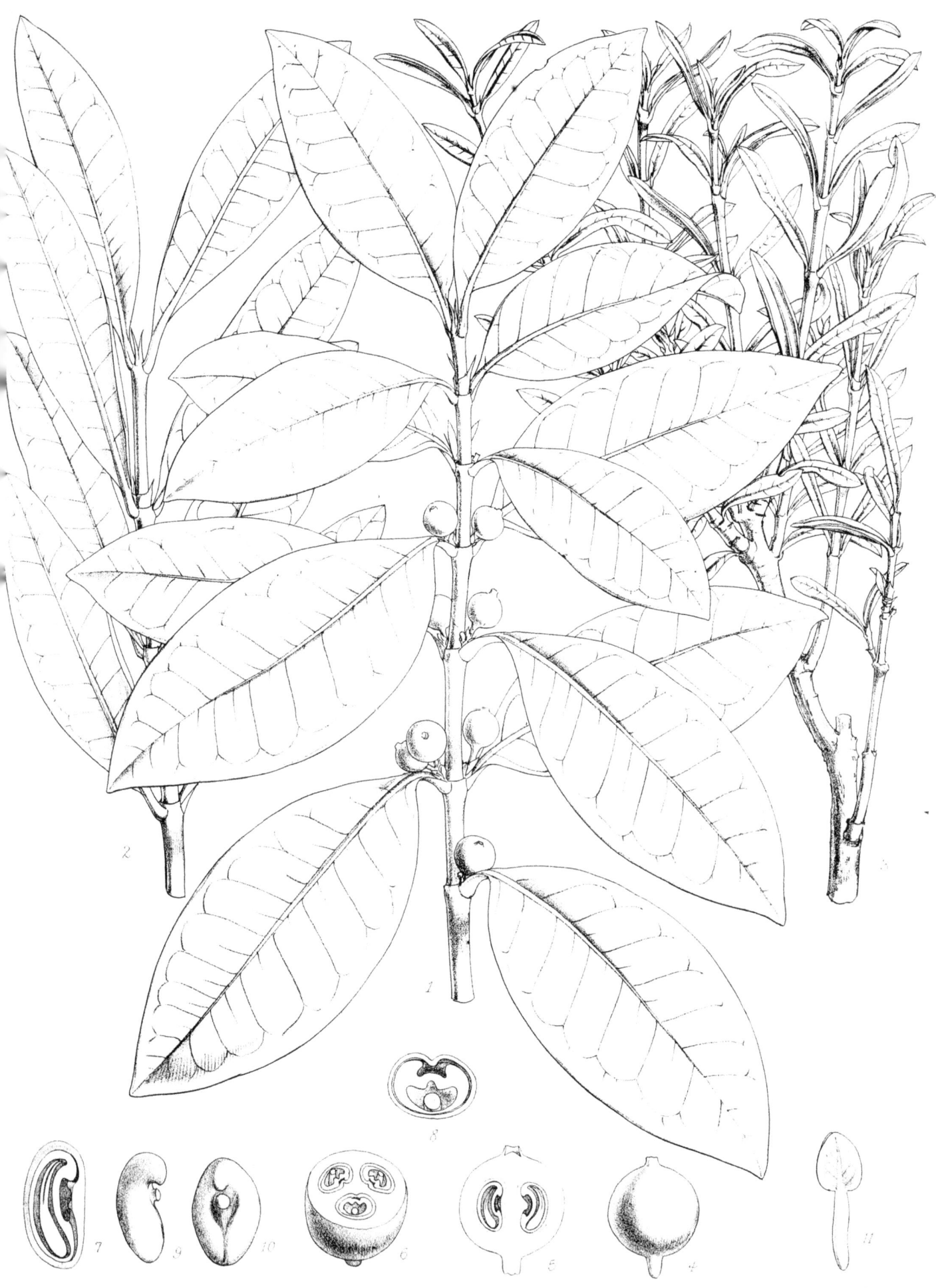

PYROSTRIA TRILOCULARIS. *Balf. fil.*

SCYPHOCHLAMYS REVOLUTA, Balf. fil.

PSIADIA CORONOPUS Benth.

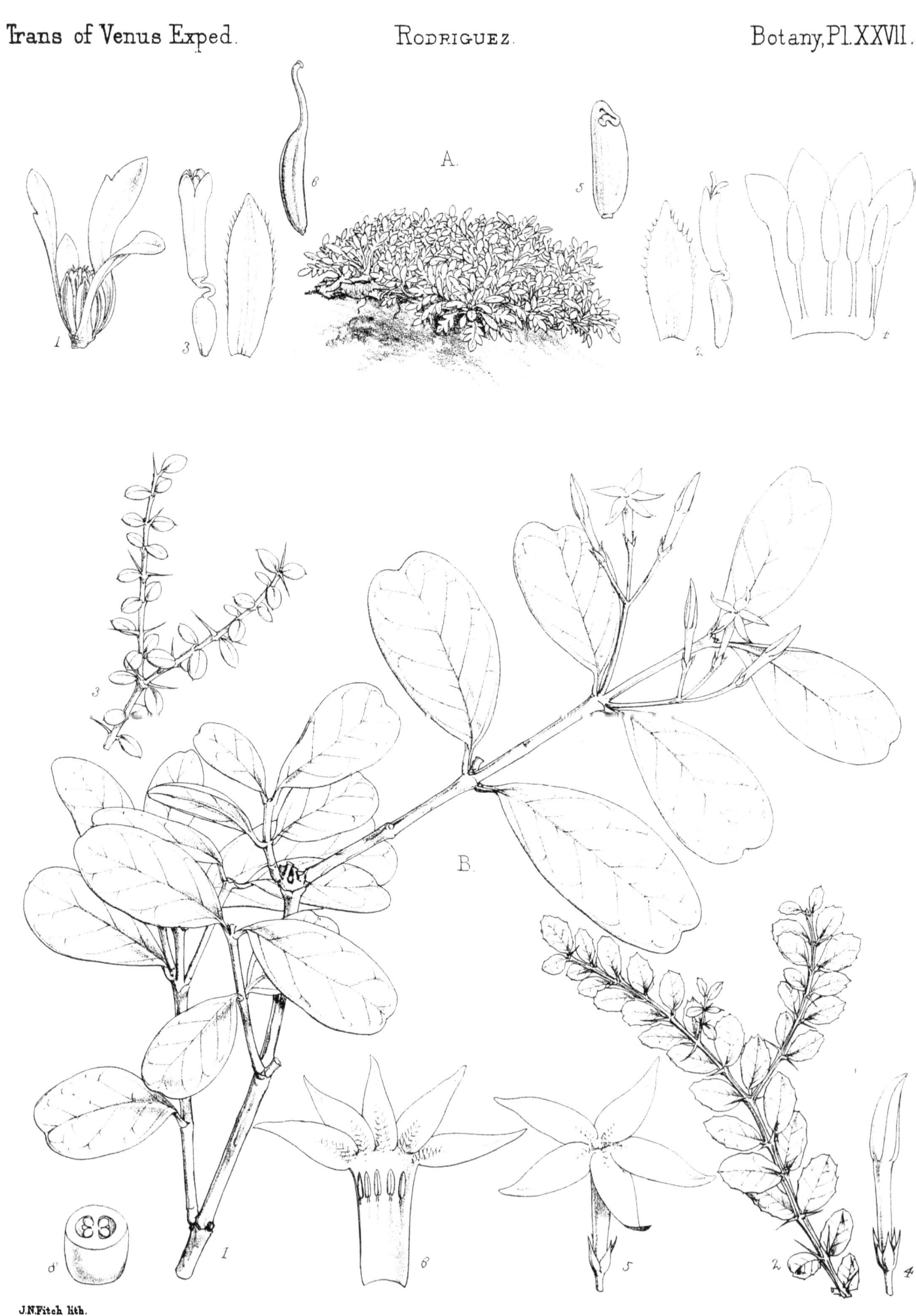

A ABROTANELLA RHYNCHOCARPA, *Balf.fil.*
B. CARISSA XYLOPICRON. *Pet Th.*

TANULEPIS SPHENOPHYLLA. *Balf. fil.*

1.3. SARCOSTEMMA VIMINALE, R Br.
4.10. " ODONTOLEPIS, Balf fil.

HYPOESTES RODRIGUESIANA, Balf. fil.

NESOGENES DECUMBENS, Balf. fil.

Trans of Venus Exped.   Rodriguez   Botany. Pl. XXXII

J.N.Fitch lith.

CLERODENDRON LACINIATUM, *Balf. fil.*

Fitch imp.

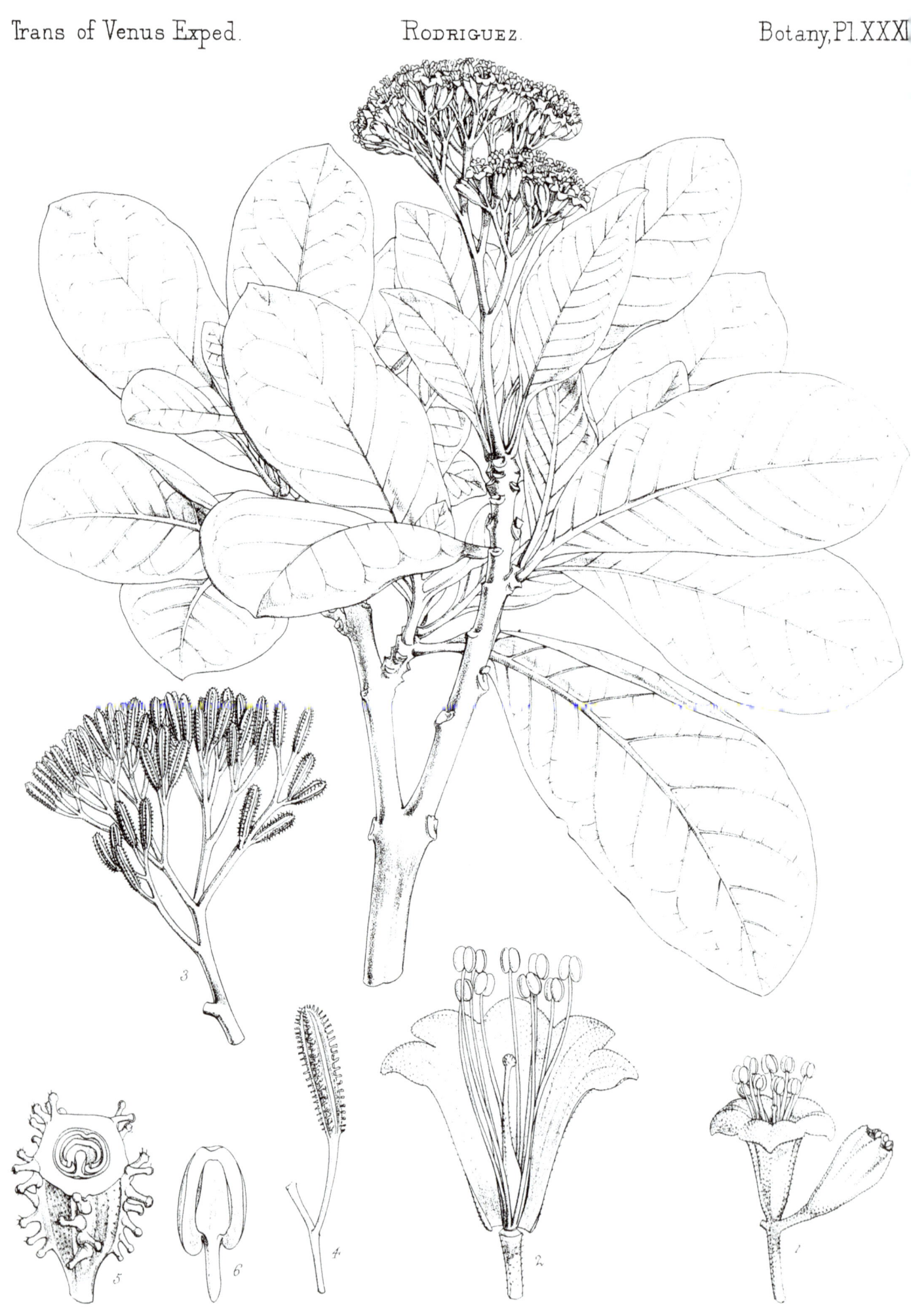

PISONIA VISCOSA, *Bulf. fil.*

PEPEROMIA RETICULATA, Balf. fil.

LISTROSTACHYS APHRODITE, *Balf.fil. & S.Moore*

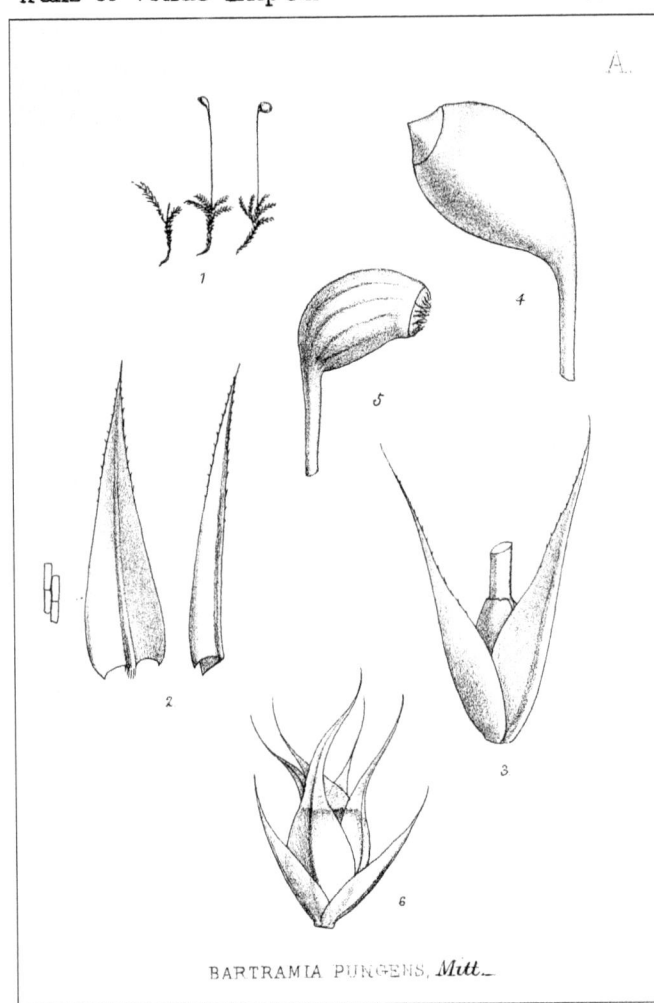
BARTRAMIA PUNGENS, Mitt.

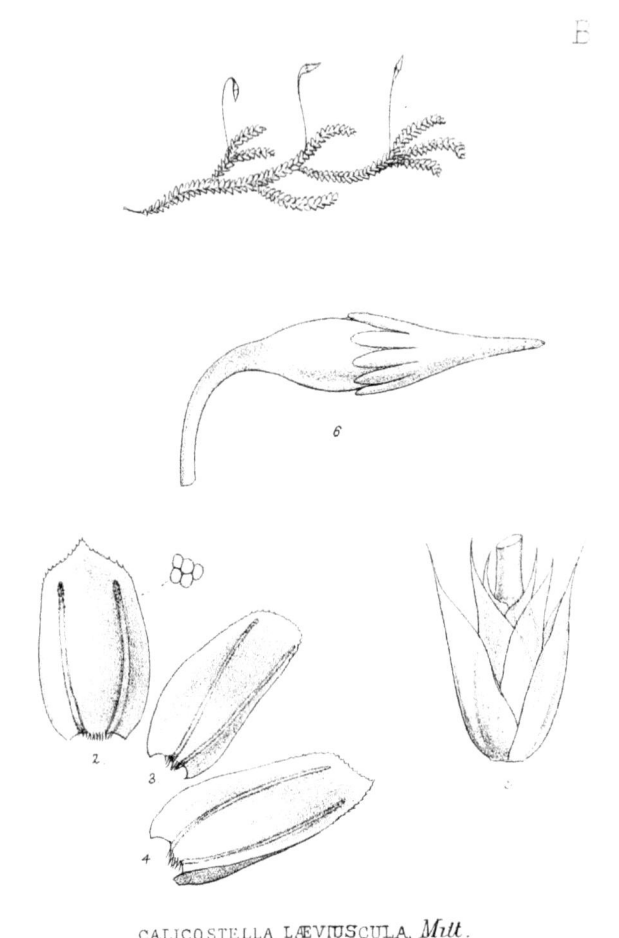
CALICOSTELLA LÆVIUSCULA, Mitt.

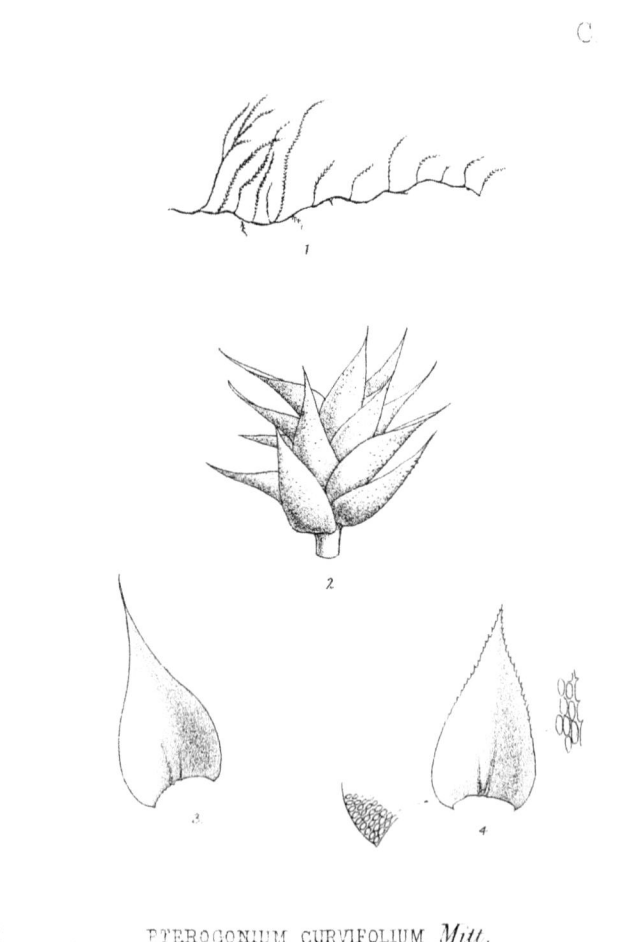
PTEROGONIUM CURVIFOLIUM, Mitt.

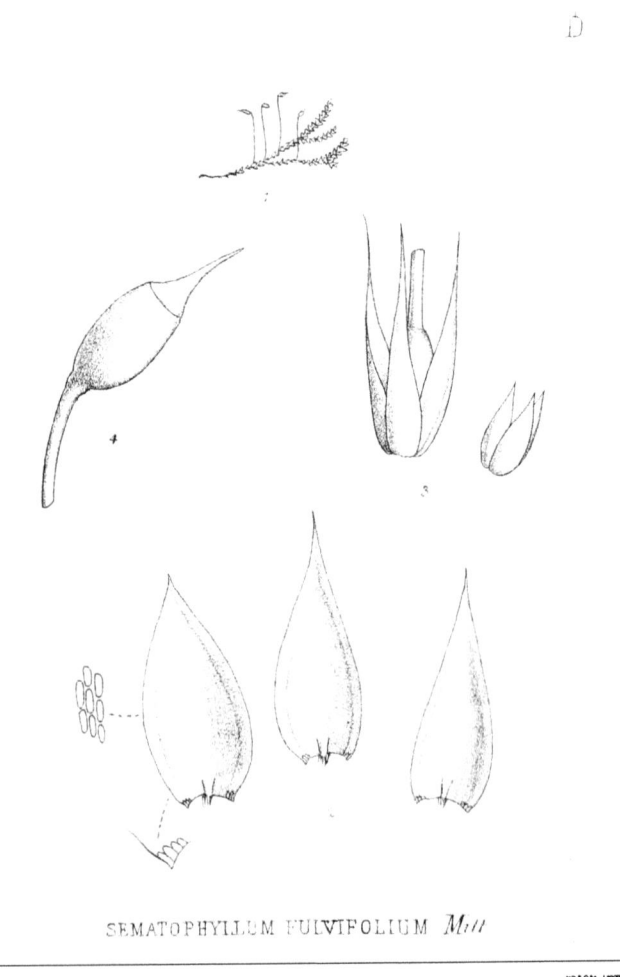
SEMATOPHYLLUM FULVIFOLIUM, Mitt.

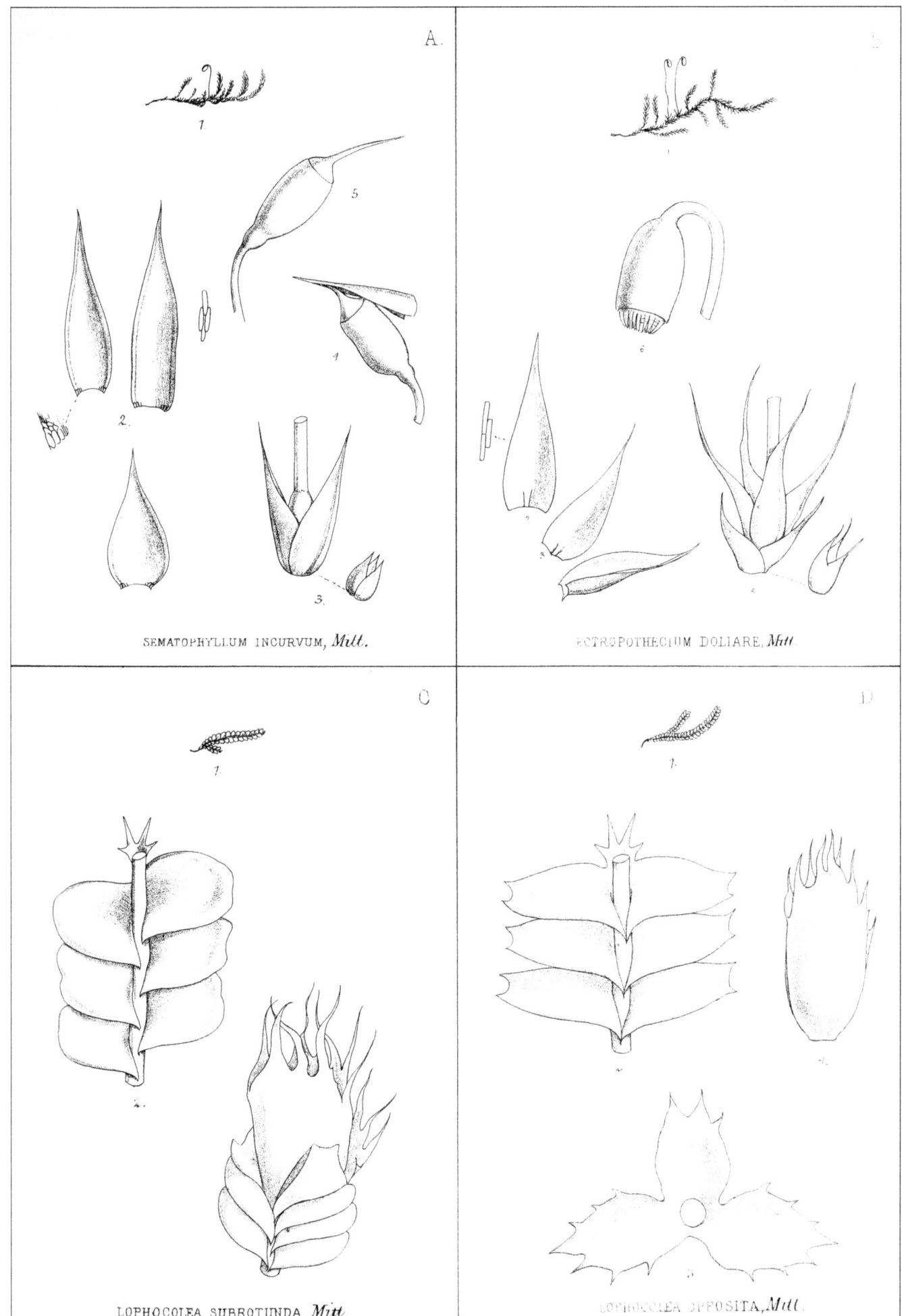

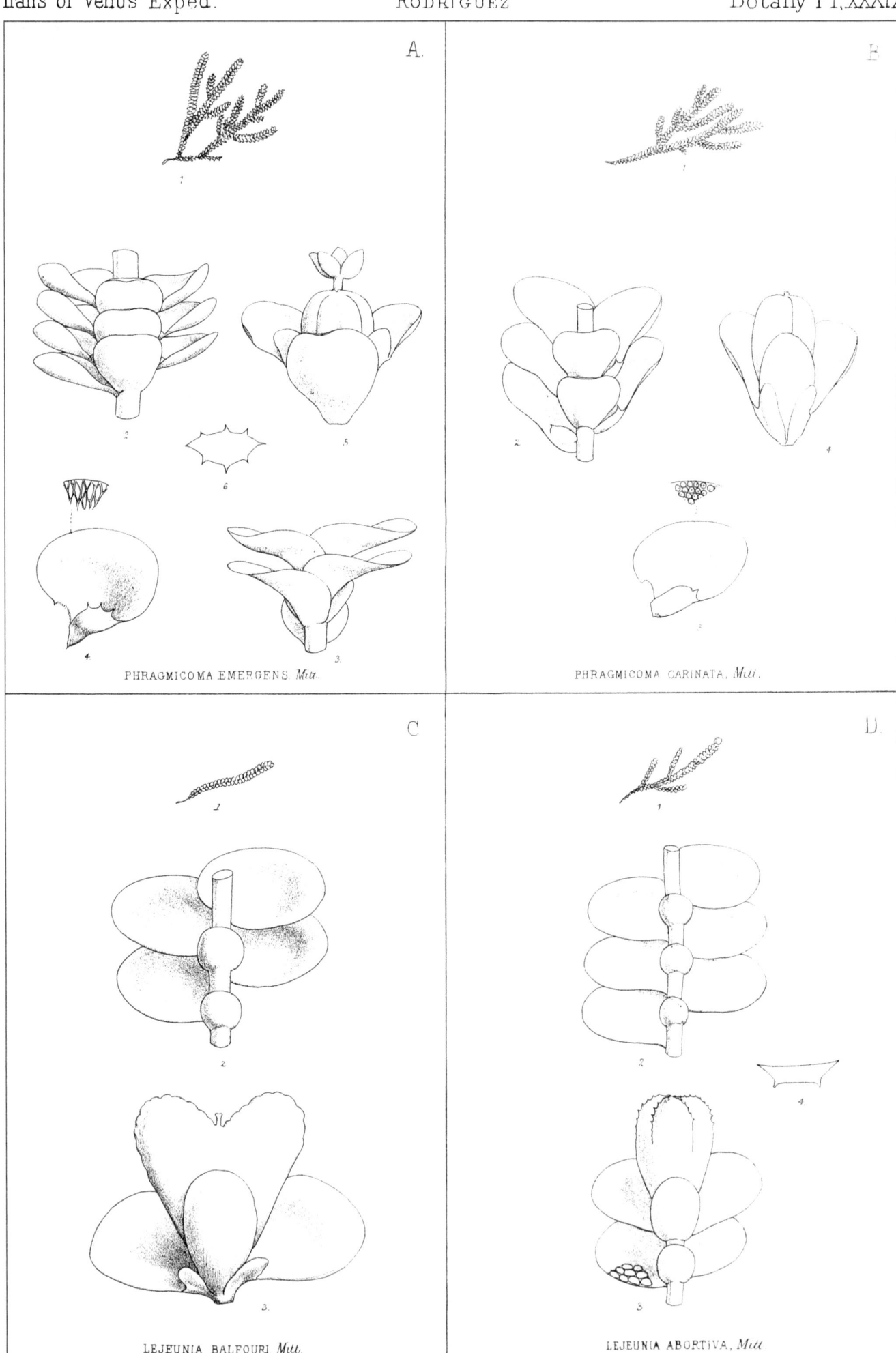

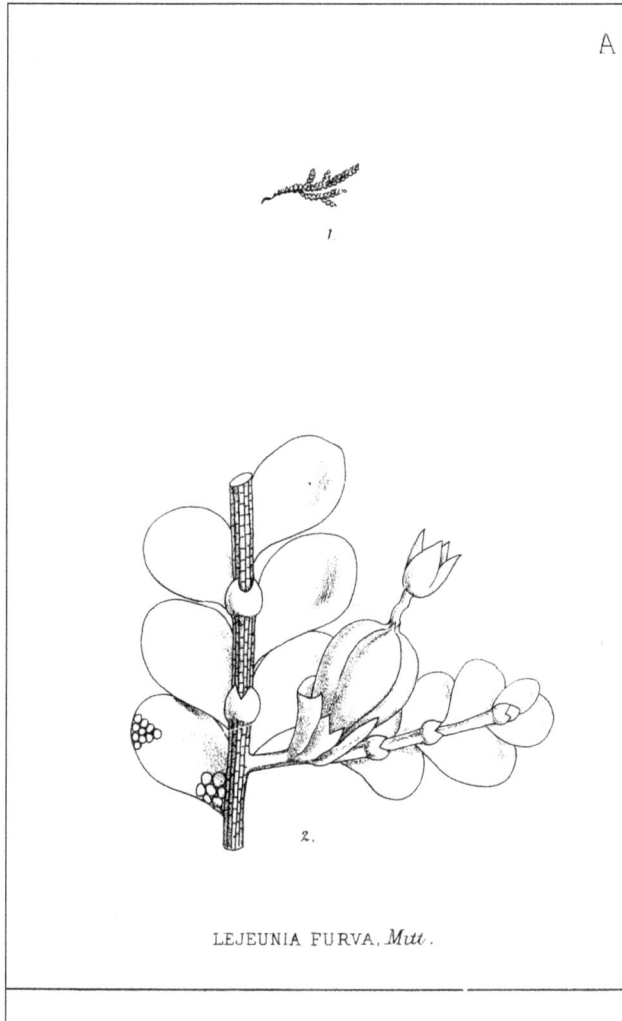

LEJEUNIA FURVA, Mitt.

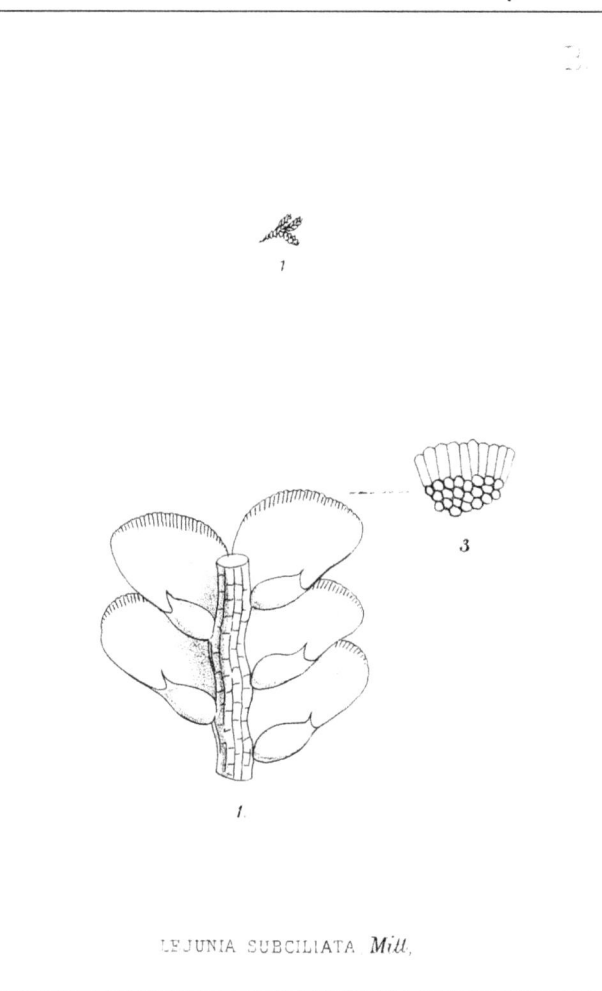

LEJUNIA SUBCILIATA, Mitt.

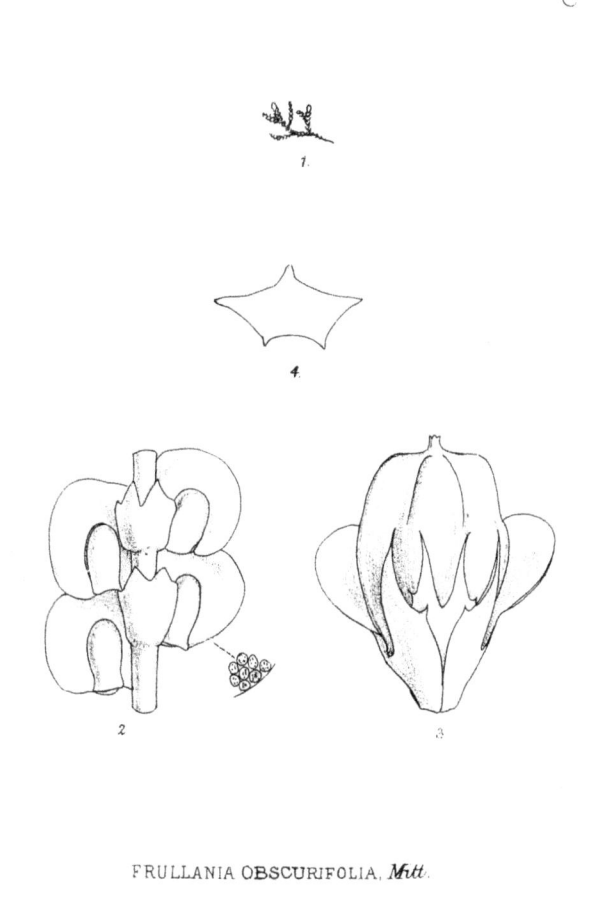

FRULLANIA OBSCURIFOLIA, Mitt.

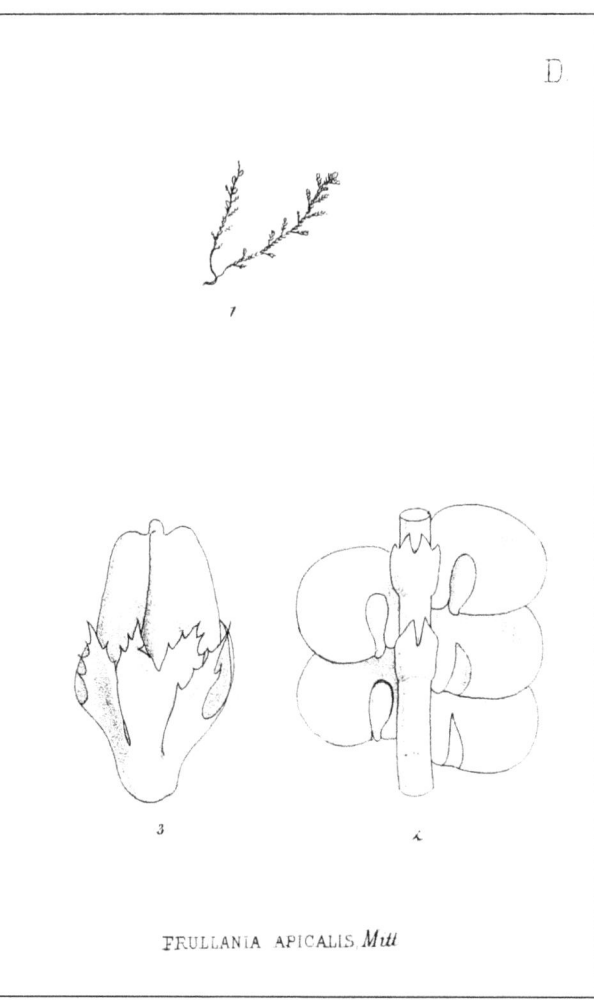

FRULLANIA APICALIS, Mitt.

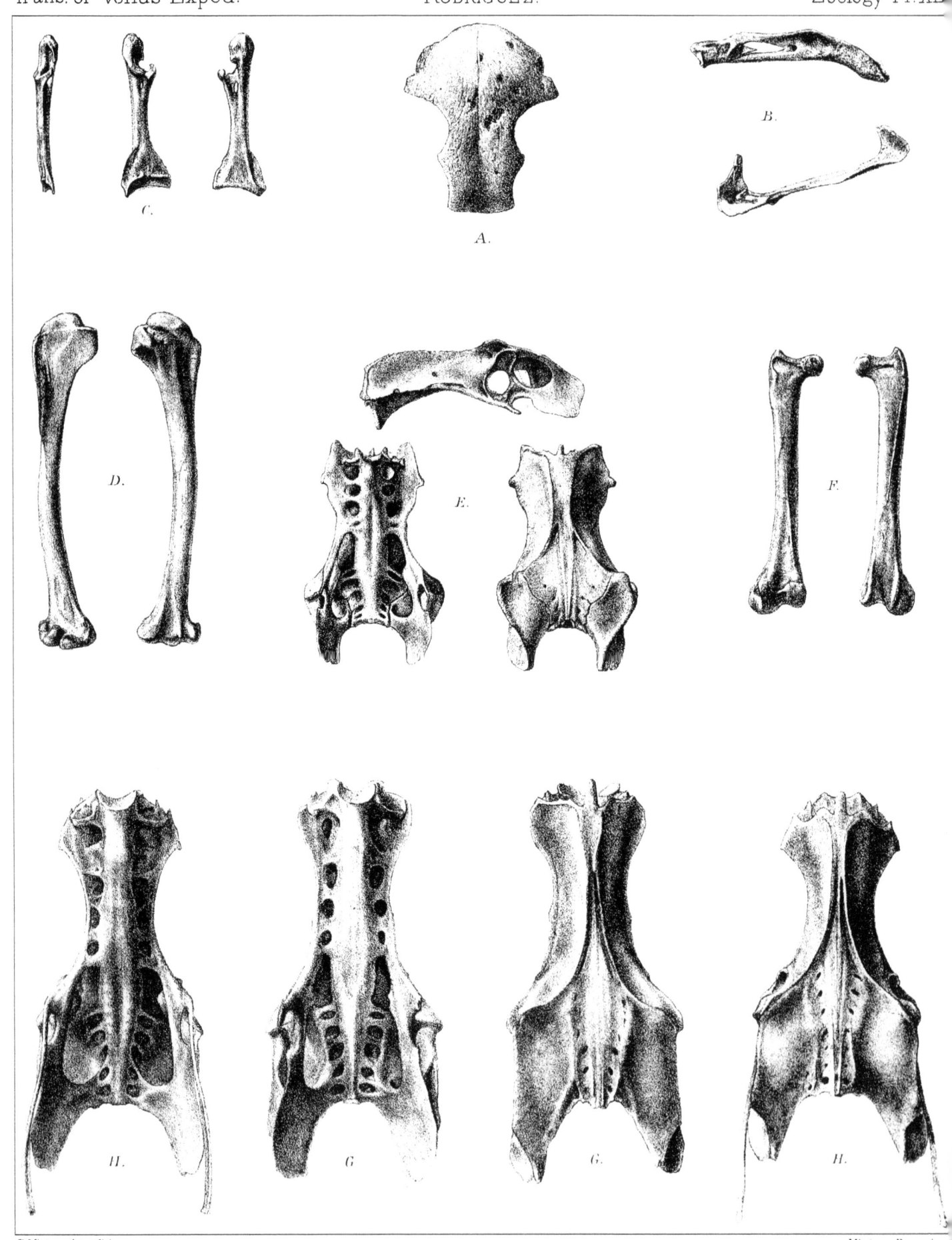

A.–F. Carine murivora. G. Nycticorax megacephalus. H. Nycticorox griseus

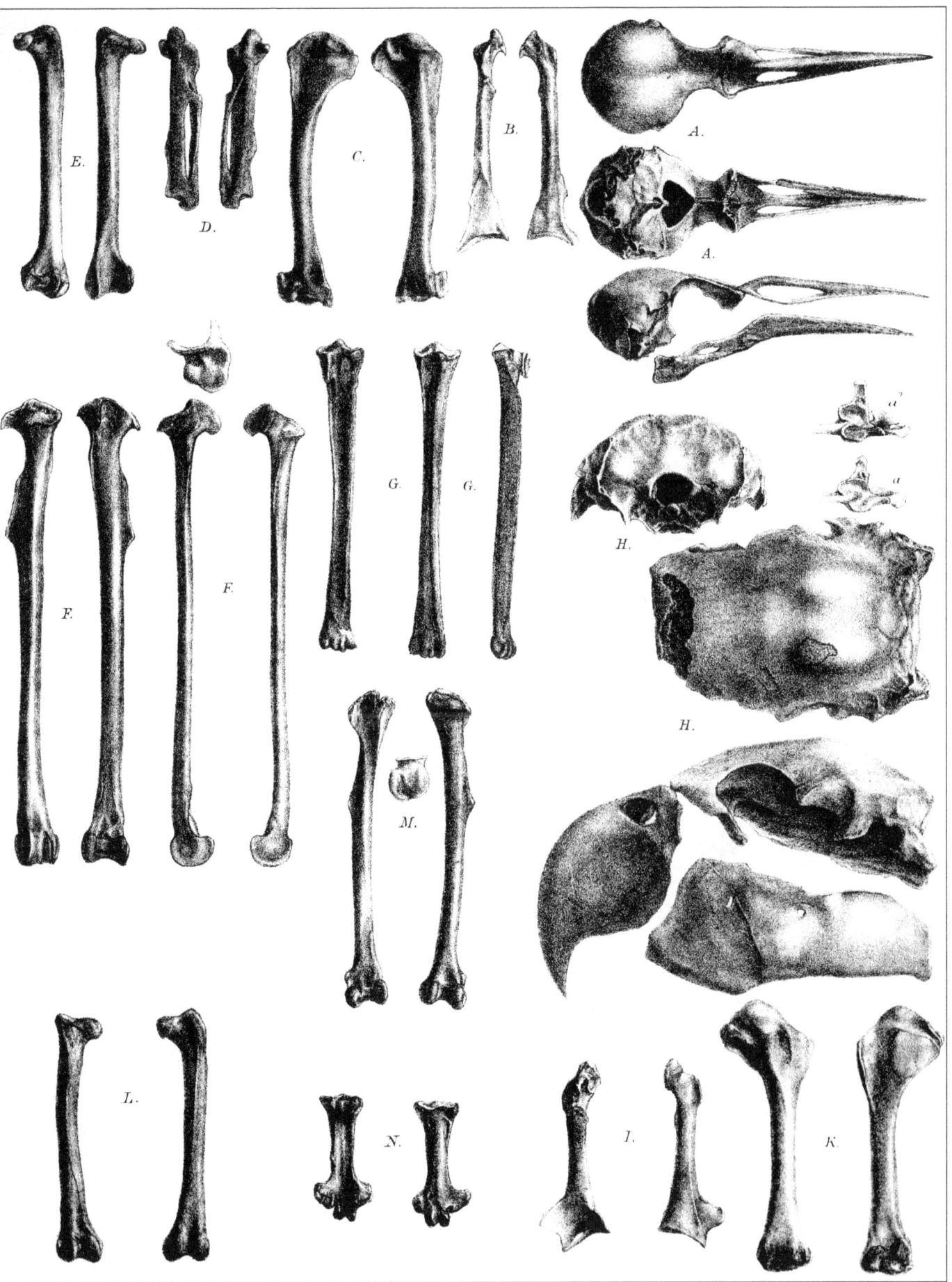

A–G. Necropsar rodericanus. H–N. Necropsittacus rodericanus.

Trans. of Venus Exped.   Rodriguez.   Zoology Pl. XLIII.

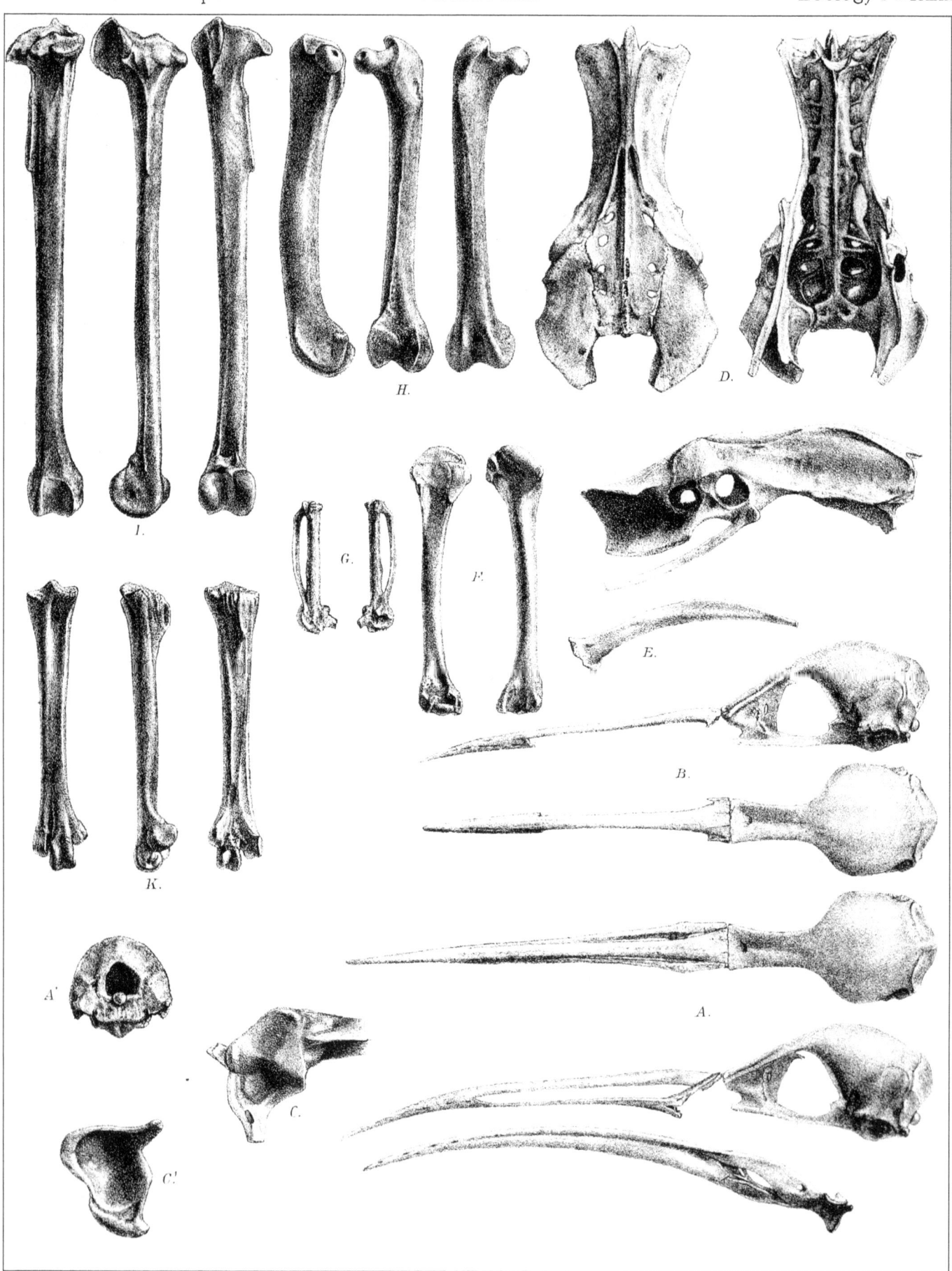

R. Mintern del. et lith.   Aphanapteryx leguati.   Mintern Bros. imp.

SOLITAIRE. — PEZOPHAPS SOLITARIA.
1–2. Eighteenth Vertebra 5–6. Twelfth Vertebra
3–4. Thirteenth Vertebra 7. Left Foot.

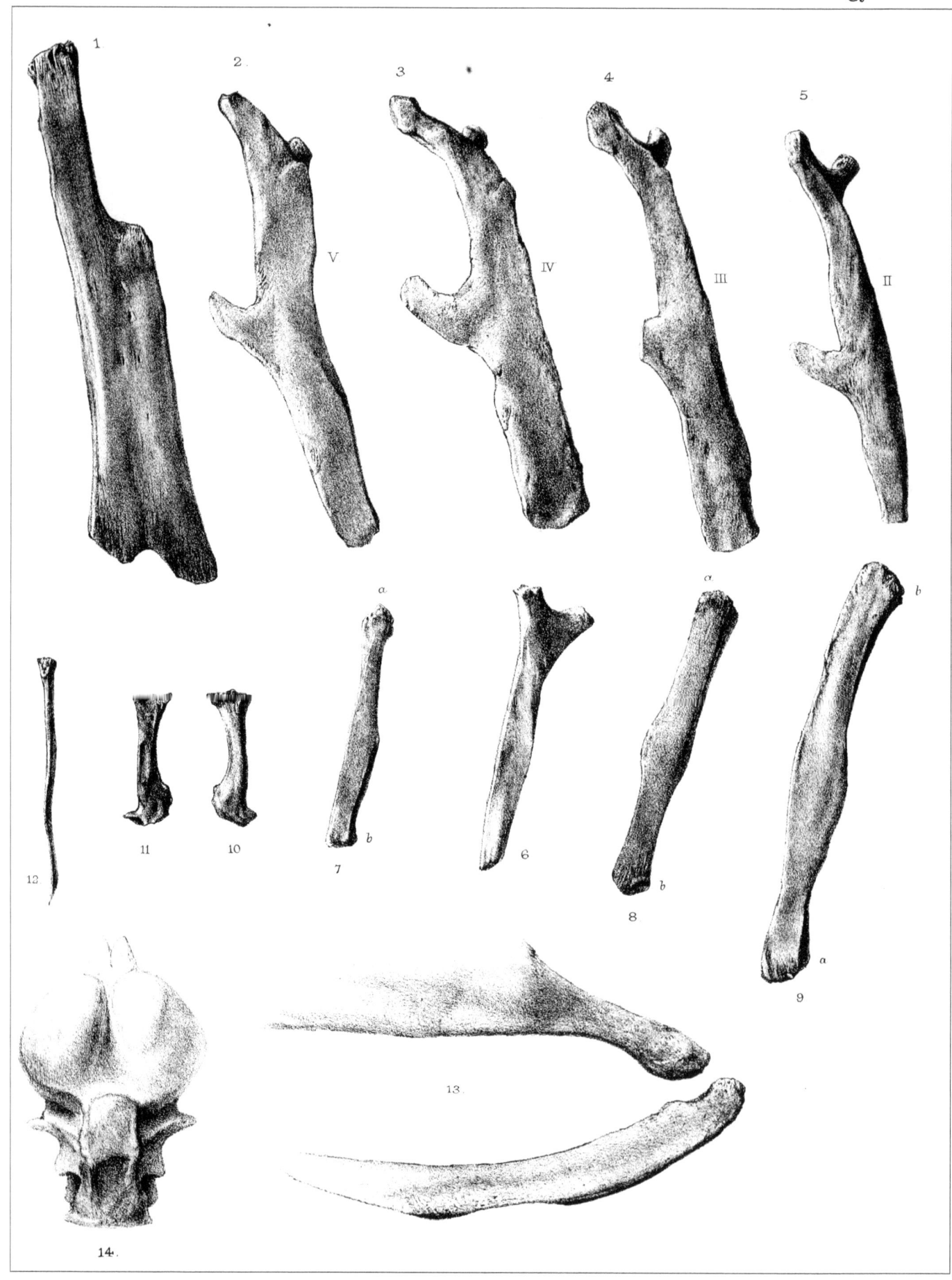

SOLITAIRE. — PEZOPHAPS SOLITARIA.

1. Sixth and seventh sternal ribs.
2-6. Vertebral ribs of right side, ♀
7-9. Sternal ribs, ♀
10, 11. Pterygoids.
12. Jugal.
13. Extremity of ischium and pubes.
14. Brain.

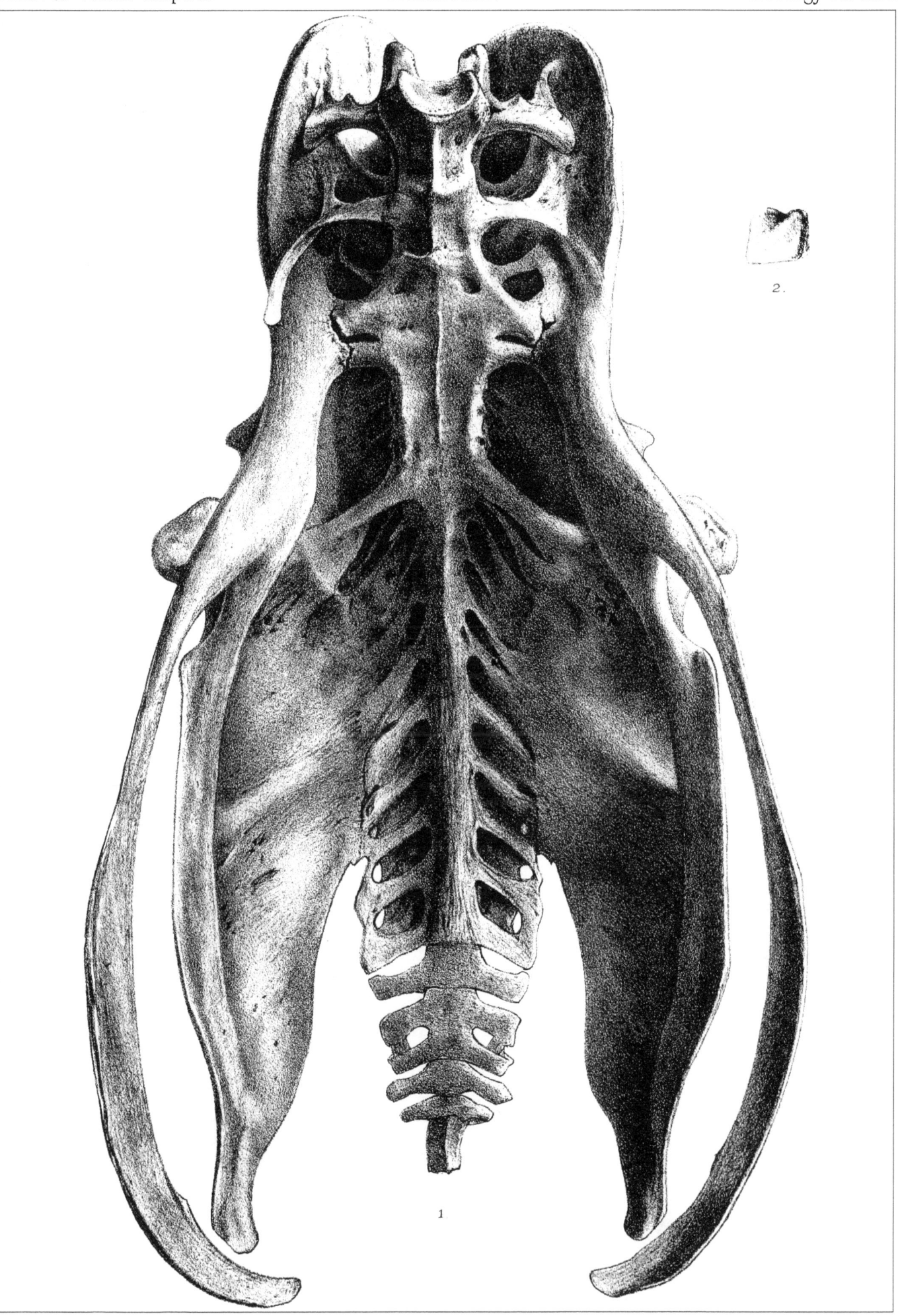

SOLITAIRE. — PEZOPHAPS SOLITARIA.
1. Pelvis of male skeleton.　　2. Side view of coccyx.

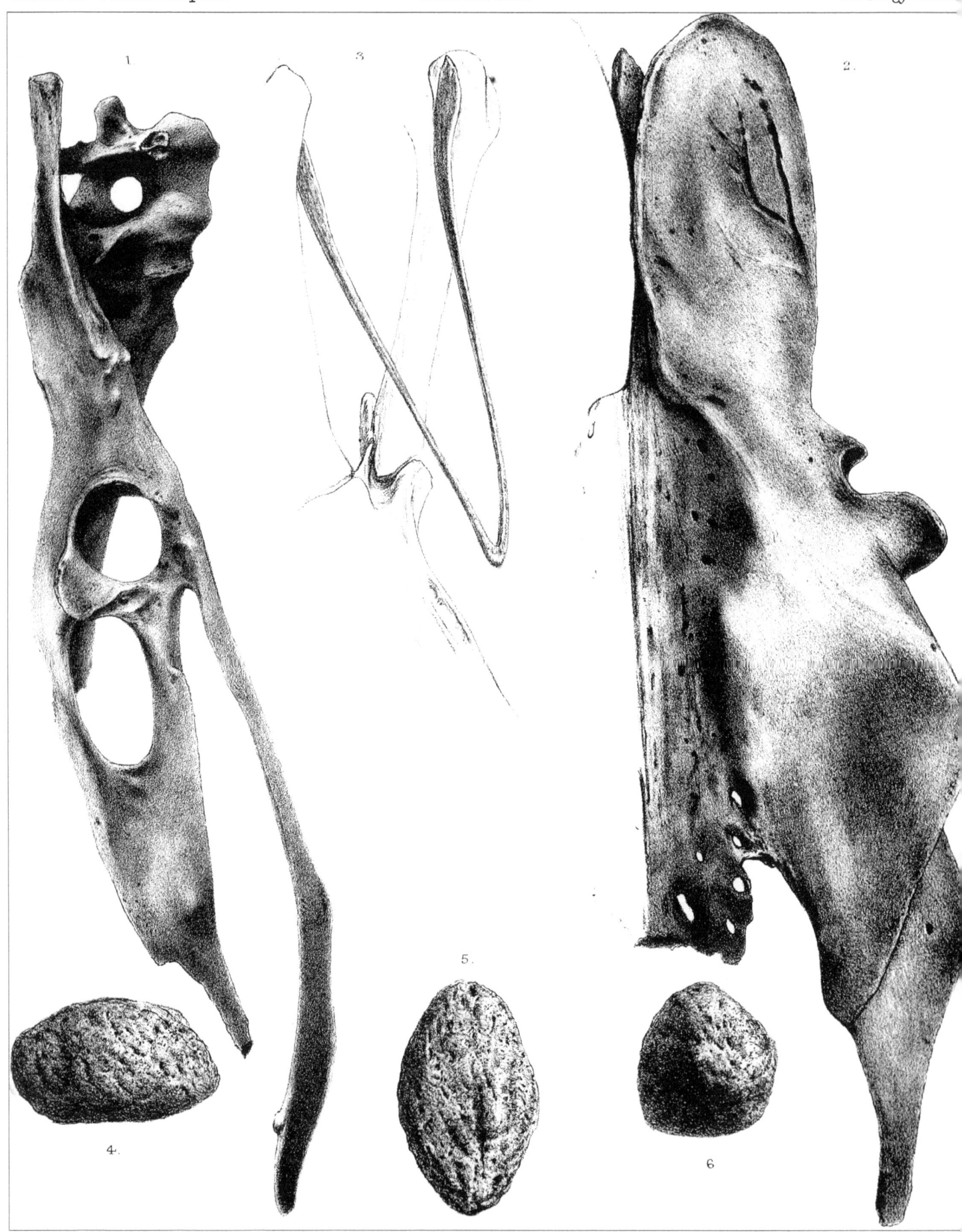

SOLITAIRE — PEZOPHAPS SOLITARIA.
1. Pelvis ♀
2. Pelvis ♂
3. Furcula.
4. Stone swallowed by the Solitaire

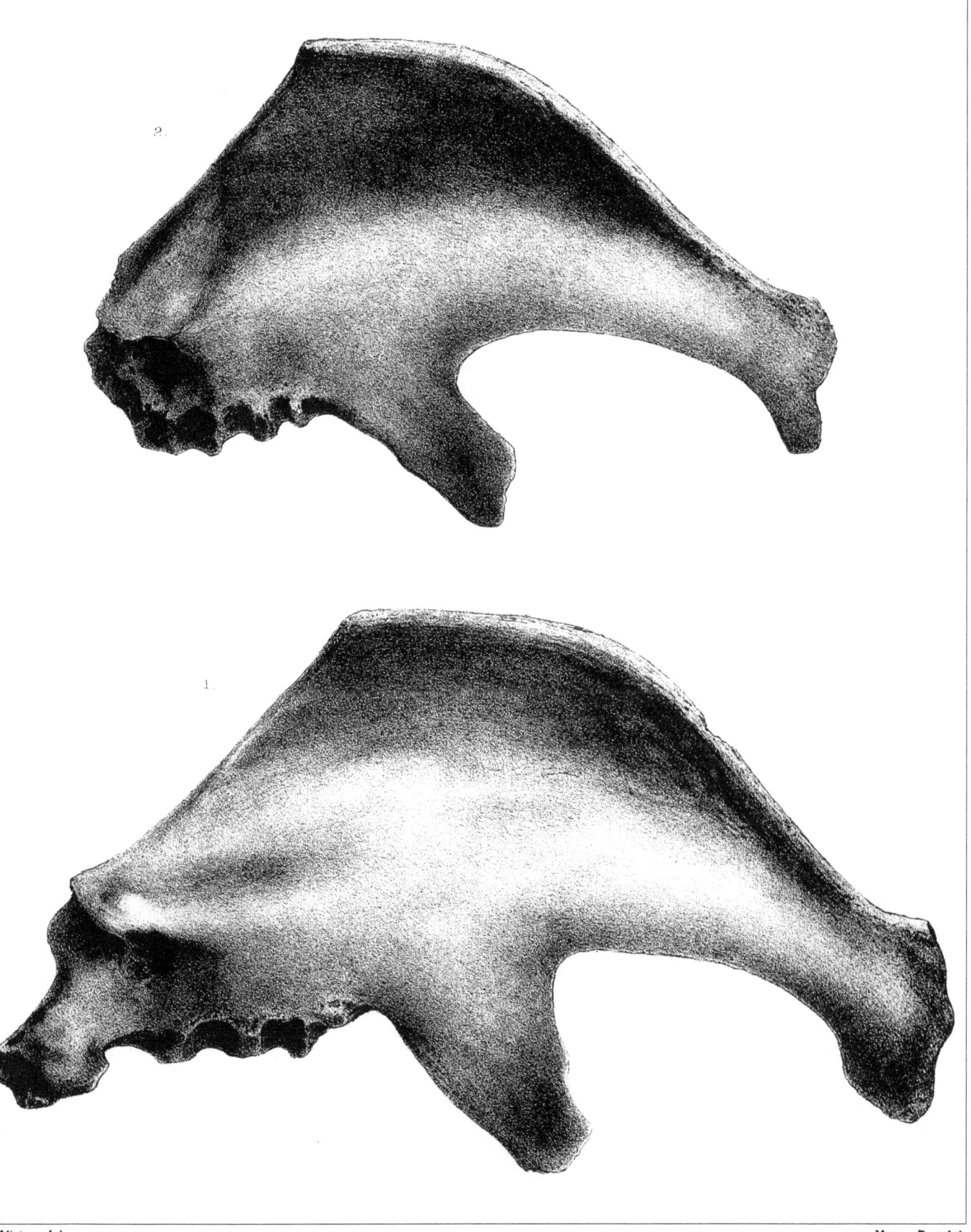

SOLITAIRE.— PEZOPHAPS SOLITARIA.
1. Sternum ♂.   2. Sternum ♀.

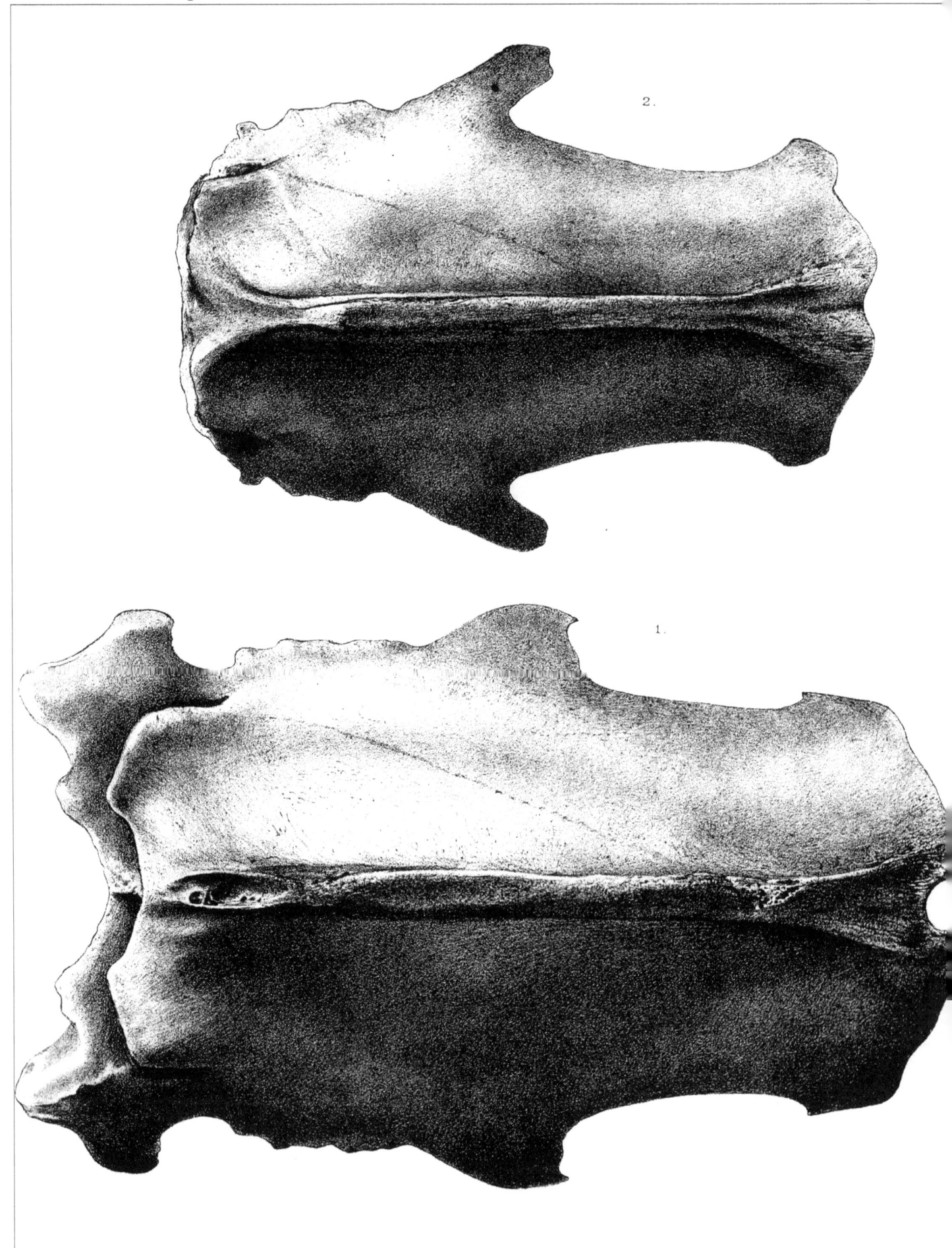

SOLITAIRE. — PEZOPHAPS SOLITARIA.
1. Sternum ♂ 2. Sternum ♀

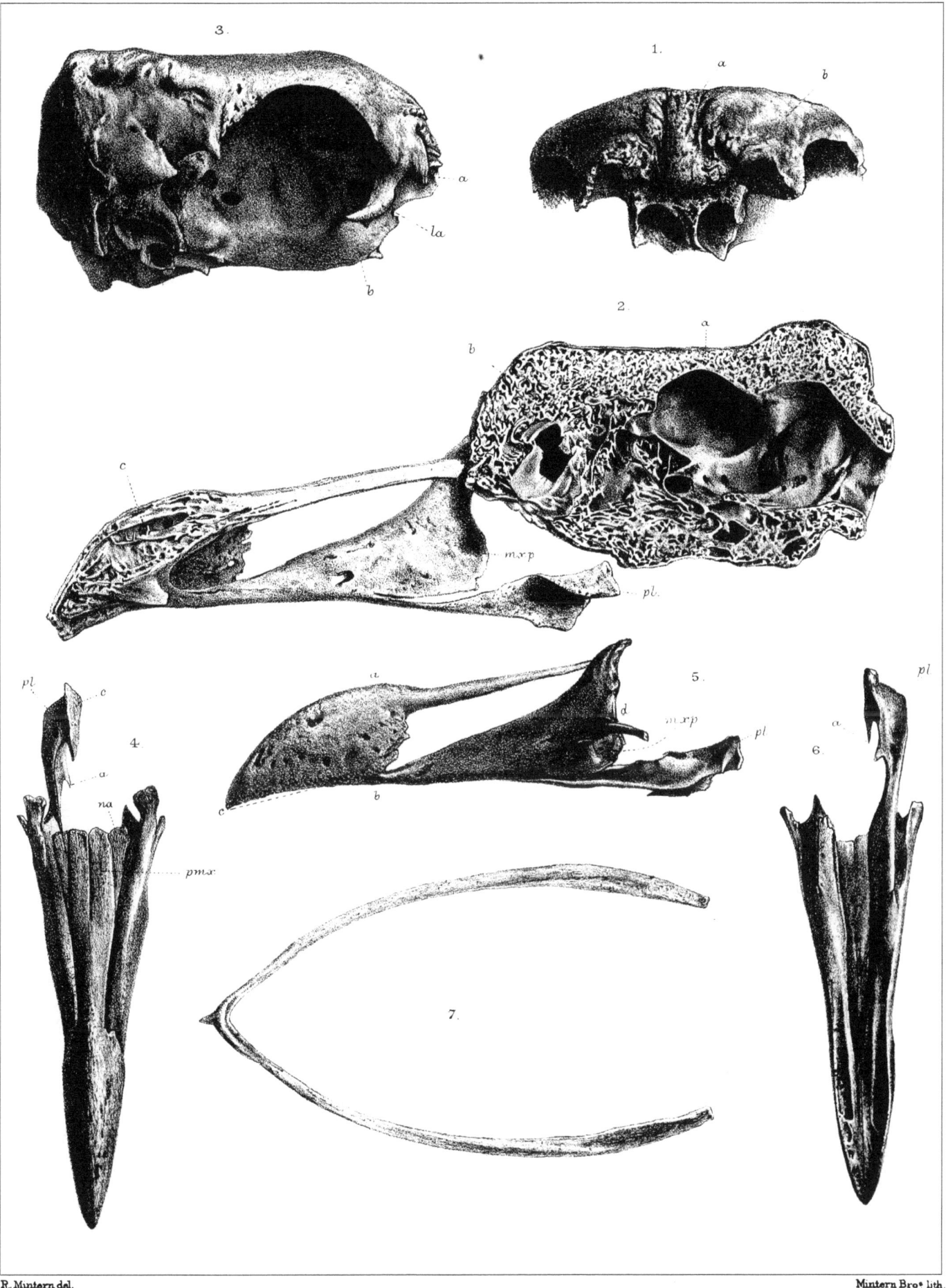

SOLITAIRE — PEZOPHAPS SOLITARIA.
1–6. Portions of Skull.   7. Furcula.

Trans. of Venus Exped. RODRIGUEZ. Zoology Pl. LI

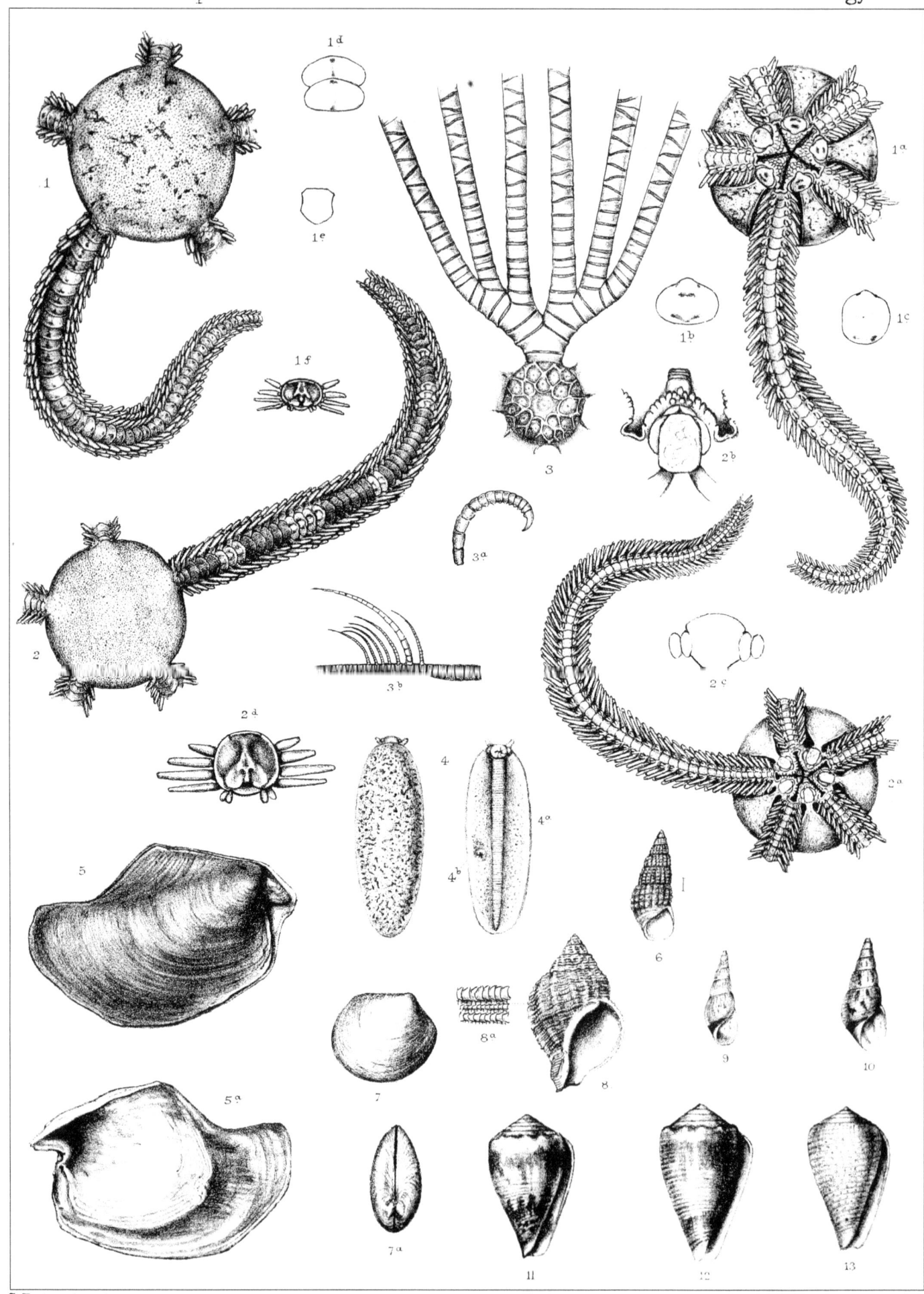

C. Berjeau del. et lith. Mintern Bros imp.

Mollusca et Echinodermata.

Trans of Venus Exped. RODRIGUEZ. Zoology. Pl. LII.

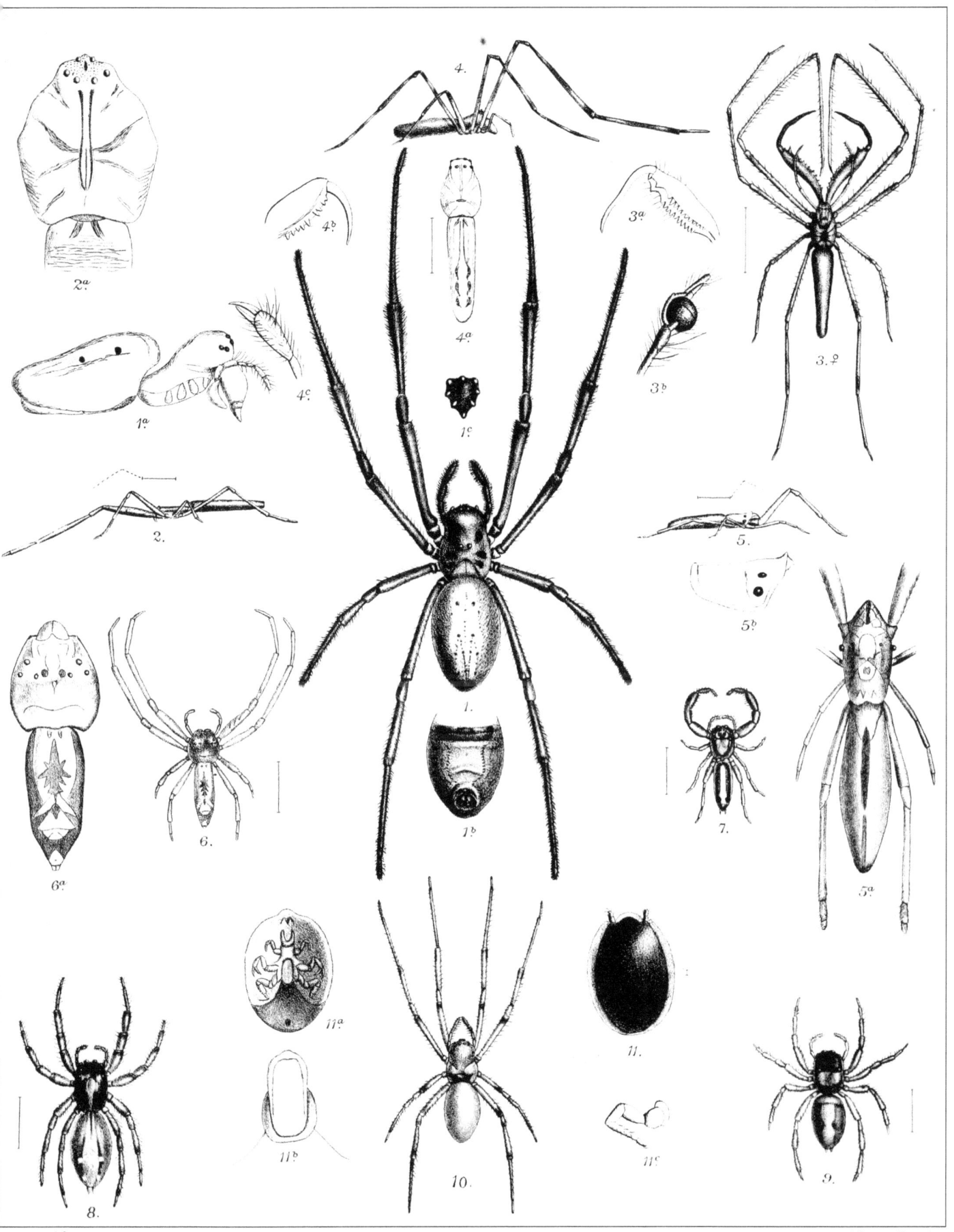

R. Mintern del. Arachnida. Mintern Bros imp.

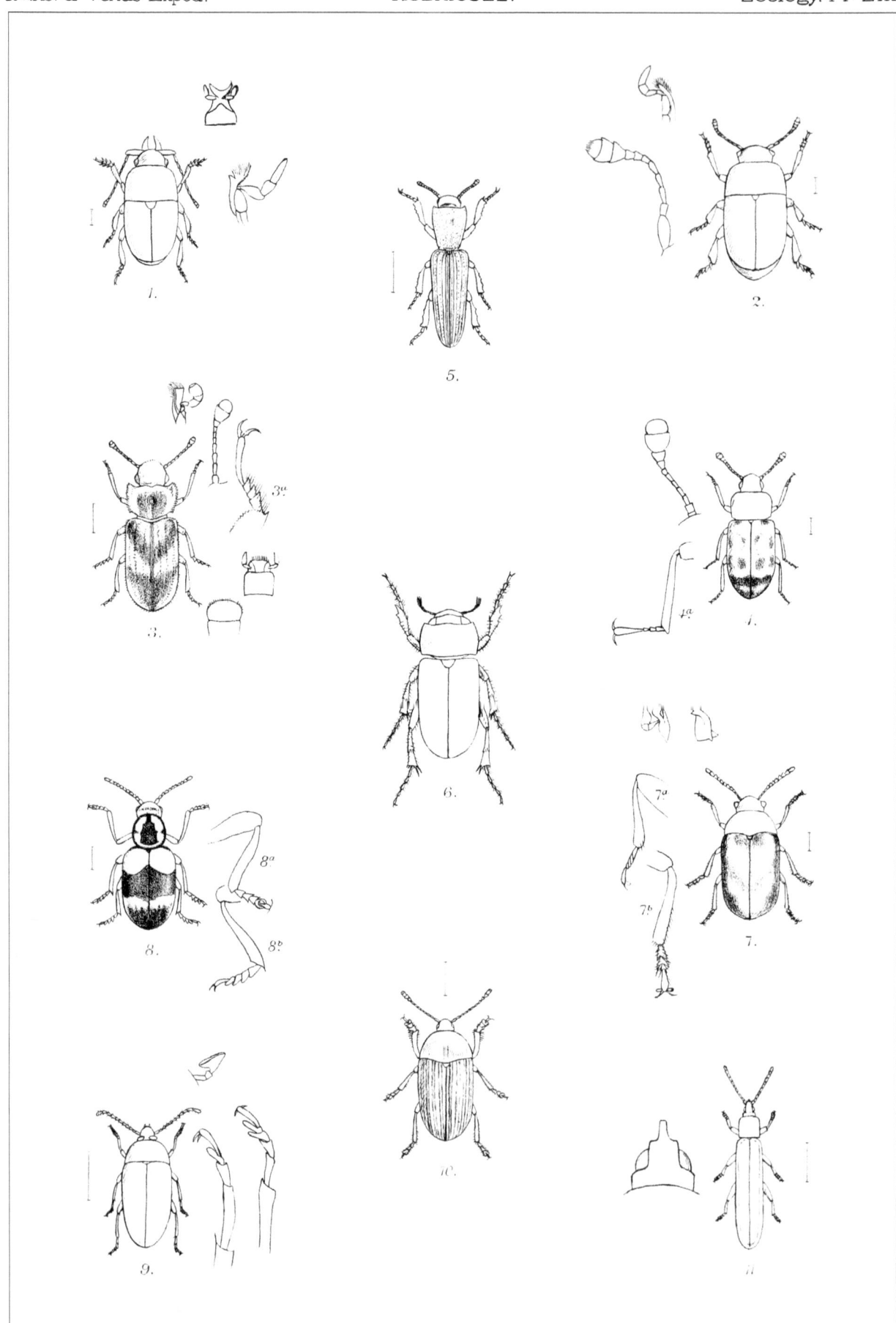

Coleoptera.

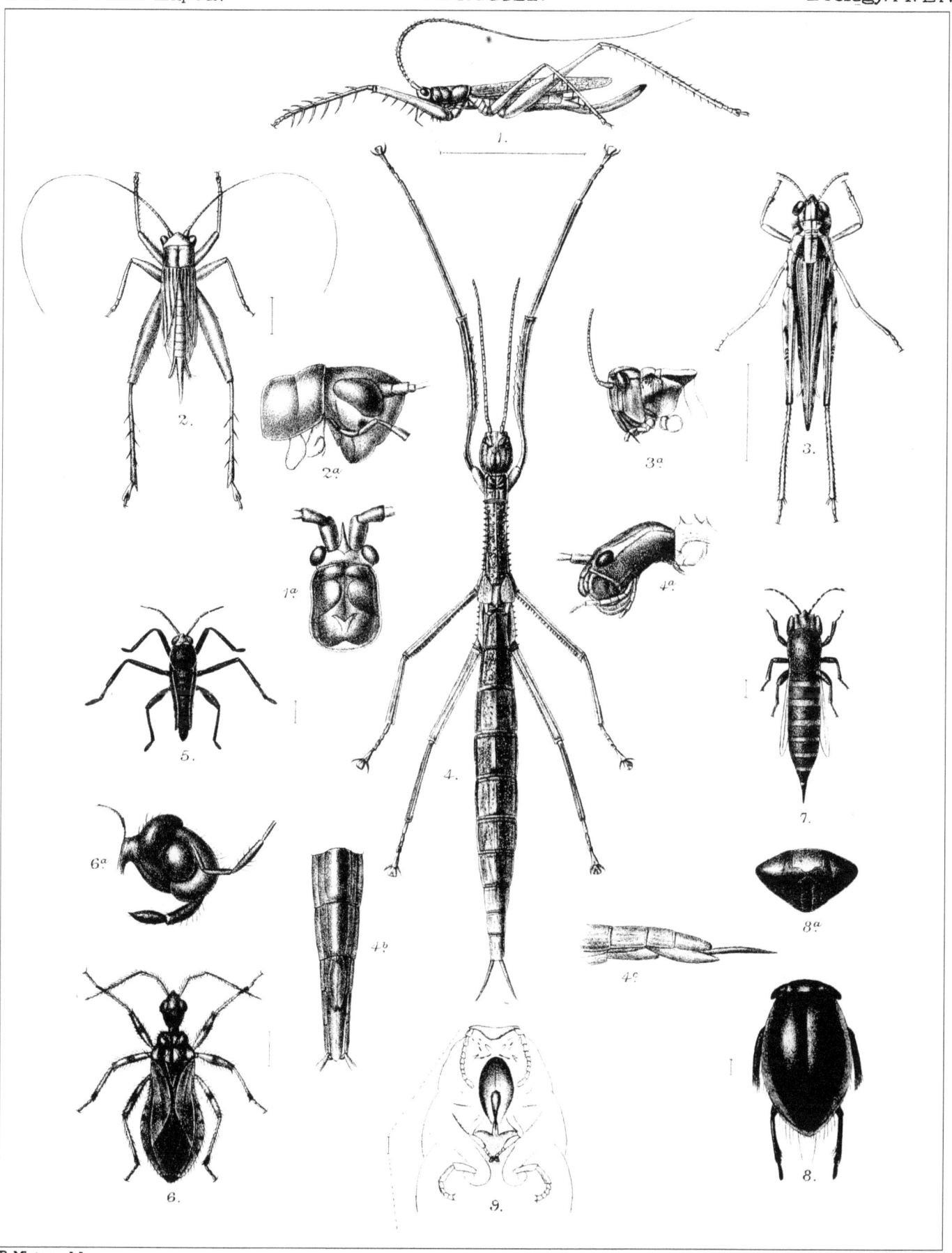

Orthoptera. & Hemiptera.

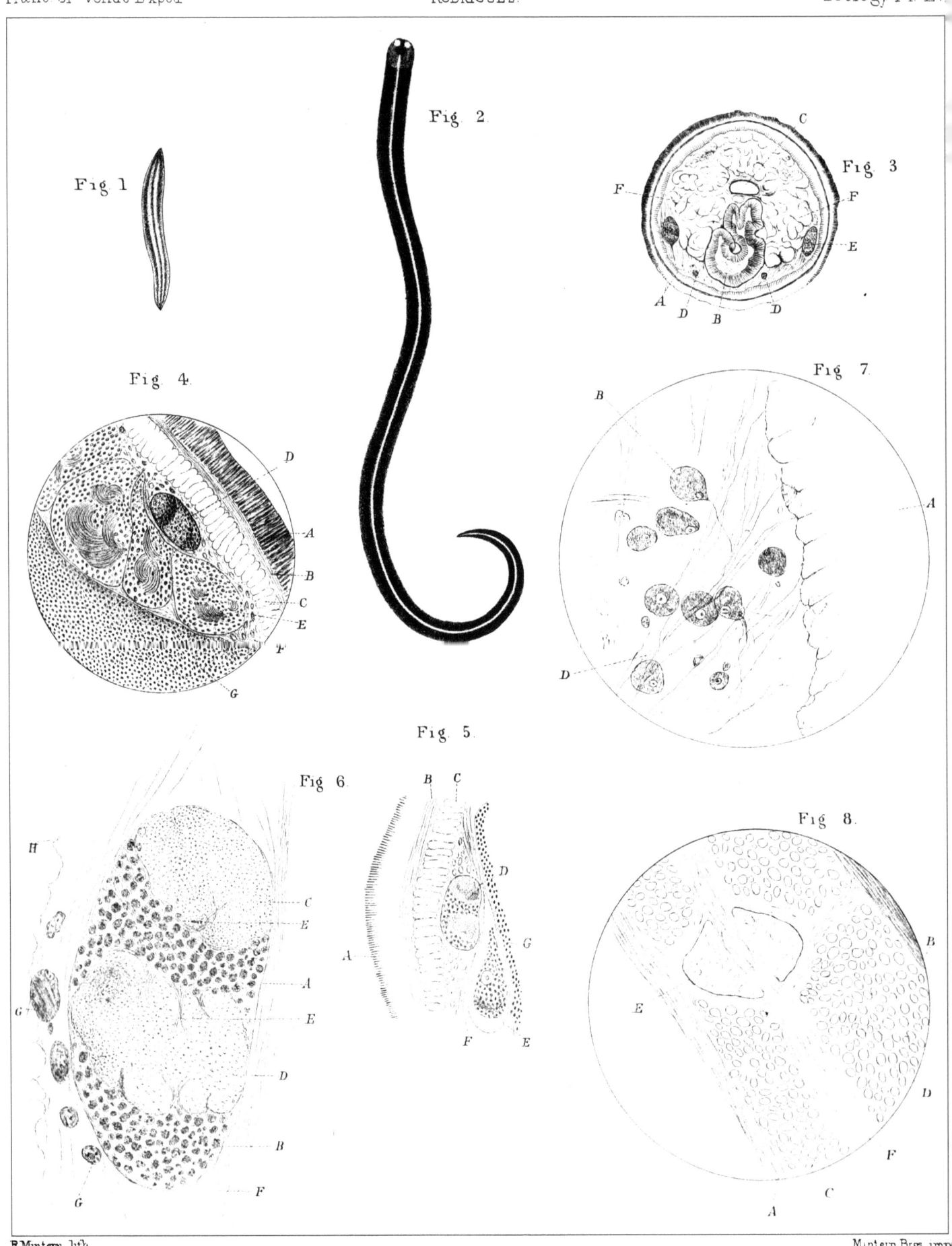

For EU product safety concerns, contact us at Calle de José Abascal, 56–1°,
28003 Madrid, Spain or eugpsr@cambridge.org.

www.ingramcontent.com/pod-product-compliance
Ingram Content Group UK Ltd.
Pitfield, Milton Keynes, MK11 3LW, UK
UKHW060049240426
12048UKWH00019B/1409